AF322930

A TALE OF DISCRETE MATHEMATICS

A Journey Through Logic, Reasoning, Structures and Graph Theory

A TALE OF DISCRETE MATHEMATICS

A Journey Through Logic, Reasoning, Structures and Graph Theory

Joseph Khoury

University of Ottawa, Canada

NEW JERSEY · LONDON · SINGAPORE · BEIJING · SHANGHAI · HONG KONG · TAIPEI · CHENNAI · TOKYO

Published by

World Scientific Publishing Co. Pte. Ltd.

5 Toh Tuck Link, Singapore 596224

USA office: 27 Warren Street, Suite 401-402, Hackensack, NJ 07601

UK office: 57 Shelton Street, Covent Garden, London WC2H 9HE

Library of Congress Cataloging-in-Publication Data

Names: Khoury, Joseph, 1968–   author.

Title: A tale of discrete mathematics : a journey through logic, reasoning, structures and
    graph theory / Joseph Khoury, University of Ottawa, Canada.

Description: New Jersey : World Scientific, [2024] | Includes index.

Identifiers: LCCN 2023049052 | ISBN 9789811285783 (hardcover) |
    ISBN 9789811285790 (ebook for institutions) | ISBN 9789811285806 (ebook for individuals)

Subjects: LCSH: Logic, Symbolic and mathematical--Textbooks. |
    Combinatorial analysis--Textbooks. | Set theory--Textbooks. | Algebraic logic--Textbooks.

Classification: LCC QA9 .K435 2024 | DDC 511.3--dc23/eng/20231221

LC record available at https://lccn.loc.gov/2023049052

British Library Cataloguing-in-Publication Data

A catalogue record for this book is available from the British Library.

For any available supplementary material, please visit
https://www.worldscientific.com/worldscibooks/10.1142/13663#t=suppl

Desk Editors: Nimal Koliyat/Kwong Lai Fun

Typeset by Stallion Press
Email: enquiries@stallionpress.com

To the one who showed me that true love is beyond discrete and continuous, it is simply out of this world. To my daughter Jo-Ann whose presence in my life fills my heart with warmth and joy every day.

Preface

In the English Cambridge dictionary, one can find the following definition of the word discrete: *clearly separate or different in shape or form.* From a mathematical perspective, this definition might not tell much about the nature of discrete mathematics, but it hints at the fact that the subject is an umbrella of many areas of mathematics that share the property of being *discrete* as opposed to *continuous.* Understanding the difference between these two concepts is key to understanding what discrete mathematics is and the scope of the topics that fall under its umbrella. A well-known example to explain the difference between these two notions is the comparison between an analog clock (i.e., a clock with three hands) and a digital one. If one has a super magnifier, then one can see that there are always two positions occupied by the seconds hand of an analogue clock so close together that we cannot distinguish them with the naked eye. We say that the seconds hand moves in a *continuous* mode (at least in theory). The digital clock, on the other hand, displays numerals which are limited in values and distinct from one another. In discrete mathematics, variables (like the numerals on digital clocks) cannot get arbitrarily small as they are just multiples of a certain unit. For example, integers are just multiples of the digit 1. We cannot choose an integer n *arbitrary* close to 0 for instance. As a consequence, notions like the instantaneous rate of change and limits of functions and sequences can only be defined in continuous mathematics.

Topics covered in a typical discrete mathematics course are certainly not new additions to mathematics. Some of them go back deep in history to the time where humans first felt the need to count, to calculate areas and to keep track of time. However, the recent advances in technology and computer science have resulted in increased interest in these topics. The basic fact that information is stored and analyzed by machines in a discrete manner made discrete mathematics a key to any advancement in computer science. Topics covered by a modern entry level discrete mathematics course include, but not limited to: propositional and predicate logic, proof techniques, recursion and induction, integers and their properties, set theory, functions, relations, combinatorics, algorithms and theory of computation, discrete probability, graph theory and discrete structures. While the book covers a wide range of these topics, it does not touch on probability theory or the classical algebraic structures like groups, rings, fields. We do however explore discrete structures like Boolean algebras and lattices.

The work on this book has been developed over many years of teaching first and second year discrete mathematics courses. During these years, I was greatly inspired by my students. Their reaction to various topics, their questions, comments and concerns allowed me to better understand how some concepts are perceived and which ones cause most difficulties.

0.1   Organization of the book

The book is organized in 11 chapters and one appendix. The content of each chapter is given briefly in what follows.

- **Chapter 1. Propositional Logic: The Foundation of Mathematical Reasoning.** Logic is at the heart of mathematical reasoning and proof techniques. Interest in the topic has seen a considerable increase with the digital revolution over the second half of the twentieth century. Chapter 1 deals with basic logical propositions, i.e., statements that are either true of false, and the logic connectives that tie them together. Although weaker than predicate logic in expressing and interpreting mathematical results, propositional logic is still a powerful tool in mathematics and computer science. The chapter covers topics like truth tables, truth trees, tautologies, contradictions, validity of logic arguments, normal forms (conjunctive and disjunctive) and others. Considering its binary nature, propositional logic offers a perfect setting to mathematically represent and design logic gates in electronic circuits.

- **Chapter 2. Set Theory and Introduction to Boolean Algebra: A Naïve Approach.** Sets are building blocks of every mathematical theory. Set theory is a vast area of mathematics and many authors like to include functions and relations under its umbrella. The book covers these topics in different chapters to allow more flexibility. In this chapter, a practical approach to set theory known in the literature as the *naïve set theory* is adopted. My intention is to present the version of set theory relevant to other topics in the book and to stay away from the formal axiomatic treatment of the subject. The chapter explores the notion of recursive definition of sets, a key concept in computer science. Boolean algebra structure is introduced in Section 6. Sum of products, product of sums and Karnaugh maps are used to minimize Boolean expressions. This is particularly important in designing efficient electronic circuits.

- **Chapter 3. Prove It: Mathematical Proof Techniques.** This chapter addresses the question of writing a sound mathematical proof. Multiple proof strategies are explored, each followed with numerous examples to enhance comprehension. Proof techniques studied include: direct proof, indirect proof, proof of an "if and only if" statements, proof by contradiction and proof by separation of cases. The chapter ends with the important technique of the proof by induction. Simple, strong and structural forms of induction. While simple and strong forms of induction are standard in any textbook of discrete mathematics, structural induction is not commonly covered and it is used to prove properties about recursively defined objects (sets, functions, ...).

- **Chapter 4. Introduction to Predicate Logic: One Step Further.** Proportional logic is a good first step to understand and analyse statements over the simple domain {True, False}. But it fails to deal with mathematical statements with truth values depending on one or more variables in a certain domain. The purpose of this chapter is to extend the language of propositional logic to capture the meaning and to analyse statements containing terms like *for all* and *there exists*. The notion of quantifiers, scope, translation to and from natural language, reasoning with quantifiers and arguments of predicate logic are explored.

- **Chapter 5. Functions: Back to the Basics.** Functions are particular type of binary relations, but this chapter introduces them as rules of correspondence between sets satisfying certain conditions. The notion of a *well-defined* function is introduced. Binary operations on sets and their properties are studied from the point of view of functions. Like sets, we consider functions defined recursively. Injective, surjective and bijective functions and their properties are studied. Invertible functions and their relation with bijective functions are also considered at the end of the chapter.

- **Chapter 6. Elementary Number Theory: The Basics, Primes, Congruences and A Bit More.** As discrete mathematics is concerned with discrete numeral system, it is only natural that integers are studied in any textbook on the subject. The chapter starts with a formal axiomatic definition of the integers. From this definition, some immediate properties of the integers are established as well as the introduction of the standard order in $\mathbb{Z}$. The chapter proceeds to explore the notions of divisibility and prime numbers. The Fundamental Theorem of Arithmetic, the gcd, the lcm and the Euclidean algorithm are studied. The chapter also touches on the notion of linear congruence, the famous Chinese remainder theorem and the famous Euler phi function. As an application, the chapter ends with the RSA algorithm for ciphering messages.

- **Chapter 7. Relations.** This chapter is an extension of Chapters 2 and 5 on set theory and functions. Relations occur very often in mathematical theories as a mean to compare elements. The emphasis in the chapter is on binary relations. Representations of relations using Boolean matrices and direct graphs are introduced and methods to create new relations from old ones are explained. The chapter revisits the notion of functions defined in Chapter 5 and a more formal definition of functions is given in terms of relations. Important properties of relations are introduced (reflexivity, symmetry, transitivity, ...) as well as their closures. In particular, Warshall's algorithm to compute the transitive closure is given. Two important classes of relations are studied in this chapter: the equivalence relation and the partial order relation. The well-ordering principal is introduced in a general setting. The chapter ends with the notion of topological sorting.

- **Chapter 8. Basic Combinatorics: The Art of Counting Without Counting.** Combinatorics is the branch of discrete mathematics that is concerned with arrangements of objects and counting outcomes. This is a vast area of mathematics and the goal of the chapter is to just give an introduction of the topic and its main results. The chapter starts with the basic counting principles: the sum rule, the product rule and the principle of inclusion–exclusion. The pigeonhole principle is then introduced and first examples are given. The chapter moves on to explore permutations and combinations of a finite number of objects, with and without repetitions. The number of surjective functions between two finite sets and derangements of objects are counted. The chapter also explores the binomial theorem and some applications.

- **Chapter 9. Basics of Graph Theory.** This chapter starts *phase 2* of this book. Started as a simple puzzle in the 18th century, graph theory has developed through the years to be an important field of study with a wide range of applications in mathematics, chemistry, computer science and engendering. Basic definitions are given including a list of basic simple graphs, the adjacency and incidence matrices associated with a graph. The first result about graph theory, the handshaking lemma, is given in Section 3. The section also looks at the question of determining if a list of natural numbers is graphical. The Havel–Hakimi algorithm to answer this question is given with several examples on how to apply it. Many important and interesting topics of graph theory are covered in the remainder of the chapter: subgraph, line graph, complement, Hamiltonian closure, connectivity, bipartite graphs, matching in bipartite graphs, Hall's marriage theorem, isomorphism of graphs. The chapter ends with the notion of a flow in a network, and the Ford–Fulkerson algorithm to solve the maximum flow problem. There is a lot of flexibility in this chapter. Sections 9.4.4, 9.5.2, 9.7, 9.8 and 9.10 can be treated as *optional* but instructors are strongly encouraged to cover some of them.

- **Chapter 10. More on Graph Theory.** The chapter continues to investigate important results and applications of graph theory. Planar graphs and Kuratowski's theorem (without proof) are given in the first section. Eulerian and Hamiltonian graphs and their properties are studied in Sections 2 and 3. The chapter finishes with the topic of graph colouring and its applications. Graph colouring ties graph theory with algebra by associating a polynomial, called chromatic polynomial, to every graph. Coefficients of the chromatic polynomial gives many interesting features of the graph.

- **Chapter 11. Trees.** Trees are special type of graphs which are very useful in computer science, chemistry and in modeling some real life problems. After establishing first results about trees (including characterizing a tree), we investigate the topic of tree traversal and its applications. Modeling with trees, prefix, infix and postfix notations and coding using trees are presented. Depth-First and Breadth-First searches are studied and the question of minimal spanning tree is tackled.

0.2 Key features of the book

There are many excellent textbooks on discrete mathematics out there, so why another one on the subject? If I have to cite couple of reasons, I would say *accessibility* and the *breadth-depth* approach. But there are certainly other features that set this textbook apart from others on the subject.

- **Accessibility.** Many students taking an introductory discrete math class have little mathematics in their final high school years. Others are returning to university after years of being in the work force and away from academia. Unlike calculus and linear algebra, topics covered in a discrete math class are not universal and can vary significantly between educational institutions, and even between different programs at the same institution. After many years of teaching logic and discrete mathematics courses, I came to realize that most textbooks on the subject fall into two categories. In the first category, the authors assume that the reader has acquired a certain level of mathematical maturity which results in a high level of abstraction and complexity in the textbook. In the second, the authors stick to the basics and students with higher mathematical maturity find themselves almost repeating what they already learned in high school. My intent in writing this book is to present the topics in a way that is accessible to students with modest mathematical maturity, but also to give all students the chance to advance gradually in acquiring a wide range of advanced topics.

- **Breadth and depth.** A key feature of this textbook is the scope of topics it covers. The following topics are explored in the book but are not standard in an introductory textbook in discrete mathematics: the method of truth tree (Chapter 1), sum of products, product of sums and Karnaugh maps (Chapter 2), recursive definitions of sets and functions and structural induction (chapters 3 and 5), the Euler phi function and the RSA algorithm (Chapter 6), closures of relations, Warshall's algorithm and topological sorting (Chapter 7), the Havel–Hakimi algorithm, Hamiltonian closure, matching in bipartite graphs, the Hall's marriage theorem, flow in a network, the Ford–Fulkerson algorithm (Chapter 9), chromatic polynomial (Chapter 10) and ruskal's and Prim's algorithms (Chapter 11). The fact is each chapter can be explored in a full textbook on its own. Efforts are made to ensure that the treatment of each topic goes as deep as possible to capture the most important features of the topic in the most direct and concise way possible.

- **A straightforward language.** The textbook is written for students from various mathematical and cultural backgrounds. Mathematical concepts can be challenging enough, and the last thing students need is a complex language to add to the challenge. Definitions are given using simple straightforward language, theorems are stated in a clear and concise matter with little room for misinterpretation. Proofs include as much details as possible to facilitate comprehension.

- **Focus on proofs.** After every semester teaching a discrete mathematics class, I survey students on a list of the most challenging components of the class. By far, mathematical proof tops the list every semester. Many students have difficulties reading and understanding a mathematical proof, let alone writing one. Mathematical proofs require a lot of reading, patience and exercise. An important feature of this book is the combination between sound mathematics and interesting applications. By sound mathematics, I mean precise results with their complete proofs. This combination is precisely what students should get from a discrete mathematics course. In this sense, the textbook is suitable for math majors as well computer science and engineering students. With the exception of very few theorems, the proof of every result is presented with all necessary details. Students are strongly encouraged to read every proof carefully and to try hard to rewrite it on their own. In every section, a certain number of exercises is dedicated to proofs of mathematical results. I always recommend to my students to start a proof by writing the first and the last sentence of the proof and then gradually fill it the lines in between.

- **Early introduction to the language of mathematics.** The book introduces logic and mathematical reasoning at an early stage. A clear and thorough understanding of mathematical logic is key to fully grasp any mathematical concept. My experience teaching some transitional courses like real analysis or group theory shows that students' difficulties are not so much in grasping the theoretical concepts, but rather in properly expressing their reasoning and understanding of statements. A typical example is the confusion between the hypothesis and the conclusion in an implication.

- **A wealth of examples.** The book contains hundreds of fully worked examples, carefully written to eliminate ambiguities and to cover as many scenarios as possible. Understanding these examples is key to fully grasp definitions and results. Many of the worked examples are also mirrored in the exercises to allow students to have a firm grasp on new concepts.

- **Extensive set of exercises.** Every section of the book ends with a set of suggested exercises designed to offer students a variety of practice problems on the topics covered in the section. Students are strongly encouraged to try as many exercises as possible. Some exercises have similarities with worked out examples in the section, others are designed to offer a bit more challenge to students. Exercises are also used in some occasions to explore some concepts not covered in the section in some informal fashion.

- **Solutions to selected exercises.** To provide some guidance to students, Appendix A contains full solutions to selected set of exercises (indicated by $\star$) from each section. These selected exercises are carefully chosen to give students enough material to practice the theory in the section. Students are strongly encouraged to try their absolute best to work these exercises before they consult the solutions. It is only by grappling with these problems that students will gain confidence and experience in the topic.

- **Flexibility.** Flexibility is desirable in any textbook, but difficult to achieve in general. This is even more challenging in a textbook on a subject of a cumulative nature like mathematics. The diversity of topics covered in the textbook offers instructors and students a comfortable level of flexibility in choosing the order in which topics are covered. For example, one can choose to completely skip the chapter on predicate logic and still have good grasp on the remaining topics of the book. Another example is the chapter on functions that can be approached from the point of view of Chapter 7 on relations. Even within the same chapter, some sections can be treated independently of other sections. In addition, some sections can be treated as optional and left to the discretion of the instructor.

- **Introduction to algorithms.** The book contains several important algorithms in various topics. The combination of abstraction and practical algorithms enriches the book and widens the range of students who could benefit from it. Although the book does not include a chapter on writing and verifying algorithms, it lays down the mathematical foundation that fuels these algorithms. We explore, among others, Warshall's algorithm to compute the transitive closure of a binary relation, the Euclidean algorithm to compute the gcd of two integers and express it as a linear combination of the integers, Havel–Hakimi algorithm to test if a sequence of natural numbers is graphical, the Ford–Fulkerson algorithm to find the maximum flow in a network, Depth-first and Breadth-first search algorithms to traverse a graph, Kruskal's and Prim's algorithms to generate a minimal spanning tree.

About the Author

 **Joseph Khoury** received his Ph.D. in Mathematics in 2001 from the University of Ottawa, Canada. He is currently an instructor and the director of the Math Help Center at the University of Ottawa. Dr. Khoury is a co-editor and a co-author of previously published World Scientific books, *Jim Totten's Problems of the Week* and *The Mathematics that Power Our World, How Does It Work?*

Contents

Chapter 1

Propositional Logic: The Foundation of Mathematical Reasoning

1.1 Introduction

Throughout history, many civilizations have realized the need to formalize rules for reasoning and proof techniques. For example, a pilar in ancient Greece education system was the *trivium* (*three roads* in Latin), which refers to the study of grammar, rhetoric, and logic to young students. With the development of new mathematical theories, the need for a precise set of rules to determine what constitutes a valid mathematical proof became paramount and preoccupied mathematicians for centuries.

Historically speaking, logic origins are hard to trace as many ancient philosophers and scientists contributed to the development of the discipline. It is however widely believed that the discipline started to take the shape we know today in ancien Greece. Aristotle, a Greek philosopher and mathematician (384–322 BC), is considered to be the author of the earliest known document on formal logic. His main goal was to develop the logic theory as a complete mathematical system to analyze human reasoning and effective arguments. While this philosophical aspect of logic remains a central component of the subject some 25 centuries after Aristotle, this area of mathematics has significantly evolved in the last century to become a key tool not only in pure and applied mathematics but also in many applications of computer science.

There are two main formal systems of mathematical logic: the propositional logic and predicate logic. Propositional logic studies statements which have truth values (either true or false) and logical operators on these statements like the words "and", "or" and others. In this context, we refer to the statements as *propositions* and to the operators on them as *logic connectives*. A unit (or atomic) statement of propositional logic is a proposition that contains no connective and as such, it is not further analyzable. Logic connectives are used to connect propositions together and form more complex ones. Propositional logic analyses statements with truth values depending uniquely and entirely on the units on which they are built. There are however mathematical statements with truth values depending not only on their underlying units but also on their interpretation within a given context. This type of statements requires a richer system of analysis and will be studied in Chapter 4 on predicate logic.

1.2 Basic terminologies and logic connectives

In the English language, the following are well-known types of sentences.

(1) *The imperative sentence.* This is a sentence that gives a command or an order. For example, *"Bring me the book"*.
(2) *The interrogative sentence.* This is a sentence that asks a direct question and which always ends with a question mark. For example, *"Is 1051 a prime number?"*.
(3) *The declarative sentence.* This is a sentence that states a fact or gives some sort of information. For example, *"1051 is not a prime number"* and "Ottawa is the capital city of Canada".

What distinguishes a declarative sentence from other types of sentences is the fact that it is either true or false but not both at the same time. From a propositional logic perspective, a declarative sentence is called a *proposition*. If a proposition p is true, we say that its *truth value* is **T** and we write $v(p) = \mathbf{T}$. If a proposition p is false, we say that its *truth value* is **F** and we write $v(p) = \mathbf{F}$. Some authors assign 1 and 0 as truth values for a true and false proposition, respectively. Propositions like " $3^3 + 3$ *is an even number*" and "*the equation $x^2 - 3x + 2 = 0$ has no real roots*" can be easily characterized as true or false. The first is true as $3^3 + 3 = 30$ is an even number and the second is false since $x = 1$ and $x = 2$ are real roots of the equation. The ease of determining the truth values of these two statements is unfortunately not an indication on the hardship of this task in general. Things can get complicated very quickly as we combine statements. For example, the task of determining the truth value of the declaration: "6851 is a prime number or π^π is a rational number if and only if $x^3 - 12354x^2 + 3x - 1200 = 0$ has a rational root" is certainly not a trivial one, even if we know the truth value of each of the individual components of the declaration.

In the syntax of propositional logic, propositions are usually referred to using letters. So we write things like A : "$2^5 > 33$", B : "31 is an even number" and p : "The domain of the function $f(x) = \sqrt{x+1}$ is the set of all real numbers."

Example 1.1. In each case, determine if the expression is a proposition. If the expression is a proposition, determine its truth value.

(a) p: "$8^{\frac{1}{3}} = 2$".
(b) q: "$8^{\frac{1}{3}}$".
(c) r: "The sum of two odd numbers".
(d) s: "The sum of two odd numbers is odd".
(e) t: "The equation $4x^2 - 4x - 3 = 0$ has at least one integer solution".

Solution. (a) p is a proposition. It states that the cubic root of 8 is equal to 2. This is true since $2^3 = 8$.
(b) q is not a proposition. It simply gives a number, namely $8^{\frac{1}{3}} = 2$.
(c) r is not a proposition. It does not declare anything.
(d) s is a proposition. It declares that when we add two odd numbers, the result is always an odd number. The truth value of s is **F** (false) since, for example, $1 + 3 = 4$ is even.
(e) t is a proposition. Its truth value is **F** since the equation has two solutions $x = -\frac{1}{2}$ and $x = \frac{3}{2}$ and neither one of them is an integer.

$\Diamond$

1.2.1 *Logic connectives*

In any natural language, a *word* is a sequence of symbols chosen from a set called the *alphabet* of the language. Sentences are formed by connecting words together with special connectors like "or", "and", a comma, etc. Propositional logic is no different from any other language. its alphabet consists of a set of propositional units, the special symbols $\wedge$, $\vee$, $\rightarrow$, $\leftrightarrow$ and $\neg$, commas and parentheses: "(" and ")". The role of the special symbols is to connect propositions together to create new ones. For example, the following (compound) proposition: "*The sun is shining or the lion is roaring if and only if it is pouring in the forest and the hunter hides behind the trees unless the leaves are falling"* is formed by connecting the (atomic) propositions *"The sun is shining"*, *"The lion is roaring"*, *"It is pouring in the forest"*, *"The hunter hides behind the trees"* and *"The leaves are falling"* using the words *"or"*, *"if and only if"*, *"and"*, and *"unless"* that we call the *logic connectives* or the *logic operators* . There are five basic logic connectives. Table 1.1 gives the names and symbols of these basic connectives. Note that, with the exception of the negation connective $\neg$, each of these connectives is a *binary operation* which means it requires two input propositions. For instance, the connective $\vee$ takes two input propositions p and q and produces the new proposition *"p or q"*.

Table 1.1   Basic logic connectives

Connective Name	Symbol	English Expressions
Conjunction	$\wedge$	and
Disjunction	$\vee$	or
Implication	$\rightarrow$	implies
Biconditional	$\leftrightarrow$	if and only if
Negation	$\neg$	not

A proposition with no logic connectives in it is called an *atomic proposition* or simply an *atom*. A logic proposition containing one or more logic connectives is called a *compound proposition*. For example, *"Ottawa is the capital city of Canada unless the sun rises from the west"* is a compound proposition formed by the disjunction of the two atoms *"Ottawa is the capital city of Canada"* and *"The sun rises from the west"*.

1.2.2   *Truth table*

The *truth table* of a logic proposition p is a table listing the truth values of p for all possible truth values of the components of p (also called the *logical variables* of p). Each list of truth values of the logical variables of p is called a *valuation* of p (see Definition 1.3 below). For example, if p is a proposition with two logical variables A and B, then there are four possible valuations of p, namely $(\mathbf{T}, \mathbf{T})$, $(\mathbf{T}, \mathbf{F})$, $(\mathbf{F}, \mathbf{T})$ and $(\mathbf{F}, \mathbf{F})$.

Example 1.2. The truth table of a logic proposition p is given below.

A	B	p
T	**T**	**F**
T	**F**	**T**
F	**T**	**F**
F	**F**	**T**

The table can be interpreted as follows: the proposition p is true when A is true and B is false or when both A and B are false. It is false when both A and B are true or when A is false and B is true.

We look next at each of the five basic connectives in a bit more detail. Table 1.2 at the end gives a summary that will serve as a basis for the rest of the discussion on propositional logic.

1.2.2.1   *The conjunction connective*

If p and q are two propositions, their *conjunction* is the compound proposition "p and q" that we denote by $(p \wedge q)$. The conjunction $(p \wedge q)$ is true if and only if both propositions p and q are true. The truth table of the conjunction connective is the following.

p	q	$(p \wedge q)$
T	T	T
T	F	F
F	T	F
F	F	F

Example 1.3. The proposition *Washington is the capital city of Canada and $3^{125689} + 11$ is a prime number* is false since the first component is false. Note that we did not need to know if $3^{125689} + 11$ is a prime number.

Example 1.4. The proposition "$8^{-\frac{1}{3}} = \frac{1}{2}$ *and* $\sqrt{7} \leq 3$" is true since it is the conjunction of two true statements.

1.2.2.2   *The disjunction connective*

If p and q are two propositions, their *disjunction* is the compound proposition "p or q" that we denote by $(p \vee q)$. The disjunction $(p \vee q)$ is always true except in the unique case when both p and q are false. The truth table of the disjunction connective is as follows.

p	q	$(p \vee q)$
T	T	T
T	F	T
F	T	T
F	F	F

Example 1.5. The proposition *Ottawa is the capital city of Canada or the sun rises from the west* is true since the first component is true. Note that it does not matter what the truth value of the second component is in this case since we know that the proposition *Ottawa is the capital city of Canada* is true.

Example 1.6. The proposition "$8^{\frac{1}{3}} = \frac{1}{2}$ *or* $\sqrt{7} > 3$" is false since it is the disjunction of two false propositions.

Example 1.7. The proposition "*Ottawa is the capital city of Canada and the sun rises from the west, or* $\sqrt[3]{-8} = -2$" is true since the second component is clearly true. Note that the proposition "*Ottawa is the capital city of Canada and the sun rises from the west*" is false in this case.

Remark 1.1. The symbol $\vee$ is used to represent the disjunction "or" in the inclusive sense. To explain this, imagine you ask a friend about his plan to get to Toronto today and he answers with the following disjunction: "*I will drive or I will take the train*". Note that exactly one of the two components of this disjunction can be true in this context. We say in this case that the disjunction "or" is used in an *exclusive sense*. On the other hand, the disjunction "or" in the proposition "*I drink my coffee or I read my book*" is used in an *inclusive sense* since the components "I drink my coffee" and "I read my book" can happen simultaneously (can be true at the same time). The symbol $\oplus$ is usually used to specify the exclusive disjunction. So $(p \oplus q)$ is true when *exactly* one of p, q is true and it is false when both are true or both are false.

1.2.2.3 *The implication connective*

A big part of human interactions comes in the form of conditional and sequential statements. Sentences like *If you walk under the rain, you will catch a cold* or *Reading this book is sufficient to have a basic understanding of Discrete Mathematics* are called conditional statements. Such statements give some information (conclusion) based on a certain assumption (hypothesis). In propositional logic, a conditional statement is referred to as an *implication*. If p and q are two propositions, the proposition "*p implies q*" (or equivalently "*if p then q*") is represented with the notation $(p \rightarrow q)$. For the implication $(p \rightarrow q)$, p is called the *hypothesis* and q is called the *conclusion*.

Example 1.8. The statement *If the stars are shining in the sky at night, then it is going to be sunny in the morning* is an implication with *The stars are shining in the sky at night* as the hypothesis and *It is going to be sunny in the morning* as the conclusion .

Example 1.9. The proposition "the sum of two rational numbers is a rational number" does not sound at first like an implication but rather like one atomic proposition. But a bit of deflexion shows that the proposition is actually stating the following: "If x and y are two rational numbers, then $x + y$ is a rational number". This makes it an implication with "x and y are two rational numbers" as the hypothesis and "$x + y$ is a rational number" as the conclusion.

Unlike the conjunction and the disjunction connectives, the truth table of the implication connective is not obvious and requires some explanation. A classic scenario used to explain the truth table of the implication connective is the following. During an election campaign, the candidate Joe makes the following statement: *"If I am elected, there will be income tax cut for the middle class in the first hundred days after the elections"* (sounds familiar?). When can you accuse Joe of lying? If Joe was not elected, then no one can accuse him of lying regardless of wether there was an income tax cut for the middle class in the first hundred days or not. In this case, his initial statement would still be valid since we don't know what would have happened had he been elected. If Joe is elected and he managed to reduce income tax for middle class in the first hundred days after the elections, then he spoke no fallacy during the campaign and his statement is true. Now, Joe would be in hot water (one would hope!) if he was elected but no income tax reduction followed for the middle class in the first hundred days after the elections. So, there is *only one* scenario where we can confirm that Joe's statement is false: he is elected but no income tax reduction follows for the middle class in the first hundred days after the elections. This suggests that an implication $(p \rightarrow q)$ is always true except in the unique case where p (the hypothesis) is true and q (the conclusion)is false. In particular, an implication with a false hypothesis is always true. The truth table of the implication connective is the following.

p	q	$(p \rightarrow q)$
T	T	T
T	F	F
F	T	T
F	F	T

Example 1.10. The proposition "*If Washington is the capital city of Canada, then $2^4 = -1$*" is true since it is an implication with a false hypothesis.

Example 1.11. The proposition "*If $8^{\frac{1}{3}} = 2$ and the equation $x^2 + 1 = 0$ has no real roots, then $\sqrt{7} > 3$*" is false since it is an implication with a true hypothesis (a conjunction of two true propositions) and a false conclusion.

Definition 1.1. For the implication $(p \to q)$: the implication $(\neg q \to \neg p)$ is called the *contrapositive* implication and the implication $(q \to p)$ is called the *converse* implication.

The reader should be very careful not to confuse the contrapositive and the converse of an implication. They are different propositions with different meanings. The following example explains the difference in a natural language context.

Example 1.12. Consider the proposition ϕ: "If there is a storm, then the school bus is cancelled". This is clearly an implication with "There is a storm" as the hypothesis and "The school bus is cancelled" as the conclusion. The contrapositive of ϕ is "If the school bus is not cancelled, then there is no storm" while the converse of ϕ is: "If the school bus is cancelled, then there is a storm".

Example 1.13. For the implication "If n is an odd integer, then $n^3 + 1$ is an even integer", the contrapositive is "If $n^3 + 1$ is an odd integer, then n is an even integer", and the converse is "If $n^3 + 1$ is an even integer, then n is an odd integer".

1.2.2.4 *The biconditional connective*

If p and q are two propositions, the compound proposition "p if and only if q" (or p is a necessary and sufficient condition for q) is called the *biconditional* of p and q that we denote with $(p \leftrightarrow q)$. The biconditional $(p \leftrightarrow q)$ is true when p and q have the same truth value (both are true or both are false) and is false when p and q have different truth values (one is true and the other is false). The truth table of the biconditional connective is the following.

p	q	$(p \leftrightarrow q)$
T	T	T
T	F	F
F	T	F
F	F	T

Example 1.14. The proposition *Washington is the capital city of Canada if and only if* $2^4 = 16$ is false since its first component is false but its second is true.

Example 1.15. The proposition $\sqrt[4]{8} = 2$ *if and only if* $\sqrt{7} > 3$ is true since both components of the proposition are false.

Example 1.16. The proposition $\sqrt[4]{8} = 2$ *or* $8^{\frac{1}{3}} = 2$ *if and only if* $\sqrt{7} < 3$ *and* $\sqrt{3} < 2$ is true since both components of the proposition are true.

1.2.2.5 *The negation connective*

Unlike the other four logic connectives, the negation connective is a unary operation (as opposed to binary for the other four). It takes one proposition as its input and produces another at the input. If p is a logic statement, then the statement *"not p"* is called the *negation of p* that we denote with the symbol $\neg p$. The negation of p has the opposite truth value than p. The truth table of the negation connective is the following.

p	$\neg p$
T	F
F	T

Example 1.17. If p is the proposition $\sqrt{7} < 3$, then $\neg p$ is the proposition $\sqrt{7} \geq 3$.

Example 1.18. The negation of the proposition *There exists a real number x that satisfies* $x^2 + 1 = 0$ (which is false) is *No real number satisfies* $x^2 + 1 = 0$ (which is true).

1.2.2.6 *Summary*

Table 1.2 gives a summary of the truth tables of the five basic logic connectives. Propositional logic calculus is mainly based on this table.

Table 1.2   Basic logic connectives

p	q	$(p \wedge q)$	$(p \vee q)$	$(p \rightarrow q)$	$(p \leftrightarrow q)$	$\neg p$	$\neg q$
T	T	T	T	T	T	F	F
T	F	F	T	F	F	F	T
F	T	F	T	T	F	T	F
F	F	F	F	T	T	T	T

Example 1.19. In each case, determine the truth value of the compound proposition.

(a) 23 is odd or the equation $\ln(2x+3)+e^{x+2} = 37$ has an integer solution.
(b) If $1 + 2 = 4$, then pigs can fly.
(c) $3^{123456} + 1$ is a prime number and the equation $x^2 + 1 = 0$ has at least one real root.
(d) The equation $x^2 + 1 = 0$ has at least one real root if and only if pigs can fly.
(e) If $1 - 2 = -1$, then $2^{12456790} + 3$ is even.

Solution. (a) The proposition is true since it is the disjunction of two propositions with the first one true. Note that it does not matter what the truth value of the second component is in this case.

(b) The proposition is true since it is an implication with a false hypothesis.

(c) The proposition is false since it is a conjunction with the second component (pigs can fly) is false. Note that it does not matter what the truth value of the first component is in this case.

(d) The proposition is true since it is a biconditional with both components have the same truth value (both are false in this case).

(e) The proposition is false since it is an implication of the form $(\mathbf{T} \rightarrow \mathbf{F})$.

$\Diamond$

Using Table 1.2, we can determine the truth value of any complex logic proposition as a function of the truth values of its components.

Example 1.20. Construct the truth table of the compound proposition $\phi : ((p \wedge \neg q) \vee \neg (q \rightarrow p))$ where p and q are logic propositions. In particular give all truth values of p and q for which ϕ is true.

Solution. The proposition ϕ is the disjunction of the two propositions $(p \wedge \neg q)$ and $\neg(q \rightarrow p)$. First, we determine the truth value of $(p \wedge \neg q)$ and $\neg(q \rightarrow p)$ and then we use the truth table for the disjunction connective. All the steps are included in the following table.

p	q	$\neg q$	$(q \to p)$	$\neg(q \to p)$	$(p \wedge \neg q)$	$((p \wedge \neg q) \vee \neg(q \to p))$
T	**T**	**F**	**T**	**F**	**F**	**F**
T	**F**	**T**	**T**	**F**	**T**	**T**
F	**T**	**F**	**F**	**T**	**F**	**T**
F	**F**	**T**	**T**	**F**	**F**	**F**

In particular, ϕ is true when p and q have opposite truth values. $\Diamond$

1.2.3 *Operations on binary strings*

Information is usually processed by computers using *bit strings*: a sequence of 0's and 1's. This type of sequences plays a central role in mathematics and computer science. We give a formal definition of binary strings as we need to refer to them in many places in this book.

Definition 1.2. A binary string is an (ordered) sequence of 0's and 1's, each of which is called a *bit* (short for binary digit). The *length* of a binary string is the number of bits in the sequence. A string of length zero is called the *empty string* (the string with no bits in it), that we denote by λ.

Because of their binary nature, bits can be used to represent truth values of logic propositions with 1 and 0 representing true and false, respectively. Logic connectives can then be applied to 1 and 0 the same way they apply to **T** and **F**. From a computer science perspective, connectives $\wedge$, $\vee$ and $\oplus$ are of particular importance. In this context, $\wedge$, $\vee$ and $\oplus$ are referred to as the AND, OR and XOR operators, respectively. Actions of these operators on *Boolean variables* p and q (these are variables which take values over the set $\{0, 1\}$) are summarized in Table 1.3 below. The bit operations in Table 1.3 can be generalized to operations on binary strings having the same length. If $s_1 = a_1 a_2 \cdots a_n$ and $s_2 = b_1 b_2 \cdots b_n$ are two binary strings of the same length, then $s_1 \wedge s_2$ is the string $c_1 c_2 \cdots c_n$ where $c_i = a_i \wedge b_i$ for each i. Strings $s_1 \vee s_2$ and $s_1 \oplus s_2$ are defined similarly.

Example 1.21. Let $s_1 = 110100111$ and $s_2 = 010100100$. Then $s_1 \wedge s_2 = 010100100$, $s_1 \vee s_2 = 110100111$ and $s_1 \oplus s_2 = 100000011$.

Another important operation on binary strings is the *concatenation*. Given two binary strings $s_1 = a_1 a_2 \cdots a_n$ and $s_2 = b_1 b_2 \cdots b_n$, the concatenation of s_1 and s_2 is the string $s_1 s_2 = a_1 a_2 \cdots a_n b_1 b_2 \cdots b_n$. In particular, if n is a natural number and s is a binary string, then s^n is the string obtained by concatenating n copies of s. In this definition, it is understood that s^0 is the empty string λ.

Example 1.22. Let $s = 11010$ and $t = 1010100100$. Then $st = 110101010100100$, $s^2 = 1101011010$ and $s^3 = 110101101011010$.

Table 1.3   AND, OR and XOR

p	q	$p \wedge q$	$p \vee q$	$p \oplus q$
1	1	1	1	0
1	0	0	1	1
0	1	0	1	1
0	0	0	0	0

1.2.4   *Exercises*

(1) In each case, determine if the expression is a logic proposition.

    (a) Is it raining now?
    ⋆(b) The sum of two even numbers is even.
    ⋆(c) $2^{1012} + 5$.
    ⋆(d) $2^{1012} + 5$ is odd.
    (e) Finish your assignment!
    (f) No prime number is even except 2.
    (g) $n^2 + 3n + 1$ is odd for some integer n.
    (h) A permutation of five objects.
    (i) There are 120 ways to permute five objects.
    (j) If Oxygen is a perfect gas, then so is Nitrogen.
    ⋆(k) An infinite number of odd numbers.

(2) In each case, the given expression is the start of a logic proposition. Complete the expression to form a: (i) true proposition (ii) false proposition. There are many possible answers for each part.

    ⋆(a) The sum of the squares of two integers.
    (b) $2^{1012} + 5$.
    (c) The product of two positive integers.
    (d) $(-64)^{\frac{1}{3}}$.
    ⋆(e) There exists a real number.
    (f) There exists no integer.

(3) In each case, a compound logic proposition is given. Determine the main logic connective of the statement. That is, write the proposition

under the form $(p \; \theta \; q)$ where p and q are two (possibly compound) propositions and θ is a logic connective.

(a) 3 is an odd and a prime number.

(b) Either $\sqrt{2}$ is a rational number, or $\sqrt{2}^{\sqrt{2}}$ is a rational number and a root of the equation $x^4 - 1 = 0$.

(c) If I fail the exam, then either I did not prepare or I was sick.

⋆(d) Joe fails the course if and only if he does not do the effort and the professor does not warn him.

(e) Joe is a smart and an honest person, or he can be really hard to deal with if and only if he senses dishonesty on your part.

(4) In each case, the statement is the disjunction of two propositions. Determine if the disjunction is used in an inclusive or an exclusive sense.

(a) You study or you fail.

⋆(b) I will either swim or take the boat to get to the island today.

(c) You can choose English or French to answer the question.

(d) You can either dial 0 or 1 on you phone to speak to a costumer service representative.

(e) Joe can speak either English or French.

(5) Determine the truth value of each of the following logic propositions.

(a) $2 - 3 = -1$ or $3^{1080} + 7$ is a multiple of 17.

(b) The equation $x^4 + 1 = 0$ has no real roots and $\sqrt{9} = 3$.

(c) If $2 - 3 = -1$, then $\sqrt{10} > 3$.

(d) If Washington is the capital city of the US and $2^3 = 8$, then $1 + 2 = 5$.

(e) If Boston is the capital city of the US and $2^3 = 8$, then $1 + 2 = 5$.

⋆(f) If 3345687609876651 is a prime number and every dog is a cat, then $\sqrt{9} = 3$.

⋆(g) $8^{23432} - 1$ is even if and only if the equation $x^3 - 4x^2 = 0$ has three distinct real roots.

(h) There is no integer n such $n^3 - 2 = 0$ if and only if $\sqrt[3]{64} = 8$ or $\sqrt[3]{7} > 2$.

(6) A and B are two atomic propositions such that the proposition $(B \rightarrow \neg A)$ is false. Determine the truth value of each of the following propositions.

(a) $(A \lor B)$ (f) $((A \lor B) \to (\neg A \land B))$
(b) $(A \land B)$ (g) $\neg(\neg A \to B)$
(c) $(\neg A \lor \neg B)$ (h) $(\neg A \leftrightarrow \neg B)$
$\star$(d) $(\neg(A \lor B) \to A)$ $\star$(i) $(\neg A \leftrightarrow (A \land B))$
(e) $((A \land B) \to B)$ $\star$(j) $(\neg B \leftrightarrow (\neg A \land B))$

$\star$(7) Give an example of an implication which is true but its converse is false.

(8) Each of the following statements is an implication. State: (i) the contrapositive and (ii) the converse of the implication.

   (a) If you snooze, you loose.

   $\star$(b) If a function is differentiable on the interval $[a, b]$, then it is continuous on that interval.

   (c) For an even integer n, $5n + 2$ is even.

   (d) The product of two even integers is even.

(9) Let ϕ and ψ be two propositions with the same set of logic variables. We say that ϕ and ψ have the same truth table if every valuation of the logic variables assigns the same truth value for ϕ and ψ. In each case, verify that the two compound propositions have the same truth table.

   (a) $(p \oplus q)$, $((p \lor q) \land \neg(p \land q))$ (c) $(p \to q)$, $(\neg p \lor q)$
   (b) $\neg(p \lor q)$, $(\neg p \land \neg q)$ $\star$(d) $(p \leftrightarrow q)$, $((p \to q) \land (q \to p))$

(10) Show that the logic formula $\phi : ((A \land \neg B) \lor \neg(B \to A)) \to (\neg A \lor \neg B)$ is true for any valuation of the variables A and B (we call such a proposition as we will see later).

(11) Consider the logic proposition $\phi : ((p \to \neg q) \lor \neg(q \to r))$ in three logic variables p, q and r. Construct the truth table of ϕ and give all valuations that satisfy it. Note that in this case, there are eight different valuations of ϕ.

(12) In each case, two binary strings s and t are given. Find: $s \land t$, $s \lor t$, $s \oplus t$ and st (the concatenation).

   (a) $s = 10$, $t = 11$ $\star$(c) $s = 110100$, $t = 110011$
   (b) $s = 1001$, $t = 1100$ (d) $s = 00011001$, $t = 10010111$

(13) Let s, t and v be three binary strings of the same length. In each case, determine if the statement is true or false. Justify your answer.

(a) $s \wedge t = t \wedge s$

(b) $s \vee t = t \vee s$

$\star$(c) $(s \wedge t) \wedge v = s \wedge (t \wedge v)$

(d) $(s \vee t) \vee v = s \vee (t \vee v)$

(e) $s \oplus (t \wedge v) = (s \oplus t) \wedge (s \oplus v)$

(f) $s \oplus (t \vee v) = (s \oplus t) \vee (s \oplus v)$

$\star$(g) $(s \oplus t)v = (s \oplus v)(t \oplus v)$

(h) $s\lambda = s$

1.3 Propositional logic: Formal point of view

The logic expression $P \rightarrow Q \vee R$ can be interpreted in two different ways. One can look at it as the disjunction of $P \rightarrow Q$ and R, or as an implication with P as hypothesis and $Q \vee R$ as conclusion. In other words, it is not clear what the main connective in the expression is. In order to determine the truth value of a logic proposition, the latter should present no ambiguity and its main connective should be clear. A proposition that is written with no ambiguity about its meaning is called a *well-formed logic formulas* or simply *formulas* of propositional logic, that we abbreviate with wff (plural: wffs). To formally define wffs, we start by fixing a set $\mathcal{A} = \{A, B, \ldots, A_1, B_1, \ldots\}$ of letters and indexed letters whose elements are called *propositional variables*. Then the set of wffs of propositional logic over $\mathcal{A}$ is defined recursively using the following rules.

(1) Every propositional variable is a wff that we call an *atomic formula*.

(2) If ϕ is a wff, then $\neg\phi$ is also a wff.

(3) If ϕ and ψ are wffs, then

(3.1) $(\phi \wedge \psi)$ is a wff.

(3.2) $(\phi \vee \psi)$ is a wff.

(3.3) $(\phi \rightarrow \psi)$ is a wff.

(3.4) $(\phi \leftrightarrow \psi)$ is a wff.

(4) The symbols **T** and **F** (True and False) are wffs.

(5) Nothing else is a wff. That is to say, **only** expressions that can be generated using the above rules are wffs.

Remark 1.2. Like the grammar of any natural language, one has to follow the above rules to the letter in forming a wff. For instance, the expression $\phi \rightarrow \psi$ does not constitute a wff since it cannot be generated following the above rules (parentheses are missing), even if there is no ambiguity in its meaning. There are, however, some instances where the rules can be relaxed a bit as we will see later. We also point out that it is sometimes convenient to use square brackets as outer denominators instead of round parentheses to improve readability of certain formulas. For example, one can write $[(\phi \to \psi) \vee (\rho \wedge \epsilon)]$ instead of $((\phi \to \psi) \vee (\rho \wedge \epsilon))$ without violating the above rules.

Example 1.23. The expression $\neg A$ is a wff by rules (1) and (2) above. The expressions $\neg(A)$ and $(\neg A)$, on the other hand, are not wffs since they cannot be constructed using the above rules.

Example 1.24. The following provides a step by step proof of the fact that the expression: $\neg((\neg A \vee B) \leftrightarrow \neg(A \to (C \wedge B)))$ is indeed a wff.

1. A, B and C are wffs: Rule (1).
2. $\neg A$ is a wff: Line 1 and Rule (2).
3. $(\neg A \vee B)$ is a wff: Lines 1, 2 and Rule (3.2).
4. $(C \wedge B)$ is a wff: Line 1 and Rule (3.1).
5. $(A \to (C \wedge B))$ is a wff: Lines 1, 4 and Rule (3.3).
6. $(A \to (C \wedge B))$ is a wff: Line 5 and Rule (2).
7. $((\neg A \vee B) \leftrightarrow \neg(A \to (C \wedge B)))$ is a wff: Lines 3, 6 and Rule (3.4).
8. $\neg((\neg A \vee B) \leftrightarrow \neg(A \to (C \wedge B)))$ is a wff: Line 7 and Rule (2).

Note that if ϕ is a wff over the set $\mathcal{A}$ of logic variables and $\mathcal{A}$ is a subset of $\mathcal{B}$, then ϕ is also a wff over the set $\mathcal{B}$. For example, the expression $\phi : ((\neg A \vee B) \leftrightarrow \neg(A \to (C \wedge B)))$ is a wff over the set $\mathcal{A} = \{A, B, C\}$ but it can also be considered as a wff over the set $\mathcal{B} = \{A, B, C, D, X, Y, Z, T\}$ containing $\mathcal{A}$.

1.3.1 *Valuations and truth tables of logic formulas*

We have seen already how to construct the truth table of a logic proposition in the first section. In this section, we use a more formal approach.

Definition 1.3. Given a set $\mathcal{E} = \{e_1, \ldots, e_n\}$ of logic variables, a *valuation* of $\mathcal{E}$ is a list $(v_1, \ldots, v_n)$ where $v_i = v(e_i)$ is a truth value for the logic variable e_i for $i = 1, 2, \ldots n\}$.

Since there are only two choices for every truth value, there are 2^n different valuations for a set $\mathcal{E}$ containing n propositional variables. The formal proof of this fact will be given in Chapter 8.

Example 1.25. If $\mathcal{E} = \{A, B, C\}$, then there are $2^3 = 8$ possible valuations of the set $\mathcal{E}$, namely $(\mathbf{T}, \mathbf{T}, \mathbf{T})$, $(\mathbf{T}, \mathbf{T}, \mathbf{F})$, $(\mathbf{T}, \mathbf{F}, \mathbf{T})$, $(\mathbf{T}, \mathbf{F}, \mathbf{F})$, $(\mathbf{F}, \mathbf{T}, \mathbf{T})$, $(\mathbf{F}, \mathbf{T}, \mathbf{F})$, $(\mathbf{F}, \mathbf{F}, \mathbf{T})$, and $(\mathbf{F}, \mathbf{F}, \mathbf{F})$.

Given a set $\mathcal{E}$ of logic variables and a wff ϕ over $\mathcal{E}$, every valuation α of $\mathcal{E}$ determines a unique truth value for ϕ. For example, the valuation $\alpha = (\mathbf{T}, \mathbf{F}, \mathbf{F})$ of the set $\mathcal{E} = \{A, B, C\}$ assigns the truth value $\mathbf{T}$ for the formula $\phi : (C \leftrightarrow (A \wedge B))$ (since in this case, ϕ has the form $\mathbf{F} \leftrightarrow \mathbf{F}$ which is true). The valuation $(\mathbf{T}, \mathbf{F}, \mathbf{T})$, on the other hand, assigns $\mathbf{F}$ as a truth value for ϕ (for this valuation, ϕ has the form $\mathbf{T} \leftrightarrow \mathbf{F}$ which is false). We say that the valuation α *satisfies* the formula ϕ if the truth value of ϕ determined by α is $\mathbf{T}$. Hence, the valuation $\alpha = (\mathbf{T}, \mathbf{F}, \mathbf{F})$ satisfies the formula $\phi : (C \leftrightarrow (A \wedge B))$ while $(\mathbf{T}, \mathbf{F}, \mathbf{T})$ does not. The truth table of ϕ is the listing of all truth values of ϕ, in a tabular form, in relation with all possible valuations of $\mathcal{E}$.

Example 1.26. The truth table of the formula $\phi : (C \leftrightarrow (A \wedge B))$ is given below. It is important to notice that in the table on the left, the column of $(A \wedge B)$ is not really a part of the truth table of ϕ. It is included to give us a better understanding how the last column is formed. The truth table of ϕ can be restricted to the table on the right. From the table, we see that the valuations that satisfy ϕ are: $(\mathbf{T}, \mathbf{T}, \mathbf{T})$, $(\mathbf{T}, \mathbf{F}, \mathbf{F})$, $(\mathbf{F}, \mathbf{T}, \mathbf{F})$ and $(\mathbf{F}, \mathbf{F}, \mathbf{F})$.

A	B	C	$(A \wedge B)$	$\phi : (C \leftrightarrow (A \wedge B))$
T	T	T	T	T
T	T	F	T	F
T	F	T	F	F
T	F	F	F	T
F	T	T	F	F
F	T	F	F	T
F	F	T	F	F
F	F	F	F	T

A	B	C	ϕ
T	T	T	T
T	T	F	F
T	F	T	F
T	F	F	T
F	T	T	F
F	T	F	T
F	F	T	F
F	F	F	T

Example 1.27. For the formula $\psi : (((\neg C \vee A) \to B) \to (A \wedge \neg C))$, the truth table is given below. Each component of the formula is analyzed separately.

A	B	C	$\neg C$	$(\neg C \vee A)$	$((\neg C \vee A) \to B)$	$(A \wedge \neg C)$	ψ
T	T	T	F	T	T	F	F
T	T	F	T	T	T	T	T
T	F	T	F	T	F	F	T
T	F	F	T	T	F	T	T
F	T	T	F	F	T	F	F
F	T	F	T	T	T	F	F
F	F	T	F	F	T	F	F
F	F	F	T	T	F	F	T

Once again, the column of $\neg C$, $(\neg C \vee A)$, $((\neg C \vee A) \to B)$ and $(A \wedge \neg C)$ are included to give us a better understanding of how the last column is formed. The valuations that satisfy ψ are: $(\mathbf{T}, \mathbf{T}, \mathbf{F})$, $(\mathbf{T}, \mathbf{F}, \mathbf{T})$, $(\mathbf{T}, \mathbf{F}, \mathbf{F})$ and $(\mathbf{F}, \mathbf{F}, \mathbf{F})$.

1.3.2 *Special types of logic formulas and coherency*

Let ϕ be a wff of propositional logic. If ϕ is true then $(\phi \vee \neg\phi)$ is true. If ϕ is false, then $\neg\phi$ is true and so $(\phi \vee \neg\phi)$ is true. This means that $(\phi \vee \neg\phi)$ is true regardless of what the truth value of ϕ is. On the other extreme, we can prove similarly that $(\phi \wedge \neg\phi)$ is false regardless of what the truth value of ϕ is. Formulas which are always true (for any valuation of their logic variables) play a central role in propositional logic. This type of formulas is directly linked to other concepts in logic, like the validity of arguments and the notion of logic equivalence.

Definition 1.4. We say that the logic formula ϕ is a *tautology* if its truth value is $\mathbf{T}$ for any valuation of its logic variables. We say that ϕ is a *contradiction* if its truth value is $\mathbf{F}$ for any valuation of its logic variables. If ϕ is neither a tautology nor a contradiction, we say that it is a *contingency* (or a *contingent formula*). The truth table of a tautology consists of $\mathbf{T}$'s only, that of a contradiction consists of $\mathbf{F}$'s only and the truth table of a contingency has both $\mathbf{T}$'s and $\mathbf{F}$'s in it.

Example 1.28. In each case, use a truth table to determine if the formula is a tautology, a contradiction or a contingency.

(a) $\phi : ((P \vee (Q \wedge R)) \to ((P \vee Q) \wedge (P \vee R)))$
(b) $\psi : (((\neg C \vee A) \to B) \to (A \wedge \neg C))$
(c) $\eta : \neg((A \vee B) \vee (\neg A \wedge \neg B))$

Solution. (a) The truth table of ϕ is the following:

P	Q	R	$(Q \wedge R)$	$(P \vee (Q \wedge R))$	$(P \vee Q)$	$(P \vee R)$	$((P \vee Q) \wedge (P \vee R))$	ϕ
T	T	T	T	T	T	T	T	T
T	T	F	F	T	T	T	T	T
T	F	T	F	T	T	T	T	T
T	F	F	F	T	T	T	T	T
F	T	T	T	T	T	T	T	T
F	T	F	F	F	T	F	F	T
F	F	T	F	F	F	T	F	T
F	F	F	F	F	F	F	F	T

The formula is a tautology.

(b) The truth table of ψ was constructed in Example 1.27 above. The table contains both **T**'s and **F**'s. The formula is a contingency.

(c) The truth table of η is the following.

A	B	$\neg A$	$\neg B$	$(A \vee B)$	$(\neg A \wedge \neg B)$	$((A \vee B) \vee (\neg A \wedge \neg B))$	η
T	**T**	**F**	**F**	**T**	**F**	**T**	**F**
T	**F**	**F**	**T**	**T**	**F**	**T**	**F**
F	**T**	**T**	**F**	**T**	**F**	**T**	**F**
F	**F**	**T**	**T**	**F**	**T**	**T**	**F**

The table consists of **F**'s only, η is a contradiction. $\Diamond$

Definition 1.5. A set $\mathcal{E}$ of logic formulas over a set $\mathcal{F}$ of logic variables is called *coherent* (or *Consistent*) if there exists at least one valuation of $\mathcal{F}$ that satisfies all formulas in $\mathcal{E}$. The set $\mathcal{E}$ is called *incoherent* otherwise.

Example 1.29. The set $\mathcal{E} = \{(A \vee B), (A \to B), (\neg A \wedge B)\}$ is coherent. The valuation $(\mathbf{F}, \mathbf{T})$ satisfies all the formulas in $\mathcal{E}$ as shown in the following truth table of the formulas in $\mathcal{E}$.

A	B	$\neg A$	$(A \vee B)$	$(A \to B)$	$(\neg A \wedge B)$
T	**T**	**F**	**T**	**T**	**F**
T	**F**	**F**	**T**	**F**	**F**
F	**T**	**T**	**T**	**T**	**T**
F	**F**	**T**	**F**	**T**	**F**

Example 1.30. The set $\mathcal{E} = \{(A \wedge B), (A \to B), (\neg A \wedge B)\}$ is incoherent. From the truth table of each of the formulas in $\mathcal{E}$ given below, we see that none of the four possible valuations of $\{A, B\}$ satisfies all the formulas in $\mathcal{E}$ at the same time.

A	B	$\neg A$	$(A \wedge B)$	$(A \to B)$	$(\neg A \wedge B)$
T	**T**	**F**	**T**	**T**	**F**
T	**F**	**F**	**F**	**F**	**F**
F	**T**	**T**	**F**	**T**	**T**
F	**F**	**T**	**F**	**T**	**F**

1.3.3 *Exercises*

(1) In each case, determine if the expression is a wff of propositional logic. If you say that the expression is a wff, determine its main logic connective.

(a) $\neg\neg\neg A$

$\star$(b) $(\neg(\neg A \wedge B) \to (A \leftrightarrow (\neg C \vee B)))$

   (c) $(A \vee B) \rightarrow \neg(C \rightarrow \neg B)$
   (d) $\neg((A \vee \neg(\neg B \rightarrow C)) \leftrightarrow ((\neg A \wedge B) \vee (C \rightarrow \neg B)))$
 $\star$(e) $(((A \rightarrow B \wedge C) \rightarrow \neg(A \vee B))$
   (f) $(((A \rightarrow (B \rightarrow C)) \leftrightarrow ((A \rightarrow B) \rightarrow C)) \rightarrow (A \rightarrow C))$

(2) Prove that the following expression ϕ is a wff:

$$((A \vee \neg(\neg B \rightarrow D)) \rightarrow ((\neg A \wedge B) \vee (C \rightarrow \neg E))).$$

How many rows are there in the truth table of ϕ?

(3) In each case, form the truth table of the logic formula. Determine if the formula is a tautology, a contradiction or a contingency. If you say that the formula is a contingency, give all valuations that satisfy the formula.

   (a) $((A \rightarrow B) \rightarrow (A \wedge \neg B))$
   (b) $((A \rightarrow \neg B) \leftrightarrow (\neg C \rightarrow (A \rightarrow B)))$
 $\star$(c) $((P \rightarrow (Q \rightarrow R)) \rightarrow ((P \rightarrow Q) \rightarrow (P \rightarrow R)))$
   (d) $((P \rightarrow Q) \rightarrow R)) \rightarrow (P \rightarrow (Q \rightarrow (P \rightarrow R)))$
   (e) $((A \vee B) \rightarrow \neg(C \rightarrow \neg B))$
   (f) $((A \vee \neg(\neg B \rightarrow C)) \leftrightarrow ((\neg A \wedge B) \vee (C \rightarrow \neg B)))$
 $\star$(g) $((A \vee B) \leftrightarrow \neg(\neg B \rightarrow A))$

(4) In each case, determine if the given set of logic formulas is coherent. If you say it is, give at least one valuation of the logic variables that satisfies all the formulas in the set.

   (a) $\{\neg(A \rightarrow \neg B), (\neg A \vee B), \neg(A \vee \neg B)\}$
   (b) $\{(\neg A \rightarrow B), (\neg B \rightarrow \neg A), (\neg A \wedge B)\}$
 $\star$(c) $\{\neg(Q \wedge R), ((\neg P \rightarrow Q) \wedge (\neg R \rightarrow P)), (P \vee \neg R), (P \rightarrow (Q \rightarrow \neg R))\}$

(5) Consider the valuation $\alpha : (v(A) = \mathbf{T}, v(B) = \mathbf{T}, v(C) = \mathbf{F}, v(D) = \mathbf{F})$ of the set $\{A, B, C, D\}$ of logic variables. In each case, determine if α satisfies the given formula.

   (a) $(\neg A \rightarrow (B \leftrightarrow (C \vee D)))$
 $\star$(b) $(((A \wedge B) \rightarrow \neg C) \rightarrow (A \wedge (C \vee D)))$
   (c) $((C \vee \neg D) \leftrightarrow \neg B) \vee (B \rightarrow \neg A))$
 $\star$(d) $(((A \rightarrow B) \rightarrow C) \rightarrow D)$
   (e) $(A \rightarrow ((B \rightarrow C) \rightarrow D))$

(6) Consider the wff $\phi : (D \vee (B \wedge \neg C)) \wedge (B \rightarrow (A \rightarrow C) \wedge \neg D)))$. If $\alpha = (v(A).v(B), v(C), v(D))$ is a valuation of the set $\{A, B, C, D\}$ that satisfies ϕ and such that $v(B) = \mathbf{T}$, determine the values of $v(A)$, $v(C)$ and $v(D)$.

1.4 Logic formulas in natural language: Translation between English and propositional logic

Automated translation softwares have been around for many years now and have came a long way in achieving good quality translations. Still, you have probably noticed how silly some translated sentences could sound using a computer software. The task of capturing the exact meaning in two different languages is certainly not an easy one, even for professional translators. The same is true when it comes to translation between natural languages and the language of propositional logic. The ability of manipulating formal logic formulas and laws is very important but has little value if one cannot express a natural language statement into a formal expression of propositional logic and vice a versa. In most cases, the difficulty of translating a natural language text into logic arises from the complexity of the text itself and the way it is written. For example, saying: *"Failing a course is a necessary condition for not studying"* sounds much harder to understand than saying :*"If you don't study, you fail"*, although the two sentences say the same thing in two different ways.

Table 1.1 above gives one way of expressing each of the basic logic connectives in English. There are, however, many other expressions in English that capture the same meaning for these connectives. In what follows, we explore the most common ways of expressing logic connectives in English. This is key in analysing statements of propositional logic.

- <u>Conjunction</u> $((p \wedge q))$: p and q; p but q; p even though q; p, moreover q; p although q.
- <u>Disjunction</u> $((p \vee q))$: p or q; p unless q; either p or q; p, otherwise q; p except if q.
- <u>Implication</u> $((p \rightarrow q))$: p implies q; If p, then q; If p, q; q if p; p only if q; q is a necessary condition for p; p is a sufficient condition for q; For p, it is necessary that q; For q, it is sufficient that p; For q, it suffices that p.
- <u>Biconditional</u> $((p \leftrightarrow q))$: p if and only if q; p is equivalent to q; p is a necessary and a sufficient condition for q; for q, it is necessary and sufficient that p.

Example 1.31. Translate each of the following sentences into a wff of propositional logic using the following atoms:

- J: Joe passes the course
- C: Joe attends all classes
- N: Joe takes notes in class
- E: Joe does well on the exam

(a) To pass the course, it is necessary that Joe attends all classes and that he takes notes in class.

(b) Doing well on the exam is sufficient for Joe to pass the course.

(c) Neither attending all classes nor taking notes in class guarantees that Joe passes the course.

(d) Joe does not do well on the exam if and only if he skips some classes unless or does not take notes in class.

(e) Joe passes the course only if he does well on the exam or he does not take notes in class.

Solution. (a) The sentence has the form "For p, it is necessary that q" which translates to $(p \rightarrow q)$ in propositional logic language. Note that the conjunction is main connective of the conclusion q. The translation of the sentence is $(J \rightarrow (C \wedge N))$.

(b) The sentence is of the form "E is sufficient for J" which translates as $(E \rightarrow J)$ in propositional logic.

(c) the sentence can be rephrased as follows: *It is not true that attending all classes implies that Joe will pass course and it is not true that taking notes in class implies that Joe will pass the course.* The translation into a wff of propositional logic is: $(\neg(C \rightarrow J) \wedge \neg(N \rightarrow J))$.

(d) This is a biconditional with a disjunction as the main connective in the second component: $(\neg E \leftrightarrow (\neg C \vee \neg N))$.

(e) This is a implication (only if connective) where the conclusion is a disjunction of two propositions: $(J \rightarrow (E \vee \neg N))$. $\Diamond$

Example 1.32. Translate the following sentence using symbols of propositional logic: *If Joe goes to class only if he does not go to work or Fred is working, then he will not pass the course unless he does not go to work and Fred does not go to class.*

Solution. We identify the following atomic propositions in the sentence:

- J: Joe goes to class
- W: Joe goes to work
- F: Fred goes to work
- C: Joe will pass the course
- D: Fred goes to class

A quick look at the statement shows that it is of the form "If p, then q" with:

- p: *Joe goes to class only if he does not go to work or Fred is working;*
- q: *Joe will not pass the course unless he does not go to work and Fred does not go to class.*

Each of p and q is a compound proposition on its own and it can be formed using the above atomic sentences as follows: $p : (J \to (\neg W \vee F))$ and $q : (\neg C \vee (\neg W \wedge \neg D))$. We conclude that the translation of the sentence is the wff: $((J \to (\neg W \vee F)) \to (\neg C \vee (\neg W \wedge \neg D)))$. $\diamond$

1.4.1   *Logic arguments*

Consider the following paragraph:

"If the arms race continues, the world is heading to a war. Joe is elected as a president only if the arms race continues. Joe will never be elected unless the steel industry endorses him. The steel industry endorses Joe for presidency if and only if the arms race continues and the world is heading to a war. Therefore, the world is heading to a war."

The paragraph sounds like a statement made by some political opponent of Joe to make the case why he should not be elected as a president. The paragraph draws a conclusion based on several facts (or premises). From the propositional logic perspective, the above paragraph is called an *argument*. We all make arguments almost on a daily basis to push an opinion forward or to persuade others about an opinion. The more "coherent" the argument, the more convincing it is. Propositional logic gives us a way to assess and measure an argument in terms of its *validity* that we will introduce later.

Definition 1.6. An *argument* of propositional logic is a list of the following form:

$$
\begin{array}{c}
\phi_1 \\
\vdots \\
\phi_n \\
\hline
\therefore \psi
\end{array}
$$

where $\phi_1, \phi_2, \ldots, \phi_n$ and ψ are logic formulas. The formulas $\phi_1, \phi_2, \ldots, \phi_n$ are called the *premises* and the formula ψ is called the *conclusion* of the argument. A small horizontal line separates the premises from the conclusion of an argument and the symbol $\therefore$ stands for *"therefore"*.

In most cases, logic arguments are given in natural languages sentences which makes them hard to analyze. As we will see later, the first step to determine the validity of an argument is to break it down to the form given in Definition 1.6 above.

Example 1.33. Write the logic argument given at the beginning of this section in standard form given in Definition 1.6.

Solution. The following propositions are the atoms in the argument:

- A: The arms race continues
- B: The world is heading to a war
- C: Joe is elected as a president
- D: The steel industry endorses Joe

The first premise translates as $(A \to B)$, the second as $(C \to A)$, the third is $(\neg C \lor D)$ and the fourth as $(D \leftrightarrow (A \land B))$. The conclusion is simply B. The translation of the argument into logic symbols is:

$$
\begin{array}{c}
(A \to B) \\
(C \to A) \\
(\neg C \lor D) \\
\underline{(D \leftrightarrow (A \land B))} \\
\therefore B
\end{array}
$$

1.4.2   *Exercises*

(1) Translate the following statement into a formula of propositional logic. Start by identifying the atomic components of the statement: *"If the European union wants to standardize its economy, then it must follow the French or the Swedish model but not both at the same time."*

(2) In each case, translate the given sentence into a wff of propositional logic using the following atoms.

- P: *The wind is blowing*
- Q: *The temperature drops below the freezing degree*
- R: *A snow storm is coming*

  (a) If the wind is blowing or the temperature drops below the freezing degree, then a snow storm is coming.

   ⋆(b) A snow storm is coming unless the wind is not blowing and the temperature drops below the freezing degree.

   (c) Neither the drop of the temperature below the freezing degree nor the fact that wind is blowing is a sufficient condition for a snow storm.

   (d) If the wind is not blowing only if the temperature drops below the freezing degree, then no snow storm is coming.

(3) Use the atoms defined in Exercise (2) to give an English translation to each of the following wffs.

   (a) $((\neg R \rightarrow P) \vee \neg Q)$ ⋆(c) $((R \rightarrow Q) \leftrightarrow \neg P)$

   (b) $(R \rightarrow (P \wedge \neg Q))$ (d) $(R \rightarrow (\neg P \wedge Q))$

(4) Translate each of the following statements into a wff of propositional logic. Start by specifying the atoms you use.

   (a) Solving suggested exercises is a necessary condition for Joe to pass his discrete math exam.

   (b) Winning the lottery is not a sufficient condition for Joe to lead a happy life.

   ⋆(c) Joe wears his hat only if he enters the construction site or he changes the roof, but he does not wear his safety glasses on site.

   (d) Joe buys the house if and only if the flooring is changed and plumping is fixed.

(5) Consider the following atomic propositions:

- P: Joe buys a new car
- Q: Joe gets the job
- R: Joe will goes on a trip
- S: The exchange rate is not favorable

Give an English translation of the following argument.

$$((P \rightarrow Q) \rightarrow R)$$
$$(R \vee S)$$
$$\neg S$$
$$\underline{(R \wedge \neg S)}$$
$$\therefore (\neg S \rightarrow \neg P)$$

⋆(6) Translate the following argument into symbols of propositional logic. Start by identifying the atomic propositions in the argument. *"If there is a power outage, then the emergency lights turn on. The emergency*

*exist is clear only if there is no power outage or the emergency lights are
turned on. A clear emergency exist is a necessary but not a sufficient
condition for all students to leave the auditorium safely. All students
leave the auditorium safely unless the emergency lights do not turn on.
Therefore, the emergency exist is clear or there is no power outage."*

⋆(7) Translate the following argument into symbols of propositional logic.
Start by identifying the atomic propositions in the argument. *"If the
user does not enter a valid password, then no access is granted unless
the user pays a fee. The user account is activated and the user is granted
access only if a fee is paid by the user except if the user account is locked.
A fee is paid by the user if and only if the user does not enter a valid
password and the user account is not locked. Therefore, paying a fee is
a necessary condition for the user to enter a valid password."*

1.5   The island of knights and knaves

In his book *"What Is the Name of This Book?"*, Raymond Smullyan (an
American mathematician and logician) proposed a series of logic puzzles
where the action takes place on a fictional island he called the *Island of
Knights and Knaves*. On that island live two tribes of inhabitants. The
Knights who always tell the truth (every sentence pronounced by a knight
is true) and the Knaves who always lie (every sentence pronounced by a
knight is false). It is assumed that every inhabitant of the island is either
a knight or a knave and that the two types are indistinguishable by sight.
While visiting the island, you meet some inhabitants who give you some
information. Your task is to analyze the statements of the inhabitants from
the propositional logic perspective and try to come up with answers to some
questions, like for instance what is the nature of each inhabitant you meet.
Simple knights and knaves puzzles can be analyzed using truth tables, but
for more complicated ones require careful case analysis. We explain this
technique in the following examples.

Example 1.34. On the island of knights and knaves you meet two
inhabitants, A and B. Inhabitant A says: *"If I am a knight, then so is
B"*. Can you determine the nature of each of the two inhabitants?

Solution. Let us start by using the method of truth table. Define the two
propositions: p: " A is a Knight" and q: " B is a Knight". Then inhabitant
A says ($p \to q$). Since every Knight speaks true, the logic formulas p and

$(p \to q)$ must have the same truth value. From the truth table below, we see that this happens only when both p and q are true, which means when both inhabitants are knights.

p	q	$(p \to q)$
T	T	T
T	F	F
F	T	T
F	F	T

Another way to solve this puzzle is to consider all possible cases and to eliminate cases that lead to a contradiction.

- **Case 1.** <u>A is a Knave.</u> In this case A lies. So the implication $(p \to q)$ is false which is only possible when p is true and q is false. So A is a knight and B is a knave, This leads to a contradiction since we assumed in this case that A is a knave.
- **Case 2.** <u>A is a Knight.</u> In this case, A speaks the truth and so $(p \to q)$ is true. Since p is true in this case (A is a knight), we conclude that q is true and so B is a knight as well. This case leads to no contradiction so it is the only possible scenario. $\Diamond$

Example 1.35. On the island of knights and knaves you meet two inhabitants, A and B. Suppose A says: *"We are both knights"* and B says *"Exactly one of us is a knight."* What can be concluded from these statements?

Solution. Define the two propositions: p: "A is a knight" and q: "B is a knight". Then inhabitant A says $(p \wedge q)$ and inhabitant B says $((p \vee q) \wedge \neg(p \wedge q))$. As in the previous example, the logic formulas p and $(p \wedge q)$ must have the same truth value and the same is true for the formulas q and $((p \vee q) \wedge \neg(p \wedge q))$. From the truth table below, we see that this happens for the two valuations $(\mathbf{F}, \mathbf{T})$ and $(\mathbf{F}, \mathbf{F})$. This means that A is a knave but B can be either a knight or a knave. Hence we cannot determine the nature of inhabitant B.

p	q	$(p \vee q)$	$(p \wedge q)$	$\neg(p \wedge q)$	$((p \vee q) \wedge \neg(p \wedge q))$
T	T	T	T	F	F
T	F	T	F	T	T
F	T	T	F	T	T
F	F	F	F	F	F

We now proceed by case analysis.

- **Case 1.** <u>*A* is a knight.</u> In this case *A* speaks the truth. So both *A* and *B* are knights. In particular, *B* speaks true which means that exactly one of the two inhabitants is a knight (as *B* says). This is a contradiction. So this case is not possible.
- **Case 2.** <u>*A* is a knave.</u> In this case, *A* lies and $(p \wedge q)$ is false. There are two possible subcases.

  - **Case 2.1.** <u>*B* is a knight.</u> There is no contradiction since *B*'s statement is true in this case.
  - **Case 2.2.** <u>*B* is a knave.</u> There is no contradiction since *B*'s statement is false in this case.

We conclude, like before, that *A* is a knave but we cannot determine the nature of *B*. $\qquad \Diamond$

Example 1.36. On the island of knights and knaves you meet a person *A* who says, "*If I am a knight, then $\sqrt{5} = 2$.*" What can you conclude?

Solution. Note first that *A* speaks the implication "I am a knight, then $\sqrt{5} = 2$". Note also that the statement "$\sqrt{5} = 2$" is False. There are two possible cases:

- **Case 1.** <u>*A* is a Knight.</u> In this case *A* speaks the truth. So the implication "*If I am a knight, then $\sqrt{5} = 2$*" is true. Since the conclusion "$\sqrt{5} = 2$" is false, the hypothesis "*I am a knight*" must be also false. This is a contradiction.
- **Case 2.** <u>*A* is a Knave.</u> In this case, *A* lies and the implication "*If I am a knight, then $\sqrt{5} = 2$*" is false. This is only possible if "*I am a knight*" is true and "$\sqrt{5} = 2$" is false. This leads also to a contradiction.

We conclude that the person you met is actually not an inhabitant of the island. $\qquad \Diamond$

Example 1.37. On the island of knights and knaves you meet three inhabitants, *A*, *B* and *C*. *A* says: "*All of us are knaves*", *B* says: "*Exactly one of us is a knight.*" *C* remains silent. Can you determine the nature of each of the three inhabitants?

Solution. Consider the two cases:

- **Case 1.** <u>*A* is a Knight.</u> In this case the proposition "*All of us are knaves*" is true. So A is a knave. This is a contradiction. This case is not possible.

- **Case 2.** <u>A is a Knave.</u> In this case, A lies and so either B or C must be a knight. We consider two subcases.

  - **Case 2.1.** <u>B is a Knight.</u> In this case B speaks the truth and C must be a knave since exactly one of the three must be a knight. No contradiction arises in this case.
  - **Case 2.2.** <u>B is a Knave.</u> In this case B lies and so C must be a knave since otherwise the statement (pronounced by B) "*Exactly one of us is a knight*" would be true. But then this means that all of them are knaves and A speaks the truth. This is a contradiction since A is a knave. This case is not possible.

We conclude that A and C are knaves, B is knight. $\Diamond$

1.5.1 *Exercises*

In the following exercises, the word "inhabitant" refers to an inhabitant of the island of knights and knaves. THe word "nature" refers to the type of the inhabitant: a knight or a knave. The exercises below are inspired by the work of Raymond Smullyan in the book "*What Is the Name of This Book?*".

$\star$(1) Give a proposition that any inhabitant can say.

(2) Give a proposition that no inhabitant can say.

(3) You meet two inhabitants A and B. B says: "*A is a knave*". Prove that A and B cannot be of the same nature. Is it possible to determine the nature of A and B?

$\star$(4) You meet two inhabitants A and B. A says: "*B is a knight but I am not.*" Can you determine the nature of each of the two inhabitants?

(5) You ask an inhabitant A: "*Is this highway 417 West?*" He replies, "*It is unless I am a knave*". Prove that it is highway 417 West.

(6) Prove that no inhabitant can say: "*I am a knight only if $\sqrt{5} \leq 2$.*"

$\star$(7) You meet three inhabitants A, B and C. A says: "*B is a knave*", B says: "*A and C are of the same nature.*" Can you tell if C is a knight or a knave?

(8) You ask an inhabitant A whether there is gold on the island. A answers with the following statement: "*There is gold on this island if and only if I am a knight.*" Can you determine whether A is a knight or a knave? Can you determine if there is gold on the island?

(9) A trial is taking place on the island of knights and knaves. The accused are the two visitors Joe and Annie of the island. The court heard from two eye witnesses A and B who are both inhabitants of the island. A says: *"If Joe is guilty, then so is Annie."* B says: *"Either Joe is innocent or A is a knight".* Show that A and B have the same nature.

(10) You are on the island of knights and knaves looking for a hotel. You meet two inhabitants A and B and you ask them the if there is a hotel on the island. A answers: *"B is a knight and there is no hotel on this island."* B answers: *"There is no hotel on this island only if A is a knave."* Can you determine the nature of each of the two inhabitants? Is there a hotel on the island?

(11) Two inhabitants A and B of the island of knights and knaves made the following statements: A: *"At least one of us is a knave but there are no caves on this island"*, B: *"A speaks the truth."* Are there caves on the island? What is the nature of each of A and B?

★(12) On the island of knights and knaves, you meet three inhabitants A, B and C. A says: *"B is a knight"*, B says: *"A is a knave unless C is knight".* C remains silent. Can you determine the nature of each of the inhabitants?

(13) On the island of knights and knaves, you meet three persons A, B and C. You know for sure that one of them is a knight, one is a knave and one is neither (like you and me). A says: *"I am a knight"*, B says: *"A is not a knave"*, and C says: *"B is not a knight".* Can you determine which of the three people is the knight, which is the knave and which is neither?

1.6 Logical equivalence

Mathematics is rich with examples of objects that look very different at the surface but in reality they share similar properties. Formulas of propositional logic can have different "appearances" but still have the exact same truth value for every valuation of their logic variables.

Definition 1.7. We say that two formulas ϕ and ψ are *logically equivalent* (or simply *equivalent*) and we write $\phi \equiv \psi$, if they have the same truth value for every valuation of their logic variables.

One way to show that two formulas are equivalent is to verify that they have the same truth table.

Example 1.38. Show that the formulas $((A \to B) \vee \neg C)$, $(C \to (\neg A \vee B))$ are equivalent.

Solution. We form the truth table of each of the two formulas. Note that we have eight valuations of the set $\{A, B, C\}$ of logic variables.

A	B	C	$\neg A$	$\neg C$	$(A \to B)$	$(\neg A \vee B)$	$((A \to B) \vee \neg C)$	$(C \to (\neg A \vee B))$
T	T	T	F	F	T	T	T	T
T	T	F	F	T	T	T	T	T
T	F	T	F	F	F	F	F	F
T	F	F	F	T	F	F	T	T
F	T	T	T	F	T	T	T	T
F	T	F	T	T	T	T	T	T
F	F	T	T	F	T	T	T	T
F	F	F	T	T	T	T	T	T

The last two columns show the two formulas have the same truth table and hence they are logically equivalent. $\Diamond$

A simple but very useful equivalence is given in the following example. This result is the base of one mathematical proof technique that we will study later.

Example 1.39. Show that a logic implication and its contrapositive are equivalent.

Solution. We have to prove that $(p \to q) \equiv (\neg q \to \neg p)$ for any given formulas p and q. The truth table of both formulas are given in the following table.

p	q	$\neg p$	$\neg q$	$(p \to q)$	$(\neg q \to \neg q)$
T	T	F	F	T	T
T	F	F	T	F	F
F	T	T	F	T	T
F	F	T	T	T	T

Since $(p \to q)$ and $(\neg q \to \neg p)$ have the same truth value for any valuation, they are equivalent. $\Diamond$

The result of Example 1.39 tells us in particular that proving an implication is the same as proving its contrapositive. For instance, proving the statement: "*If n^2 is an even integer, then so is n*" directly (by assuming n^2 even and then proving n is even) is more challenging than proving the contrapositive of the theorem, namely "If n is odd, then so is n^2".

Remark 1.3. Let p, q be two logic formulas. It is not hard to see that $p \equiv q$ if and only if the formula $(p \leftrightarrow q)$ is a tautology (see Exercise (7) below).

The notion of logic equivalence is the basis of all laws of propositional logic. These laws are used in simplifying expressions, understanding electronic circuit performance and to perform logic calculus. Table 1.4 below gives the basic equivalences in propositional logic, together with their known names. The symbols **T** and **F** in the table represent respectively a tautology and a contradiction. Each of these equivalences can be easily verified with a truth table as we did in the previous examples. The logic equivalences of Table 1.4 will be referred to as the *basic logic equivalences*.

Table 1.4   Basic equivalences of propositional logic

Name	Equivalence
Commutativity of the conjunction	$(p \wedge q) \equiv (q \wedge p)$
Commutativity of the disjunction	$(p \vee q) \equiv (q \vee p)$
Commutativity of the biconditional	$(p \leftrightarrow q) \equiv (q \leftrightarrow p)$
Associativity of the conjunction	$(p \wedge (q \wedge r)) \equiv ((p \wedge q) \wedge r)$
Associativity of the disjunction	$(p \vee (q \vee r)) \equiv ((p \vee q) \vee r)$
Distributivity	$(p \vee (q \wedge r)) \equiv ((p \vee q) \wedge (p \vee r))$
Distributivity	$(p \wedge (q \vee r)) \equiv ((p \wedge q) \vee (p \wedge r))$
De Morgan's law	$\neg(p \wedge q) \equiv (\neg p \vee \neg q)$
De Morgan's law	$\neg(p \vee q) \equiv (\neg p \wedge \neg q)$
Negation of the implication	$\neg(p \rightarrow q) \equiv (p \wedge \neg q)$
Negation of the biconditional	$\neg(p \leftrightarrow q) \equiv ((p \wedge \neg q) \vee (\neg p \wedge q))$
Double negation	$\neg\neg p \equiv p$
Excluded middle	$(p \vee \neg p) \equiv \mathbf{T}$
Contradiction	$(p \wedge \neg p) \equiv \mathbf{F}$
Implication elimination	$(p \rightarrow q) \equiv (\neg p \vee q)$
Material equivalence (1)	$(p \leftrightarrow q) \equiv ((p \rightarrow q) \wedge (q \rightarrow p))$
Material equivalence (2)	$((p \leftrightarrow q) \equiv ((p \wedge q) \vee (\neg q \wedge \neg p))$
Or-simplification	$(p \vee p) \equiv p$
Or-simplification (true)	$(p \vee \mathbf{T}) \equiv \mathbf{T}$
Or-simplification (false)	$(p \vee \mathbf{F}) \equiv p$
And-simplification	$(p \wedge p) \equiv p$
And-simplification (true)	$(p \wedge \mathbf{T}) \equiv p$
And-simplification (false)	$(p \wedge \mathbf{F}) \equiv \mathbf{F}$

Remark 1.4.

(1) The commutativity and associativity laws of the conjunction and the disjunction operators allow to us to "relax" a bit the parenthesis used

in wffs. For instance, we can write $(p \wedge q \wedge r)$ instead of $((p \wedge q) \wedge r)$ and $(p \vee q \vee r)$ instead of $((p \vee q) \vee r)$.

(2) The associativity laws for the conjunction and the disjunction operators work with any number of logic formulas. So one can write expressions like $(p_1 \wedge p_2 \wedge \cdots \wedge p_n)$ or $(p_1 \vee p_2 \vee \cdots \vee p_n)$ for any number of logic formulas without the need of parenthesis in the middle.

(3) The distributive laws are also valid for any number of logic formulas:

$$
\begin{aligned}
(p \wedge (p_1 \vee p_2 \vee \cdots \vee p_n)) &\equiv ((p \wedge p_1) \vee (p \wedge p_2) \vee \cdots \vee (p \wedge p_n)) \\
(p \vee (p_1 \wedge p_2 \wedge \cdots \wedge p_n)) &\equiv ((p \vee p_1) \wedge (p \vee p_2) \wedge \cdots \wedge (p \vee p_n))
\end{aligned}
$$

Example 1.40. Use the basic logic equivalences to show each of the following equivalences.

(a) $\phi : (p \vee \neg(p \wedge q)) \equiv \mathbf{T}$
(b) $\psi : (\neg(\neg p \rightarrow q) \vee \neg(p \wedge \neg q)) \equiv (\neg p \vee q)$
(c) $\eta : (\neg(p \vee q) \vee (\neg p \vee q)) \equiv (p \rightarrow q)$

Solution.

(a)

$$
\begin{aligned}
\phi &\equiv (p \vee (\neg p \vee \neg q)) \text{ (De Morgan's law)} \\
&\equiv ((p \vee \neg p) \vee \neg q) \text{ (Associativity of the disjunction)} \\
&\equiv (\mathbf{T} \vee \neg q) \text{ (Excluded middle)} \\
&\equiv \mathbf{T} \text{ (Or-simplification (true))}
\end{aligned}
$$

(b)

$$
\begin{aligned}
\psi &\equiv ((\neg p \wedge \neg q) \vee \neg(p \wedge \neg q)) \text{ (Negation of } \rightarrow) \\
&\equiv ((\neg p \wedge \neg q) \vee (\neg p \vee \neg\neg q)) \text{ (De Morgan)} \\
&\equiv ((\neg p \wedge \neg q) \vee (\neg p \vee q)) \text{ (Double negation)} \\
&\equiv ((\neg p \vee \neg p \vee q) \wedge (\neg q \vee \neg p \vee q)) \text{ (Distributivity)} \\
&\equiv ((\neg p \vee q) \wedge (\neg q \vee \neg p \vee q)) \text{ (Or-simplification)} \\
&\equiv ((\neg p \vee q) \wedge (\neg p \vee \mathbf{T})) \text{ (Excluded middle)} \\
&\equiv ((\neg p \vee q) \wedge \mathbf{T}) \text{ (Or-simplification (true))} \\
&\equiv (\neg p \vee q) \text{ (And-simplification (true))}
\end{aligned}
$$

(c)

$$\eta \equiv ((\neg p \wedge \neg q) \vee (\neg p \vee q)) \text{ (De Morgan)}$$
$$\equiv ((\neg p \vee \neg p \vee q) \wedge (\neg q \vee \neg p \vee q)) \text{ (Distributivity)}$$
$$\equiv ((\neg p \vee q) \wedge (\mathbf{T} \vee \neg p)) \text{ (Or-simplification, Excluded-middle)}$$
$$\equiv ((\neg p \vee q) \wedge \mathbf{T}) \text{ (Or-simplification (true))}$$
$$\equiv (\neg p \vee q) \text{ (And-simplification (true))}$$
$$\equiv (p \to q) \text{ (Material implication)}$$

$\Diamond$

1.6.1   *Using a truth table to write an expression for a logic formula*

In many instances, a propositional logic formula is given by its truth table and one would like to derive an expression for the formula. For example, suppose you are told that a formula ϕ in two logic variables A and B has the following truth table:

A	B	ϕ
T	T	F
T	F	F
F	T	F
F	F	T

We would like to write a possible formula for ϕ. The task is not hard in this case since ϕ is always false except for the valuation $(\mathbf{F}, \mathbf{F})$ of $\{A, B\}$. This suggests that formula is the conjunction $(\neg A \wedge \neg B)$ of $\neg A$ and $\neg B$. This is only one possible expression for ϕ. Note that since $(\neg A \wedge \neg B) \equiv \neg(A \vee B)$ (DeMorgan's law), $\neg(A \vee B)$ is another expression equivalent to ϕ.

Things are not always that simple when the formula contains a large number of logic variables. Consider for instance the formula ψ given by the following truth table:

A	B	C	ψ
T	T	T	F
T	T	F	F
T	F	T	T
T	F	F	F
F	T	T	F
F	T	F	T
F	F	T	F
F	F	F	T

You can certainly try to guess a logic formula with the same truth table as ψ, but you will quickly realize how cumbersome this task can be. Surprisingly enough, there is an easy and straightforward method to solve this problem. First, we need some terminologies. A *literal* is defined to be a propositional variable (atomic formula) or the negation of propositional variable. So A, $\neg B$, $\neg Z_{12}$ are literals. A *minterm* on the set $\{x_1, x_2, \ldots, x_n\}$ of logic variables is a logic formula of the form $(y_1 \wedge y_2 \wedge \cdots \wedge y_n)$ where y_i is either x_i or $\neg x_i$. In other words, a minterm is the conjunction of literals in which every logic variable appears exactly once in a non-negated or negated form. For the set $\{A, B, C\}$ of logic variables, $(A \wedge \neg B \wedge C)$, $(\neg A \wedge B \wedge \neg C)$ and $(\neg A \wedge \neg B \wedge \neg C)$ are examples of minterms. A minterm is true for exactly one valuation of the logic variables. The converse also holds: given any valuation α of the logic variables, there exists a unique minterm satisfied by α. For example, the minterm $(A \wedge \neg B \wedge C)$ is true if and only if A is true, B is false and C is true which corresponds to the valuation $(\mathbf{T}, \mathbf{F}, \mathbf{T})$ of $\{A, B, C\}$. For the valuation $\alpha = (\mathbf{F}, \mathbf{F}, \mathbf{F})$ of $\{A, B, C\}$, $(\neg A \wedge \neg B \wedge \neg C)$ is the unique minterm satisfied by η.

Going back to the formula ψ given in the last table, notice that ψ is true only on rows 3, 6 and 8 of the truth table. These rows correspond to the minterms $(A \wedge \neg B \wedge C)$, $(\neg A \wedge B \wedge \neg C)$ and $(\neg A \wedge \neg B \wedge \neg C)$ respectively. A valuation α of $\{A, B, C\}$ satisfies the formula ψ if and only if α is one of the valuations $(\mathbf{T}, \mathbf{F}, \mathbf{T})$, $(\mathbf{F}, \mathbf{T}, \mathbf{F})$ or $(\mathbf{F}, \mathbf{F}, \mathbf{F})$. The same is true for any valuation satisfying the formula $\phi = ((A \wedge \neg B \wedge C) \vee (\neg A \wedge B \wedge \neg C) \vee (\neg A \wedge \neg B \wedge \neg C))$. We conclude that $\psi \equiv \phi$.

The above reasoning can be applied to any logic formula. If a logic formula ϵ is given by its truth table, locate the rows where $v(\epsilon) = \mathbf{T}$ and for each one of these rows, write the corresponding minterm. The disjunction of all these minterms gives a formula logically to ϵ.

1.6.2 *Exercises*

(1) Use truth tables to prove DeMorgan's laws.
(2) Prove the equivalence: $\neg (p \to q) \equiv (p \wedge \neg q)$.
$\star$(3) Prove the distributive laws:

    (a) $(p \vee (q \wedge r)) \equiv ((p \vee q) \wedge (p \vee r))$
    (b) $(p \wedge (q \vee r)) \equiv ((p \wedge q) \vee (p \wedge r))$

$\star$(4) If P, Q and R are three propositions, use a truth table to prove the equivalence $((P \vee Q) \to R) \equiv ((P \to R) \wedge (Q \to R))$.

(5) In each case, determine if $\phi \equiv \psi$.

  (a) $\phi = (\neg(A \to \neg B))$, $\psi = (A \leftrightarrow (\neg B \to A))$

  (b) $\phi = ((\neg P \to (Q \wedge R))$, $\psi = ((\neg P \to Q) \wedge (\neg R \to P))$

$\star$(c) $\phi = (A \to (B \wedge C))$, $\psi = ((A \to B) \wedge (A \to C))$

  (d) $\phi = ((P \wedge Q) \to R)$, $\psi = (P \to (Q \to R))$

  (e) $\phi = (X \to (Y \vee Z))$, $\psi = ((X \to Y) \vee (X \to Z))$

(6) Prove the following equivalences using the basic logic equivalences.

  (a) $\neg(\neg p \to q) \equiv \neg(p \vee q)$

  (b) $\neg(A \leftrightarrow B) \equiv (A \leftrightarrow \neg B)$

$\star$(c) $((A \leftrightarrow B) \wedge \neg A) \equiv \neg(A \vee B)$

  (d) $(\neg(P \vee (Q \wedge \neg R)) \wedge Q) \equiv (\neg P \wedge Q \wedge R)$

  (e) $(\neg((p \vee q) \to r) \wedge \neg(r \to q)) \equiv \mathbf{F}$

$\star$(f) $((S \to (\neg P \to S)) \to P) \equiv P$

  (g) $((A \vee B) \leftrightarrow \neg(A \vee \neg C)) \equiv ((\neg A \wedge B \wedge C) \vee (\neg A \wedge \neg B \wedge \neg C))$

(7) Let p, q be two logic formulas. Prove that $p \equiv q$ if and only if $(p \leftrightarrow q)$ is a tautology.

$\star$(8) The symbol $\downarrow$ stands for "nor" in the literature, so $(p \downarrow q)$ means "neither p nor q". Form the truth table for $\downarrow$. Find a formula equivalent to $(p \vee q)$ which uses $\downarrow$ and $\neg$ only.

(9) The notation $(p \uparrow q)$ stands for "not both p and q". Form the truth table for $\uparrow$. Find a formula equivalent to $(p \wedge q)$ and which uses $\uparrow$ and $\neg$ only.

(10) Use the notations and your answers to Exercises (8) and (9).

  (a) Express $(p \to q)$ using:

    (i) $\downarrow$ and $\neg$ only   (ii) $\uparrow$ and $\neg$ only

  (b) Prove each of the following logic equivalences:

    (i) $(\neg p \downarrow \neg q) \equiv \neg(p \uparrow q)$ (ii) $(\neg p \uparrow \neg q) \equiv \neg(p \downarrow q)$.

$\star$(11) Use basic logic equivalences to find a logic formula ϕ equivalent to $(p \to (p \leftrightarrow q))$ such that the only connectives in ϕ are $\wedge$ and $\neg$.

(12) Use basic logic equivalences to carefully prove the following extensions of DeMorgan's laws:

(a) $\neg(p \vee q \vee r) \equiv (\neg p \wedge \neg q \wedge \neg r)$
(b) $\neg(p \wedge q \wedge r) \equiv (\neg p \vee \neg q \vee \neg r)$

$\star$(13) The truth table of a logic formula ψ is given.

A	B	C	ψ
T	T	T	F
T	T	F	F
T	F	T	F
T	F	F	F
F	T	T	T
F	T	F	T
F	F	T	F
F	F	F	F

(a) Find a logic formula equivalent to ψ.
(b) Find a logic formula equivalent to ψ that does not contain the connective $\wedge$.
(c) Find a logic formula equivalent to ψ that does not contain the connective $\vee$.

(14) In each case, use basic logic equivalences to prove that the formula is a tautology.

(a) $(\neg B \rightarrow (A \vee \neg(A \vee B)))$
$\star$(b) $(((A \rightarrow B) \wedge (B \rightarrow C)) \rightarrow (A \rightarrow C))$
(c) $(((A \vee B) \wedge (A \rightarrow C) \wedge (B \rightarrow C)) \rightarrow C)$

(15) Use DeMorgan's laws to give the negation of each of the following statements.

(a) The function f is not continuous but $f(0) = 0$.
(b) The function f is continuous unless $f(0) \neq 0$.
(c) The function f is not continuous nor differentiable.
$\star$(d) Either the function f is continuous or $f(0) \neq 0$.
(e) The function f is both continuous and logarithmic.

(16) Negate each of following logic formulas. Use the basic logic equivalences to simplify your answer so that the only connectives present in the final answer are $\wedge$ and $\neg$.

(a) $((p \rightarrow (p \wedge q)) \rightarrow q)$ (c) $((p \vee q) \leftrightarrow r)$
(b) $(\neg p \rightarrow (\neg q \rightarrow (\neg(p \vee q) \wedge r)))$

1.7   The method of truth trees

Constructing the truth table of a logic formula becomes unmanageable as the number of logic variables grows. For instance, if a formula has six logic variables, its truth table consists of $2^6 = 64$. With this large number of rows, you can imagine that mistakes in setting up the valuations of the logic variables and the corresponding truth values of the formula are almost inevitable. In using a truth table to check the nature of a logic formula or the coherency of a set of formulas, a lot of effort is waisted on listing valuations (or rows of the truth table) that are actually not needed. To explain this, let us revisit Example 1.29 above where we studied the coherency of the set $\mathcal{E} = \{(A \vee B), (A \rightarrow B), (\neg A \wedge B)\}$ of logic formulas. The truth tables of the formulas in the set $\mathcal{E}$ is as follows.

A	B	$\neg A$	$(A \vee B)$	$(A \rightarrow B)$	$(\neg A \wedge B)$
T	T	F	T	T	F
T	F	F	T	F	F
F	T	T	T	T	T
F	F	T	F	T	F

Note that only the highlighted row matters for the purpose of determining the coherency of the set. The method of *truth tree* is designed to eliminate unnecessary rows in a truth table and get straight to the rows that matter for the problem at hand. With truth trees, we explore all valuations for which all the formulas in a certain set are true at the same time without listing the other valuations. The method is given below as an algorithm to solve a variety of problems in propositional logic: testing tautology, coherency of a set of logic formulas and the validity of a logic argument. In the context of propositional logic, a truth tree consists of an upside down tree where the root (or trunk) is drawn on top and leaves growing downward. The method is based on the nine *branching rules* given in Figure 1.1 below. There is a branching rule for every logic operator (connective) and another for its negation. Each branching rule is designed to give the valuations for which the formula at the root is true. For example, in order for $(p \wedge q)$ to be true, both p and q must be true which explain the same branch on which p and q are linked in the first rule in Figure 1.1. Similarly, $(p \vee q)$ is true when either p or q is true which explain the two branches for that rule in Figure 1.1. In fact, the branching rules are just a way to represent pictorially some of the basic equivalences of Table 1.4.

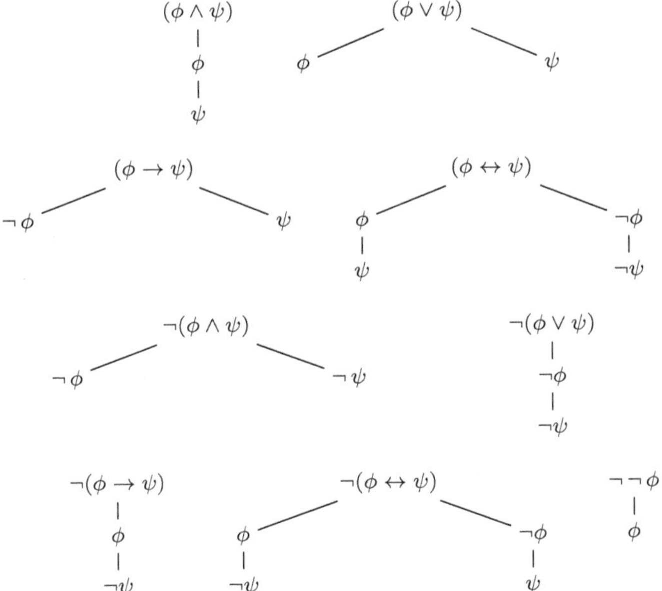

Fig. 1.1   The nine branching rules.

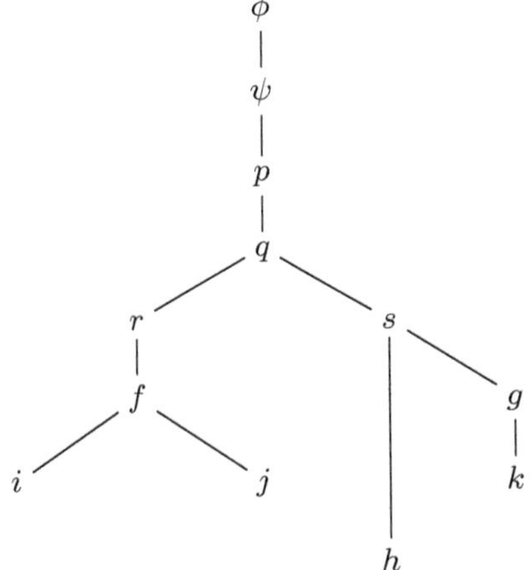

In the tree on the left, formulas $\phi, \psi, p, q, r, s, f, g, i, j, k, h$ are called the *nodes* and lines joining the formulas are called the *edges* of the tree. Formulas $\phi, \psi, p, q, r, s, f, g$ are called the *internal nodes* and i, j, k, h are called the *leaves* of the tree.

1.7.1   *Description of the method*

Let $\mathcal{E} = \{p_1, p_2, \ldots, p_n\}$ be a set of logic formulas. To determine the coherency of $\mathcal{E}$ using the method of truth tree, we follow the steps blow.

(1) We start by listing all the formulas in $\mathcal{E}$, each on a separate line. This part forms the *root* of the tree.
(2) Pick a formula in the root and apply the appropriate branching rule to decompose that formula. This step requires a clear understanding of what the main connective of the formula is. Although the choice of the first formula to branch does not change the end result, there is usually an advantage in starting with a formula requiring a one-branch rule.
(3) If a certain formula and its negation appear on the same branch, the branch represents a logical inconsistency (a contradiction). We "close" the branch in this case. This is indicated by the symbol ✖.
(4) If all branches are closed, then the process stops.
(5) If there are open branches, go to step (2) and continue to branch all compound formulas remaining including the ones at the root. *A compound formula must be branched at every open branch extending from it.*
(6) The branching process ends when all the branches are closed or when there are no more compound formulas to branch on any of the branches.

At the end of the branching process, we draw the following **conclusion**: if there exists an open branch, then following all the atoms present on that branch gives one or more valuations that satisfy *all* the formulas at the root. The existence of an open branch means in particular that the set $\mathcal{E}$ is coherent. If all branches are closed, then there exists no valuation that satisfy all the formulas at the root and the set $\mathcal{E}$ is incoherent.

Example 1.41. Use the method of truth tree to determine if the set $\mathcal{E} = \{(A \vee B), (A \to B), (\neg A \wedge B)\}$ of logic formulas is coherent. If $\mathcal{E}$ is coherent, give at least one valuation that satisfies the set.

Solution. The first step is to write all the formulas in $\mathcal{E}$ at the root of a truth tree and numbering the corresponding rows for future references:

$$(A \vee B) \qquad\qquad 1$$

$$(A \to B) \qquad\qquad 2$$

$$(\neg A \wedge B) \qquad\qquad 3$$

The next step is to choose a formula at the root and branch it using the corresponding branching rule. As mentioned before, we can start with any of the three formulas but it could be more beneficial in the process to start with a one-branch formula. So, we start with the formula on line 3 since it is a conjunction. We check it off (to keep track of branched formulas) and branch it using branching rule 1 from Figure 1.1. We also write "from 3" in lines 4 and 5 to indicate that these new lines come from the branching of the formula on line 3.

$$
\begin{array}{lll}
(A \vee B) & & 1 \\
(A \to B) & & 2 \\
(\neg A \wedge B) \; \checkmark & & 3 \\
\quad | & & \\
\neg A & \text{from 3} & 4 \\
B & \text{from 3} & 5 \\
\end{array}
$$

In the latest tree above, formulas on lines 1 and 2 are still compound and we can proceed with either one. We check formula $(A \vee B)$ from line 1 and we branch it at the bottom of the tree using branching rule 2 from Figure 1.1. We only have one open branch extending from the formula $(A \vee B)$ in this case. We write "from 1" in line 6 to indicate that it originates from branching the formula on line 1.

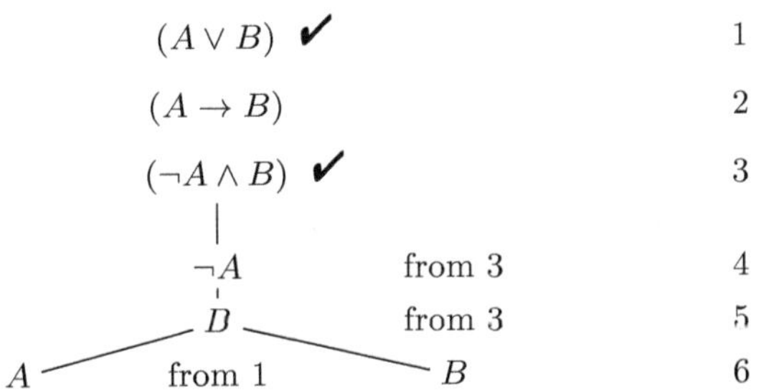

The latest tree ends with two branches. Note that the left branch contains A and $\neg A$ at the same time, so we close it since it corresponds to a contradiction. One more compound formula left to branch from line 2. Since the left branch is closed, we only have to branch it on the right one using rule 3 from Figure 1.1 (again, a compound formula must be branched on *all* open branches extending from it).

$$
\begin{array}{lll}
(A \vee B) \;\checkmark & & 1 \\
(A \to B) \;\checkmark & & 2 \\
(\neg A \wedge B) \;\checkmark & & 3 \\
\neg A & \text{from 3} & 4 \\
B & \text{from 3} & 5 \\
A \quad\quad B & \text{from 1} & 6 \\
\times \quad \neg A \quad B & \text{from 2} & 7
\end{array}
$$

Since no more compound formulas are left to branch, the branching process ends. Every open branch in the final tree gives at least one valuation for which *all* the formulas at the roots are true. The first open branch from the left corresponds to the valuation $(\mathbf{F}, \mathbf{T})$ of the logic variables $\{A, B\}$ (since $\neg A$ and B appear on that branch). The second open branch corresponds to the same valuation. We conclude that the set $\mathcal{E}$ is coherent and that $(\mathbf{F}, \mathbf{T})$ is a valuation (in fact the only one) that satisfy the set. $\quad\diamond$

Remark 1.5. It is very useful to close a branch as soon as we notice it contains a contradiction (a formula together with its negation) but in some cases, we don't notice that early enough. This could lengthen the process a bit but it does change the end result.

Example 1.42. Use the method of truth tree to determine if the set $\mathcal{E} = \{((C \vee D) \leftrightarrow \neg B), (A \to B), (B \leftrightarrow C)\}$ of logic formulas is coherent. If you say it is, give at least one valuation that satisfies the set.

Solution. Here, we will skip the details of every step and draw the final picture of the truth tree. Note that lines 9 and 10 in the tree result from branching the formula on line 3.

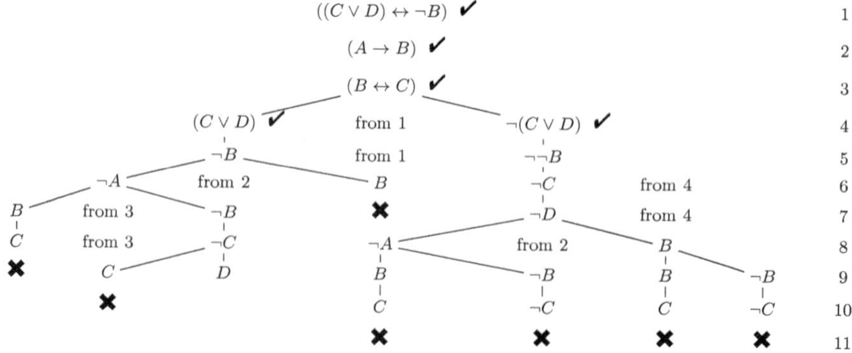

There is an open branch corresponding to the valuation $\alpha = (\mathbf{F}, \mathbf{F}, \mathbf{F}, \mathbf{T})$ of $\{A, B, C, D\}$. This means that the set is coherent and α is a valuation that satisfies $\mathcal{E}$. $\Diamond$

1.7.2 Tautologies and the method of truth tree

The method of a truth tree can be effective in determining if a wff is a tautology. Given a formula ϕ, we consider the set $\mathcal{E} = \{\neg\phi\}$. If $\mathcal{E}$ is coherent, then there a valuation α of the logic variables of ϕ that satisfies $\neg\phi$. In other words, α does not satisfy ϕ and so ϕ is not a tautology. Conversely, if ϕ is not a tautology then the set $\mathcal{E} = \{\neg\phi\}$ is coherent since there exists a valuation that does not satisfy ϕ. This suggests that the method of truth tree can be used to check if ϕ is a tautology as follows. Start a truth tree with the *negation* $\neg\phi$ of ϕ at the root. Branch the tree completely. If every branch is closed at the end of the branching process, the formula ϕ is a tautology as there exists no valuation satisfying $\neg\phi$. If there exists an open branch, the formula is not a tautology and every open branch gives at least one valuation for which the formula ϕ is false.

Example 1.43. For the wff: $\phi : (((A \to B) \wedge (B \to C)) \to (A \to C))$, use the method of truth tree to determine if it is a tautology. If you say that ϕ is not a tautology, give all valuations of $\{A, B, C\}$ that do not satisfy it.

Solution. We start a truth tree with the negation of the original formula at the root. We completely branch the tree as we did previously.

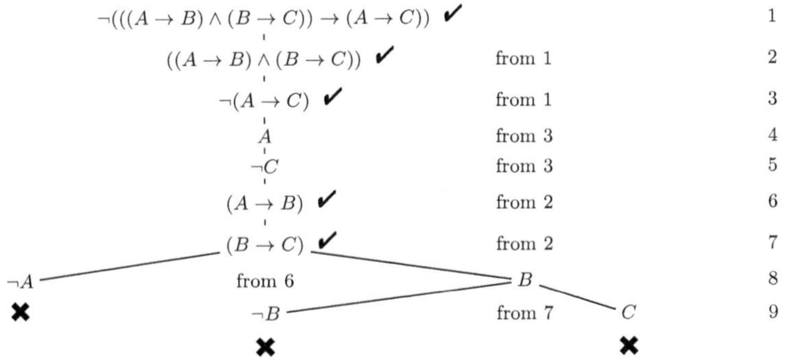

$$
\begin{array}{lll}
\neg(((A \to B) \wedge (B \to C)) \to (A \to C)) ~\checkmark & & 1 \\
((A \to B) \wedge (B \to C)) ~\checkmark & \text{from 1} & 2 \\
\neg(A \to C) ~\checkmark & \text{from 1} & 3 \\
A & \text{from 3} & 4 \\
\neg C & \text{from 3} & 5 \\
(A \to B) ~\checkmark & \text{from 2} & 6 \\
(B \to C) ~\checkmark & \text{from 2} & 7 \\
\neg A \qquad\qquad B & \text{from 6} & 8 \\
\bigstar \qquad \neg B \qquad\quad C & \text{from 7} & 9 \\
\qquad\quad \bigstar \qquad\quad \bigstar & &
\end{array}
$$

All the branches are closed. There is no valuation that makes $\neg\phi$ true (or ϕ false). The formula $(((A \to B) \wedge (B \to C)) \to (A \to C))$ is then a tautology. $\Diamond$

Example 1.44. Use the method of truth tree to determine if the formula $\psi : ((\neg A \to (B \to C)) \to (\neg B \to A))$ is a tautology. If you say that ψ is not a tautology, give all valuations of $\{A, B, C\}$ that do not satisfy it.

Solution. The following is the completed truth tree of $\neg\psi$.

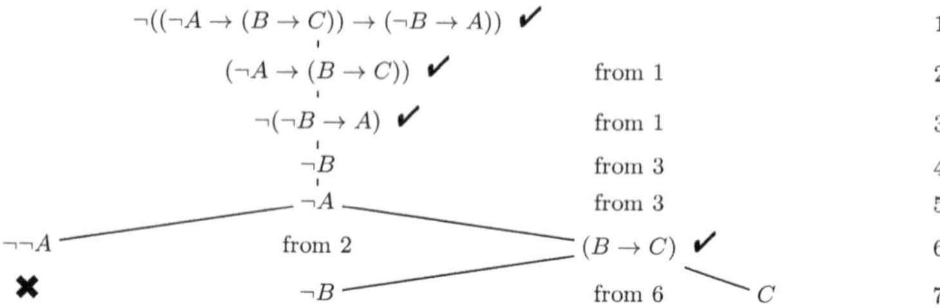

There are open branches. The formula ψ is not a tautology. Following the first open branch, we notice that A and B are false (they appear as $\neg A$ and $\neg B$) and C can take either one of the truth values (since C does not appear on that branch). On the second open branch, we notice that A and B are false and C is true (they appear as $\neg A$ and $\neg B$ and C on that branch). The valuations of $\{A, B, C\}$ that make the formula ψ false are $(\mathbf{F}, \mathbf{F}, \mathbf{T})$ and $(\mathbf{F}, \mathbf{F}, \mathbf{T})$. Note that we can actually verify our result. For instance, for the valuation $(\mathbf{F}, \mathbf{F}, \mathbf{T})$, the formula $(B \to C)$ is true and $(\neg B \to A)$ is false. So $(\neg A \to (B \to C))$ is true (since it is of the form $(\mathbf{T} \to \mathbf{T})$) and consequently $((\neg A \to (B \to C)) \to (\neg B \to A))$ is false (since it is of the form $(\mathbf{T} \to \mathbf{F})$). We can verify similarly that the formula is false for the valuation $(\mathbf{F}, \mathbf{F}, \mathbf{F})$. $\Diamond$

1.7.3   *Exercises*

(1) Let ϕ be a logic formula and let T (respectively S) be the completely branched truth trees with ϕ (respectively $\neg\phi$) as the only formula at the root. For each of the following statements, determine if it is true or false. Justify your answer.

    ⋆(a) If T has all its branches open, then ϕ is a contradiction.

    (b) If S has all its branches open, then ϕ is a contradiction.

    (c) If T has all its branches closed, then ϕ is a tautology.

    ⋆(d) If S only has all its branches closed, then ϕ is a tautology.

    (e) If the set $\{\phi\}$ is coherent, then ϕ is a tautology.

    (f) If the set $\{\neg\phi\}$ is coherent, then ϕ is a not a tautology.

    ⋆(g) If the set $\{\phi\}$ is incoherent, then ϕ is a contradiction.

(2) In each case, use the method of truth tree to determine if the set $\mathcal{E}$ is coherent. If you say it is, give all valuations that satisfy the set.

　$\star$(a) $\mathcal{E} = \{(A \to (\neg A \to B)), (A \vee C), (C \to A)\}$
　　(b) $\mathcal{E} = \{(A \leftrightarrow (B \to C)), \neg(A \to C), ((B \wedge C) \to \neg A), (\neg A \vee C)\}$
　　(c) $\mathcal{E} = \{((Y \wedge Z) \to X), (X \to \neg(Y \to Z)), (X \to \neg Z), ((X \wedge Y) \to Y)\}$

(3) In each case, use the method of truth tree to determine if the formula is a tautology. If the formula is not a tautology, give all valuations that do not satisfy it.

　　(a) $((A \to (\neg A \to B)) \leftrightarrow B)$.
　　(b) $((A \leftrightarrow (B \to C)) \to \neg(A \wedge B \wedge \neg C))$.
　$\star$(c) $((X \to (Y \to Z)) \to ((\neg X \to Y) \to Z))$.
　　(d) $(((X \to (\neg Y \to Z)) \to ((X \vee Y) \to Z)) \to \neg X)$.

(4) So far we have seen two method to determine if a given formula is a tautology, namely the method of truth table and the method of truth tree. A third method, called the *contradiction method*, consists of assuming that ϕ is not a tautology and proving that this leads to a contradiction. We then can conclude that ϕ is a tautology. In each case, use the contradiction method to prove that the given formula is a tautology.

　　(a) $(((A \to B) \wedge (B \to C)) \to (A \to C))$.
　$\star$(b) $(((A \to B) \wedge (C \to D)) \to ((A \to D) \vee (C \to B)))$.

1.8　Validity of logic arguments

Unlike logic propositions, arguments are neither true nor false. They are either valid or invalid. The definition of logic arguments does not stipulate anything about the truth values of the premises and the conclusion. However, it is clear that if we want to analyze the validity of an argument, we have to relate somehow the truth values of the premises with that of the conclusion. For instance, it does not take a lot of analysis to see that the following argument

The safe was open

Joe is the only person who has the keys for the safe

We are going on a trip next year

is not valid. The reason is simple: the premises do not necessarily imply the conclusion. We still have to formally define what validity means in this context.

Definition 1.8. A logic argument is called *valid* if it is not possible for the premises to be all true at the same time while the conclusion is false. The argument is called *invalid* otherwise. In other words, an argument is invalid only when there exists a valuation that satisfies every premise but not the conclusion. Such a valuation is called a *counter-example* of the argument. Only invalid arguments have counter-examples.

The following theorem makes the connection between the validity of an argument and the notion of a tautology.

Theorem 1.1. *Let $\mathcal{A}$ be a logic argument with premises $p_1, \ldots, p_n$ and conclusion p. Then $\mathcal{A}$ is valid if and only if the implication $\phi : ((p_1 \wedge \cdots \wedge p_n) \to p)$ is a tautology.*

Proof. Assume that $\mathcal{A}$ is valid and let α be a valuation of all the logic variables in the premises and the conclusion of the argument. If all the premises $p_1, \ldots, p_n$ are satisfied by α at the same time, then p is also satisfied by α by the validity of $\mathcal{A}$. The implication ϕ is true in this case. If there exists at least one premise not satisfied by α, then the conjunction $(p_1 \wedge \cdots \wedge p_n)$ is false and the implication ϕ is also true in this case. This shows that if $\mathcal{A}$ is valid then ϕ is a tautology. Conversely, assume that ϕ is a tautology and let α be a valuation that satisfies $p_1, \ldots, p_n$ at the same time, then α satisfies also their conjunction $(p_1 \wedge \cdots \wedge p_n)$ and hence it satisfies the conclusion p since ϕ is a tautology. This shows that the argument is valid. $\qquad\square$

To check the validity of a logic argument with premises $p_1, \ldots, p_n$ and conclusion p, we can use either a truth table or a truth tree.

(a) With a truth table, we only focus on the lines where $p_1, \ldots, p_n$ are true at the same time. If p happens to be also true for each of these lines, then the argument is valid. If there exists a line for which $p_1, \ldots, p_n$ are all true and p is false, the corresponding valuation on this line is a counter-example and the argument is invalid. If there is no line on which all the premises are true at the same time, the argument is still valid (since it has no counter-example in this case).

(b) With a truth tree, we write all the premises and the *negation* of the conclusion at the root of the tree and we branch the tree as we did in the previous section. If all branches are closed at the end, then there exists no valuation that makes all premises and the negation of the conclusion true at the same time. In this case, the argument is valid since no counter-example exists. If there exists an open branch, the argument is invalid since any valuation along such a branch makes all premises true and the conclusion false (hence a counter-example).

Example 1.45. In each case, determine if the argument is valid using a truth table. If you say that the argument is invalid, give at least one counter-example.

$$
(a) \quad \frac{\begin{array}{c} (A \to B) \\ (\neg A \vee C) \end{array}}{B}
\qquad
(b) \quad \frac{\begin{array}{c} (P \to \neg Q) \\ (P \wedge R) \\ (\neg Q \to (P \to R)) \\ R \end{array}}{(P \wedge \neg Q)}
\qquad
(c) \quad \frac{\begin{array}{c} (A \to (B \vee C)) \\ (\neg A \wedge B) \\ A \end{array}}{(A \to C)}
$$

Solution. For the first argument, the truth table of the premises and the conclusion is the following:

A	B	C	$\neg A$	$(A \to B)$	$(\neg A \vee C)$	B
T	T	T	F	T	T	T
T	T	F	F	T	F	T
T	F	T	F	F	T	F
T	F	F	F	F	F	F
F	T	T	T	T	T	T
F	T	F	T	T	T	T
F	F	T	T	T	T	F
F	F	F	T	T	T	F

The last two rows (in gray) show valuations for which the premises are true but the conclusion is false. The argument is then *invalid*. Counter-examples of the argument are the valuations $(\mathbf{F}, \mathbf{F}, \mathbf{T})$ and $(\mathbf{F}, \mathbf{F}, \mathbf{F})$ of $\{A, B, C\}$. For the second argument, the table of the premises and the conclusion is:

P	Q	R	$\neg Q$	$(P \to R)$	$(P \to \neg Q)$	$(P \wedge R)$	$(\neg Q \to (P \to R))$	$(P \wedge \neg Q)$
T	T	T	F	T	F	T	T	F
T	T	F	F	F	F	F	T	F
T	F	T	T	T	T	T	T	T
T	F	F	T	F	T	F	F	T
F	T	T	F	T	T	F	T	F
F	T	F	F	T	T	F	T	F
F	F	T	T	T	T	F	T	F
F	F	F	T	T	T	F	T	F

From the table, there is only one valuation that satisfies all the premises (highlighted in grey in the table) and for this valuation, the conclusion is also true. The argument is then *valid*. Finally, the table of the premises and the conclusion of the third argument is:

A	B	C	$\neg A$	$(B \wedge C)$	$(A \to (B \wedge C))$	$(\neg A \wedge B)$	A	$(A \to C)$
T	T	T	F	T	T	F	T	T
T	T	F	F	F	F	F	T	F
T	F	T	F	F	F	F	T	T
T	F	F	F	F	F	F	T	F
F	T	T	T	T	T	T	F	T
F	T	F	T	F	T	T	F	T
F	F	T	T	F	T	F	F	T
F	F	F	T	F	T	F	F	T

There is no valuation that makes all the premises true at the same time (the set of premises is incoherent). The argument is valid regardless of the truth value of the conclusion. Remember, a counter-example exists only when all the premises are true *and* the conclusion is false.   ◇

Example 1.46. In each case, determine if the argument is valid using a truth tree. If you say that the argument is invalid, give at least one counter-example.

$$(a) \quad \frac{\begin{array}{c} (A \to (B \vee C)) \\ (\neg A \wedge B) \\ (B \to \neg C) \end{array}}{(A \vee C)} \qquad (b) \quad \frac{\begin{array}{c} (\neg A \to (C \to B)) \\ (B \leftrightarrow C) \\ (B \vee A) \\ (C \to D) \end{array}}{(A \vee D)}$$

Solution. In each case, we start a truth tree with all the premises and the negation of the conclusion at the root. We branch all the formulas at the root using the nine branching rules. The order in which the formulas are branched does not affect the final outcome, but it helps to choose formulas the branching of which results in early closing of branches in the tree.

(a)

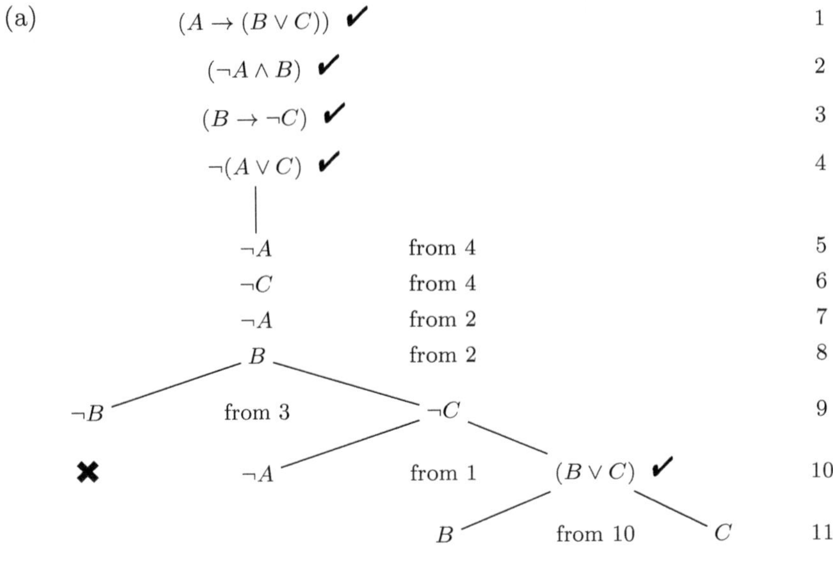

Since there are open branches, the argument is *invalid*. Following the open branches, we see that the only counter-example for the argument is the following valuation: $(v(A) = \mathbf{F}, v(B) = \mathbf{T}, v(C) = \mathbf{F})$.

(b)

All branches are closed. Notice that we didn't have to branch the formula $(C \to B)$ on line 8 as we were able to close all branches before the need to break it up. There are no counter-examples and the argument is therefore *valid*. $\diamond$

1.8.1   *Exercises*

(1) Let $\mathcal{A}$ be a logic argument with premises $p_1, \ldots, p_n$ and conclusion p. For each of the following statements, determine if it is true or false. Justify your answer.

    $\star$(a) If there exists a valuation that satisfies p_i for every i, then $\mathcal{A}$ is valid.

    (b) If p is a tautology, then $\mathcal{A}$ is valid.

    $\star$(c) If p is a contradiction, then $\mathcal{A}$ is invalid.

    $\star$(d) If there exists a valuation that satisfies p and p_i for every i, then $\mathcal{A}$ is valid.

    (e) If there exists a valuation that makes p and p_i false for all i, then $\mathcal{A}$ is invalid.

    (f) If the set $\{p_1, \ldots, p_n\}$ is coherent, then $\mathcal{A}$ is valid.

    $\star$(g) If the set $\{p_1, \ldots, p_n\}$ is incoherent, then $\mathcal{A}$ is valid.

    (h) If the set $\{p_1, \ldots, p_n, \neg p\}$ is coherent, then $\mathcal{A}$ is invalid.

$\star$(2) Let $\mathcal{E}$ be the set of the logic formulas $((B \to A) \to C)$, $((\neg C \to B) \to A)$, $(((A \to B) \land (B \to C)) \to (A \to C))$ and $\neg A$.

    (a) Determine if $\mathcal{E}$ is coherent.

    (b) Use the part (a) to determine the validity of the following argument:

$$\frac{\begin{array}{c} ((B \to A) \to C) \\ ((\neg C \to B) \to A) \\ (((A \to B) \land (B \to C)) \to (A \to C)) \end{array}}{A}$$

(3) In each case, determine if the argument is valid using a truth table. If you say that the argument is invalid, give at least one counter-example.

$$(a) \quad \frac{\begin{array}{c} (\neg A \lor B) \\ (A \to C) \end{array}}{B} \qquad \star(b) \quad \frac{\begin{array}{c} (B \to \neg A) \\ \neg(\neg A \lor \neg C) \\ (\neg C \to (B \to \neg C)) \\ (A \to C) \end{array}}{(A \land \neg B)} \qquad (c) \quad \frac{\begin{array}{c} (A \to (\neg C \to B)) \\ (\neg A \land B) \\ (A \to C) \end{array}}{A}$$

(4) In each case, determine if the argument is valid using a truth tree. If you say that the argument is invalid, give at least one counter-example.

$$(a) \quad \frac{\begin{array}{c} (\neg A \to (\neg B \to C)) \\ \neg(A \lor \neg B) \\ (C \to \neg B) \end{array}}{(\neg A \to C)}$$

$$\star(b) \quad \frac{\begin{array}{c} (A \lor \neg(C \to B)) \\ (B \leftrightarrow D) \\ (\neg B \to A) \\ (\neg D \to \neg C) \end{array}}{(\neg D \lor A)}$$

$\star$(5) Sometime it is easier to prove the validity of a logic argument using the *contradiction method:* assume that there exists a counter-example and prove that this leads to a contradiction. Use the contradiction method to prove that the following argument is valid.

$$\frac{\begin{array}{c} ((A \leftrightarrow B) \to D) \\ (\neg D \leftrightarrow C) \\ \neg(B \lor A) \\ (C \to D) \end{array}}{(D \to \neg A)}$$

(6) In each case, start by translating the argument into symbols of propositional logic. Then determine the validity of the argument using a truth table or a truth tree.

   (a) The interest rate increases if the unemployment rate exceeds 6%. The unemployment rate exceeds 6%. Then, the interest rate increases.

   $\star$(b) If Joe calls for a meeting, then Phil or Amanda attends. If Phil and Amanda attend the meeting, there is a huge argument. Therefore, there is a huge argument if Joe calls for a meeting.

   (c) If Joe goes to France, he visits the Eiffel tour. If Joe goes to Egypt, he visits the Pyramids. Joe goes to France or to Egypt. Therefore, Joe visits the Eiffel tour or the Pyramids.

   $\star$(d) If Joe wants to get his lifeguard certificate, it is necessary that he finishes his course. In order to finish his course, it is sufficient that Joe reduces his gym time. Reducing the gym time is sufficient to lower Joe's self-esteem. If Joe has a low self-esteem, he will not finish his course. Therefore, Joe will not get his lifeguard certificate.

$\star$(7) Consider the following argument: *If the dog barks, then there is a stranger at the door. No stranger is at the door unless the lights are off. The lights are off only if the sensor stops working or there is a power outage. Having a power outage is sufficient for the censor to work. the dog barks if there is power outage. The dog does not bark. Then, there is a stranger at the door only if there is a power outage.*

(a) Using the following atoms, translate the argument into symbols of propositional logic.

- B: The dog barks
- S: There is a stranger at the door
- L: The lights are on
- P: There is a power outage
- W: The censor is working

(b) Determine the validity of the argument. If you say that the argument is not valid, give a counter-example.

(8) Consider the following argument: *The Maple Leafs are defeated if either the Canadians or the Senators make it to the playoffs. In order for the Maple leafs to change their manager, it suffices that they are defeated or the Senators make it to the playoffs. The Maple leafs will keep their manager only if they claim bankruptcy. The Maple leafs are not defeated. Then, the Maple leafs change their manager.*

(a) Translate the argument into symbols of propositional logic. Start by identifying the atomic statements.
(b) Determine the validity of the argument. If you say that the argument is not valid, give a counter-example.

(9) Using the atoms:

- S: The system is upgraded
- U: Users can access the data
- F: Users can save files
- B: The system has a bug

translate the argument below into symbols of propositional logic. Determine the validity of the argument. If you say that the argument is not valid, give a counter-example.

Whenever the system is upgraded, users cannot access the data. Users cannot access the data unless they can save files. If users cannot save files, then the system is not upgraded. Files cannot be saved by users if and only of the system has a bug. Therefore, the system has a bug only if users cannot access data.

1.9 Formal proof of validity of an argument: Rules of inference for propositional logic

A mathematical proof is, in essence, a form of a valid argument that establishes a certain conclusion from a set of statements (premises). Starting with the premises, we add a small "valid" step at a time until we finally reach the conclusion. The challenge is to properly insert a new statement from others already established in the proof without "breaking" any law of logic. The inference rules are designed exactly for that purpose. If the argument is valid, the inference rules allow us to deduce the conclusion using the given premises or newly inferred ones (like the basic logic equivalence we studied before). As an example, if propositions $(p \rightarrow q)$ and p are already established, then we can *infer* proposition q. This is one of the inference rules named as Modus Ponens. The rule is represented in the form of a logic argument:

$$(p \rightarrow q)$$
$$p$$
$$\overline{}$$
$$\therefore q$$

The Modus Ponens inference rule is another way of saying that the wff $(((p \rightarrow q) \wedge p) \rightarrow q)$ is a tautology. Every inference rule is associated with a tautology of propositional logic. Table 1.5 gives the inference rules of propositional logic together with their known English names and the tautologies associated with them. Each of these rules is given in the form of simple valid argument the validity of which can be easily verified by a truth table or a truth tree. From these simple arguments, we can construct more complex valid ones. Together with the main logic equivalences of Table 1.4 above, the inference rules form a system to provide a *formal proof* of a valid argument. Such a proof consists of a finite sequence of numbered lines (with a justification for each line) with the conclusion of the argument at the last line. On each new line, either a hypothesis or a newly inferred formula (from Table 1.4 or Table 1.5) is introduced.

Example 1.47. The following is a valid argument:

$$(s \rightarrow t)$$
$$(\neg r \rightarrow s)$$
$$(\neg p \wedge q)$$
$$(r \rightarrow p)$$
$$\overline{}$$
$$\therefore t$$

Table 1.5   Inference rules propositional logic

Rule of Inference	Name	Associated Tautology
$(p \rightarrow q)$ p $\therefore q$	Modus ponens	$(((p \rightarrow q) \wedge p) \rightarrow q)$
$(p \rightarrow q)$ $\neg q$ $\therefore \neg p$	Modus tollens	$(((p \rightarrow q) \wedge \neg q) \rightarrow \neg p)$
$(p \vee q)$ $(\neg p \vee r)$ $\therefore (q \vee r)$	Resolution	$(((p \vee q) \wedge (\neg p \vee r)) \rightarrow (q \vee r))$
$(p \vee q)$ $\neg p$ $\therefore q$	Disjunctive Syllogism	$(((p \vee q) \wedge \neg p) \rightarrow q)$
$(p \rightarrow q)$ $(q \rightarrow r)$ $\therefore (p \rightarrow r)$	Hypothetical Syllogism	$(((p \rightarrow q) \wedge (q \rightarrow r)) \rightarrow (p \rightarrow r))$
$(p \wedge q)$ $\therefore p$	or $\wedge$-elimination	$((p \wedge q) \rightarrow p)$
p $\therefore (p \vee q)$	$\vee$-introduction	$(p \rightarrow (p \vee q))$
p q $\therefore (p \wedge q)$	$\wedge$-introduction	$((p \wedge q) \rightarrow (p \wedge q))$
$(p \rightarrow \mathbf{F})$ $\therefore \neg p$	Contradiction	$((p \rightarrow \mathbf{F}) \rightarrow \neg p)$
$(p \rightarrow r)$ $(q \rightarrow r)$ $\therefore ((p \vee q) \rightarrow r)$	Proof by cases	$(((p \rightarrow r) \wedge (q \rightarrow r)) \rightarrow ((p \vee q) \rightarrow r))$

The formal proof of the argument goes as follows.

1. $(s \rightarrow t)$ Premise
2. $(\neg r \rightarrow s)$ Premise
3. $(\neg p \wedge q)$ Premise
4. $(r \rightarrow p)$ Premise
5. $\neg p$ $\wedge$-elimination, 3
6. $\neg r$ Modus tollens, 4, 5
7. s Modus ponens, 2, 6
8. t Modus ponens, 1, 7.

Example 1.48. The argument with premises $(A \vee B)$, $((A \vee \neg C) \vee \neg B)$, $((B \to C) \wedge D)$ and conclusion A is valid. Its proof goes as follows.

   1. $(A \vee B)$ Premise
   2. $((A \vee \neg C) \vee \neg B)$ Premise
   3. $((B \to C) \wedge D)$ Premise
   4. $(B \to C)$ $\wedge$-elimination, 3
   5. $(A \vee (A \vee \neg C))$ Resolution, 1, 2
   6. $((A \vee A) \vee \neg C)$ Logic equivalence (Associativity of $\vee$), 5
   7. $(A \vee \neg C)$ Logic equivalence (Or-simplification), 6
   8. $(C \to A)$ Logic equivalence (Material implicaion), 7
   9. $(B \to A)$ Hypothetical syllogism, 4, 8
 10. $(\neg B \vee A)$ Logic equivalence (Material implicaion), 9
 11. $(A \vee A)$ Resolution, 1, 01
 12. A Logic equivalence (Or-simplification), 11

Remark 1.6. If we want to prove an implication $(p \to q)$ at some point of a formal proof (in particular, if the conclusion of the argument is $(p \to q)$), then we can add p as a premise with the justification: *"Hypothesis to prove $(p \to q)$"*. To insert a formula ϕ in a formal proof, we can also proceed by contradiction assuming $\neg \phi$ as a premise with the justification: *"Hypothesis to use contradiction"*. If this leads to a contradiction (something like $(\psi \wedge \neg\psi)$ for some formula ψ) in the proof, then we can infer ϕ.

Example 1.49. The following is a formal proof of the (valid) argument with premises $(\neg F \vee C)$, $((C \wedge D) \to (\neg A \to E))$, $(B \to (D \wedge F))$, $\neg E$ and conclusion $(B \to A)$.

   1. $(\neg F \vee C))$ Premise
   2. $((C \wedge D) \to (\neg A \to E))$ Premise
   3. $(B \to (D \wedge F))$ Premise
   4. $\neg E$ Premise
   5. B Hypothesis, to prove $(B \to A)$
   6. $(D \wedge F)$ Modus pollens, 3, 5
   7. F $\wedge$-elimination, 6
   8. $(F \to C)$ Logic equivalence (Material implication), 1
   9. C Modus pollens, 7, 8
 10. D $\wedge$-elimination, 6
 11. $(C \wedge D)$ $\wedge$-introduction, 9, 10
 12. $(\neg A \to E)$ Modus pollens, 2, 11
 13. $\neg\neg A$ Modus tollens, 4, 12
 14. A Logic equivalence (Double negation), 13

1.9.1   *Exercises*

(1) Consider the argument with $(C \to (A \to B))$, $\neg(A \to B)$ as premises and with $\neg C$ as conclusion. Prove that the argument is valid by proving that the associated implication is a tautology. Give a formal proof of the argument.

(2) In each case, indicate the inference rule used in the (valid) argument.

    $\star$(a) Joe will go to France or to Egypt. Joe cannot go to Egypt. Therefore, Joe will go to France.

    (b) Joe cannot go to Egypt. If Joe goes to France, he must go to Egypt. Therefore, Joe cannot go to France.

    (c) Joe goes to Egypt. Joe goes to France. Therefore, will goes to Egypt and to France.

    $\star$(d) Joe can program in Java. Therefore, Joe can program either in Java or in C++.

    (e) Joe can program in Java and in C++. Therefore, Joe can program in Java.

    (f) If Joe can program in Java, then he can program in C++. If Joe can program in C++, then he can program in C. Therefore, if Joe can program in Java then he can program in C.

    $\star$(g) If it rains, Joe goes fishing. If it snows, Joe goes fishing. Therefore, if it rains or it snows, Joe goes fishing.

$\star$(3) Consider the logic argument with premises $(\neg p \to r)$, $(p \to q)$, $(\neg r \lor s)$ and conclusion $(q \lor s)$.

    (a) Use a truth tree to prove that the argument is valid.

    (b) The following is a formal proof of the argument using the inference rules. Complete the proof by adding justification to each step in the proof.

Step	Justification
1. $(\neg p \to r)$	
2. $(p \to q)$	
3. $(\neg r \lor s)$	
4. $(r \to s)$	
5. $(\neg p \to s)$	
6. $(\neg q \to \neg p)$	
7. $(\neg q \to s)$	
8. $(q \lor s)$	

(4) For each of the following arguments, use a truth table or a truth tree to verify that the argument is valid. Then, give a formal proof of the

validity of the argument using basic logic equivalences and the rules of inference.

<table>
<tr><td>(a)</td><td>⋆(b)</td><td>(c)</td></tr>
</table>

$$
\begin{array}{ccc}
\begin{array}{c}
(D \to C) \\
(A \to (B \wedge C)) \\
\underline{(C \to A)} \\
\therefore (D \to B)
\end{array}
&
\begin{array}{c}
(r \to (s \vee t)) \\
(t \to u) \\
((\neg p \vee q) \to r) \\
\underline{(\neg s \wedge \neg u)} \\
\therefore p
\end{array}
&
\begin{array}{c}
(C \vee D) \\
(D \leftrightarrow (\neg C \to \neg A)) \\
\underline{(A \vee (B \wedge D))} \\
\therefore (C \to D)
\end{array}
\end{array}
$$

⋆(5) Consider the following argument: *The war is over only if the Klingon Kingdom surrenders. The Klingon Kingdom surrenders or the war is over if and only if King Namnar dies. For the Klingon Kingdom not to surrender, it is necessary and sufficient that either King Namnar dies or prince Shawra does not take the thrown. Therefore, the Klingon Kingdom surrenders unless prince Shawra does not take the thrown.*

  (a) Translate the argument into symbols of propositional logic.
  (b) Verify the validity of the argument using a truth tree.
  (c) Give a formal proof of the validity of the argument using the inference rules.

(6) Repeat the questions in the previous exercise for the following argument: *We go hunting today only if it is not cold but sunny. It is cold today but sunny. If it is sunny, we take the dirt road. Taking the dirt road is sufficient to get us to the cabin before sunset. Therefore, we don't go hunting today but we get to the cabin before sunset.*

1.10   Normal forms

Table 1.4 above shows in particular the following two logic equivalences: $(p \to q) \equiv (\neg p \vee q)$ and $(p \leftrightarrow q) \equiv ((p \wedge q) \vee (\neg p \wedge \neg q))$. This suggests that we can eliminate any appearance of the connectives $\to$ and $\leftrightarrow$ from a compound logic formula and restrict our logic connectives to elements of the set $\{\neg, \wedge, \vee\}$. This is important from a practical point of view as it allows an easy interpretation of logic expressions by computer programs. For the next definition, recall that a *literal* is either a propositional variable or the negation of a propositional variable. If P is a propositional variable, then P and $\neg P$ are called *complementary literals*.

Definition 1.9. A *disjunctive clause* (respectively *conjunctive clause*) is a disjunction (respectively a conjunction) of literals in which no complementary literals appear.

Example 1.50. $(P \vee \neg Q \vee R)$ is a disjunctive clause, $(\neg A \wedge \neg B \wedge \neg C)$ is a conjunctive clause. On the other hand, $\neg(A \wedge \neg B \wedge \neg C)$, $(A \wedge B \wedge C \wedge B \wedge \neg A)$ are neither since the first one is the negation of a conjunctive clause (hence not a clause) and the second contains complementary literals.

Remark 1.7. A clause (disjunctive or conjunctive) is completely determined by its set of literals. For example, in the context of disjunctive clauses (respectively, conjunctive clauses) we can look at the set $\{\neg A, B, C, \neg D\}$ as being the disjunctive clause $(\neg A \vee B \vee C \vee \neg D)$ (respectively $(\neg A \wedge B \wedge C \wedge \neg D)$). In particular, the empty set is a clause: this is an empty disjunction (or conjunction) of literals. Also, one literal can be considered as a disjunctive clause and a conjunctive clause at the same time.

Definition 1.10. A *Disjunctive Normal Form* (DNF for short) is a disjunction of a finite number of conjunctive clauses. A *Conjunctive Normal Form* (CNF for short) is a conjunction of a finite number of disjunctive clauses.

Example 1.51. The formula $((A \vee \neg B \vee C) \wedge (\neg A \vee \neg B \vee C) \wedge (A \vee B))$ is a CNF. The formula $((A \wedge B \wedge C) \vee (\neg A \wedge C))$ is a DNF.

Remark 1.8. Any nonempty clause is at the same time a DNF and a CNF. For instance, $(A \vee B \vee C)$ can be seen as one disjunctive clause and as such, it is a CNF. But since each of the literals A, B and C is a conjunctive clause on its own (Remark 1.7), the formula $(A \vee B \vee C)$ is also a DNF. It is also important to note that in a DNF or a CNF, not all logic variables need to appear in every clause. If every logic variable appears in every clause in a normal form, we say that the form is *complete*.

Example 1.52. The formula $\phi : ((A \rightarrow \neg B) \rightarrow C)$ is clearly not a DNF nor a CNF. But we can use the basic logic equivalences to find a DNF and a CNF equivalent to ϕ.

$$\begin{aligned}
((A \to \neg B) \to C) &\equiv (\neg(A \to \neg B) \vee C) \quad &&\text{(since } (p \to q) \equiv (\neg p \vee q)) \\
&\equiv (\neg(\neg A \vee \neg B) \vee C) \quad &&\text{(since } (p \to q) \equiv (\neg p \vee q)) \\
&\equiv ((\neg\neg A \wedge \neg\neg B) \vee C) \quad &&\text{(De Morgan law)} \\
&\equiv ((A \wedge B) \vee C) \quad &&\text{(since } \neg\neg p \equiv p).
\end{aligned}$$

The last formula $((A \wedge B) \vee C)$ is a DNF equivalent to ϕ. Applying the distributivity law, we get that $\phi \equiv ((A \vee C) \wedge (B \vee C))$ which gives a CNF equivalent to ϕ.

There is nothing special about the formula ϕ treated in the previous example. The following theorem guarantees that we can always find a DND and a CNF equivalent to any given logic formula. We state the theorem without giving the proof.

Theorem 1.2. *Every formula of propositional logic is logically equivalent to a formula in disjunctive normal form and to a formula in Conjunctive normal form.*

In general, there is more than one DNF (respectively CNF) equivalent to a logic formula ϕ but there exists a unique *complete* DNF (respectively CNF) equivalent to ϕ.

The first step in obtaining a normal form equivalent to a logic formula ϕ is to get rid of any occurrence of the connectives $\to$ and $\leftrightarrow$ appearing ϕ using the two logic equivalences given at the beginning of this section. Basic logic equivalences are then used to rearrange the formula into the required form. In the literature, one can find detailed algorithms to find a DNF and a CNF equivalent to a given logic formula. We will not be concerned with the details of these algorithms but rather, we explore different practical ways to get these normal forms through various examples.

Example 1.53. Find a DNF equivalent to the formula:

$$\phi : (((\neg A \leftrightarrow B) \to (C \vee A)) \wedge B).$$

Solution. For a DNF, we can use one of the following three methods.

(1) **The method of a truth table.** We have used this method before. It consists of locating all the valuations that satisfy the formula from the truth table. Each of these valuations corresponds to a conjunctive clause. The disjunction of all these conjunctive clauses forms a DNF

equivalent to the original formula. It is worth noting here that a DNF obtained from a truth table is always complete as every clause would contain every logic variable exactly once.

A	B	C	$\neg A$	$(\neg A \leftrightarrow B)$	$(C \vee A)$	$((\neg A \leftrightarrow B) \rightarrow (C \vee A))$	ϕ
T	T	T	F	F	T	T	T
T	T	F	F	F	T	T	T
T	F	T	F	T	T	T	F
T	F	F	F	T	T	T	F
F	T	T	T	T	T	T	T
F	T	F	T	T	F	F	F
F	F	T	T	F	T	T	F
F	F	F	T	F	F	T	F

The formula is true in rows 1, 2 and 5 which correspond to the conjunctive clauses $(A \wedge B \wedge C)$, $(A \wedge B \wedge \neg C)$ and $(\neg A \wedge B \wedge C)$ respectively. A DNF equivalent to ϕ is given by making the disjunction of the three clauses: $((A \wedge B \wedge C) \vee (A \wedge B \wedge \neg C) \vee (\neg A \wedge B \wedge C))$.

(2) **The method of algebraic manipulations.** Using the basic logic equivalences, this method usually starts by eliminating the connectives $\rightarrow$ and $\leftrightarrow$ from the formula. In the following, we show the algebraic manipulations to come up with a DNF for ϕ but we leave it to the reader to fill in the justification for each step.

$$
\begin{aligned}
\phi &\equiv ((\neg(\neg A \leftrightarrow B) \vee (C \vee A)) \wedge B) \\
&\equiv ((((\neg A \wedge \neg B) \vee (\neg\neg A \wedge B)) \vee (C \vee A)) \wedge B) \\
&\equiv ((\neg A \wedge \neg B \wedge B) \vee (\neg\neg A \wedge B \wedge B) \vee (C \wedge B) \vee (A \wedge B)) \\
&\equiv ((\neg A \wedge \mathbf{F}) \vee (A \wedge B) \vee (C \wedge B) \vee (A \wedge B)) \\
&\equiv (\mathbf{F} \vee (A \wedge B) \vee (C \wedge B) \vee (A \wedge B)) \\
&\equiv ((A \wedge B) \vee (C \wedge B) \vee (A \wedge B)) \\
&\equiv ((A \wedge B) \vee (C \wedge B))
\end{aligned}
$$

So $((A \wedge B) \vee (C \wedge B))$ is a DNF equivalent to ϕ. Note that this DNF is not complete.

(3) **The method of a truth tree.** A DNF of ϕ can be obtained from the truth table by locating the valuations that satisfy the formula. It should then come as no surprise that a DNF can also be found by following all open branches in a completed truth tree with ϕ at the root. Each open

branch at the end gives a conjunctive clause and a DNF equivalent to ϕ is obtained by forming the disjunction of all these clauses.

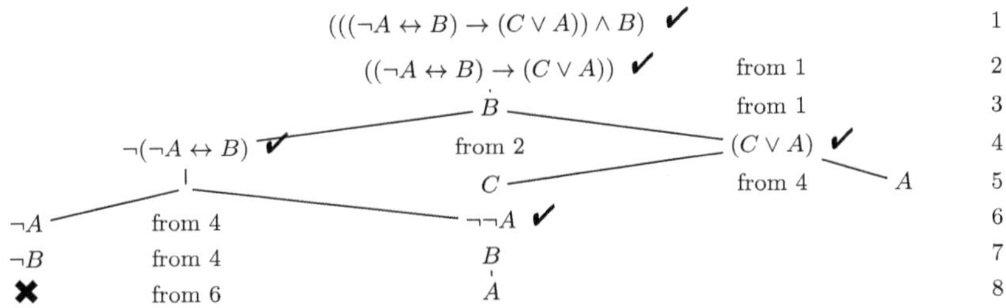

There are three open branches in the tree. The first one contributes with the clause $(A \wedge B)$, the second one with the clause $(B \wedge C)$ and the third one with the clause $(A \wedge B)$ once more (so it is irrelevant). We conclude that a DNF equivalent to ϕ is $((A \wedge B) \vee (B \wedge C))$. $\qquad \Diamond$

To find a CNF equivalent to a logic formula ϕ, only the methods of truth tables or algebraic manipulations are suitable. Using the truth table of ϕ, locate the rows where the formula is false. Each one of these rows contributes with a disjunctive clause to a CNF as follows. If the truth value of a logic variable A is **F** on that row, then A appears in the clause. If the truth value of A is **T** on the row, then $\neg A$ appears in the clause. Conjunction of all the corresponding disjunctive clauses gives a CNF equivalent to the formula. This gives a hint on why the truth tree method cannot be employed to find a CNF. The truth tree, when completed, reveals all the valuations for which the formula is true while we are interested in those that do not satisfy the formula. Let us use these two methods to find a CNF equivalent to the formula ϕ of the previous example.

(1) **The method of truth table.** We locate the rows where the formula is false, namely rows 3, 4, 6, 7 and 8. In row 3, A is true, B is false and C is true. The disjunctive clause corresponding to the third row is then $(\neg A \vee B \vee \neg C)$. Doing the same for rows 4, 6, 7 and 8 gives the following CNF equivalent to ϕ: $((\neg A \vee B \vee \neg C) \wedge (\neg A \vee B \vee C) \wedge (A \vee \neg B \vee C) \wedge (A \vee B \vee \neg C) \wedge (A \vee B \vee C)$. Like in the case of DNF, this is a complete CNF equivalent to ϕ.

(2) **The method of algebraic manipulations.** In the previous example, the method of algebraic manipulations gave us $((A \wedge B) \vee (C \wedge B))$ as a DNF equivalent to ϕ. We can use this form to find a CNF equivalent to ϕ: $((A \wedge B) \vee (C \wedge B)) \equiv ((A \vee C) \wedge B)$ (by the distributivity law). The expression $((A \vee C) \wedge B)$ is a (not complete) CNF equivalent to ϕ.

Example 1.54. Find a DNF and a CNF equivalent to the formula:

$$\psi : (((A \to \neg B) \vee (A \leftrightarrow C)) \to (\neg B \vee C)).$$

Solution. The truth table of ψ is the following.

A	B	C	$\neg B$	$(A \to \neg B)$	$(A \leftrightarrow C)$	$((A \to \neg B) \vee (A \leftrightarrow C))$	$(\neg B \vee C)$	ψ
T	T	T	F	F	T	T	T	T
T	T	F	F	F	F	F	F	T
T	F	T	T	T	T	T	T	T
T	F	F	T	T	F	T	T	T
F	T	T	F	T	F	T	T	T
F	T	F	F	T	T	T	F	F
F	F	T	T	T	F	T	T	T
F	F	F	T	T	T	T	T	T

and the completed truth tree of ψ is

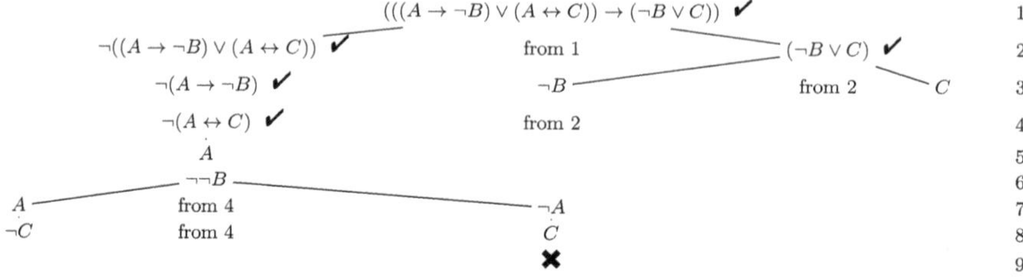

Using the truth table, a (complete) DNF equivalent to ψ is: $((A \wedge B \wedge C) \vee (A \wedge B \wedge \neg C) \vee (A \wedge \neg B \wedge C) \vee (A \wedge \neg B \wedge \neg C) \vee (\neg A \wedge B \wedge C) \vee (\neg A \wedge \neg B \wedge C) \vee (\neg A \wedge \neg B \wedge \neg C))$. For a CNF, the truth table of ψ shows only one valuation for which the formula is false and the corresponding disjunctive clause is $(A \vee \neg B \vee C)$. This is a (complete) CNF equivalent to ψ. A DNF equivalent to ψ can be formed from the truth tree of ψ. Note that the tree has three open branches, each corresponding to a conjunction clause. Forming the disjunction of all open branches yields the following (incomplete) DNF of ψ: $((A \wedge B \wedge \neg C) \vee \neg B \vee C)$. Algebraic manipulations can also be used to form a DNF and a CNF of the formula:

$$\psi \equiv (\neg((A \to \neg B) \vee (A \leftrightarrow C)) \vee (\neg B \vee C)) \text{(Material implication)}$$
$$\equiv ((\neg(A \to \neg B) \wedge \neg(A \leftrightarrow C)) \vee (\neg B \vee C)) \text{ (De Morgan)}$$
$$\equiv (((A \wedge \neg\neg B) \wedge ((A \wedge \neg C) \vee (\neg A \wedge C))) \vee (\neg B \vee C)) \text{ (Negation of } \to, \leftrightarrow)$$
$$\equiv (((A \wedge B) \wedge ((A \wedge \neg C) \vee (\neg A \wedge C))) \vee (\neg B \vee C)) \, (\neg\neg B \equiv B)$$
$$\equiv (((A \wedge B \wedge A \wedge \neg C) \vee (A \wedge B \wedge \neg A \wedge C)) \vee (\neg B \vee C)) \text{(Distributivity)}$$
$$\equiv (((A \wedge B \wedge \neg C) \vee (\mathbf{F} \wedge B \wedge \neg C)) \vee (\neg B \vee C)) (p \wedge p \equiv p,\ p \wedge \neg p \equiv \mathbf{F})$$
$$\equiv ((A \wedge B \wedge \neg C) \vee \mathbf{F} \vee (\neg B \vee C)) (p \wedge \mathbf{F} \equiv \mathbf{F})$$
$$\equiv ((A \wedge B \wedge \neg C) \vee \neg B \vee C) \text{(Since } p \vee \mathbf{F} \equiv p)$$

The formula $((A \wedge B \wedge \neg C) \vee \neg B \vee C)$ is a DNF equivalent to ψ. We can manipulate this DNF to get a CNF as follows: $\psi \equiv ((A \wedge B \wedge \neg C) \vee \neg B \vee C) \equiv ((A \vee \neg B \vee C) \wedge (B \vee \neg B \vee C) \wedge (\neg C \vee \neg B \vee C)) \equiv ((A \vee \neg B \vee C) \wedge (\mathbf{T} \vee C) \wedge (\mathbf{T} \vee \neg B)) \equiv ((A \vee \neg B \vee C) \wedge \mathbf{T} \wedge \mathbf{T}) \equiv (A \vee \neg B \vee C).$ $\Diamond$

1.10.1 *Exercises*

$\star$(1) The truth table of a formula ϕ containing three logic variables A, B and C is given below. Find a DNF and a CNF equivalent to ϕ.

A	B	C	ϕ
T	T	T	F
T	T	F	F
T	F	T	T
T	F	F	T
F	T	T	T
F	T	F	T
F	F	T	F
F	F	F	T

(2) For each of the following logic formulas, find a DNF equivalent to the formula using: (i) a truth table; (ii) a truth tree; (iii) algebraic manipulations.

   (a) $(A \to (B \leftrightarrow C))$
$\star$ (b) $((A \to B) \leftrightarrow (\neg A \to (B \wedge C)))$
   (c) $(\neg A \to (B \to \neg(A \to (B \wedge C))))$
   (d) $((A \to (B \leftrightarrow C)) \to (C \leftrightarrow (A \to \neg B)))$

(3) For each of the following logic formulas in the previous exercise, find a CNF equivalent to the formula using: (i) a truth table; (ii) algebraic manipulations.

$\star$(4) It is known that $((A \wedge \neg B) \vee (B \wedge \neg C))$ is a DNF equivalent to a logic formula ϕ containing three logic variables A, B and C. Find the complete DNF equivalent to ϕ.

 (5) It is known that $((A \vee \neg B) \wedge (B \wedge \neg C))$ is a CNF equivalent to a logic formula ψ containing three logic variables A, B and C. Find the complete CNF equivalent to ψ.

$\star$(6) Give a formula ϕ in DNF on three logic variables A, B and C such that ϕ is true for exactly three valuations of the logic variables.

 (7) Give a formula ϕ in CNF on three logic variables A, B and C such that ϕ is true for exactly three valuations of the logic variables.

Chapter 2

Set Theory and Introduction to Boolean Algebra: A Naïve Approach

2.1 Introduction

A building block of every mathematical theory is the concept of a *set*. Mathematical models and structures are generally based on a "well-defined" collection of objects called the *elements* or the *members* of that collection. The term *well-defined* in this context refers to the fact that there is no ambiguity in determining if a given object is a member of the collection or not. For example, if $\mathcal{C}$ is the collection of all real numbers that satisfy the equation $x^2 - 3x + 2 = 0$, then clearly 1 and 2 are members of $\mathcal{C}$ (in fact the only members) since they satisfy the equation but -2 and $\sqrt{2}$ are not. Unfortunately, things are not always that clear. Consider for example the collection $\mathcal{D}$ of all "good" students in a Discrete Math course. If Joe is a student in this course, it is not well-defined how to determine if he is a member of $\mathcal{D}$ unless we give a clear and precise definition of what a "good student" means. Another example is the collection $\mathcal{W}$ of all wealthy families in a certain country. What threshold of wealth determines if a family belongs to $\mathcal{F}$?

Set theory was founded in the late nineteenth century by German mathematician Georg Cantor (1845–1918). Cantor was the first to give a formal definition of a set, and his approach was based on the fact that the collection of all "objects" satisfying a certain condition P is a well-defined set. Cantor's approach was strongly opposed by some of his contemporary mathematicians as they saw in it some ambiguities that could lead to paradoxes, like the famous Russell paradox. To explain this paradox, consider the following scenario (known as the barber paradox). In a certain town, Joe is the only barber and he shaves everyone who does not shave

65

himself. Let S be the collection of all people of the town who do not shave themselves. Then Joe is either an element of S or he is not. If Joe is an element of S, then he does not shave himself (by definition of S) and so he must shave himself since he shaves everyone in S (as being the barber). This is clearly a contradiction. If Joe is not an element of S, then he shaves himself and so he (as the barber) does not shave himself. This also leads to a contradiction. To avoid such situations leading to paradoxes, an *axiomatic approach* to set theory was developed by Ernst Zermelo (1871–1953) in 1908 and later modified by Abraham Fraenkel (1891–1965) to form what is now known as Zermelo-Fraenkel set theory.

In this book, we look at set theory from Cantor's perspective and we will not be concerned with the axiomatic point of view. In this framework, the term "object" is used loosely and intuitively. The good news is that from a practical point of view, this "naive" approach to set theory is all what we need to build on for the other topics in the coming chapters.

2.2   Basic definitions and terminology

A **set** is a *well-defined* collection of *distinct* objects called the *elements* of the set. Some applications allow repeating elements in the same collection in which case the collection is called a *multiset*. In what follows, we assume (unless otherwise specified) that no repetition of elements occur in a given set. Sets are usually denoted by capital letters A, B, C, ... and their elements by lowercase letters a, b, c, ... If A is a set and x is an element of A, we say that x *belongs* to A and we write $x \in A$ for short. If x is not an element of A, we write $x \notin A$. If in a certain context all sets are parts of a set $\mathcal{U}$, we say that $\mathcal{U}$ is a *universal set* (or simply a *universe*). In other words, a universal set is a set containing all required elements in a particular situation. There exists a set that does not contain any element that we call the *empty set* and we denote by $\emptyset$ or $\{\}$. So the statement "$x \in \emptyset$" is always false for any object x. A set that contains only one element is called a *singleton*.

Example 2.1.

(a) A standard deck of playing cards is a set with every card as an element.

(b) The collection of all prime numbers between 10 and 20 is a set. Its consists of the elements 11, 13, 17, and 19.

(c) The collection A of all integers n such that $n^2 \leq 16$ is a set. The elements of A are -4, -3, -2, -1, 0, 1, 2, 3 and 4.

(d) The collection of all large cities in the world is not a set since the term "large city" is not well-defined.

Number sets appear frequently in many areas of mathematics. The following are the universal notations used for these sets.

- $\mathbb{N}$ is the set of all natural numbers: $0, 1, 2, \ldots$.
- $\mathbb{Z}$ is the set of all integers: $\ldots, -3, -2, -1, 0, 1, 2, 3 \ldots$.
- $\mathbb{Q}$ is the set of all rational numbers: $\frac{a}{b}$ where $a, b \in \mathbb{Z}$ with $b \neq 0$.
- $\mathbb{R}$ is the set of all real numbers.
- (For the readers familiar with this class of numbers) $\mathbb{C}$ is the set of all complex numbers: $a + ib$ where $a, b \in \mathbb{R}$ and i is the imaginary number satisfying $i^2 = -1$.

2.2.1 *Describing sets*

There are several ways to describe a set:

(1) **The Roster notation.** Using this notation, elements of the set are listed ia row, inside curly brackets.

Example 2.2. The set A of all prime numbers between 10 and 20 can be expressed in roster notation as $A = \{11, 13, 17, 19\}$.

Example 2.3. The set of all integers n such that $n^2 \leq 16$ can be expressed as $Q = \{-4, -3, -2, -1, 0, 1, 2, 3, 4\}$ in roster notation.

Example 2.4. The set of all English vowels is $\{a, e, i, o, u\}$.

The order in which the elements of a set appear in roster notation is not important. For instance $\{1, 2, 3\}$ and $\{3, 2, 1\}$ represent the same set.

Remark 2.1. Elements of a set can be any objects, including sets. For example, $A = \{1, \{1\}, \{\{1\}\}, \{1, \{1\}\}\}$ is a set containing four elements: 1, $\{1\}$, $\{\{1\}\}$ and $\{1, \{1\}\}$.

(2) **The set-builder notation.** Obviously, it is not always possible to list all the elemnts of a set. Another way to represent a set is to specify the rules (or properties) that determine the set membership. For example, the notation $\{x; \ x \in \mathbb{R}$ and $x \geq 2\}$ represents the set of

all real numbers that are greater than or equal to 2. This is simply the interval $[2, \infty[$ (see Section 2.3.1 below). The semicolon ";" used in this notation (just after the variable x) sands for "such that". Sometimes a vertical line "|" is used for the same purpose.

Example 2.5. The set $S = \{(x, y); \; x, y \in \mathbb{R} \text{ and } y - 2x + 1 = 0\}$ represents the line $y = 2x - 1$ in the Cartesian plane.

Example 2.6. The set $S = \{x \in \mathbb{Z}, x = n^2 \text{ for some integer } n\}$ is the collection of all numbers which are squares of integers. So $1, 4, 16, 25, 36$ are elements of S, while $3, 8, 11$ are not.

Example 2.7. Consider the set $A = \{x \in \mathbb{N}; \; 2x^2 + x - 1 = 0\}$. The roots of the quadratic equation $2x^2 + x - 1 = 0$ are -1 and $\frac{1}{2}$. Since -1 and $\frac{1}{2}$ are not elements of $\mathbb{N}$, $A = \emptyset$ (the empty set). If we change the definition of the set to $A = \{x \in \mathbb{Z}; \; 2x^2 + x - 1 = 0\}$, then A is singleton $\{-1\}$.

Remark 2.2. There are often different ways to describe a given set using set builder notation. For example, the set $\{1, 2, 3, 6\}$ can be described as $\{n \in \mathbb{N}; n \text{ is a divisor of } 6\}$. The same set can also be described as $\{x \in \mathbb{R}; (x - 1)(x - 2)(x - 3)(x - 6) = 0\}$.

(3) **Recursive definition.** Very often, sets are too large to list all their elements in a roster manner or there are no convenient or clear properties that determine their membership. A recursive approach could be useful to define the set in this case. Given a set A (part of a universal set $\mathcal{U}$), a recursive definition of A consists of a procedure that allows the generation new elements of A using a list of initial elements of A and a set of rules. More formally, a recursive definition of a set A consists of three clauses: (1) The *Basis clause*, (2) The *Recursive clause* and (3) The *Terminal clause*. In the basis clause, we declare some elements of the universe $\mathcal{U}$ to be elements of A. This will ensure in particular that A is not empty. In the recursive clause, we give a set of one or more rules to generate new elements of A from elements already known to be in the set. Finally, in the terminal clause, we declare that the first two clauses are the only way to generate the elements of A. That is, an element x of the universe $\mathcal{U}$ is in A if and only if x is one of the initial elements given in the basis clause or x can be obtained by applying a finite number of rules given in the recursive clause to elements already known to be in A.

Example 2.8. In the universe $\mathcal{U} = \mathbb{Z}$ of all integers, consider the set A defined recursively as follows:

(1) **Basis clause.** $0 \in A$.
(2) **Recursive clause.** If $x \in A$, then $x + 1 \in A$.
(3) **Terminal clause.** Nothing else is in A unless it is obtained from the Basis and Recursive Clauses.

Since $0 \in A$ (basis clause), $0 + 1 = 1 \in A$ by the recursive clause. Now that we know that $1 \in A$, $2 = 1 + 1 \in A$, $3 = 2 + 1 \in A, \ldots$ and hence every non-negative integer is in A. Of course, we still need a formal way to prove this observation. This requires a technique we will study later.

Example 2.9. The set S of all binary strings can be defined recursively as follows.

(1) **Basis clause.** $\lambda \in S$, where λ the empty string.
(2) **Recursive clause.** If $x \in S$, then $x0$, $0x$, $x1$ and $1x$ are in S.
(3) **Terminal clause.** Nothing else is in S unless it is obtained from the Basis and Recursive Clauses.

Example 2.10. In the universe $\mathcal{U} = \mathbb{N}$, consider the set A defined recursively as follows:

(1) **Basis clause.** $3 \in A$.
(2) **Recursive clause.** If $x, y \in A$, then $x - y \in A$ and $x + y \in A$.
(3) **Terminal clause.** Nothing else is in A unless it is obtained from the Basis and Recursive Clauses.

Since $3 \in A$, $3 - 3 = 0 \in A$ and $3 + 3 = 6 \in A$. Since $3 \in A$ and $6 \in A$, $3 + 6 = 9 \in A$. Similarly, we can prove that 12, $15, \ldots$ are also elements of A. It looks like A is the set of all multiple of 3 in $\mathbb{N}$. But that isjust an observation, not a formal proof.

When defining a set recursively, we assume that the terminal clause holds even if it is not included in the definition.

Remark 2.3. For a set A defined recursively, to prove that $x \in A$ we must show that it is either one of the elements listed in the basis clause or it can be generated from already known elements of A using a finite number of applications of the rules given in the recursive step. Proving that $x \notin A$ is, however, more difficult.

(4) **The Venn diagram.** This is a visual way of representing sets, which can be helpful to understand set operations like intersection, union and

others. In a typical Venn diagram, the universal set $\mathcal{U}$ is represented by the interior of a rectangle $\mathcal{R}$ and other sets are represented by the interior of regions (often circles) inside $\mathcal{R}$.

Example 2.11. The following represents the Venn diagram of three sets A, B and C inside a universal set $\mathcal{U}$ with the area in gray representing the elements in A and B but not in C:

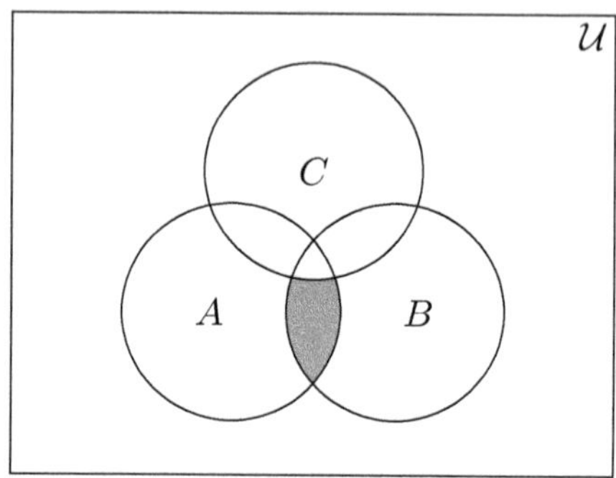

While Venn diagrams are useful to visualize properties of sets, they should not be used as proofs of these properties.

2.2.2   *Set equality: Finite and infinite sets*

Since a set is completely determined by its elements. two sets are the same (or equal) if and only if they have the same elements. The equality of sets is expressed formally with the *Axiom of Extensionality*:

Axiom of Extensionality. Two sets A and B are *equal,* and we write $A = B$, if and only if A contains every element of B and B contains every element of A (the two sets have exactly the same elements).

As a consequence of the above axiom, sets A and B are *different,* and we write $A \neq B$, if at least one element of A is not an element of B or at least one element of B is not an element of A.

Example 2.12. The sets $\{a, b, c\}$, $\{b, a, c\}$, $\{c, a, b\}$ are all equal since all three have the same elements a, b and c. The sets $A = \{\{1\}, 2, 3\}$ and $B = \{1, 2, 3\}$ are not equal since $\{1\}$ is an element of A but not of B (or $1 \in B$ but $1 \notin A$).

Example 2.13. The sets $A = \{x \in \mathbb{R}; x^2 - 3x + 2 = 0\}$ and $B = \{n \in \mathbb{N}; n$ is a divisor of $3\}$ are equal. They have the same elements, namely 1 and 3.

We often need to know the "size" of various sets in discrete mathematics. That is, we need to know if a set contains a finite number of elements and if it does, we need to compute that number.

Definition 2.1. A set A is called *finite* if it contains a finite number of elements. If a set is not finite, it is called *infinite*. For a finite set A, the number of elements of A is called the *cardinality* of A and is denoted by $|A|$ or by $\#A$.

It should noted that the notion of cardinality can be extended to infinite sets. To have a proper definition of infinite cardinality, one needs to notion of a bijective function. In this chapter, the word "cardinality" refers to the number of elements in a finite set.

Example 2.14.

(a) Number sets $\mathbb{N}$, $\mathbb{Z}$, $\mathbb{Q}$ and $\mathbb{R}$ are all infinite.
(b) The set $A = \{x \in \mathbb{R}; x^5 - x = 0\}$ is finite with cardinality 3 since the only real numbers satisfy the equation $x^5 - x = 0$ are $x = -1$, $x = 0$ and $x = 1$.
(c) The empty set is a finite set of cardinality zero.
(d) The set A of all three-digit integers with sum of digits equals to 3 is finite with cardinality 6 since $A = \{102, 111, 120, 201, 210, 300\}$.

2.2.3   *Exercises*

(1) In each case, determine if the collection represents a well-defined set.

   (a) All tall buildings in New York city.
 ⋆(b) All experts.
   (c) All professors at the University of Ottawa.
   (d) All old citizens in Canada.
 ⋆(e) All integers with absolute value less than or equal to 30.
   (f) All words in the English language dictionary having at most three vowels.
 ⋆(g) All large integers.

(2) In each case, write the set using the roaster notation.

   (a) The set of all colors of a rainbow.

  ⋆(b) The set of all integers between $-\frac{21}{2}$ and $\frac{5}{2}$.

   (c) The set of all real roots of the equation $x^4 - 1 = 0$.

   (d) The set of all planets in the solar system.

   (e) The set of all integer multiples of 7 between -22 and 82.

  ⋆(f) The set of all natural numbers that are less than or equal to 100, divisible by 3 and by 5 at the same time.

(3) In each case, write the set A using a set-building notation: $\{x \in \mathcal{U};\ P(x)\}$. Specify the Universe $\mathcal{U}$ and the property $P(x)$. Note that there is more than one possible answer in each case.

   (a) $A = \{0, 1, 4, 9, 16, 25, 36, \ldots\}$.

  ⋆(b) $A = \{-\sqrt{2}, -1, 0, 1, \sqrt{2}\}$.

   (c) $A = \{2, 3, 5, 7, 11, 13, 17, 19\}$.

  ⋆(d) $A = \{1, 2, 4, 8, 16, 32, \ldots\}$.

   (e) A is the set of all real numbers strictly between -2 and 2.

(4) In each case, determine if the statement "$x \in A$" is true or false. Justify your answer.

   (a) $x = 231$, $A = \{n \in \mathbb{N};\ n$ is an odd multiple of 3$\}$.

   (b) $x = 53361$, $A = \{n \in \mathbb{N};\ n$ is a perfect square$\}$.

  ⋆(c) $x = \sqrt[3]{2}$, $A = \{y \in \mathbb{R};\ y^n \in \mathbb{N}$ for some integer $n\}$.

   (d) $x = \frac{2}{3}$, $A = \{a \in \mathbb{R};\ p(a) = 0$ for some polynomial of degree 2$\}$.

   (e) $x = 117$, A is the set of all natural numbers either divisible par 11 or have a remainder of 3 upon division by 7.

  ⋆(f) $x = -2$, $A = \{y \in \mathbb{R};\ |y^2 - 10| \leq 5\}$

(5) In each case, determine if the statement is true or false. Justify your answer.

  ⋆(a) $1 \in \{-1, \{1\}, \{1, \{1\}\}\}$ (e) $5 \notin \{1, 3, 5\}$

   (b) $\{1\} \in \{-1, \{1\}, \{1, \{1\}\}\}$ (f) $0 \in \emptyset$

   (c) $\{1, 2\} \in \{1, 2, 3, 4\}$ (g) $J \in \{Joe\}$

  ⋆(d) $\{5\} \notin \{1, 3, 5\}$ ⋆(h) $1011 \in \{10, 11, 1001, 0011, 1001, 1111\}$

(6) In each case, determine if the set equality is true or false. Justify your answer.

   (a) $\{1, \{1\}\} = \{\{1\}, 1\}$.

  ⋆(b) $\{1, \{1\}\} = \{1\}$.

   (c) $\{1, \{1\}, 2\} = \{\{2\}, 1, \{1\}\}$
   (d) $\{x \in \mathbb{R};\ x = x^2\} = \{x \in \mathbb{R};\ x^4 = x\}$.
   (e) $\{\emptyset\} = \{0\}$.
$\star$(f) $\{x \in \mathbb{R};\ x^2 - 4x + 5 = 0\} = \{x \in \mathbb{Z};\ x^2 + x - 1 = 0\}$.

(7) Which of the following are empty sets.

   (a) A is the set of all European countries sharing a border with Mexico.
$\star$(b) $A = \{n \in \mathbb{R};\ 2n^2 - n + 3 = 0\}$.
   (c) $A = \{n \in \mathbb{N};\ 2n^2 - n - 3 = 0\}$.
   (d) $A = \{n \in \mathbb{N};\ n$ is odd number and divisible by $6\}$.
   (e) A is the set of all intersection points of the two lines $y = x$ and $y = x + 1$ in the Cartesian plan.
$\star$(f) $A = \{\{\emptyset\}\}$.
   (g) $A = \{x \in \mathbb{R};\ ||x + 2| - 2| \leq 1\}$.

(8) In each case, determine if the set is finite or infinite. If the set is finite, give its cardinality.

   (a) $\{n \in \mathbb{N};\ n > 34\}$
$\star$(b) $\{n \in \mathbb{Z};\ -5 \leq n < 5\}$
$\star$(c) $\{n \in \mathbb{R};\ -5 \leq n < 5\}$
   (d) $\{n \in \mathbb{Q};\ -5 \leq n < 5\}$
   (e) $\{n \in \mathbb{Z};\ |n| \leq 10\}$
   (f) $\{n \in \mathbb{R};\ |n| \leq 10\}$
$\star$(g) The set of all binary strings starting with 0.
   (h) The set of all binary strings of length 4 starting with 0
   (i) The set of all ordered pairs (x, y) where $x, y \in \mathbb{N}$ with $1 \leq x < 7$ and $3 \leq y \leq 12$

(9) In each case, the set A is defined recursively by providing: (i) a basis clause and (ii) a recursive clause. Give four elements of A different from those given at the basis clause.

$\star$(a) (i) $1 \in A$
     (ii) If $x \in A$, then $3x^2 - x + 2 \in A$.

  (b) (i) $1, 2 \in A$
     (ii) If $x, y \in A$, then $2x + 3y + 2 \in A$.

$\star$(c) (i) $(0, 0) \in A$
     (ii) If $(a, b) \in A$, then $(a+1, b) \in A$, $(a, b+1) \in A$, $(a+2, b+1) \in A$

  (d) (i) $01, 10, 11 \in A$
     (ii) If $x \in A$, then $x01x \in A$.

(10) In each case, give a recursive definition of the set A. You do not need to prove that your definition works.

  (a) A is the set of all *even* natural numbers.
  ⋆(b) A is the set of all *odd* natural numbers.
  (c) A is the set of all binary strings ending with 0.
  (d) A is the set of all binary strings with odd length.

(11) In each case, a set A of *natural numbers* is defined recursively (with the terminal clause assumed true but omitted). Describe, using words, the elements of A. No justification is required.

  (a)   (i) $5 \in A$
    (ii) If $x, y \in A$, then $x + y \in A$.

  ⋆(b)   (i) $1, 2, 3, 4 \in A$
    (ii) If $x, y \in A$, then $x + 5 \in A$.

  (c)   (i) $1 \in A$
    (ii) If $x \in A$, then $2x \in A$.

  (d)   (i) $2 \in A$
    (ii) If $x \in A$, then $x + 3 \in A$.

(12) In each case, a set A of *integers* is defined recursively (with the terminal clause assumed true but omitted). Describe, using words, the elements of A. No justification is required.

  (a)   (i) $1, 2 \in A$
    (ii) If $x \in A$, then $x + 2 \in A$ and $x - 2 \in A$.

  ⋆(b)   (i) $0 \in A$
    (ii) If $x \in A$, then $x - 3 \in A$.

  (c)   (i) $-1 \in A$
    (ii) If $x \in A$, then $x^2 - 1 \in A$.

(13) In each case, a set A of binary strings is defined recursively (with the terminal clause assumed true but omitted). As usual, if $x, y \in A$ then xy means the concatenation of x and y (the string x followed by the string y) and the symbol λ stands for the empty string. Give a description of the elements of A. No justification is required.

  ⋆(a)   (i) $\lambda \in A$
    (ii) If $x \in A$, then $1x0 \in A$.

  (b)   (i) $0, 1 \in A$
    (ii) If $x \in A$, then $0x, x0, 1x, x1 \in A$.

  (c)   (i) $0 \in A$
    (ii) If $x \in A$, then $x0, x1 \in A$.

  ⋆(d)   (i) $\lambda \in A$
    (ii) If $x \in A$, then $1x, x1, 00x, x00 \in A$.

(14) A set A of integers is defined recursively as follows: (i) $-1, 1 \in A$; (ii) If $x \in A$, then $2x + 1 \in A$ with the assumption that nothing else is in A. True or False: A is the set of all *odd* integers.

★(15) In the universe $\mathcal{U} = \{(x, y); x, y \in \mathbb{R}\}$ of all ordered pairs of real numbers, a set A is defined recursively below. Give ten distinct elements of the set A. Give a description of the elements of A as a subset of the Cartesian plane.

   (i) $(0, 0) \in A$
   (ii) If $(m, n) \in A$, then $(m + 1, n) \in A$ and $(m, n + 1) \in A$
   (iii) Nothing else is in A.

(16) Well-parenthesized expressions are important in mathematics and computer science. For example, (), ()() and ((())) are well-defined parentheses sequences whereas ((, (()(and ()) are not (do you see why?). Give a recursive definition of the collection of all well-defined and non-empty parentheses sequences.

2.3 Subsets and power set

In this section, all sets considered are parts of a universal set $\mathcal{U}$.

Definition 2.2. We say that a set B is a *subset* of a set A, and we write $B \subseteq A$, if every element of B is at the same time an element of A. We say that B is a *proper* subset of A, and we write $B \subset A$, if B is a subset of A which is not equal to A.

The following is a Venn diagram representation of the set inclusion $B \subseteq A$:

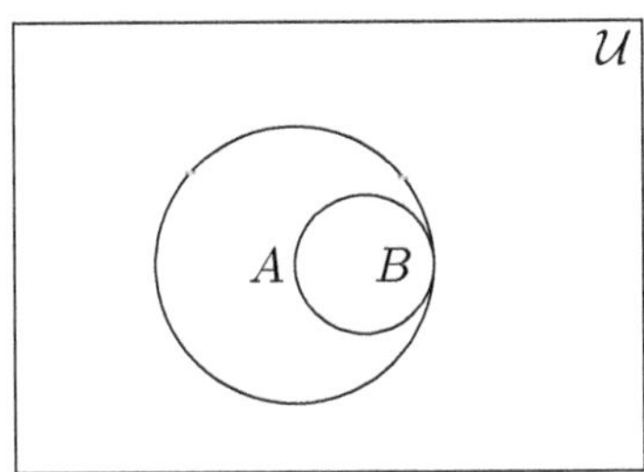

Remark 2.4.

(1) If $B \subset A$ is true, then $B \subseteq A$ is also true. We often use the notation $B \subseteq A$ for any subset B of A, proper or not. The notation $B \subset A$

is only used when we need to emphasis the fact that B is a proper subset of A.

(2) Every set is subset of itself.

(3) Since the statement "$x \in \emptyset$" is a contradiction (always false) for any $x \in \mathcal{U}$, the implication ($x \in \emptyset \Rightarrow x \in A$) is true for any set A. This shows that the empty set $\emptyset$ is a subset of any given set.

(4) The empty set has no proper subsets.

(5) If $B \subseteq A$ and A is a finite set, then B is also finite and $|B| \leq |A|$.

Example 2.15. Clearly $\mathbb{N} \subset \mathbb{Z} \subset \mathbb{Q} \subset \mathbb{R} \subset \mathbb{C}$.

Example 2.16. Let $A = \{\{1\}, 1, 2\}$. Give the truth value of each of the following statements.

(a) $\{1\} \in A$	(d) $\{2\} \subseteq A$	(g) $\emptyset \in A$
(b) $\{1\} \subseteq A$	(e) $\{1,2\} \in A$	(h) $\{\emptyset\} \subseteq A$
(c) $\{2\} \in A$	(f) $\{1,2\} \subseteq A$	(i) $\{\{1\}, 1\} \subseteq A$

Solution.   (a) True: $\{1\}$ is an element of the set A.

(b) True: $\{1\}$ has a unique element (1) which is also an element of A.

(c) False: $\{2\}$ is not an element of A (2 is but not $\{2\}$).

(d) True: $\{2\}$ has a unique element (2) which is also an element of A.

(e) False: $\{1,2\}$ is not an element of A.

(f) True: $1 \in A$ and $2 \in A$, so $\{1,2\} \subseteq A$.

(g) False: the empty set is not an element of A.

(h) False: if $\{\emptyset\} \subseteq A$, then $\emptyset \in A$ which is false.

(i) True: $\{1\} \in A$, $1 \in A$, so $\{\{1\}, 1\} \subseteq A$. $\Diamond$

Set equality can be expressed in terms of inclusion as indicated in the following theorem.

Theorem 2.1. *Given two sets A and B, then $A = B$ if and only if $A \subseteq B$ and $B \subseteq A$.*

Proof.   Assume first that $A = B$. Then for any $x \in A$, $x \in B$ and so $A \subseteq B$. Similarly, $B \subseteq A$. Conversely, if $A \subseteq B$ and $B \subseteq A$ then any element of A is an element of B and vice versa. This proves $A = B$. $\square$

Example 2.17. Let A be the subset of $\mathbb{Z}$ of all multiples of 12, B the subset of $\mathbb{Z}$ of all multiples of 4 and of 6 at the same time. Prove that $A = B$.

Solution. We prove that $A \subseteq B$ and $B \subseteq A$. Let $x \in A$ be an arbitrary element. Then $x = 12k$ for some $k \in \mathbb{Z}$. Note that $x = 12k = 4(3k) = 6(2k)$ with both $3k$, $2k \in \mathbb{Z}$. This shows in particular that x is a multiple of 4 and 6 at the same time and so $x \in B$. This proves that $A \subseteq B$. For the other inclusion, let $x \in B$ be an arbitrary element. Then $x = 4s$ and $x = 6t$ for some $s, t \in \mathbb{Z}$. This implies that $4s = 6t$ or $2s = 3t$. In particular, $3t$ must be even (as $2s$ is even). This is only possible if t is even (since 3 is odd). Write $t = 2n$ for some $n \in \mathbb{Z}$. Then $x = 6t = 6(2n) = 12n$: a multiple of 12. We conclude that $x \in A$ and hence $B \subseteq A$. $\diamond$

Definition 2.3. The *powerset* of a set A, denoted by $\wp(A)$ (or sometimes 2^A), is defined as the set of all subsets of A.

Elements of the powerset $\wp(A)$ of a set A are precisely all subsets of A:

$$(\mathcal{X} \in \wp(A) \Leftrightarrow \mathcal{X} \subseteq A).$$

If A is any set, then $\emptyset$ and A are both elements of $\wp(A)$ (by Remark 2.4 above) and the other elements of $\wp(A)$ are simply the proper subsets of A.

Example 2.18. The power set of $A = \{a, b, c\}$ is:

$$\wp(A) = \{\emptyset, \{a\}, \{b\}, \{c\}, \{a, b\}, \{a, c\}, \{b, c\}, \{a, b, c\}\}.$$

In the last example, the cardinality of A is 3 and that of $\wp(A)$ is $8 = 2^3$. This is no coincidence as shown in the following theorem. There are different ways to prove the theorem. We will present one using the principle of mathematical induction in the next chapter. For now, we just state the result.

Theorem 2.2. *If A is a finite set with $|A| = n$, then $|\wp(A)| = 2^n$.*

Example 2.19. For each of the following statements, determine if it is true or false. Justify your answer.

(a) $\{1\} \in \wp(\{\{1\}, 1, 2\})$
(b) $\{\{1, 2\}\} \subseteq \wp(\{\{1\}, 1, 2\})$
(c) $\{\{1\}, \{1, \{2\}\}\} \subseteq \wp(\{\{1\}, 1, 2\})$
(d) $\{\{\{1\}\}\} \in \wp(\wp(\{\{1\}, 1, 2\}))$
(e) If $A = \{0, \{0\}, \{\{0\}\}, \{0, \{0\}\}\}$, then $|\wp(A)| = 32$
(f) There exists a set A such that $|\wp(A)| = 4094$.

Solution.   (a) True since $\{1\} \subseteq \{\{1\}, 1, 2\}$.

(b) True since $\{1, 2\} \in \wp(\{\{1\}, 1, 2\})$ and the inclusion $\{\{1, 2\}\} \subseteq \wp(\{\{1\}, 1, 2\})$ is true.

(c) False since $\{1, \{2\}\}$ is not a subset $\{\{1\}, 1, 2\}$.

(d) True. The statement is equivalent to $\{\{\{1\}\}\} \subseteq \wp(\{\{1\}, 1, 2\})$ which in turns is equivalent to $\{\{1\}\} \in \wp(\{\{1\}, 1, 2\})$. The last statement is equivalent to $\{\{1\}\} \subseteq \{\{1\}, 1, 2\}$ which is true since $\{1\} \in \{\{1\}, 1, 2\}$.

(e) False since $|\wp(A)| = 2^{|A|} = 2^4 = 16$.

(f) False. If such a set A exists, then $2^{|A|} = 4094$ which implies that $|A| = \frac{\ln(4094)}{\ln(2)}$. Since $\frac{\ln(4094)}{\ln(2)} \approx 11.99$ is not an integer, such a set A does not exist. $\Diamond$

2.3.1   *Intervals in $\mathbb{R}$*

The set $\mathbb{R}$ of real numbers is at the foundation of many mathematical theories. Notions like domain, range, continuity, differentiability, convergence and others are often defined on certain subsets of $\mathbb{R}$ called intervals. This type of subsets of $\mathbb{R}$ draws its properties from the fact that there exists a natural order on real numbers. If $x, y \in \mathbb{R}$, the notation $x < y$ means x is *strictly less* than y and $y > x$ means that y is *strictly greater* than x. The notation $x \leq y$ means that x is less than or equal to y and $x \geq y$ means that x is greater than or equal to y. This ordering on the reals allows us to conveniently think of $\mathbb{R}$ as a (continuous) set of points on a straight line, known as the **real line**. We normally choose 0 as a reference point on the line, so any number to the right (respectively, to the left) of 0 is referred to as positive (respectively, negative). If $x < y$, then that would translate to x being to the left of y on the real line (or y is to the right of x). We also make the distinction between non-negative and positive numbers. A real number is non-negative (respectively, non-positive) if it is either 0 or positive (respectively, 0 or negative). The following is an illustration of the set $\mathbb{R}$ as a line together with some elements (points) on it:

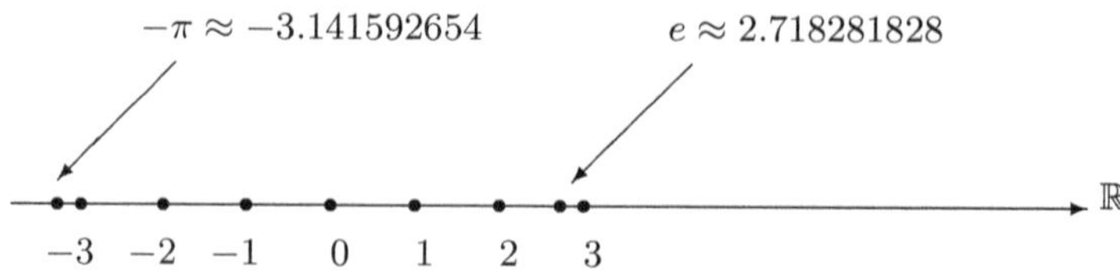

For real numbers a, b with $a < b$, the subset $\{x \in \mathbb{R};\ a < x < b\}$ of $\mathbb{R}$ is called an *open interval* that we denote by $]a, b[$ or (a, b) (in this book, we use square brackets for intervals). Similarly, the subsets $\{x \in \mathbb{R};\ a \leq x < b\}$, $\{x \in \mathbb{R};\ a < x \leq b\}$ and $\{x \in \mathbb{R};\ a \leq x \leq b\}$ of $\mathbb{R}$ are referred to using the interval notations as $[a, b[$, $]a, b]$ and $[a, b]$ respectively. The subset $[a, b]$ (that includes the two endpoints) is called a *closed interval*. A standard way of graphically representing intervals on the real line is to use a thick line segment with filled or/and empty circles at its ends. Empty circles are used to exclude points while filled ones are used to include points in the interval. This is illustrated in the following diagrams.

- $]a, b[= \{x \in \mathbb{R};\ a < x < b\}$:

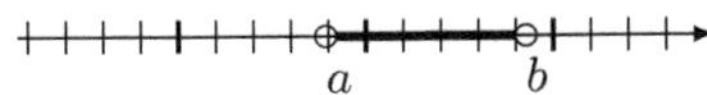

- $[a, b] = \{x \in \mathbb{R};\ a \leq x \leq b\}$:

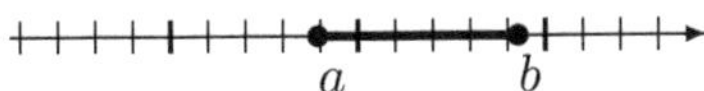

- $[a, b[= \{x \in \mathbb{R};\ a \leq x < b\}$:

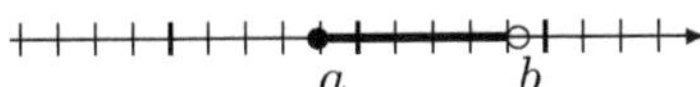

- $]a, b] = \{x \in \mathbb{R};\ a < x \leq b\}$:

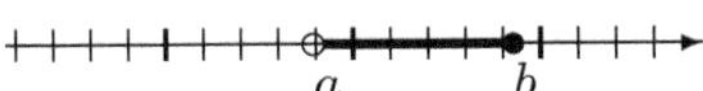

The following are some special (unbounded) intervals:

- $]a, \infty[= \{x \in \mathbb{R};\ x > a\}$:

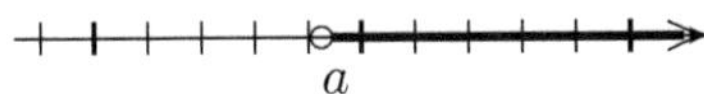

- $[a, \infty[= \{x \in \mathbb{R};\ x \geq a\}$:

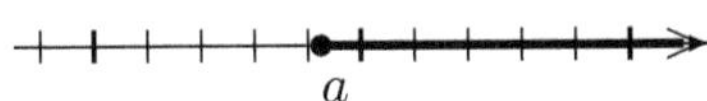

- $]-\infty, a[= \{x \in \mathbb{R};\ x < a\}$:

- $]-\infty, a] = \{x \in \mathbb{R};\ x \le a\}$:

Clearly, $]-\infty, +\infty[= \mathbb{R}$.

Example 2.20. For $x \in \mathbb{R}$, the notation $|x|$ stands for the absolute value of x. So $|x| = x$ if $x \ge 0$ and $|x| = -x$ if $x < 0$. To write the subset $A = \{x \in \mathbb{R};\ |2x - 5| \le 4\}$ of $\mathbb{R}$ using interval notations, we start by noticing that the condition $|2x - 5| \le 4$ is equivalent to $-4 \le 2x - 5 \le 4$. Simplifying gives: $\frac{1}{2} \le x \le \frac{9}{2}$. So $A = \left[\frac{1}{2}, \frac{9}{2}\right]$.

2.3.2 *Exercises*

$\star$(1) Can a power set be empty? Explain.

(2) Let $A = \{1, 2, 3, 4, \{a\}, \{b\}\}$. Which of the following are subsets of A?

$\star$(a) $\emptyset$	$\star$(d) $\{a, \{a\}, 2\}$	(g) $\{1, 2, 3, \{3\}, 4, \{a\}\}$
(b) $\{1, \{a\}\}$	(e) $\{4, 3, \{b\}\}$	$\star$(h) $\{\{b\}, 3, 4, 2, 1, \{a\}\}$
(c) $\{\{1\}, a\}$	(f) $\{\{a, b\}, 3, 4\}$	(i) $\{\{b\}, 3, 4, 2, 1, 0, \{a\}\}$

(3) In each case, prove that the sets A and B are equal.

   (a) A is the set of all integers which are multiple of 6 and 8 at the same time, B is the set of all integers which are multiple of 24.

  $\star$(b) $A = \{3^n;\ n \text{ is an even integer}\}$, $B = \{9^k;\ k \in \mathbb{Z}\}$.

   (c) $A = \{x \in \mathbb{R};\ |2x + 3| \le 2\}$, $B = \left[-\frac{5}{2}, -\frac{1}{2}\right]$.

  $\star$(d) $A = \{12s + 3t;\ s, t \in \mathbb{Z}\}$, $B = \{3k;\ k \in \mathbb{Z}\}$.

   (e) $A = \{x \in \mathbb{R};\ x^2 - 5x \le -4\}$, $B = [1, 4]$.

(4) Let $A = \{\{2\}, \{1, 2\}, 3, 4\}$. Determine the truth value of each of the following statements. Justify your answer.

(a) $\{1\} \in A$	$\star$(e) $\{1, 2\} \in A$	$\star$(i) $\{1, 2, 3\} \subseteq A$
$\star$(b) $\{1\} \subseteq A$	(f) $\{\{1\}, 2\} \subseteq A$	(j) $\{\{1, 2\}, \{2\}\} \subseteq A$
(c) $\{2\} \in A$	(g) $\{3, 4\} \in A$	(k) $\{\{2\}, 3, 4\} \subseteq A$
(d) $\{2\} \subseteq A$	(h) $\{3\} \in A$	(l) $\{\{\emptyset\}\} \subseteq A$

$\star$(5) Let $A = \{a, b, c, d\}$. Give all elements of $\wp(A)$.

(6) Let $A = \{a, b\}$.

  (a) Give all elements of $\wp(A)$.

  (b) Give all elements of $\wp(\wp(A))$.

(7) If $|A| = 2^3$, what is $|\wp(A)|$?

$\star$(8) Is there a set A with $|\wp(A)| = 32766$? If yes, what is $|A|$?

(9) Is there a set A with $|\wp(\wp(A))| = 4294967296$? If yes, what is $|A|$?

(10) Determine the truth value of each of the following statements. Justify your answer.

  $\star$(a) $\wp(\emptyset) = \emptyset$

  (b) $\wp(\emptyset) = \{0\}$

  $\star$(c) $\emptyset \subseteq \wp(\emptyset)$

  (d) $\{\emptyset\} \subseteq \wp(\emptyset)$

  (e) $\{\emptyset, \{\emptyset\}\} \subseteq \wp(\emptyset)$

  $\star$(f) $|\wp(\wp(\{\emptyset, \{\emptyset\}\}))| = 16$

(11) Let $A = \{3n^3 + 2;\ n \in \mathbb{Z}\}$. Determine the truth value of each of the following statements. Justify your answer.

  (a) $\frac{1}{2} \in A$

  (b) $-1 \in A$

  (c) $1 \in A$

  $\star$(d) $\{-22, 2, 26\} \in \wp(A)$

  (e) $\{-79, 0, 5, 26\} \in \wp(A)$

  $\star$(f) $\wp(A) \subseteq \wp(\mathbb{N})$

(12) In each case, determine the truth value of the statement. Justifier your answer.

  (a) $\{\{1, 2, 3\}, \{1\}, \{\frac{9}{2}\}\} \in \wp(\wp(\mathbb{Q}))$

  (b) $\emptyset \in \wp(\wp(\wp(\mathbb{N})))$

  $\star$(c) $\{\mathbb{Z}, [-3, 3]\} \subseteq \wp(\mathbb{Q})$.

  (d) If X is a set satisfying $\wp(\mathbb{Z}) \subseteq X$, then $\{x \in \mathbb{R};\ x^6 - 4x^4 = 0\} \in X$.

(13) In each case determine if the given set can be realized as the power set of a set A. If you say it is, determine the set A.

  (a) $\{\emptyset, \{a\}, \{a, b\}\}$

  $\star$(b) $\{\{1\}, \{2\}, \{1, 2\}, \{a\}, \{b\}, \{a, b\}, \{1, a\}, \{2, b\}\}$

  (c) $\{\emptyset, \{\{a\}\}, \{\{a\}, \{b\}\}, \{\{b\}\}, \{2\}, \{\{a\}, 2\}, \{\{b\}, 2\}, \{\{a\}, 2, \{b\}\}\}$

(14) Write each of the following intervals using a set-builder notation.

  (a) $\,]-1, 2]$ (c) $[-1, 3]$ $\star$(e) $\,]-\infty, -1]$ $\star$(g) $[2, \infty[$

  $\star$(b) $[2, 3[$ (d) $\,]2, 5[$ (f) $\,]1, \infty[$ (h) $\,]-\infty, 3[$

(15) Represent each of the following sets as intervals on the real line.

  (a) $A = \{x \in \mathbb{R}; -2 \le x < 4\}$.
  (b) B is the set of all real numbers greater than -2.
  (c) $C = \{x \in \mathbb{R}; |x + 2| \le \frac{7}{2}\}$.
  (d) $D = \{x \in \mathbb{R}; x^2 - 4x < -x - 2\}$.

(16) Give two sets A and B such that $A \in B$ and $A \subseteq B$ are true at the same time.

(17) Let A, B be two subsets of a universe $\mathcal{U}$. Prove that $A = B \Leftrightarrow \wp(A) = \wp(B)$.

(18) Let A, B and C be three subsets of a universe $\mathcal{U}$.

  $\star$(a) Prove that if $A \subseteq B$ and $B \subseteq C$, then $A \subseteq C$.
  (b) Prove that if $A \subseteq B$, $B \subseteq C$ and $C \subseteq A$, then $A = C$.

(19) (**Representation of finite sets in computers**) While the order of elements in a finite set is not important, a certain order on the elements is needed when representing a set in a computer. Let $\mathcal{U}$ be a finite universal set. For a computer representation of subsets of $\mathcal{U}$, we start by fixing a certain order on the elements of $\mathcal{U}$, say $a_1, a_2, \ldots, a_n$. A subset B of $\mathcal{U}$ is represented by a binary string of length n with the ith term is 0 if $a_i \in B$ and 1 if $a_i \in B$. Represent each of the following subsets of the universe $\mathcal{U} = \{a, b, c, d, e, f, g, h\}$ as a binary string of length 8.

(a) $\emptyset$	$\star$(c) $\{e, h\}$	$\star$(e) $\{a, b, d, e, h\}$
(b) $\{d\}$	(d) $\{b, c, g\}$	(f) $\{a, b, c, d, e, g\}$

(20) Using the notations of the previous exercise, find the subset of $\mathcal{U}$ corresponding to each of the following binary strings.

$\star$(a) 10001110	$\star$(c) 11011110	(e) 00001000
(b) 00001010	(d) 01001000	(f) 11000001

2.4   Operations on sets

Just like formulas of propositional logic can be combined using logic connectives, set operations can be used to create new sets from already known ones. Basic operations on sets include intersection, union, difference, complement and (cartesian) product. In this section, we define these operations and give their main properties. We also make the connection with propositional logic connectives.

As usual, $\mathcal{U}$ denotes the universe and all sets considered are subsets of $\mathcal{U}$. If E is a set and $x \in \mathcal{U}$, then $E(x)$ denotes the proposition "$x \in E$".

2.4.1 *Intersection*

Definition 2.4. The *intersection* of two sets A and B, denoted by $A \cap B$, is the set of all elements of $\mathcal{U}$ belonging to A and B at the same time.

$$A \cap B = \{x \in \mathcal{U};\ x \in A \text{ and } x \in B\}.$$

Sets A and B are called *disjoint* if $A \cap B = \emptyset$. So disjoint sets have no common elements.

The following Venn diagram shows (in grey) the subset of $\mathcal{U}$ that represents the intersection of A and B:

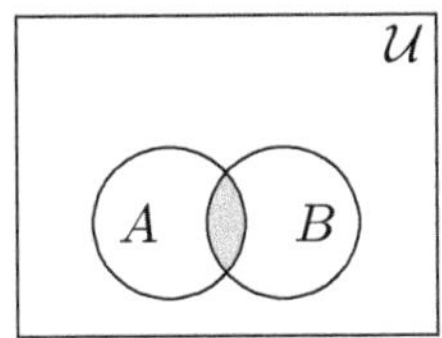

Example 2.21. Let $A = \{2, 4, 6, 8, 10\}$, $B = \{2, 6, 9, 13\}$ and $C = \{0, 9, 13\}$, then $A \cap B = \{2, 6\}$, $A \cap C = \emptyset$ (so sets A and C are disjoint), $B \cap C = \{9, 13\}$.

Example 2.22. For the intervals $I =]-1, 3]$ and $J =]1, 5]$ of $\mathbb{R}$, $I \cap J =]1, 3]$. This is illustrated in the following diagram where the interval I is drawn in grey and the interval J in black. The two colors overlap in $]1, 3]$ giving the intersection of the two intervals.

Using logic terminology, intersection of sets corresponds to the conjunction connective $\wedge$:

$$A \cap B = \{x \in \mathcal{U};\ (A(x) \wedge B(x))\}.$$

Properties of the intersection operation are given in the following theorem.

Theorem 2.3. *Let A, B and C be three sets. The following hold.*

(a) $A \cap B = B \cap A$ (Commutativity of $\cap$).
(b) $(A \cap B) \cap C = A \cap (B \cap C)$ (Associativity of $\cap$).
(c) $A \cap A = A$ (Idempotence of intersection).
(d) $A \cap \emptyset = \emptyset$ ($\emptyset$ is absorbent for the intersection).
(e) $A \cap \mathcal{U} = A$ ($\mathcal{U}$ is neutral for the intersection).
(f) $A \cap B \subseteq A$ and $A \cap B \subseteq B$.

Proof. Every property in the above theorem follows from a corresponding property for the conjunction connective $\wedge$. For instance, properties (a) and (b) follow from the commutativity and the associativity properties of $\wedge$, respectively. We leave the details of the other parts for the reader. $\qquad\square$

Example 2.23. If A and B are two sets, prove that $A \cap B = B \Leftrightarrow B \subseteq A$.

Solution. This is an "if and only if" statement. We need to prove two directions. First, assume that $A \cap B = B$. If $x \in B$, then $x \in A \cap B$ and so $x \in A$. This proves that $B \subseteq A$. Conversely, assume that $B \subseteq A$. If $x \in B$, then $x \in A$ and so $x \in A \cap B$. This shows that $B \subseteq A \cap B$. The inclusion $A \cap B \subseteq B$ is given by part (f) of Theorem 2.3 above. We conclude that $A \cap B = B$. $\qquad\qquad\diamond$

We can generalize the intersection operation to any number of sets. For example, if A, B and C are three sets, then $A \cap B \cap C$ is the set of all elements of $\mathcal{U}$ that belong to the three sets at the same time. By the associativity of the intersection, this is the same set as $(A \cap B) \cap C$ or $A \cap (B \cap C)$. More generally, given a set I and a collection A_i ($i \in I$) of (indexed) subsets of $\mathcal{U}$, we define the intersection of sets A_i, denoted by $\bigcap_{i \in I} A_i$, as being the subset of $\mathcal{U}$ consisting of all elements that belong to each set A_i[1]:

$$\bigcap_{i \in I} A_i = \{x \in \mathcal{U}; x \in A_i \text{ for every } i \in I\}. \tag{2.1}$$

If I is a finite set of cardinality n, the intersection in (2.1) is $A_1 \cap A_2 \cap \cdots \cap A_n$ that we denote by $\bigcap_{i=1}^{n} A_i$. The infinite intersection $A_1 \cap A_2 \cap A_3 \cap \cdots$ is denoted by $\bigcap_{i=1}^{\infty} A_i$.

Example 2.24. For the sets $A = \{1, 2, 3, 6, 7\}$, $B = \{0, 1, 2, 7, 8\}$, $C = \{1, 2, 7, 9, 10\}$, $A \cap B \cap C = \{1, 2, 7\}$.

Example 2.25. Consider the intervals $I =]-1, 3]$, $J =]1, 5]$ and $W = [4, 7]$ of $\mathbb{R}$. Then $I \cap J \cap W = \emptyset$ since there are no common elements between all three intervals.

---

[1] The set I is called an index set.

Example 2.26. A well-known property of real numbers is the Archimedean property. One version of this property states that for any real number x and any positive real number y, there exists $n \in \mathbb{N}$ such that $ny > x$. In this example, we use the property to prove that the infinite intersection of a family of intervals of $\mathbb{R}$ could be reduced to a single element. For integer $i \geq 1$, let A_i be the interval $[0, \frac{1}{i}[$ of $\mathbb{R}$. So $A_1 = [0, 1[$, $A_2 = [0, \frac{1}{2}[$, $A_3 = [0, \frac{1}{3}[$, and so on. Clearly $\ldots \subset A_4 \subset A_3 \subset A_2 \subset A_1$. Since $0 \in A_i$ for each i, we have that $0 \in \bigcap_{i=1}^{\infty} A_i$. Conversely, if $x \in \bigcap_{i=1}^{\infty} A_i$ is an arbitrary element, then $0 \leq x < \frac{1}{i}$ for any integer $i \geq 1$. This implies that $x = 0$ since otherwise we can find an integer $i \geq 1$ such that $x \geq \frac{1}{i}$ (by the Archimedean property). We conclude that $\bigcap_{i=1}^{\infty} A_i = \{0\}$.

Example 2.27. In this example, we look at an empty intersection of an infinite number of sets. For each integer $k \geq 1$, let A_k be the open interval $]k, \infty[$ of $\mathbb{N}$. That is, $A_k = \{n \in \mathbb{N}; n > k\}$. If there exits an integer n in $\bigcap_{k=1}^{\infty} A_k$, then $n > k$ for any $k \geq 1$. Since no such integer exists (clearly, $\mathbb{N}$ is not bounded above), we conclude that $\bigcap_{k \in \mathbb{N}} A_k = \emptyset$.

2.4.2   Union

Definition 2.5. The *union* of two sets A and B, denoted by $A \cup B$, is the subset of $\mathcal{U}$ of all elements belonging to either A *or* B (or both):

$$A \cup B = \{x \in \mathcal{U}; \ x \in A \text{ or } x \in B\}.$$

The following shows the Venn diagram of $A \cup B$:

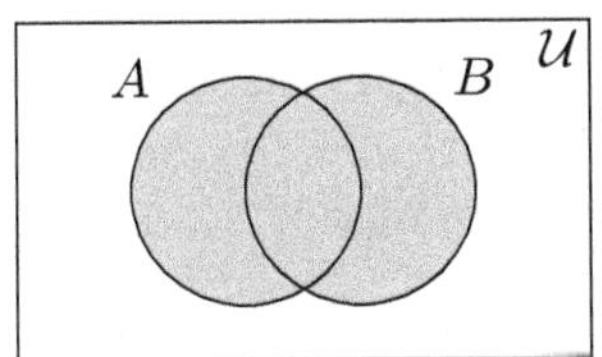

Example 2.28. Using the sets in Example 2.21, we have $A \cup B = \{2, 4, 6, 8, 9, 10, 13\}$, $A \cup C = \{0, 2, 4, 6, 8, 9, 10, 13\}$, $B \cup C = \{0, 2, 6, 9, 13\}$.

Example 2.29. For the intervals $I =]-1, 3]$ and $J =]1, 5]$ of $\mathbb{R}$, $I \cup J =]-1, 5]$. This is illustrated in the following diagram where the interval I is drawn in grey and the interval J in black.

Using logic terminology, union of sets corresponds to the disjunction connective $\vee$:

$$A \cup B = \{x \in \mathcal{U};\ (A(x) \vee B(x))\ \text{is true}\}.$$

Properties of the union operation are given in the following theorem. Again, all these properties follow directly from properties of logic connectives. We omit the formal proof.

Theorem 2.4. *Let A, B and C be three sets. The following hold.*

(1) $A \cup B = B \cup A$ (Commutativity of $\cup$).
(2) $(A \cup B) \cup C = A \cup (B \cup C)$ (Associativity of $\cup$).
(3) $A \cup A = A$ (Idempotence of union).
(4) $A \cup \emptyset = A$ ($\emptyset$ is neutral for the union).
(5) $A \cup \mathcal{U} = \mathcal{U}$ ($\mathcal{U}$ is absorbent for the union).
(6) $A \subseteq A \cup B$ and $B \subseteq A \cup B$.
(7) $A \cap (B \cup C) = (A \cap B) \cup (A \cap C)$ (Distributivity of $\cap$ over $\cup$).
(8) $A \cup (B \cap C) = (A \cup B) \cap (A \cup C)$ (Distributivity of $\cup$ over $\cap$).

Example 2.30. If A and B are two sets, prove that $A \cup B = B \Leftrightarrow A \subseteq B$.

Solution. Assume first that $A \cup B = B$. If $x \in A$, then $x \in A \cup B$ and so $x \in B$. This proves that $A \subseteq B$. Conversely, assume that $A \subseteq B$ and let $x \in A \cup B$. Then $x \in A$ or $x \in B$. In both cases, $x \in B$. This shows that $A \cup B \subseteq B$. The inclusion $B \subseteq A \cup B$ is also true by Theorem 2.4. This proves that $A \cup B = B$. $\Diamond$

Like in the case of intersection, the union can be generalized to an arbitrary family of sets. If I is an index set and $(A_i)_{i \in I}$ is a family of subsets of $\mathcal{U}$, we define the union of sets A_i, denoted by $\bigcup_{i \in I} A_i$, as follows:

$$\bigcup_{i \in I} A_i = \{x \in \mathcal{U};\ x \in A_i\ \text{for some}\ i \in I\}.$$

We also use the notation $\bigcup_{i=1}^{n} A_i$ to denote the finite union $A_1 \cup A_2 \cup \cdots A_n$ and $\bigcup_{i=1}^{\infty} A_i$ to denote the infinite union $A_1 \cup A_2 \cup A_3 \cup \cdots$

Example 2.31. Let us find $\bigcup_{i=1}^{\infty} A_i$ where $A_i = [0, \frac{1}{i}[$ as in Example 2.26 above. As we have noticed before: $\ldots \subset A_4 \subset A_3 \subset A_2 \subset A_1$. If $x \in \bigcup_{i=1}^{\infty} A_i$, then there exists at least one $i \geq 1$ such that $x \in A_i$. Since $A_i \subseteq A_1$, $x \in A_1$ and so $\bigcup_{i=1}^{\infty} A_i \subseteq A_1$. The inclusion $A_1 \subseteq \bigcup_{i=1}^{\infty} A_i$ is also true by Theorem 2.4. We conclude that $\bigcup_{i=1}^{\infty} A_i = A_1 = [0, 1[$.

2.4.3 *Difference, complement and symmetric difference*

In the department of Mathematics and Statistics at a certain university, there are two groups of faculty members: group A of pure mathematicians and group B of statisticians. A faculty member in the department who is not a pure mathematician must be statistician. We say that group B is the *complement* of group A in the department (also group A is the *complement* of group B.)

Definition 2.6. Let A and B be two sets in a universe $\mathcal{U}$.

- The *set difference* of A and B, denoted by $A\backslash B$, is the subset of $\mathcal{U}$ of those elements that belong to A but not to B:

$$A\backslash B = \{x \in \mathcal{U};\ x \in A \text{ and } x \notin B\}.$$

- The *complement* of A, denoted by $\overline{A}$ (other notations include A', A^c and $C(A)$), is the subset $\mathcal{U}\backslash A$. In other words, the complement of A is the set of all elements of $\mathcal{U}$ that are not in A:

$$\overline{A} = \{x \in \mathcal{U};\ x \notin A\}.$$

- The *symmetric difference* of A and B, denoted by $A \oplus B$, is the subset of $\mathcal{U}$ of those elements that belong to A or to B but not to both. In other words,

$$A \oplus B = (A \cup B)\backslash(A \cap B).$$

An alternative way of defining $A \oplus B$ is given in Exercise (10) below.

The regions in grey in the following diagrams show the Venn diagrams of $A\backslash B$, $\overline{A}$ and $A \oplus B$.

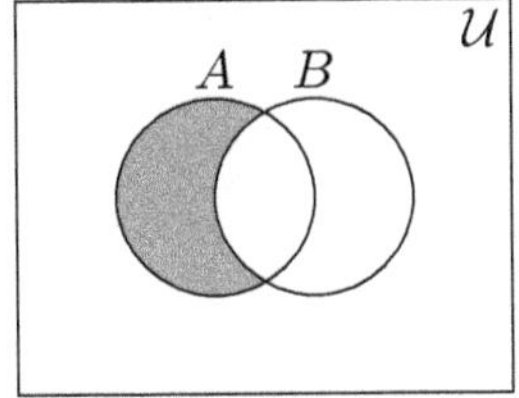
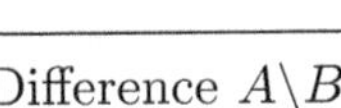
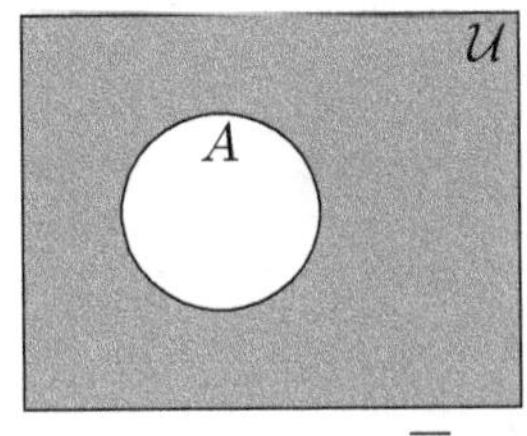

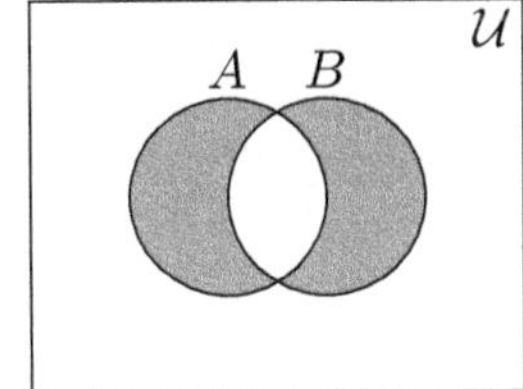

Difference $A\backslash B$ Complement $\overline{A}$ Symmetric difference $A \oplus B$

Using propositional logic terminology, the difference, complement and symmetric difference are interpreted as follows:

- $A \backslash B = \{x \in \mathcal{U};\ (A(x) \wedge \neg B(x))\ \text{is true}\}$.
- $\overline{A} = \{x \in \mathcal{U};\ \neg A(x)\ \text{is true}\}$.
- $A \oplus B = \{x \in \mathcal{U};\ ((A(x) \vee B(x)) \wedge \neg (A(x) \wedge B(x)))\ \text{is true}\}$.

Example 2.32. Consider the subsets

$$A = \{1,2,3,6,7\},\ B = \{0,1,2,7,8\},\ C = \{1,2,7,9,10\}$$

of the universe $\mathcal{U} = \{0,1,2,3,4,5,6,7,8,9,10\}$. In each case, list the elements of the given set.

(a) $A \backslash B$ (e) $\overline{A} \backslash \overline{B}$ (i) $A \oplus B$

(b) $B \backslash A$ (f) $A \cap \overline{B \cup C}$ (j) $B \oplus C$

(c) $C \backslash A$ (g) $\overline{A \cap C}$ (k) $(A \oplus B) \oplus C$

(d) $A \backslash (B \backslash C)$ (h) $\overline{A} \backslash (\overline{B} \backslash \overline{C})$ (l) $A \oplus (B \oplus C)$

Solution.   (a) $A \backslash B = \{3,6\}$.

(b) $B \backslash A = \{0,8\}$.

(c) $C \backslash A = \{9,10\}$.

(d) $B \backslash C = \{0,8\}$. So $A \backslash (B \backslash C) = \{1,2,3,6,7\}$.

(e) $\overline{A} = \{0,4,5,8,9,10\}$, $\overline{B} = \{3,4,5,6,9,10\}$, so $\overline{A} \backslash \overline{B} = \{0,8\}$.

(f) $B \cup C = \{0,1,2,7,8,9,10\}$, so $\overline{B \cup C} = \{3,4,5,6\}$ and $A \cap \overline{B \cup C} = \{3,6\}$.

(g) $A \cap C = \{1,2,7\}$, so $\overline{A \cap C} = \{0,3,4,5,6,8,9,10\}$.

(h) $\overline{A} = \{0,4,5,8,9,10\}$, $\overline{B} = \{3,4,5,6,9,10\}$, $\overline{C} = \{0,3,4,5,6,8\}$. So $\overline{B} \backslash \overline{C} = \{9,10\}$ and $\overline{A} \backslash (\overline{B} \backslash \overline{C}) = \{0,4,5,8\}$.

(i) $A \oplus B = (A \cup B) \backslash (A \cap B) = \{0,1,2,3,6,7,8\} \backslash \{1,2,7\} = \{0,3,6,8\}$.

(j) $B \oplus C = (B \cup C) \backslash (B \cap C) = \{0,1,2,7,8,9,10\} \backslash \{1,2,7\} = \{0,8,9,10\}$

(k) Using part (i), $(A \oplus B) \cup C = \{0,1,2,3,6,7,8,9,10\}$ and $(A \oplus B) \cap C = \emptyset$. So $(A \oplus B) \oplus C = \{0,1,2,3,6,7,8,9,10\}$.

(l) Using part (j), $A \cup (B \oplus C) = \{0,1,2,3,6,7,8,9,10\}$ and $A \cap (B \oplus C) = \emptyset$. So $A \oplus (B \oplus C) = \{0,1,2,3,6,7,8,9,10\}$. $\Diamond$

Example 2.32 shows in particular that the difference of sets is not a commutative operation as $A \backslash B \neq B \backslash A$ for sets A and B given in the example.

Example 2.33. For the intervals $I = [0,1]$ and $J = {]}0,1{[}$ of $\mathbb{R}$, $I \backslash J = \{0,1\}$ and $J \backslash I = \emptyset$.

Main properties of the difference, complement and symmetric difference operations are given in the following theorem.

Theorem 2.5. *Let A, B and C be three sets. The following hold.*

(a) $A \backslash B = A \cap \overline{B}$

(b) $A \backslash A = \emptyset$

(c) $A \backslash \emptyset = A$

(d) $\overline{\overline{A}} = A$

(e) $A \oplus A = \emptyset$

(f) $A \oplus \emptyset = A$

(g) $A \oplus B = B \oplus A$

(h) $(A \oplus B) \oplus C = A \oplus (B \oplus C)$

(i) $\overline{A \cap B} = \overline{A} \cup \overline{B}$ *(De Morgan)*

(j) $\overline{A \cup B} = \overline{A} \cap \overline{B}$ *(De Morgan)*

Proof. We prove parts (a) and (i) only. The reader is encouraged to prove the other parts. Let x be an arbitrary element of the universe $\mathcal{U}$. For (a): $x \in A \backslash B \Leftrightarrow (x \in A) \wedge (x \notin B) \Leftrightarrow (x \in A) \wedge (x \in \overline{B}) \Leftrightarrow x \in A \cap \overline{B}$. For (i): $x \in \overline{A \cap B} \Leftrightarrow x \notin A \cap B \Leftrightarrow x \notin A$ or $x \notin B$ (since otherwise $x \in A \cap B$) $\Leftrightarrow x \in \overline{A}$ or $x \in \overline{B} \Leftrightarrow x \in \overline{A} \cup \overline{B}$. $\square$

Remark 2.5. Propositional logic equivalences we saw in the first chapter can be used to prove the properties given in Theorem 2.5. For example, the expression $\overline{A \cap B}$ corresponds to the logic formula $\neg (A(x) \wedge B(x))$ which corresponds (by De Morgan's law of propositional logic) to the statement $(\neg A(x) \vee \neg B(x))$. In terms of set theory, this last statement corresponds to $\overline{A} \cup \overline{B}$.

The following table gives a summary of the basic properties of set operations. In the table, A, B and C are subsets of a universe $\mathcal{U}$. The third column gives an interpretation of each of the properties in terms of propositional logic equivalences.

Name	Set Theory	Logic
Commutativity of $\cap$	$A \cap B = B \cap A$	$(A(x) \wedge B(x)) \equiv (B(x) \wedge A(x))$
Commutativity of $\cup$	$A \cup B = B \cup A$	$(A(x) \vee B(x)) \equiv (B(x) \vee A(x))$
Associativity of $\cap$	$(A \cap B) \cap C = A \cap (B \cap C)$	$(A(x) \wedge B(x)) \wedge C(x)) \equiv ((A(x) \wedge B(x)) \wedge C(x))$
Associativity of $\cup$	$(A \cup B) \cup C = A \cup (B \cup C)$	$(A(x) \vee B(x)) \vee C(x)) \equiv ((A(x) \vee B(x)) \vee C(x))$
Distributivity	$A \cup (B \cap C) = (A \cup B) \cap (A \cup C)$	$(A(x) \vee (B(x) \wedge C(x))) \equiv ((A(x) \vee B(x)) \wedge (A(x) \vee C(x)))$
Distributivity	$A \cap (B \cup C) = (A \cap B) \cup (A \cap C)$	$(A(x) \wedge (B(x) \vee C(x))) \equiv ((A(x) \wedge B(x)) \vee (A(x) \wedge C(x)))$
Elimination of set difference	$A \backslash B = A \cap \overline{B}$	$(A(x) \wedge \neg B(x))$
De Morgan law	$\overline{A \cap B} = \overline{A} \cup \overline{B}$	$\neg (A(x) \wedge B(x)) \equiv (\neg A(x) \vee \neg B(x))$
De Morgan law	$\overline{A \cup B} = \overline{A} \cap \overline{B}$	$\neg (A(x) \vee B(x)) \equiv (\neg A(x) \wedge \neg B(x))$
Double complementation	$\overline{\overline{A}} = A$	$\neg \neg A(x) \equiv A(x)$
Complementation law	$A \cup \overline{A} = \mathcal{U}$	$(A(x) \vee \neg A(x)) \equiv \mathbf{T}$ (a tautology)
Exclusion law	$A \cap \overline{A} = \emptyset$	$(A(x) \wedge \neg A(x)) \equiv \mathbf{F}$ (a contradiction)
Identity law	$A \cup \emptyset = A$	$(A(x) \vee \mathbf{F}) \equiv A(x)$
Identity law	$A \cap \mathcal{U} = A$	$(A(x) \wedge \mathbf{T}) \equiv A(x)$
Domination law	$A \cup \mathcal{U} = \mathcal{U}$	$(A(x) \vee \mathbf{T}) \equiv \mathbf{T}$
Domination law	$A \cap \emptyset = \emptyset$	$(A(x) \wedge \mathbf{F}) \equiv \mathbf{F}$
Idempotent law	$A \cap A = A$	$(A(x) \wedge A(x)) \equiv A(x)$
Idempotent law	$A \cup A = A$	$(A(x) \vee A(x)) \equiv A(x)$

Example 2.34. A, B, C are three subsets of a universal set $\mathcal{U}$. In each case, use set operations properties to prove the set equality.

(a) $A \setminus (B \setminus C) = (A \setminus B) \cup (A \setminus \overline{C})$ (b) $(A \cap B) \cup (A \cap \overline{B}) = A$

Solution.

(a)

$$
\begin{aligned}
A \setminus (B \setminus C) &= A \cap \overline{(B \setminus C)} && \text{(Elimination of set difference)} \\
&= A \cap \overline{B \cap \overline{C}} && \text{(Elimination of set difference)} \\
&= A \cap (\overline{B} \cup \overline{\overline{C}}) && \text{(De Morgan law)} \\
&= (A \cap \overline{B}) \cup (A \cap \overline{\overline{C}}) && \text{(Distributivity law)} \\
&= (A \setminus B) \cup (A \setminus \overline{C}) && \text{(Elimination of set difference)}
\end{aligned}
$$

(b)

$$
\begin{aligned}
(A \cap B) &\cup (A \cap \overline{B}) \\
&= A \cap (B \cup \overline{B}) && \text{(Distributivity law)} \\
&= A \cap \mathcal{U} && \text{(Complementation law)} \\
&= A && \text{(Identity law)}
\end{aligned}
$$

$\Diamond$

2.4.4   *Cartesian product*

Given n objects $a_1, \ldots, a_n$, we can form the set $\{a_1, \ldots, a_n\}$ and the ordered n-tuple $(a_1, \ldots, a_n)$. The difference between the two is the order. In the set $\{a_1, \ldots, a_n\}$, the order in which we list the elements is not important and does not change the set. This is not the case for an n-tuple. If we change the order in which the components in $(a_1, \ldots, a_n)$ appear, we obtain a different n-tuple. An example you have probably seen before is the notion of a point determined by its two coordinates in the Cartesian plane. Ordered pairs $(1, 2)$ and $(2, 1)$ in the plane represent two different points.

Definition 2.7. The ordered n-tuple $(a_1, \ldots, a_n)$ is an **ordered** listing of the objects $a_1, \ldots, a_n$ where a_1 appears as a first component, a_2 as a second component and so on. Two ordered n-tuples $(a_1, \ldots, a_n)$ and $(b_1, \ldots, b_n)$ are equal if and only if $a_i = b_i$ for every $i \in \{1, \ldots, n\}$.

The collection of all n-tuples with component a_i is a set A_i is called the Cartesian product of the sets A_i's, named after the French mathematician René Descartes.

Definition 2.8. The *Cartesian product* of the sets $A_1, A_2, \ldots, A_n$, denoted by $A_1 \times A_2 \times \cdots \times A_n$, is the set of all ordered n-tuples $(a_1, a_2, \ldots, a_n)$ with $a_i \in A_i$ for $i \in \{1, 2, \ldots, n\}$. In some cases, it is more convenient to denote an element $(a_1, a_2, \ldots, a_n)$ of the Cartesian product by a finite sequence (or a list) $a_1 a_2 \ldots a_n$. If $A_1 = A_2 = \cdots = A_n = A$, then the Cartesian product $A \times A \times \cdots \times A$ is denoted by A^n.

Example 2.35. Consider the two sets $A = \{\bigcirc, \diamond, \triangle\}$, $B = \{\oplus, \oslash\}$. Then:

- $A \times B = \{(\bigcirc, \oplus), (\diamond, \oplus), (\triangle, \oplus), (\bigcirc, \oslash), (\diamond, \oslash), (\triangle, \oslash)\}$
- $B \times A = \{(\oplus, \bigcirc), (\oplus, \diamond), (\oplus, \triangle), (\oslash, \bigcirc), (\oslash, \diamond), (\oslash, \triangle)\}$
- $A \times A = \{(\bigcirc, \bigcirc), (\bigcirc, \diamond), (\bigcirc, \triangle), (\diamond, \bigcirc), (\diamond, \diamond), (\diamond, \triangle),$ $(\triangle, \bigcirc), (\triangle, \diamond), (\triangle, \triangle)\}$
- $B \times B = \{(\oplus, \oplus), (\oplus, \oslash), (\oslash, \oplus), (\oslash, \oslash)\}$.

The above example shows in particular that $A \times B \neq B \times A$ in general.

Example 2.36. Every 10 digits phone number (with no restriction) can be represented by a 10-tuple $(x_1, x_2, \ldots, x_{10})$ where $x_i \in \mathbb{N}$ for all i. So the set of all 10 digits phone numbers is a subset of $\mathbb{N}^{10}$.

Example 2.37. The set of all binary strings of a given length n is simply A^n where $A = \{0, 1\}$. In particular, the set of all bytes is A^8 (a *byte* is defined to be a binary string of length 8).

Theorem 2.6. *Let A, B, C and D be four subsets of a universal set $\mathcal{U}$. Then the following properties hold.*

(a) $A \times (B \cap C) = (A \times B) \cap (A \times C)$.

(b) $(B \cap C) \times A = (B \times A) \cap (C \times A)$.

(c) $A \times (B \cup C) = (A \times B) \cup (A \times C)$.

(d) $(B \cup C) \times A = (B \times A) \cup (C \times A)$.

(e) $(A \backslash B) \times C = (A \times C) \backslash (B \times C)$.

(f) $(A \cap B) \times (C \cap D) = (A \times C) \cap (B \times D)$.

Proof. We only prove part (e) leaving the other parts as exercises to the reader. If $(x, y) \in (A \backslash B) \times C$, then $x \in A \backslash B$ and $y \in C$. So $x \in A$, $x \notin B$ and $y \in C$. We conclude that $(x, y) \in A \times C$ and $(x, y) \notin B \times C$. In other words, $(x, y) \in (A \times C) \backslash (B \times C)$. This proves that $(A \backslash B) \times C \subseteq (A \times C) \backslash (B \times C)$. Conversely, let $(x, y) \in (A \times C) \backslash (B \times C)$. Then $(x, y) \in A \times C$ and $(x, y) \notin B \times C$. So $x \in A$, $y \in C$ and either $x \notin B$ or $y \notin C$. Since we know that $y \in C$, we must have that $x \notin B$. We conclude that $x \in A$, $x \notin B$ and

$y \in C$. In other words, $x \in A \backslash B$ and $y \in C$ and so $(x, y) \in (A \backslash B) \times C$. This proves that $(A \times C) \backslash (B \times C) \subseteq (A \backslash B) \times C$. $\square$

2.4.5   *Exercises*

(1) Consider the following three subsets $A = \{1, 2, 4, 5\}$, $B = \{4, 5, 6, 8\}$ and $C = \{0, 1, 3, 4, 8, 9\}$ of the universe $\mathcal{U} = \{0, 1, 2, 3, 4, 5, 6, 7, 8, 9\}$. In each case, determine the truth value of the statement.

(a) $1 \in A \backslash B$

(b) $1 \in A \cap C$

(c) $1 \in A \oplus B$

$\star$(d) $\{0, 4, 8\} \subseteq A \oplus C$

(e) $9 \in \overline{A}$

(f) $7 \in \overline{A \cup C}$

(g) $\{2, 5\} \subseteq A \backslash C$

$\star$(h) $0 \in A \backslash (B \backslash C)$

(i) $8 \in C \backslash \overline{A \cup B}$

(j) $2 \in A \oplus (B \oplus C)$

$\star$(k) $\{1, 2\} \subseteq A \oplus \overline{B}$.

(2) Consider the following three subsets $A = \{a, b, e, f, g\}$, $B = \{a, d, h, i, j\}$ and $C = \{c, e, g, h, i\}$ of the universe $\mathcal{U} = \{a, b, c, d, e, f, g, h, i, j\}$. In each case, list all elements of the given set.

$\star$(a) $A \cap C$

(b) $A \cup B$

(c) $B \cup \overline{C}$

(d) $A \cap B \cap C$

$\star$(e) $\overline{A \cup C} \cap B$

(f) $A \backslash C$

$\star$(g) $A \backslash (B \backslash C)$

(h) $C \backslash \overline{A \cup B}$

(i) $A \oplus C$.

(3) The following represents the Venn diagram of three subsets A, B, C of a universal set $\mathcal{U}$. In each case, shade the region in the diagram that corresponds to the given set.

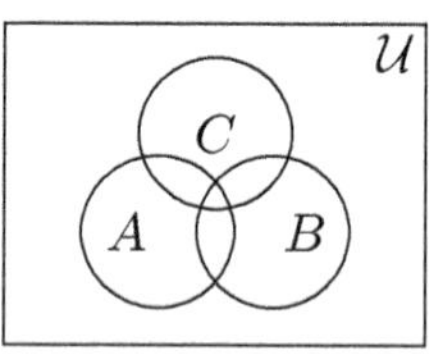

(a) $A \cap B$

(b) $A \cup C$

(c) $A \cap \overline{B}$

(d) $A \cap (B \cup C)$

(e) $B \cup \overline{A \cap C}$

(f) $\overline{A} \cap \overline{B}$

(g) $\overline{A} \cap C$

(h) $\overline{A \cap B \cap C}$

(i) $\overline{A} \cap \overline{B} \cap \overline{C}$

(j) $A \oplus (B \oplus C)$.

(4) In a certain department, there are four groups of professors: group A of all pure mathematicians, group B of all applied mathematicians, group C of all statisticians and group D of all computer scientists.

Some professors belong to more than one group. In each case, express the given set in terms of sets A, B, C and D and the set operations. In this context, the universe is the set of all professors in the department.

  (a) All professors who are pure and applied mathematicians.

$\star$(b) All professors who are pure and applied mathematicians but not computer scientists.

  (c) All professors who are computer scientists but not applied mathematicians.

  (d) All professors who are neither pure nor applied mathematicians.

$\star$(e) All professors who are either pure or applied mathematicians but not statisticians.

(5) A, B, C are subsets of a universal set $\mathcal{U}$. In each case, give an expression for the subset of $\mathcal{U}$ determined by the region in grey in the given Venn diagram. Write your answer in terms of the three sets A, B, C and the set operations.

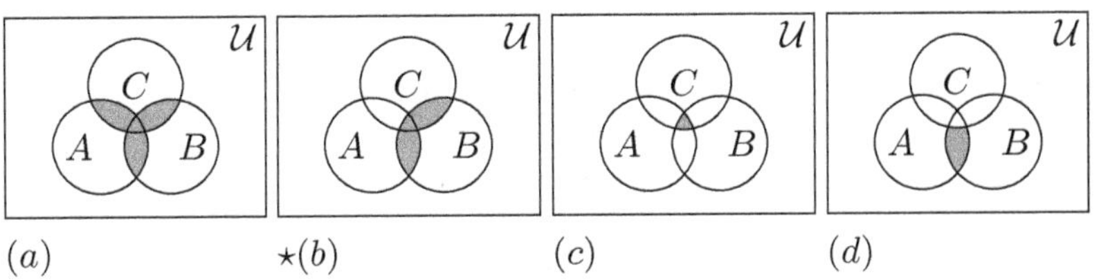

     (a) $\star(b)$ (c) (d)

(6) Let $\mathcal{U} = \mathbb{N}\backslash\{0\}$ be the universe of all positive integers. Consider the following subsets of $\mathcal{U}$: A is the set of all even numbers, B is the set of all odd numbers, C is the set of all prime numbers, D is the set of all powers of 3 and E is the set of all multiples of 3. Express each of the following statements using A, B, C, D, E and the symbols $\cap$, $\cup$, $\backslash$, $\overline{}$, $\in$, $\subseteq$, $\notin$, $\nsubseteq$, and $\emptyset$.

  (a) 2 is the only even number which is also prime.

$\star$(b) 3 is prime but not even.

  (c) 9 is odd, a power of 3 and a multiple of 3.

  (d) Every power of 3 is also a multiple of 3.

$\star$(e) Every positive integer is either even or odd.

  (f) No positive integer is even and odd at the same time.

  (g) there exist even multiples of 3.

$\star$(h) there exist multiples of 3 which are not powers of 3.

$\star$(7) Give an example of three sets A, B and C such that each of the sets $A \cap B$, $A \cap C$ and $B \cap C$ is non-empty but $A \cap B \cap C = \emptyset$.

(8) Let A and B be two subsets of a universal set $\mathcal{U}$. what can be said about the two sets in each of the following cases?

    (a) $A \cup B = \emptyset$ (b) $A \oplus B = \emptyset$ (c) $A \oplus B = A \cup B$

(9) Let A be a subset of a universal set $\mathcal{U}$. In each case, prove the set equality.

    (a) $A \oplus \emptyset = A$ (b) $A \oplus A = \emptyset$ $\star$(c) $A \oplus \mathcal{U} = \overline{A}$ (d) $A \oplus \overline{A} = \mathcal{U}$

$\star$(10) Let A and B be two subsets of a universal set $\mathcal{U}$. We defined the symmetric difference of A and B as being $(A \cup B) \setminus (A \cap B)$. In some textbooks, the symmetric difference is defined as being the subset $(A \setminus B) \cup (B \setminus A)$ of $\mathcal{U}$. Prove that $(A \cup B) \setminus (A \cap B) = (A \setminus B) \cup (B \setminus A)$.

(11) Prove De Morgan's law: $\overline{A \cup B} = \overline{A} \cap \overline{B}$ (the other De Morgan's law was proven in this section).

$\star$(12) Let A, B be two subsets of a universe $\mathcal{U}$.

    (a) Use set operations properties to prove that $(A \cap B) \cup \overline{(\overline{A} \cup B)} = A$.

    (b) Prove the set equality of the previous part by setting and proving the corresponding tautology of propositional logic.

(13) A, B, C are three subsets of a universal set $\mathcal{U}$. Each of the set equalities below is incorrect in general. In each case, give examples of sets A, B and C for which the set equality does not hold.

    (a) $A \setminus (B \setminus C) = (A \setminus B) \setminus C$ (c) $A \cap (B \cup C) = (A \cap B) \cup C$

 $\star$ (b) $A \setminus (B \setminus C) = (A \setminus B) \setminus (A \setminus C)$ (d) $(A \cap B) \setminus C = A \cap (B \setminus C)$

(14) A, B and C are subsets of a universal set $\mathcal{U}$. In each case, determine if the statement is true or false. If the statement is true, give a proof. If the statement is false, give an example to show that it does not hold.

   $\star$(a) If $A \setminus B = \emptyset$, then $A = B$.

   (b) If $A \subseteq B$, then $A \oplus B = B \setminus A$.

   (c) If $A \setminus C = B \setminus C$, then $A = B$.

   (d) $A \setminus \overline{A} = A$.

  $\star$ (e) If $A \subseteq A \cap B$, then $A \cup B \subseteq B$.

   (f) If $A \cup B \subseteq C$ and $C \setminus B \subseteq C \setminus A$, then $A \subseteq B$.

(15) Let A, B be two subsets of a universal set $\mathcal{U}$. Prove each of the following statements.

$\star$(a) $A \subseteq B \Leftrightarrow A \cap B = A$

(b) $A \subseteq B \Leftrightarrow A \cup B = B$

(c) $A \cup B = A \cap B \Leftrightarrow A = B$

(d) $A \backslash B = B \backslash A \Leftrightarrow A = B$

$\star$(e) $A \backslash B = A \backslash (A \cap B)$

(f) $A \backslash (A \oplus B) = A \cap B$

(16) Let A, B and C be three subsets of a universal set $\mathcal{U}$.

(a) Prove that $A \cap (B \oplus C) = (A \cap B) \oplus (A \cap C)$ using properties of set operations.

(b) Give an example to show that $A \cup (B \oplus C) \neq (A \cup B) \oplus (A \cup C)$ in general.

(17) Given three subsets A, B and C of a universal set $\mathcal{U}$:

(a) Write the corresponding logic proposition for each of the subsets $A \oplus (B \oplus C)$ and $(A \oplus B) \oplus C$ of $\mathcal{U}$.

(b) Prove that $A \oplus (B \oplus C) = (A \oplus B) \oplus C$ by showing that the logic propositions in part (a) are logically equivalent.

(18) Given three subsets A, B and C of a universal set $\mathcal{U}$, prove that:

$\star$ (a) $A \cap (B \cup C) = (A \cap B) \cup C \Leftrightarrow C \subseteq A$

(b) $A \cup (B \cap C) = (A \cup B) \cap C \Leftrightarrow A \subseteq C$

(19) Let A, B and C be three subsets of a universal set $\mathcal{U}$. In each case, use properties of set operations to prove the set equality.

(a) $A \cap (\overline{A} \cup B) = A \cap B$

$\star$(b) $A \backslash (B \cap C) = (A \backslash B) \cup (A \backslash C)$

(c) $C \backslash (\overline{A} \cap B) = (C \backslash \overline{A}) \cup (C \backslash B)$

$\star$(d) $\overline{A \cap (\overline{A} \cup B)} \cup B = \mathcal{U}$

(e) $\overline{(A \cup B) \cap C} = \overline{A} \cap \overline{C} \cap \overline{B} \cap \overline{C}$

$\star$(20) For each integer $i \geq 0$, consider the subset $A_i = \{0, 1, \ldots, i\}$ of $\mathbb{N}$. Find $\bigcap_{i=0}^{\infty} A_i$ and $\bigcup_{i=0}^{\infty} A_i$.

(21) For each $i \in \mathbb{N}$, consider the subset $A_i = \{k \in \mathbb{Z};\ -i \leq k \leq i\}$ of $\mathbb{Z}$.

(a) Write the sets A_0, A_1, A_2 and A_5 using roster notation.

(b) Find $\bigcap_{i=0}^{\infty} A_i$ and $\bigcup_{i=0}^{\infty} A_i$.

(22) Let A, B and C be three sets with $B \cap C = \emptyset$. X and Y are two sets such that $X \subseteq A \cup B$ and $Y \subseteq A \cup C$. Show that $X \cap Y \subseteq A$.

(23) Let A, B be two subsets of a universal set $\mathcal{U}$.

$\star$ (a) Prove that if $A \subseteq B$, then $\wp(A) \subseteq \wp(B)$.

(b) Prove that $\wp(A \cap B) = \wp(A) \cap \wp(B)$.

(c) Prove that $\wp(A) \cup \wp(B) \subseteq \wp(A \cup B)$.

(d) Give an example of two sets A and B such that $\wp(A) \cup \wp(B) \neq \wp(A \cup B)$.

(24) If A is a set, what is $A \times \emptyset$?

$\star$(25) What conclusion can you draw about sets A and B if $A \times B = \emptyset$?

(26) What conclusion can you draw about sets A and B if $A \times B = B \times A$?

(27) In each case, write the given cross product in roster notations.

    (a) $A \times B$ where $A = \{a, b, c\}$, $B = \{1, 2\}$.

    (b) $A \times B$ where $A = \{a, b\}$, $B = \{1, 2, 3, 4\}$.

  $\star$ (c) $B \times A$ where $A = \{(a, 1), (b, 2), (c, 3)\}$, $B = \{\alpha, \beta\}$.

  $\star$ (d) $A \times B \times C$ where $A = \{a, b\}$, $B = \{1, 2\}$ and $C = \{\alpha, \beta\}$.

    (e) $A \times B \times C \times D$ where $A = \{0, 1\}$, $B = \{2, 3\}$, $C = \{a, b\}$ and $D = \{c\}$.

(28) Joe wants to go to campus but he wants to stop at his favorite coffee shop first. From his house to the coffee shop, Joe can take either walk or take a street car. From the coffee shop to campus, Joe can take a bus, a train or a taxi. List the set of all possible transport modes that Joe can take from his house to campus.

$\star$(29) Give an example to show that the Cartesian product is not associative. That is, give an example of three sets A, B and C such that $A \times (B \times C) \neq (A \times B) \times C$.

(30) Consider the two intervals $I = [-1, 4[$ and $J =]2, 5]$ of $\mathbb{R}$. In each case, determine if the given point in the Cartesian plane $\mathbb{R}^2$ is an element of $I \times J$.

    (a) $(-1, 0)$ (c) $(0, 2)$ $\star$ (e) $(1, 3)$ $\star$ (g) $(\frac{9}{2}, 5)$

 $\star$ (b) $(0, 5)$ (d) $(-\frac{1}{2}, \frac{5}{2})$ (f) $(\pi, 5)$

(31) Draw the region of the plane $\mathbb{R}^2$ that represents the Cartesian product $I \times J$ where I and J are the intervals given in Exercise (2).

(32) Find two sets A and B such that $A \times B$ is the given set.

    (a) $\{(a, 0), (a, 1), (a, 2)\}$

    (b) $\{(0, 0), (1, 0), (2, 0)\}$

  $\star$ (c) $\{(a, 0), (a, 1), (a, 2), (b, 0), (b, 1), (b, 2)\}$

    (d) $\{(-1, (a, 0)), (0, (a, 1)), (1, (a, 2)), (2, (b, 0)), (3, (b, 1)), (4, (b, 2))\}$

(33) Determine the truth value of each of the following statements.

(a) $(c, d) \in \{a, c\} \times \{b, d\}$

$\star$ (b) $\{(a, b), (c, d)\} = \{a, b, c, d\}$

(c) $\{(a, b), (c, d)\} \subseteq \{a, c\} \times \{b, d\}$

$\star$ (d) $\{(a, b), (c, d)\} = \{(a, c), (b, d)\}$

(e) $|\{a, c\} \times \{b, d\}| = 4$

$\star$ (f) $\{(a, c), (b, d)\} \in \wp(\{a, c\} \times \{b, d\})$

(34) If $A = \{a, b\}$ and $B = \{0\}$, write the set $\wp(A) \times \wp(B)$ using roster notation.

(35) Let A, B and C are three sets such that $C \neq \emptyset$ and $A \times C = B \times C$. Prove that $A = B$. Does the result remain true if we drop the assumption that $C \neq \emptyset$?

(36) In each case, give possible sets A and B such that the cardinality of $A \times B$ is the given integer.

(a) 0

$\star$ (b) 1

(c) 2

(d) 3

(e) 7

(f) 9

(g) 12

$\star$ (h) 36

(37) If A and B are two finite sets with $|A| = m$ and $|B| = n$, what is $|A \times B|$?

$\star$(38) Let $A = \{a, b\}$, $B = \{b, c\}$ and $C = \{c, d\}$. List all the elements of $A \times B \times C$.

(39) Prove part (a) of Theorem 2.6.

$\star$(40) Prove part (c) of Theorem 2.6.

(41) Prove part (f) of Theorem 2.6.

2.5 Partition of a set

A family $(X_i)_{i \in I}$ of subsets of a universe $\mathcal{U}$ is called *pairwise disjoint* if $X_i \cap X_j = \emptyset$ whenever $i \neq j$. That is, any two distinct sets of the family are disjoint. In this section, we look at a pairwise disjoint family of non-empty subsets of a set A such that their union covers A. Such a family is called a *partition* of A.

Definition 2.9. Let A be a non-empty set. A *partition* of A is a family $\{A_i; \ i \in I\}$ of subsets A_i of A indexed by a set I (finite or infinite) satisfying the following conditions.

(a) $A_i \neq \emptyset$ for each i.

(b) The family is pairwise disjoint: $A_i \cap A_j = \emptyset$ whenever $i \neq j$.

(c) $\cup_{i \in I} A_i = A$ (this means in particular that every element of A belongs to *exactly* one of the subsets A_i's).

If $\mathcal{F} = \{A_i;\ i \in I\}$ is a partition of A, then each member A_i of $\mathcal{F}$ is called a *block* of the partition. The following diagram is an illustration of a partition of a set A into 8 blocks.

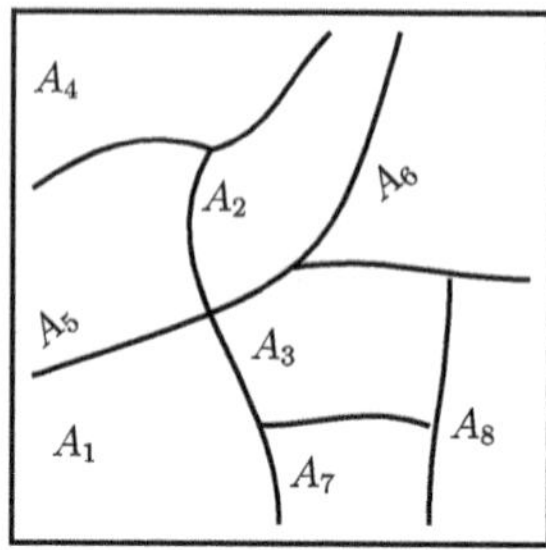

Example 2.38. Let $A = \{1, 2, 4, 6, 9, 13, 20\}$. In each case, determine if the given collection of subsets of A forms a partition of A.

(a) $S_1 = \{\{1, 4, 6\}, \{2\}, \{9, 13, 20\}\}$
(b) $S_2 = \{\{1, 9, 13\}, \{2, 20\}, \{1\}, \{4, 6\}\}$
(c) $S_3 = \{\{9, 13\}, \{2, 20\}, \{1\}, \{4, 6\}\}$
(d) $S_4 = \{\{1, 2\}, \{9, 20\}, \{4, 13\}\}$

Solution. Clearly S_1 and S_3 are partitions of A since they satisfy all three conditions of Definition 2.9. S_2 is not a partition since the two subsets $\{1, 9, 13\}$ and $\{1\}$ are not disjoint. S_4 is not a partition since the union of all elements of A does not equal to A (note that the element 6 of A does not belong to any of the subsets in S_4). $\Diamond$

Example 2.39. Let E and O denote the subsets of $\mathbb{N}$ of all even and odd integers, respectively. Clearly E and O are non-empty and $E \cap O = \emptyset$ since no integer is even and odd at the same. On the other hand, every integer is either even or odd (so an element of $E \cup O$). We conclude that $\{E, O\}$ is a partition of $\mathbb{N}$.

Given a non-empty set A, $\{A\}$ and $\{\{a\};\ a \in A\}$ are partitions of A. We refer to the partition $\{A\}$ of A as *trivial*.

Definition 2.10. Let A be a non-empty set, $\mathcal{F} = \{A_i;\ i \in I\}$ and $\mathcal{G} = \{B_j;\ j \in J\}$ two partitions of A. We say that $\mathcal{F}$ is a *refinement* of $\mathcal{G}$ if for every $i \in I$, $A_i \subseteq B_j$ for some $j \in J$. If in addition, $A_i \subset B_j$ (strict inclusion) for some $i \in I$ and for some $j \in J$, then $\mathcal{F}$ is called a *proper refinement* of $\mathcal{G}$.

Example 2.40. Let $A = \{a, b, c, d, e\}$. Consider the partitions $\mathcal{F} = \{\{a, b\}, \{c, d, e\}\}$, $\mathcal{F}_1 = \{\{a\}, \{b\}, \{c, d\}, \{e\}\}$, $\mathcal{F}_2 = \{\{a\}, \{b\}, \{c, d, e\}\}$ and $\mathcal{F}_3 = \{\{a\}, \{b, c, d, e\}\}$. Then both $\mathcal{F}_1$ and $\mathcal{F}_2$ are refinements of partition $\mathcal{F}$, but $\mathcal{F}_3$ is not.

2.5.1 *Exercises*

(1) In each case, give all possible partitions of the given set A.

    (a) $A = \{1, 2\}$

    ⋆(b) $A = \{1, 2, 3\}$

    (c) A is the set of all binary strings of length 2

(2) Let A be the set of all binary strings of length 3. Give a partition $\mathcal{F}$ of A such that each element of $\mathcal{F}$ is of cardinality 2.

(3) Let $A = \{\alpha, \beta, \gamma, \epsilon, \rho, \phi, \xi\}$. In each case, determine if the given collection of subsets of A forms a partition of A.

    ⋆(a) $\{\{\alpha, \beta\}, \{\gamma, \epsilon\}, \emptyset, \{\rho, \phi, \xi\}\}$

    (b) $\{\{\rho, \phi, \xi\}, \{\alpha\}, \{\beta, \epsilon\}, \{\gamma\}\}$

    (c) $\{\{\alpha, \beta, \epsilon\}, \{\rho, \phi, \xi\}, \{\alpha, \gamma\}\}$

    ⋆(d) $\{\{\rho, \phi, \xi\}, \{\alpha, \beta\}, \{\gamma\}\}$

(4) Let $A = \{1, 2, 3, 4, 5, 6, 7\}$.

    (a) Give a partition of A containing three blocks.

    ⋆(b) Give a partition of A containing five blocks.

    (c) Give two different partitions of A with the same number of blocks.

(5) Let $A = \{1, 2\}$ and $B = \{a, b\}$. Give two different partitions of $A \times B$.

(6) Let A, B be two non-empty sets such that $A \subset B$ (strict inclusion). Prove tat $\{A, B \backslash A\}$ is a partition of B.

⋆(7) Prove that the family $\{\{1, 2, 3\}, \{4, 5\}, \{6, 7\}\}$ is a partition of the set $A = \{1, 2, 3, 4, 5, 6, 7\}$. Give a refinement of $\mathcal{F}$ consisting of five blocks.

⋆(8) Consider the intervals $I =]-\infty, 0[$, $J =]0, \infty[$ of $\mathbb{R}$. Prove that the family $\{I, J, \{0\}\}$ of subsets of $\mathbb{R}$ is a partition of $\mathbb{R}$. Give a refinement of $\mathcal{F}$ consisting of five blocks.

(9) Let $\mathbb{Z}_0$, $\mathbb{Z}_1$ and $\mathbb{Z}_2$ denote the subsets of $\mathbb{Z}$ of integers having remainder 0, 1 and 2 upon division by 3, respectively. Prove that $\{\mathbb{Z}_0, \mathbb{Z}_1, \mathbb{Z}_2\}$ is a partition of $\mathbb{Z}$.

(10) If $r \geq 0$ is a real number, let A_r be the circle centered at the origin $(0, 0)$ and of radius r in the Cartesian plane $\mathbb{R}^2$. Prove that the family $\mathcal{F} = \{A_r; r \geq 0\}$ is a partition of the plane $\mathbb{R}^2$. (*Hint.* Given a point

(x, y) of $\mathbb{R}^2$, consider the circle centered at $(0,0)$ and with radius $r = \sqrt{x^2 + y^2}$).

$\star(11)$ Suppose that the family $\mathcal{F} = \{A_i;\ i \in I\}$ is a partition of a non-empty set A. Let B be a non-empty subset of A. Prove that the family $\mathcal{G} = \{B \cap A_i;\ i \in I \text{ and } B \cap A_i \neq \emptyset\}$ is a partition of B.

(12) If $\mathcal{F} = \{A_i;\ i \in I\}$ is a non-trivial partition of a set A, is it true that the family $\{A \backslash A_i;\ i \in I\}$ is also a partition of A?

(13) Let A and B be two non-empty subsets of a universe $\mathcal{U}$.

   (a) Assume that A and B are disjoint and let $\mathcal{F} = \{A_i;\ i \in I\}$ and $\mathcal{G} = \{B_j;\ i \in J\}$ be partitions of A and B, respectively. Prove that the family $\{A_i \cup B_j;\ (i, j) \in I \times J\}$ is a partition of $A \cup B$.

   (b) Prove (by providing a counter-example) that if we drop the assumption that A and B are disjoint, then the result in part (a) is not true in general.

$\star(14)$ A and B are two non-empty sets. Prove that if $\{A_i;\ i \in I\}$ is a partition of A and $\{B_j;\ j \in J\}$ is a partition of B, then the family $\{A_i \times B_j;\ (i, j) \in I \times J\}$ is a partition of $A \times B$.

(15) Let $\mathcal{F} = \{A_i;\ i \in I\}$ be a partition of a set A. For each $i \in I$, let $\{A_{ij}; j \in J_i\}$ be a partition of the set A_i. Prove that $\mathcal{G} = \{A_{ij}; i \in I,\ j \in J_i\}$ is a partition of A which is a refinement of $\mathcal{F}$.

2.6   Introduction to Boolean algebra

The close connection between the laws of propositional logic and the properties of set operations is just an illustration of a more general mathematical structure called *Boolean algebra*. In the mid 1800's, the British mathematician George Boole came up with a mathematical system to model the logic laws in an algebraic way. After almost a century of refinements and improvements on the system by various mathematicians and algorithm designers, the model finally found its way into real world applications and became an important tool in many areas of science and engineering. For instance, the Boolean algebra system is ideal in representing a two-state device, like an electrical transistor.

Definition 2.11. A *Boolean algebra* is a set B with at least two elements and two binary operations (i.e., operations requiring two variables as input) denoted as $+$ and $\cdot$ such that the following axioms hold for all elements x, y and z of B.

(B1) (Internal operations) $x + y$ and $x \cdot y$ are elements of B for all $x, y \in B$.

(B2) (Commutativity laws): $x + y = y + x$ and $x \cdot y = y \cdot x$ for all $x, y \in B$.

(B3) (Distributivity laws): $x + (y \cdot z) = (x + y) \cdot (x + z)$ and $x \cdot (y + z) = (x \cdot y) + (x \cdot z)$ for all $x, y, z \in B$.

(B4) (Identity laws) There exist distinct elements 0 and 1 of B such that: $x + 0 = x$ and $x \cdot 1 = x$ for every $x \in B$. The element 0 (respectively 1) is called the *identity element* of $+$ (respectively of $\cdot$).

(B5) (Complement laws) For every $x \in B$, there exists an element x', called the *complement* of x, satisfying $x + x' = 1$ and $x \cdot x' = 0$.

The operation $\cdot$ is usually denoted by a juxtaposition of operands, so we write xy instead of $x \cdot y$. Also, the identity element of the operation $+$ (respectively, of the operation $\cdot$) of a Boolean algebra B is referred to as the *zero element* of B (respectively, the *unit element* of B).

Notice that each of the axioms in Definition 2.11 comes in pair. In a Boolean algebra, the dual of a property P is the property obtained from P by interchanging $+$ and $\cdot$, 0 and 1. So each axiom consists of a property and its dual. It is also useful to notice that for any $x \in B$, $0 + x = x$, $1x = x$, $x' + x = 1$ and $x'x = 0$ by the commutativity laws. The same laws also allow us to write $(yz) + x = (y + x)(z + x)$ and $(y + z)x = (yx) + (zx)$ for all $x, y, z \in B$.

Example 2.41. The simplest example of a Boolean algebra is the set $B_2 = \{0, 1\}$ containing exactly two elements (the identity elements), together with the operations $+$ and $\cdot$ given by the following tables:

$+$	0	1
0	0	1
1	1	1

$\cdot$	0	1
0	0	0
1	0	1

and the complements $0' = 1$, $1' = 0$. Axioms (B1)–(B5) of Definition 2.11 can be easily verified from the tables above. The reader is encouraged to verify the details (see Exercise (2) below).

Example 2.42. Given a set A, the power set $\wp(A)$ of A with the operations $\cup$ and $\cap$ (union and intersection of sets) and the complement $\overline{X}$ of a subset X of A is a Boolean algebra. The identity element of $\cup$ is $\emptyset$ and that of $\cap$ is A itself. The verification that axioms (B1)-(B5) hold is left to the reader (see Exercise (3) below).

Example 2.43. Considering that our first encounter with the laws of Boolean algebra was through propositional logic, it is only natural to assume that set Σ of all wffs of propositional logic is a Boolean algebra with the operations $\vee$, $\wedge$ and $\neg$ (as addition, multiplication and complementation respectively). There is, however, a catch. In Σ, we don't have the notion of "equality" of wffs but rather "equivalence" of these objects. That causes some issues with the axioms of Boolean algebra. Given a wff ϕ, we consider the subset $[\phi]$ of Σ of all wffs which are logically equivalent to ϕ (called the class of ϕ). So all wffs equivalent to ϕ are simply represented by ϕ. On the set $\pi = \{[\phi]; \phi \in \Sigma\}$, we define the operations: $[\phi] \vee [\psi] = [\phi \vee \psi]$, $[\phi] \wedge [\psi] = [\phi \wedge \psi]$ and $\neg[\phi] = [\neg\phi]$. Then the set π with these new operations is a Boolean algebra with $[\mathbf{F}]$ (the class of a contradiction) as the 0 element and $[\mathbf{T}]$ (the class of a tautology) as the unit element. We will revisit this type of construction when we study equivalence relations in Chapter 7.

Example 2.44. Consider the set $D = \{1, 3, 5, 15\}$ of all positive integers divisors of 15. On the set D, we define the following operations: $x + y = \mathrm{lcm}(x, y)$ (the least common multiple of x and y), $xy = \gcd(x, y)$ (the greatest common divisor of x and y) and $x' = \frac{15}{x}$. The tables of values for the operations $+$, $\cdot$ and $'$ are as follows:

<table>
<tr><td>

$+$	1	3	5	15
1	1	3	5	15
3	3	3	15	15
5	5	15	5	15
15	15	15	15	15

</td><td>

$\cdot$	1	3	5	15
1	1	1	1	1
3	1	3	1	3
5	1	1	5	5
15	1	3	5	15

</td><td>

x	x'
1	15
3	5
5	3
15	1

</td></tr>
</table>

From the tables, we see $x + y$, xy and x' are all elements of D for any $x, y \in D$. We also see that $x + 1 = x$ and $x \cdot 15 = x$ for any $x \in D$. In other words, D has 1 and 15 as the zero and the unit element, respectively. The tables show also that $+$ and $\cdot$ are commutative operations. The distributivity laws $x(y + z) = xy + xz$ and $x + yz = (x + y)(x + z)$ can be easily verified from the tables. Finally, note that $xx' = 1$ and $x + x' = 15$ for each $x \in D$. This shows that $(D, +, \cdot, ')$ is a Boolean algebra. It is worth mentioning here that the result of this example cannot be generalized to the set of divisors of any natural number n (see Exercise (17) below).

Using axioms (B1)-(B5) of Definition 2.11, we derive a series of interesting results that apply to any Boolean algebra. These results play an important role in the applications of this structure.

Theorem 2.7. *Given a Boolean algebra B, the following hold:*

(a) *The zero element of B is unique.*
(b) *The unit element of B is unique.*
(c) *For any element $b \in B$, the complement b' is unique.*
(d) *For any element $b \in B$, $(b')' = b$ (the complement of the complement of $b \in B$ is b).*

Proof. We only prove part (a) and leave the other parts as exercises for the reader. Assume that z_1 and z_2 are both zero elements of B. Then $z_2 + z_1 = z_2$ (since z_1 is a zero element) and $z_1 + z_2 = z_1$ (since z_2 is a zero element). But since $z_1 + z_2 = z_2 + z_1$ (commutativity of $+$), $z_1 = z_2$. $\square$

Theorem 2.8. *Given a Boolean algebra B, the following hold:*

(a) *(Idempotent laws) $b + b = b$ and $bb = b$ for any $b \in B$.*
(b) *(Boundless laws) $b + 1 = 1$ and $b0 = 0$ for any $b \in B$.*
(c) *(Absorption laws) $a + ab = a$ and $a(a + b) = a$ for any $a, b \in B$.*
(d) *(Associative laws) $a + (b + c) = (a + b) + c$ and $a(bc) = (ab)c$ for any $a, b, c \in B$.*
(e) *(De Morgan's laws) $(a + b)' = a'b'$ and $(ab)' = a' + b'$ for any $a, b \in B$.*

Proof. Each part of the theorem consists of a property and its dual. In each case, we prove one property and leave the proof of its dual as an exercise.

(a) For any $b \in B$,

$$
\begin{aligned}
b &= b + 0 && \text{(Axiom (B4))} \\
&= b + (bb') && \text{(Axiom (B5))} \\
&= (b + b)(b + b') && \text{(Axiom (B3))} \\
&= (b + b)(1) && \text{(Axiom (B5))} \\
&= b + b && \text{(Axiom (B4))}
\end{aligned}
$$

(b) For any $b \in B$:

$$
\begin{aligned}
b0 &= b0 + 0 && \text{(Axiom (B4))} \\
&= b0 + bb' && \text{(Axiom (B5))} \\
&= b(0 + b') && \text{(Axiom (B3))} \\
&= bb' && \text{(Axiom (B4))} \\
&= 0 && \text{(Axiom (B5))}
\end{aligned}
$$

(c) For any $a, b \in B$: $a + ab = a1 + ab$ (Axiom (B4)) $= a(1 + b)$ (Axiom (B3)) $= a1$ (Boundless laws, part (b)) $= a$ (Axiom (B4)).

(d) Let $x = a + (b + c)$ and $y = (a + b) + c$. We need to prove that $x = y$. Note that $ax = a\,(a + (b + c)) = a$ by the absorption law. Also, $ay = a\,((a + b) + c) = a(a + b) + ac$ (by the distributivity laws) $= a + ac$ (absorption law) $= a$ (by the absorption law again). On the other hand,

$$
\begin{aligned}
a'x &= a'\,(a + (b + c)) \\
&= a'a + a'(b + c) & \text{(Axiom (B3))} \\
&= 0 + a'(b + c) & \text{(Axiom (B5))} \\
&= a'(b + c) & \text{(Axiom (B4))}
\end{aligned}
$$

and

$$
\begin{aligned}
a'y &= a'\,((a + b) + c) \\
&= a'(a + b) + a'c & \text{(Axiom (B3))} \\
&= a'a + a'b + a'c & \text{(Axiom (B3))} \\
&= 0 + a'b + a'c & \text{(Axiom (B5))} \\
&= a'b + a'c & \text{(Axiom (B4))} \\
&= a'(b + c) & \text{(Axiom (B3))}.
\end{aligned}
$$

We conclude that $ax = ay$ and $a'x = a'y$. Therefore:

$$
\begin{aligned}
x &= 1x & \text{(Axiom (B4))} \\
&= (a + a')x & \text{(Axiom (B5))} \\
&= ax + a'x & \text{(Axiom (B3))} \\
&= ay + a'y & (\text{Since } ax = ay \text{ and } a'x = a'y) \\
&= (a + a')y & \text{(Axiom (B3))} \\
&= 1y & \text{(Axiom (B5))} \\
&= y & \text{(Axiom (B4))}.
\end{aligned}
$$

(e) We prove that $(a + b)' = a'b'$. By the uniqueness of the complement element (Theorem 2.8), we need to show two things: $(a+b)+(a'b') = 1$

and $(a + b)(a'b') = 0$.

$$
\begin{aligned}
(a + b) + (a'b') &= a + (b + a'b') &&\text{(Associative law)} \\
&= a + (b + a')(b + b') &&\text{(Axiom (B3))} \\
&= a + (b + a')(1) &&\text{(Axiom (B5))} \\
&= a + (b + a')(\text{Axiom (B4)}) \\
&= a + (a' + b) &&\text{(Axiom (B2))} \\
&= (a + a') + b &&\text{(Associativity laws)} \\
&= 1 + b &&\text{(Axiom (B5))} \\
&= 1 &&\text{(Boundless laws).}
\end{aligned}
$$

Also:

$$
\begin{aligned}
(a + b)(a'b') &= a(a'b) + b(a'b') &&\text{(Axiom (B3))} \\
&= a(a'b) + (a'b')b &&\text{(Axiom (B2))} \\
&= (aa')b + a'(b'b) &&\text{(Associativity of $\cdot$)} \\
&= 0b + a'0 &&\text{(Axiom (B5))} \\
&= 0 + 0 &&\text{(Boundless laws)} \\
&= 0 &&\text{(Axiom (B4)).} \quad\square
\end{aligned}
$$

Remark 2.6. Some authors include the associativity of the operations $+$ and $\cdot$ of a Boolean algebra as axioms for simplicity. Theorem 2.8 shows that this is not necessary as this can be proven using the other axioms of Boolean algebra as given in Definition 2.11.

2.6.1 *Boolean expressions*

When designing an electronic device, engineers make an effort to minimize the number of gates (switches) used in the circuit. Boolean algebra laws are useful for that purpose. Since each switch is either off or on, the two-state Boolean algebra B_2 of Example 2.41 is of particular importance in circuit designs. By a *Boolean variable* over a Boolean algebra B, we mean a symbol that can only take values in B. For example, a Boolean variable over the algebra B_2 is a variable x that can only assume one of the values 0 or 1. A *Boolean expression* (also known as *Boolean form*) in n Boolean variables $x_1, \ldots, x_n$ over a Boolean algebra B is a correspondence rule f (or a function) that assigns to every element $(x_1, x_2, \ldots, x_n)$ of B^n (the Cartesian product $B \times B \times \cdots \times B$) a unique element $f(x_1, x_2, \ldots, x_n)$ of the algebra B. Simply put, a Boolean expression is one containing some Boolean variables, the operations $+$, $\cdot$ and $'$. For example, $f(x, y, z, t) = x'yz + yzt' + xyzxt + x'z'$ is a Boolean expression in four Boolean variables x, y, z, t. The value $f(\alpha_1, \ldots, \alpha_n)$ of a Boolean expression f at the element $(\alpha_1, \ldots, \alpha_n)$ of B^n is obtained by substituting the variable x_i with the value α_i in the expression. For example, the following are some values of the Boolean expression $f(x, y) = x'y + y'$ over the algebra B_2: $f(1, 1) = 1'(1) + 1' = 0$, $f(1, 0) = 1'(0) + 0' = 1$, $f(0, 1) = 0'(1) + 1' = 1$ and $f(0, 0) = 0'(1) + 0' = 1$. Two Boolean expressions f and g of n Boolean variables are called *equivalent*, and we write $f \equiv g$, if $f(\alpha_1, \ldots, \alpha_n) = g(\alpha_1, \ldots, \alpha_n)$ for every $(\alpha_1, \ldots, \alpha_n) \in B^n$.

Like in the case of the standard arithmetic, there is a hierarchy in applying the operations $+$, $\cdot$ and $'$ in a Boolean expression. The complementation takes precedence over the product and the product takes precedence over addition. For example, the expression $xy + z$ means $(xy) + z$ rather than $x(y + z)$.

Example 2.45. Prove that the expressions $f(x, y, z) = x'y'z + x'yz' + xy'z + xyz + xyz'$ and $g(x, y, z) = y'z + yz' + xy$ in the Boolean variables x, y and z over the algebra B_2 are equivalent.

Solution. We proceed in two ways. First, we form the table of values of f and g at each of the eight elements of B^3:

x	y	z	x'	y'	z'	$f(x, y, z)$	$g(x, y, z)$
1	1	1	0	0	0	1	1
1	1	0	0	0	1	1	1
1	0	1	0	1	0	1	1
1	0	0	0	1	1	0	0
0	1	1	1	0	0	0	0
0	1	0	1	0	1	1	1
0	0	1	1	1	0	1	1
0	0	0	1	1	1	0	0

we see that f and g have the same value for any triplet (x, y, z) of Boolean variables and we conclude that $f \equiv g$. You can imagine that if the number of the Boolean variables is four or more, then verifying the equivalence of two Boolean expressions using a table is not practical (the same way it was not for logic formulas). A more efficient way is through *algebraic manipulations* of Boolean algebra laws. In what follows we show the manipulations with brief justifications leaving the full details for the reader.

$$f(x, y, z) = x'y'z + x'yz' + xy'z + xyz + xyz'$$
$$= (x'y'z + xy'z) + (x'yz' + xyz') + xyz \qquad \text{(Commutativity)}$$
$$= (x' + x)y'z + (x' + x)yz' + xyz \qquad \text{(Distributivity)}$$
$$= 1y'z + 1yz' + xyz \qquad \text{(Complementation laws)}$$
$$= y'z + yz' + xyz \qquad \text{(Identity element)}$$
$$= y'z + y(z' + xz) \qquad \text{(Distributivity)}$$
$$= y'z + y(z' + x)(z' + z) \qquad \text{(Distributivity)}$$
$$= y'z + y(z' + x)1 \qquad \text{(Complementation laws)}$$
$$= y'z + y(z' + x) \qquad \text{(Identity law)}$$
$$= y'z + yz' + xy \qquad \text{(Distributivity)}$$
$$= g(x, y, z) \qquad \Diamond$$

Example 2.46. In this example, we show that the expressions $f(a, b, c) = a'b'c' + ab'c' + abc'$ and $g(a, b, c) = ac' + b'c'$ are equivalent:

$$f(a, b, c) = a'b'c' + ab'c' + abc'$$
$$= (a' + a)b'c' + abc' \qquad \text{(Distributivity)}$$
$$= 1b'c' + abc' \qquad \text{(Complementation laws)}$$
$$= b'c' + abc' \qquad \text{(Identity element)}$$
$$= (b' + ab)c' \qquad \text{(Distributivity)}$$
$$= (b' + a)(b' + b)c' \qquad \text{(Distributivity)}$$
$$= (b' + a)1c' \qquad \text{(Complementation laws)}$$
$$= (b' + a)c' \qquad \text{(Identity law)}$$
$$= b'c' + ac' \qquad \text{(Distributivity)}$$
$$= ac' + b'c' \qquad \text{(Commutativity)}$$

2.6.2 *Exercises*

⋆(1) Is the set $\mathbb{R}$ of all real numbers with the standard addition and multiplication a Boolean algebra? Justify your answer.

(2) Verify that axioms (B1)-(B5) of Boolean algebra are satisfied for the set $B_2 = \{0, 1\}$ with the binary operations defined in Example 2.41 above.

$\star$(3) Given a set A, prove that the power set $\wp(A)$ of A is a Boolean algebra with the union, intersection and complement of subsets of A.

$\star$(4) Prove that the unit element of a Boolean algebra is unique.

(5) Let B be a Boolean algebra, and let $b \in B$.

    (a) Prove that the complement b' of b is unique.

  $\star$ (b) Prove that $(b')' = b$.

(6) Let B be a Boolean algebra with unit element 1 and let $b \in B$. Prove that $bb = b$.

(7) Let B be a Boolean algebra with unit element 1 and let $b \in B$. Prove that $b + 1 = 1$.

(8) Let B be a Boolean algebra and let $a, b \in B$. Prove that $a' + b = 1$ if and only if $ab' = 0$.

$\star$(9) Let B be a Boolean algebra and let $a, b \in B$. Prove that $a + b = b$ if and only if $ab = a$.

(10) Prove that in any Boolean algebra B, $bc + acb = cb$ for any $a, b, c \in B$.

$\star$(11) Prove that there is no Boolean algebra containing exactly three pairwise distinct elements.

(12) Let B_1 and B_2 be two Boolean algebras. On the Cartesian product $B_1 \times B_2$, we define two operations, also denoted by $+$ and juxtaposition as follows. For any $x, y \in B_1$ and $a, b \in B_2$: $(x, y) + (a, b) = (x + a, y + b)$ and $(x, y)(a, b) = (xa, yb)$ where $x + a$ and xa are the addition and the multiplication (respectively) in B_1 while $y + b$ and yb are the addition and the multiplication (respectively) in B_2. Prove that $B_1 \times B_2$ is a Boolean algebra with respect to these operations. Clearly state what the zero and the unit elements of $B_1 \times B_2$ are and what is the complement of an element $(a, b) \in B_1 \times B_2$.

(13) The cancellation law for addition of real numbers states that if $x, y, z \in \mathbb{R}$ such that $x + y = x + z$, then $y = z$. Does that law hold in a Boolean algebra?

$\star$(14) Prove De Morgan law: $(ab)' = a' + b'$ in any Boolean algebra.

(15) Find the dual of each of the following Boolean expressions.

    $\star$(a) $abc' + a'bc'$ $\star$(c) $(ab)' + abc'$

    (b) $a(bc + a'b'c)$ (d) $(a + b)(ac + ac') + ab' + c$

(16) Let M be a positive real number. On the interval $I = [0, M]$ of $\mathbb{R}$, we define the following operations: $x + y = \max\{x, y\}$, $xy = \min\{x, y\}$ and $x' = M - x$. Prove that these operations on I satisfy axioms

$(B1) - (B4)$. Show that there are only two elements of I satisfy axiom $(B5)$.

(17) Using the notations of the Example 2.44 above, is the set $B = \{1, 2, 3, 4, 6, 12\}$ of all positive divisors of 12 a Boolean algebra? Justify your answer.

(18) In each case, prove that the given Boolean expressions in the logic variables a, b, c over a certain Boolean algebra B are equivalent.

- $\star$ (a) $(a + (ab)')'ab'$ and 0
- (b) $(ab)' + a$ and 1
- (c) $(a + b)(ac + ac') + ab' + c$ and $a + c$
- (d) $a'(a + b) + (b + a)(a + b')$ and $a + b$
- $\star$ (e) $abc' + abc + a'bc + a + c$ and $a + c$
- (f) $a'b'c' + a'bc' + ab'c + abc'$ and $a'c' + bc' + ab'c$

(19) Let B be a Boolean algebra and let C be non-empty subset C of B. We say that C is a *subalgebra* of B if for any $x, y \in C$ we have that $x + y$, xy and x' are elements of C.

(a) Prove that any Boolean algebra B other than B_2 contains at least two distinct subalgebras.

(b) Prove that C is a subalgebra of B if and only if C is a Boolean algebra with the inherited operations from B.

(c) Prove that C is a subalgebra of B if and only if for any $x, y \in C$ we have that $x + y$, and x' are elements of C.

2.7 Sum of products (SOP), product of sums (POS), and Karnaugh maps

In Chapter 1, we studied the notions of Disjunctive Normal Form (DNF) and Conjunctive Normal Form (CNF) of wffs of propositional logic. In this section, we generalize these notions to expressions in any Boolean algebra using a slightly different terminology. In what follows, B is a Boolean algebra, $S = \{x_1, \ldots, x_n\}$ is a set of Boolean variables over B with an assumed ordering $x_1, \ldots, x_n$ on the variables. A *literal* over S is either a variable x_i or the complement x_i' of a variable.

2.7.1 *Sum of products (SOP)*

A *fundamental product* over S is either a literal or the product of two or more literals such that no two literals involve the same variable. For example, $x_1 x_2' x_3'$, $x_1 x_2 x_3'$ and $x_1' x_2' x_3'$ are fundamental products while $x_1 x_2' x_2$, $x_1 x_2 x_1 x_3$ and $x_1' x_2' x_1$ are not. A fundamental product P is said to be a *part* of another fundamental product Q if every literal appearing in P also appears in Q. For example, the fundamental product xyt' is a part of $xyz't'$ but not a part of $x'yt'$. A Boolean expression $f(x_1, \ldots, x_n)$ is said to be in *sum of products* form (SOP for short) if it is one fundamental product or the sum of two or more fundamental products none of which is a part of another. For example, $f(x_1, x_2, x_3) = x_1 x_2 x_3' + x_1' x_2 + x_1 x_3$ is in SOP form but $g(x_1, x_2, x_3) = x_1 x_2 x_3' + x_1' x_2 + x_1 x_2 x_3 + x_1 x_3'$ is not (since the last term is part of the first term). A *minterm* is a fundamental product in which every variable x_i appears *exactly once* (either as x_i or as its complement x_i' but not both). For example, the following are all the minterms over the set of variables $\{x, y, z\}$: $x'y'z'$, $x'y'z$, $x'yz'$, $x'yz$, $xy'z'$, $xy'z$, xyz' and xyz. A Boolean expression is said to be in *complete sum of products form* (CSOP for short) if it is in SOP form and each fundamental product appearing in it is a minterm. For example, the expressions $xy'z + x'y'z'$ and $x'yz + xy'z' + xy'z + xyz + xyz'$ are in CSOP forms over the variables x, y and z.

Every Boolean expression can be written in a SOP and CSOP forms using laws of Boolean algebra. We omit the proof of this result but we show some examples of how it works.

Example 2.47. Write each of the following Boolean expressions over the variables x, y, z, t in SOP and CSOP forms.

  (a) $f(x, y, z, t) = xy'z + xyzt + x'y'zt' + xy'zt$
  (b) $g(x, y, z, t) = x'y'z + xy + x'yzt'$

Solution.   (a) Expression f is not in SOP form as the term $xy'z$ is part of the term $xy'zt$. Note that $xy'z + xy'zt = xy'z(1 + t) = xy'z$ since $1 + t = 1$ by the absorption law. So, $f(x, y, z, t) = xy'z + xyzt + x'y'zt'$ is a SOP form of f. To find a CSOP form of f, every variable has to appear in every fundamental product. This is the case for the products $xyzt$ and $x'y'zt'$ but not for the first product $xy'z$. To this end, we use the identity $a + a' = 1$ in any Boolean algebra and we write $xy'z = xy'z(t + t') = xy'zt + xy'zt'$. So $f(x, y, z, t) = xy'zt + xy'zt' + xyzt + x'y'zt'$ is the CSOP form of f.

  (b) Expression g is in SOP form but not in CSOP form. The first term of g is missing the variable t and the second term is missing the variables z

and t:

$$x'y'z + xy + x'yzt' = x'y'z(t + t') + xy(z + z')(t + t') + x'yzt'$$
$$= x'y'zt + x'y'zt' + (xyz + xyz')(t + t') + x'yzt'$$
$$= x'y'zt + x'y'zt' + xyzt + xyzt' + xyz't + xyz't'$$
$$+ x'yzt'$$

So an expansion of g as a CSOP form is $x'y'zt + x'y'zt' + xyzt + xyzt' + xyz't + xyz't' + x'yzt'$. $\Diamond$

2.7.2 *Product of sums (POS)*

The dual of the sum of product notion is the *product of sums*. We start with some terminology. A *fundamental sum*, is either a literal or the sum of two or more literals such that no two literals involve the same variable. For example, $x_1 + x_2' + x_3'$, $x_1 + x_2 + x_3'$ and $x_1' + x_2' + x_3'$ are fundamental sums while $x_1 + x_2' + x_2$, $x_1 + x_2 + x_1'$ and $x_1' + x_2' + x_1$ are not. A fundamental sum S is said to be a *part* of another fundamental sum T if every literal appearing in S also appears in T. For example, the fundamental sum $x + y + t'$ is a part of $x + y + z' + t'$ but not a part of $x' + y + t$. A Boolean expression $f(x_1, \ldots, x_n)$ is said to be in *product of sums* form (POS for short) if it is one fundamental sum or the product of two or more fundamental sums none of which is a part of another. For example, $f(x_1, x_2, x_3) = x_2'(x_1 + x_2 + x_3')(x_1' + x_2)(x_1 + x_3)$ is in POS form but $g(x_1, x_2, x_3) = (x_1 + x_2 + x_3')(x_1' + x_2)(x_1 + x_2)$ is not. A *maxterm* is a fundamental sum in which every variable x_i appears *exactly once* (either as x_i or as its complement x_i' but not both). A Boolean expression is said to be in *complete product of sums form* (CPOS for short) if it is in POS form and each fundamental sum appearing in it is a maxterm. For example, the expression $(x + y' + z)(x' + y' + z')$ is in CPOS form over the variables x, y and z.

Like the SOP form, every Boolean expression can be expressed in a POS and CPOS forms using Boolean algebra laws. The following example illustrates such conversion.

Example 2.48. In this example, we write the Boolean expression $f(x, y, z) = x'y + z(x'y' + xyz')$ over the variables x, y, z in POS and CPOS forms. Justifications of the steps are omitted.

$$f(x, y, z) = x'y + z(x'y' + xyz')$$
$$= x'y + x'y'z + xy \underbrace{z'z}_{=0}$$
$$= x'y + x'y'z$$
$$= x'(y + y'z)$$
$$= x' \underbrace{(y + y')}_{=1}(y + z)$$
$$= x'(y + z).$$

We conclude that $x'(y + z)$ is a POS form for f. For a CPOS form, each variable must appear in every fundamental sum. If a variable α is missing from a sum S, it is easy to insert it back by writing: $S = S + 0 = S + \alpha\alpha' = (S + \alpha)(S + \alpha')$ using distributivity. This process is repeated until every variable appears in every sum:

$$x'(y + z) = (x' + yy')(y + z + xx')$$
$$= (x' + y)(x' + y')(y + z + x)(y + z + x')$$
$$= (x' + y + zz')(x' + y' + zz')(y + z + x)(y + z + x')$$
$$= (x' + y + z)(x' + y + z')(x' + y' + z)(x' + y' + z')$$
$$(y + z + x)(y + z + x')$$

So $f(x, y, z) = (x' + y + z)(x' + y + z')(x' + y' + z)(x' + y' + z')(y + z + x)(y + z + x')$ is an expansion of f in CPOS form.

2.7.3   *Karnaugh maps*

Boolean expression are used in designing electronic logic circuits as explained in Section 2.8 below. It is therefore important to be able to simplify a Boolean expression as much as possible and obtain a minimal form of such an expression in order to optimize the cost of making the circuit. By a minimal form, we mean an equivalent Boolean expression that contains the smallest possible number of terms. The method of Karnaugh maps was first introduced by Edward Westbrook Veitcheitch in 1952 and then modified to its actual form by Maurice Karnaugh in 1953. For a Boolean expression in n variables $f(x_1, \ldots, x_n)$ written in CSOP form, the Karnaugh map gives a schematic representation of f in the form of a rectangular area divided in 2^n cells. Each cell represents one of the 2^n possible minterms in the variables $x_1, x_2 \ldots, x_n$. We place 1 in each cell representing a minterm present in the CSOP of f. The allocation of minterms in the cells is not random and must be done in a particular way for the map to be effective in simplifying the expression. First a definition.

Definition 2.12. Two minterns are called *adjacent* if they differ in exactly one position. For instance, the two minterms (in four variables) $xyz't$ and $xy'z't$ are adjacent since they differ at the second position only.

For the adjacent minterms $xyz't$ and $xy'z't$, notice the following:

$$xyz't + xy'z't = xz't \underbrace{(y + y')}_{=1} = xz't. \tag{2.2}$$

The variable y which represents the different position in the minterms has completely disappeared when the minterms are added together. This observation is not proper to the example in equation (2.2): *every time two adjacent minterms are added, the sum is obtained by simply dropping the variable at the different position.* Exploiting this property of adjacent minterms, Karnaugh maps assign adjacent minterms to adjacent cells (cells sharing a common edge, not just a single point). If in a Karnaugh map representing a Boolean expression f there is 1 in two adjacent cells, then the sum of the corresponding minterms in f can be simplified by eliminating the variable at the different position.

2.7.4 *Two-variable Karnaugh maps*

A Karnaugh map in two variables x and y is a 2×2 array of 4 cells, one for each of the four possible minterms in x and y. The rows represent one variable (so a variable and its complement) and the columns the other. As mentioned above, the cells are labeled strategically so that two adjacent cells correspond to two adjacent minterms as shown in the following diagram.

	y	y'
x	xy	xy'
x'	$x'y$	$x'y'$

Example 2.49. The diagram below represents the Karnaugh maps for the Boolean expressions $f(x,y) = xy' + x'y$, $g(x,y) = xy + xy' + x'y'$, and $h(x,y) = xy + xy' + x'y + x'y'$.

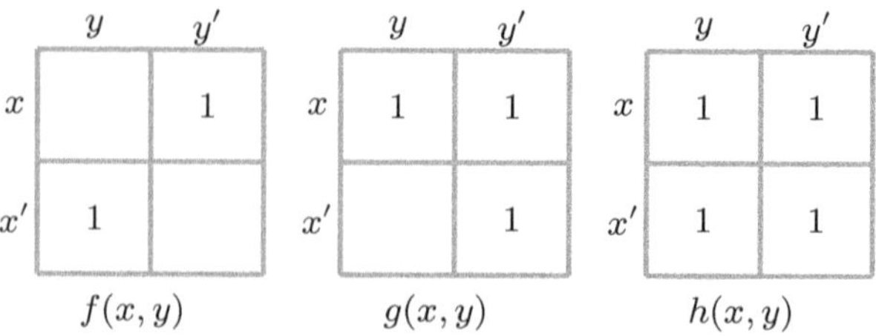

$$f(x,y) \qquad\qquad g(x,y) \qquad\qquad h(x,y)$$

2.7.5   *Three-variable Karnaugh maps*

A Karnaugh map in three variables x, y and z is a 2×4 (or a 4×2) array of $2^3 = 8$ cells, one for each of the eight possible minterms in x, y and z. One way to construct a three-variable Karnaugh map is to let the rows represent one variable and the columns all possible minterms in the other two variables. Like in the case of two variables, the cells are labeled so that two adjacent cells correspond to two adjacent minterms as shown in the following diagram. Note that the first and the last cells in each row are adjacent as they differ by only one position. The physical adjacency of these extreme cells may also be realized by imagining the Karnaugh map wrapped around a cylinder.

	yz	yz'	$y'z'$	$y'z$
x	xyz	xyz'	$xy'z'$	$xy'z$
x'	$x'yz$	$x'yz'$	$x'y'z'$	$x'y'z$

Example 2.50. The diagram below represents the Karnaugh maps for the given Boolean expressions in three variables: $f(x,y,z) = x'yz + x'yz' + xy'z + x'y'z'$, $g(x,y,z) = xyz + x'y'z' + x'yz' + xy'z'$, and $h(x,y,z) = xyz + x'yz + xyz' + x'yz' + xy'z' + x'y'z' + x'y'z$.

	yz	yz'	$y'z'$	$y'z$
x				1
x'	1	1	1	

$$f(x,y,z)$$

	yz	yz'	$y'z'$	$y'z$
x	1		1	
x'		1	1	

$$g(x,y,z)$$

	yz	yz'	$y'z'$	$y'z$
x	1	1	1	
x'	1	1	1	1

$$h(x, y, z)$$

2.7.6 *Four-variable Karnaugh maps*

A Karnaugh map in four variables x, y, z and t is a 4×4 array of $2^4 = 16$ cells, one for each of the sixteen possible minterms in x, y, z and t. The rows represent all possible minterms in two of the four variables and the columns represent all possible minterms in the other two variables. Like before, the cells are labeled so that two adjacent cells correspond to two adjacent minterms as shown in the following diagram. The extreme cells in each row are adjacent as they differ by one position, just like before.

	zt	zt'	$z't'$	$z't$
xy	$xyzt$	$xyzt'$	$xyz't'$	$xyz't$
xy'	$xy'zt$	$xy'zt'$	$xy'z't'$	$xy'z't$
$x'y'$	$x'y'zt$	$x'y'zt'$	$x'y'z't'$	$x'y'z't$
$x'y$	$x'yzt$	$x'yzt'$	$x'yz't'$	$x'yz't$

Example 2.51. The diagram below shows the Karnaugh maps of the Boolean expressions in four variables $f(x, y, z, t) = xyzt + xyz't + x'y'zt + x'y'zt' + x'y'z't' + x'y'z't + x'yzt$ and $g(x, y, z, t) = xyzt + xyzt' + xy'zt + xy'zt' + xy'z't' + x'y'z't' + x'y'z't$.

	zt	zt'	$z't'$	$z't$
xy	1			1
xy'				
$x'y'$	1	1	1	1
$x'y$	1			

$$f(x,y,z,t)$$

	zt	zt'	$z't'$	$z't$
xy	1	1		
xy'	1	1	1	
$x'y'$			1	1
$x'y$				

$$g(x,y,z,t)$$

2.7.7 *Using Karnaugh to minimize a Boolean expression*

Once the Karnaugh map is drawn from the CSOP form of a Boolean expression f, it can be used to minimize the number of terms in f using the following steps.

(a) **Grouping of 1's in the map.** Grouping the 1's in a Karnaugh map consists of enclosing a rectangular area of cells in the map containing 1's. Each cell in a group must be *adjacent* to one or more cells in the same group. It is important to remember that cells containing 1's on the extreme columns in a row are adjacent. For example, the two cells containing 1 in the first row of the Karaugh map representing the expression $f(x,y,z,t)$ of Example 2.51 above are adjacent and can be enclosed in the same group. The goal of this step is to minimize the number of groups using the following rules.

(1) Every group must enclose a rectangular area containing 2^k, $k = 0, 1, 2, \ldots$ cells with 1 in each cell. So, a group in a two-variable Karnaugh map can enclose 1, 2 or 4 cells. A group in a three-variable Karnaugh map can enclose 1, 2, 4 or 8 cells and a group in a four-variable Karnaugh map can enclose 1, 2, 4, 8 or 16 cells.

(2) The number of groups in a Karnaugh map must be *as small as possible*. This is achieved by enclosing the largest possible number of 1's in a group while respecting Rule (1) at the same time.

(3) Every 1 in a Karnaugh map must be enclosed in at least one group. Groups can have overlapping 1's but they must also include non-common 1's.

(b) Once the grouping of 1's is done on a Karnaugh map according to the above rules, minimizing a Boolean expression is done as follows:

(1) Each group of 1's is simplified to a single product term. In this single term, we drop the variables that change status (x to x' or vice versa) and keep those that remain unchanged in the group.

(2) The minimization of the Boolean expression is obtained by adding all the product terms from all the groups in the map.

Example 2.52. In this example, we minimize each of the following CSOP forms in three variables using the technique of Karnaugh maps.

(a) $f(x, y, z) = xyz + xy'z + xy'z' + x'yz' + x'y'z'$
(b) $g(x, y, z) = xyz + xyz' + xy'z' + xy'z + x'yz' + x'y'z'$
(c) $h(x, y, z) = xyz' + xy'z' + xy'z + x'yz + x'y'z' + x'y'z.$

The Karnaugh maps for the expressions together with the appropriate groupings in each map are given in the following diagram:

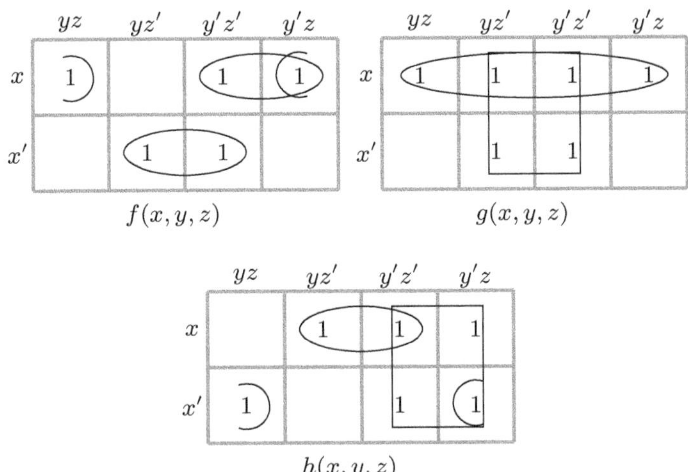

(a) The minimal number of groups that will cover all the 1's in the Karnaugh map of $f(x, y, z)$ is 3. The first group consists of the two 1's in the extreme columns of the first row. This group contributes with the term xz to the minimal form. The second group consists of the two 1's in the last two cells in the first row. This group contributes with the term xy'. The last group consists of the two 1's in the second row. It contributes with $x'z'$ to the minimal form. Summing up the contributions of all groups, the minimal form of $f(x, y, z)$ is $xz + xy' + x'z'$.

(b) From the Karnaugh map of $g(x, y, z)$, only two groups suffice to cover all the 1's in the map. The first consists of the four 1's in the first row. This group contributes with the term x to the minimal form. The second group consists of the square containing the four 1's in the middle of the map. It contributes with z' to the minimal form. We conclude that the minimal form of $g(x, y, z)$ is $x + z'$.

(c) Following the Karnaugh map of $h(x, y, z)$ given in the diagram above, three groups are needed to cover the 1's in the map. The minimal form of $h(x, y, z)$ is $xz' + y' + z'z$.

Example 2.53. The diagram

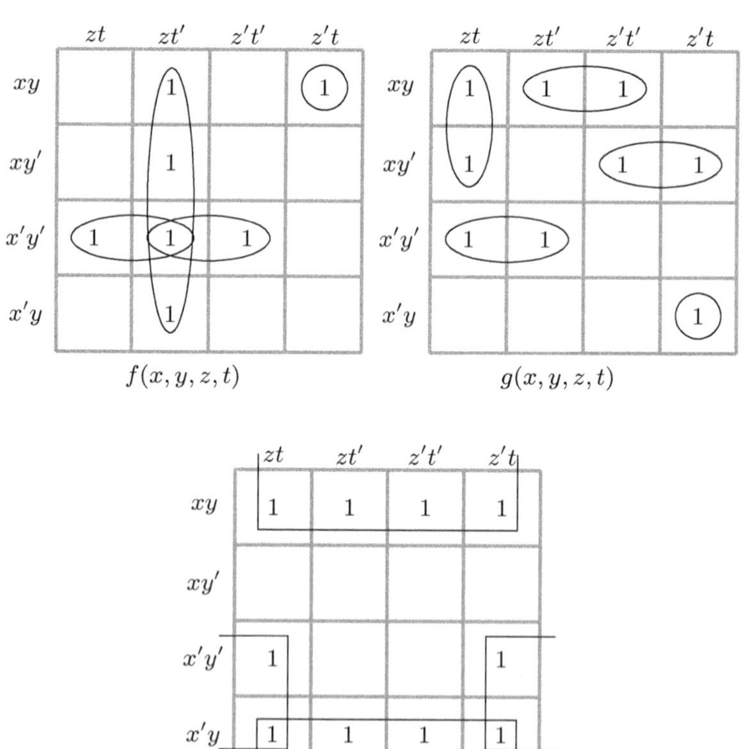

$$f(x, y, z, t) \qquad\qquad g(x, y, z, t)$$

$$h(x, y, z, t)$$

represents the Karnaugh maps (with the appropriate groupings in each map) for the CSOP forms in four variables $f(x, y, z, t) = xyzt' + xyz't + xy'zt' + x'y'zt + x'y'zt' + x'y'z't' + x'yzt'$, $g(x, y, z, t) = xyzt + xyzt' + xyz't' + x'y'zt + x'y'zt' + x'yz't + xy'zt + xy'z't' + xy'z't$ and $h(x, y, z, t) = xyzt + xyzt' + xyz't' + xyz't' + x'y'zt + x'y'z't' + x'yzt + x'yzt' + x'yz't' + x'yz't$.

(a) The minimal number of groups of 1's in the Karnaugh map of $f(x, y, z, t)$ is 4. The first consits of just the single 1 in the last cell of the first row. This contributes with $xyz't$ to the minimal form. The second group consists of the four 1's in the second column. This group contributes with the term zt'. The third group consists of the first two 1's in the third row. This group contributes with $x'y'z$. The last grouping of 1's consists of the two 1's in the second and third cells of the third row. It contributes with $x'y't'$ to the minimal form. Summing up the contributions of all groups, we conclude that the minimal form of $f(x, y, z, t)$ is $xyz't + zt' + x'y'z + x'y't'$.

(b) From the Karnaugh map of $g(x, y, z, t)$, the minimal number of groupings that cover all the 1's is 4, as shown in the diagram above. The minimal form of $g(x, y, z, t)$ is $xzt + xyt' + xy'z' + x'y'z + x'yz't$.

(c) Following the Karnaugh map of $h(x, y, z, t)$ given in the diagram above, the minimal number of groups that cover all the cells containing 1's is 2. The first consists of eight 1's connecting the first and last rows. The contribution of this grouping to the minimal form is just y. The second grouping consists of the four 1's at the first and fourth columns. This second grouping contributes with $x't$ to the minimal form. From this, we get that the minimal form of $h(x, y, z)$ is $y + x't$.

2.7.8 *Exercises*

(1) In each case, determine if the given Boolean expression is in SOP form, POS form or neither.

    (a) $(x + y)z$ (c) $xyz + z(y + z)$ ⋆ (e) $x(y + z) + (y + zt)w$
    (b) $xy + zt + w$ ⋆ (d) $x + y + z$

(2) In each case, convert each of the SOP forms into CSOP form over the set $\{x, y, z, t\}$ of logic variables.

    (a) $xyt + xy'z'$ ⋆ (c) $xyz' + yzt + xyzt' + xt$
    (b) $xyz + xzt + yz'$ (d) $x'y' + xy'zt + xzt$

(3) In each case, write the given Boolean expression of the logic variables x, y and z in SOP and CSOP forms using Boolean algebra laws.

    (a) $xy + xz + yz$ $\star$ (c) $x(yz + xy'z')' + (yz)(x + y')$
    (b) $(((x + yz)')' + (x'z))'$ (d) $xy(y + z)' + (x + yz + y')(x + y')'$

(4) In each case, write the given Boolean expression of the logic variables x, y and z in POS and CPOS forms using Boolean algebra laws.

    (a) $x'y + xz + y'z$ (c) $x' + y'(x + z)z + xz$
$\star$ (b) $xy(x'yz + y'z')'$ (d) $x(yz + xy'z)' + (x' + yz)(x' + y)'$

(5) The Karnaugh map of a Boolean expression $f(x, y, z, t)$ in four variables is given. Write a CSOP form of f and find a minimal form of the expression.

	zt	zt'	$z't'$	$z't$
xy	1	1		1
xy'		1		1
$x'y'$			1	
$x'y$		1	1	1

$\star(a)$

	zt	zt'	$z't'$	$z't$
xy	1			1
xy'	1			
$x'y'$	1	1	1	1
$x'y$	1			1

(b)

(6) What is the minimal form of a Boolean expression f if the Karnaugh map of a CSOP form of f has 1 in every cell?

$\star$(7) The CSOP form of a Boolean expression $f(x, y, z, t)$ contains the minterm $x'yzt'$. List all minterms that are represented by cells adjacent to the cell representing $x'yzt'$ in the Karnaugh map of f.

(8) In each case, minimize each of the given CSOP form in two variables using the technique of Karnaugh maps.

    (a) $x'y + x'y'$ $\star$ (b) $xy' + x'y$ (c) $xy + xy' + x'y$

(9) Minimize each of the following CSOP forms in three variables using the technique of Karnaugh maps.

(a) $x'yz + xyz' + xy'z + xy'z'$

$\star$ (b) $x'yz + x'yz' + x'y'z' + x'y'z + xyz' + xy'z'$

(c) $x'yz' + x'y'z' + x'y'z + xyz + xy'z' + xy'z.$

(10) Minimize each of the following CSOP forms in four variables using the technique of Karnaugh maps.

$\star$ (a) $x'yzt + x'yz't + x'y'zt' + x'y'zt + xy'zt' + xy'z't' + xyzt'$

(b) $x'yzt' + xy'zt' + x'yz't' + xy'zt + xyzt' + xyz't + xyzt + xy'z't'$

(c) $x'yzt + x'yzt' + x'yz't' + x'yz't + xy'zt' + xy'z't + xyzt + xyzt' + xyz't' + x'y'z't'.$

2.8 Logic gates

In this section, we only consider the simple Boolean algebra B_2 consisting of two elements given in Example 2.41 above.

Logic gates are building blocks in any electronic design as they have the power to allow or block a binary signal through parts of the circuit. You can think of them as being the "decision makers" in electronic circuits. The circuits we consider in this section are *combinational* which means that their outputs depend solely on the inputs and not on the current state of the circuit. Logic gates implement Boolean algebra laws and as such, there are three basic types of such gates. The first one is the *OR gate*. This gate receives two or more binary Boolean variables $x_1, x_2 \ldots, x_n$ (each is either 0 or 1) as inputs and produces their sum $x_1 + x_2 + \cdots + x_n$ as the output. The second gate is the *AND gate* which takes the inputs $x_1, x_2 \ldots, x_n$ and produces their product $x_1 x_2 \cdots x_n$. The third basic gate is the NOT gate. Unlike the first two, this is unary gate: it takes one input x (binary Boolean variable) and produces its complement x'.

The following diagram gives graphical illustrations of each of the basic gates using two binary inputs for the OR and AND gates:

Depending on the technology used in building electronic circuits, there are more logic gates used in the industry than the three basic ones described above. For instance, the XOR gate corresponds to the Exclusive OR operator: it takes two inputs x and y and produces $x \oplus y = (x + y)(xy)'$. Other widely used gates include the combination of the OR and AND gates with the NOT gate. The NOR gate takes inputs $x_1, x_2, \ldots, x_n$ and produces the complement $(x_1 + x_2 + \cdots + x_n)'$ of their sum. The NAND gate is an AND gate followed by a NOT gate. It takes binary inputs $x_1, x_2, \ldots, x_n$ and produces the complement $(x_1 x_2 \cdots x_n)'$ of their product. The graphical representations of the XOR, NOR and the NAND gates are given in the diagram below.

XOR gate   NOR gate   NAND gate

A typical electronic circuit is usually a complex network of logic gates carefully wired to perform specific tasks. Figure 2.1 shows an example of a logic circuit with four binary inputs x_0, x_1, x_2 and x_3 and with logic gates having multiple input entries. The network is designed to produce a binary output function $f(x_0, x_1, x_2, x_3)$ in response to various combinations of binary inputs. The filled little circles in the network ($\bullet$) indicate that the wires are connected at that junction and electrons can flow in both directions to the corresponding gates. Unconnected wires are represented by intersecting lines ($\perp$) with no filled circle at the intersection point.

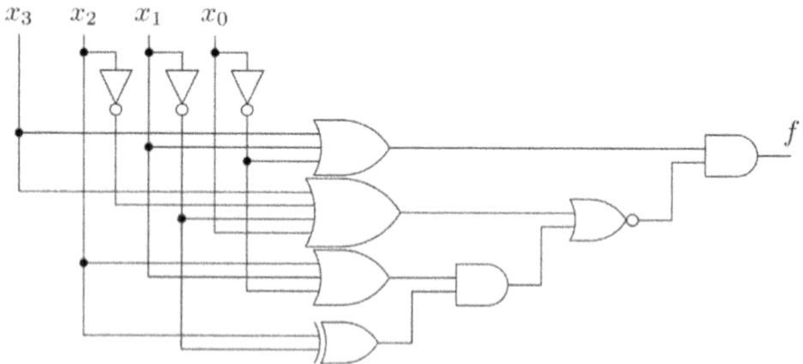

Fig. 2.1   A logic circuit with four binary inputs.

A quick look at the circuit in Figure 2.1 shows that it could be a real challenge to determine what the final output value is for a given list of binary inputs in a circuit. Another challenge to deal with in electronic circuit design is to ensure that the design is cost effective in terms of the number of gates used considering that a given binary output can be achieved using several designs. In previous sections, we have seen two ways to simplify a Boolean expression. The first one is algebraic using the laws of Boolean algebra and the second is graphical using Karnaugh maps (after expanding the expression in CSOP form).

Example 2.54. Consider the following logic circuit in three binary inputs x, y and z. Find the output of the circuit as a function $f(x, y, z)$ of the three inputs. Simplify the expression $f(x, y, z)$ and replace the logic circuit with another that uses fewer gates and produces the same output.

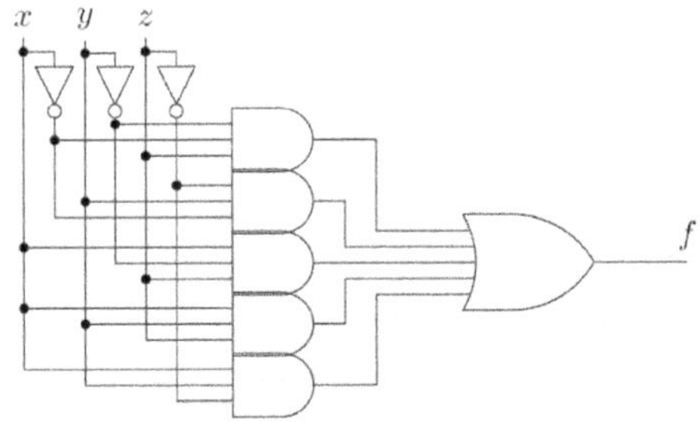

Solution. Following the first AND gate from the top, it is easy to see that its input is $x'y'z$. Similarly, the inputs for the other AND gates (starting from the second one on top) are $x'yz'$, $xy'z$, xyz and xyz' respectively. All five AND gates are connected to an OR gate making the final circuit output $f(x, y, z) = x'y'z + x'yz' + xy'z + xyz + xyz'$. We obviously wish to come up with a simpler circuit that gives exactly the same output for every list of binary values for x, y and z. This can be achieved using Boolean algebra laws to simply the expression $f(x, y, z)$. Since $f(x, y, z)$ is already in CSOP form, a faster way to get to a minimal expression of $f(x, y, z)$ is to use a Karnaugh map with appropriate groupings:

	yz	yz'	$y'z'$	$y'z$
x	1	1		1
x'		1		1

Summing the contribution of each group, we find that the minimal form of the output of the logic circuit is the expression $y'z + yz' + xy$. We can then replace the original circuit with the following equivalent, but more efficient one.

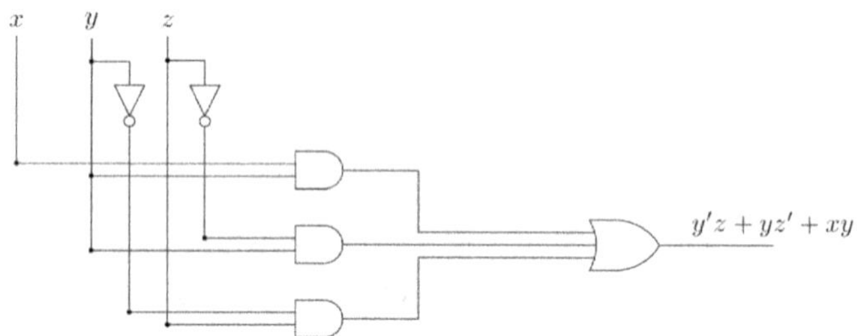

$\Diamond$

2.8.1   *Exercises*

(1) In each case, draw a possible logic circuit with two inputs x and y and with the given expression as an output: (a) $xy' + x'$; (b) $(xy)' + xy'$.

(2) In each case, draw a possible logic circuit with three inputs x, y and y and with the given expression as an output: (a) $xy' + x'z + yz'$; (b) $(x + z)' + xy'$; (c) $(xz')'(y + z') + x$

(3) Consider the following logic circuit with three (binary) inputs x, y and z. Find the output of the circuit as a function $f(x, y, z)$ of the two inputs. Use Boolean algebra laws or Karnaugh maps to simplify the expression $f(x, y, z)$ and replace the logic circuit with another that uses fewer gates and produces the same output.

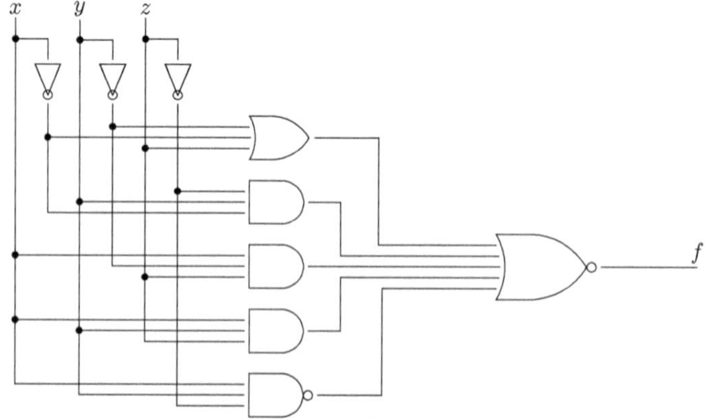

(4) Repeat the previous exercise for the following logic circuit.

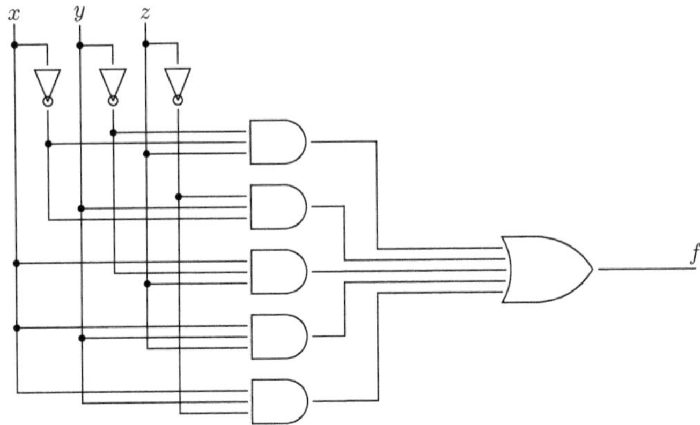

(5) Consider the following logic circuit with four (binary) inputs x, y, z and t. Find the output of the circuit as a function $f(x, y, z, t)$ of the four inputs. Use Boolean algebra laws or Karnaugh maps to simplify the expression $f(x, y, z, t)$ and replace the logic circuit with another that uses fewer gates and produces the same output.

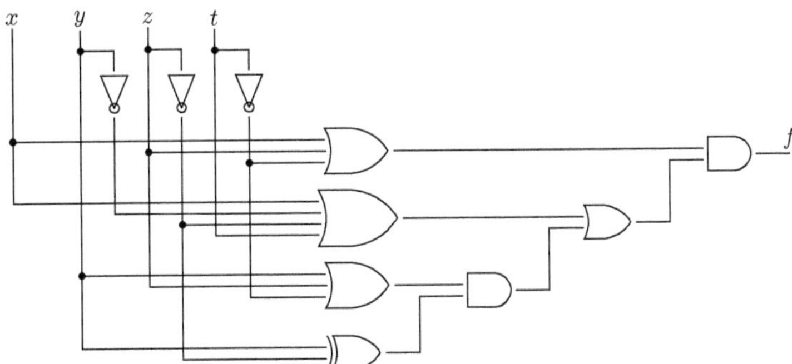

Chapter 3

Prove It: Mathematical Proof Techniques

3.1 Introduction

Since the dawn of history, ancient civilizations realized the need to develop mathematical tools to help solve their daily problems. From finding the area of a circular field to estimating square roots, ancient scholars developed practical "procedures" with very little interest in proving why these procedures actually work. Mathematics started to take the shape we know today at the hands of ancient Greeks who benefited from the knowledge of ancient Egyptian and various mesopotamian civilizations and added a key component of their own: the art of mathematical proof. But what are the elements of a sound mathematical proof? and why do we actually need a proof in the first place? The first question has been a subject of debate and research for many centuries among mathematicians and philosophers. To justify an argument by saying "it is clear" or "it is evident" is certainly not going to convince many people. Mathematicians established a precise framework to define what constitutes a valid proof, starting with the rules of the mathematical language and then declaring some mathematical results to be true (axioms). As for the question *"why do we need proofs?"*, maybe the answer is in a result proposed by Euclid more than 2300 years ago. In the first textbook ever written (*Elements*) on Mathematics, Euclid proposed (and proved) that there are infinitely many prime numbers. Today, having a sufficiently large supply of prime numbers is crucial for cyber security. Admittedly, most of us agree with this result with not much argument, but Euclid wanted to tackle a more fundamental question: *"How can we be sure of this fact beyond any reasonable doubt?"* Obviously, we can test Euclid result by listing the first prime numbers we know (2, 3, 5, 7, ...) and try to add one prime at a time. Two major problems with that approach. First, no matter how many integers we have tested, there will always be untested ones. Second, the task of testing for primality of integers becomes harder and harder as we progress in our list of tested integers. For example, is the integer 80629 prime?

In an ideal world, a jury cannot deliver a conviction on "suspicious grounds" only. A verdict must always to be backed by hard evidence "beyond any reasonable doubt". Mathematics is indeed an ideal world where *almost* nothing is accepted without evidence and proof (I say *almost* since we still have to accept some axioms without proof).

3.2   Proving an implication and a biconditional

It is not an overstatement to say that a large family of mathematical results are of the form "If p, then q" for some statements p and q. In Chapter 1, we learnt that such an expression is the implication $(p \to q)$. Sometimes a statement does not look at all like an implication at first glance and one has to think of its true meaning in order to write it as such. For example, the statement *"The sum of two even integers is even"* can be rewritten as *"If a and b are two even integers, then $a + b$ is even"* which shows that the original statement is indeed an implication. We discuss two techniques to prove the validity of an implication.

3.2.1   *Direct proof*

We have seen in Chapter 1 that the implication $(p \to q)$ is always true except in the unique case where the hypothesis p is true and the conclusion q is false. The method of direct proof is designed to prove the implication by ruling out this unique case. Namely, a *direct proof* of $(p \to q)$ consists of assuming that the hypothesis p is true and then proving that the conclusion q must be also true. Formally, the technique of direct proof is based on the following valid argument of propositional logic:

$$\begin{array}{c} (p \to q) \\ p \\ \hline q \end{array}$$

Example 3.1. Give a direct proof of the following result: *"The sum of two even integers is even."*

Solution. As indicated above, the statement can be rewritten as: "*If a and b are even integers, then a + b is even.*" To give a direct proof of this implication, we assume that a and b are even integers and we prove (under this assumption) that $a + b$ must be even. Since a and b are even, we can write $a = 2s$ and $b = 2t$ for some integers s and t. Then $a + b = 2s + 2t = 2(s + t)$ is even. $\Diamond$

Example 3.2. Prove that the square of an even integer is even.

Solution. We need to prove the following implication: "*If n is even, then n^2 is even.*" Assume that n is an even integer, then $n = 2k$ for some integer k. So $n^2 = (2k)^2 = 4k^2 = 2(2k^2)$ is even. $\Diamond$

Example 3.3. For $x \in \mathbb{R}$, prove that if $x^2 < 9$, then $-3 < x < 3$.

Solution. Assume that $x^2 < 9$. Then $x^2 - 9 < 0$ and so $(x - 3)(x + 3) < 0$. This implies that $x - 3$ and $x + 3$ have opposite signs. So either $x - 3 < 0$ and $x + 3 > 0$ or $x - 3 > 0$ and $x + 3 < 0$. Note that the second case is a contradiction since it says that $x > 3$ and $x < -3$ at the same time. The first case implies that $-3 < x < 3$. $\Diamond$

Example 3.4. If m, n are two integers with $n \neq 0$, we say that n divides m if there exists an integer k such that $m = kn$. Let a, b and c be three integers with $a \neq 0$. Prove the following result: *If a divides b and a divides c, then a divides mb + nc for any integers m and n.*

Solution. Assume a divides b and c. Then $b = sa$ and $c = ta$ for some integers s and t. For any integers m and n, $mb + nc = msa + nta = (ms + nt)a$. We conclude that a divides $mb + nc$. $\Diamond$

Example 3.5. Let x be a real number. Prove that "$x^8 - 1 = 0$ *only if x is either* -1 *or* 1".

Solution. Assume that $x^8 - 1 = 0$. Since $x^8 - 1 = (x^4 - 1)(x^4 + 1) = (x^2 - 1)(x^2 + 1)(x^4 + 1) = (x - 1)(x + 1)(x^2 + 1)(x^4 + 1)$ and $(x^2 + 1)$, $(x^4 + 1)$ have no real roots, the only real solutions of the equation $x^8 - 1 = 0$ are $x = -1$ and $x = 1$. $\Diamond$

Example 3.6. Let $\mathcal{U}$ be a universal set, A and B subsets of $\mathcal{U}$. Recall that the symmetric difference of A and B is $A \oplus B = (A \backslash B) \cup (B \backslash A)$. Prove that if $A \oplus B = \emptyset$, then $A = B$.

Solution. Assume $A \oplus B = \emptyset$. We need to prove that $A = B$. If $x \in A$, then $x \in B$ since otherwise $x \in A \backslash B \subseteq A \oplus B$ which is absurd by the assumption that $A \oplus B = \emptyset$. This shows that $A \subseteq B$. The proof that $B \subseteq A$ isdone similarly. We conclude that $A = B$. $\Diamond$

3.2.2   *Indirect proof*

In Chapter 1, we established the fact that an implication is logically equivalent to its contrapositive: $(p \rightarrow q) \equiv (\neg q \rightarrow \neg p)$. An *indirect proof* of the implication $(p \rightarrow q)$ consists of a direct proof of the contrapositive $(\neg q \rightarrow \neg p)$.

Example 3.7. Let n be an integer. Prove the following statement using an indirect proof: "*If n^2 is even, then n is even*".

Solution. The contrapositive of this implication is: "*If n is odd, then n^2 is odd*". We prove this implication using a direct proof. Assume that n is odd and write $n = 2k + 1$ for some integer k. Then $n^2 = 4k^2 + 4k + 1 = 2(2k^2 + 2k) + 1$ is odd. $\Diamond$

Example 3.8. Let n, a and b be three integers with $n \neq 0$. Prove the following: "*If n is not a divisor of $a + b$, then n is not a common divisor of a and b*".

Solution. The contrapositive of this implication is: "*If n is a common divisor of a and b, then it is a divisor of $a + b$*". Assume that n is a common divisor of a and b and write $a = sn$ and $b = tn$ for some integers s and t. Then $a + b = sn + tn = (s + t)n$ which implies that n is a divisor of $a + b$. It is worth mentioning that this is a particular case of Example 3.4 with $m = n = 1$. $\Diamond$

Example 3.9. Let m and n be two integers. Give an indirect proof of the following: "*If the product of two integers is even, then at least one of them must be even.*"

Solution. The statement can be rewritten as follows: "*If m and n are two integers such that mn is an even, then either m or n is even*". The contrapositive of this implication is: "*If m and n are two odd integers, then mn is odd*". We prove this implication using a direct proof. Assume m and n are both odd and write $m = 2s + 1$ and $m = 2t + 1$ for some integers s

and t. Then $mn = (2s+1)(2t+1) = 4st + 2s + 2t + 1 = 2(2st + s + t) + 1$ is also odd. $\Diamond$

Example 3.10. Prove the following using indirect proof: "*If the sum of two integers is even, then they have the same parity*".

Solution. The statement can be formulated as follows: "*If m and n are two integers such that $m + n$ is even, then m and n have the same parity*". The contrapositive of this implication is: "*If m and n are two integers of different parities, then $m + n$ is odd*". We prove this implication using a direct proof. Assume m and n are two integers of different parities. Without loss of generality, we may assume that m is even and n is odd and write $m = 2s$, $n = 2t + 1$ for some integers s and t. Then $m + n = 2(s + t) + 1$ is odd. $\Diamond$

Example 3.11. Let A, B be two subsets of a universal set $\mathcal{U}$. Use an indirect proof to show the following: "*If $B \not\subseteq A$, then $A \cup B \neq A$*".

Solution. The contrapositive of this implication is: "*If $A \cup B = A$, then $B \subseteq A$*". Assume that $A \cup B = A$ and let $x \in B$ be an arbitrary element. Then $x \in A \cup B$ and so $x \in A$ by the assumption that $A \cup B = A$. We conclude that $B \subseteq A$. $\Diamond$

At this point, you might be wondering how to decide when to use a direct or an indirect approach to prove an implication. While there are no general rules to follow, the following are some tips that could help with the decision.

(1) If the hypothesis contains "heavier" computational components that the conclusion, an indirect approach is probably appropriate. It is usually more useful to have the computational components in the conclusion. For example, to prove the implication "*If n^3 is odd, then n is odd*", staring with the hypothesis "*n^3 is odd*" does not leave us with much to work with to prove that "*n is odd.*" On the other hand, an indirect proof of the implication allows us to assume that n is even and then "compute" the value of n^3.

(2) If the negation of the conclusion and the hypothesis can be easily stated, this could be an indication that an indirect proof is appropriate.

(3) If a direct approach leads to a dead end or requires complex and advanced techniques, it is probably worth considering an indirect

proof. For example, if we adopt a direct approach to prove the implication "*If n^3 is odd, then n is odd*" then we start by assuming that n^3 is odd and we write $n^3 = 2k + 1$ for some integer k. From this point on, the reasoning gets complicated to prove that n is odd.

3.2.3 *Exercises*

(1) Let n be an integer. Prove that if n^3 is odd, then n is odd.

(2) Prove that the sum of two odd integers is even.

$\star$(3) Let n be an integer. Prove: "*If n is a multiple of 3, then $n^2 + 2n$ is a multiple of 3*".

(4) Prove that the sum of two integers of different parities is odd.

$\star$(5) Prove that the product of two integers of different parities is even.

(6) Prove the following: *If a, b and c are three integers such that $a \neq 0$ and a is a common divisor of b and c, then a divides $b^2 - c^2$.*

$\star$(7) Let n be an integer. Prove: $n^5 + 7$ *is even only if n is odd.*

(8) Let m be an integer. Prove: *If $m^3 + 5m + 3$ is even, then m is odd.*

(9) Prove that the sum of four consecutive integers is always even.

(10) Let m, n be two integers. Prove: *If $m^2 = n^2$, then $m = n$ or $m = -n$.*

(11) Let A, B be two subsets of a universal set $\mathcal{U}$. In each case, prove the given statement.

    (a) If $A \cap B \neq A$, then $A \nsubseteq B$.

    (b) If $A \cup B \neq B$, then $A \cap B \neq A$.

$\star$(12) Prove: *If x and y are two rational numbers, then so is $mx + ny$ for any integers m and n.*

(13) Prove: *If x is an irrational number, then $\frac{1}{x}$ is irrational.*

(14) Prove: *If x is a real number with $e^x - e^{-x} = \sqrt{5}$, then $e^x + e^{-x} = 3$.*

$\star$(15) Prove: *If $x \geq 0$ and $y \geq 0$ are real numbers, then $\frac{x+y}{2} \geq \sqrt{xy}$.*

(16) For natural numbers x and y, we write $x \Diamond y$ if and only if $3x + 2y$ is a multiple of 5. For example, $1 \Diamond 1$ and $5 \Diamond 10$. Prove: *If $x \Diamond y$ and $a \Diamond b$, then $(x + a) \Diamond (y + b)$.*

(17) Let a and b be two positive integers. Verify that

$$2^{ab} - 1 = (2^a - 1)\left(2^{a(b-1)} + 2^{a(b-2)} + \cdots + 2^a + 1\right).$$

Recall that a positive integer $m \geq 2$ which is not prime is called composite. In other words, the integer n is composite if and only if $n = ab$ for some integers a, b such that $1 < a \leq b < n$. Use part (a) to prove the following: "*If m is a composite integer, then $2^m - 1$ is also composite*".

$\star$(18) Let x be a positive real number. Prove: *If x is not a rational number, then $\sqrt{x}$ is not a rational number.*

(19) Let n, a and b be three integers such that $n \neq 0$. Prove: *If n is not a divisor of $3a + 2b$, then n is not a common divisor of a and b.*

(20) Let x be a real number. we define $\lceil x \rceil$ (respectively $\lfloor x \rfloor$) as being the smallest integer greater than or equal to x (respectively, the greatest integer less than or equal to x). For example, $\lceil \frac{4}{3} \rceil = 2$, $\lfloor \frac{16}{3} \rfloor = 5$. Prove each of the following statements:

   (a) If $x \in \mathbb{Z}$, then $\lceil x \rceil - \lfloor x \rfloor = 0$.
   (b) If $x \notin \mathbb{Z}$, then $\lceil x \rceil - \lfloor x \rfloor = 1$.

$\star$(21) Two positive integers a and b are said to be relatively prime (or coprime) if and only if the only positive common divisor of a and b is 1. Prove the following for positive integers a and b: *If a^2 and b^2 are relatively prime then so are a and b.*

(22) Let m, n and v be three non-negative real numbers. Prove: *If $mn = v$, then $m \leq \sqrt{v}$ or $n \leq \sqrt{v}$.*

$\star$(23) Natural numbers like $0, 1, 4, 9, 16, 25, \ldots$ are called perfect squares. Prove the following for natural numbers m and n: *If mn is not a perfect square, then at least one of the integers m and n is not a perfect square.*

(24) Let $m \geq 2$ and $n \geq 1$ be two integers. Prove the following: *If $m^n - 1$ is a prime number, then n is odd.*

(25) Let a, b be two real numbers. Prove that if $ab = 0$, then $a = 0$ or $b = 0$.

(26) Prove the following: *If the sum of two real numbers is irrational, then at least one of them must be irrational.*

(27) Let a be a positive integer. Prove: *If n is an integer other than -1 and 1, then n is not a common divisor of a and $a + 1$.*

(28) Prove that if the average of three distinct integers is 11, then at least one of the integers must be greater than or equal to 12.

3.3   Proving a biconditional

Proving a biconditional statement $(p \leftrightarrow q)$ is equivalent to proving both implications $(p \rightarrow q)$ and $(q \rightarrow p)$. This is based on the fact that $(p \leftrightarrow q)$ and $((p \rightarrow q) \wedge (q \rightarrow p))$ are logically equivalent.

Example 3.12. Prove that for any integer n, n is even if and only if n^2 is even.

Solution. The implication n is even implies n^2 is even was proved in Example 3.2. The converse implication n^2 *is odd implies n is odd* was proved in Example 3.7. $\Diamond$

Example 3.13. Prove that for any integer n, $3n + 5$ is even if and only if n odd.

Solution. Assume first that $3n + 5$ is even. Then n cannot be even since otherwise, $n = 2k$ for some $k \in \mathbb{Z}$ and so $3n + 5 = 6k + 5 = 2(3k + 2) + 1$ is odd. So n is odd. Conversely, assume n is odd and write $n = 2k + 1$ for some $k \in \mathbb{Z}$. Then $3n + 5 = 3(2k + 1) + 5 = 6k + 8 = 2(3k + 4)$ is even. $\Diamond$

Example 3.14. Prove that for any real number x, $4x^2 + 5x - 8 = 2x^2 + 3x + 16$ if and only if $x = -4$ or $x = 3$.

Solution. The result follows from the fact that $4x^2 + 5x - 8 = 2x^2 + 3x + 16 \Leftrightarrow x^2 + x - 12 = 0 \Leftrightarrow (x + 4)(x - 3) = 0$ for any real number x. This can be easily verified by arranging terms and factoring . $\Diamond$

3.3.1   *Exercises*

  (1) Let a, b be two positive integers. Prove that a divides b and b divides a if and only if $a = b$.

$\star$(2) Let n be an integer. Prove that n is a multiple of 6 if and only if n is even and n is a multiple of 3.

  (3) Let a, b be two integers. Prove that ab is odd if and only if neither a nor b is even.

  (4) For $n \in \mathbb{Z}$, prove that $5n + 3$ is even if and only if $3n + 2$ is odd.

  (5) For $n \in \mathbb{Z}$, prove that $3n^2 + 5n + 2 = 0$ if and only if $2n^2 + 5n + 3 = 0$.

$\star$(6) For any real numbers x and y, prove that $x < y$ if and only if $x < \frac{x+y}{2}$.

  (7) Let A, B be two subsets of a universal set $\mathcal{U}$. Prove that $\wp(A) \subseteq \wp(B)$ if and only if $A \subseteq B$.

$\star$(8) Let A, B be two subsets of a universal set $\mathcal{U}$. Prove that $A \backslash (A \backslash B) = B$ if and only if $B \subseteq A$.

3.4   Proof by contradiction

For a logic proposition p, either p or its negation $\neg p$ is true but not both. One way to prove that p is true is to establish the impossibility of having $\neg p$ true. The method of the proof by contradiction consists of assuming that $\neg p$

is true and then proving that this leads to a contradiction. In other words, a proof by contradiction of p consists of proving the validity of an implication of the form $(\neg p \to (r \wedge \neg r))$ for some proposition r. Since $(r \wedge \neg r)$ is always false (a contradiction), the fact that the implication $(\neg p \to (r \wedge \neg r))$ is true implies that $\neg p$ must be false and hence p must be true.

Example 3.15. Prove that there exist no positive integers m and n such that $m^2 - n^2 = 10$.

Solution. Assume by contradiction that there exist two positive integers m and n such that $m^2 - n^2 = 10$. Then $(m - n)(m + n) = 10$. Since $10 = 1 \times 10 = 2 \times 5$, either $m - n = 1$ and $m + n = 10$ or $m - n = 2$ and $m + n = 5$. From the first pair of equations, we get that $2m = 11$ which leads to a contradiction since $m = \frac{11}{2}$ is not an integer. A Similar reasoning leads to a contradiction from the second pair of equations. We conclude that no positive integers m and n exist such that $m^2 - n^2 = 10$. $\qquad \Diamond$

Example 3.16. A classic example of a proof by contradiction is the proof that $\sqrt{2}$ is not rational. Assume by contradiction that $\sqrt{2}$ is rational and write $\sqrt{2} = \frac{a}{b}$ for some positive integers a and b. Furthermore, we may assume that $\frac{a}{b}$ is simplified to its lowest terms so that a and b are coprime (they have no common divisor except 1). Squaring both sides of the equation $\sqrt{2} = \frac{a}{b}$, we get that $a^2 = 2b^2$. This implies in particular that a^2 is even. Example 3.7 above shows that n must be even. Write $n = 2k$ for some integer k, then the equation $a^2 = 2b^2$ becomes $4k^2 = 2b^2$ which simplifies to $b^2 = 2k^2$. This shows in particular that b^2 is even and consequently, b is even. We conclude that a and b must be even. This is a contradiction to the assumption that a and b are coprime.

Example 3.17. Prove that there is no largest even integer.

Solution. Suppose by contradiction that E is the largest even integer. Write $E = 2n$ for some integer n. Then the integer $N = 2n + 2 = 2(n + 1)$ is even and greater than E. This is a contradiction to the assumption that E is the largest even integer. $\qquad \Diamond$

It is often the case that we have to combine proof techniques to prove some mathematical results. In the following example, a direct proof and a proof by contradiction are combined.

Example 3.18. For an integer n, prove the following: *If $3n + 2$ is odd, then n is odd.*

Solution. We prove this implication using a direct proof. Assume that $3n + 2$ is odd. We need to prove that n is odd. To this end, We proceed by contradiction. Assume that n is even, then $n = 2k$ for some integer k. This implies that $3n + 2 = 3(2k) + 2 = 2(3k + 1)$ is even: a contradiction to the initial assumption that $3n + 2$ is odd. We conclude that n is odd. $\Diamond$

A word of caution. Before trying a proof by contradiction of an implication, one must decide if this is the most efficient strategy. For instance, the proof of the result in Example 3.18 can be done more efficiently using an indirect proof of the implication.

Recall that an integer $p \geq 2$ is called prime if the only positive divisors of p are 1 and p itself. For example, 2, 3, 5 and 7 are prime numbers whereas 4, 6 and 8 are not prime. One of the oldest mathematical results (found in Euclid's Elements around 300 BC) is the fact that there are infinitely many prime numbers. We show this result in the following example using a proof by contradiction. We also use the fact that any integer $n \geq 2$ has a prime divisor p (this will be established in Chapter 6).

Example 3.19. Prove that there are infinitely many prime numbers.

Solution. Assume by contradiction that there exists a finite number of prime numbers $p_1, p_2, \ldots, p_n$. Let $N = p_1 p_2 \ldots p_n + 1$. For any i, the remainder of the division of N by p_i is 1, which means that N is not divisible by any of the p_i's. Since N must have a prime divisor, there exists a prime number p which is not in the set $\{p_1, p_2, \ldots, p_n\}$. This is a contradiction to the assumption that $\{p_1, p_2, \ldots, p_n\}$ is the set of all the prime numbers. $\Diamond$

3.4.1 *Exercises*

(1) Prove that there exist no integers a and b such that $6a + 2b = 1$.
(2) Prove that there is no smallest odd integer.
$\star$(3) Prove that there is no smallest positive rational number.
(4) Recall that an angle θ is called *obtuse* if $\theta > 90°$. Prove that a triangle cannot have more than one obtuse angle.

(5) Prove that $\sqrt{3}$ is irrational (*Hint.* Use the following fact: *If p is a prime number which divides the product mn of two integers, then p must divide at least one of them*).

(6) Prove that $\sqrt[3]{2}$ is irrational (Use the same hint as the previous exercises).

$\star$(7) Let n be a positive integer. Prove that if $n+1$ objects are placed in n boxes, then at least one box must contain at least two objects.

(8) Prove that there exists no integer n such that $n^2 - 2$ is a multiple of 4.

(9) Prove that the graphs of $y = x^4 + 2x^2 + 2$ and $y = x^2 + 1$ do not intersect.

(10) Prove that no positive integers m and n exist such that $m^2 - n^2 = 14$.

$\star$(11) Recall that a natural number of the form k^2 for some integer k is called a *perfect square*. Prove that there exists no perfect square of the form $4n + 2$ where n is an integer.

(12) Prove that if n is a perfect square, then $n + 6$ is not a perfect square (see Exercise (11)).

(13) You meet three inhabitants A, B and C on the island of knights and knaves (see Chapter 1). Inhabitant A tells you: "*B is a knight and C is a knave.*" Inhabitant B tells you: "*If A is a knight, then C is a knight.*" Use a proof by contradiction to prove that B is a knight.

(14) Let a, b and c be three positive integers such that $c^2 = a^2 + b^2$. Prove that at least one of the integers a and b must be even.

$\star$(15) Let A, B and C be three subsets of a universal set $\mathcal{U}$ such that $A \cap B \subseteq C$. Use a proof by contradiction to show that $B \cap (A \backslash C) = \emptyset$.

(16) Let a, b be two real numbers satisfying $a^5 b - 3a - 4b = 0$. Prove that if $b \neq 0$ then $a \neq 0$.

$\star$(17) Prove that $\frac{2+3\sqrt{2}}{3-2\sqrt{2}}$ is irrational.

(18) For an integer n, prove that at least one of the integers n and $n+1234567$ is not divisible by 3.

(19) Consider the following (incorrect) result: "*If x, y are real numbers satisfying $2x + 3y = 12$, then $x \neq 3$ and $y \neq 4$.*" Someone proposes a proof of the result as follows. *Assume that $2x + 3y = 12$. If $x = 3$ and $y = 4$, then $2x + 3y = 2(3) + 3(4) = 18 \neq 12$. This is a contradiction. We conclude that the result is true.* What is wrong with this proof?

$\star$(20) For a real number x, prove that if $2 \leq x \leq 3$ then $x^2 - 5x + 6 \leq 0$.

3.5   Proof by separation of cases

A friend tells you: *"If it rains today then I am going fishing. If it does not rain today then I am going fishing."* Clearly, your friend is going fishing today. This conclusion is true because it is true in all "possible cases". More formally, given logic propositions $p_1, p_2, \ldots, p_n$ and p, one can prove (see Exercise (19) of Section 3.6.3 below) that the propositions $((p_1 \vee p_2 \vee \cdots \vee p_n) \to p)$ and $((p_1 \to p) \wedge (p_2 \to p) \wedge \cdots \wedge (p_n \to p))$ are logically equivalent. So, if each one of the implications $(p_1 \to p)$, $(p_2 \to p)$, $\cdots$, $(p_n \to p)$ is true then the implication $((p_1 \vee p_2 \vee \cdots \vee p_n) \to p)$ is also true. In particular, if $(p_1 \vee p_2 \vee \cdots \vee p_n)$ is a tautology and each of the implications $(p_1 \to p)$, $(p_2 \to p)$, $\cdots$, $(p_n \to p)$ is true, then p must be true. A proof by separation of cases is based on the following valid argument:

$$
\begin{array}{c}
(p_1 \to p) \\
(p_2 \to p) \\
\vdots \\
(p_n \to p) \\
\underline{(p_1 \vee p_2 \vee \cdots \vee p_n)} \\
\\
p
\end{array}
$$

To prove a proposition p using a proof by separation of cases, the first step is to identify a complete and exhaustive list of possible cases for the problem (or the variables) at hand. These cases must be mutually exclusive. This is probably the hardest part of this proof strategy. Once all the possible cases are clearly identified, we prove p in each one of these cases. In principle, it should be easier to deal with p separately in each case. Once p is proven in each possible case, p is true.

Example 3.20. Prove that $n^3 - n + 1$ is odd for any integer n.

Solution. Given an integer n, there are two possibilities: either n is even or odd. We prove the statement in each of these two cases. If n is even, write $n = 2k$ for some integer k. Then $n^3 - n + 1 = (2k)^3 - 2k + 1 = 2\left(4k^3 - k\right) + 1$: this is odd and the result is true in this case. If n is odd, write $n = 2k + 1$ for some integer k. Then $n^3 - n + 1 = (2k + 1)^3 - (2k + 1) + 1 = (8k^3 + 12k^2 + 6k + 1) - (2k + 1) + 1 = 8k^3 + 12k^2 + 4k + 1 = 2(4k^3 + 6k^2 + 2k) + 1$: this is odd and the result is also true in this case. Since the result is true in each of the two possible cases, it is true for any integer n. $\Diamond$

Example 3.21. Recall that the absolute value of a real number x, denoted by $|x|$, is defined as follows:

$$|x| = \begin{cases} x & \text{if } x \geq 0 \\ -x & \text{if } x < 0 \end{cases}$$

Prove that for any real number x, $|x + 3| + |x - 5| \geq 8$.

Solution. First note that $x + 3 = 0$ for $x = -3$ and $x - 5 = 0$ for $x = 5$. Given a real number x, there are three possible positions of x with respect to the values -3 and 5: $x \leq -3$, $-3 < x \leq 5$ and $x > 5$. We prove the inequality $|x + 3| + |x - 5| \geq 8$ in each of these three cases. If $x \leq -3$, then $x+3 \leq 0$ and $x-5 < 0$. This implies that $|x+3|+|x-5| = -(x+3)-(x-5) = -2x+2 = -2(x-1)$. Since $x \leq -3$ in this case, $x-1 \leq -4$ and $-2(x-1) \geq 8$. The inequality is true in this case. If $-3 < x \leq 5$, $x + 3 > 0$ and $x - 5 \leq 0$. This implies that $|x + 3| + |x - 5| = (x + 3) - (x - 5) = 8 \geq 8$. The inequality is true in this case. For $x > 5$, $x + 3 > 0$ and $x - 5 > 0$ and so $|x + 3| + |x - 5| = (x + 3) + (x - 5) = 2x - 2 = 2(x - 1)$. Since $x > 5$ in this case, $x - 1 > 4$ and $2(x - 1) > 8$. The inequality is also true in this case. Since the inequality is true in each of the three possible cases, it is true for any real number x. $\diamond$

Example 3.22. In this example, we use a proof by separation of cases to prove the following: *for any real numbers x and y, $|x + y| \leq |x| + |y|$.* This is an important inequality in Mathematics known as *the triangular inequality*. Depending on the positions of x and y with respect to 0, there are four possible cases to consider. We prove the inequality in each one of these cases.

(1) $x \geq 0$ and $y \geq 0$. In this case, $x + y \geq 0$, $|x| = x$ and $|y| = y$. So $|x + y| = x + y$ and $|x| + |y| = x + y$. The result is true in this case.

(2) $x < 0$ and $y < 0$. In this case, $x + y < 0$, $|x| = -x$, and $|y| = -y$. So $|x + y| = -x - y$ and $|x| + |y| = -x - y$. The result is true in this case.

(3) $x > 0$, $y < 0$. In this case, $|x| = x$ and $|y| = -y$. Consider the following subcases:

    (3.1) $x+y \geq 0$. In this case, $|x+y| = x+y$ and $|x|+|y| = x+(-y) > x+y$ since $-y > y$. The inequality is true.

    (3.2) $x+y < 0$. In this case, $|x+y| = -(x+y)$ and $|x|+|y| = x+(-y) > -x - y = -(x+y)$ since $x > -x$. The inequality is also true in this case.

(4) $x < 0$, $y > 0$. A similar argument to the previous case shows that the inequality is true in this case.

We conclude that $|x + y| \leq |x| + |y|$ for any real numbers x and y.

Example 3.23. Prove that $n^3 - n$ is divisible par 3 for any integer n.

Solution. Given an integer n, let r be the remainder of the division of n by 3. There are three possible values of r: 0, 1 and 2. We prove the result for each one of these three possibilities. If $r = 0$, then n is divisible by 3 and can be written as $n = 3k$ for some integer k. So $n^3 - n = (3k)^3 - 3k = 3\left(9k^3 - k\right)$ is divisible by 3. If $r = 1$, then $n = 3k + 1$ for some integer k. So $n^3 - n = (3k+1)^3 - (3k+1) = \left(27k^3 + 27k^2 + 9k + 1\right) - (3k+1) = 27k^3 + 27k^2 + 6k = 3\left(9k^3 + 9k^2 + 2k\right)$: divisible by 3. Finally, if $r = 2$, then $n = 3k+2$ for some integer k and $n^3 - n = (3k + 2)^3 - (3k + 2) = \left(27k^3 + 54k^2 + 36k + 8\right) - (3k + 2) = 27k^3 + 54k^2 + 33k + 6 = 3\left(9k^3 + 18k^2 + 11k\right)$ is divisible by 3. Since the result is true in each of the possible cases, it is true for any integer n. $\Diamond$

Example 3.24. On the island of Knights and Knaves you meet two inhabitants A and B. You ask the question: *"Is either one of you a Knight?"* Inhabitant A answers: *"I am a knight if and only if my friend is a Knave."* Inhabitant B says nothing. Use a proof by separation of cases to prove that B is a Knave.

Solution. There are two possible cases.

(1) A is a knight. In this case, the statement *"I am a knight if and only if my friend is a Knave"* is true. Since the proposition *"I am a knight"* is true (in this case), so is *"my friend is a Knave"*. We conclude that B must be a knave in this case.

(2) A is a knave. In this case, the statement *"I am a knight if and only if my friend is a Knave"* is false. Since the proposition *"I am a knight"* is false (in this case), *"my friend is a Knave"* must be true. We conclude that B must be a knave in this case.

We conclude that B is a Knave. $\Diamond$

3.5.1   *Exercises*

(1) Prove that $n^2 - n$ is even for any integer n.

$\star$(2) Prove that $n^3 + 3n + 5$ is odd for any integer n.

(3) Prove that if n is an integer not divisible by 5, then the remainder of the division of n^2 by 5 is either 1 or 4.

$\star$(4) Prove that $n^2 + n + 8$ is not divisible by 3 for any integer n.

(5) Prove that for any real number x, $|-x| = |x|$.

(6) Prove that for any real numbers x and y, $|xy| = |x||y|$.

(7) Prove that for any real number x, $|x + 3| + |x - 4| \geq 7$.

(8) On the of Knights and Knaves you meet two inhabitants A and B. You ask the question: *"Is either one of you a Knave?"* Inhabitant A answers: *"I am a knave if and only if my friend is a Knight."* Inhabitant B says nothing. Use a proof by cases to prove that B is a Knave.

$\star$(9) On the of Knights and Knaves you meet three inhabitants A, B and C. Inhabitant A says: *"B is a knight."* Inhabitant B says: *"Either A is a knave and C is a knight, or A is a knight and C is a knave."* Inhabitant C remains silent. Can you determine the nature of C? Use a proof by cases to justify your answer.

(10) On a certain island, there are three kind of inhabitants: the Knights who always speak the truth, the Knaves who always lie and the Normals who, like most of us, sometimes lie and sometimes speak the truth. You meet three inhabitants A, B and C of the island, each of whom knows the type of the other two and you know for sure that one is a knight, one is a knave and one is normal. A says *"I am the Knight"*, B says *"A is not a knave"* and C says: *"B is not a knave"*. Prove that A is the Knave, B is the Normal and C is the Knight.

(11) Prove that for any real numbers x and y, $\max\{x, y\} + \min\{x, y\} = x + y$.

$\star$(12) A binary operation $\oplus$ is defined on the real numbers as follows: $x \oplus y = \max\{x, y\}$. For example $\pi \oplus 4 = 4$ and $\frac{3}{2} \oplus 2 = 2$. Prove that the operation $\oplus$ is associative. That is, prove that $x \oplus (y \oplus z) = (x \oplus y) \oplus z$ for any real numbers x, y and z.

(13) Let A be a positive real number. Prove the following: *"If $|x| \leq A$, then $-A \leq x \leq A$ for any real number x"* (*Hint.* Consider the cases $x \geq 0$ and $x < 0$)

(14) Let A, B and C be three subsets of a universal set $\mathcal{U}$ such that $A \oplus B = A \oplus C$. Use a proof by separation of cases to show $B = C$.

3.6   Proof by induction

Let $n_0 \geq 0$ be an integer and let $P(n)$ be a property concerning the natural numbers. In this section, we present a powerful method to prove a statement of the form "*$P(n)$ is true for any integer $n \geq n_0$*". The following scenario is often used to illustrate how an inductive proof works. Imagine an infinite row of domino pieces numbered $0, 1, 2, \ldots, n, \ldots$ placed vertically in a row. You are given the following information: (i) Domino piece number 0 falls over; (ii) For every $n \geq 0$, if domino piece number n falls over, then so does domino piece number $n + 1$. What can you conclude? Since domino piece number 0 falls over (from (i)), piece number 1 falls (from (ii)). Now, since piece number 1 falls over, the same will happen to piece number 2 (from (ii)). Continuing this reasoning, we can say with reasonable confidence that all the domino pieces will fall. The problem here is that "a reasonable" confidence is not sufficient to prove beyond any reasonable doubt the observation we made about the domino pieces. The principle of mathematical induction provides the framework for a sound and effective proof technique for statements concerning natural numbers. The principle is based on an axiom known as the *well-ordering property*. It is an intuitively simple property of the natural numbers but has powerful consequences. The axiom of well-ordering will be revisited in Chapter 7.

Axiom (The well-ordering principle). Every non-empty subset of the set $\mathbb{N}$ of natural numbers has a smallest element.

For example, the set A of all positive integers which are multiple of 3 has a least element, namely 3. It is very important to keep in mind that the well-ordering principle applies only for *non-empty* subsets of $\mathbb{N}$. For instance, the subset B of $\mathbb{Z}$ of all multiple of 3 has no smallest element.

In this section, we explore three (equivalent) versions of the principle of mathematical induction.

3.6.1   *The principle of weak induction*

Theorem 3.1 (The principle of weak induction). *Let $P(n)$ be a property concerning the natural numbers. If the following two statements are satisfied, then $P(n)$ is true for any integer $n \geq n_0$:*

*(i) **Base step.** $P(n_0)$ is true;*

*(ii) **Inductive step.** For an integer $n \geq n_0$: if $P(n)$ is true then $P(n+1)$ is true.*

Proof. The proof uses the well-ordering principle of natural numbers. Assume that $P(n)$ satisfies conditions (i) and (ii) of Theorem 3.1. Let A be the set of all integers $n \geq n_0$ such that $P(n)$ is false. If A is not empty, then A must have a least element k by the well-ordering principle. Since $P(n_0)$ is true, $k > n_0$ and so $k - 1 \geq n_0$. The integer $k - 1$ is not in A as it is strictly smaller than the least element of A. In particular, $P(k-1)$ is true. Condition (ii) of Theorem 3.1 implies that $P(k)$ is true. This is a contradiction to the fact that $k \in A$. We conclude that A is empty and $P(n)$ is true for any $n \geq n_0$. $\qquad\square$

The inductive step of Theorem 3.1 consists of proving the implication $P(n) \to P(n+1)$. This means that we assume $P(n)$ is true (*induction hypothesis*) and then we prove $P(n+1)$ is true (*conclusion*).

Example 3.25. Prove that the sum of the first n positive integers is equal to $\frac{n(n+1)}{2}$.

Solution. Using the sigma notation for the sum, we need to prove that $\sum_{i=1}^{n} i = 1 + 2 + 3 + \cdots + n = \frac{n(n+1)}{2}$ for $n \geq 1$. We use the principal of weak induction.

Base step. $P(1)$ is true since $1 = \frac{1(1+1)}{2}$.

Inductive step. Let $n \geq 1$ be an integer.

Induction hypothesis. Assume $P(n)$ is true: $\sum_{i=1}^{n} = \frac{n(n+1)}{2}$.

Conclusion. We need to prove that $P(n+1)$ is true. That is, we need to prove that $\sum_{i=1}^{n+1} i = \frac{(n+1)(n+2)}{2}$:

$$
\begin{aligned}
\sum_{i=1}^{n+1} i &= \left(\sum_{i=1}^{n} i \right) + (n+1) \\
&= \frac{n(n+1)}{2} + (n+1) \qquad \text{(By the induction hypothesis)} \\
&= \frac{n(n+1) + 2(n+1)}{2} \\
&= \frac{(n+1)(n+2)}{2}.
\end{aligned}
$$

$\diamondsuit$

The next example is an important result which is widely used in many area of Mathematics.

Example 3.26. Prove that a finite set of cardinality n has 2^n subsets.

Solution. We use mathematical induction on the cardinality n of A.

Base step. If $n = 0$, then $A = \emptyset$. In this case A has only one subset, namely $\emptyset$. Since $2^0 = 1$, $P(0)$ is true.

Inductive step. Let $n \geq 0$ be an integer.

Induction hypothesis. Assume $P(n)$ is true. That is, assume that every finite set of cardinality n has 2^n subsets.

Conclusion. We need to prove that $P(n + 1)$ is true. Let A be a finite set containing $n + 1$ elements. Fix an element a of A (this is possible since $A \neq \emptyset$ in this case) and let $B = A \backslash \{a\}$. Then B is of cardinality n and $A = B \cup \{a\}$. By the induction hypothesis, the set B has 2^n subsets. Each subset X of A is either a subset of B (if $a \notin X$) or of the form $Y \cup \{a\}$ (if $a \in X$) for some subset Y of B. In other words, for each subset X of B, we can associate exactly two subsets of A, namely X and $X \cup \{a\}$. Since B has 2^n subsets, A has $2(2^n) = 2^{n+1}$ subsets. By the principle of induction, the result is true for any $n \geq 0$. $\Diamond$

Example 3.27. Show that $5^n - 1$ is divisible by 4 for any integer $n \geq 0$.

Solution. We prove the result using weak induction.

Base step. For $n = 0$, $5^0 - 1 = 0$ is divisible by 4. So $P(0)$ is true.

Inductive step. Let $n \geq 0$ be an integer.

Induction hypothesis. Assume that the result is true for n. That is assume that $5^n - 1 = 4k$ for some integer k.

Conclusion. We need to prove that $5^{n+1} - 1$ is divisible by 4.

$$
\begin{aligned}
5^{n+1} - 1 &= 5(5^n) - 1 \\
&= 5(4k + 1) - 1 \qquad \text{(By the induction hypothesis)} \\
&= 20k + 4 \\
&= 4(5k + 1): \qquad \text{This is a multiple of 4.}
\end{aligned}
$$

We conclude that $5^n - 1$ is divisible by 4 for any integer $n \geq 0$. $\Diamond$

Example 3.28. Recall that for a natural number n, the factorial of n, denoted by $n!$, is defined to be the integer $n \cdot (n - 1) \cdots 2 \cdot 1$ with the convention that $0! = 1$. For what values of the natural number n, the inequality $3^n < n!$ holds? Justify your answer using the principle of induction.

Solution. The following table shows the truth value of the inequality for the first possible values of n.

n	$3^n < n!$	True/False
0	$3^0 < 0!$	False
1	$3^1 < 1!$	False
2	$3^2 < 2!$	False
3	$3^3 < 3!$	False
4	$3^4 < 4!$	False
5	$3^5 < 5!$	False
6	$3^6 < 6!$	False
7	$3^7 < 7!$	True
8	$3^8 < 8!$	True

The table suggests that the inequality is true for $n \geq 7$. We prove this observation using weak induction.

Base step. For $n = 7$, $3^7 = 2187$ and $7! = 5040$. So $P(7)$ is true.

Inductive step. Let $n \geq 7$ be an integer.

Induction hypothesis. Assume that $3^n < n!$.
Conclusion. We need to prove that $3^{n+1} < (n+1)!$.

$$
\begin{aligned}
3^{n+1} &= 3(3^n) \\
&< 3(n!) &&\text{(By the induction hypothesis)} \\
&< (n+1)n! &&\text{(Since } 3 < n+1 \text{ as } n \geq 7) \\
&= (n+1)!
\end{aligned}
$$

We conclude that the inequality $3^n < n!$ holds for any $n \geq 7$. $\Diamond$

Example 3.29. Prove that

$$
\left(1 - \frac{1}{2^2}\right)\left(1 - \frac{1}{3^2}\right) \cdots \left(1 - \frac{1}{n^2}\right) = \frac{n+1}{2n} \tag{3.1}
$$

for any integer $n \geq 2$.

Solution. We use the principle of induction, the weak form.

Base step. For $n = 2$, both left and right sides of (3.1) are equal to $\frac{3}{4}$. The result is true in this case.

Inductive step. Let $n \geq 2$ be an integer.

Induction hypothesis. Assume that the equality is true for n.

Conclusion. We need to prove that the equality remains true for $n+1$. That is, we need to prove that

$$\left(1 - \frac{1}{2^2}\right)\left(1 - \frac{1}{3^2}\right)\cdots\left(1 - \frac{1}{n^2}\right)\left(1 - \frac{1}{(n+1)^2}\right) = \frac{n+2}{2(n+1)}.$$

We compare the left-hand side (LHS) and the right-hand side (RHS) of this equation.

$$
\begin{aligned}
LHS &= \left(1 - \frac{1}{2^2}\right)\left(1 - \frac{1}{3^2}\right)\cdots\left(1 - \frac{1}{n^2}\right)\left(1 - \frac{1}{(n+1)^2}\right) \\
&= \frac{n+1}{2n}\left(1 - \frac{1}{(n+1)^2}\right) \quad \text{(By the induction hypothesis)} \\
&= \frac{n+1}{2n}\left(\frac{n^2 + 2n}{(n+1)^2}\right) = \frac{(n+1)n(n+2)}{2n(n+1)^2} = \frac{n+2}{2(n+1)} \\
&= RHS
\end{aligned}
$$

We conclude that the equality is true for any $n \geq 2$. $\Diamond$

Example 3.30. A puzzle invented by the French mathematician Edouard Lucas (1842–1891) is widely known as the *Towers of Hanoi*. The puzzle consists of three long pegs (labeled as A, B and C in the diagram below) and a number of disks that can fit on any peg. No two disks are of the same size. To start, place all the disks on peg A in a decreasing order of sizes with the largest disk at the bottom. The goal is to move all the disks and place them on a different peg, say C. There are two rules to follow: you can only move one disk at a time and you can never put a bigger disk on top of a smaller disk. Let a_n be the *minimal number of moves* needed to solve the puzzle with n discs, $n \geq 1$. Use mathematical induction to prove that $a_n = 2^n - 1$.

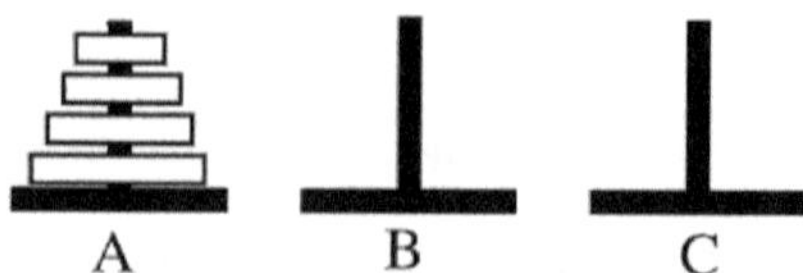

Solution. We use induction on the number n of disks used in the puzzle.

Base step. It is clear that the minimal number of moves required to solve the puzzle with one disc is 1. Since $2^1 - 1 = 1$, the result is true in this case.

Inductive step. Let $n \geq 1$ be an integer.

Induction hypothesis. Assume that $a_n = 2^n - 1$.

Conclusion. We need to prove that $a_{n+1} = 2^{n+1} - 1$. Note that the minimal number of moves to transfer the $n+1$ disks from peg A to peg C is achieved as follows. First, move the top n disks on peg A to peg B leaving the larger disk on A. By the induction hypothesis, the minimal number of moves to achieve this step is $2^n - 1$. Then move the largest disk to the empty peg C. This requires only one move. Finally, move the n disks from peg B to peg C on top of largest disk we placed on peg C. By the induction hypothesis, the minimal number of moves to achieve this step is $2^n - 1$. We conclude that the minimal number of moves needed to solve the puzzle with $n+1$ discs is $2\left(2^n - 1\right) + 1 = 2^{n+1} - 1$ as required. $\Diamond$

3.6.2 *The principle of strong induction*

Given a property $P(n)$ concerning the natural numbers, sometimes the assumption that $P(n)$ holds is not enough to prove that $P(n+1)$ holds as in the weak induction version. There is another, but equivalent version, of Theorem 3.1 known as the *principle of strong induction*. In this new version, we use the fact that $P(k)$ is true for any k with $n_0 \leq k \leq n$ to prove that $P(n+1)$ is true. Like the weak version, a proof by strong induction consists of a base step and an inductive step. However, for the base step, we may have to prove the validity of $P(n)$ for several initial values and in the inductive step, the induction hypothesis assumes the result is true for all integers *up to* n and not just for n itself.

Theorem 3.2 (The principle of strong induction). *Let $P(n)$ be a property concerning the natural numbers. If the following two statements are true, then $P(n)$ is true for any integer $n \geq n_0$:*

Base step. *$P(n_0)$ is true;*

Inductive step. *For an integer $n \geq n_0$, if $P(k)$ is true for any integer k with $n_0 \leq k \leq n$, then $P(n+1)$ is also true.*

Proof. The proof follows the same argument as in the proof of Theorem 3.1 above with few minor modifications (see Exercise (37) below). $\square$

Weak induction, strong induction and the well-ordering principal are all equivalent results. It is a matter of choosing the most convenient form to prove the problem at hand. Strong induction is useful when the result $P(n+1)$ depends not only on the immediate previous value n but also on other smaller values. A typical example is when we prove properties of recursively defined sequences.

Remark 3.1. In the base step of a proof by strong induction, you might have to prove that the result is true for multiple values n_0 through m_0 and not just for n_0. It is sometimes preferable to do a rough work starting with the inductive step before the base step. This can reveal the values of n that must be included in the base step.

Example 3.31. Consider the sequence a_n; $n \geq 1$ of real numbers defined recursively as follows: $a_1 = 3$, $a_2 = 5$ and $a_n = 3a_{n-1} - 2a_{n-2}$ for all $n \geq 3$. Show that $a_n = 2^n + 1$ for any $n \geq 1$.

Solution. We prove this result using strong induction.

> **Base step.** Note that the recurrence relation $a_n = 3a_{n-1} - 2a_{n-2}$ is true only for $n \geq 3$. This means that we must verify the result for $n = 1$ and $n = 2$ before we can use this relation. For $n = 1$, $2^1 + 1 = 3 = a_1$ and for $n = 2$, $2^2 + 1 = 5 = a_2$.
>
> **Inductive step.** Let $n \geq 2$ be an integer.
>
> > *Induction hypothesis.* Assume the result is true for any integer k with $1 \leq k \leq n$. That is $a_k = 2^k + 1$ for $1 \leq k \leq n$.
> >
> > *Conclusion.* We need to prove that $a_{n+1} = 2^{n+1} + 1$:

$$\begin{aligned}
a_{n+1} &= 3a_n - 2a_{n-1} \\
&= 3\left(2^n + 1\right) - 2\left(2^{n-1} + 1\right) \text{ (By the induction hypothesis)} \\
&= 3(2^n) + 3 - 2^n - 2 \\
&= 2(2^n) + 1 \text{ (Since } 3(2^n) - 2^n = 2(2^n)) \\
&= 2^{n+1} + 1.
\end{aligned}$$

We conclude that $a_n = 2^n + 1$ for any $n \geq 1$. $\Diamond$

Example 3.32. Prove that any integer greater than or equal to 8 can be expressed as the sum of 3's and/or 5's.

Solution. We start with some "rough work" to see what is needed for the base case. Since $n + 1 = (n - 2) + 3$, the induction hypothesis can be used to deduce that $n - 2$ is a sum of 3's and/or 5's and conclude that n is also a sum of 3's and/or 5's. But there is one small catch: to use the induction hypothesis, we must have $n - 2 \geq 8$ and so $n \geq 10$. If we include the cases $n = 8, 9, 10$ in the Base step, we can then assume that $n \geq 10$.

Base step. Since $8 = 3 + 5$, $9 = 3 + 3 + 3$ and $10 = 5 + 5$, the property is true for $n = 8, 9$ and 10.

Inductive step. Let $n \geq 10$ be an integer.

Induction hypothesis. Assume that the result is true for any integer k with $8 \leq k \leq n$. That is $k = 3s + 5t$ for some non-negative integers s and t.

Conclusion. We need to prove that $n + 1$ is a sum of 3's and/or 5's. Since $n + 1 = (n - 2) + 3$ and $8 \leq n - 2 \leq n$ (remember that $n \geq 10$, so $n - 2 \geq 8$), the induction hypothesis implies that $n - 2$ is a sum of 3's and 5's and consequently, $n + 1$ is also a sum of 3's and 5's. $\Diamond$

Example 3.33. Consider the sequence a_n; $n \geq 1$ of real numbers defined recursively as follows: $a_1 = 1$, $a_2 = 7$ and $a_n = 4a_{n-1} - 3a_{n-2}$ for $n \geq 3$. Give the values of a_3, a_4 and a_5. Guess a formula for the general term a_n and prove your guess using strong induction.

Solution. Using the recursive relation $a_n = 4a_{n-1} - 3a_{n-2}$, we get that $a_3 = 4a_2 - 3a_1 = 25$, $a_4 = 4a_3 - 3a_2 = 79$ and $a_5 = 4a_4 - 3a_3 = 241$. From these first terms, it looks like $a_n = 3^n - 2$ for any $n \geq 1$. We prove this observation using the principal of strong induction.

Base step. The assertion is clearly true for $n = 1$ and $n = 2$.

Inductive step. Let $n \geq 2$ be an integer.

Induction hypothesis. Assume that $a_k = 3^k - 2$ for any integer k with $1 \leq k \leq n$.

Conclusion. We need to prove that $a_{n+1} = 3^{n+1} - 2$:

$$
\begin{aligned}
a_{n+1} &= 4a_n - 3a_{n-1} \text{ (By the given recursive relation and since } n + 1 \geq 3) \\
&= 4\left(3^n - 2\right) - 3\left(3^{n-1} - 2\right) \text{ (By the induction hypothesis)} \\
&= 4 \cdot 3^n - 8 - 3^n + 6 \\
&= 3 \cdot 3^n - 2 \\
&= 3^{n+1} - 2.
\end{aligned}
$$

We conclude that $a_n = 3^n - 2$ for any $n \geq 1$, as guessed. $\Diamond$

Example 3.34. Consider the sequence a_n; $n \geq 1$ of real numbers defined recursively as follows:

$$a_n = \begin{cases} 3 & \text{if } n \text{ is not divisible by 3} \\ \left(a_{\frac{n}{3}}\right)^3 & \text{if } n \text{ is divisible by 3} \end{cases}$$

(a) Give the values of the first nine terms $a_1, a_2, \ldots, a_9$ of the sequence.

(b) Show that $a_n \leq 3^n$ for any $n \geq 1$.

Solution.   (a) We have: $a_1 = a_2 = 3$, $a_3 = \left(a_{\frac{3}{3}}\right)^3 = (a_1)^3 = 27$, $a_4 = a_5 = 3$, $a_6 = \left(a_{\frac{6}{3}}\right)^3 = (a_2)^3 = 27$, $a_7 = a_8 = 3$, $a_9 = \left(a_{\frac{9}{3}}\right)^3 = (a_3)^3 = 27^3 = 19683$.

(b) We use strong induction.

> **Base step.** Since $a_1 = 3 \leq 3^1$, the result is true in this case.
>
> **Inductive step.** Let $n \geq 1$ be an integer.
>
> > *Induction hypothesis.* Assume that $a_k \leq 3^k$ for all integers k with $1 \leq k \leq n$.
> >
> > *Conclusion.* We need to prove that $a_{n+1} \leq 3^{n+1}$. There are two possibilities: either 3 divides $n+1$ or not. If 3 divides $n+1$, then $a_{n+1} = \left(a_{\frac{n+1}{3}}\right)^3$. Since $1 \leq \frac{n+1}{3} \leq n$, the induction hypothesis implies that $a_{\frac{n+1}{3}} \leq 3^{\frac{n+1}{3}}$ and so $a_{n+1} = \left(a_{\frac{n+1}{3}}\right)^3 \leq \left(3^{\frac{n+1}{3}}\right)^3 = 3^{n+1}$ and the result is true in this case. If 3 does not divide $n+1$, then $a_{n+1} = 3 \leq 3^{n+1}$ and the result is also true in this case. $\diamond$

Example 3.35. Consider the sequence a_n; $n \geq 1$ of real numbers defined recursively as follows:

$$a_1 = 1, \ a_2 = 1; \ a_n = a_{n-1} + a_{n-2} \ \text{ for all } n \geq 3.$$

Show that $a_n \geq \left(\frac{3}{4}\right)^n$ for any $n \geq 1$.

Solution. We use strong induction.

> **Base step.** Since $a_1 = 1 \geq \left(\frac{3}{4}\right)^1$ and $a_2 = 1 \geq \frac{9}{16} = \left(\frac{3}{4}\right)^2$, the result is true for $n = 1$ and $n = 2$.
>
> **Inductive step.** Let $n \geq 2$ be an integer.
>
> > *Induction hypothesis.* Assume that $a_k \geq \left(\frac{3}{4}\right)^k$ for all integers k with $1 \leq k \leq n$.

Conclusion. We need to prove that $a_{n+1} \geq \left(\frac{3}{4}\right)^{n+1}$:

$$
\begin{aligned}
a_{n+1} &= a_n + a_{n-1} \\
&\geq \left(\frac{3}{4}\right)^n + \left(\frac{3}{4}\right)^{n-1} \quad \text{(By the induction hypothesis)} \\
&= \left(\frac{3}{4}\right)^{n-1} \left(\frac{3}{4} + 1\right) \\
&= \left(\frac{3}{4}\right)^{n-1} \left(\frac{7}{4}\right) \\
&\geq \left(\frac{3}{4}\right)^{n-1} \left(\frac{3}{4}\right)^2 \quad \text{(Since } \tfrac{7}{4} \geq \tfrac{9}{16} = \left(\tfrac{3}{4}\right)^2 \text{)} \\
&= \left(\frac{3}{4}\right)^{n+1}.
\end{aligned}
$$

The result is true by the principle of induction. $\diamond$

In the following Example, we prove one part of an important result known as the Fundamental Theorem of Arithmetic.

Example 3.36. Use strong induction to prove that every integer $n \geq 2$ can be written as a product of primes: $n = p_1 p_2 \cdots p_k$ (each p_i is prime number).

Solution. We use strong induction.

Base step. Since $n = 2$ is a prime number, the result is true in this case.
Inductive step. Let $n \geq 2$ be an integer.

Induction hypothesis. Assume that each integer k with $2 \leq k \leq n$ is the product of primes.
Conclusion. We need to prove that $n + 1$ is the product of primes. If $n + 1$ is prime, the result holds and there is nothing to prove. If $n + 1$ is not prime, then we can write $n + 1 = ab$ for some integers a and b with $2 \leq a \leq n$ and $2 \leq b \leq n$. The induction hypothesis implies that each of a and b is the product of primes. This shows that $n + 1$ is also a product of primes in this case. $\diamond$

Note that the Fundamental Theorem of Arithmetic states that every integer $n \geq 2$ can be written as the product of primes in a *unique* way, up to the order of the primes.

3.6.3   *Exercises*

(1) Prove that $\sum_{i=1}^{n}(2i-1) = 1 + 3 + 5 + \cdots + (2n-1) = n^2$ for any integer $n \geq 1$.

$\star$(2) Prove that $\sum_{k=1}^{n}(2k-1)^2 = 1^2 + 3^3 + 5^2 + \cdots + (2n-1)^2 = \frac{n(2n-1)(2n+1)}{3}$ for any integer $n \geq 1$.

(3) Prove that $1^3 + 2^3 + 3^3 + \cdots + n^3 = \left(\frac{n(n+1)}{2}\right)^2$ for any integer $n \geq 1$.

(4) Prove that:

$$(1 \times 2) + (2 \times 3) + (3 \times 4) + \cdots + n \times (n+1) = \frac{n(n+1)(n+2)}{3}$$

for any integer $n \geq 1$.

(5) Prove that $2 \cdot 2 + 3 \cdot 2^2 + 4 \cdot 2^3 + \cdots + (n+1) \cdot 2^n = n \cdot 2^{n+1}$ for any integer $n \geq 1$.

(6) Prove that $1 + 5 + 9 + 13 + \cdots + (4n-3) = 2n^2 - n$ for any integer $n \geq 1$.

(7) Given an integer $n \geq 1$, guess a formula for the expression $\left(1 - \frac{1}{2}\right)\left(1 - \frac{1}{3}\right) \cdots \left(1 - \frac{1}{n+1}\right)$ and then prove your guess using mathematical induction.

$\star$(8) Let a, r be two real numbers with $r \neq 1$. For an integer $n \geq 0$, the sum $S_n = a + ar + ar^2 + \cdots + ar^n$ is called a finite geometric sum of ratio r with $n+1$ terms. Use mathematical induction to prove that $S_n = a\frac{1-r^{n+1}}{1-r}$. Deduce the value of the expression $1 - \frac{1}{3} + \frac{1}{9} - \frac{1}{27} + \cdots + \frac{(-1)^n}{3^n}$.

(9) Prove the following generalized version of the triangular inequality established in Example 3.22 above: $|x_1 + x_2 + \cdots + x_n| \leq |x_1| + |x_2| + \cdots + |x_n|$ for any real numbers $x_1, x_2, \ldots, x_n$ using induction on the integer $n \geq 1$.

(10) Show that $n^3 - n$ is divisible by 3 for any integer $n \geq 0$.

(11) Show that $7^n - 2^n$ is divisible by 5 for any integer $n \geq 0$.

$\star$(12) Show that $2^{3n+1} + 5$ is divisible by 7 for any integer $n \geq 0$.

(13) Show that $3^{2n} + 3$ is divisible by 4 for any integer $n \geq 0$.

$\star$(14) Show that $10^{n+2} + 11^{2n+1}$ is divisible by 111 for any integer $n \geq 0$.

(15) For what values of the natural number n does the inequality $2^n < n!$ hold? Justify your answer using the principle of induction.

(16) For what values of the natural number n does the inequality $n^2 > 2n + 3$ hold? Justify your answer using induction.

$\star$(17) For what values of the natural number n does the inequality $2n+1 < 2^n$ hold? Justify your answer using the principle of induction.

(18) Prove that the constant term (i.e., the term that contains no x in it) in the expansion of $(2x+5)^n$ is equal to 5^n for any integer $n \geq 0$.

(19)(a) Given three logic formulas p, q and r, use a truth table to prove that $((p \vee q) \to r) \equiv ((p \to r) \wedge (q \to r))$.

   (b) Let $n \geq 1$ be an integer. Given logic formulas $p_1, p_2, \ldots, p_n$ and p, use induction on n and part (a) to prove that $((p_1 \vee p_2 \vee \cdots \vee p_n) \to p) \equiv ((p_1 \to p) \wedge (p_2 \to p) \wedge \cdots \wedge (p_n \to p))$.

$\star$(20) (**Bernoulli's inequality**) Let $x > -1$ be a real number. Prove that $(1+x)^n \geq 1 + nx$ for any integer $n \geq 1$.

(21) (*Linear Algebra is required*) Consider the 2×2 matrix $A = \begin{bmatrix} 1 & 1 \\ 0 & 1 \end{bmatrix}$.

Prove that $A^n = \begin{bmatrix} 1 & n \\ 0 & 1 \end{bmatrix}$ for any integer $n \geq 0$. Note that A^0 is defined

to be the identity matrix $I_2 = \begin{bmatrix} 1 & 0 \\ 0 & 1 \end{bmatrix}$.

(22) (*Linear Algebra is required*) Consider the 2×2 matrix $A = \begin{bmatrix} 1 & 1 \\ 1 & 1 \end{bmatrix}$.

Prove that $A^n = \begin{bmatrix} 2^{n-1} & 2^{n-1} \\ 2^{n-1} & 2^{n-1} \end{bmatrix}$ for any integer $n \geq 1$.

(23) Consider the sequence a_n; $n \geq 0$ of real numbers defined recursively as follows: $a_0 = 1$ and $a_n = na_{n-1}$ for $n \geq 1$. Show that $a_n = n!$ for any $n \geq 0$.

(24) Recall that if x is a real number, $\lfloor x \rfloor$ denotes the largest integer less than or equal to x. For example, $\lfloor \pi \rfloor = 3$, $\lfloor \frac{17}{3} \rfloor = 5$. Consider the sequence a_n; $n \geq 1$ of real numbers defined recursively as follows: $a_1 = 1$ and $a_n = a_{\lfloor \frac{n}{2} \rfloor}$ for $n \geq 2$. Give the values of the terms $a_1, a_2, \ldots, a_9$ of the sequence. Use induction to show that $a_n \leq n$ for any $n \geq 1$.

(25) Consider the sequence $a_0, a_1, a_2, \ldots$ defined by $a_0 = 0$ and for $n \geq 1$, $a_n = 2^n + 2a_{n-1}$. Prove that $a_n = n2^n$ for any $n \geq 0$.

(26) Consider the sequence a_n; $n \geq 1$ of real numbers defined recursively as follows: $a_1 = 1$, $a_2 = 3$ and $a_n = 3a_{n-1} - 2a_{n-2}$ for all $n \geq 3$. Show that $a_n = 2^n - 1$ for any $n \geq 1$.

$\star$(27) Consider the sequence a_n; $n \geq 1$ of real numbers defined recursively as follows: $a_1 = 1$, $a_2 = 2$ and $a_n = 2a_{n-1} - a_{n-2}$ for all $n \geq 3$. Give the values of a_3, a_4 and a_5. Guess a formula for the general term a_n in terms of n and prove your guess using induction.

(28) A *convex polygon* is a polygon with at least three sides such that any two of its vertices can be joined by a straight line that lies in the interior of the polygon. Use induction on n to prove that the sum of the interior angles of any n-sided convex polygon is equal to $180(n-2)$ degrees.

(29) Consider the sequence a_n; $n \geq 1$ of real numbers defined recursively as follows: $a_1 = 3$, $a_2 = 5$ and $a_n = a_{n-1} + 2a_{n-2}$ for all $n \geq 3$. Show that a_n is odd for any $n \geq 1$.

(30) Consider the sequence a_n; $n \geq 0$ of real numbers defined recursively as follows: $a_1 = 1$, $a_2 = 8$ and $a_n = a_{n-1} + 2a_{n-2}$ for all $n \geq 3$. Show that $a_n = 3(2^{n-1}) + 2(-1)^n$ for any $n \geq 1$.

$\star$(31) Consider the sequence a_n; $n \geq 0$ of real numbers defined recursively as follows: $a_0 = 0$, $a_1 = 1$ and $a_n = 7a_{n-1} - 12a_{n-2}$ for all $n \geq 2$. Show that $a_n = 4^n - 3^n$ for any $n \geq 0$.

(32) Consider the sequence a_n; $n \geq 1$ of real numbers defined recursively as follows:

$$a_n = \begin{cases} 2 & \text{if } n \text{ is odd} \\ \left(a_{\frac{n}{2}}\right)^2 & \text{if } n \text{ is even} \end{cases}$$

Give the values of the first eight terms $a_1, a_2, \ldots, a_8$ of the sequence. Show that $a_n \leq 2^n$ for any $n \geq 1$.

(33) The well-known Fibonacci sequence f_n, $n \geq 1$ is defined recursively as follows: $f_0 = 0$, $f_1 = 1$ and $f_n = f_{n-1} + f_{n-2}$ for all $n \geq 2$. Show that $f_n = \frac{1}{\sqrt{5}}\left[\left(\frac{1+\sqrt{5}}{2}\right)^n - \left(\frac{1-\sqrt{5}}{2}\right)^n\right]$ for any $n \geq 0$.

$\star$(34) For the Fibonacci sequence defined in Exercise (33), prove that $f_n \geq \left(\frac{1+\sqrt{5}}{2}\right)^{n-2}$ for any $n \geq 3$.

(35) Define a variation of the Fibonacci sequence as follows: $a_1 = 1$, $a_2 = 2$ and $a_n = a_{n-1} + a_{n-2}$ for all $n \geq 3$. Show that $a_n \geq \left(\frac{4}{5}\right)^n$ for all $n \geq 1$.

(36) Let $\mathcal{U}$ be a universal set, $n \geq 2$ an integer and $A, B_1, \ldots, B_n$ be subsets of $\mathcal{U}$. Prove that:

    (a) $A \cap (B_1 \cup \cdots \cup B_n) = (A \cap B_1) \cup \cdots \cup (A \cap B_n)$.
    (b) $A \cup (B_1 \cap \cdots \cap B_n) = (A \cup B_1) \cap \cdots \cap (A \cup B_n)$.

(37) Modify the proof of Theorem 3.1 above to prove the strong version of mathematical induction (Theorem 3.2).

(38) Define a sequence $a_0, a_1, \ldots, a_n, \ldots$ of real numbers as follows: $a_0 = 0$, $a_1 = 1$ and $a_n = \frac{a_{n-1} + (n-1)a_{n-2}}{n}$ for all $n \geq 2$.

    (a) Prove that a_n is a rational number for any $n \geq 0$.
    (b) Prove that $0 \leq a_n \leq 1$ for any $n \geq 0$.

(39) Prove that any integer greater than or equal to 18 can be expressed as the sum of 3's and/or 10's.

(40) Prove that every amount of postage of 12 cents or more can be formed using just 4 cents and 5 cents stamps.

$\star$(41) What amounts of postage can be formed using just 5-cent and 7-cent stamps? Prove your answer by using strong induction.

(42) Show that every integer $n \geq 1$ can be written as the sum of distinct powers of two: $n = 2^{k_1} + 2^{k_2} + \cdots + 2^{k_r}$ where the k_i's are distinct non-negative integers. (*Hint.* For the inductive step, consider two cases: $n + 1$ is even and $n + 1$ is odd).

(43) Let a be a positive real number. Someone is trying to convince you that $a^n = a$ for any integer $n \geq 1$. He proposed the following proof of this (clearly false) fact using strong induction.

Base step. For $n = 1$, $a^1 = a$ is clearly true.

Inductive step. Let $n \geq 1$ be an integer.

Induction hypothesis. Assume that $a^k = a$ for any integer k with $1 \leq k \leq n$.

Conclusion. We prove that $a^{n+1} = a$:

$$a^{n+1} = \frac{a^n \cdot a^n}{a^{n-1}}$$

$$= \frac{a \cdot a}{a} \quad \text{(By the induction hypothesis)}$$

$$= a.$$

Can you point out the flaw in the above proof?

3.7 Recursive definition of sets and the principle of structural induction

In the previous section, we considered sequences defined recursively by giving some initial terms and a rule to generate new terms from known ones. Recursive definitions are not exclusive to sequences, they occur in others mathematical structures like sets, functions, relations and others. In some cases, a mathematical object cannot be defined explicitly by a simple expression but rather with a set of rules that can be used as "pieces" that we can put together to have a bigger picture of the object. In this case, we say (similar to sequences) that the object is defined *recursively*. The process of defining a mathematical object recursively is called *recursion*. In this section, our main focus is on the recursive definitions of sets.

Definition 3.1. Let S be a subset of a universal set $\mathcal{U}$. A recursive definition of S consists of the following three steps:

Base step. We declare that some elements $s_0, \ldots, s_n$ of $\mathcal{U}$ to be in S.

Recursion step. A set of rules is given to generate new elements of S from elements already known to be in the set.

Restriction step. We declare that no elements of $\mathcal{U}$ are in S other than those obtained from the previous two steps.

Example 3.37. Consider the subset S of $\mathbb{N}$ defined recursively as follows:

Base step. $0 \in S$.

Recursion step. If $x \in S$, then $x + 2 \in S$.

Recursion step. There are no elements of $\mathbb{N}$ that belong to S other than those obtained from the previous two steps.

Since $0 \in S$, the Recursion rule implies that $2 = 0 + 2 \in S$. Since $2 \in S$, we have that $4 = 2 + 2 \in S$. Similarly, $6, 8, 10$ are in S. Looks like S is the set of all even natural numbers.

Example 3.37 begs the questions: While it is obvious that the elements of S are precisely the even natural numbers, how can we prove that beyond any doubt? A version of mathematical induction, known as *structural induction*, can be used to that end. The underlying idea of this version of induction reasoning is very similar to that of weak or natural induction. We state the principle of structural induction for sets but we omit the proof.

Theorem 3.3 (Principle of structural induction). *Let $\mathcal{U}$ be a universal set and let S be a subset of $\mathcal{U}$ defined recursively as above. Let P be a property that elements of $\mathcal{U}$ can satisfy. If the following two statements are satisfied, then every element of S satisfies the property P:*

Base step. *All elements of the Base step of S satisfy P.*

Recursive step. *For any rule R in the Recursion step of S, if $x_1, x_2, \ldots, x_n$ are elements of S satisfying P, then the new element $R(x_1, x_2, \ldots, x_n)$ obtained by applying R to $x_1, x_2, \ldots, x_n$ also satisfies the property P.*

Example 3.38. In this example, we prove our observation about the set S of Example 3.37 using the principal of structural induction. Let E be the set of all even natural numbers. We need to prove that $S = E$. Note that every element of E is of the form $2n$ for some $n \in \mathbb{N}$, so to prove that E is a subset of S, we can use (weak) induction on n.

Base step. For $n = 0$, $2n = 0$ which is in S by the basic step of the recursive definition of S.

Inductive step. Let $n \geq 0$ be an integer.

Induction hypothesis. Assume that $2n$ is in S.

Conclusion. We need to prove that $2(n+1)$ is an element of S. By the Recursive step in the definition of S and the induction hypothesis, we have that $2(n+1) = 2n + 2$ is in S.

This proves that E is a subset of S. For the converse, we use structural induction.

Base step. The only element in the Base step of the recursive definition of S is 0, which is in E.

Recursive step. Assume x is an element of S which is already in E. The (only) recursive rule defining S applied to x produces $x + 2$ which belongs to E since it is even (sum of two even numbers.)

We conclude that S is a subset of E. Thus $S = E$.

Example 3.39. Consider the subset E of $\mathbb{N}$ defined recursively as follows:

Base step. 3 is in E.

Recursion step. If $x, y \in E$, then $x + y \in E$.

Restriction step. Nothing else is in E.

Prove that E is the set of all positive multiples of 3.

Solution. Let $S = \{3k; \, k \geq 1 \text{ is an integer }\}$ be the set of all positive multiples of 3. We need to prove that $E = S$. We prove the inclusion $S \subseteq E$ by (weak) induction on k:

Base step. For $k = 1$, $3k = 3$ is in E by the Base step of the recursive definition of E.

Inductive step. Let $k \geq 1$ be an integer.

Induction hypothesis. Assume that $3k$ is an element of E.

Conclusion. We need to prove that $3(k + 1)$ is an element of E. Since $3k$ is assumed to be in E (by the induction hypothesis), the Recursive step in the definition of E implies that $3(k+1) = 3k + 3$ is in E.

This proves that S is a subset of E. For the converse, we use the principle of structural induction.

Base step. The base element 3 of E is a positive multiple of 3, and so it is an element of S.

Inductive step. Let x and y be two elements of E and assume that they are already elements of S. So $x = 3m$ and $y = 3n$ for some positive integers m and n. Then $x + y = 3(m + n)$ is a positive multiple of 3 and hence an element of S.

This proves that E is a subset of S and hence $E = S$. $\Diamond$

Example 3.40. Let $\mathcal{U}$ be a universal set and let $a \in \mathcal{U}$. A subset E of $\mathcal{U}$ is defined recursively as follows:

Base step. $a \in E$.
Recursion step. If $x \in E$, then $(x) \in E$.
Restriction step. Nothing else is in E.

Prove that every element in E contains an equal number of right and left parentheses.

Solution. Let P be the property *"contains an equal number of right and left parentheses"*. Since E is defined recursively, we can use structural induction to prove that every element of E satisfies P.

Base step. The base element a has the same number of right and left parentheses, namely 0. So a satisfies P.
Recursive step. Consider an element x of E. So x has the same number of right and left parentheses, say k. Then (x) also has the same number of right and left parentheses, namely $k + 1$.

By the principle of Structural induction, every element in E contains an equal number of right and left parentheses. $\Diamond$

Example 3.41. Give a recursive definition of the subset of $\mathbb{N}$ of integers not divisible by 5. Prove that your definition is correct.

Solution. Let us start by noticing that $1, 2, 3, 4$ are all positive integers not divisible by 5. Moreover, if we add 5 to each of these integers we get the integers $6, 7, 8$ and 9 which are also not divisible by 5. Let E be the set of all positive integers not divisible by 5 and let S be the subset of $\mathbb{N}$ defined recursively as follows.

Base step. $1 \in S, 2 \in S, 3 \in S, 4 \in S$.
Recursive step. If $x \in S$, then $x + 5 \in S$.
Restriction step. Nothing else is in S except integers in the base step or produced by the recursion step.

We need to prove that $E = S$. If $x \in E$, then $x = 5q + r$ for some integers k and r with $q \geq 0$ and $r \in \{1, 2, 3, 4\}$ (the remainder of the division of x by 5 cannot be zero since x is not divisible by 5). we prove that $x \in S$ using induction on q. If $q = 0$, then $x = r \in S$ by the Base step of the definition of S. Assume that $5q + r \in S$ for some $q \geq 0$, then $5(q+1)+r = (5q+r)+5 \in S$ by the fact that $5q + r \in S$ and the recursion step of the definition of S. By the principle of (weak) induction, we conclude that $5q + r \in S$ for any $q \geq 0$ and any $r \in \{1, 2, 3, 4\}$. This proves that E is a subset of S. We use Structural induction to prove that S is a subset of E. First note that every element of the base step in the definition of S is in E (as being not divisible by 5). Next, assume that x is an element of S which is already in E. Then $x + 5$ is also in E since otherwise $x + 5$ is divisible by 5 and x would be divisible by 5. This proves that $S \subseteq E$. We conclude that $S = E$. $\Diamond$

For the next set of examples, we introduce some terminologies.

Definition 3.2. An *alphabet* is a finite non-empty set of symbols that we usually denote by Σ. A *string* or a *word* over the alphabet Σ is a finite ordered sequence of symbols from the alphabet. The set of all strings on an alphabet Σ is denoted by Σ^*. The *length*, $l(s)$, of a string s is the number of symbols used in s. The *empty string*, denoted by λ, is the string with no symbols in it, so $l(\lambda) = 0$. A *language* over an alphabet Σ is a subset of Σ^*. If s and t are two strings, the *concatenation* of s and t is the string, denoted by st, obtained by writing s followed by t. For example, if $s = abacd$ and $t = xyztu$, then $st = abacdxyztu$ and $ts = xyztuabacd$.

Example 3.42. Given an alphabet Σ, the following is a recursive definition of the set Σ^* of all possible words over Σ:

Base step. The empty string λ is in Σ^*.
Recursion step. If $s \in \Sigma^*$ and $x \in \Sigma$, then $sx \in \Sigma^*$.
Restriction step. Nothing else is in S unless it is obtained from the Basic and Recursive steps. $\Diamond$

An alphabet of particular interest is $\Sigma = \{0, 1\}$. A word over this alphabet is called a *binary string*. Each of the two characters 0 and 1 is called a bit. Binary strings are widely used in computer science.

Example 3.43. Consider the set S of all nonempty binary strings defined recursively as follows:

Base step. 0 is in S.

Recursion step. If $x \in E$, then $0x \in S$ and $1x \in S$.

Restriction step. Nothing else is in S unless it can be obtained from the previous two steps.

Prove that any string in S must end with a 0.

Solution. Let P be the property "*ends with* 0" that can be satisfied by binary strings. We use structural induction to prove that every element of S satisfies P.

Base step. The base element 0 clearly satisfies P.

Recursion step. Assume that x is a string that ends with 0. The two rules given in the recursion step applied of the definition of S applied to x produce the strings $0x$ and $1x$, each of which ends with 0.

By the principle of structural induction, every string in S ends with a 0. $\Diamond$

Example 3.44. In Chapter 1, we defined the notion of a well-formed formula of proposition logic (wff). If L is the set of all wffs of proposition logic, then L can be defined recursively as follows.

Base step. Every logic variable (letter and indexed letter of the English alphabet) is in L. Also, $\mathbf{T}$ and $\mathbf{F}$ are in L.

Recursion step. If $x, y \in S$, then $\neg x \in L$, $(x \wedge y) \in L$, $(x \vee y) \in L$, $(x \rightarrow y) \in L$ and $(x \leftrightarrow y) \in L$.

Restriction step. The only wffs of proposition logic are those obtained from the previous two steps.

Example 3.45. The following is a recursive definition of the set S of all binary strings containing an odd number of 0's:

Base step. $0 \in S$.

Recursion step. If $x \in S$, then $x1 \in S$, $1x \in S$, $00x \in S$, $x00 \in S$.

Restriction step. Nothing else is in S except strings in the base step or produced by the recursion step.

Recursive definitions are also used in defining some geometrical structures that involve repetitive patterns. The infinite repetition of these geometric patterns usually leads a complex and self-similar objects known as *fractals*. In addition to their fascinating and artistic shapes, fractals hold a wealth of interesting properties that are of great interest to mathematics. The following is an example of such a structure.

Example 3.46. Consider the set S of geometrical shapes K_n, $n \geq 1$ defined recursively as follows.

Base step. K_1 is an equilateral triangle of a certain fixed side length a.

Recursion step. If $n \geq 2$, then K_n is the figure obtained from K_{n-1} by replacing every straight edge in K_{n-1} with an edge that has the shape shown in the following diagram.

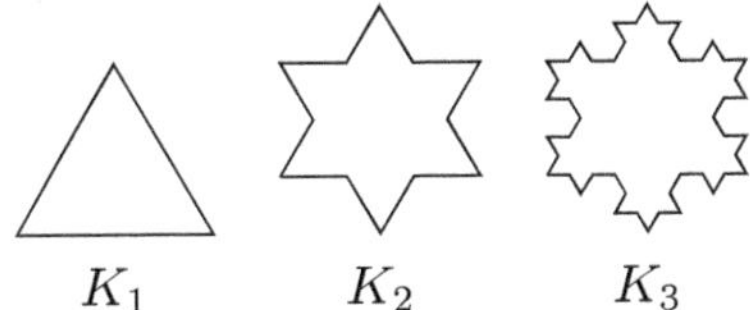

Edge in K_{n-1} The transformed edge in K_n

This transformation is done as follows: divide a straight edge in K_{n-1} of length x into three equal segments, construct an equilateral triangle with side length $\frac{x}{3}$ with one side being the middle segment and then remove that middle segment.

Restriction step. No other shapes are in S except those defined at the base and recursive Steps.

The following are representations of K_1, K_2 and K_3:

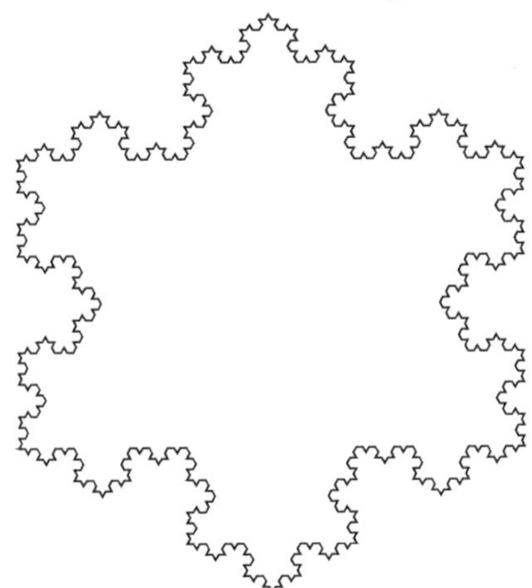

K_1 K_2 K_3

A successive application of the above process leads eventually to the following shape in S that is often referred to as Koch snowflake (first described by the Swedish mathematician Helge von Koch in 1904):

3.7.1 *Exercises*

(1) In each case, give five elements of the subset S of $\mathbb{Z}$ defined recursively. The base step is labeled by (i), the recursion step by (ii). The restriction step is omitted but it is understood that nothing else is in S except elements defined in the base step and the recursion step.

    (a) (i) $0 \in S$; (ii) If $x \in S$, then $3x + 2 \in S$.

    (b) (i) $-1 \in S$, $1 \in S$; (ii) If $x, y \in S$, then $x + y \in S$ and $x - y \in S$.

    (c) (i) $1, 2, 3 \in S$; (ii) If $x \in S$, then $x + 4 \in S$.

    $\star$(d) (i) $0 \in S$, $2 \in S$. (ii) If $x, y \in S$, then $x + 2^y \in S$ and $x^2 + y \in S$.

(2) In each case, give five elements of the subset S of the set of all binary strings defined recursively. The base step is labeled by (i), the recursion step by (ii). The restriction step is omitted but it is understood that nothing else is in S except elements defined in the base step and the recursion step.

    (a) (i) $\lambda \in S$; (ii) If $x \in S$, then $x00 \in S$.

    $\star$(b) (i) $0 \in S$; (ii) If $x \in S$, then $x0 \in S$ and $1x \in S$.

    (c) (i) $\lambda \in S$; (ii) If $x \in S$, then $x0 \in S$ and $1x \in S$.

    (d) (i) $1 \in S$; (ii) If $x, y \in S$, then $xy \in S$ and $0x1 \in S$.

(3) Consider the subset S of $\mathbb{N} \times \mathbb{N}$ defined recursively as follows:

Base step. $(0, 1) \in S$.

Recursion step. Let $(x, y) \in S$.

    (i) If $x \leq 5$ and $y \leq 12$, then $(x + 1, y + 2) \in S$.

    (ii) If $x > 5$ and $y \leq 12$, then $(x - 2, y + 2) \in S$.

Restriction step. Nothing else is in S except pairs from the base step or produced by the Recursion step.

Give 10 pairs in S other than the pair defined in the base step.

(4) Give a recursive definition of the following subset of $\mathbb{N}$:

$$S = \{1, 2, , 7, 11, 12, 16, 17, 21, 22, \ldots\}.$$

You don't need to justify your answer.

(5) Consider the set S of binary strings defined recursively as follows:

Base step. $01 \in S$, $1 \in S$.

Recursion step. If $x, y \in S$, then $xyx \in S$ and $xxyxyy \in S$.

Restriction step. Nothing else is in S except strings in the base step or produced by the recursion step.

Prove that each of the following strings belongs to S. Justify your answer.

(a) 010110111   ⋆(b) 1011011011   (c) 101101010100

(6) Consider the subset S of $\mathbb{N}$ defined recursively as follows:

Base step. $3 \in S, 7 \in S$.

Recursion step. If $x, y \in S$, then $5x \in S$ and $x + y \in S$.

Restriction step. Nothing else is in S except elements in the base step or can be produced by the recursion step.

Which of the following are elements of S? Justify your answer.

(a) 1   (c) 5   (e) 30   (g) 35046.
(b) 4   (d) 15   (f) 50   (h) 37621

(7) In each case give a recursive definition of the given subset of the set of binary strings. You don't need to justify your answer.

(a) $S = \{1, 010, 00100, 0001000, 000010000, \ldots\}$
(b) $S = \{\lambda, 00, 0000, 0000000, 00000000, 0000000000, \ldots\}$
(c) $S = \{00, 000, 0000, 00000, 000000, \ldots\}$
(d) $S = \{\lambda, 01, 0011, 000111, 00001111, \ldots\}$

For Exercises (8)-(16), give a recursive definition of the set S. Make sure your definition covers all the elements of the set. No proof of the correctness of the definition is required.

⋆(8) S is the set of strings over an alphabet Σ.

(9) S is the set of all even integers (both positive and negative).

(10) S is the set of binary strings that have the same numbers of 0's as the number of 1's.

⋆(11) S is the set of binary strings with even length.

(12) S is of the set of binary strings with even length that begin with a 0.

(13) S is of the set of binary strings with odd length that begin with a 1.

(14) S is of the set of binary strings which contain exactly one 0.

⋆(15) S is of the set of non-empty binary strings that start with 1.

(16) S is the set POLY of all polynomials in one variable x with integer coefficients.

⋆(17) Consider the set subset S of $\mathbb{Z}$ defined recursively as follows:

Base step. $0 \in S$.

Recursion step. If $x \in S$, then $x + 3 \in S$ and $x - 3 \in S$.

Restriction step. No elements in S other than those obtained from the previous two steps.

Prove that every integer in S is divisible by 3.

(18) Consider the subset S of $\mathbb{Z}$ defined recursively as follows:

Base step. $3 \in S$, $11 \in S$.

Recursion step. If $x, y \in S$, then $x + y \in S$.

Restriction step. No elements in S other than those obtained from the previous two steps.

Prove that every integer $n \geq 20$ is an element of S (*Hint.* Use strong induction on n).

(19) Let S be the subset of $\mathbb{N} \times \mathbb{N}$ defined recursively as follows:

Base step. $(0,0)$ is an element of S.

Recursion step. If (a, b) is in S, then $(a + 3, b + 2)$ and $(a + 1, b + 4)$ is in S.

Restriction step. Nothing else is in E other than elements from the previous two steps.

Prove that for any pair (a, b) in S, $a + b$ is a multiple of 5.

$\star$(20) Let S be the subset of $\mathbb{N} \times \mathbb{N}$ defined recursively as follows:

Base step. $(0,0) \in S$.

Recursion step. If (a, b) is in S, then $(a, b + 1) \in S$, and $(a + 2, b + 1) \in S$.

Restriction step. Nothing else is in S other than those obtained from the previous two steps.

Show that if $(a, b) \in S$, then $a \leq 2b$.

(21) Let S be the subset of $\mathbb{Z}$ defined recursively as follows:

Base step. $5 \in S$.

Recursion step. If $a \in S$, then $a - 15 \in S$ and $a^2 \in S$.

Restriction step. Nothing else is in S other than those obtained from the previous two steps.

Show that every element S is a multiple of 5.

(22) A *palindrome* is a word (over some alphabet) that reads the same forwards and backwards. For Example *mom, dad* and *civic* are palindromes over the English alphabet. Give a recursive definition of the set P all non-empty binary strings which are palindromes.

(23) Give a recursive definition of the set of all positive integers which are multiples of 5. Prove that your definition is correct.

(24) Give a recursive definition of the set of all positive integers *not divisible* by 3. Prove that your definition is correct.

$\star$(25) Give a recursive definition of the set of all non-negative integers having a remainder of 2 upon division by 3. Prove that your definition is correct.

$\star$(26) Let $a \geq 2$ be an integer. Give a recursive definition of the set $E = \{a^{a^n}; n \in \mathbb{N}\}$. Prove that your definition is correct.

(27) Refer to the set $S = \{K_1, K_2, \ldots, K_n, \ldots\}$ of geometric shapes leading to Koch snowflake of Example 3.46. Use induction on n to prove that shape K_n has $3 \cdot 4^{n-1}$ line segments.

(28) The following shape belongs to a recursively defined set S of plane figures. The base step of set S states that a square of a certain side length a belongs to S.

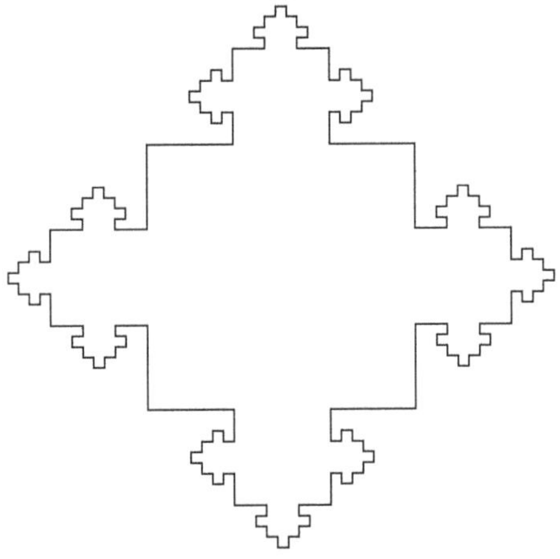

(a) Give a possible recursion step defining S.

(b) Starting with a square, how many applications of the recursion step does it take to produce the above given shape?

Chapter 4

Introduction to Predicate Logic: One Step Further

4.1 Introduction

The following is an old logic argument known since the time of ancient Greece: "*All men are mortal. Socrates is a man. Therefore Socrates is mortal.*" While it is obvious that the argument is valid (unless we dispute the fact that every man is mortal), propositional logic techniques we saw in Chapter 1 fail to prove its validity. Of course, one can treat the statements P: "*All men are mortal*", Q: "*Socrates is a man*" and R: "*Socrates is mortal*" as atomic propositions and proceed to analyze the argument with P and Q as premises and R as conclusion. But one realizes quickly that the argument is invalid. For instance, $v(P) = \mathbf{T}$, $v(Q) = \mathbf{T}$ and $v(R) = \mathbf{F}$ is a counterexample. At first glance, it looks like there is a gap between the propositional logic interpretation of a valid argument and what our mind tells us what such an argument should be. The fact is neither propositional logic interpretation is misleading, nor our mind is playing tricks on us. The problem in the use of the word *All* in the first premise of the argument. Propositional logic is ill-equipped to capture the true meaning of expressions like *for all* and *there exists*. To express the statement *All men are mortal* using propositional logic, we need to use a huge number of conjunctions of propositions ranging over the set of all men ever lived: "*Joe is mortal and Moe is mortal and Boe is mortal $\cdots$*" (in particular, *All men are mortal* is not atomic). In other words, we need to identify every man ever lived and verify the predicate *is mortal* for all men.

An expression like $P(n)$: "*n is evenly divisible by 3*" where n is an arbitrary integer is called an *open statement* since its truth value depends on the choice of n in the set of all integers. An open statement is not a logic proposition unless we specify a value for each of the variables it contains. For the open statement $P(n)$: "*n is evenly divisible by 3*", $P(1)$ is false but $P(9)$ is true. We can turn $P(n)$ into a logic proposition by writing "*n is evenly divisible by 3 for all integers n*" or "*There exists an integer n which is evenly divisible by 3*". The first one is clearly false but the second is true. The added expressions *for all* and *there exists* are called *quantifiers* in this context. If P represents the expression "*is evenly divisible by 3*", then we say that P is a *unary* predicate symbol (the term *unary* refers to the fact that P can be applied to a single variable at a time). The symbol P can be applied to any integer n and the expression $P(n)$ means "*The integer n is evenly divisible by 3.*" Predicate symbols can handle any number of variables depending in the context of the problem at hand. For example, let Q denote the predicate "*is less than or equal to*" on the set of all real numbers. Then Q takes two reals numbers x and y as inputs and returns the proposition $Q(x, y)$: "*x is less than or equal to y*". Hence, $Q(1, 2)$ is true while $Q(-1, -2)$ is false. We say in this case that Q is a binary predicate symbol.

The goal of this chapter is to extend the language of propositional logic to deal with statements containing predicates and quantifiers. This extension is known in the literature as *predicate logic* or *First-order logic*. Our goal is to keep the treatment of this topic as simple and as informal as possible.

4.2   The basics of predicate logic

Given an open statement containing some variables, the set from which the variables can take values is called the *universe of discourse* (or simply the *universe*). Throughout this chapter, the universe of discourse is denoted by the symbol $\mathcal{U}$.

Definition 4.1. A *constant* (also known as an *individual*) refers to a specific element of $\mathcal{U}$. A *variable* is a symbol that can range over the elements of $\mathcal{U}$. A variable is used to denote an arbitrary and unspecified element of the universe and can be thought of as a "placeholder" for objects in $\mathcal{U}$. A *term* (also called an *argument*) refers to either a constant or a variable. A *predicate* refers to a property that objects in $\mathcal{U}$ can satisfy or a relationship between objects in $\mathcal{U}$. A predicate that depends on one variable (respectively two variables, three variables) is called a *unary* (respectively *binary*, *ternary*). In general, if a predicate depends on n variables, we call it an *n-ary* predicate or a predicate of *arity n*. A 0-ary predicate is just a formula of propositional logic.

Example 4.1. In the universe of all living things, the predicate M: *"is a mammal"* is a unary predicate. So $M(cow)$ is true but $M(tulip)$ is false. In the universe of all students and professors at the University of Ottawa, the predicate Q: *"is a student of"* is a binary predicate and $Q(x, y)$ reads *"x is a student of y"*. In the universe of all real numbers, the statement $B(x, y, z)$: *"The number x belongs to the interval $[y, z]$"* represents a ternary predicate.

Example 4.2. Let the universe of discourse be the set $\mathbb{N}$ of all natural numbers. The following are predicates over $\mathbb{N}$ with different arities:

(a) P: *"is a prime number"*: this is a unary predicate. The open statement $P(n)$ reads *"n is a prime number"*. So $P(4)$ and $P(27)$ are false while $P(3)$ and $P(131)$ are true.

(b) Q: *"is a prime divisor of"* is a binary predicate. The open statement $Q(m, n)$ reads *"m is a prime divisor of n"*. Hence $Q(2, 84)$, $Q(7, 84)$ are true but $Q(4, 84)$ and $Q(13, 84)$ are false.

(c) G: *"is the gcd of"* (here the expression gcd stands for the *greatest common divisor*). We look at G in this context as a ternary predicate. The open statement $G(d, a, b)$ reads *"d is the greatest common divisor of a and b"* or simply *"$d = \gcd(a, b)$"*. So $G(2, 4, 6)$ is true (as 2 is the gcd of 4 and 6) whereas $G(2, 8, 12)$ is false (as $\gcd(8, 12) = 4$).

Another important class of objects in predicate logic is that of *functions*. Like predicates, functions have arities that correspond to the number of arguments they take. If n is a natural number, then an *n-arity* function f on the universe $\mathcal{U}$ takes n terms $x_1, \ldots, x_n$ of $\mathcal{U}$ as inputs and produces a unique output which is another element of $\mathcal{U}$ that we denote by $f(x_1, \ldots, x_n)$. We often denote an *n-ary* function f by $f(x_1, \ldots, x_n)$. For example, the standard addition on the natural numbers is a binary function $A(m, n)$ that takes two natural numbers m and n and produces another natural number that we denote by $m + n$. The substraction, on the other hand, is not a function on the universe of natural numbers as the output of two natural numbers is not necessarily a natural number. It should be noted here that not every element of $\mathcal{U}$ is necessarily an output of a given function. For instance, the real number -1 cannot be an output of the ternary function $f(x, y, z) = x^2 + y^2 + z^2$ on the universe of all real numbers. Like in the case of predicates, the variables of a function can be thought of as "placeholders" that can be evaluated using constants of the universe. For example, for the ternary function $f(x, y, z) = -2x + xy + z$ on the universe $\mathbb{R}$, $f(-1, 0, z) = 2 + z$, $f(-1, y, 1) = 3 - y$ and $f(-1, -2, -4) = 0$. Note that a 0-ary function is a constant function that assigns to every variable the same element (output) of the universe.

The notion of a *term* as given in Definition 4.1 can now be extended as follows. If f is an n-ary function and $t_1, t_2, \ldots, t_n$ are terms (individual constants or individual variables), then $f(t_1, t_2, \ldots, t_n)$ is also called a *term*.

Both functions and predicates take a finite number of arguments in the universe as inputs, but the two notions are different and they serve different purposes. For an n-ary function f, an n-ary predicate P and n terms $t_1, \ldots, t_n$ of $\mathcal{U}$, $f(t_1, \ldots, t_n)$ is an element of the universe whereas $P(t_1, \ldots, t_n)$ is a "claim" (or a statement) about the terms $t_1, \ldots, t_m$. For example, in the universe of all citizen in a certain country, we can look at "*mother*" as a function symbol that assigns to every citizen his or her mother. So if x is a citizen, then $mother(x)$ is the (unique) citizen who is the mother of x. Now, consider the binary predicate M: "*is the mother of*". If x, y are two citizens, then $M(x, y)$ is the statement: "x is the mother of y" which could be either true or false depending on the choice of x and y. If P is an n-ary predicates and $x_1, \ldots, x_n$ are variables in $\mathcal{U}$, the expression $P(x_1, \ldots, x_n)$ is called a *propositional function*. We often don't distinguish between the predicate symbol P and the propositional function $P(x_1, \ldots, x_n)$. We use the term "predicate" to refer to both.

Typically, variables (and terms in general) are denoted by lower case letters or indexed letters: $a, b, n, x, y, z, x_1, y_2, \ldots$. Capital letters $P, Q, R, \ldots$ are used to denote predicate symbols. Function symbols are usually denoted by lower case letter like f, g, h, etc... If P is an n-ary predicate symbol, the order in which the terms appear in $P(t_1, \ldots, t_n)$ is important. For example, if P is the binary predicate "*is the father of*", then $P(Joe, Bob)$ has a totally different meaning than $P(Bob, Joe)$.

Definition 4.2. For an n-ary predicate P on the universe $\mathcal{U}$, the *truth set* of P is the collection of all n-tuples $(a_1, \ldots, a_n)$ with each a_i in $\mathcal{U}$ such that the proposition $P(a_1, \ldots, a_n)$ is true.

Example 4.3. In each case, determine the truth set T of the unary predicate $P(x)$: "$x^2 \leq 4$" on the universe $\mathcal{U}$: (a) $\mathcal{U} = \mathbb{N}$; (b) $\mathcal{U} = \mathbb{Z}$; (c) $\mathcal{U} = \mathbb{R}$.

Solution. Note first that the statement $x^2 \leq 4$ is equivalent to $-2 \leq x \leq 2$. In the universe $\mathcal{U} = \mathbb{N}$, T is the set $\{0, 1, 2\}$. In the universe $\mathcal{U} = \mathbb{Z}$, T is the $\{-2, -1, 0, 1, 2\}$. In the universe of real numbers, T is the closed interval $[-2, 2]$ which, unlike the other two cases, is an infinite set. $\diamondsuit$

Example 4.4. In each case, determine the truth set T of the binary predicate $P(x, y)$: "$x^2 - 2x + y^2 = 3$" on the universe $\mathcal{U}$: (a) $\mathcal{U} = \mathbb{N}$; (b) $\mathcal{U} = \mathbb{Z}$; (c) $\mathcal{U} = \mathbb{R}$.

Solution. The statement $P(x, y)$ is equivalent to $(x - 1)^2 + y^2 = 4$ by completion of the square. In the universe $\mathcal{U} = \mathbb{N}$, the equation $(x-1)^2 + y^2 = 4$ is only possible if $x - 1 = 0$ and $y = 2$ or $x - 1 = 2$ and $y = 0$. So $T = \{(1, 2), (3, 0)\}$ in this case. In the universe $\mathcal{U} = \mathbb{Z}$, the equation $(x - 1)^2 + y^2 = 4$ is only possible if $x - 1 = 0$ and $y = \pm 2$ or $x - 1 = \pm 2$ and $y = 0$. So $T = \{(1, -2), (1, 2), (-1, 0), (3, 0)\}$. In the universe $\mathcal{U} = \mathbb{R}$, T consists of the circle in the Cartesian plane centered at the point $(1, 0)$ and of radius 2. In this case, T is an infinite set. $\diamondsuit$

Connectives of propositional logic carry over in a natural way to predicate logic. If $P(x)$, $Q(x)$ are unary predicates, it should be clear to the reader what it is meant by expressions like $\neg P(x)$, $(P(x) \wedge Q(x))$, $(P(x) \vee Q(x))$, $(P(x) \to Q(x))$ and $(P(x) \leftrightarrow Q(x))$. In a similar fashion, we can extend these notations to include predicates with any arity.

Example 4.5. Using the notations of Example 4.2 above, the proposition $\neg P(n)$ reads "n *is a composite number*" (that is a number with at least one positive divisor d other than 1 and n). The statement $(Q(m, n) \to Q(m, n^2))$ reads "*If m is a prime divisor of n, then it is a prime divisor of n^2*" (which is true).

Example 4.6. Let the universe of discourse be the set of positive integers. Consider the unary predicate $P(x)$: "x *is a prime number*" and the binary predicate $D(x, y)$: "x *is a divisor of y*". In each case, determine the truth value of the given proposition.

(a) $(P(123) \vee D(7, 91))$
(b) $(P(7) \to D(5, 91))$
(c) $((D(7, 91) \wedge \neg P(3)) \leftrightarrow \neg P(11))$

Solution. Since $D(7, 91)$ is true (7 is a divisor of 91), proposition (a) is true regardless the truth value of $P(123)$ (truth value of a disjunction). Proposition (b) is false since $P(7)$ is true but $D(5, 91)$ is false (truth value of an implication). Proposition (c) is true since $(D(7, 91) \wedge \neg P(3))$ is false and $\neg P(11)$ is also false (truth value of a biconditional). $\Diamond$

4.2.1   *Exercises*

(1) Let the universe of discourse be the set of all students at a certain university. Identify the constants and the predicates in each of the given English sentences. Determine the arity of each predicate.

   (a) Joe is a student in pure math.

 ⋆ (b) Joe finished his first year with a higher average than Patrick.

   (c) Joe and Jane are classmates.

   (d) Joe is a common friend of Jane and Rochelle.

(2) With the notations of Example 4.2 above, determine the truth value of each of the following propositions:

   (a) $P(43)$ (c) $Q(9, 153)$ (e) $G(4, 16, 32)$

   (b) $P(121)$ (d) $Q(23, 2875)$ (f) $G(8, 48, 616)$

(3) For a positive real number a, an *integer power* of a is a real number of the form a^k for some $k \in \mathbb{Z}$. Consider the open statement $P(z)$: "z *is an integer power of* 3". Find the truth value of each of the following.

   (a) $P(1)$ (c) $P(243)$ (e) $P\left(\frac{1}{27}\right)$

 ⋆ (b) $P(91)$ (d) $P(2181)$ ⋆ (f) $P\left(\frac{1}{2187}\right)$

(4) Consider the unary predicate $P(x)$: "$x > \frac{1}{x^2}$" over the universe $\mathcal{U} = \mathbb{R}$.

   (a) Give the truth value of each of the propositions: $P(-2)$, $P(-1)$, $P\left(-\frac{1}{3}\right)$, $P(1)$, $P\left(\frac{1}{3}\right)$, $P\left(\frac{3}{2}\right)$ and $P(2)$.

   (b) Determine the truth set of P (*Hint.* $x^3 - 1 = (x - 1)(x^2 + x + 1)$).

(5) In the universe of all real numbers, determine the truth set T of each of the following predicates.

   (a) $A(x)$: "$|x| \geq 6$" (here $|x|$ means the absolute value of x).

 ⋆ (b) $B(x, y)$: "$x^2 + 2x + y^2 - 2y = 7$".

$\star$(6) Consider the unary predicate $P(x)$: "$x^2 + 3x \le 2x + 6$". In each case, determine the truth set of P in the given universe $\mathcal{U}$: (a) $\mathcal{U} = \mathbb{N}$; (b) $\mathcal{U} = \mathbb{Z}$; (c) $\mathcal{U} = \mathbb{R}$.

(7) Consider the ternary predicate $S(x, y, z)$: "$z^2 = x^2 + y^2$" on the universe of all real numbers.

   (a) Determine the truth value of each of the propositions: $S(1, \sqrt{2}, 3)$, $S(1, \sqrt{2}, \sqrt{3})$ and $S(-1, \sqrt{2}, -\sqrt{3})$.
   (b) Describe the set of points (x, y) of real numbers (i.e., points in the Cartesian plane) such that the statement $S(x, y, 2)$ is true.

(8) With the notations of Example 4.6 above, determine the truth value of each the following propositions.

   (a) $(D(3, 51) \wedge P(31))$
   (b) $(D(1, 0) \rightarrow (D(5, 3454345660) \vee P(3123457)))$
   $\star$ (c) $\neg(\neg D(11, 1) \leftrightarrow D(2, 23))$
   (d) $((D(11, 1) \rightarrow D(2, 23)) \rightarrow \neg(D(34, 78956434) \vee P(2)))$

(9) (requires trigonometry) Let the universe of discourse be the interval $[0, \pi]$ of $\mathbb{R}$. Consider the predicates $A(x)$: "$\sin x \cos x > 0$", $C(x, y)$: "$\cos x < \sin y$" and $S(x, y)$: "$\sin^2 x + \cos^2 y \le \frac{3}{2}$". In each case, determine the truth value of the given proposition.

   (a) $A\left(\frac{\pi}{3}\right)$
   (b) $A\left(\frac{5\pi}{6}\right)$
   $\star$ (c) $C(0, \pi)$
   (d) $S\left(\frac{\pi}{2}, \frac{\pi}{3}\right)$

   (e) $S\left(0, \frac{\pi}{3}\right)$
   (f) $(C(0, 0) \vee S(0, 0))$
   (g) $\neg\left(A(0) \rightarrow S\left(\frac{\pi}{2}, \frac{\pi}{2}\right)\right)$
   $\star$ (h) $\left(\neg A\left(\frac{\pi}{4}\right) \leftrightarrow \left(\neg S(0, 0) \vee C\left(\frac{\pi}{6}, \frac{\pi}{3}\right)\right)\right)$

(10) Let the universe of discourse be the set of all real numbers. Consider the predicates $P(x)$: "$x > 0$", $Q(x)$: "$|x - 1| \ge 2$" and $R(x)$: "$x^2 - 3x + 2 \ge 0$". In each case, determine the truth set of the given open statement.

   (a) $(P(x) \wedge R(x))$
   (b) $(P(x) \wedge Q(x) \wedge R(x))$

   (c) $\neg(P(x) \vee R(x))$
   $\star$ (d) $(\neg P(x) \rightarrow Q(x))$

4.3   Quantifiers: The language of predicate logic

If P is a unary predicate and x is a variable in the universe of discourse, we have seen that one way to turn the open statement $P(x)$ into a logic proposition is by attributing a constant value to the variable. Another way of making a logic proposition out of $P(x)$ is to add words like *"for all"* and *"there exits"* as a prefix before the open statement. These two expressions are called *quantifiers* and the resulting expression is called a *quantified statement.*. The process of adding a quantifier to an open statement is called *quantification*. The role of a quantifier is to clarify wether a statement is true for all or for only some elements of the universe. There are two types of quantifiers.

- **The universal quantifier.** If $P(x)$ is a unary predicate, the *universal quantification* of $P(x)$ is the logic proposition *"$P(x)$ is true for all x in the universe of discourse"*. In predicate logic language, the universal quantification of $P(x)$ is represented by the expression $(\forall x)\, P(x)$.
- **The existential quantifier.** If $P(x)$ is a unary predicate, the *existential quantification* of $P(x)$ is the logic proposition *"there exists at least one element x in the universe of discourse such that $P(x)$ is true"* or *"$P(x)$ is true for some x in the universe of discourse"*. In predicate logic language, the existential quantification of $P(x)$ is represented by the expression $(\exists x)\, P(x)$.

Mathematics is rich with results that can be expressed as quantified statements. It is therefore important to be clear on what it means for the propositions $(\forall x)P(x)$ and $(\exists x)P(x)$ to be true. Of course, this widely depends on the universe of discourse.

Definition 4.3. Let P be a unary predicate over a universe $\mathcal{U}$.

- The universal statement $(\forall x)P(x)$ is *true* if and only if the proposition $P(\alpha)$ is true for every element α of $\mathcal{U}$. The universal statement $(\forall x)P(x)$ is *false* if and only if the proposition $P(\alpha)$ is false for at least one element α of $\mathcal{U}$. In this case, we say that the element $\alpha \in \mathcal{U}$ is a *counterexample* of the universal statement.
- The existential statement $(\exists x)P(x)$ is *true* if and only if the proposition $P(\alpha)$ is true for at least one element α in $\mathcal{U}$. The existential statement $(\exists x)P(x)$ is *false* if and only if the proposition $P(\alpha)$ is false for every element α of $\mathcal{U}$.

Example 4.7. Consider the unary predicates $P(x)$: "$6x^2 - 7x - 3 = 0$", $Q(x)$: "$x^2 - 1 \leq 0$. In each case, determine the truth value of the quantified statement in the given universe $\mathcal{U}$.

(a) $(\exists x)\,P(x)$, $\mathcal{U} = \mathbb{Z}$.
(b) $(\exists x)\,P(x)$, $\mathcal{U} = \mathbb{Q}$.
(c) $(\forall x)\,P(x)$, $\mathcal{U} = \mathbb{R}$.

(d) $\neg\,(\forall x)\,Q(x)$, $\mathcal{U} = \mathbb{Z}$.
(e) $(\forall x)\,(P(x) \to Q(x))$, $\mathcal{U} = \mathbb{R}$.
(f) $(\exists x)\,(\neg P(x) \leftrightarrow Q(x))$, $\mathcal{U} = \mathbb{N}$.

Solution. First note that the equation $6x^2 - 7x - 3 = 0$ has two roots: $x = -\frac{1}{3}$ and $x = \frac{3}{2}$ and the inequality $x^2 - 1 \leq 0$ is valid if and only if $-1 \leq x \leq 1$.

(a) False. None of the roots of the equation $6x^2 - 7x - 3 = 0$ is an integer.
(b) True. Both roots of $6x^2 - 7x - 3 = 0$ are rational numbers.
(c) False. The statement says that $6x^2 - 7x - 3 = 0$ for any real number x, which is clearly false. For example, $P(1) = -4 \neq 0$.
(d) True. The statement $(\forall x)\,Q(x)$ says that $x^2 - 1 \leq 0$ for any real number x, which is false (for example, $Q(2) = 3 > 0$). So, $\neg\,(\forall x)\,Q(x)$ is true.
(e) False. The statement says that the implication $(P(x) \to Q(x))$ is true for all real numbers. This is false since $P\left(\frac{3}{2}\right)$ is true but $Q\left(\frac{3}{2}\right)$ is false.
(f) True. For the element $0 \in \mathbb{N}$, both $\neg P(0)$ and $Q(0)$ are true and so the biconditional $(\neg P(0) \leftrightarrow Q(0))$ is true. $\qquad\qquad \Diamond$

We can define the notion of equivalence for open statements using the universal quantifiers.

Definition 4.4. We say that the two open statements $P(x)$ and $Q(x)$ on the universe $\mathcal{U}$ are *logically equivalent*, and we write $P(x) \equiv Q(x)$, if the universal statement $(\forall x)\,(P(x) \leftrightarrow Q(x))$ is true. In other words, $P(x) \equiv Q(x)$ if and only if the biconditional $(P(a) \leftrightarrow Q(a))$ is true for each element a in the universe.

Example 4.8. The statements $P(x)$: "$x^4 - 1 > 0$" and $Q(x)$: "$x^2 - 1 > 0$" are logically equivalent over the universe $\mathbb{R}$. To see this, notice that $x^4 - 1 = (x^2 + 1)(x^2 - 1)$ and $x^2 + 1 > 0$ for any real number x. So $x^2 - 1 > 0$ if and only if $x^4 - 1 > 0$ and this implies that $(\forall x)\,(P(x) \leftrightarrow Q(x))$ is true in $\mathbb{R}$.

When the universe of discourse is specified and every symbol (predicate, function, term, ...) in a quantified statement is assigned an interpretation in the universe, the statement becomes a logic proposition. For quantified statements over the same universe and sharing the same interpretations of symbols, it make sense to talk about logic equivalences of such statements. The following theorem gives some basic equivalences of predicates logic.

Theorem 4.1. *Let $P(x)$, $Q(x)$ be two predicates over a universe $\mathcal{U}$. Then:*

(a) $(\forall x)\,(P(x) \wedge Q(x)) \equiv ((\forall x)P(x) \wedge (\forall x)Q(x))$ *(the universal quantifier distributes over conjunction).*

(b) $(\exists x)\,(P(x) \vee Q(x)) \quad \equiv \quad ((\exists x)P(x) \vee (\exists x)Q(x))$ *(the existential quantifier distributes over disjunction).*

(c) *If R is a proposition not containing any quantifier, then:*

$\quad$ (i) $(\forall x)(P(x) \vee R) \equiv ((\forall x)P(x) \vee R)$
$\quad$ (ii) $(\forall x)(P(x) \wedge R) \equiv ((\forall x)P(x) \wedge R)$
$\quad$ (iii) $(\exists x)(P(x) \vee R) \equiv ((\exists x)P(x) \vee R)$
$\quad$ (iv) $(\exists x)(P(x) \wedge R) \equiv ((\exists x)P(x) \wedge R)$

Proof. We only prove parts (a) and (c)-(i), leaving the proof of the other parts to the reader. For part (a), notice that $(\forall x)\,(P(x) \wedge Q(x))$ says that $(P(x) \wedge Q(x))$ is true for any $x \in \mathcal{U}$. This is equivalent to say that $P(x)$ is true for any $x \in \mathcal{U}$ and $Q(x)$ is true for any $x \in \mathcal{U}$, which translates to $((\forall x)P(x) \wedge (\forall x)Q(x))$. For part $(c) - (i)$, We argue according to the truth value of R. If R is true, then so is $(P(x) \vee R)$ for any $x \in \mathcal{U}$ and the statement $(\forall x)(P(x) \vee R)$ is true. On the other hand, $((\forall x)(P(x) \vee R)$ is also true in this case. If R is false, then $(\forall x)(P(x) \vee R)$ and $((\forall x)P(x) \vee R))$ are each equivalent to $(\forall x)P(x)$ and hence they are equivalent. $\qquad\square$

Note that in general, $(\forall x)\,(P(x) \vee Q(x)) \not\equiv ((\forall x)P(x) \vee (\forall x)Q(x))$ and $(\exists x)\,(P(x) \wedge Q(x)) \not\equiv ((\exists x)P(x) \wedge (\exists x)Q(x))$. To see this, it suffices to come up with counter-examples where each pair of these propositions have different truth values (see Exercise (12) below).

4.3.1 *Negation of a quantified statement*

Let P be a unary predicate over a universe $\mathcal{U}$. The negation, $\neg(\forall x)P(x)$, of the statement $(\forall x)P(x)$ is the statement *"It is not the case that $P(x)$ is true for every x in $\mathcal{U}$."* This is equivalent to say that *"There exists an element x of $\mathcal{U}$ that does not satisfy P"*. Using symbols of predict logic, the last statement translates to $(\exists x)\neg P(x)$. Similarly, the negation of $(\exists x)P(x)$ is $(\forall x)\neg P(x)$.

Negation rules for predicate logic:

- $\neg(\exists x)P(x) \equiv (\forall x)\neg P(x)$
- $\neg(\forall x)P(x) \equiv (\exists x)\neg P(x)$

Example 4.9. Let A, B and C be three unary predicate symbols over a universe $\mathcal{U}$. Find the negation of each of the following quantified statements.

(a) $(\exists x)\,(A(x) \vee \neg B(x))$
(b) $(\forall x)\,(\neg A(x) \vee (B(x) \rightarrow C(x)))$

Solution. For part (a), $\neg(\exists x)\,(A(x) \vee \neg B(x)) \equiv (\forall x)\neg\,(A(x) \vee \neg B(x)) \equiv (\forall x)\,(\neg A(x) \wedge B(x))$. For part (b), $\neg(\forall x)\,(\neg A(x) \vee (B(x) \rightarrow C(x))) \equiv (\exists x)\neg\,(\neg A(x) \vee (B(x) \rightarrow C(x))) \equiv (\exists x)\,(A(x) \wedge \neg(B(x) \rightarrow C(x))) \equiv (\exists x)\,(A(x) \wedge B(x) \wedge \neg C(x))$. $\Diamond$

4.3.2 *Multiple quantifiers*

There are many instances where we need to use multiple quantifiers in a statement. Expressions like *"For every x and for every y, $P(x,y)$"* and *"There exists x such that for every y, $P(x,y)$"* are examples of statements containing multiple quantifiers.

Definition 4.5. Let P be a binary predicate over a universe $\mathcal{U}$. In what follows, an "element" means "an element of $\mathcal{U}$".

(a) The statement $(\forall x)(\forall y)P(x,y)$ reads *For all elements x and y, $P(x,y)$ is true.* The statement is true if and only if for arbitrary elements a and b, the proposition $P(a,b)$ is true.
(b) The statement $(\exists x)(\exists y)P(x,y)$ reads *There exist elements x and y such that $P(x,y)$ is true.* The statement is true if and only if the proposition $P(a,b)$ is true for some elements a and b.
(c) The statement $(\forall x)(\exists y)P(x,y)$ reads *For every element x, there exists an element y such that $P(x,y)$ is true.* The statement is true if and only if for every $a \in \mathcal{U}$, we can find at least one element $b \in \mathcal{U}$ such that $P(a,b)$ is true.
(d) The statement $(\exists x)(\forall y)P(x,y)$ reads *There exists an element x such that $P(x,y)$ is true for every element y.* The statement is true if and only if an element a can be found such that $P(a,b)$ is true for every element b.

Remark 4.1. Let P be a binary predicate symbol on the universe $\mathcal{U}$. The order in which the quantifiers occur in the statements $(\forall x)(\forall y)P(x,y)$ and $(\exists x)(\exists y)P(x,y)$ does not matter. In other words, $(\forall x)(\forall y)P(x,y)$ and $(\forall y)(\forall x)P(x,y)$ have the same meaning, and the same is true for $(\exists x)(\exists y)P(x,y)$ and $(\exists y)(\exists x)P(x,y)$. The same is *not* true for the statements $(\forall x)(\exists y)P(x,y)$ and $(\exists x)(\forall y)P(x,y)$. In these statements, switching quantifiers changes completely the meaning of the statement. The next example illustrates that (compare parts (a) and (d) of the example).

Example 4.10. Consider the binary predicate $P(x,y)$: "$x+y=1$" over the universe $\mathbb{R}$. In each case, determine the truth value of the given statement. Start by interpreting the meaning of the statement in English.

(a) $(\forall x)\,(\exists y)\,P(x,y)$ (c) $(\exists x)\,(\exists y)\,P(x,y)$

(b) $(\forall x)\,(\forall y)\,P(x,y)$ (d) $(\exists x)\,(\forall y)\,P(x,y)$

Solution. (a) The statement reads *For any real number x, there exists a real number y such that $x + y = 1$*. This is true: given a real number x, the real number $y = 1 - x$ satisfies the property.

(b) The statement reads *For all real numbers x and y, $x+y = 1$* or simply *The sum of any two real numbers is 1*. This is clearly false. For instance $P(-1,0)$ is false.

(c) The statement reads *There exist two real numbers with sum equals to 1*. This is true as the sum of -2 and 3 is equal to 1.

(d) The statement reads *There exists a real number x such that $x + y = 1$ for any real number y*. If such an element x exists, then in particular $x+0 = 1$ (choosing $y = 0$) and $x+1 = 1$ (choosing $y = 1$). This means that $x = 1$ and $x = 0$, which is absurd. The statement is false. $\Diamond$

Quantified statements can include any number of quantifiers as shown in the following example.

Example 4.11. In the universe of real numbers, consider the open statement in three variables: $A(x,y,z)$: "$x + y + z = 0$". In each case, determine the truth value of the given statement. Start by interpreting the meaning of the statement in English.

(a) $(\forall x)(\forall y)(\forall z)A(x,y,z)$ (c) $(\exists x)(\forall y)(\forall z)A(x,y,z)$

(b) $(\forall x)(\forall y)(\exists z)A(x,y,z)$ (d) $(\exists x)(\exists y)(\exists z)A(x,y,z)$

Solution. (a) The statement reads *For all real numbers x, y and z, we have that $x + y + z = 0$* or simply: *The sum of any three real numbers is equal to 0*. This is false since, for example, $1 + 2 + 3 = 6 \neq 0$.

(b) The statement reads *For all real numbers x and y, there exists a real number z such that $x + y + z = 0$*. This is true since $z = -x - y$ is a real number satisfying $x + y + z = 0$.

(c) The statement reads *There exists a real number x such that $x + y + z = 0$ for all real numbers y and z*. If such a real number x exists then in particular $x+0+0 = 0$ (choosing $y = z = 0$) and $x+1+0 = 0$ (choosing $y = 1$, $z = 0$). This means that $x = 0$ and $x = -1$, which is absurd. The statement is false.

(d) The statement reads *There exist real numbers x, y and z such that $x + y + z = 0$*. This is true. For instance $x = y = z = 0$ satisfy the property. $\Diamond$

4.3.3 *Syntax of the predicate logic language*

As an extension of propositional logic, predicate logic language combines logic connectives, predicates and quantified statements to create a more powerful mathematical language. Similar to what we did in propositional logic, we formally define the notion of a *well-formed formula* of predicate logic. We start with the definition of the *syntax* of the language.

Definition 4.6. The *syntax* of the predicate logic language consists of:

- *Individual constants*: typically denoted by lower case letters from the beginning of the alphabet $a, b, c, a_0, a_1, \ldots$
- *Individual variables*: typically denoted by lower case letters from the end of the alphabet $x, y, z, x_0, x_1, \ldots$
- *Function symbols*: lower case letters $f, g, h, p, q, f_0, f_1, \ldots$ Every function has an arity $n \in \mathbb{N}$. Functions with 0-arity are just individual constants.
- *Predicate symbols*: capital letters $A, B, P, Q, \ldots, A_1, B_1, \ldots$ Like functions, every predicate symbol has an arity $n \in \mathbb{N}$ and a 0-arity predicate is just a propositional logic formula.
- *Connectives of propositional logic*: $\wedge, \vee, \rightarrow, \leftrightarrow, \neg$.
- The *universal and existential quantifiers*: $\forall, \exists$.
- The symbols "$=$" and "$\neq$".
- *Parentheses*: $(,), [,], \{, \}$.
- *Comma*: ,

Like wffs of propositional logic, well-formed formulas of predicate logic are defined recursively using a set of rules.

Definition 4.7. The following set of rules defines recursively what constitutes a *well-formed formula* (wff for short) of predicate logic.

(1) For any $n \in \mathbb{N}$ and any n-ary predicate P, $P(t_1, \ldots, t_n)$ is a wff (called an *atomic formula*) for any terms $t_1, \ldots, t_n$ in the universe.
(2) If ϕ and ψ are wffs then each of the following is also a wff: (2.1) $\neg\phi$; (2.2) $(\phi \wedge \psi)$; (2.3) $(\phi \vee \psi)$; (2.4) $(\phi \rightarrow \psi)$; (2.5) $(\phi \leftrightarrow \psi)$.
(3) If ϕ is a wff and x is a variable, then the following are wffs: (3.1) $(\forall x)\phi$; (3.2) $(\exists x)\phi$.
(4) A string of symbols is a wff if and only if it is obtained by a finite number of applications of the above rules.

Before we look at some examples, some remarks are in order.

(1) Every wff of propositional logic is in particular a wff of predicate logic.
(2) If t_1, t_2 are two terms, then the expression $t_1 = t_2$ is considered a wff as it stands for the binary predicate $E(t_1, t_2)$ where P: "*is the same element as*". Similarly, $t_1 \neq t_2$ is also a wff as it stands for $\neg(t_1 = t_2)$.
(3) In rule (3) of Definition 4.7, we don't require that the wff ϕ contains an occurrence of the variable x. There is, however, another school of thought in the literature that requires x to appear in ϕ to avoid quantifying expressions not containing the variable of quantification.
(4) While expressions like $(\forall x)(\exists x)P(x)$ and $(\exists x)(\forall x)P(x)$ are wffs according to Definition 4.7, they don't make a lot of sense in terms on actual meaning and we will stay away from using them in practice.

Example 4.12. Let P, Q be unary predicates, R a binary predicate over a universe $\mathcal{U}$ and let a, b be two constants in $\mathcal{U}$. In each case, determine if the expression is a wff of predicate logic. If you say that the expression is not a wff, propose changes to make it as such.

(a) $(P(a) \rightarrow R(a, b))$
(b) $(R(a, b) \rightarrow (\exists x)P(x))$
(c) $(\forall x)(\exists y)P(y) \wedge R(y, x)$
(d) $(\exists x)(\forall y)\,(R(x, y) \wedge (\forall z)Q(z))$
(e) $(\forall x)(\forall y)R(y, x) \rightarrow (\exists x)P(x) \rightarrow Q(y)$

Solution. (a) The expression is a wff as it is a wff of propositional logic.
(b) The expression is a wff: $R(a, b)$ and $P(x)$ are wffs (Rule (1)). Since $P(x)$ is a wff, $(\exists x)\,P(x)$ is a wff by Rule (3.2). Rule (2.4) implies that $(R(a, b) \rightarrow (\exists x)\,P(x))$ is a wff.
(c) The expression is not a wff. It cannot be generated by the rules of Definition 4.7. Some parentheses are missing to make the expression a wff. For instance, both $(\forall x)\,((\exists y)P(y) \wedge R(y, x))$ and

$(\forall x)(\exists y)\,(P(y) \wedge R(y, x))$ are wffs obtained from the given expression by adding suitable parentheses.

(d) The expression is a wff. The details are left to the reader.

(e) The expression is not a wff as it cannot be generated by applying the rules of Definition 4.7. It can be made a wff by adding appropriate parentheses. For instance, $((\forall x)\,(\forall y)\,R(y, x) \to ((\exists x)P(x) \to Q(y)))$ and $(\forall x)\,((\forall y)\,R(y, x) \to ((\exists x)P(x) \to Q(y)))$ are wffs obtained from the original expression. $\diamond$

Definition 4.8. If ψ is a wff that occurs as part of another wff ϕ, we say that ψ is a *subformula* of ϕ.

Example 4.13. Consider the wff ϕ:

$$(\forall x)(\forall y)\,((\exists z)(P(x) \to Q(x, z)) \vee (\forall z)(\neg Q(y, z) \leftrightarrow P(y)))\,.$$

Then $(\exists z)(P(x) \to Q(x, z))$ and $(\forall z)(\neg Q(y, z) \leftrightarrow P(y))$ are two subformulas of ϕ.

4.3.4 *Exercises*

(1) Let the universe of discourse be the set of real numbers. Let P be the binary predicate: $P(x, y) :$ "$x \geq y$". In each case, determine the truth value of the given statement:

$\star$ (a) $(\exists x)(\forall y)P(x, y)$ (c) $(\forall x)(\forall y)P(x, y)$
   (b) $(\exists x)(\exists y)P(x, y)$ $\star$(d) $(\forall x)(\exists y)P(x, y)$

(2) Let the universe of discourse be the set $\mathcal{U} = \{-1, 0, 1, 2, 6\}$. Given the predicates $P(x)$: "$x > 0$", $Q(x)$: "$|x - 1| \geq 2$" and $R(x)$: "$x^2 - 3x + 2 < 0$", determine the truth value of each of the following quantified statements.

   (a) $(\exists x)\,(P(x) \wedge R(x))$ (d) $(\exists x)(\exists y)\,(Q(x) \vee R(y))$
$\star$ (b) $(\forall x)\,(Q(x) \to P(x))$ (e) $(\forall x)(\forall y)\,(Q(x) \vee R(y))$
   (c) $(\exists x)(\forall y)\,(Q(x) \to \neg R(y))$ $\star$ (f) $(\forall x)(\exists y)\,((\neg P(x) \vee Q(y)) \to R(x))$

(3) Let the universe of discourse be the set $\mathbb{N}$ of natural numbers. Consider the unary predicate $P(x)$: "x *is a prime number*" and the binary predicate $D(x, y)$: "x *is a divisor of* y". In each case, determine the truth value of the given quantified statement.

$\star$ (a) $(\exists x)D(1,x)$
   (b) $(\exists x)(P(x) \wedge D(x,14))$
   (c) $(\forall y)D(1,y)$
$\star$ (d) $(\forall y)(D(y,2) \leftrightarrow P(y))$

   (e) $(\exists x)(\exists y)D(x,y)$
$\star$ (f) $(\forall x)((x \neq 0) \rightarrow (\exists y)D(x,y))$
   (g) $(\exists x)(\forall y)D(x,y)$

(4) Using the notations of the previous exercise, determine the truth value of the quantified statement $(\forall z)(P(z) \rightarrow (\exists t)((t \neq z) \wedge (P(t) \wedge D(z,t))))$.

(5) Let the universe of discourse be the set of all natural numbers. Consider the unary predicates $P(m)$: "*m is odd*", $Q(m)$: "*m > 2*", $R(m)$: "*m is a prime number*" and the binary $L(m,n)$: "*m < n − 1*". In each case, determine the truth value of the given statement. Remember that 1 is not considered a prime number.

   (a) $(\exists m)\,L(1,m)$
   (b) $(\forall m)\,(\neg P(m) \vee Q(m))$
$\star$ (c) $(R(3) \rightarrow (\exists x)(\neg P(x) \wedge R(x)))$
   (d) $\left(P\left(2^5 + 1\right) \rightarrow (\forall m)(\exists n)L(m,n)\right)$
$\star$ (e) $\left(L\left(2^5 + 1, 3^3 + 6\right) \leftrightarrow (\exists m)(\neg Q(m) \wedge R(m))\right)$
   (f) $(\forall m)(\forall n)\left((R(m) \wedge L(n,m)) \rightarrow R(n)\right)$

(6) Let the universe of discourse be the set of all integers. In each case, determine the truth value of the quantified statement.

   (a) $(\exists m)(\exists n)\left(m^2 - n^2 = m + n\right)$
   (b) $(\forall m)(\forall n)\left((m^2 = mn) \rightarrow (m = n)\right)$
$\star$ (c) $(\forall m)(\exists n)\left(m^2 + n^2 = 3\right)$
   (d) $(\forall m)(\forall n)\left((m < n) \rightarrow (m^2 < n^2)\right)$

(7) Let the universe of discourse be the set of real numbers. In each case, determine the truth value of the quantified statement.

   (a) $(\forall x)(\forall y)(\forall z)\left((x + y) + z = x + (y + z)\right)$
$\star$ (b) $(\exists x)(\forall y)(\forall z)\left(xy + z = x + yz\right)$
   (c) $(\forall x)(\forall y)\,(\exists z)\left(x^2 z - xy^3 + 2z = 0\right)$
   (d) $(\forall x)(\forall y)(\forall z)\left((xy = xz) \rightarrow (y = z)\right)$

(8) If A is a subset of a universe $\mathcal{U}$, we sometimes need to attach a quantifier to only elements of A and not to the entire universe. Given a unary predicate P, the notations $(\forall_{x \in A})\,P(x)$ and $(\exists_{x \in A})\,P(x)$ are used to indicate the sentences :"*Every element of the universe which is in A satisfies P*" and "*There exists an element of the universe which is in A and which satisfies P*", respectively. In the following, $\mathbb{R}^+$ and $\mathbb{N}^+$ denote the set of all positive real numbers and the set of all positive integers, respectively. Determine the truth value of each of the following quantified statements over the universe $\mathcal{U} = \mathbb{R}$. No formal proof is required, just state what you think the truth value should be.

(a) $\left(\exists_{n\in\mathbb{N}+}\right)\left(\exists_{m\in\mathbb{N}+}\right)\left(n < \frac{1}{m}\right)$

$\star$ (b) $\left(\exists_{x\in\mathbb{R}+}\right)\left(\forall_{n\in\mathbb{N}+}\right)\left(x < \frac{1}{n}\right)$

(c) $\left(\forall_{x\in\mathbb{R}+}\right)\left(\exists_{n\in\mathbb{N}+}\right)\left(\frac{1}{n} < x\right)$

(d) $\left(\forall_{x\in\mathbb{R}+}\right)\left(\forall_{n\in\mathbb{N}+}\right)\left(x > \frac{1}{n}\right)$

$\star$ (e) $\left(\forall_{n\in\mathbb{N}+}\right)\left(\exists_{x\in\mathbb{R}+}\right)\left(x < \frac{1}{n}\right)$

(f) $\left(\exists x\right)\left(\exists_{n\in\mathbb{N}+}\right)\left(x^2 < \frac{1}{n} \leq \frac{1}{2}\right)$

(9) Let P and Q be two unary predicates on a universe $\mathcal{U}$. In each case, give the negation of the given statement. In your final answer, the negation symbol must be to the right of all quantifiers appearing in the statement.

(a) $(\forall x)\left((P(x) \to (\exists y)Q(y)) \to \neg P(x)\right)$
(b) $(\exists x)\left((\forall x)P(x) \vee (\exists x)\neg Q(x)\right)$
(c) $(\forall x)(\exists y)\left(P(y) \to (\exists x)\neg Q(x)\right)$
$\star$ (d) $(\exists x)(\exists y)(\forall z)\left((\forall u)(P(u) \vee Q(x)) \wedge (P(y) \to Q(z))\right)$

(10) Let P and Q be two binary predicates on a universe $\mathcal{U}$. In each case, give the negation of the given statement. In your final answer, the negation symbol must to the right of all quantifiers appearing in the statement.

(a) $(\forall x)(\exists y)\left(P(x, y) \to (\exists z)Q(z, x)\right)$
$\star$ (b) $(\exists x)(\forall y)\left((\forall t)P(x, t) \vee Q(y, t)\right)$
(c) $(\forall x)(\exists y)\left(P(x, y) \to \neg(Q(x, y) \to (\exists z)P(y, z))\right)$
(d) $(\forall x)(\forall y)(\exists z)\left(P(x, z) \vee (\exists t)(P(x, t) \to Q(y, t))\right)$

(11) Let the universe of discourse be $\mathcal{U} = \mathbb{N}$. In each case, find a wff equivalent to the given expression. Your final answer should not contain the negation sign in it.

(a) $\neg(\forall n)\left((\exists k)(n = 2k) \vee (\exists m)(n = 2m + 1)\right)$
$\star$ (b) $\neg(\exists n)(\forall m)\left((m < n) \to (\exists k)(n = m + k)\right)$
(c) $\neg(\forall n)(\forall m)(\exists k)\left(((k > 0) \wedge (km = kn)) \to (m = n)\right)$

(12) Let P and Q be two unary predicates on a universe $\mathcal{U}$.

(a) Prove that $(\forall x)\left(P(x) \vee Q(x)\right) \not\equiv \left((\forall x)P(x) \vee (\forall x)Q(x)\right)$ by providing a counter-example.
(b) Prove that $(\exists x)\left(P(x) \wedge Q(x)\right) \not\equiv \left((\exists x)P(x) \wedge (\exists x)Q(x)\right)$ by providing a counter-example different from the one you gave in the previous part.
(c) Give a proof of part (b) assuming part (a) is true.

(13) Let A and B be two unary predicates on a universe $\mathcal{U}$. In each case, determine if the given expression represents a well-formed formula of predicate logic.

(a) $A(x)$
(b) $(A(x))$
(c) $(\forall x)A(x) \to B(a)$

(d) $((\forall x)A(x) \to (\exists y)(A(a) \vee B(y)))$
(e) $((\forall x)(\forall y)(\exists z)A(x) \to B(y))$
(f) $(\forall y)(\exists z)A(x) \to B(y)$

⋆(14) Let A, B be two unary predicates, C and D two binary predicates on a universe $\mathcal{U}$. Consider the expression ϕ:

$$((\exists x)(\forall y)\,((\forall z)C(x,z) \to B(y)) \wedge (\exists z)(A(z) \leftrightarrow (\forall x)D(x,z)))\,.$$

  (a) Prove that ϕ is a wff of predicate logic.
  (b) Give four *non-atomic* subformulas of ϕ.

(15) (Calculus required) The notion of continuity of a real-valued function is a topic usually approached informally in the first calculus course. Unless you have taken some advanced Calculus or a real analysis class, you were probably told that a real-valued function f is *continuous* if its graph can be drawn without lifting your pen from the paper. Of course, this is far from being a valid mathematical definition. Formally, a real-valued function f with domain D is said to be continuous at a point $x = a \in D$ if the following condition holds: "Given any positive real number ϵ, there exists a positive real number δ such that for any element x in the domain of f, if $|x - a| < \delta$ then $|f(x) - f(a)| < \epsilon$". We say that the function is continuous on its domain if it is continuous at every point of its domain.

  (a) Express the fact that the function f is continuous at a point $x = a \in D$ of its domain using a wff of predicate logic.
  (b) Express the fact that the function f is continuous on its domain D using a wff of predicate logic.
  (c) Express the fact that the function f is not continuous at the point $x = a$ of its domain using a wff of predicate logic that does not contain the negation symbol.

4.4   Translation from and to predicate logic

There is certainly no shortage of rules and tips on proper translation from and to predicate logic in the literature as the topic is of great interest for mathematicians and linguists alike. Our intention is not to go in depth in the philosophy of translation, but rather explore a variety of scenarios through examples. First, few guidance tips.

(1) The following table gives the translations of some standard sentences that occur frequently in various forms. In the table, A, B are unary predicates over a certain universe $\mathcal{U}$. The sentence "*All A's are B's*" is a short for "*For any $x \in \mathcal{U}$, if $A(x)$ is true then $B(x)$ is also true*".

A similar interpretation is given for each of the other expressions in the first column of the table.

English	Predicate Logic
All A's are B's	$(\forall x)\,(A(x) \rightarrow B(x))$
Some A's are B's	$(\exists x)\,(A(x) \wedge B(x))$
No A's are B's	$(\forall x)\,(A(x) \rightarrow \neg B(x))$
Only A's are B's	$(\forall x)\,(B(x) \rightarrow A(x))$
Not all A's are B's	$\neg(\forall x)\,(A(x) \rightarrow B(x)) \equiv (\exists x)(A(x) \wedge \neg B(x))$

(2) The universal quantifier is typically (but not always) followed by the implication connective (as in the first entry in the above table).

(3) The existential quantifier is typically (but not always) followed by the conjunction connective. There are however instances where we combine the existential quantifier with an implication. A typical example is when $\exists$ is part of the hypothesis of the implication like in the sentence: *"If there exists a person who is expert in Mathematics and international diplomacy, then there is a chance to achieve world peace."*

(4) Negated statements are usually not easy to interpret and should be avoided when possible. The negation rules of quantifiers can be used for that purpose. For example, the table above gives the translation of *"Not all A's are B's"* in two equivalent ways: $\neg(\forall x)\,(A(x) \rightarrow B(x))$ and $(\exists x)(A(x) \wedge \neg B(x))$. Clearly, the second one is easier to interpret.

(5) A common mistake to avoid is to translate the sentence *"Some A's are B's"* as $(\exists x)\,(A(x) \rightarrow B(x))$.

Example 4.14. Consider the statement ϕ: *"All lions are carnivores"* over the universe of all animals. To translate the statement into a wff of predicate logic, we could simply use the first entry in the above table but we expand a bit more to explain the reasoning behind that entry. Consider the unary predicates L: *"is a lion"* and C: *"is a carnivore"*. For $x \in \mathcal{U}$, the implication $(L(x) \rightarrow C(x))$ translates in English as: *"If x is a lion, then x is carnivore"*. So in order for ϕ to be true, the implication $(L(x) \rightarrow C(x))$ must be satisfied by every element of the universe. This can be formulated as follows: *"For any $x \in \mathcal{U}$: if x is a lion then x must be carnivore"*. In predicate logic notations: $(\forall x)(L(x) \rightarrow C(x))$. Note that this is equivalent to the formula $(\forall x)(\neg C(x) \rightarrow \neg L(x))$. This last version says that if an animal is not a carnivore, then it is not a lion.

Example 4.15. Consider the English statement ψ: *"There is a Canadian hockey player who is admired by every other Canadian hockey player"*. Here

the universe is the set of all Canadian hockey players. Consider the unary predicate H: "*is a Canadian hockey player*" and the binary predicate A: "*is admired by*". The statement ψ can be interpreted as follows: "*There is a Canadian hockey player x such that if y is any other Canadian hockey player, then y admires x*". The last part of this sentence (after "such that") is the implication $((H(y) \wedge (y \neq x)) \rightarrow A(x,y))$. The expression $(y \neq x)$ is used to indicate that y is a Canadian hockey player "other than" x. So the complete translation of ψ is the following wff of predicate logic:
$$(\exists x)\,(H(x) \wedge (\forall y)((H(y) \wedge (y \neq x)) \rightarrow A(x,y)))\,.$$

Example 4.16. Let the universe of discourse be the set of all animals. Consider the unary predicates C: "*is a cat*", D: "*is a dog*", S: "*is a snake*", T: "*has a tail*", W: "*has wings*", the binary predicate B: "*bites*", the constants a and c representing *Adler* (a snake) and *Cane* (a dog) respectively. Translate each of the following sentences from English to wff of predicate logic using the given predicates and constants.

(a) All dogs and all cats have tails.
(b) No snake has a tail.
(c) If Adler bites Cane, then it bites any other dog.
(d) Either Cane has no wings or every other dog has no wings.
(e) Some dogs bite every cat.
(f) Only cats have tails and no snake bites a dog.

Solution. (a) $((\forall x)(D(x) \rightarrow T(x)) \wedge (\forall y)(C(y) \rightarrow T(y)))$.

(b) $\neg(\exists x)(S(x) \wedge T(x)) \equiv (\forall x)\neg(S(x) \wedge T(x)) \equiv (\forall x)(\neg S(x) \vee \neg T(x)) \equiv (\forall x)(S(x) \rightarrow \neg T(x))$.

(c) $(B(a,c) \rightarrow (\forall x)\,((D(x) \wedge (x \neq c)) \rightarrow B(a,x)))$.

(d) $(\neg W(c) \vee (\forall x)\,((D(x) \wedge (x \neq c)) \rightarrow \neg W(x)))$.

(e) $(\exists x)\,(D(x) \wedge (\forall y)\,(C(y) \rightarrow B(x,y)))$.

(f) $((\forall x)\,(T(x) \rightarrow C(x)) \wedge (\forall y)\,(S(y) \rightarrow (\forall z)\,(D(z) \rightarrow \neg B(y,z))))$. $\Diamond$

4.4.1 *Exercises*

(1) Let the universe of discourse be the set of all human beings ever lived. Use the predicate symbol S: "*is the son of*" and the constants j: "*Joe*", m: "*Mich*" to translate in English each of the following wffs of predicate logic.

$\star$ (a) $(\exists x)S(x,j)$

(b) $(\forall x)(\exists y)S(y,x)$

(c) $(\exists x)(\forall y)S(y,x)$

$\star$ (d) $(S(m,j) \to (\exists x)S(x,j))$

(e) $(S(m,j) \to (\exists x)(\exists y)S(y,x))$

(2) Let the universe of discourse be the set of all members of a certain family. Use the predicate symbol $R(x,y)$: "*x respects y*" and the constants j: "Joe", c: "Cathy" to translate each of the following English statements to a wff formula of predicate logic.

(a) Joe respects someone.

$\star$ (b) If Joe respects Cathy, then Joe respects everyone.

(c) There is someone who respects someone else.

$\star$ (d) Everyone respects someone.

(e) No one respects Cathy unless she is respected by Joe.

(3) Let the universe of discourse be $\mathbb{R}$. Use the predicates: $I(x)$: "*x is an integer*", $R(x)$: "*x is a rational number*", $C(x,y)$: "$x = y^3$", $G(x,y)$: "$x > y$" to translate each of the given English sentence into a wff of predicate logic.

$\star$ (a) There exists a rational number which is not an integer.

(b) Every integer is a rational number.

(c) Every real number is either a rational number or the cube of a rational number.

(d) The cube of any integer is an integer.

$\star$ (e) The cube of some rational number is less than to -2.

(f) No real number is greater than the cube of an integer.

(g) Some rational number is between any two distinct integers.

$\star$ (h) No cube of an integer is greater than the cube of a rational number.

(4) Let the Universe of discourse be the set of all animals. Consider the predicates: $B(x)$: "*x is a bear*", $D(x)$: "*x is a dear*", $L(x)$: "*x is a lion*", $H(x,y)$: "*x can hunt y*", $R(x,y)$: "*x outruns y*", and the two constants b: "Yogi" (a bear), s: "Simba" (a lion). In each case, translate the given wff of predicate logic into English using the above notations.

$\star$ (a) $\neg(\exists x)\,(D(x) \wedge H(x,s))$

(b) $\neg(\forall x)\,(D(x) \to R(x,s))$

(c) $(\forall x)\,((B(x) \wedge H(x,s)) \to (\forall y)\,(L(y) \to H(x,y)))$

$\star$ (d) $(\forall x)\,((B(x) \wedge (\forall z)(L(z) \wedge H(x,z))) \to (x = b))$

(e) $(\forall x)\,(L(x) \to (\forall y)\,(D(y) \to R(x,y)))$

$\star$ (f) $(\exists x)(\forall y)\,((L(y) \wedge R(y,x)) \to H(y,x))$

(g) $(\exists x)\,(L(x) \wedge (\forall y)\,(\neg H(y,x) \wedge \neg(\exists z)(B(z) \wedge R(z,x))))$

$\star$ (h) $(\exists x)\,(B(x) \wedge (x \neq b) \wedge (\forall t)(R(x,t) \to H(x,t)))$

(i) $(\forall x)\,(L(x) \to (\forall y)\,(R(x,y) \vee H(x,y)))$

(5) In the universe $\mathbb{N}$, translate each of the following statements into a wff of predicate logic. Clearly specify the predicate symbols used.

  $\star$ (a) Only odd numbers are prime.

     (b) There are odd numbers which are not prime.

     (c) Every natural number is either even, odd or prime.

  $\star$ (d) All prime numbers are odd except for 2.

     (e) If all prime numbers are odd, then there exists an odd number divisible by 2.

     (f) Some numbers are neither prime nor even.

  $\star$ (g) The only divisors of a prime numbers are 1 and itself.

For the questions (6)-(10) below, use the following predicate symbols: $S(x)$: "x is a student", $B(x)$: "x is a book", $R(x,y)$: "x has read y" and the two constants: j : "Joe" (a student) and l: "No logic" (a book).

(6) Determine which of the following wffs of predicate logic is the correct translation of the English sentence: *"Not all students have read the book No logic"*.

  (a) $(\exists x)\,(S(x) \to \neg R(x,l))$

  (b) $(\exists x)\,(S(x) \wedge \neg R(x,l))$

  (c) $(\exists x)\,(S(x) \to (\exists y)(B(y) \wedge \neg R(x,y)))$

  (d) $\neg(\forall y)\,(\neg S(y) \vee \neg R(y,l))$

(7) Determine which of the following wffs of predicate logic is the correct translation of the English sentence: *"If Joe has read No logic, then at least one student has not read it"*.

  (a) $(R(j,l) \to (\exists x)(S(x) \wedge \neg R(x,l)))$

  (b) $(R(j,l) \to (\forall x)(S(x) \to \neg R(x,l)))$

  (c) $(R(j,l) \to \neg(\exists x)(S(x) \wedge \neg R(x,l)))$

  (d) $(R(j,l) \to \neg(\forall x)(S(x) \vee R(x,l)))$

$\star$(8) Determine which of the following wffs of predicate logic is the correct translation of the English sentence: *"There is at least one student who has not read any book"*.

  (a) $(\exists x)\,(S(x) \to (\forall y)(B(y) \wedge R(x,y)))$

  (b) $(\exists x)\,(S(x) \wedge (\forall y)(B(y) \to \neg R(x,y)))$

  (c) $(\exists x)\,(S(x) \to (\exists y)(B(x) \wedge \neg R(x,l)))$

  (d) $\neg(\forall x)\,(S(x) \to (\forall y)\,(B(y) \to \neg R(x,y)))$

$\star$(9) Which of the following sentences is the correct translation of the wff of predicate logic: $(\exists x)\,(B(x) \wedge (x \neq l) \wedge (\forall y)(S(y) \to R(y,x)))$?

(a) No book other than *No logic* has been read by any student.
(b) Any book other than *No logic* has been read by at least one student.
(c) There is a student who has read every book other than *No logic*.
(d) There is a book other than *No logic* that has been read by every student.

(10) Which of the following sentences is the correct translation of the wff of predicate logic: $(R(j,l) \rightarrow (\forall y)\,((B(y) \wedge (y \neq l)) \rightarrow R(j,y)))$?

(a) If Joe has read *No logic*, then so has every other student.
(b) Joe has read *No logic* only if he has read every other book.
(c) Joe has read all books only if he has read *No logic*.
(d) Neither Joe nor any other student has read a book other than *No logic*.

$\star$(11) An $m \times n$ matrix is a rectangular arrangement of real numbers in m rows and n columns. The entry on the ith row and the jth column of a matrix A is denoted by a_{ij} and the matrix

$$A = \begin{bmatrix} a_{11} & a_{12} & \cdots & a_{1n} \\ \vdots & \vdots & \cdots & \vdots \\ a_{m1} & a_{m2} & \cdots & a_{mn} \end{bmatrix}$$

is usually denoted by $A = [a_{ij}]$. Which of the following six 3×3 matrices satisfy the following property:

$$(\forall i)\,(\exists j)\,\big((a_{ij} \leq -5) \vee (\exists_{n \in \mathbb{N}})(a_{ij} = n^2)\big)\,.$$

$$A = \begin{bmatrix} 1 & 4 & -7 \\ -1 & 2 & 3 \\ 2 & 0 & 3 \end{bmatrix}, \quad B = \begin{bmatrix} 0 & 1 & -1 \\ 2 & 6 & 8 \\ 2 & 1 & 0 \end{bmatrix}, \quad C = \begin{bmatrix} 1 & 1 & -10 \\ 0 & 16 & 0 \\ 49 & -12 & 36 \end{bmatrix}$$

$$D = \begin{bmatrix} -1 & 2 & 3 \\ 5 & -2 & 7 \\ 0 & 0 & 1 \end{bmatrix}, \quad E = \begin{bmatrix} 1 & 0 & -6 \\ -7 & -4 & 5 \\ 0 & 0 & 9 \end{bmatrix}, \quad F = \begin{bmatrix} -1 & 5 & -6 \\ 0 & 4 & 2 \\ -1 & -9 & 0 \end{bmatrix}$$

(12) With the notations of Exercise (11), which of the following six matrices satisfy the following property:

$$(\exists i)(\forall j)\,((i \neq j) \rightarrow (a_{ij} + a_{ji} > 0))\,.$$

$$A = \begin{bmatrix} 1 & 4 & -7 \\ -1 & 2 & -1 \\ 2 & 0 & 3 \end{bmatrix}, \quad B = \begin{bmatrix} 0 & 1 & -1 \\ 2 & 0 & 1 \\ 2 & 1 & 0 \end{bmatrix}, \quad C = \begin{bmatrix} 1 & 1 & -10 \\ 0 & 16 & 0 \\ 9 & -12 & 36 \end{bmatrix}$$

$$D = \begin{bmatrix} 1 & 0 & 0 \\ 0 & 1 & 0 \\ 0 & 0 & 1 \end{bmatrix}, \quad E = \begin{bmatrix} 1 & 0 & -6 \\ -7 & 4 & 25 \\ 0 & 0 & 9 \end{bmatrix}, \quad F = \begin{bmatrix} -1 & 0 & 7 \\ 0 & 4 & 2 \\ -1 & 0 & 0 \end{bmatrix}$$

(13) Consider the following rectangular array of symbols:

$$\begin{array}{ccccc} \bigstar & \star & \circ & \times & \bullet \\ \checkmark & \div & \spadesuit & \square & \blacksquare \\ * & \bigcirc & \spadesuit & \blacktriangle & \heartsuit \\ \diamond & \infty & \$ & \clubsuit & \blacklozenge \end{array}$$

Shapes in the above array could come in different forms and sizes. For instance, a circle can be full, empty, large or small. Use the notations: $C(x)$: "x is a circle", $S(x)$: "x is a square", $T(x)$: "x is a triangle", $D(x)$: "x is a symbol of a deck suit" (in a standard deck, there are four suits: clubs ($\clubsuit$), diamonds ($\diamond$), hearts ($\heartsuit$) and spades ($\spadesuit$)), $E(x)$: "x is a star", $F(x,y)$: "x and y are adjacent in the same column", $G(x,y)$: "x and y are adjacent in the same row" to translate each of the following formulas in English. In each case, determine the truth value of the sentence.

$\star$ (a) $(\exists x)(\exists y)\,(E(x) \wedge E(y) \wedge G(x,y))$

   (b) $(\forall x)\,(C(x) \to (\exists y)(D(y) \wedge F(y,x)))$

   (c) $(\forall x)(\exists y)\,((x \neq y) \wedge (G(y,x) \vee F(x,y)))$

$\star$ (d) $(\forall x)\,(T(x) \to \neg(\exists z)(S(z) \wedge F(z,x)))$

   (e) $(\forall x)(\forall y)\,((F(x,y) \wedge (x = y)) \to (D(x) \vee S(y)))$

4.5 Scope of a quantifier, bound and free variables

A wff of predicate logic can get complicated very quickly with the sequence of symbols and punctuation. One reason why we impose strict rules on writing wffs is to make sure there is no ambiguity on what the *scope* of each quantifier is within the formula and whether an occurrence of a variable is *free* or *bound*. Understanding these terms is essential in interpreting and manipulating wffs.

Definition 4.9. Let ϕ be a wff of predicate logic containing the expression (Qx) where Q is a quantifier ($\forall$ or $\exists$) and x is a variable. The *scope* of

(Qx) in ϕ is the expression $(Qx)\psi$ where ψ is the *smallest* subformula of ϕ occurring immediately to the right of (Qx). Here the term "smallest subformula" refers to the wff with the smallest number of symbols. An occurrence of x in ϕ is called *bound* by the quantifier Q if it appears in the scope of (Qx). We say that x is a *bound variable* in ϕ if at least one occurrence of x is bound by a certain quantifier in ϕ. If an occurrence of x in ϕ is not bound, we say that it is *free*. We also say in this case that x is a *free variable*. A wff with no free occurrence of any of its variables is called a *closed formula* or a *sentence*. A wff which is not closed is called *open*. Every wff of predicate logic is either closed (a sentence) or open.

Remark 4.2. Closed wffs are statements with well determined truth values: they are either true or false. Open wffs, on the other hand, can be true for some variables and false for others in the universe of discourse.

Example 4.17. Consider the following wffs:

(a) $\phi_1 : ((\exists x)A(x) \to B(y))$.
(b) $\phi_2 : (\exists x)\,((\forall x)A(x) \to (\forall y)(\exists z)(B(z) \to (C(y) \vee D(z))))$.

In ϕ_1, the scope of the $(\exists x)$ is the subformula $(\exists x)A(x)$. The two occurrences of the variable x in ϕ_1 are bound and the only occurrence of the variable y is free. The formula ϕ_1 is open since it contains a free variable. Formula ϕ_2 is a bit more complex. The following table displays the quantifiers and their scopes in ϕ_2.

Quantifier	Scope
$(\exists x)$	$(\exists x)\,((\forall x)A(x) \to (\forall y)(\exists z)(B(z) \to (C(y) \vee D(z))))$
$(\forall x)$	$(\forall x)A(x)$
$(\forall y)$	$(\forall y)(\exists z)(B(z) \to (C(y) \vee D(z)))$
$(\exists z)$	$(\exists z)(B(z) \to (C(y) \vee D(z)))$

Since all occurrences of the variables in ϕ_2 are bound, ϕ_2 is closed (a sentence).

Remark 4.3. An occurrence of a variable x in a wff can be within the scope of more than one quantifier. For example, in the wff

$$(\exists x)\,(A(x) \to (\forall x)B(x))$$

the occurrence of the variable x in $B(x)$ is in the scopes of $(\exists x)$ and $(\forall x)$ at the same time. However, the predicate occurrence of x in $B(x)$ is bound *only* by the quantifier closest to x on the left, namely $(\forall x)$. In this case, we call this quantifier the *binder* of the variable.

Example 4.18. Consider the following formulas of predicate logic:

(a) $\psi_1 : ((\exists x)A(x) \to (B(x) \vee C(y)))$
(b) $\psi_2 : (\exists x)(A(x) \to (\forall x)(B(x) \vee (\forall y)C(y)))$
(c) $\psi_3 : (\forall x)(A(x, y) \vee (\exists y)\neg B(y, x))$

In ψ_1, the first two occurrences of the variable x are bound by $(\exists x)$, the third occurrence of x and the occurrence of the variable y are free. The formula ψ_1 is open since it has free variables. In ψ_2, the first two occurrences of x are bound by $(\exists x)$ while the third and fourth occurrences of x are bound by $(\forall x)$. The two occurrences of the variable y in ψ_2 are bound by $(\forall y)$. The formula ψ_2 is closed since no occurrence of any of its variables is free. In ψ_3, all three occurrences of the variable x are bound by $(\forall x)$. The first occurrence of the variable y is free while the second and third are bound by $(\exists y)$. The formula ψ_3 is open as it contains a free variable.

Remark 4.4. A variable can be free and bound at the same time in a wff as it is the case for the variable y in formula ψ_3 of Example 4.18.

Often, one has to change variable names in a wff to eliminate potential confusion or to determine logical equivalences. If done arbitrarily, this could change the meaning of the formula. For example, consider the wff $(\forall x)(\exists y)A(y, x)$. The universal quantifier applies to the second place in $A(y, x)$ while the existential quantifier applies to the first place. If we interchange the roles of the variables in $(\forall x)(\exists y)$, we obtain the wff $(\forall y)(\exists x)A(y, x)$. In spite of the apparent similarity, the two wffs are very different. To see this, let us interpret both formulas in the universe of all people with $A(z, t)$ representing "*z admires t*". The formula $(\forall x)(\exists y)A(y, x)$ is translated as "*Everyone is admired by someone*" while the formula $(\forall y)(\exists x)A(y, x)$ reads "*Everyone admires someone*". The meanings of theses two sentences are completely different.

Example 4.19. Consider the wffs

- $\epsilon_1 : ((\forall x)A(x) \vee (\forall x)B(x))$
- $\epsilon_2 : (\forall x)(A(x) \vee B(x))$.

In ϵ_1, the first two occurrences of the variable x are bound by the first universal quantifier while the third and fourth occurrences are bound by the second universal quantifier. However, the occurrences of the variable x in the subformulas $(\forall x)A(x)$ and $(\forall x)B(x)$ are completely independent of each other. If we change the name of the variable x in one of the subformulas, we do not change the meaning of the original formula.

In other words, the wffs $((\forall x)A(x) \vee (\forall y)B(y))$, $((\forall y)A(y) \vee (\forall x)B(x))$, $((\forall z)A(z) \vee (\forall w)B(w))$ and ϵ_1 are all equivalent. Variables in ϵ_2 behave a bit differently. All occurrences of the variable x in ϵ_2 are bound by the universal quantifier, and these occurrences are not independent of each other. If we change the name of the variable x in one of these occurrences, we get a different formula. For instance, the formulas $(\forall x)\,(A(x) \vee B(x))$ and $(\forall x)\,(A(x) \vee B(y))$ are not equivalent. Note that $(\forall x)\,(A(x) \vee B(x))$ is a closed formula while $(\forall x)\,(A(x) \vee B(y))$ is open since the occurrence of the variable y in it is free.

When we change the name of a bound variable x by a quantifier (Qx) to a different name y in a wff, we must:

- make sure that we change all bound occurrences of x to y within the scope of (Qx);
- make sure *no* free variable in ϕ can become bound after renaming the variable x.

Example 4.20. Let P and Q be two binary predicates and consider the wff $\phi:\ ((\forall x)(\exists y)P(x,y) \rightarrow (\exists x)(\forall y)Q(x,y))$. The formula has occurrences of two different variables x and y. But a closer look shows that ϕ has in fact four independent variables that happen to use only two different names. The use of the same name for independent variables in a formula is a bit misleading. It makes sense to change the names of some of these variables to avoid this confusion. Note first that ϕ has the form $(p \rightarrow q)$ with $p:\ (\forall x)\,(\exists y)\,P(x,y)$ and $q:\ (\exists x)\,(\forall y)\,Q(x,y)$. Both p and q are wffs and none of the quantifiers in p binds any of the variables in q. It is this property that makes variables in p and q independent of each others. In the operand q, the two occurrences of x are bound by $(\exists x)$ and the two occurrences of y are bound by $(\forall y)$. Changing the names of the (bound) occurrences of x and y in q to u and v respectively gives the wff $(\exists u)\,(\forall v)\,Q(u,v)$ which is equivalent to q. Hence ϕ is equivalent to the formula $((\forall x)\,(\exists y)\,P(x,y) \rightarrow (\exists u)(\forall v)Q(u,v))$.

If ϕ is a wff of predict logic containing a variable x and if t is a term, the notation x/t stands for the *substitution* of every *free* occurrence of the variable x in ϕ by the term t. The resulting wff obtained by applying the substitution x/t to ϕ is denoted by $\phi(x/t)$. If there are no free occurrences of x in ϕ, then $\phi(x/t) = \phi$.

Example 4.21. Let A, B be two binary predicates, x, y two variables and t a term in the universe of discourse. Consider the two wffs $\phi_1:(\exists x)A(x,y)$ and $\phi_2:(A(x,y) \wedge (\forall y)B(x,y))$. Then $\phi_1(x/t) = \phi_1$ (since there are no free occurrences of x in ϕ_1), $\phi_1(y/t) = (\exists x)A(x,t)$, $\phi_2(x/t) = (A(t,y) \wedge (\forall y)B(t,y))$ and $\phi_2(y/t) = (A(x,t) \wedge (\forall y)B(x,y))$.

4.5.1   *Exercises*

(1) Let A, B be two unary predicates over a universe $\mathcal{U}$. In each case, give the scope of each occurrence of a quantifier in the given wff. For each occurrence of a variable in the formula, determine if it is free or bound and determine if the formula is closed (statement) or open.

   (a) $(\exists y)(\forall x)(A(x) \wedge B(y))$
   (b) $(\exists x)(A(x) \wedge (\forall x)B(x))$
   (c) $(\forall x)(A(x) \rightarrow (\exists y)B(y))$
   (d) $((\forall x)A(x) \rightarrow B(x))$
 $\star$ (e) $(\exists x)(\forall y)(A(x) \vee (\exists z)(A(y) \vee B(z)))$
   (f) $((\forall x)A(x) \leftrightarrow (\exists y)(A(a) \vee B(y)))$
 $\star$ (g) $((\forall x)(\forall y)(\neg B(x) \wedge A(y)) \rightarrow ((\exists x)A(x) \rightarrow \neg B(y)))$

(2) Let A, B be two binary predicates over a universe $\mathcal{U}$ and let a be a constant in $\mathcal{U}$. In each case, give the scope of each occurrence of a quantifier in the given wff. For each occurrence of a variable in the formula, determine if it is free or bound and determine if the formula is closed (statement) or open.

   (a) $(\forall x)(\forall y)(\forall z)\,((A(x,y) \wedge B(y,z)) \rightarrow (\exists w)B(w,z))$
   (b) $(\forall x)\,(A(x,y) \rightarrow (\forall z)\,((\exists x)B(z,x) \wedge A(x,z)))$
   (c) $((A(x,y) \vee (\forall z)\,(B(a,x) \rightarrow A(a,z))) \wedge (\exists x)B(x,a))$.
   (d) $((\forall x)A(a,x) \rightarrow ((\forall y)B(y,y) \wedge (\forall z)\,(A(a,a) \rightarrow B(z,y))))$

(3) Give a wff ϕ of predicate logic containing three variables x, y and z each of which appears four times in ϕ such that only one occurrence of x is free and only two occurrences of each of y and z are free. The wff must also contain three unary predicate symbols A, B and C and one binary predicate symbol D.

(4) In each case, a wff ϕ of predicate logic over a universe $\mathcal{U}$ is given. Given a term t and a constant a of $\mathcal{U}$, write each of the expressions: (i) $\phi(x/t)$; (ii) $\phi(y/t)$; (iii) $\phi(z/t)$.

   (a) $(P(a) \rightarrow Q(a,a))$
 $\star$ (b) $(P(a) \rightarrow (\forall x)(Q(x,z) \rightarrow Q(y,z)))$
   (c) $(P(a) \rightarrow (\exists x)(\forall y)(R(x,y,z)))$
 $\star$ (d) $((\forall x)(A(x) \rightarrow B(x,y)) \rightarrow ((\exists x)A(x) \rightarrow B(x,z)))$
   (e) $(\exists z)(P(x,z) \vee ((\forall y)Q(x,y) \rightarrow (\exists x)R(x,y,z)))$

4.6 Interpretation

For a wff ϕ of predicate logic to have a meaning, a universe of discourse must be specified and an interpretation of each of the symbols in ϕ in the universe must be given. Changing the universe or any of the interpretations of the symbols can completely change the meaning of the wff. Let us consider, for instance, the wff of predicate logic $\phi : (\forall x)(\exists y)P(x, y)$. If the universe of discourse is $\mathbb{R}$ and $P(x, y)$ stands for: "$x^2 + y^2 = 1$", then ϕ is false since it states that "*for every real number x, there exists a real number y such that $x^2 + y^2 = 1$.*" This is false since no such y exists for $x = 2$. If the universe of discourse is the set of all human beings and $P(x, y)$ is the binary predicate "*y is a parent of x*", then ϕ is true since it states that "*Everyone has a parent*".

Definition 4.10. Let ϕ be a wff of predicate logic. An *interpretation* of ϕ consists of the following components:

(1) A non-empty universe of discourse $\mathcal{U}$ from which values for the variables and constants are drawn.
(2) For every predicate symbol in ϕ, a relation over $\mathcal{U}$ is assigned.
(3) For every n-ary function symbol f in ϕ, a rule of correspondence that associates to every n-tuple $(x_1, \ldots, x_n)$ of $\mathcal{U}^n$ a unique element $f(x_1, \ldots, x_n)$ of $\mathcal{U}$ (i.e., a function from $\mathcal{U}^n$ to $\mathcal{U}$) is assigned.
(4) Every occurrence of a free variable x in ϕ is assigned a unique constant $x_{\mathcal{U}}$ in $\mathcal{U}$.
(5) Every occurrence of a constant c in ϕ is assigned a designated element $c_{\mathcal{U}}$ in $\mathcal{U}$.

It should be noted here that occurrences of the same free variable (or same constant) in a wff are assigned the same element of $\mathcal{U}$ by an interpretation.

Example 4.22. Consider the wff $\phi : (\forall x)(\forall y)\left(P(x, y) \to Q(f(x, y), y)\right)$. We consider two interpretations I_1 and I_2 of ϕ. For I_1, the universe of discourse is $\mathbb{R}$, $P(x, y) : $ "$x + y = 1$", $Q(x, y) : $ "$2x + 3y = -2$" and $f(x, y) = xy$. In this interpretation, ϕ reads: "*For all real numbers x and y, if $x + y = 1$ then $2xy + 3y = -2$.*" This is clearly false. For I_2, we use the same notations as in I_1 but with the set $\mathcal{U} = \{1, 2\}$ as the universe of discourse. Then ϕ is true in this case since $P(x, y)$ is false for any $x, y \in \mathcal{U}$ (no sum of two elements of $\mathcal{U}$ is equal to 1).

Example 4.23. Consider the wff $\phi : (\exists x)\,(P(x) \wedge (\forall y)(Q(x,y) \to C(y)))$. Let I_1 be the interpretation of ϕ with universe $\mathbb{Z}$, $P(x)$: "$x > 0$", $Q(x,y)$: "$x > y$" and $C(x)$: "$x \leq 3$". Then ϕ states the following: "*There exists a positive integer x such that if y is any integer strictly smaller than x then y is less than equal to 3*". This statement is true. For example, $x = 1$ satisfies ϕ. Now consider a second interpretation I_2 with universe being the set of all *living* humans (at a certain moment of time), $P(x)$: "*x is a parent*", $Q(x,y)$: "*y is a child of x*" and $C(x)$: "*x is an orphan*". Then with respect to I_2, ϕ is the following statement: "*There exists a living person who is a parent and any child of whose is an orphan.*" This is clearly false.

Definition 4.11. Let ϕ be a wff of predicate logic. An interpretation with respect to which ϕ is true (respectively false) is called a *model* (respectively, a *countermodel*) of ϕ. The wff ϕ is called *valid* if every interpretation of ϕ is a model. Otherwise, ϕ is called *invalid*. We say that ϕ is *satisfiable* if there exits at least one model of ϕ. Otherwise, ϕ is called *unsatisfiable*.

From the definition, it is clear that every valid wff is in particular satisfiable and every unsatisfiable wff is also invalid.

The notion of a valid wff of predicate logic corresponds to that of a tautology in propositional logic. But unlike the method of the truth table of proportional logic, there is no straightforward method to verify if a wff of predicate logic is valid. in theory, one has to prove that the wff is true with respect to any of its interpretations. Of course, this is not practical as it is impossible to consider all possible interpretations and verify the truth value of the wff for each of them. Validity in predicate logic is a hard problem in general, but there are some situations where the problem can be dealt with reasonably easily. One of these situations is the case of an implication. To prove the validly of a wff of the form $(\phi_1 \to \phi_2)$ (where ϕ_1 and ϕ_2 are wffs of predicate logic), we consider an arbitrary interpretation I which is a model for ϕ_1 and we prove that I is also a model for ϕ_2.

Example 4.24. The wff $\phi : ((\forall x)P(x) \to (\exists x)P(x))$ is valid. To see this, consider an arbitrary interpretation I of ϕ with universe $\mathcal{U}$ such that I is a model of $(\forall x)P(x)$. Then for any element $a \in \mathcal{U}$, $P(a)$ is true. This implies that there exists $a \in \mathcal{U}$ such that $P(a)$ is true and I is also a model for $(\exists x)P(x)$. We conclude that ϕ is valid.

The proof of the validity of the wff of Example 4.24 was done using a *direct approach*: assume that I is a model for the hypothesis of the implication and then prove that I must be a model for the conclusion. An *indirect approach* to prove the validity is to proceed by contradiction: assume that the wff is invalid and then prove that this leads to a contradiction. We illustrate this approach in the following example.

Example 4.25. Let $\phi : ((\forall x)(P(x) \to Q(x)) \to ((\forall x)P(x) \to (\forall x)Q(x)))$. In this example we use an indirect approach to prove the validity of the wff. Assume (using contradiction) that ϕ is not valid and let I be a countermodel of ϕ. Then $(\forall x)(P(x) \to Q(x))$ is true and $((\forall x)P(x) \to (\forall x)Q(x))$ is false with respect to I. The latter assumption implies that $(\forall x)P(x)$ is true and $(\forall x)Q(x)$ is false with respect to I. Then there exists an element a in the universe such that $Q(a)$ is false but $P(a)$ is true. This means in particular that the implication $(P(a) \to Q(a))$ is false and therefore $(\forall x)(P(x) \to Q(x))$ is false. This is a contradiction. We conclude that ϕ has no countermodel and it is in fact valid.

The results of the previous two examples are part of a set of basic valid wffs of predicate logic that we give in the following theorem.

Theorem 4.2. *The following are valid wffs of predicate logic.*

(a) $((\forall x)P(x) \to (\exists x)P(x))$
(b) $((\forall x)(P(x) \to Q(x)) \to ((\forall x)P(x) \to (\forall x)Q(x)))$
(c) $(((\forall x)P(x) \vee (\forall x)Q(x)) \to (\forall x)(P(x) \vee Q(x)))$
(d) $((\exists x)(P(x) \wedge Q(x)) \to ((\exists x)P(x) \wedge (\exists x)Q(x)))$
(e) $((\exists y)(\forall x)P(x, y) \to (\forall x)(\exists y)P(x, y))$

The proof of parts (a) and (b) were given in the previous two examples. The proofs of the other parts are left as exercises (see Exercise (4) below).

4.6.1 *Exercises*

(1) Consider the wff $\phi : (\forall x)\,((\exists y)P(x, y) \to Q(f(x, z), c))$. Find the truth value of ϕ with respect to each of the following interpretations.

$\star$ (a) The universe is $\mathcal{U} = \mathbb{R}$, $P(x, y)$: "$x + y = 1$", $f(x, z) = xz$, $Q(x, y)$: "$x^2 + y = 3$", $z = 2$ and $c = -1$.
  (b) The universe is $\mathcal{U} = \{1, 3\}$ and same other notations as part (a).

(2) Consider the wff $\phi : (\exists x)(\forall y)\,((P(x, y) \wedge E(x, y)) \to Q(y, c))$. Find the truth value of ϕ with respect to each of the following interpretations.

  (a) The universe is the set of all prime numbers $(2, 3, 5, 7, \ldots)$, $P(x, y)$: "x is a divisor of y", $E(x, y)$: "$x \neq y$", $Q(x, y)$: "$x \leq y$", $c = 1$.

★ (b) The universe is the set of all countries of the world, $P(x,y)$: "x and y are in the same continent", $E(x,y)$: "x and y share a land border", $Q(x,y)$: "x has a smaller area than y", c stands for Canada.

(3) Let ϕ be a wff of predicate logic. Determine the truth value of each of the following statements.

★ (a) If ϕ is satisfiable, then it is valid.
 (b) If ϕ is valid, then it is satisfiable.
 (c) If ϕ is unsatisfiable, then it is invalid.
 (d) If ϕ is invalid, then it is unsatisfiable.
★ (e) If ϕ is unsatisfiable, then every interpretation of ϕ is a countermodel.
 (f) If ϕ is satisfiable, then every interpretation of ϕ is a model.

(4) In each case, prove that the wff is valid.

 (a) $(((\forall x)P(x) \vee (\forall x)Q(x)) \to (\forall x)(P(x) \vee Q(x)))$
★ (b) $((\exists x)(P(x) \wedge Q(x)) \to ((\exists x)P(x) \wedge (\exists x)Q(x)))$
 (c) $((\exists y)(\forall x)P(x,y) \to (\forall x)(\exists y)(P(x,y))$

(5) In each case, prove that the wff is satisfiable by giving a model.

 (a) $(\forall x)(P(x) \to Q(x))$
 (b) $(P(a) \wedge (\forall x)\neg Q(x))$
 (c) $(\forall x)(P(x) \wedge (\forall y)((P(y) \wedge (x \neq y)) \to \neg D(x,y)))$
★ (d) $(\forall x)(\neg P(f(x,x),y) \to Q(x,y))$

(6) In each case, prove that the wff is invalid but satisfiable by providing a model and countermodel.

★ (a) $(\forall x)\,(M(x) \to (\exists y)P(x,y))$.
 (b) $(((\forall x)A(x) \to (\forall x)B(x)) \to (\forall x)(A(x) \to B(x)))$.
 (c) $(\exists x)(P(x) \wedge (\forall y)(Q(y) \to R(x,y)))$
 (d) $(\exists x)\,(P(x) \wedge (\forall y)\,(D(y,x) \to E(x,y)))$.

(7) In each case, prove that the wff is unsatisfiable. Proceed by contradiction by assuming that the wff has a model and get a contradiction.

 (a) $(\exists x)(P(x) \wedge \neg P(x))$
 (b) $(\exists x)(\forall y)(P(x,y) \wedge \neg P(x,y))$.

(8) Consider the wff $\phi : (\forall x)(\forall y)\,((P(x) \wedge Q(x)) \to P(y))$. Prove that ϕ is invalid. Prove that ϕ is true with respect to any interpretation with a universe consisting of a single element.

(9) Let ϕ be a wff and suppose that $x_1, \ldots, x_n$ are all the free variables in ϕ. The wff $(\forall x_1)(\forall x_2) \cdots (\forall x_n)\phi$ is called a *universal closure* of ϕ. The

wff $(\exists x_1)(\exists x_2)\cdots(\exists x_n)\phi$ is called an *existential closure* of ϕ. In each case, give a universal and an existential closure of the given wff.

(a) $(\exists x)(\forall y)P(x,y)$
(b) $(\forall x)(\forall y)P(x,z)$
(c) $(\exists x)(P(x,y) \rightarrow (\forall z)(P(z,y) \wedge P(u,v)))$

(10) Let ϕ be a wff containing one free variable x. Prove that:

(a) ϕ is valid if and only if the universal closure of ϕ is valid;
(b) ϕ is valid if and only if the existential closure of ϕ is valid.

4.7 Reasoning with quantifiers: Arguments of predicate logic

There are four principles governing the reasoning process in predicate logic: the Universal Instantiation (UI), the Universal Generalization (UG), the Existential Instantiation (EI) and the Existential Generalization (EG). Instantiation principles allow us to remove quantifiers while generalizing ones allow us to introduce them.

4.7.1 *The universal instantiation (UI)*

The *universal instantiation principle* (UI) states that if the unary predicate P is true for any element of the universe $\mathcal{U}$, then it is true for a specific element a of $\mathcal{U}$: *From* $(\forall x)P(x)$, *we can infer* $P(a)$ *for any* a *in* $\mathcal{U}$.

Example 4.26. Recall that the Archimedean property of $\mathbb{R}$ states that for any real number x and any positive real number y, there exists an integer n such that $ny > x$. Using symbols of predict logic, the Archimedean property can be expressed as follows: $(\forall x)(\forall_{y>0})(\exists_{n\in\mathbb{Z}})(ny > x)$. Using the universal instantiation principle, the statements $(\forall_{y>0})(\exists_{n\in\mathbb{Z}})(ny > \pi)$ and $(\exists_{n\in\mathbb{Z}})(n\sqrt{2} > \pi)$ are true. The first is derived from the original statement using $x = \pi$ and the second using $x = \pi$ and $y = \sqrt{2}$.

4.7.2 *The universal generalization (UG)*

The universal generalization principle (UG) allows the introduction of the universal quantifier, i.e., the validation of the universally quantified statement $(\forall x)P(x)$: *To introduce* $(\forall x)P(x)$, *we must prove that* $P(x)$ *is true for an arbitrary element* x *of* $\mathcal{U}$. In applying this principle, it is very important that we don't assume anything special about the element x, except the fact that it belong to the universe of discourse $\mathcal{U}$.

Example 4.27. Let $\mathcal{U} = \mathbb{R}$. Prove each of the following statements.

(a) $(\forall x)\left(x^2 - 2x + 2 \neq 0\right)$.
(b) $(\forall_{x>0})\left(x + \frac{9}{x} \geq 6\right)$.
(c) $(\forall_{x\geq 0})(\exists_{y\leq 0})\left(x = y^2\right)$.

Solution. (a) Let $x \in \mathbb{R}$ be arbitrary. Assume (by contradiction) that $x^2 - 2x + 2 = 0$ for some $x \in \mathbb{R}$. Then $x^2 - 2x = -2$. Adding 1 on both sides, we get that $x^2 - 2x + 1 = -2$. Since the left hand side is $(x-1)^2$ and the right side is negative, this leads to a contradiction. We conclude that $x^2 - 2x + 2 \neq 0$ for any real number x.

(b) Let x be an arbitrary positive real number. We know that $(x-3)^2 \geq 0$. By expanding, we get $x^2 - 6x + 9 \geq 0$. Since $x > 0$, the inequality remains true if we divide both sides by x: $x - 6 + \frac{9}{x} \geq 0$. Adding 6 to both sides we get $x + \frac{9}{x} \geq 6$.

(c) Let x be an arbitrary non-negative real number. We need to prove the existence of a non-positive real number y such that $x = y^2$. Let $y = -\sqrt{x}$, then clearly $y \leq 0$ and $x = y^2$. $\Diamond$

The (UG) principle works with any n-ary predicate in general.

Example 4.28. Let $\mathcal{U} = \mathbb{Z}$ and let $M(x, y)$ be the binary predicate: "*x is a multiple of y*". Prove that $(\forall x)(\forall y)(\forall z)\left(M(x, y) \rightarrow M(xz, y)\right)$.

Solution. We have to show the following: "*For any integers x, y and z, if x is a multiple of y then xz is a multiple of y.*" Start by considering three arbitrary integers x, y and z such that x is a multiple of y. Then $x = ky$ for some integer k. Multiplying both sides by z, we get that $xz = kyz = (kz)y$ and so xz is a multiple of y (since kz is an integer). $\Diamond$

To refute the statement $(\forall x)\,P(x)$ with respect to a certain interpretation, it suffices to show that $P(a)$ is false for at least one element a. Such an element is called a *counterexample* of $(\forall x)\,P(x)$.

Example 4.29. In the universe $\mathcal{U} = \mathbb{R}$, consider the two predicates: $P(x)$: "$x \leq 3$" and $Q(x)$: "$x^2 \leq 9$. Prove that the statement ϕ : $(\forall x)\left(P(x) \rightarrow Q(x)\right)$ is false. Give a different universe in which ϕ is true.

Solution. Let $a = -4$, then $P(-4)$ is true but $Q(-4)$ is false. So the implication $(P(-4) \rightarrow Q(-4))$ is false. We conclude that $a = -4$ is a counterexample of ϕ and ϕ is false. There are many possibilities for a different universe in which ϕ is true. For instance, if we take $\mathcal{U} = \{-1, 0, 1\}$ then clearly both $P(x)$ and $Q(x)$ are true for any $x \in \mathcal{U}$ and consequently $(P(x) \to Q(x))$ is also true for any $x \in \mathcal{U}$. Another universe we could choose is $\mathcal{U} = \{5, 6, 7\}$. In this universe $P(x)$ is false for any x and so the implication $(P(x) \to Q(x))$ is true for any x. $\Diamond$

Example 4.30. In the universe of all integers, consider the two predicates: $E(x)$: "*x is even*" and $D(x, y)$: "*x is a divisor of y*". Prove that the statement $(\forall x)\, (\forall y)\, ((D(x, y) \wedge E(y)) \to E(x))$ is false.

Solution. Let $x = 3$, $y = 12$. Then x is a divisor of y and y is even so $(D(3, 12) \wedge E(12))$ is true. The implication $((D(3, 12) \wedge E(12)) \to E(3))$ is, however, false since $E(3)$ is false. $\Diamond$

4.7.3 *The existential instantiation (EI)*

The principle of existential instantiation states the following: *If $(\exists x)P(x)$ is true, then $P(a)$ holds for some element a of the universe.* Here, we stress the fact that a is not an arbitrary element of the universe but rather an element for which $P(a)$ is true. We might not know what elements of $\mathcal{U}$ satisfy P, but we know that at least one exists and hence we can give it a name. A consequence of the EI principle is the following: Let Q be a statement not involving the variable x and suppose that the existential statement $(\exists x)P(x)$ and universal statement $(\forall x)(P(x) \to Q)$ hold. Then we can conclude that Q is true. Here is the justification: first we know that $P(a)$ holds for some a in the universe. From $(\forall x)(P(x) \to Q)$, we get that $(P(a) \to Q)$ is true. Since both $P(a)$ and $(P(a) \to Q)$ are true, Q is true.

4.7.4 *The existential generalization (EG)*

To introduce $(\exists x)P(x)$, it suffices to prove that $P(a)$ is true for some element a of $\mathcal{U}$. Such an element is called a *witness* of the statement $(\exists x)\, P(x)$. In other words, the existential generalization (EG) states that if $P(a)$ holds for some element a, then we can infer $(\exists x)\, P(x)$.

Example 4.31. In the universe $\mathbb{N}$, consider the two predicates: $P(x)$: "*x is prime*", $E(x)$: "*x is even*" and $M(x)$: "*x is a multiple of* 3". Prove each of the following statements.

(a) $(\exists x)\, (P(x) \wedge E(x))$

(b) $(\exists x)\,(\neg P(x) \wedge \neg E(x))$

(c) $(\exists x)\,(\exists y)\,(\neg P(x+y) \wedge P(x) \wedge M(y))$.

Solution. (a) $a = 2$ is a witness since it is prime and even.

(b) $a = 15$ is a witness since it is not prime and not even.

(c) Here we need to find two natural numbers a and b one of which is prime, the other is a multiple of 3 and such that $a + b$ is not prime. Integers $a = 3$, $b = 3$ satisfy these requirements. $\Diamond$

It is worth mentioning here that there could exist many elements of the universe satisfying the statement $(\exists x)P(x)$. If there exists a *unique* element x satisfying $P(x)$, we often use the expression $(\exists! x)\,P(x)$. For example, $x = 2$ is the only natural number which is even and prime at the same time, so we could replace the expression in part (a) of Example 4.31 with $(\exists! x)\,(P(x) \wedge E(x))$. This cannot be done for parts (b) and (c) of the same example. Clearly, if we have $(\exists! x)\,P(x)$, we can infer $(\exists x)\,P(x)$ or equivalently, the wff $((\exists! x)\,P(x) \rightarrow (\exists x)\,P(x))$ is valid.

4.7.5   *Arguments of predicate logic*

Like in propositional logic, an argument of predicate logic is a list of wffs each called a *premise*, followed by a wff called the *conclusion* of the argument. An argument is called *valid* if and only if every common model for all the premises is also a model for the conclusion. So, one way to prove the validity of an argument in predicate logic is to assume that I is an interpretation with respect to which every premise is true and then prove that the conclusion of the argument is also true with respect to I. A more formal way is to use the inference rules of propositional logic we saw in Chapter 1 together with the four principles of reasoning in predicate logic.

Example 4.32. Let the universe be the set of all countries in the world. Translate the following argument using predicate logic symbols and then prove its validity: *Only scandinavian countries are affected by the hurricane. Canada is not a scandinavian country. Therefore Canada is not affected by the hurricane.*

Solution. Consider the following predicates $S(x)$: "x is a scandinavian country", $H(x)$: "x is affected by the hurricane", c: "Canada". Then the argument can be translated as follows:

$$(\forall x)\,(H(x) \to S(x))$$
$$\frac{\neg S(c)}{\neg H(c)}$$

For the validity of the argument, assume that I is a model for both $(\forall x)\,(H(x) \to S(x))$ and $\neg H(c)$, so both $(\forall x)\,(H(x) \to S(x))$ and $\neg H(c)$ are true with respect to I. The (UI) principle tells that $(H(c) \to S(c))$ is true. Since $\neg S(c)$ is also true, the Modus tollens rule of propositional logic tells us that $\neg H(c)$ is true. The formal proof of the validity of the argument is shown below with a justification of each row.

1. $(\forall x)\,(H(x) \to S(x))$ Premise
2. $(H(c) \to S(c))$ (UI), Line 1
3. $\neg S(c)$ Premise
4. $\neg H(c)$ Modus tollens, Line 2, Line 3 $\Diamond$

Example 4.33. In the universe $\mathbb{N}$, consider the following argument: *There exists a prime number which is even. Any prime number is strictly greater than 1. Therefore, there exists an even number strictly greater than 1.* Of course, the conclusion of that argument is true as we can choose the integer $n = 2$ that satisfies it. But, we want to deduce this conclusion in a formal way from the two premises. For this end, consider the unary predicates $P(n)$: "n is prime", $E(n)$: "n is even" and the binary predicate $G(m, n)$: "$m > n$." Using symbols of predicate logic, the argument is translated as follows:

$$(\exists n)\,(P(n) \wedge E(n))$$
$$\frac{(\forall n)\,(P(n) \to G(n, 1))}{(\exists n)\,(E(n) \wedge G(n, 1))}$$

The proof of the validity of the argument goes as follows:

1. $(\exists n)\,(P(n) \wedge E(n))$ Premise
2. $(\forall n)\,(P(n) \to G(n, 1))$ Premise
3. $(P(a) \wedge E(a))$ for some integer a Line 1, (EI)
4. $(P(a) \to G(a, 1))$ Line 2, (UI)
5. $P(a)$ Line 3, $\wedge$-Elimination
6. $G(a, 1)$ Line 4, Line 5, Modus ponens
7. $(P(a) \wedge G(a, 1))$ Line 5, Line 6, $\wedge$-introduction
8. $(\exists n)\,(E(n) \wedge G(n, 1))$ Line 7, (EG)

Example 4.34. Consider the following argument of predicate logic:

$$(\forall x)\,(P(x) \to Q(x))$$
$$\underline{(\forall x)\,(Q(x) \to R(x))}$$
$$(\forall x)\,(P(x) \to R(x)))$$

The proof of the validity of the argument is as follows:

1. $(\forall x)\,(P(x) \to Q(x))$	Premise
2. $(\forall x)\,(Q(x) \to R(x))$	Premise
3. $(P(c) \to Q(c))$, $c \in \mathcal{U}$ is arbitrary	(UI), Line 1
4. $(Q(c) \to R(c))$	(UI), Line 2
5. $(P(c) \to R(c))$, $c \in \mathcal{U}$ is arbitrary	Hypothetical syllogism, Lines 3, 4
6. $(\forall x)\,(P(x) \to R(x)))$	(UG), Line 5

When the conclusion of an argument is a universally quantified implication like $(\forall x)(A(x) \to B(x))$, we can introduce $A(c)$ as a premise on some line in the proof where c is an arbitrary element of the universe. The purpose of this premise is to prove $B(c)$ and consequently the implication $(A(c) \to B(c))$. From this, we can conclude $(\forall x)(A(x) \to B(x))$ using the (UG) principle. The next example illustrates this technique.

Example 4.35. Consider the following argument of predicate logic:

$$(\forall x)\,(\neg P(x) \to Q(x))$$
$$\underline{(\forall x)\,((\neg P(x) \wedge Q(x)) \to R(x))}$$
$$(\forall x)\,(\neg P(x) \to R(x)))$$

The proof of the validity of the argument is as follows:

1. $(\forall x)\,(\neg P(x) \to Q(x))$	Premise
2. $(\forall x)\,((\neg P(x) \wedge Q(x)) \to R(x))$	Premise
3. $\neg P(c)$, c is an arbitrary element	Premise
4. $(\neg P(c) \to Q(c))$	(UI), Line 1
5. $Q(c)$	Modus ponens, Lines 3, 4
6. $(\neg P(c) \wedge Q(c))$	$\wedge$-Introduction, Lines 3, 5
7. $((\neg P(c) \wedge Q(c)) \to R(c))$	(UI), Line 2
8. $R(c)$	Modus ponens, Lines 6, 7
9. $(\neg P(c) \to R(c))$, c is an arbitrary element	Lines 3–8
10. $(\forall x)\,(\neg P(x) \to R(x)))$	(UG), Line 9

4.7.6　*Exercises*

(1) Let the universe be the set $\mathbb{N}$ of all natural numbers. If $m, n \in \mathbb{N}$, then $\gcd(m, n)$ and $\operatorname{lcm}(m, n)$ stand for the greatest common divisor and the smallest common multiple of m and n, respectively. Consider the

unary predicate $P(m)$: "*m is a prime number*" and the three binary predicates: $D(m, n)$: "*m is a divisor of n*", $G(m, n)$: "$\gcd(m, n) = 1$" and $L(m, n)$: "$\text{lcm}(m, n) = 1$". Prove each of the following statements.

  (a) $(\exists x)P(x)$
  (b) $(\forall x)D(1, x)$
$\star$ (c) $(\exists x)(\exists y)L(x, y)$
  (d) $(\exists x)(\forall y)G(x, y)$
$\star$ (e) $(\exists x)(\forall y)D(x, y)$
  (f) $(\forall x)(\forall y)\left((P(x) \wedge D(y, x)) \to ((y = 1) \vee (y = x))\right)$
$\star$ (g) $(\forall x)(\forall y)\left((P(x) \wedge \neg D(x, y)) \to G(x, y)\right)$
  (h) $(\forall x)(\forall y)(\forall z)\left((D(x, y) \wedge D(y, z)) \to D(x, z)\right)$

(2) Let the universe of discourse be the interval $\left[0, \frac{\pi}{3}\right]$ of $\mathbb{R}$. Knowing that the property $(\forall x)(\sin x \le x)$ is true, prove that $\frac{\pi^2}{36}\left(2 + \sqrt{3}\right) \ge 1$ (*Hint.* $\sin\left(\frac{\pi}{12}\right) = \frac{\sqrt{6}-\sqrt{2}}{4}$).

$\star$(3) Let the universe be the set $\mathbb{N}$ of all natural numbers. Consider the predicate $P(x)$:"*x is prime*". Knowing that the statement:

$$(\forall a)\left((\exists k)\left(a = 3^{2^k}\right) \to (\exists b)\left((a < b) \wedge P(b)\right)\right)$$

is true, prove that there exists a prime number greater than 1853020188851841.

(4) Prove each of the following statements over the universe $\mathcal{U} = \mathbb{R}$.

  (a) $(\exists x)(\forall y)\left((y \le 0) \vee (xy < y)\right)$
  (b) $(\forall x)\left((x \ne 0) \to (\exists y)\left(x^2 y + x = 0\right)\right)$
$\star$ (c) $(\exists x)(\forall y)\left(((y > 0) \wedge (x \ge y)) \to (y > xy^2 + e^{xy})\right)$

(5) Consider the predicates: $P(m, n)$: "$m < n^2$" and $Q(m, n)$: "*n is a multiple of m*" over the universe $\mathcal{U} = \mathbb{Z}$. Refute each of the following statements.

$\star$ (a) $(\exists! x)P(100, x)$
  (b) $(\forall x)\left(P(0, x) \wedge Q(2, x)\right)$
  (c) $(\forall x)\left(Q(-1, x) \to \left(P(x^2, x) \vee \neg Q(x, 2x)\right)\right)$

(6) Let P, Q and R be three unary predicates over a universe $\mathcal{U}$. Use the rules of inferences to prove the validity of the following argument.

$$\begin{array}{c} (\exists x)\left(P(x) \wedge Q(x)\right) \\ (\forall x)\left(P(x) \to R(x)\right) \\ \hline (\exists x)\left(Q(x) \wedge R(x)\right) \end{array}$$

(7) Let A, B and C be three unary predicates over a universe $\mathcal{U}$ and let $\alpha \in \mathcal{U}$ be a constant. Prove the validity of the following argument.

$$
\begin{array}{c}
(\forall x)\,(A(x) \to B(x)) \\
(\forall x)\,(B(x) \to C(x)) \\
\neg C(\alpha) \\
\hline
\neg A(\alpha)
\end{array}
$$

$\star$(8) Let P, Q, R and S be four unary predicates over a universe $\mathcal{U}$. Prove the validity of the following argument.

$$
\begin{array}{c}
(\forall x)\,(\neg P(x) \to Q(x)) \\
(\exists x)\,\neg P(x) \\
(\forall x)\,(Q(x) \to R(x)) \\
(\forall x)\,(S(x) \to \neg R(x)) \\
\hline
(\exists x)\,\neg S(x)
\end{array}
$$

(9) Consider the following argument of predicate logic: *"No buffalo has a night vision. Carnivores cannot dig holes. Only animals who can dig holes don't have night vision. Therefore, buffalos are not carnivores."* Translate the argument using symbols of predicate logic. Clearly specify each predicate. Determine if the argument is valid.

Chapter 5

Functions: Back to the Basics

5.1 Introduction

Functions are building blocks in almost every mathematical theory. The transition to a higher level of mathematics is only achieved when the notion of functions is fully understood. It goes without saying that the whole topic of modern Calculus is based on the study of functions and their behaviour. Functions are also used as "measuring tools" for things like the atmospheric pressure in terms of altitude, the temperature when the windchill is factored or the time that a computer takes to execute a task in terms of the complexity of the task. Functions can also appear in some unexpected places like in matrix algebra where an $m \times n$ matrix with real entries is often treated as a (linear) function from $\mathbb{R}^n$ to $\mathbb{R}^m$.

There are two directions one can take to introduce the concept of functions. One direction is to start by introducing the more general concept of *binary relations* and then talk about functions as particular types of such relations. The other direction is to develop the concept as a topic on its own and then relate it to that of binary relations. In this book, we choose the second direction for two main reasons. First, it allows a softer introduction to the topic as most readers have studied functions in some context before. Second, it allows the reader to have all the advantages and properties of functions in the tool box "early" enough and without having to indulge in the theory of relations which can be overwhelming. In Chapter 7, we make a clear and precise link between functions and relations.

5.2 Basic definitions and terminology

Definition 5.1. Let A and B be two sets. A *function* f from A to B, denoted by $f : A \to B$, is a rule of correspondence that assigns to *every* element $a \in A$ a *unique* element $b \in B$ that we denote by $b = f(a)$. If $b = f(a)$, then b is called the *image* of a and a is called a *pre-image* of b. The set A is called the *domain* of the function f and the set B is called the *codomain* of f. Two functions f and g are said to be *equal* if they have the same domain, the same codomain and if $f(x) = g(x)$ for every element x in the domain. The set of all functions $f : A \to B$ is denoted by B^A.

For a correspondence rule f from a set A to a set B to define a function, we emphasis that the following two conditions must be satisfied:

(1) Every element of A has an image;
(2) The image of every element of A is unique.

There are many ways to represent a function. If A, B are subsets of $\mathbb{R}$, a function $f : A \to B$ can (but not always) be defined in terms of an explicit formula, like $f(x) = 2^{3x+2} + 3$ or $\frac{\ln x}{x}$. In case of finite sets A and B, a function $f : A \to B$ can be represented by a diagram showing the two sets and arrows assigning to each element $a \in A$ its image $f(a) \in B$ as shown in the following example.

Example 5.1. Let $A = \{a, b, c, d\}$ and $B = \{1, 2, 3, 4, 5\}$. The following diagram represents the function from A to B defined by $f(a) = 2$, $f(b) = 1$, $f(c) = 3$ and $f(d) = 1$.

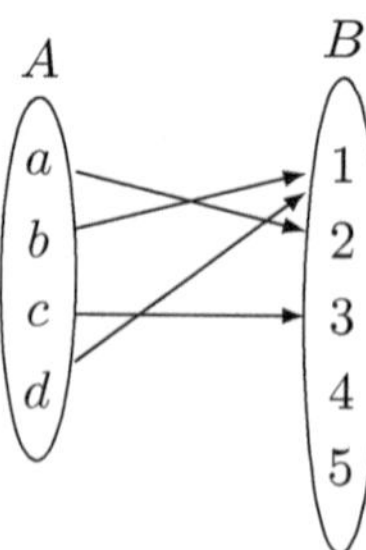

Functions can also be defined "piecewise" where multiple expressions are used to define the output of the function over different pieces of the domain.

Example 5.2. Consider the function $f : \mathbb{R} \to \mathbb{R}$ defined as follows:

$$f(x) = \begin{cases} \frac{1}{x^2-1} & \text{for } x \leq -2 \\ \frac{x}{x^2+1} & \text{for } -2 < x < 2 \\ x^2 + 1 & \text{for } 2 \leq x \leq 5 \\ 2x + 3 & \text{for } x > 5 \end{cases}$$

For this function, the domain $\mathbb{R}$ is divided in four "pieces": $]-\infty, -2]$, $]-2, 2[$, $[2, 5]$ and $]5, \infty[$ with different output on each piece. For instance, to find $f(-1)$ we use the expression $\frac{x}{x^2+1}$ since $-1 \in]-2, 2[$ and we get that $f(-1) = \frac{-1}{(-1)^2+1} = -\frac{1}{2}$. Similarly, $f(-3) = \frac{1}{(-3)^2-1} = \frac{1}{8}$ (since $-3 \leq -2$) and $f(7) = 2(7) + 3 = 17$ (since $7 > 5$).

Piecewise defined functions are often useful to model real life situations as shown in the following example.

Example 5.3. A repair company charges fees according to the following scheme:

- A technician visit up to 60 minutes costs \$75;
- A technician visit over 60 minutes and up to 90 minutes costs \$100;
- A technician visit over 90 minutes costs \$100 plus \$4 per minute above 90 minutes.

The company fee can then be modeled by the following piecewise function $f(t)$ where t represents the time of the visit, in minutes:

$$f(t) = \begin{cases} 75 & \text{for } 0 < t \leq 60 \\ 100 & \text{for } 60 < t \leq 90 \\ 100 + 4(t - 90) & \text{for } t > 90 \end{cases}$$

Hence, a visit of 45 minutes costs $f(45) = \$75$, a visit of 83 minutes costs $f(83) = \$100$ and a visit of 115 minutes costs $f(110) = 100 + 4(115 - 90) = \$200$.

For some sets, there are different ways to represent the elements of the set. A typical example is the set $\mathbb{Q}$ of all rational numbers. In $\mathbb{Q}$, $\frac{1}{2}$, $\frac{2}{4}$ and $\frac{128}{256}$ all represent the same rational number. In this case, one has to check that the rule defining a function on $\mathbb{Q}$ does not depend on the way we represent the elements in this set. If that is the case, we say that the function is *well-defined*. In general, a function $f : A \to B$ is well-defined if the following condition holds for all elements x_1, x_2 of A:

$$x_1 = x_2 \Rightarrow f(x_1) = f(x_2). \tag{5.1}$$

Example 5.4. The function $f : \mathbb{Q} \to \mathbb{Z}$ defined by $f\left(\frac{a}{b}\right) = a + b$ is not well-defined on $\mathbb{Q}$ since $\frac{2}{4} = \frac{3}{6}$ but $f\left(\frac{2}{4}\right) = 6 \neq f\left(\frac{3}{6}\right) = 9$.

Example 5.5. The function $f : \mathbb{Q} \to \mathbb{Q}$ defined by $f\left(\frac{a}{b}\right) = \frac{2a}{3b}$ is well-defined. To see this, let $\frac{a}{b}, \frac{c}{d} \in \mathbb{Q}$ such that $\frac{a}{b} = \frac{c}{d}$. Then $ad = bc$ and so $6ad = 6bc$ (multiplying by 6). This last equation can be rewritten in the form $(2a)(3d) = (3b)(2c)$ which implies that $\frac{2a}{3b} = \frac{2c}{3d}$ or equivalently, $f\left(\frac{a}{b}\right) = f\left(\frac{c}{d}\right)$. This proves that condition (5.1) above is satisfied and f is well-defined.

Definition 5.2. Let $f : A \to B$ be a function, $S \subseteq A$ and $T \subseteq B$. The *image* of S under f, denoted by $f(S)$, is the subset $f(S) = \{f(s); \; s \in S\}$ of B. When $S = A$, $f(A)$ is denoted by $\mathrm{Im}(f)$ and is called the *image* of the function f. If $\mathrm{Im}(f) = \{b\}$, then f is called a *constant function*. The *inverse image* of T under f, denoted by $f^{-1}(T)$, is the subset of A of all elements with images in T: $f^{-1}(T) = \{x \in A; \; f(x) \in T\}$.

It is important to distinguish between the image and the codomain of a function. The image is a subset of the codomain which is not necessarily equal to the codomain. When the equality holds, the function will be called onto as we will see later.

Example 5.6. With the notations of Example 5.1, consider the subsets $S_1 = \{b, d\}$ and $S_2 = \{b, c, d\}$ of A and the subsets $T_1 = \{1, 2\}$, $T_2 = \{4, 5\}$ of B. Then $f(S_1) = \{1\}$, $f(S_2) = \{1, 3\}$, $f^{-1}(T_1) = \{a, b, d\}$ and $f^{-1}(T_2) = \emptyset$ (no element of A has its image in the set T_2). The image of the function f is $\mathrm{Im}(f) = \{1, 2, 3\}$.

Example 5.7. Consider the function $f : \mathbb{R} \to \mathbb{R}$ defined by the rule $f(x) = x^2 + 1$. Since $x^2 + 1 \geq 1$, $\mathrm{Im}(f) \subseteq [1, \infty[$. Conversely, if $y \in [1, \infty[$ let $x = \sqrt{y - 1}$. Then $x \in \mathbb{R}$ since $y - 1 \geq 0$ in this case. Moreover, $f(x) = (\sqrt{y - 1})^2 + 1 = y$. This shows that $y \in \mathrm{Im}(f)$. We conclude that $\mathrm{Im}(f) = [1, \infty[$. For the subset $S_1 = \{-1, 1\}$, $f(S_1) = \{2\}$ since $f(-1) = f(1) = 2$. For the interval $S_2 = [-1, 1]$, $f(S_2) = [1, 2]$. To see this let $x \in [-1, 1]$, then $-1 \leq x \leq 1$ which implies that $0 \leq x^2 \leq 1$ and so $1 \leq x^2 + 1 = f(x) \leq 2$. This shows that $f(S_2) \subseteq [1, 2]$. For the converse, assume that $y \in [1, 2]$. Then $1 \leq y \leq 2$ and so $0 \leq \sqrt{y - 1} \leq 1$. Let $x = \sqrt{y - 1}$, then $x \in [0, 1] \subseteq [-1, 1]$ and $f(x) = (\sqrt{y - 1})^2 + 1 = y$. This shows that $y \in f(S_2)$ and thus $[1, 2] \subseteq f(S_2)$. For the subset $T_1 = \{1\}$ of the codomain, $f^{-1}(T_1) = \{0\}$. For the subset $T_2 = [5, 10]$ of the codomain, $f^{-1}(T_2) = \{x \in \mathbb{R}; \; 5 \leq x^2 + 1 \leq 10\} = \{x \in \mathbb{R}; \; 4 \leq x^2 \leq 9\}$. Note that the solution set of the inequality $x^2 \geq 4$ is $]-\infty, -2] \cup [2, \infty[$ and that of the inequality $x^2 \leq 9$ is $-3 \leq x \leq 3$. Thus, $f^{-1}(T_2)$ the intersection of the two solution sets, namely $[-3, -2] \cup [2, 3]$.

Example 5.8. Consider the function $f : \mathbb{Z} \times \mathbb{Z} \to \mathbb{Z}$ defined by the rule $f(m, n) = 3m + 2n$ (in theory, we should write $f((m, n))$ instead of $f(m, n)$ but we use the latter for simplicity). The domain of f is clearly $\mathbb{Z} \times \mathbb{Z}$. For $\mathrm{Im}(f)$, notice that for any $m \in \mathbb{Z}$ we have that $m = 3m + 2(-m) = f(m, -m)$. Since $(m, -m) \in \mathbb{Z} \times \mathbb{Z}$, $m \in \mathrm{Im}(f)$ and so $\mathrm{Im}(f) = \mathbb{Z}$. For the subset $S = \{(-1, 1), (0, 0), (-3, 4), (1, 1)\}$ of the domain, $f(S) = \{-1, 0, 5\}$. For the subset $T = \{0\}$ of $\mathbb{Z}$, $(0, 0)$, $(-2, 3)$, $(4, -6)$ are elements of $f^{-1}(T)$.

Theorem 5.1. *Let $f : A \to B$ be a function, S_1, S_2 two subsets of A and T_1, T_2 two subsets of B. Then:*

(a) $f(S_1 \cap S_2) \subseteq f(S_1) \cap f(S_2)$.
(b) $f(S_1 \cup S_2) = f(S_1) \cup f(S_2)$.
(c) $f^{-1}(T_1 \cap T_2) = f^{-1}(T_1) \cap f^{-1}(T_2)$.
(d) $f^{-1}(T_1 \cup T_2) = f^{-1}(T_1) \cup f^{-1}(T_2)$.

Proof.

(a) Let $b \in f(S_1 \cap S_2)$. Then there exists $a \in S_1 \cap S_2$ such that $b = f(a)$. Since $a \in S_1$ and $a \in S_2$, $b \in f(S_1)$ and $b \in f(S_1)$ and thus $b \in f(S_1) \cap f(S_2)$.
(b) Left as an exercise (see Exercise (18) below).
(c) Let $a \in A$. Then:

$$a \in f^{-1}(T_1 \cap T_2) \Leftrightarrow f(a) \in T_1 \cap T_2$$
$$\Leftrightarrow f(a) \in T_1 \text{ and } f(a) \in T_2$$
$$\Leftrightarrow a \in f^{-1}(T_1) \text{ and } a \in f^{-1}(T_2)$$
$$\Leftrightarrow a \in f^{-1}(T_1) \cap f^{-1}(T_2).$$

This shows that $f^{-1}(T_1 \cap T_2) = f^{-1}(T_1) \cap f^{-1}(T_2)$.
(d) Left as an exercise (see Exercise (10) below).

$\square$

Definition 5.3. Let $f : A \to B$ be a function, $S \subseteq A$. The *restriction* of f to S, denoted by $f\mid_S$, is the function $f\mid_S : S \to B$, defined by $f\mid_S(x) = f(x)$ for any $x \in S$. The function f is called an *extension* of a function $g : S \to B$ to A if $f\mid_S = g$ (so $f(x) = g(x)$ for any $x \in S$).

5.2.1   *Combining real-valued functions*

Real-valued functions on a set A (i.e., functions $f : A \to \mathbb{R}$) can be combined to form new ones by using operations on reals numbers like addition, subtraction, multiplication and division. Later on, we look at another important operation on functions called *composition*.

Definition 5.4. Let A be any set and let f and g be two functions from A to $\mathbb{R}$. We define the following new functions from A to $\mathbb{R}$.

(a) The *addition* of f and g, denoted by $f + g$, is defined by: $(f + g)(x) = f(x) + g(x)$ for any $x \in A$.

(b) The *subtraction* of f and g, denoted by $f - g$, is defined by: $(f - g)(x) = f(x) - g(x)$ for any $x \in A$.

(c) The *multiplication* of f and g, denoted by fg, is defined by: $(fg)(x) = f(x)g(x)$ for any $x \in A$.

(d) The *division* of f and g, denoted by $\frac{f}{g}$ (or simply f/g), is defined by: $\left(\frac{f}{g}\right)(x) = \frac{f(x)}{g(x)}$ for any $x \in A$ with $g(x) \neq 0$.

(e) Let $k \in \mathbb{R}$. The *scaling* of the function f by a factor of k is the function, denoted by kf, defined by $(kf)(x) = kf(x)$ for any $x \in A$.

Given real-valued functions $f, f_1, \ldots, f_n$ on a set A, we say that f is a *linear combination* of $f_1, \ldots, f_n$ if there exists real numbers $k_1, \ldots, k_n$ such that $f = k_1 f_1 + \cdots + k_n f_n$.

Example 5.9. The function $f(x) = \frac{1}{x^2 - 4}$ is a linear combination of $f_1(x) = \frac{1}{x+2}$ and $f_2 = \frac{1}{x+2}$ since one can easily verify that $\frac{1}{x^2-4} = \frac{-\frac{1}{4}}{x+2} + \frac{\frac{1}{4}}{x-2}$ which implies that $f = -\frac{1}{4}f_1 + \frac{1}{4}f_2$.

Example 5.10. In each case, determine if the function f can be written as a linear combination of the functions f_i's on $\mathbb{R}$. Justify your answer.

(a) $f(x) = -2x^2 + 4x - 2$, $f_1(x) = -x^2 + 2x + 1$, $f_2(x) = -x^2 + 4x + 2$, $f_3(x) = -2x + 3$.

(b) $f(x) = 6x + 1$, $f_1(x) = x^2 + 1$, $f_2(x) = 2x - 2$, $f_3(x) = 2x^2 + x + 1$.

(c) (Trigonometry) $f(x) = 1$ (constant function), $f_1(x) = \cos^2 x$, $f_2(x) = \sin^2 x$, $f_3(x) = x^2$, $f_4(x) = 2^x$

Solution.   (a) We look for real numbers a_1, a_2 and a_3 such that $f = a_1 f_1 + a_2 f_2 + a_3 f_3$. This means in particular that $f(x) = a_1 f(x) +$

$a_2 f_2(x) + a_3 f_3(x)$ for any $x \in \mathbb{R}$:

$$-2x^2 + 4x - 2 = a_1(-x^2 + 2x + 1) + a_2(-x^2 + 4x + 2) + a_3(-2x + 3)$$
$$\Rightarrow -2x^2 + 4x - 2 = (-a_1 - a_2)x^2 + (2a_1 + 4a_2 - 2a_3)x$$
$$+ a_1 + 2a_2 + 3a_3$$

Comparing the coefficients of x^2, x and the constant terms on both sides of the last equation, we get the following system of equations: $-a_1 - a_2 = -2$, $2a_1 + 4a_2 - 2a_3 = 4$ and $a_1 + 2a_2 + 3a_3 = -2$. Solving this linear system (details left to the reader), we find that $a_1 = 3$, $a_2 = -1$ and $a_3 = -1$. We conclude that $f = 3f_1 - f_2 - f_3$ is a linear combination of f_1, f_2 and f_3.

(b) Again, We look for real numbers a_1, a_2 and a_3 such that $f = a_1 f_1 + a_2 f_2 + a_3 f_3$:

$$6x + 1 = a_1(x^2 + 1) + a_2(2x - 2) + a_3(2x^2 + x + 1)$$
$$\Rightarrow 6x + 1 = (a_1 + 2a_3)x^2 + (2a_2 + a_3)x + a_1 - 2a_2 + a_3.$$

Comparing the coefficients of x^2, x and the constant terms on both sides of the last equation, we get the following system of equations: $a_1 + 2a_3 = 0$, $2a_2 + a_3 = 6$ and $a_1 - 2a_2 + a_3 = 1$. From the first equation, we get that $a_1 = -2a_3$. Substituting in the third equation, we get that $-2a_2 - a_3 = 1$ or $2a_2 + a_3 = -1$. This is a contradiction to the second equation. We conclude that f is not a linear combination of f_1, f_2 and f_3.

(c) Using the trigonometric identity $\sin^2 x + \cos^2 x = 1$, we get that $f = f_1 + f_2$ which implies that $f = f_1 + f_2 + 0 \cdot f_3 + 0 \cdot f_4$. We conclude that f is a linear combination of f_1, f_2, f_3 and f_4. $\diamond$

Given an expression $f(x)$ where x is a real number, the largest subset of $\mathbb{R}$ for which $f(x)$ is a real number (so $f(x)$ is defined) is called the domain of f that we denote by D_f. Note that $f : D_f \to \mathbb{R}$ is a function. If $f(x)$ and $g(x)$ are two expressions with $x \in \mathbb{R}$, the domain of each of the expressions $f(x) + g(x)$, $f(x) - g(x)$ and $f(x)g(x)$ is $D_f \cap D_g$. In the case of the expressions $\frac{f(x)}{g(x)}$, we additionally require that $g(x) \neq 0$ so the domain of $\frac{f(x)}{g(x)}$ is $\{x \in D_f \cap D_g; \; g(x) \neq 0\}$.

Example 5.11. Consider the following two expressions: $f(x) = \sqrt{x^2 - 4}$ and $g(x) = \frac{x-3}{\sqrt{x-1}}$ where x is a real number. Find the domain of each of the following expressions: $f(x) + g(x)$, $f(x) - g(x)$, $f(x)g(x)$ and $f(x)/g(x)$.

Solution. The domain D_f of $f(x)$ is the set of all real numbers x such that $x^2 - 4 \geq 0$ or equivalently $D_f =]-\infty, -2] \cup [2, \infty[$. The domain D_g of $g(x)$ is the set of all real numbers x such that $x - 1 > 0$ or equivalently $D_g =]1, \infty[$. The intersection of the two domains is $D_f \cap D_g = [2, \infty[$. We know then that the domain of each of the functions $f(x) + g(x)$, $f(x) - g(x)$, $f(x)g(x)$ is $[2, \infty)$. For the expression $\frac{f(x)}{g(x)}$, the domain is the interval $[2, \infty)$ from which we remove the values of x satisfying $g(x) = 0$. Notice that $x = 3$ is the only value with $g(x) = 0$, and since $3 \in [2, \infty[$, the domain of $\frac{f(x)}{g(x)}$ is $[2, 3[\cup]3, \infty[$. $\qquad\qquad \Diamond$

5.2.2   *Exercises*

(1) For each of the following rules, determine if it defines a function. If the rule is a function, determine its domain and codomain. If the rule is not a function, propose changes (restriction of the domain or codomain, change the rule, ...) to make it a function.

   $\star$ (a) The rule that assigns a student number to each student at Oxford University.

     (b) The rule that assigns to each Swiss citizen the language he or she speaks. Note that Switzerland has four official languages and some of its citizens can speak all four languages.

     (c) The rule that assigns the name of a parent to every student at Oxford University.

     (d) The rule that assigns the value $\frac{x}{x^2+1}$ to every real number x.

   $\star$ (e) The rule that assigns the value $\frac{x}{x^2-1}$ to every real number x.

     (f) The rule that assigns the absolute value $|n|$ to every integer n.

   $\star$ (g) The rule that assigns the value $\pm 2^x$ to every real number x.

     (h) The rule that assigns to every real number x the value y such that $y^2 - x = 4$.

(2) Consider the expression $f(x) = \frac{x}{x^2-3}$ where $x \in \mathbb{R}$. For each of the following statements, determine if it is true or false. Justify your answer.

   $\star$ (a) f represents a function from $\mathbb{R}$ to $\mathbb{R}$.

     (b) f represents a function from $\mathbb{Z}$ to $\mathbb{R}$.

     (c) f represents a function from $\mathbb{N}$ to $\mathbb{R}$.

   $\star$ (d) f represents a function from $\mathbb{Z}$ to $\mathbb{Z}$.

     (e) f represents a function from $\mathbb{R} \setminus \{-\sqrt{3}, \sqrt{3}\}$ to $\mathbb{R}$.

   $\star$ (f) f represents a function from $] - \sqrt{3}, \sqrt{3}]$ to $\mathbb{R}$.

(3) For the piecewise function defined in Example 5.2 above, find: $f(-5)$, $f(-2)$, $f(2)$, $f(3)$, $f(5)$ and $f(6)$.

(4) Prove that the functions $f(x) = \sqrt{9x^2 + 2} - 3x$ and $g(x) = \frac{2}{\sqrt{9x^2+2}+3x}$ from $\mathbb{R}$ to $\mathbb{R}$ are equal.

(5) In each case, determine if the given rule is a well-defined function.

  (a) $f : \mathbb{Q} \to \mathbb{Q}$ defined by $f\left(\frac{a}{b}\right) = \frac{1}{b}$.

$\star$ (b) $f : \mathbb{Q} \to \mathbb{Q}$ defined by $f\left(\frac{a}{b}\right) = \frac{a^3}{b^3}$.

  (c) Let X be a non-empty finite set with at least two elements and let $\wp(X)$ be the powerset of X. Let $f : \wp(X) \to X$ be the rule that assigns to every subset Y of X its "first element". For instance, $f(\{a, b\}) = a$.

$\star$ (d) Let F be the set of all members of a certain family, and let H be the set of all hobbies. A rule $f : F \to H$ is defined by $f(x)$ is a hobby of the family member x.

(6) Let M be the set of all math professors at a certain university. Let S be the set of all math fields of research (algebra, analysis, discrete math, topology, ...). Consider the rule of correspondence f from M to S that assigns to each professor his or her research interest. What condition each mathematician must satisfy in order for f to be a well-defined function from M to S?

(7) In your Calculus classes, you were probably introduced to the idea of a "function" $f : A \to B$ a little differently in the sense that not every element of A needs to have an image. In fact, any function analysis you did in a Calculus class started by finding the "domain" of the function. A *partial function* from A to B is a rule of correspondence in which $f(a)$ is not necessarily defined for every $a \in A$. In this case, we use the notation $f : A \rightharpoonup B$ and we say that A is the *source* of f. The domain of a partial function f is the subset D of A of all elements a of A for which $f(a)$ is defined (as an element of B). The notion of function as given in Definition 5.1 is sometimes referred to as *total function*. In each case, determine if the given rule is a total or a partial function. In case of partial function, clearly determine what the domain is.

  (a) $f : \mathbb{Z} \to \mathbb{Z}$, $f(n) = \frac{n+1}{2}$

  (b) $f : \mathbb{R} \to \mathbb{R}$, $f(x) = \frac{x}{x^2+1}$

$\star$ (c) $f : \mathbb{N} \to \mathbb{N}$, $f(n) = n - 4$

  (d) $f : \mathbb{Z} \times \mathbb{Z} \to \mathbb{Z}$, $f(m, n) = m + n - 1$

$\star$ (e) $f : \mathbb{Z} \to \mathbb{Z} \times \mathbb{Z}$, $f(n) = \left(\frac{n}{3}, \frac{n}{4}\right)$

(8) Prove that the function $f : \mathbb{N}\backslash\{0\} \to \mathbb{N}\backslash\{0\}$ defined by:

$$f(n) = \begin{cases} 1 & \text{if } n = 1 \\ 1 + f\left(\frac{n}{2}\right) & \text{if } n \text{ is even} \\ f(5n - 3) & \text{if } n \text{ is odd and } n \geq 3 \end{cases}$$

is not well-defined (*Hint.* Try to compute $f(3)$).

$\star$(9) Determine the image of the function $f : \mathbb{R}\backslash\{\frac{2}{3}\} \to \mathbb{R}$ defined by $f(x) = \frac{-x+2}{3x-2}$. Carefully prove that your answer is correct.

(10) Give a function $f : \mathbb{R} \to \mathbb{Z}$ such that $\text{Im}(f) = \mathbb{Z}$.

(11) Give a function $f : \mathbb{Z} \times \mathbb{Z} \to \mathbb{Z}$ such that $\text{Im}(f) \neq \mathbb{Z}$.

(12) Give a function $f : \mathbb{Z} \times \mathbb{Z} \to \mathbb{Z}$ such that $\text{Im}(f) = \mathbb{N}$.

$\star$(13) Give a function $f : \mathbb{R} \to \mathbb{Z}$ such that $\text{Im}(f) = \{1, 2, 3\}$.

(14) Let $A = \{-2, -1, 0, 3, 4\}$, $S = \{-1, 0, 3\} \subseteq A$. Consider the following four functions: $f : A \to \mathbb{Z}$ defined by $f(-2) = 10$, $f(-1) = 7$, $f(0) = 4$, $f(3) = -5$ and $f(4) = -8$; $g : S \to \mathbb{Z}$ defined by $g(-1) = 7$, $g(0) = 4$, $g(3) = -5$; $h : \mathbb{Z} \to \mathbb{Z}$ defined by $h(n) = -3n + 4$ and $k : \mathbb{Z} \to \mathbb{Z}$ defined by $k(x) = -x^2 + 4$. In each case, determine if the statement is true or false. Justify your answer.

(a) $f\mid_S = g$	(d) k is an extension of g to $\mathbb{Z}$
$\star$ (b) $h\mid_S = g$	$\star$ (e) k is an extension of f to $\mathbb{Z}$
(c) $h\mid_A = f$	(f) $h = k$

$\star$(15) Let $A = \{a, b, c, d, e\}$ and $B = \{1, 2, 3, 4, 5, 6\}$, $S = \{a, b, c\} \subseteq A$. Consider the function $g : S \to B$ defined by $g(a) = 3$, $g(b) = 1$ and $g(c) = 4$. How many extensions of g to a function $f : A \to B$ are there?

(16) A certain company offers the following payment scheme for its employees. If an employee works 40 hours or less per week, the regular hourly rate is \$22. For every additional hour between 41 and 50 hours weekly (inclusive), the hourly rate is \$33. For every additional hour between 51 and 60 hours weekly (inclusive), the hourly rate increases to \$40. No employee is allowed to work more than 60 hours per week.

    (a) Set up a piecewise function $P(h)$ that gives an employee weekly pay in terms of h, the number of hours worked per week.

    (b) Determine the weekly pay if an employee works: (i) 32 hours a week (ii) 48 hours a week; (iii) 55 hours a week.

(17) Give an example where the inclusion of part (a) of Theorem 5.1 is strict.

(18) Prove part (b) of Theorem 5.1.

$\star$(19) Prove part (d) of Theorem 5.1.

(20) Consider the two functions: $f(x) = \sqrt{9 - x^2}$ and $g(x) = \frac{x-2}{\sqrt{x-1}}$ where x is a real number.

    (a) Find the domain D_f of f and the domain D_g of g.

    (b) Find the domain of each of the functions $f + g$, $f - g$, fg and $\frac{f}{g}$.

(c) Find the value of each of the following expressions, if the expression is defined.

(i) $(f+g)(-2)$ (iii) $(fg)(3)$ (v) $(g-f)\left(\frac{3}{2}\right)$

(ii) $(f-g)(3)$ (iv) $\left(\frac{f}{g}\right)(\sqrt{2})$ (vi) $\left(\frac{f}{g}\right)(0)$.

(21) Consider the function $f(x) = x + \frac{1}{x}$ from $\mathbb{R}\backslash\{0\}$ to $\mathbb{R}$. In each case, determine the value of the given expression. Simply your expression as much as possible.

(a) $f(x+1)$.

(b) $2f(x-1) - 3f(x+2)$.

(c) $\frac{f(x+h)-f(x)}{h}$, $h \neq 0$.

(22) For the three real-valued functions $f(x) = \frac{1}{\sqrt{x^3-4x}}$, $g(x) = \frac{1}{x+1}$ and $h(x) = \sqrt{9-x^2}$, find the domain of the function $f+g+h$.

(23) Consider the two functions from $\mathbb{R}$ to $\mathbb{R}$ defined as following:

$$f(x) = \begin{cases} x^2 & \text{for } x < 1 \\ x-1 & \text{for } x \geq 1 \end{cases}, \quad g(x) = \begin{cases} -2x+3 & \text{for } x < -1 \\ x+1 & \text{for } x \geq -1 \end{cases}$$

(a) Find the value of each of the following expressions: $(f+g)(-1)$, $(f-g)(0)$, $(fg)(1)$ and $(f/g)(3)$.

(b) Write each of the functions $f+g$, $f-g$, fg and f/g as a piecewise function on $\mathbb{R}$.

(24) Let a be a non-negative real number. The *Heaviside function* at a (also known as the *unit step function*) is the function, denoted by $U(x-a)$, from $\mathbb{R}$ to $\{0,1\}$ defined by:

$$U(x-a) = \begin{cases} 0 \text{ for } x < a \\ 1 \text{ for } x \geq a \end{cases}$$

Express each of the following functions in terms of one or more Heaviside functions and possibly other real-valued functions.

(a) $f(x) = \begin{cases} 1 & \text{for } x \leq \pi \\ 0 & \text{for } x > \pi \end{cases}$

(c) $h(x) = \begin{cases} 0 & \text{for } x < \frac{\pi}{2} \\ \cos(2x) & \text{for } x \geq \frac{\pi}{2} \end{cases}$

(b) $g(x) = \begin{cases} 0 & \text{for } x < 1 \\ 1 & \text{for } 1 \leq x < 3 \\ 0 & \text{for } x \geq 3 \end{cases}$

$\star$ (d) $k(x) = \begin{cases} 0 & \text{for } x < 1 \\ x-1 & \text{for } 1 \leq x < 3 \\ x & \text{for } x \geq 3 \end{cases}$

(25) Let $f : \mathbb{R} \to \mathbb{R}$ be a function. We say that f is *even* if $f(-x) = f(x)$ for every $x \in \mathbb{R}$. We say that f is *odd* if $f(-x) = -f(x)$ for every $x \in \mathbb{R}$. For each of the following functions, determine if it is even, odd or neither. Justify your answer.

(a) $f_1(x) = x^2 - x$

(b) $f_2(x) = x^4 + 2x^2 + 1$

$\star$ (c) $f_3(x) = -2x^3 + 4x$

(d) $f_4(x) = x^2 2^x$

(e) $f_5(x) = \frac{x^2}{x^2+1}$

(f) $f_6(x) = \frac{x^3}{x^3+1}$

$\star$ (g) $f_7(x) = \frac{|x|}{|x|^3+1}$

(h) $f_8(x) = x^{-5}|x|$

(i) $f_9(x) = \begin{cases} 0 & \text{for } x < 1 \\ 1 - x^2 & \text{for } 1 \le x < 3 \\ -1 & \text{for } x \ge 3 \end{cases}$

(26) (See Exercise (25) for the definition of even and odd functions) Let $f : \mathbb{R} \to \mathbb{R}$ and $g : \mathbb{R} \to \mathbb{R}$ be two functions. For each of the following statements, determine if it is true or false. Justify your answer.

$\star$ (a) If f and g are even functions, then $f + g$ is even.

(b) If f and g are odd functions, then $f + g$ is odd.

(c) If f and g are even functions, then fg is even.

(d) If f and g are odd functions, then fg is odd.

(e) If f and g are even functions, then $\frac{f}{g}$ is even.

(f) If f and g are odd functions, then $\frac{f}{g}$ is odd.

$\star$ (g) If f is even and g is odd, then $f + g$ is even.

(h) If f is even and g is odd, then $f + g$ is odd.

(i) If f is even and g is odd, then fg is even.

(j) If f is even and g is odd, then fg is odd.

(27) In each case, functions f, $f_1, \ldots, f_n$ from $\mathbb{R}$ to $\mathbb{R}$ are given. If possible, write the function f as a linear combination of the functions f_i's.

(a) $f(x) = 2x^2 - 9x$, $f_1(x) = x^2 - x$, $f_2(x) = -3x + 2$, $f_3(x) = x^2 - 1$.

(b) $f(x) = -3x^2 + 5$, $f_1(x) = x^2 + 2x - 1$, $f_2(x) = x + 1$, $f_3(x) = x^2 - x$.

(c) $f(x) = \cos(2x)$, $f_1(x) = \cos^2 x$, $f_2(x) = \sin^2 x$, $f_3(x) = x^2$, $f_4(x) = x^3 + 2x^2 + 1$ (*Hint.* Use the identity $\cos(2x) = \cos^2 x - \sin^2 x$).

5.3   Some special functions

In this section, we explore some special types of functions that appear frequently in mathematics and computer science. In particular, we look at the notion of sequences and the related notion of summation from a functional point of view.

Definition 5.5. Let A be a set and $B \subseteq A$ a subset of A. the *identity function* of A, denoted by 1_A or id_A, is the function from A to A defined by $1_A(x) = x$ for any $x \in A$. The inclusion function of B (also called the embedding or the injection), denoted by i_B, is the function $i_B : B \to A$ defined by $i_B(x) = x$ for any $x \in B$.

The identity function of a set A leaves every element of A fixed. It plays an important role in defining the notion of the inverse of a function.

Definition 5.6. Let A be a subset of a universal set $\mathcal{U}$. The *characteristic function* of A is the function $\chi_A : \mathcal{U} \to \{0, 1\}$ defined by:

$$\chi_A(x) = \begin{cases} 1 & \text{if } x \in A \\ 0 & \text{if } x \notin A \end{cases}$$

Characteristic functions have some interesting properties that we explore in the following theorem.

Theorem 5.2. *Let A, B be two subsets of a universal set $\mathcal{U}$. Then:*

(a) $\chi_A = \chi_B \Leftrightarrow A = B$
(b) $\chi_A \leq \chi_B \Leftrightarrow A \subseteq B$
(c) $\chi_{A \cap B} = \chi_A \cdot \chi_B$
(d) $\chi_{A \cup B} = \chi_A + \chi_B - \chi_A \cdot \chi_B$
(e) $\chi_{A \backslash B} = \chi_A - \chi_{A \cap B}$
(f) $\chi_{\overline{A}} = 1 - \chi_A$

Proof. We only prove parts (a) and (c), leaving the other parts as exercises for the reader.

(a) If $\chi_A = \chi_B$, then for any $x \in \mathcal{U}$: $x \in A \Leftrightarrow \chi_A(x) = 1 \Leftrightarrow \chi_B(x) = 1 \Leftrightarrow x \in B$. This proves that $A = B$. Conversely, if $A = B$ then for any $x \in \mathcal{U}$: $\chi_A(x) = 1 \Leftrightarrow x \in A \Leftrightarrow x \in B \Leftrightarrow \chi_B(x) = 1$ which proves that $\chi_A = \chi_B$.

(c) Let $x \in \mathcal{U}$. If $x \in A \cap B$, then $\chi_{A \cap B}(x) = 1$ and $\chi_A(x) = \chi_B(x) = 1$ since $x \in A$ and $x \in B$. The result is true in this case. If $x \notin A \cap B$, then $\chi_{A \cap B}(x) = 0$ and either $x \notin A$ or $x \notin B$ which implies that $\chi_A(x) = 0$ or $\chi_B(x) = 0$. Thus $\chi_{A \cap B}(x) = \chi_A(x) \cdot \chi_B(x) = 0$ in this case. We conclude that $\chi_{A \cap B}(x) = \chi_A(x) \cdot \chi_B(x)$ for any $x \in \mathcal{U}$ and the functions $\chi_{A \cap B}$ and $\chi_A \cdot \chi_B$ are then equal. $\qquad\square$

Definition 5.7. Let A, B be two sets. The function $\pi_A : A \times B \to A$ defined by $\pi_A(a, b) = a$ is called the *projection* from $A \times B$ onto A. We define π_B similarly.

Example 5.12. Let $A = \{a, b, c, d\}$, $B = \{1, 2, 3, 4, 5\}$. Then $\pi_A(a, 2) = a$, $\pi_B(a, 2) = 2$, $\pi_A(d, 3) = d$ and $\pi_B(d, 3) = 3$. For the subset $\Gamma = \{(a, 5), (b, 1), (b, 4), (c, 3), (c, 5)\}$ of $A \times B$, $\pi_A(\Gamma) = \{a, b, c\}$ and

$\pi_B(\Gamma) = \{1, 3, 4, 5\}$. For the subset $C = \{a\}$ of A, $\pi_A^{-1}(C) = \{(a, 1), (a, 2), (a, 3), (a, 4), (a, 5)\}$.

Example 5.13. Consider the Cartesian plane $\mathbb{R} \times \mathbb{R}$ and the subset $\Gamma = \{(x, y) \in \mathbb{R} \times \mathbb{R}; y = x^2\}$. Using the notations of Definition 5.7, $A = B = \mathbb{R}$. We have that $\pi_A(\Gamma) = \mathbb{R}$ since every real number is the first coordinate of a point on the parabola $y = x^2$ (the domain of the function $y = x^2$ is $\mathbb{R}$). However, $\pi_B(\Gamma) = [0, \infty[$ since every second coordinate of a point on the parabola $y = x^2$ is a non-negative real number.

The next function plays a central role in combinatorics.

Definition 5.8. The factorial function is the function from $\mathbb{N}$ to $\mathbb{N}$ that assigns to every integer n its factorial, denoted by $n!$, defined as follows:

$$n! = \begin{cases} 1 & \text{if } n = 0 \\ n \cdot (n-1) \cdots 2 \cdot 1 & \text{if } n \geq 1 \end{cases}$$

Example 5.14. $4! = 4 \cdot 3 \cdot 2 \cdot 1 = 24$, $7! = 7 \cdot 6 \cdot 5 \cdot 4 \cdot 3 \cdot 2 \cdot 1 = 5040$.

The next two functions we define are handy tools in situations where real numbers need to be rounded to the "closet" integers.

Definition 5.9. The *ceiling* function, denoted by $\lceil \, \rceil$, is the function $\lceil \, \rceil : \mathbb{R} \to \mathbb{Z}$ defined by $\lceil x \rceil =$ the smallest integer greater than or equal to x. The *floor* function, denoted by $\lfloor \, \rfloor$, is the function $\lfloor \, \rfloor : \mathbb{R} \to \mathbb{Z}$ defined by $\lfloor x \rfloor =$ the largest integer less than or equal to x.

Note that by definition, $\lfloor n \rfloor = \lceil n \rceil = n$ for any integer n.

Example 5.15. $\lfloor \pi \rfloor = 3$, $\lceil \pi \rceil = 4$, $\lfloor -\frac{5}{2} \rfloor = -3$, $\lceil -\frac{5}{2} \rceil = -2$.

Theorem 5.3. *For any real number x and any integer n:*

(a) $\lceil x \rceil = n \Leftrightarrow n - 1 < x \leq n$

(b) $\lfloor x \rfloor = n \Leftrightarrow n \leq x < n + 1$

(c) $\lceil x \rceil < x + 1$

(d) $x < \lfloor x \rfloor + 1$

(e) $\lceil x \rceil - \lfloor x \rfloor = \begin{cases} 0 & \text{if } x \in \mathbb{Z} \\ 1 & \text{if } x \notin \mathbb{Z} \end{cases}$

(f) $\lceil x + n \rceil = \lceil x \rceil + n$

(g) $\lfloor x + n \rfloor = \lfloor x \rfloor + n$

Proof.

(a) If $\lceil x \rceil = n$, then in particular $x \leq n$. On the other hand, if $x \leq n - 1$, then $\lceil x \rceil \leq n - 1$ by definition of the ceiling function. This is a contradiction. We conclude that $x > n - 1$ and so $n - 1 < x \leq n$. Conversely, if $n - 1 < x \leq n$ then $\lceil x \rceil = n$ by definition of $\lceil x \rceil$.

(b) Left as an exercise (see Exercise (13) below).

(c) This is a direct consequence of part (a).

(d) This is a direct consequence of part (b).

(e) If $x \in \mathbb{Z}$, $\lceil x \rceil = \lfloor x \rfloor = x$ and so $\lceil x \rceil - \lfloor x \rfloor = 0$. If $x \notin \mathbb{Z}$, let $m = \lfloor x \rfloor$. By definition of the floor function and since x is not an integer, we must have that $m < x < m + 1$. This implies in particular that $\lceil x \rceil = m + 1$ by definition of the ceiling function. We conclude that $\lceil x \rceil - \lfloor x \rfloor = m + 1 - m = 1$ and the result follows.

(f) Let $x \in \mathbb{R}$ and let $m = \lceil x \rceil$. By part (a), $m - 1 < x \le m$. Adding n to all sides we get that $m + n - 1 < x + n \le m + n$ and so $\lceil x + n \rceil = m + n = \lceil x \rceil + n$ by part (a) once again.

(g) Left as an exercise (see Exercise (14) below). $\square$

Given two integers n and d with $d \ge 1$, the division algorithm that we will investigate in Chapter 6 states that there exist *unique* integers q and r such that $n = qd + r$ and $0 \le r < d$. Integers q and r are called the quotient and the remainder (respectively) of the division of n by d.

Definition 5.10. Let d be a positive integer. We define two functions associated with d. The function $\mathrm{mod}_d : \mathbb{Z} \to \mathbb{N}$ defined by $\mathrm{mod}_d(n) = $ the remainder of the division of n by d. If $\mathrm{mod}_d(n) = r$, we write $n \ (\mathrm{mod}\ d) = r$. The function $\mathrm{div}_d : \mathbb{Z} \to \mathbb{Z}$ defined by $\mathrm{div}_d(n) = $ the quotient of the division of n by d.

Example 5.16. $22 \ (\mathrm{mod}\ 6) = 4$, $\mathrm{div}_6(22) = 3$ since $22 = 3 \cdot 6 + 4$, $-5 \ (\mathrm{mod}\ 3) = 1$ and $\mathrm{div}_3(-5) = -2$ since $-5 = (-2)(3) + 1$.

Example 5.17. If today is a Friday, what day of the week will it be in 128 days from today?

Solution. Since every week consists of seven days, we know that it will be a Friday every multiple of seven days from today. So it is a Friday again in seven days, in 14 days and so on. Since $128 \ (\mathrm{mod}\ 7) = 2$, we know then that it will a Sunday 128 days from today.

Example 5.18. What is the 100th digit in the decimal expansion of $\frac{2}{7}$?

Solution. The decimal expansion of $\frac{2}{7}$ is $0.2857142857\ldots$ with 285714 as the periodical part. Since the periodical part contains 6 digits, we compute $100 \ (\mathrm{mod}\ 6) = 4$. This means that the 100th digit in the decimal expansion of $\frac{2}{7}$ is 7.

5.3.1   *Exercises*

(1) Let A be a set and B a subset of A. What is the relation between the identity function 1_A of A and the inclusion function i_B of B?

(2) Set $A = B = \mathbb{R}$. Determine $\pi_A(\Gamma)$ and $\pi_B(\Gamma)$ where $\Gamma = \{(x, y) \in A \times B; \ 2x^2 + 3y^2 = 12\}$.

(3) Let $e = 2.71828$ (approximation for the base of the natural logarithm) and $\phi = 1.618$ (approximation for the golden ratio). Find the value of each of the following.

   (a) $\lceil e \rceil$ (d) $\lfloor \phi \rfloor$ (g) $\lfloor \frac{e+\phi}{e} \rfloor$ $\star$ (i) $\lfloor \frac{e+\phi}{\phi} \rfloor$

   (b) $\lfloor e \rfloor$ $\star$ (e) $\lceil e + \phi \rceil$ (h) $\lceil \frac{e+\phi}{e} \rceil$ (j) $\lceil \frac{e+\phi}{\phi} \rceil$

   (c) $\lceil \phi \rceil$ (f) $\lfloor e + \phi \rfloor$

(4) Let $n \geq 1$ be an integer. Find the value of each of the following:

  $\star$ (a) $\lceil \frac{2n+1}{2} \rceil$ (b) $\lfloor \frac{2n+1}{2} \rfloor$ (c) $\lceil \frac{n+1}{n} \rceil$ (d) $\lfloor \frac{n+1}{n} \rfloor$

(5) Prove part (b) of Theorem 5.2.

$\star$(6) Prove part (d) of Theorem 5.2.

(7) Prove part (e) of Theorem 5.2.

(8) Prove part (f) of Theorem 5.2.

(9) Let $\mathcal{U} = \{a, b, c, d, e, f, g, h, i\}$. Given a subset A of $\mathcal{U}$, the characteristic function of A can be represented by a binary string of length 9 (since $|A| = 9$) by assigning 1 or 0 to each element of $\mathcal{U}$ following the (natural) order $a, b, \ldots, i$ of these elements. For example, if $A = \{b, d, h\}$ then χ_A is represented by the binary string 010100010. Conversely, every binary string of length 9 represents the characteristic function of a certain subset A of $\mathcal{U}$, and hence the set A. For example, the binary string 111000000 represents the set $A = \{a, b, c\}$.

   (a) In each case, give the characteristic function of the given subset as a binary string: (i) $\{b, e, f, i\}$; (ii) $\{a, c, e, g, h\}$; (iii) $\{c, d, e, f, g, h, i\}$.

   (b) In each case, give the subset of $\mathcal{U}$ with the characteristic function is represented by the given binary string of length 9: (i) 111110000; (ii) 101010100; (iii) 001100111.

$\star$(10) If x is a real number, what is the value of $\lfloor x \rfloor + \lfloor -x \rfloor$? Justify your answer (*Hint.* Consider two cases: $x \in \mathbb{Z}$ and $x \notin \mathbb{Z}$).

(11) Prove that $\lfloor \frac{n}{2} \rfloor = \begin{cases} \frac{n}{2} & \text{if } n \text{ is even} \\ \frac{n-1}{2} & \text{if } n \text{ is odd} \end{cases}$

(12) Let $a, b \geq 1$ be two integers. Prove that the number of integers less than or equal to a and divisible by b is $\lfloor \frac{a}{b} \rfloor$.

(13) Prove part (b) of Theorem 5.3.

$\star$(14) Prove part (g) of Theorem 5.3.

(15) Find each of the following integers.

$\quad$ (a) $26 \pmod 4$ $\qquad$ (d) $2771 \pmod{47}$ $\qquad$ (g) $\text{div}_{61}(2222)$

$\star$ (b) $-85 \pmod{13}$ $\quad$ $\star$ (e) $\text{div}_{41}(2211)$ $\qquad$ (h) $\text{div}_{98}(-21459)$

$\quad$ (c) $-228 \pmod{23}$ $\quad$ (f) $\text{div}_{13}(-221)$

(16) Today is Tuesday. In each case, determine what day of the week it is after n days.

$\quad$ (a) $n = 70$ $\qquad$ (b) $n = 254$ $\qquad$ (c) $n = 1543$ $\qquad$ (d) $n = 11149$

(17) In each case, find the nth digit after the decimal point in the decimal representation of $\frac{3}{7}$.

$\quad$ (a) $n = 70$ $\qquad$ (b) $n = 254$ $\qquad$ $\star$ (c) $n = 1543$ $\qquad$ (d) $n = 11149$

5.4 Binary operations

In this section, we explore an important class of functions. Addition, subtraction and multiplication of real numbers are examples of binary operations on $\mathbb{R}$. Properties of binary operations on a set A are the foundation of any algebraic structure on A.

Definition 5.11. A binary operation on a set A is a function $\lambda : A \times A \to A$. It is customary to denote a binary operation by $\star$ so we write $a \star b$ to denote the element $\lambda(a, b)$ of A.

Example 5.19. The standard addition of natural number is a binary operation on $\mathbb{N}$ while subtraction is not. Both addition and subtraction are binary operations on the set $\mathbb{Z}$ of integers. Division is not a binary operation on $\mathbb{Z}$ since, for example, $\frac{3}{4}$ does not result in an integer. We say in this case that $\mathbb{Z}$ is not closed under the division operation.

Example 5.20. Let A be a set, $\wp(A)$ the power set of A. Then $\cap$ and $\cup$ are examples of binary operations on $\wp(A)$.

Definition 5.12. Let $\star$ be a binary operation on a non-empty set A.

(a) We say that $\star$ is *commutative* if $a \star b = b \star a$ for any $a, b \in A$.

(b) We say that $\star$ is *associative* if $(a \star b) \star c = a \star (b \star c)$ for any $a, b, c \in A$.

(c) If $\circ$ is another binary operation on A, we say that $\star$ is *distributive* over $\circ$ if $a \star (b \circ c) = (a \star b) \circ (a \star c)$ for any $a, b, c \in A$.

(d) We say that the element $e \in A$ is an *identity element* for the binary operation $\star$ if $e \star x = x \star e = x$ for any $x \in A$.

Before we look at some examples, we prove the uniqueness of the identity element, when it exists.

Theorem 5.4. *Let $\star$ be a binary operation on a set A. If an identity element exists for $\star$, then it is unique.*

Proof. Assume that e, e' are two identity elements of A with respect to $\star$. On one hand, $e \star e' = e$ (since e' is an identity element). On the other hand, $e \star e' = e'$ (since e is an identity element). We conclude that $e = e'$ since they are both equal to $e \star e'$. $\qquad\qquad\square$

Example 5.21. On the set $\mathbb{R}$ of all real numbers, the binary operation of standard addition is both commutative and associative. The binary operation of standard subtraction is neither commutative ($1 - 2 \neq 2 - 1$) nor associative ($1 - (2 - 3) = 2$ but $(1 - 2) - 3 = -4$). The binary operation of standard multiplication of real numbers is both commutative and associative. Moreover, the multiplication is distributive over the addition: $x(y + z) = xy + xz$ for every $x, y, z \in \mathbb{R}$. The identity element of $\mathbb{R}$ with respect to the addition is 0, and with respect to the multiplication is 1. There is, however, no identity element of $\mathbb{R}$ with respect to the subtraction.

Example 5.22. Let A be a set, $\wp(A)$ the power set of A. The binary operation $\cap$ is both commutative and associative with identity element A (since for any $X \subseteq A$, $X \cap A = A \cap X = X$). The binary operation $\cup$ is also commutative and associative with identity element $\emptyset$ (since for any $X \subseteq A$, $X \cup \emptyset = \emptyset \cup X = X$). Moreover, $\cap$ is distributive over $\cup$ and vice a versa.

Example 5.23 (linear algebra). *Let M_{22} be the set of all 2×2 matrices with standard addition and multiplication of matrices:*

$$\begin{bmatrix} a & b \\ c & d \end{bmatrix} + \begin{bmatrix} x & y \\ z & t \end{bmatrix} = \begin{bmatrix} a+x & b+y \\ c+z & d+t \end{bmatrix}, \quad \begin{bmatrix} a & b \\ c & d \end{bmatrix} \cdot \begin{bmatrix} x & y \\ z & t \end{bmatrix} = \begin{bmatrix} ax+bz & ay+bt \\ cx+dz & cy+dt \end{bmatrix}.$$

Matrix addition is commutative and associative, with the zero matrix $\begin{bmatrix} 0 & 0 \\ 0 & 0 \end{bmatrix}$ as identity element. Matrix multiplication is not commutative, but *associative (see Exercise (7) below). The identity element of the matrix multiplication is* $\begin{bmatrix} 1 & 0 \\ 0 & 1 \end{bmatrix}$ *(see Exercise (8) below).*

Definition 5.13. Let $\star$ be a binary operation on a non-empty set A with identity element e. We say that the element $a \in A$ is *invertible* with respect to $\star$ if there exists an element $b \in A$ such that $a \star b = b \star a = e$. The element b is called an inverse of a with respect to $\star$.

We prove that the inverse element, when it exists, is unique provided that the binary operation is associative.

Theorem 5.5. *Let $\star$ be an associative binary operation on a non-empty set A having an identity element e. If $a \in A$ is invertible, then its inverse is unique.*

Proof. Assume that $b, c \in A$ are two inverses of an invertible element $a \in A$. Then $a \star b = b \star a = e$ and $a \star c = c \star a = e$. Notice that

$$
\begin{aligned}
b &= b \star e && \text{(Since } e \text{ is the identity element of } \star) \\
&= b \star (a \star c) && \text{(Since } a \star c = e) \\
&= (b \star a) \star c && \text{(Since } \star \text{ is associative)} \\
&= e \star c && \text{(Since } b \star a = e) \\
&= c && \text{(Since } e \text{ is the identity element of } \star)
\end{aligned}
$$

We conclude that $b = c$ and the inverse of a in unique. $\square$

Remark 5.1. If we drop the assumption that the operation $\star$ is associative, then there is no guaranty that the inverse is unique (see Exercise (17) below).

Notation 5.1. If $a \in A$ is invertible with respect to the associative binary operation $\star$, then the (unique) inverse of a is denoted by a^{-1}. So a^{-1} is the unique element of A such that $a \star a^{-1} = e$ and $a^{-1} \star a = e$ where e is the identity element of A. The reader should be careful not to confuse a^{-1} with $\frac{1}{a}$ as we the set A is not necessarily a subset of $\mathbb{R}$ and the element a could be a matrix, a vector or any other object.

Example 5.24. For $a \in \mathbb{R}$, the inverse of a with respect to the standard addition is $-a$ since $a + (-a) = -a + a = 0$ (the identity element of the addition). If a is a non-zero real number, then the inverse of a with respect to the standard multiplication is $a^{-1} = \frac{1}{a}$ since $a \cdot \frac{1}{a} = \frac{1}{a} \cdot a = 1$ (the identity element of the multiplication).

Example 5.25. Let $A = \begin{bmatrix} a & b \\ c & d \end{bmatrix} \in \mathbb{M}_{22}$. The inverse of A with respect to the standard addition of matrices is $\begin{bmatrix} -a & -b \\ -c & -d \end{bmatrix}$ since $\begin{bmatrix} a & b \\ c & d \end{bmatrix} + \begin{bmatrix} -a & -b \\ -c & -d \end{bmatrix} = \begin{bmatrix} 0 & 0 \\ 0 & 0 \end{bmatrix}$ (the identity element of the addition in $\mathbb{M}_{22}$). For matrix multiplication, the story is differient as not all elements of $\mathbb{M}_{22}$ are invertible. For example, $\begin{bmatrix} 1 & 2 \\ 2 & 5 \end{bmatrix} \in \mathbb{M}_{22}$ is invertible with respect to matrix multiplication but $\begin{bmatrix} 1 & 2 \\ 2 & 4 \end{bmatrix}$ is not (see Exercise (14) below).

5.4.1   *Exercises*

(1) In each case, determine if $\star$ determines a binary operation on the given set A. Justify your answer.

     $\star$ (a) $A = \mathbb{N}$, $x \star y = x + y - xy$.
         (b) $A = \mathbb{Z}$, $x \star y = x + y - xy$.
         (c) $A = [1, 4]$, $x \star y = \sqrt{x + y}$.
         (d) $A = \mathbb{Z}$, $x \star y = x^y$.
     $\star$ (e) $A = \mathbb{R} \backslash \mathbb{Q}$ (the set of all irrational numbers), $x \star y$ is the standard multiplication of real numbers.
         (f) $A = \mathbb{Q}^{+}$ (positive rational numbers), $x \star y = x^{\sqrt{y}}$.

(2) Given a binary operation $\star$ on a set A, we say that the subset S of A is *closed under* $\star$ if $x \star y \in S$ for any $x, y \in S$. In each case, determine if the subset S of A is closed under the given binary operation on A.

         (a) $A = \mathbb{Z}$, $S = \mathbb{N}$, $x \star y = x + y$.
         (b) $A = \mathbb{Z}$, $S = \mathbb{N}$, $x \star y = x - y$.
     $\star$ (c) $A = \mathbb{N}$, $S = \{1, 2, 3, 4, 6, 12, 39\}$, $x \star y =$ the smallest element of $\mathbb{N}$ which is a common multiple of x and y.
         (d) $A = \mathbb{Z}$, $S = \{1, 3, 4, 6, 8, 12, 36\}$, $x \star y =$ the greatest element of $\mathbb{N}$ which is a common divisor of x and y.
     $\star$ (e) $A = \mathbb{R} \times \mathbb{R}$, $S = \mathbb{Z} \times \mathbb{Z}$, $(x, y) \star (a, b) = (x + a, y + b)$.

(3) In each case, determine if the binary operation $\star$ on the set A is commutative, associative and if it has an identity element. Justify your answers.

$\star$ (a) $A = \mathbb{Q}$, $a \star b = \frac{ab}{3}$

(b) $A = \mathbb{Q}$, $a \star b = ab + 1$

(c) $A = \mathbb{N}\backslash\{0\}$, $a \star b = a^b + 1$

(d) $A = \mathbb{Z}$, $a \star b = 3^{ab}$

(e) $A = \mathbb{R}\backslash\{1\}$, $a \star b = \frac{b}{a-1}$

$\star$ (f) $A = \mathbb{R}$, $a \star b = a + b + 2$

(4) Give a binary operation on $\mathbb{R}$ which is:

    (a) commutative but not associative

    (b) associative but not commutative

(5) Let $\star$ be a binary operation on a set A. Prove that if $\star$ has an identity element e, then e is invertible. What is e^{-1}?

(6) In each case, determine the number of *different* binary operations on a finite set A of cardinality n: (i) $n = 1$, (ii) $n = 2$ (iii) $n = 3$.

(7) Prove that matrix multiplication of 2×2 real matrices is not commutative but it is associative.

(8) Prove that the identity element of the multiplication of 2×2 matrices is $\begin{bmatrix} 1 & 0 \\ 0 & 1 \end{bmatrix}$.

(9) Let A be a non-empty set and let $\wp(A)$ its power set.

    (a) What elements of $\wp(A)$, if any, are invertible with respect to the binary operation $\cap$ on $\wp(A)$?

    $\star$ (b) What elements of $\wp(A)$, if any, are invertible with respect to the binary operation $\cup$ on $\wp(A)$?

(10) Recall that if B and C are two subsets of a universal set $\mathcal{U}$, then the symmetric difference of B and C is the subset $B \oplus C = (B \cup C)\backslash(B \cap C)$ of $\mathcal{U}$. This defines a binary relation on $\wp(\mathcal{U})$. In Chapter 2, we have seen that $\oplus$ is both commutative and associative.

    (a) Show that $\wp(\mathcal{U})$ has an identity element with respect to $\oplus$.

    (b) Show that every element of $\wp(\mathcal{U})$ has an inverse with respect to $\oplus$. Give the inverse X^{-1} of an element $X \in \wp(\mathcal{U})$.

$\star$(11) Let $\star$ be an associative binary operation on a set A and suppose that A has an identity element e. Give a careful proof that if $a \in A$ is invertible, then so is its inverse a^{-1}. What is $(a^{-1})^{-1}$?

$\star$(12) Let $A = \{a, b, c, d, e\}$. A binary operation $\star$ on A is defined by the table below, known as Cayley Table. In the table, the value of each composition $x \star y$ is given in the cell where the row of x and the column of y intersect. For example, $a \star c = c$, $b \star d = e$ and $e \star b = b$.

$\star$	a	b	c	d	e
a	a	b	c	d	e
b	b	d	a	e	b
c	c	a	a	c	e
d	d	e	c	e	a
e	e	b	e	a	c

(a) Show that $\star$ is commutative.

(b) Show that $\star$ is not associative.

(c) Show that A has an identity element with respect to $\star$.

(d) For each element of A, determine if it is invertible and if it is give one inverse.

(13) Let $A = \{1, 2, 3, 4, 5\}$. Consider the binary operation $\star$ on A defined by $x \star y = \max\{x, y\}$, where $\max\{x, y\}$ stands for the maximum of x and y.

$\star$ (a) Form the Cayley Table (see Exercise (12) above) showing the composition $x \star y$ for every $x, y \in A$.

(b) Show that $\star$ is commutative.

$\star$ (c) Show that $\star$ is associative.

(d) Determine the identity element of A with respect to $\star$ if it exists.

(e) For each element of A, determine if it is invertible and if you say it is, give its inverse.

(14) Let $A = \begin{bmatrix} a & b \\ c & d \end{bmatrix} \in M_{22}$. Prove that if $ad - bc \neq 0$, then A is invertible with respect to matrix multiplication with inverse

$$A^{-1} = \begin{bmatrix} \frac{a}{ad-bc} & \frac{b}{ad-bc} \\ \frac{c}{ad-bc} & \frac{d}{ad-bc} \end{bmatrix}.$$

(15) Consider the binary operation $\star$ on the cartesian product $\mathbb{R} \times \mathbb{R}$ defined by: $(a, b) \star (c, d) = (a + c, b + d + 1)$.

(a) Is $\star$ commutative? associative? Justify your answer.

(b) Does $\mathbb{R} \times \mathbb{R}$ have an identity element with respect to $\star$? Justify your answer.

(16) Give an example of a binary operation on a non-empty set A that has an identity element e but no element other than e is invertible.

$\star$(17) Let $A = \{1, 2, 3\}$ and consider the binary relation $\star$ on A given by the following table of operations:

$\star$	1	2	3
1	1	2	3
2	2	1	1
3	3	1	1

(a) Show that $\star$ is not associative.

(b) Show that $\star$ has an identity element.

(e) Give an element $a \in A$ with more than one inverse element.

$\star$(18) Let $n \geq 2$ be an integer. On the set $A = \{0, 1, 2, \ldots, n-1\}$, consider the binary operation $\star$ defined as follows:

$$x \star y = \begin{cases} x + y & \text{if } x + y < n \\ x + y - n & \text{if } x + y \geq n \end{cases}$$

(a) Show that $\star$ is a binary operation on A.

(b) Show that $\star$ has an identity element.

(e) Show that every element $a \in A$ is invertible and give an inverse for an arbitrary element $a \in A$.

(19) Let Σ be an alphabet, Σ^* be the set of all finite strings on Σ. Recall that if $s, t \in \Sigma^*$, then st represents the concatenation of s and t (the string s followed by the string t). Consider the binary operation on Σ^* defined by the concatenation of strings. Is this operation commutative? associative? does it have an identity element?

5.5 Functions defined recursively

In Chapter 2, the notion of recursively defined sets was introduced. In Chapter 3, we dealt with real-valued recursively defined sequences in the context of proof by induction. In this section, we extend the notion of recursively defined procedures to functions. Note that any real-valued sequence can be considered as a real-valued function with the domain being some subset X of $\mathbb{N}$. In simple terms, a recursively defined function is a function defined in terms of "itself" in the following sense: to construct $f(x)$, we use values of the same function f at some previously computed images $f(y)$ with $y \neq x$.

Definition 5.14. Let S be a recursively defined set. Recall that this means that S is defined by a base step, a recursive step and a restriction step. We can construct a function f having S as the domain using the following steps.

Base step. For each element x in the base step of S, a value $f(x)$ is given.

Recursion step. A set of rules is given that allows, for any recursively defined element x of S, to define $f(x)$ in terms of previously constructed values of f.

Restriction step. Only values obtained from the previous two steps are valid images of the function f.

A function defined this way is said to be *recursively defined* on S.

While Definition 5.14 applies to any function on a recursively defined domain, we will mainly consider functions with $\mathbb{N}$ (or a subset of $\mathbb{N}$) as domain. In this case, the base step consists of giving the value of $f(0)$ and the recursion step consists of giving the value of $f(n)$ in terms of $f(0),\ldots,$ $f(k)$ for some $k < n$. For simplicity, we also assume that the restriction step is satisfied and we will not bother to mention it every time.

Remark 5.2. When defining a function recursively, one has to make sure that using the base and the recursion steps, the image $f(x)$ of any element x in the domain can be computed. One also must be careful not to create a rule in the recursive step relating $f(x)$ with $f(y)$ where y is not in the domain of f.

Example 5.26. Determine if the following rule represents a valid recursively defined function from $\mathbb{N}$ to $\mathbb{N}$.

Base step. $f(0) = 1$.
Recursion step. If $n \geq 1$, then $f(n) = 3f(n-2) + 2$.

Solution. This is not a valid recursively defined function since $f(1) = 3f(-1) + 2$ and $f(-1)$ is not defined. In other words, both steps are not enough to compute $f(1)$ and so f is not a well-defined function on $\mathbb{N}$. $\Diamond$

Example 5.27. A function $f : \mathbb{N} \to \mathbb{N}$ is defined recursively as follows:

Base step. $f(0) = 1$, $f(1) = 1$.
Recursion step. If $n \geq 2$, then $f(n) = 3f(n-1) + 2f(n-2)$.

Find the value of $f(3)$, $f(5)$ and $f(7)$.

Solution. Using the rule of the recursion step, we get that $f(2) = 3f(1) + 2f(0) = 5$ (by the base step). Now that we know the value of $f(2)$, we can compute: $f(3) = 3f(2) + 2f(1) = 17$ and $f(4) = 3f(3) + 2f(2) = 61$. We conclude that $f(5) = 3f(4) + 2f(3) = 217$. We still need $f(6)$ to compute $f(7)$: $f(6) = 3f(5) + 2f(4) = 773$. Finally, $f(7) = 3f(6) + 2f(5) = 2753$.

Example 5.28. Recall that if $n \geq 0$ is an integer, then the factorial of n, denoted by $n!$, is defined by $0! = 1$ and $n! = n(n-1)(n-2)\cdots(3)(2)(1)$ for $n \geq 1$. Give a recursive definition of the function $f : \mathbb{N} \to \mathbb{N}$ defined by $f(n) = n!$. Prove the correctness of your definition.

Solution. Note that $0! = 1$ and $n! = n \times (n-1)!$ for $n \geq 1$. Let $f : \mathbb{N} \to \mathbb{N}$ be the function defined recursively as follows:

Base step. $f(0) = 1$.
Recursion step. If $n \geq 1$, then $f(n) = nf(n-1)$.

We prove that $f(n) = n!$ using induction on n. For the base step, note that $f(0) = 1 = 0!$ by definition. For the inductive step, assume that $f(n) = n!$ for $n \geq 0$. We need to prove that $f(n+1) = (n+1)!$. By the recursive definition of f, $f(n+1) = (n+1)f(n) = (n+1) \cdot n!$ (by the induction hypothesis) $= (n+1)!$ (by definition of the factorial of an integer). By the principle of induction, $f(n) = n!$ for any $n \geq 0$ and the above recursive definition of f is indeed that of the factorial function. $\Diamond$

Example 5.29. Give a recursive definition of the function f on the set of positive integers defined by $f(n) = n(n+1)$ for any integer $n \geq 1$. Prove that your definition is correct.

Solution. Note that $f(1) = 2$ and for $n \geq 2$, $f(n) = n(n+1) = (n-1)n + 2n = f(n-1) + 2n$. Let $g : \mathbb{N}\backslash\{0\} \to \mathbb{N}$ defined recursively as follows:

Base step. $g(1) = 2$.
Recursion step. If $n \geq 2$, then $g(n) = g(n-1) + 2n$.

We prove that $f(n) = g(n)$ for any $n \geq 1$ using induction on n. For $n = 1$, $g(1) = 2$ by definition and $f(1) = 1 \cdot (1+1) = 2$ so the base step is true. For the inductive step, let $n \geq 1$ and assume that $f(n) = g(n)$. We need to prove that $f(n+1) = g(n+1)$. By the recursive definition of g, $g(n+1) = g(n) + 2(n+1) = f(n) + 2(n+1)$ (by the induction hypothesis) $= n(n+1) + 2(n+1) = (n+1)(n+2) = f(n+1)$. By the principle of induction, $f(n) = g(n)$ for any $n \geq 1$ and the above recursive definition of f is indeed correct. $\Diamond$

Example 5.30. Give a recursive definition of the function $f : \mathbb{N} \to \mathbb{N}$ defined by $f(n) = 3^n$ for $n \geq 0$. Prove that your definition is correct.

Solution. Start by noticing that $f(0) = 3^0 = 1$ and for $n \geq 1$, $f(n) = 3^n = 3 \cdot 3^{n-1} = 3f(n-1)$. Consider then the function $g : \mathbb{N} \to \mathbb{N}$ defined recursively as follows.

Base step. $g(0) = 1$.
Recursion step. If $n \geq 1$, then $g(n) = 3g(n-1)$.

We prove that $f(n) = g(n)$ for $n \geq 0$ using induction on n. For $n = 0$, $f(0) = 3^0 = 1$ and $g(0) = 1$ by definition of g. The base step is true. For the inductive step, let $n \geq 1$ and assume that $f(n) = g(n)$. We need to prove that $f(n+1) = g(n+1)$. By the recursive definition of g, $g(n+1) = 3g(n) = 3f(n)$ (by the induction hypothesis) $= 3 \cdot 3^n = 3^{n+1} = f(n+1)$. By the principle of induction, $f(n) = g(n)$ for any $n \geq 0$ and the above recursive definition of f is indeed correct. $\Diamond$

Example 5.31. Let Σ be an alphabet and let Σ^* be the set of all finite strings (words) on Σ. If $w \in \Sigma^*$ then the length of w, denoted by $|w|$, is the number of characters appearing in w. Recall that the empty string λ is the one with length zero. Give a recursive definition of the function $l : \Sigma^* \to \mathbb{N}$ defined by $l(w) = |w|$.

Solution. Recall that a recursive definition of the set Σ^* is given as follows.

Base step. $\lambda \in \Sigma^*$.
Recursion step. If $w \in \Sigma^*$ and $x \in \Sigma$, then $wx \in \Sigma^*$.

Using this recursive definition of Σ^*, we can define the function l recursively as follows:

Base step. $l(\lambda) = 0$.
Recursion step. If $w \in \Sigma^*$ and $x \in \Sigma$, then $l(wx) = l(w) + 1$. $\Diamond$

Example 5.32. Given an alphabet Σ, the *reverse* of a string $w = x_1 x_2 \cdots x_n \in \Sigma^*$ is defined as being the string $w^R = x_n x_{n-1} \cdots x_1$. For example, the reverse of the string $w = abbacd$ is $w^R = dcabba$. Give a recursive definition of the function $R : \Sigma^* \to \Sigma^*$ defined by $R(w) = w^R$ using a recursive definition of the set Σ^*.

Solution. Using the recursive definition of the set Σ^* given in Example 5.31, the function R can be defined recursively as follows:

Base step. $R(\lambda) = \lambda$ where λ is the empty string.
Recursion step. If $w \in \Sigma^*$ and $x \in \Sigma$, then $R(wx) = xR(w)$. $\diamond$

5.5.1 *Exercises*

(1) In each case, determine if the given rule represents a valid recursively defined function with domain $\mathbb{N}$.

   $\star$ (a) $f(0) = \frac{2}{3}$, $f(n) = 2f(n-2) + n$ for any $n \geq 1$.
     (b) $f(0) = \frac{2}{3}$, $f(1) = 3$, $f(n) = 2f(n-2) + n$ for any $n \geq 2$.
     (c) $f(0) = 2$, $f(n) = 3f(n-1) + 5f(n-2) + n$ for any $n \geq 2$.
     (d) $f(0) = 2$, $f(1) = 0$, $f(n) = 3f(n-2) - 2f(n-3)$ for any $n \geq 2$.
   $\star$ (e) $f(0) = 2$, $f(1) = 0$, $f(n) = 3f(n-1) - 2f(n-2)$ for any $n \geq 2$.
     (f) $f(0) = 2$, $f(1) = 0$, $f(n) = 3f(n-1)$ for any $n \geq 1$.

(2) A function $f : \mathbb{N} \to \mathbb{Z}$ is defined recursively as follows:

    Base step. $f(0) = -1$, $f(1) = 2$.
    Recursion step. If $n \geq 2$, then $f(n) = 2f(n-1) - 3f(n-2)$.

    Find the value of $f(3)$, $f(5)$ and $f(7)$.

$\star$(3) The function $f : \mathbb{N} \to \mathbb{Q}$ is defined recursively as follows:

    Base step. $f(0) = 1$, $f(1) = 3$.
    Recursion step. If $n \geq 2$, $f(n) = \frac{f(n-1)}{f(n-2)} + 3n$.

    Find the values of $f(2)$, $f(4)$, $f(6)$ and $f(7)$.

(4) In each case, give a recursive definition of the function $f : \mathbb{N} \to \mathbb{N}$ defined by an explicit formula. Prove that your definition is correct using mathematical induction.

    (a) $f(n) = 5n$ (c) $f(n) = n^2$ (e) $f(n) = \frac{n(n+1)}{2}$
  $\star$ (b) $f(n) = 5^n$ (d) $f(n) = n^3$ $\star$ (f) $f(n) = 3^n - 2^n$

(5) Consider the function $f : \mathbb{N} \to \mathbb{Z}$ defined recursively as follows:

    Base step. $f(0) = 2$, $f(1) = 7$.
    Recursion step. If $n \geq 2$, then $f(n) = f(n-1) + 2f(n-2)$.

    Use mathematical induction to prove that $f(n) = 3 \cdot 2^n - (-1)^n$ for any $n \geq 0$.

$\star$(6) Let $f : \mathbb{N} \to \mathbb{N}$ be the function defined recursively as follows:

    Base step. $f(0) = 5$.
    Recursion step. If $n \geq 1$, $f(n) = 2 + f(n-1)$.

Find a formula for $f(n)$ an prove that your formula is correct using induction.

(7) Consider the function $f : \mathbb{N} \to \mathbb{N}$ defined recursively as follows:

   Base step. $f(0) = 1$.

   Recursion step. If $n \geq 1$, then $f(n) = f(n-1) + (2n-1)$

Find a formula for $f(n)$ an prove that your formula is correct using induction.

$\star$(8) Joe deposited \$5000.00 in a bank account that pays an interest rate of 11% compound annually. Give a recursive definition of the function $B(n)$ that represents the amount Joe will have in this account at the end of nth year where $n \geq 1$ is an integer.

(9) (McCarthy's 91 function) Consider the function $M : \mathbb{N}\backslash\{0\} \to \mathbb{N}$ defined recursively as follows:

$$M(n) = \begin{cases} M(M(n+11)) & \text{if } 1 \leq n \leq 100 \\ n - 10 & \text{if } n > 100 \end{cases}$$

Compute the values of each of the following expressions:

(a) $M(110)$ (c) $M(100)$ (e) $M(90)$ (g) $M(80)$
(b) $M(101)$ (d) $M(99)$ (f) $M(89)$ (h) $M(70)$

(10) (Ackermann function) Consider the function $A : \mathbb{N} \times \mathbb{N} \to \mathbb{N}$ defined recursively as follows:

$$A(m,n) = \begin{cases} n+1 & \text{if } m = 0 \\ A(m-1,1) & \text{if } m \geq 1 \text{ and } n = 0 \\ A(m-1, A(m,n-1)) & \text{if } m \geq 1 \text{ and } n \geq 1 \end{cases}$$

Compute the values of each of the following expressions:

(a) $A(0,0)$ $\star$ (c) $A(0,1)$ (e) $A(2,0)$ $\star$ (g) $A(1,3)$
(b) $A(1,0)$ (d) $A(1,1)$ (f) $A(2,1)$ (h) $A(2,2)$

$\star$(11) With the notations of Exercise (10), prove that $A(1,n) = n+2$ for $n \geq 0$ using induction on n.

(12) With the notations of Exercise (10), prove that $A(2,n) = 3 + 2n$ for $n \geq 0$ using induction on n (*Hint.* Use Exercise (11)).

(13) With the notations of Exercise (10), prove that $A(3,n) = 2^{n+3} - 3$ for $n \geq 0$ using induction on n (*Hint.* Use Exercise (12)).

(14) Let $f : \mathbb{N} \to \mathbb{N}$ be the polynomial function $f(n) = An^2 + Bn + C$ on $\mathbb{N}$ of degree 2, where A, B, C are fixed in $\mathbb{N}$ with $A \neq 0$.

(a) Prove that the function $g : \mathbb{N} \to \mathbb{N}$ defined by $g(n) = f(n+1) - f(n)$ is a linear function of n.

(b) Prove that the function $h : \mathbb{N} \to \mathbb{N}$ defined by $h(n) = g(n+1) - g(n)$ is a constant function of n.

$\star$ (c) Consider the function $f : \mathbb{N} \to \mathbb{N}$ defined recursively as follows:

$$f(n) = \begin{cases} 1 & \text{if } n = 0 \\ f(n-1) + 2n + 1 & \text{if } n \geq 1 \end{cases}$$

(i) Compute $f(n)$ for $n = 0, 1 \ldots, 8$.

(ii) Compute the values of the difference function $g(n) = f(n+1) - f(n)$ for $n = 0, 1 \ldots, 7$.

(iii) Compute the values of the difference function $h(n) = g(n+1) - g(n)$ for $n = 0, 1 \ldots, 6$. What do you notice?

(iv) Use parts (a) and (b) to suggest a general closed-form expression for $f(n)$.

(v) Find all coefficients in your expression of the previous part.

(vi) Prove (using induction) that your closed-form expression of the previous part is indeed equal to $f(n)$.

5.6 Injective, surjective and bijective functions

In this section, we look at some classes of functions that play an important roles in many areas of mathematics.

Definition 5.15. A function $f : A \to B$ is called *injective* (also called *one-to-one* or an *injection*) if for every elements $x, y \in A$, the condition $f(x) = f(y)$ implies that $x = y$. In other words, a function is injective if it assigns two distinct elements of the codomain to two distinct elements of the domain.

Example 5.33. The function $f : \mathbb{R} \to \mathbb{R}$ defined by $f(x) = \frac{1}{1+x^2}$ is not injective. To see this, we need to find two distinct real numbers with the same image. Take $x = -1$ and $y = 1$, then $x \neq y$ but $f(x) = f(y) = \frac{1}{2}$.

Example 5.34. If we restrict the domain of the function in the previous example to $[0, \infty[$, then the function becomes injective. To see this, let $x, y \in [0, \infty[$ such that $f(x) = f(y)$ then $\frac{1}{1+x^2} = \frac{1}{1+y^2}$ and so $x^2 = y^2$. This imply that $x = y$ since x, y are positive.

Example 5.35. Let $\mathbb{M}_{22}$ be the set of all 2×2 matrices with real entries. If $A = \left[\begin{smallmatrix} a & b \\ c & d \end{smallmatrix}\right] \in \mathcal{M}_{22}$ then the determinant of A, denoted by $\det(A)$, is the real number $\det(A) = ad - bc$. The function $\det : \mathbb{M}_{22} \to \mathbb{R}$ assigning each matrix to its determinant is not injective since $\det\left(\left[\begin{smallmatrix} 1 & 0 \\ 0 & 1 \end{smallmatrix}\right]\right) = \det\left(\left[\begin{smallmatrix} -1 & 0 \\ 0 & -1 \end{smallmatrix}\right]\right) = 1$.

Example 5.36. Let $f : \mathbb{R} \times \mathbb{R} \to \mathbb{R} \times \mathbb{R}$ defined by $f(x, y) = (-x + y, x + y)$ (here, for simplicity, we use the notation $f(x, y)$ instead of $f((x, y))$ to denote the image of the element (x, y)). If (x, y), (z, t) are two elements of $\mathbb{R} \times \mathbb{R}$ such that $f(x, y) = f(z, t)$, then $(-x + y, x + y) = (-z + t, z + t)$. This implies that $-x + y = -z + t$ and $x + y = z + t$. Adding these two equations, we obtain that $2y = 2t$ and therefore $y = t$. From this (using either one of the equations), we get that $x = z$. We conclude that $(x, y) = (z, t)$ and the function is injective.

Example 5.37. Given a set A and a subset B of A, it is clear that the identity map 1_A of A (assigning each element of A to itself) and the inclusion map from B to A (the restriction of 1_A to B) are injective.

Example 5.38. In this example, we prove that the function $f : \mathbb{Z} \to \mathbb{Z}$ defined by:

$$f(n) = \begin{cases} n & \text{for } n \geq 0 \\ n - 1 & \text{for } n < 0 \end{cases}$$

is injective. Let $a, b \in \mathbb{Z}$ such that $f(a) = f(b)$. We prove that $a = b$ using a proof by separation of cases. There are three possible cases: $a \geq 0$ and $b \geq 0$, $a > 0$ and $b < 0$ and $a < 0$ and $b < 0$ (the case $a < 0$ and $b > 0$ can be omitted as being similar to the case $a > 0$ and $b < 0$). We prove that $a = b$ in each case. Note that if $a \geq 0$ and $b < 0$, then $f(a) = a \neq f(b) = b - 1$ which contradicts the assumption $f(a) = f(b)$ and so this case cannot occur. If $a \geq 0$ and $b \geq 0$ then $f(a) = a$ and $f(b) = b$. The relation $f(a) = f(b)$ implies that $a = b$ in this case. If $a < 0$ and $b < 0$ then $f(a) = a - 1$ and $f(b) = b - 1$ and the relation $f(a) = f(b)$ also implies that $a = b$ in this case. We conclude that f is injective.

Definition 5.16. A function $f : A \to B$ is called *surjective* (also called *onto* or a *surjection*) if $\text{Im}(f) = B$. In other words, f is surjective if and only if for every $b \in B$, there exists $a \in A$ such that $b = f(a)$.

Example 5.39. The function $f : \mathbb{R} \to \mathbb{R}$ defined by $f(x) = \frac{1}{1+x^2}$ is not surjective. To see this, notice that $\text{Im}(f) =]0, \infty[\neq \mathbb{R}$.

Example 5.40. A closer look at the function f of Example (5.39) shows that each image $f(x) = \frac{1}{1+x^2}$ satisfies $0 < f(x) \leq 1$. If we restrict the codomain of the function to $]0, 1]$, then the function becomes surjective. The inclusion $\text{Im}(f) \subseteq]0, 1]$ is true by the above observation. To prove the inclusion $]0, 1] \subseteq \text{Im}(f)$, we fix $y \in]0, 1]$ and we prove it is possible to find a real number x such that $y = f(x)$. Solving for x in the equation $y = \frac{1}{1+x^2}$ gives us two choices $x = \pm\sqrt{\frac{1-y}{y}}$ (both are real numbers since $\frac{1-y}{y} > 0$ in this case).

Example 5.41. With the notations of Example (5.35), the function $\det : \mathcal{M}_{22} \to \mathbb{R}$ is surjective. To see this, fix a real number y. We need to find a 2×2 matrix A such that $\det(A) = y$. There are many possibilities. For example, the matrix $A = \begin{bmatrix} y & 0 \\ 1 & 1 \end{bmatrix}$ satifies $\det(A) = y$.

Example 5.42. The function $f : \mathbb{R} \times \mathbb{R} \to \mathbb{R} \times \mathbb{R}$ of Example (5.36) is surjective. To see this, we fix an element (c, d) in the codomain $(\mathbb{R} \times \mathbb{R})$ and we find an element (a, b) in the domain (also $\mathbb{R} \times \mathbb{R}$) such that $f(a, b) = (c, d)$. This last equation translates to $(-a + b, a + b) = (c, d)$ which gives us the two relations $-a + b = c$ and $a + b = d$. Solving for a and b, we get that $a = \frac{d-c}{2} \in \mathbb{R}$ and $b = \frac{c+d}{2} \in \mathbb{R}$. We conclude that $\text{Im}(f) = \mathbb{R} \times \mathbb{R}$ and f is surjective.

Example 5.43. Given a set A and a subset B of A, the identity map 1_A of A is surjective but the inclusion map from B to A is not.

Example 5.44. The function $f : \mathbb{Z} \to \mathbb{Z}$ defined by

$$f(n) = \begin{cases} n - 1 & \text{for } n \geq 0 \\ n & \text{for } n < 0 \end{cases}$$

is surjective. The proof is left as an exercise (see Exercise (24) below).

Definition 5.17. A function $f : A \to B$ is called *bijective* (or a bijection between A and B) if it is injective and surjective at the same time. In other words, f is bijective if and only if for every $b \in B$, there exists a *unique* $a \in A$ such that $b = f(a)$.

Example 5.45. The function $f : \mathbb{R} \to \mathbb{R}$ defined by $f(x) = \frac{1}{1+x^2}$ is not bijective since it not injective (nor surjective).

Example 5.46. The function $f : [0, \infty[\to]0, 1]$ defined by $f(x) = \frac{1}{1+x^2}$ is bijective by Examples (5.34) and (5.40) above.

Example 5.47. The function $f : \mathbb{R} \to \mathbb{R}$ defined by $f(x) = 3x + 2$ is bijective. For the injectivity, let $x, y \in \mathbb{R}$ such that $f(x) = f(y)$ then $3x+2 = 3y+2$ and consequently $x = y$. The function is injective. For the subjectivity, let $y \in \mathbb{R}$ be arbitrary, then it is straightforward to verify that the real number $x = \frac{y-2}{3}$ satisfies $f(x) = y$. The function is surjective.

Example 5.48. The function $f : \mathbb{R} \times \mathbb{R} \to \mathbb{R} \times \mathbb{R}$ defined by $f(x, y) = (-x + y, x + y)$ is bijective by Examples (5.36) and (5.42) above.

Example 5.49. The identity function of any set is bijective.

Given two finite sets A and B, the existence of injective or surjective functions between A and B has implications on the cardinalities $|A|$ and $|B|$ of A and B. More precisely, we have the following theorem.

Theorem 5.6. *Let A and B be two finite sets and let $f : A \to B$ be a function.*

> *(a) If f is injective, then $|A| \leq |B|$.*
> *(b) If f is surjective, then $|A| \geq |B|$.*
> *(c) If f is bijective, then $|A| = |B|$.*

Proof. Let $A = \{a_1, a_2, \ldots, a_n\}$ and $B = \{b_1, b_2, \ldots, b_m\}$. If $f : A \to B$ is injective, then the elements $f(a_1)$, $f(a_2)$, ..., $f(a_n)$ of B are pairwise distinct. This implies in particular that B contains at least n elements. This proves part (a). If $f : A \to B$ is surjective, then every element of B is an image. This implies that the list of images $f(a_1)$, $f(a_2)$, ..., $f(a_n)$ contains every element of B at least once and so $|B| \leq n$. This proves part (b). Part (c) is a direct consequence of the previous two parts. $\square$

Sometimes, a more useful way to interpret the result of Theorem 5.6 is to use the contrapositive of each part. Given two finite sets A and B:

> (a) If $|A| > |B|$, then no function $f : A \to B$ can be injective.
> (b) If $|A| < |B|$, then no function $f : A \to B$ can be surjective.
> (c) If $|A| \neq |B|$, then no function $f : A \to B$ can be bijective.

5.6.1 *Exercises*

(1) Consider the two sets $A = \{a, b, c, d\}$, $B = \{x, y, z, t\}$. Give a function $f : A \to B$ which is:

   (a) injective but not surjective.
   (b) surjective but not injective.
   (c) neither injective nor surjective.
   (d) bijective.

$\star$(2) Let $A = \{a, b, c\}$. List all bijective functions from A to itself.

(3) Let $f : A \to B$ be a function. Use a predicate logic formula to express the fact that f is:

   (a) injective   (b) surjective   (c) bijective

(4) In each case, determine if the given function from $\mathbb{N}$ to $\mathbb{N}$ is injective, surjective, bijective or if it has none of these properties. Justify your answer.

   (a) $f(n) = 2n + 3$ (c) $f(n) = 2^n$
   (b) $f(n) = n^2$ (d) $f(n) = n^2 - n$

(5) In each case, determine if the given function from $\mathbb{Z}$ to $\mathbb{Z}$ is injective, surjective, bijective or if it has none of these properties. Justify your answer.

   $\star$ (a) $f(n) = n - 5$ (c) $f(n) = 2n + 6$ (e) $f(n) = n^2 - 3n + 2$
      (b) $f(n) = n + 5$ (d) $f(n) = n^2$ $\star$ (f) $f(n) = (-1)^n + n$

(6) In each case, determine if the given function from $\mathbb{R}$ to $\mathbb{Z}$ is injective, surjective, bijective or has none of these properties. Justify your answer.

   (a) $f(x) = \lceil x \rceil$ (c) $f(x) = \lceil x \rceil - \lfloor x \rfloor$
   (b) $f(x) = \lfloor x \rfloor$ (d) $f(x) = \lceil x \rceil + \lfloor x \rfloor$

(7) For each of the following functions from $\mathbb{R}$ to $\mathbb{R}$, determine if it is injective, surjective, bijective or if it has none of these properties. Justify your answer. The notation $|x|$ denotes the absolute value of the real number x.

(a) $f(x) = \frac{x}{1+|x|}$

(b) $f(x) = -4x + 5$

$\star$ (c) $f(x) = |x| - x^2$

$\star$ (d) $f(x) = \sqrt[3]{x} + 1$

(e) $f(x) = x^3 - x$

(f) $f(x) = 2^{x^3+1}$

(8) In each case, a piecewise function $f : \mathbb{Z} \to \mathbb{Z}$ is given. Determine if the function is injective, surjective, bijective or if it has none of these properties. Justify your answer.

(a) $f(n) = \begin{cases} n & \text{if } n \geq 0 \\ n - 1 & \text{if } n \leq -1 \end{cases}$

(b) $f(n) = \begin{cases} n + 1 & \text{if } n \text{ is even} \\ n - 1 & \text{if } n \text{ is odd} \end{cases}$

$\star$ (c) $f(n) = \begin{cases} n & \text{if } n \text{ is even} \\ \frac{n-1}{2} & \text{if } n \text{ is odd} \end{cases}$

(9) Let Σ^* be the set of all binary strings. If $w = x_1 x_2 \cdots x_n \in \Sigma^*$, the reverse of w is defined as being the string $w^R = x_n x_{n-1} \cdots x_1$ and $l(w)$ is the number of bits in w. For each of the following functions, determine if it is injective, surjective, bijective or if it has none of these properties. Justify your answer.

$\star$ (a) $f : \Sigma^* \to \Sigma^*$, $f(w) = w^R$.

(b) $f : \Sigma^* \to \mathbb{N}$, $f(w) = l(w)$.

(c) $f : \Sigma^* \to \Sigma^*$, $f(w) = sw$: concatenation with a fixed string s.

$\star$ (d) $f : \Sigma^* \to \mathbb{N}$, $f(w) = $ number of 0's in w.

(10) For each of the following functions, determine if it is injective, surjective, bijective or if it has none of these properties. Justify your answer.

(a) $f : \mathbb{R} \times \mathbb{R} \to \mathbb{R}$, $f(x, y) = x + y$.

(b) $f : \mathbb{R} \times \mathbb{R} \to \mathbb{R} \times \mathbb{R}$, $f(x, y) = (3x + y, 2x - 3y)$.

(c) $f : \mathbb{Z} \to \mathbb{Z} \times \mathbb{Z}$, $f(m) = (m, m^2)$.

(d) $f : \mathbb{R} \times \mathbb{R} \to \mathbb{R}$, $f(x, y) = x^2 + y^2$.

(11) Give an example of a function $f : \mathbb{N} \to \mathbb{N}$ which is surjective but not injective (*Hint.* Think of a piecewise defined function).

$\star$(12) Give a bijective function $f : \mathbb{N} \to \mathbb{N}$ different from the identity function of $\mathbb{N}$. Justify your answer (*Hint.* Think of a piecewise defined function).

(13) Construct a bijective function $f : \mathbb{M}_{22} \to \mathbb{M}_{22}$ different from the identity function of $\mathbb{M}_{22}$.

(14) For real numbers a, b, prove that the function $f : \mathbb{R} \to \mathbb{R}$ defined by $f(x) = ax + b$ is bijective if and only if $a \neq 0$.

(15) Construct a bijection $f : \mathbb{N} \to E$ where E is the set of all non-negative even integers.

$\star$(16) Let $\mathbb{Z}^* = \mathbb{Z}\backslash\{0\}$ be the set of non-zero integers. Is the function $f : \mathbb{Z} \times \mathbb{Z}^* \to \mathbb{Q}$ defined by $f(m,n) = \frac{m}{n}$ injective, surjective? Justify.

(17) Consider the expression $f(x) = \frac{2x-1}{x-1}$ where $x \in \mathbb{R}$. Find subsets A and B of $\mathbb{R}$ such that $f : A \to B$ is a bijection. Justify your answer.

$\star$(18) Let $a < b$ be two distinct real numbers. Prove that the expression $f(t) = \frac{b-a}{2}t + \frac{b+a}{2}$ defines a bijection from $[-1, 1]$ to $[a, b]$. Start by proving that $f(t) \in [a, b]$ for any $t \in [-1, 1]$.

(19) Let $f : A \to B$ be a function and let C be a subset of A. In each case, determine the truth value of the given statement.

   (a) If f is injective, then so is the restriction $f\,|_C$ of f to C.

   (b) If f is surjective, then so is the restriction $f\,|_C$ of f to C.

(20) Let f and g be two functions from $\mathbb{R}$ to $\mathbb{R}$. Determine the truth value of each of the following statements. Justify your answer.

   (a) If f and g are injective, then $f + g$ is injective.

   (b) If $f + g$ is injective, then at least one of the functions f and g must be injective.

   (c) If $f + g$ is injective, then both functions f and g must be injective.

   (d) If f and g are surjective, then $f + g$ is surjective.

   (e) If $f + g$ is surjective, then at least one of the functions f and g must be surjective.

   (c) If $f+g$ is surjective, then both functions f and g must be surjective.

(21) Repeat Exercise 20 with $f + g$ replaced by fg.

$\star$(22) Let A be a non-empty set, $\wp(A)$ the powerset of A. Construct function from $f : A \to \wp(A)$ which is injective but not surjective.

(23) Prove that the function of Example (5.38) above is not surjective.

$\star$(24) Prove that the function of Example (5.44) above is surjective.

(25) Let A and B be two finite sets such that $|A| = 4$ and $|B| = 5$. In each case, determine the truth value the statement. Justify your answer.

   (a) A function $f : A \to B$ cannot be injective.

  $\star$ (b) A function $f : A \to B$ cannot be surjective.

   (c) A function $f : B \to A$ cannot be injective.

   (d) A function $f : B \to A$ cannot be surjective.

   (e) A function $f : \wp(A) \to A \times B$ cannot be injective.

  $\star$ (f) A function $f : \wp(A) \to A \times B$ cannot be surjective.

(26) Let A be a set, $\wp(A)$ be the powerset of A and let B be a subset of A.

    (a) Is the function $\phi : \wp(A) \to \wp(A)$ defined by $\phi(X) = X \cap B$ injective? surjective? Justify your answer.

    (b) Is the function $\psi : \wp(A) \to \wp(A)$ defined by $\psi(X) = X \cup B$ injective? surjective? Justify your answer.

(27) Let A be a non-empty set and let B be a subset of A. When is χ_B (the characteristic function of B) injective? surjective?

$\star$(28) Let A be a non-empty set with powerset $\wp(A)$. Let $\{0,1\}^A$ be the set of all function from A to $\{0,1\}$. Consider the function $\phi : \wp(A) \to \{0,1\}^A$ defined by $\phi(B) = \chi_B$ (the characteristic function of B). Prove that ϕ is bijective.

(29) Let A, B be two non-empty sets. Is the projection function $\pi_A : A \times B \to A$ (defined by $\pi_A(x,y) = x$) injective? surjective? What if one of the sets A, B is empty?

(30) Let $f : A \to B$ be a function, S_1 and S_2 be two subsets of A. Theorem 5.1 above shows that $f(S_1 \cap S_2) \subseteq f(S_1) \cap f(S_2)$. Prove that the equality holds if f is injective.

(31) Let A, B be two subsets of a universal set $\mathcal{U}$. As usual, $\wp(E)$ denotes the power set of the set E. Consider the function $f : \wp(A \cup B) \to \wp(A) \times \wp(B)$ defined by $f(X) = (X \cap A, X \cap B)$. Is f injective? surjective? Justify.

(32) Consider the function $f : \mathbb{N} \to \mathbb{N}$ defined recursively as follows:

$$f(0) = 1 \text{ and } f(n+1) = \begin{cases} \frac{1}{2} f(n) & \text{if } f(n) \text{ is even} \\ 3f(n) + 5 & \text{if } f(n) \text{ is odd} \end{cases}$$

    Is f injective? surjective?

(33) Let $f : A \to B$ be a function, X a subset of A and Y a subset of B.

    (a) Prove that $X \subseteq f^{-1}(f(X))$.

  $\star$ (b) Prove that if f is injective then $X = f^{-1}(f(X))$.

    (c) Prove that $f(f^{-1}(Y)) \subseteq Y$.

    (d) Prove that if f is surjective then $Y = f(f^{-1}(Y))$.

(34) Let A and B be two finite sets with the same cardinality and let $f : A \to B$ be a function. Prove that f is injective if and only if f is surjective.

5.7   Composition of functions and invertible functions

In Section 5.2.1, we saw how real-valued functions can be combined to create new ones. In this section, we introduce a more general way to combine functions.

Definition 5.18. Let $f : A \to B$, $g : B \to C$ be two functions such that the domain of g is the same as the codomain of f. The *composition* of g and f is the function from A to C, denoted by $g \circ f$, defined by: $(g \circ f)(x) = g\,(f(x))$ for any $x \in A$.

The image $(g \circ f)(a)$ of an element $a \in A$ is found in two steps: first the unique image of a under f (as element of B) is found and then the function g is applied to $f(a)$ to find the unique image $g\,(f(a))$ of $f(a)$ in C.

Remark 5.3. We could have a situation where $g \circ f$ is defined but $f \circ g$ is not. Even if both $g \circ f$ and $f \circ g$ are defined, they are not equal in general. It is also important to keep in mind that when dealing with the composition $g \circ f$, the function f comes first followed by the function g.

Example 5.50. Functions f and g are represented by the diagram below.

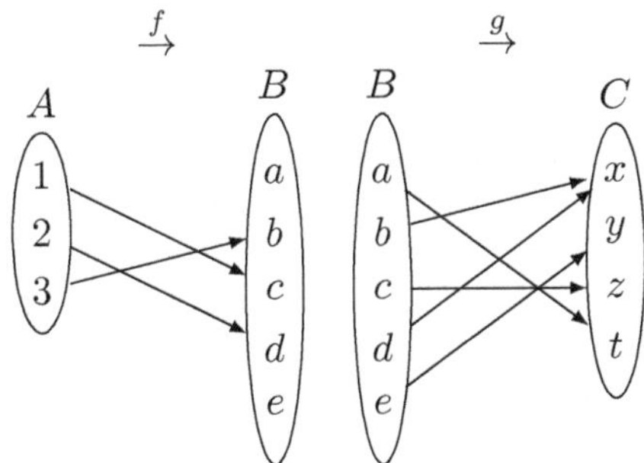

Draw a diagram representing the function $g \circ f$.

Solution. To draw the diagram representing $g \circ f$, we need to know the image of each element of A: $(g \circ f)(1) = g\,(f(1)) = g(c) = z$, $(g \circ f)(2) = g\,(f(2)) = g(d) = x$, and $(g \circ f)(3) = g\,(f(3)) = g(b) = x$. The diagram representing $g \circ f$ is the following:

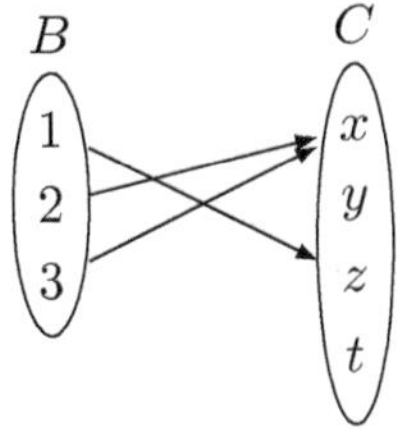

Example 5.51. Consider the functions $f : \mathbb{R} \to \mathbb{Z}$ defined by $f(x) = \lfloor x \rfloor$ (the floor function) and $g :\ ]0, \infty[\to \mathbb{R}$ defined by $g(x) = \ln(x)$ (the natural logarithm). Since the codomain of g is equal to the domain of f, the function $f \circ g$ from $]0, \infty[$ to $\mathbb{Z}$ is defined and it is given by $(f \circ g)(x) = f(g(x)) = \lfloor \ln(x) \rfloor$. The function $g \circ f$, on the other hand, is not defined as the codomain of f is not equal to the domain of g.

Example 5.52. Consider the functions $f : \mathbb{R} \times \mathbb{R} \to \mathbb{R}$ defined by $f(x, y) = -2x + 3y$ and $g : \mathbb{R} \to \mathbb{R}$ defined by $g(x) = x^3 + 1$. Since the codomain of f is equal to the domain of g, the function $g \circ f$ from $\mathbb{R} \times \mathbb{R}$ to $\mathbb{R}$ is defined and it is given by $(g \circ f)(x, y) = g(f(x, y)) = (-2x + 3y)^3 + 1$. The function $f \circ g$, on the other hand, is not defined as the codomain of g is not equal to the domain of f.

The next example shows that even if $g \circ f$ and $f \circ g$ are both defined, they need not to be equal. In other words, the composition of functions is not commutative.

Example 5.53. Consider the functions $f(x) = x^3 + 1$ and $g(x) = -2x + 3$ from $\mathbb{R}$ to $\mathbb{R}$. Clearly both $g \circ f$ and $f \circ g$ exist. For $x \in \mathbb{R}$, $(g \circ f)(x) = g(f(x)) = g(x^3 + 1) = -2(x^3 + 1) + 3 = -2x^3 + 1$, and $(f \circ g)(x) = f(g(x)) = f(-2x + 3) = (-2x + 3)^3 + 1$.

Given three functions $f : A \to B$, $g : B \to C$ and $h : C \to D$, we can form the function $g \circ f$ from A to C and then compose it with h to get the function $h \circ (g \circ f)$ from A to D. On the other hand, we can form the function $h \circ g$ from B to D and then compose it with f to get the function $f \circ (g \circ h)$ from A to D. The next theorem shows that these two ways of composing the three functions are equivalent. In other words, the theorem proves that composition of functions is associative.

Theorem 5.7. *If $f : A \to B$, $g : B \to C$ and $h : C \to D$ are functions, then $h \circ (g \circ f) = (h \circ g) \circ f$.*

Proof. We have to prove that the two functions take the same value at any given element of the domain A. For $x \in A$, $(h \circ (g \circ f))(x) = h((g \circ f)(x)) = h(g(f(x)))$. On the other hand, $((h \circ g) \circ f)(x) = ((h \circ g)(f(x)) = h(g(f(x)))$. This shows that $h \circ (g \circ f) = (h \circ g) \circ f$. $\square$

Theorem 5.7 tells us that we don't have to bother much with the parentheses in the expressions $h \circ (g \circ f)$ or $(h \circ g) \circ f$. We simply write $h \circ g \circ f$ for both.

Another important feature of Theorem 5.7 is that it allows us to define the notion of the power of a function $f : A \to A$.

Definition 5.19. Let $f : A \to A$ be a function. For any integer $n \geq 0$, define the function f^n from A to A as follows: if $n = 0$ then $f^n = 1_A$ (the identity function of A) and if $n \geq 1$, $f^n = \underbrace{f \circ f \circ \cdots \circ f}_{n \text{ times}}$.

Example 5.54. For the function $f(x) = -2x + 3$ from $\mathbb{R}$ to $\mathbb{R}$, $f^2(-1) = (f \circ f)(-1) = f(f(-1)) = f(5) = -7$ and $f^3(-1) = (f \circ f^2)(-1) = f(f^2(-1)) = f(-7) = 17$.

Example 5.55. Let A be a set, $f : A \to A$ be a function. Given a natural number n, we can define the function f^n recursively as follows:

Base step. $f^0 = 1_A$ where 1_A is the identity function of A.
Recursion step. If $n \geq 1$, then $f^n = f \circ f(n-1)$.

(The restriction step is omitted but assumed true.)

Theorem 5.8. *Let $f : A \to B$, $g : B \to C$ be two functions. Then:*

(a) If f and g are injective, then so is $g \circ f$.
(b) If f and g are surjective, then so is $g \circ f$.
(c) If f and g are bijective, then so is $g \circ f$.
(d) If $g \circ f$ is injective, then so is f.
(e) If $g \circ f$ is surjective, then so is g.

Proof.

(a) Left as an exercise (see Exercise (8) below).
(b) Suppose that both f and g are surjective. For any $z \in C$, there exists an element $y \in B$ such that $z = g(y)$ since g is surjective. The surjectivity of f implies that there exists an element $x \in A$ such that $y = f(x)$. Thus, $z = g(y) = g(f(x)) = (g \circ f)(x)$. This shows that $g \circ f$ is surjective.
(c) This is a direct consequence of parts (a) and (b).
(d) Assume that $g \circ f$ is injective, and let $a, b \in A$ such that $f(a) = f(b)$. Then $g(f(a)) = g(f(b))$ (by applying the function g) and so $(g \circ f)(a) = (g \circ f)(b)$. The injectivity of $g \circ f$ implies that $a = b$. We conclude that f is injective.
(e) Left as an exercise (see Exercise (9) below). $\qquad\square$

5.7.1   *Invertible functions*

If x is a non-zero real number, then the inverse of x with respect to the multiplication of real numbers is the (unique) real number y satisfying: $xy = 1$. We usually denote the inverse of x by $\frac{1}{x}$ or x^{-1}. We can extend this idea to functions. First notice that if $f : A \to B$ is a function, then composing f with the identity functions of A and B does not change f: $f \circ 1_A = f$ and $1_B \circ f = f$ (see Exercise (4) below). In this context, the composition of functions acts like the multiplication and the identity functions play the role of unit element 1. Using this analogy (keeping in mind that composition of functions is not commutative), we can formally define the notion of an invertible function.

Definition 5.20. A function $f : A \to B$ is called *invertible* if there exists a function $g : B \to A$ such that $g \circ f = 1_A$ and $f \circ g = 1_B$. The function g us called an *inverse* of f.

Like the multiplication of real numbers, the inverse of a function is unique when it exists as shown in the following theorem.

Theorem 5.9. *If the function $f : A \to B$ is invertible, then its inverse is unique.*

Proof.   Assume that g and h are two inverses of f. We need to prove that $f = g$. By definition, g and h are functions from B to A satisfying: $g \circ f = 1_A$, $f \circ g = 1_B$, $h \circ f = 1_A$ and $f \circ h = 1_B$. We have that:

$$
\begin{aligned}
g &= 1_A \circ g \text{ (Exercise (4) below)} \\
&= (h \circ f) \circ g \text{ (Since } h \circ f = 1_A) \\
&= h \circ (f \circ g) \text{ (Associativity of } \circ) \\
&= h \circ 1_B \text{ (Since } f \circ g = 1_B) \\
&= h \text{ (Exercise (4) below).}
\end{aligned}
$$

$\square$

Notation 5.2. If $f : A \to B$ is invertible, we denote its unique inverse by f^{-1}. In other words, f^{-1} is the unique function from B to A satisfying $f^{-1} \circ f = 1_A$ and $f \circ f^{-1} = 1_B$.

Example 5.56. The function $f : [0, \infty[\to [0, \infty[$ defined by $f(x) = x^2$ is invertible with inverse $f^{-1} : [0, \infty[\to [0, \infty[$ given by $f^{-1}(x) = \sqrt{x}$. To see this, notice that $(f \circ f^{-1})(x) = f\left(f^{-1}(x)\right) = f(\sqrt{x}) = (\sqrt{x})^2 = x$ and $(f^{-1} \circ f)(x) = f^{-1}(f(x)) = f^{-1}(x^2) = \sqrt{x^2} = x$ for any $x \in [0, \infty[$.

Example 5.57. Let a, b be real numbers with $a \neq 0$. The function $f : \mathbb{R} \to \mathbb{R}$ defined by $f(x) = ax + b$ is invertible. First notice that if $y = ax + b$, then $x = \frac{y-b}{a}$. This suggests that the inverse of f is the function $f^{-1} : \mathbb{R} \to \mathbb{R}$ given by $f^{-1}(x) = \frac{x-b}{a}$. To prove this, we verify that $f \circ f^{-1} = f^{-1} \circ f = 1_{\mathbb{R}}$: $(f \circ f^{-1})(x) = f\left(f^{-1}(x)\right) = f\left(\frac{x-b}{a}\right) = a\left(\frac{x-b}{a}\right) + b = x$ and $(f^{-1} \circ f)(x) = f^{-1}(f(x)) = f^{-1}(ax + b) = \left(\frac{(ax+b)-b}{a}\right) = x$ for any $x \in \mathbb{R}$.

Example 5.58. The function $f : \mathbb{R} \times \mathbb{R} \to \mathbb{R} \times \mathbb{R}$ defined by $f(x, y) = (x + y, 2y)$ is invertible. Simple calculation shows that if $(z, t) = f(x, y)$, then $x = z - \frac{1}{2}t$ and $y = \frac{1}{2}t$. This suggests that the inverse of f is the function $g : \mathbb{R} \times \mathbb{R} \to \mathbb{R} \times \mathbb{R}$ defined by $g(x, y) = \left(x - \frac{1}{2}y, \frac{1}{2}y\right)$. To prove this, notice that $(f \circ g)(x, y) = f(g(x, y)) = f\left(x - \frac{1}{2}y, \frac{1}{2}y\right) = \left(x - \frac{1}{2}y + \frac{1}{2}y, 2(\frac{1}{2}y)\right) = (x, y)$ and $(g \circ f)(x, y) = g(f(x, y)) = g(x + y, 2y) = \left(x + y - \frac{1}{2}(2y), \frac{1}{2}(2y)\right) = (x, y)$ for any $(x, y) \in \mathbb{R} \times \mathbb{R}$. We conclude that $f^{-1} = g$.

Like the case of real numbers with standard multiplication, not all functions are invertible. The following theorem gives a necessary and sufficient condition for a function to be invertible.

Theorem 5.10. *A function $f : A \to B$ is invertible if and only if it is bijective.*

Proof. Assume first that f is invertible. Then there exists a function $g : B \to A$ such that $g \circ f = 1_A$ and $f \circ g = 1_B$. Since $g \circ f = 1_A$ and 1_A is injective, f is injective by part (d) of Theorem 5.8. Since $f \circ g = 1_B$ and 1_B is surjective, f is surjective by part (e) of Theorem 5.8. This proves that f is bijective. Conversely, assume that f is bijective. Then for every element $y \in B$, there exists a unique element $x \in A$ such that $f(x) = y$. Define a function $g : B \to A$ as follows: $g(y) =$ the *unique* element $x \in A$ such that $f(x) = y$. Then $(g \circ f)(x) = g(f(x)) = g(y) = x$ for any $x \in A$ and $(f \circ g)(y) = f(g(y)) = f(x) = y$ for any $y \in B$. This shows that f is invertible and that $g = f^{-1}$. $\qquad\square$

Example 5.59. The function $f : \mathbb{R} \to \mathbb{R}$ defined by $f(x) = x^2$ is not invertible since it is not bijective.

Example 5.60. The function $f : \mathbb{R} \times \mathbb{R} \to \mathbb{R} \times \mathbb{R}$ defined by $f(x, y) = (-x + y, x + y)$ is invertible since it was proven to be bijective in Example 5.48 above. Moreover, if $(z, t) = f(x, y)$ then $(z, t) = (-x + y, x + y)$ and so $z = -x + y$ and $t = x + y$. Solving for x and y, we get that $x = \frac{t-z}{2}$

and $y = \frac{t+z}{2}$. This shows that the inverse function $f^{-1} : \mathbb{R} \times \mathbb{R} \to \mathbb{R} \times \mathbb{R}$ is defined by $f^{-1}(x, y) = \left(\frac{y-x}{2}, \frac{x+y}{2}\right)$.

Theorem 5.11. *Let $f : A \to B$ and $g : B \to A$ be two functions. Then:*

(a) If f is invertible, then so is f^{-1} and $(f^{-1})^{-1} = f$.

(b) If f and g are invertible, then so are the functions $f \circ g$ and $g \circ f$. Moreover, $(f \circ g)^{-1} = g^{-1} \circ f^{-1}$ and $(g \circ f)^{-1} = f^{-1} \circ g^{-1}$.

The proof of the two parts of the theorem are left as exercises (see Exercises (24) and (25) below).

5.7.2   *Exercises*

(1) Let f and g be two functions from $\mathbb{Z}$ to $\mathbb{Z}$. The following table shows the values of the two functions at some integers.

x	-2	-1	0	1	2
$f(x)$	2	6	-2	1	-1
$g(x)$	8	9	1	2	4

Find each of the following:

    $\star$ (a) $(f \circ f)(0)$ (c) $(f \circ g)(1)$ (e) $(g \circ g \circ g)(0)$

    (b) $(g \circ f)(-2)$ $\star$ (d) $(g \circ g)(1)$ $\star$ (f) $(f \circ g \circ g)(0)$

(2) Let f, g be the two functions from $\mathbb{R}$ to $\mathbb{R}$ defined by $f(x) = -2x + 3$ and $g(x) = x^2 - 3$. Let $a \in \mathbb{R}$ be an arbitrary real number. Find:

    (a) $(f \circ g)(-1)$ (d) $(g \circ g)(-1)$ (g) $(f \circ g \circ f)(-1)$

    $\star$ (b) $(g \circ f)(-1)$ (e) $(f \circ g)(a - 1)$ $\star$ (h) $(f \circ f \circ g)(a)$

    (c) $(f \circ f)(-1)$ $\star$ (f) $(g \circ f)(a - 1)$ (i) $(g \circ f \circ f)(a)$

(3) Let $\mathcal{H}$ be the set of all humans ever lived. Consider the two functions f and m from $\mathcal{H}$ to $\mathcal{H}$ defined as follows: $f(x) =$ the father of x and $m(x) =$ the mother of x. In each case, describe in words the output of the composite function.

    $\star$ (a) $f \circ f$ (c) $f \circ m$ (e) $m \circ m \circ m$

    (b) $m \circ f$ (d) $f \circ f \circ f$ $\star$ (f) $f \circ m \circ m$

(4) Let $f : A \to B$ be a function, 1_A, 1_B be the identity functions of A and B, respectively. Prove that $f \circ 1_A = f$ and $1_B \circ f = f$.

(5) Let f, g and h be the three functions from $\mathbb{R}$ to $\mathbb{R}$ defined by: $f(x) = x^2 + 1$, $g(x) = -2x + 3$ and $h(x) = \frac{x}{x^2+1}$. Give an expression (in terms of the real number x) of each of the following functions. Simply your answer as much as possible.

(a) $f \circ g$ $\star$ (c) $g \circ h$ $\star$ (e) f^2 $\star$ (g) $f \circ g \circ h$

(b) $g \circ f$ (d) $h \circ f$ (f) g^3 (h) $g \circ h \circ f$

(6) In each case, write the given function $f : \mathbb{R} \to \mathbb{R}$ as $g \circ h$ for some functions g and h.

(a) $f(x) = \frac{1}{\sqrt{1+x^2}}$ $\star$ (b) $f(x) = \frac{|x|}{1+|x|^2}$ (c) $\left(\frac{x}{1+x^2}\right)^3 + 1$

(7) We have seen that the composition of functions is not a commutative operation in general. Let f and g be the two functions from $\mathbb{R}$ to $\mathbb{R}$ defined by $f(x) = 3x + 2$ and $g(x) = ax + b$ where $a, b \in \mathbb{R}$. Find a relation that a and b must satisfy in order to have $g \circ f = f \circ g$.

(8) Let $f : A \to B$, $g : B \to C$ be two functions. Prove that if f and g are injective, then so is $g \circ f$.

$\star$(9) Let $f : A \to B$, $g : B \to C$ be two functions. Prove that if $g \circ f$ is surjective, then g must be surjective.

(10) Let $f : A \to B$, $g : B \to C$ be two functions.

   (a) Give an example to show that if $g \circ f$ is injective, then g needs not be injective.

   (b) Give an example to show that if $g \circ f$ is surjective, then f needs not be surjective.

(11) Recall that a function $f : \mathbb{R} \to \mathbb{R}$ is called *linear* if it has the form $f(x) = mx + b$ for some real numbers m and b. Prove that the composition of two linear functions is linear.

(12) A function $f : \mathbb{R} \to \mathbb{R}$ is called non-decreasing if for every $x, y \in \mathbb{R}$, the condition $x \leq y$ implies that $f(x) \leq f(y)$. If f and g are two non-decreasing functions from $\mathbb{R}$ to $\mathbb{R}$, prove that their composition (either $f \circ g$ or $g \circ f$) is also non-decreasing.

(13) Let $A = \{a, b, c, d, e, f\}$ and $B = \{1, 2, 3, 4, 5, 6\}$. In each case, a function $f_i : A \to B$ is given by a table of values for $i = 1, 2, 3, 4$. Determine if f_i is invertible and if it is, find the inverse f_i^{-1} as a table

of values.

x	a	b	c	d	e	f
$f_1(x)$	2	1	3	4	1	6

x	a	b	c	d	e	f
$f_2(x)$	2	3	1	4	6	5

x	a	b	c	d	e	f
$f_3(x)$	1	6	2	3	5	4

x	a	b	c	d	e	f
$f_4(x)$	3	6	2	1	3	5

(14) In each case, determine if the pair of functions from $\mathbb{R}$ to $\mathbb{R}$ are inverses of each other.

    (a) $f(x) = \frac{x+1}{2}$, $g(x) = 2x - 1$
    (b) $f(x) = 3 - x^{\frac{1}{5}}$, $g(x) = -(x+3)^5$
    (c) $f(x) = \frac{1}{x^2+1}$, $g(x) = x^2 + 1$
    (d) $f(x) = \sqrt[3]{x} - 5$, $g(x) = (x+5)^3$

$\star$(15) Prove that the function $f : \mathbb{R}\backslash\{\frac{3}{2}\} \to \mathbb{R}\backslash\{\frac{1}{2}\}$ defined by $f(x) = \frac{x+1}{2x-3}$ is invertible. Find an expression for the inverse f^{-1} of f.

(16) In each case, determine if the function f is invertible. If you say that it is, find its inverse.

    $\star$ (a) Let A is be a finite set of cardinality $n \geq 2$, $f : \wp(A) \to \{0, 1, \ldots, n\}$ defined by $f(X) = |X|$ where $|X|$ is the cardinality of X.
    (b) $f : \mathbb{N} \to \mathbb{N}$ defined by $f(n) = \frac{n}{2}$ if n is even and $f(n) = \frac{n+1}{2}$ if n is odd.
    (c) $f : \mathbb{R} \to \mathbb{R}$ defined by $f(x) = x^{\frac{1}{5}} + 5$.
    $\star$ (d) $f : \mathbb{M}_{22} \to \mathbb{M}_{22}$ defined by $f\left(\begin{bmatrix} a & b \\ c & d \end{bmatrix}\right) = \begin{bmatrix} a & c \\ b & d \end{bmatrix}$.
    (e) Let Σ^* be the set of all binary strings, $f : \Sigma^* \to \Sigma^*$ defined by $f(w) = 0w0$.

(17) (Calculus) Is the function $f : \left[-\frac{\pi}{2}, \frac{\pi}{2}\right] \to [-1, 1]$ defined by $f(x) = \sin x$ invertible? If you say it is, give its inverse (*Hint.* Draw the graph of $\sin x$).

(18) Let A and B be two sets. Construct a bijection $f : A \times B \to B \times A$.

(19) Let A be a non-empty set, $f : A \to A$ a function.

    (a) Prove that if f is injective, then so is f^n for any integer $n \geq 1$.
    $\star$ (b) Prove that if f is surjective, then so is f^n for any integer $n \geq 1$. (*Hint.* Use induction on n and Theorem 5.8 above).

(20) Let A, B, C be three non-empty sets, $f : A \to B$ and $g : B \to C$ two functions. Prove:

⋆ (a) If $g \circ f$ is injective and f is surjective, then g is injective.

(b) If $g \circ f$ is surjective and f is injective, then f is surjective.

(21) Let $f : A \to A$ be a function, $m, n \geq 0$ be two integers. Prove:

(a) $f^m \circ f^n = f^{m+n}$ (b) $(f^m)^n = f^{mn}$

(*Hint.* Fix one integer and use induction on the other, together with Exercise (4) above).

(22) Let f and g be two bijective functions from a set A to itself. Prove that if $f \circ g = 1_A$, then $g \circ f = 1_A$.

(23) Consider the function $f : \mathbb{R} \to \mathbb{R}$ defined piecewise as follows:

$$f(x) = \begin{cases} x + 1 & \text{for } x < -5 \\ 2x - 3 & \text{for } -5 \leq x < 5 \\ x - 5 & \text{for } x \geq 5 \end{cases}$$

Compute each of the following.

(a) $f^2(0)$ (b) $f^4(0)$ (c) $f^6(0)$ (d) $f^9(0)$

(24) Prove part (a) of Theorem 5.11 above.

(25) Prove part (b) of Theorem 5.11 above.

(26) Let A be a non-empty set, $\wp(A)$ the powerset of A. Prove that the function $\wp(A) \to \wp(A)$ defined by $f(X) = \overline{X}$ (the complement of X in A) is invertible. What is f^{-1}?

⋆(27) Let $f : A \to B$ be a function. Show that there exists a set C, a surjective function $g : A \to C$ and an injective function $h : C \to B$ such that $f = h \circ g$.

(28) Let $f : A \to B$ be a function. Prove that f is injective if and only for any set C and any two functions g, h from C to A, the condition $f \circ g = f \circ h$ implies that $g = h$.

Chapter 6

Elementary Number Theory

6.1 Introduction

Integers are fundamental objects in mathematics and computer science. Many problems arising in areas such as finite groups, finite fields and cryptography can be reduced to questions concerning properties of integers. For example, an area of great interest in discrete mathematics is *combinatorics* which consists of counting techniques that often deals with cardinalities of finite sets. The goal of this chapter is to review basic notions and results concerning integers. In particular, we look at the division algorithm, the Euclidean algorithm to compute the greatest common divisor, the Fundamental Theorem of Arithmetics and others.

6.2 Integers defined axiomatically

In mathematics, an axiom is a result that we accept as being true intuitively without a proof or deep analysis. For example, you have always used the fact that addition of integers is commutative ($m + n = n + m$ for any integers m and n) without questioning its validity. Axioms are usually the building blocks of mathematical theories; they are the stepping stones for larger and deeper discussions. In previous chapters, we talked about the set $\mathbb{Z}$ of integers and used some of its properties without giving a formal definition of this set. The purpose of this section is to introduce $\mathbb{Z}$ using an axiomatic approach.

Definition 6.1. $\mathbb{Z}$ is a set (with elements called integers), equipped with two binary operations that we call addition and multiplication denoted by $+$ and $\cdot$ (or just a juxtaposition), respectively. These operations satisfy axioms A1 to A11 below. In the following, x, y and z are integers.

A1. $x + y \in \mathbb{Z}$ ($\mathbb{Z}$ is closed under addition).

A2. $xy \in \mathbb{Z}$ ($\mathbb{Z}$ is closed under multiplication).

A3. $x + y = y + x$ (commutativity of addition).

A4. $(x + y) + z = x + (y + z)$ (associativity of addition).

A5. $xy = yx$ (commutativity of multiplication).

A6. $(xy)z = x(yz)$ (associativity of multiplication).

A7. $x(y + z) = xy + xz$ (distributivity of $\cdot$ with respect to $+$).

A8. There exists a special integer denoted by 0 that satisfies $x+0 = 0+x = x$ for any $x \in \mathbb{Z}$ (identity element for addition).

A9. There exists a special integer denoted by 1 that satisfies $1 \neq 0$ and $x \cdot 1 = 1 \cdot x = x$ for any $x \in \mathbb{Z}$ (identity element for multiplication).

A10. For any integer $x \in \mathbb{Z}$, there exists an element $y \in \mathbb{Z}$ satisfying $x+y = y + x = 0$. The element y is called an *additive inverse* of x.

A11. If $xy = xz$ and $x \neq 0$, then $y = z$ (the cancellation property).

Based on the above axioms, one can prove important properties satisfied by the addition and multiplication of integers. The first result is about the uniqueness of the identity elements and the additive inverse.

Theorem 6.1.

   (a) The identity element for the addition is unique.

   (b) The identity element for the multiplication is unique.

   (c) If x is an integer, then x has a unique additive inverse.

Proof. We only prove part (a), the other two parts are left as exercises (see Exercises (2) and (3) below). Assume that 0 and $0'$ are two identity elements for the addition. On one hand, $0 + 0' = 0'$ since 0 is an identity element for the addition. On the other hand, $0 + 0' = 0$ since $0'$ is also an identity element for the addition. We conclude that $0' = 0 = 0 + 0'$.   □

The unique additive inverse of an integer x is denoted by $-x$. If x, y are two integers then the integer $x + (-y)$, the addition of x and the additive inverse of y, is denoted by $x - y$ and called (as usual) the *subtraction* of x and y. In particular, we have that $x - x = 0$ for any integer x.

Theorem 6.2. *Let x, y, z be arbitrary integers.*

   (a) $-0 = 0$ *(d) If $x + z = y + z$, then $x = y$*

   (b) $-(-x) = x$ *(e) $xy = 0 \Leftrightarrow x = 0$ or $y = 0$*

   (c) $x0 = 0$

Property (d) is referred to as the cancellation property for the addition.

Proof.

(a) The result follows by the uniqueness of the additive inverse since $0 + 0 = 0$ (because 0 is the identity element for the addition).
(b) Left as an exercise (see Exercise (4) below).
(c) Since $0 + 0 = 0$, we have $x(0 + 0) = x0$ (multiplying both sides with x). Using distributivity, we get that $x0 + x0 = x0$. Now add $-x0$ (the additive inverse of $x0$) to both sides of the last equation: $(x0+x0)+(-x0) = x0+(-x0)$. Using the associativity of the addition and the subtraction notation, we get that $x0 + (x0 - x0) = (x0 - x0)$. Since $x0 - x0 = 0$, the last equation simplifies to $x0+0 = 0$. The result follows from the fact that 0 is the identity element for the addition.
(d) Left as an exercise (see Exercise (7) below).
(e) If either $x = 0$ or $y = 0$, then $xy = 0$ by part (c). Assume that $xy = 0$ and $x \neq 0$. We need to prove that y must be zero:

$$xy = 0 \text{ (assumption)}$$
$$\Rightarrow xy = x0 \text{ (since } x0 = 0 \text{ by part (c))}$$
$$\Rightarrow y = 0 \text{ (axiom A11).}$$

$\square$

There are more than one way to define the positive natural numbers $1, 2, \ldots$ as a subset of $\mathbb{Z}$. We use an axiomatic approach to do that by declaring that there exists a subset $\mathbb{Z}^+$ of the integers satisfying the following axioms:

N1. $\mathbb{Z}^+ \neq \emptyset$.
N2. $\mathbb{Z}^+$ is closed under addition: if $m, n \in \mathbb{Z}^+$ then, $m + n \in \mathbb{Z}^+$.
N3. $\mathbb{Z}^+$ is closed under multiplication: if $m, n \in \mathbb{Z}^+$ then, $mn \in \mathbb{Z}^+$.
N4. *(Trichotomy law)* For any integer m, exactly one of the following is true: (i) $m \in \mathbb{Z}^+$; (ii) $m = 0$; (iii) $-m \in \mathbb{Z}^+$.

It is worth mentioning here that none of the above axioms states that $1 \in \mathbb{Z}^+$ but we can actually prove that using the axioms (see Exercise (6) below).

Definition 6.2. We define the set of *natural numbers* as being the subset $\mathbb{N} = \mathbb{Z}^+ \cup \{0\}$ of $\mathbb{Z}$.

6.2.1 *Order on the integers*

The standard order on the integers can now be defined using the addition operation and the subset $\mathbb{N}$ of $\mathbb{Z}$.

Definition 6.3. Let m and n be two integers. We say that m *is greater than* n and we write $m > n$ if $m - n \in \mathbb{Z}^+$. If m is greater than n, we say that n *is less than* m and we write $n < m$. We say that m is *greater than or equal to* n and we write $m \geq n$ if $m > n$ or $m = n$. Similarly, m is *less than or equal to* n (written as $m \leq n$) if $m < n$ or $m = n$. An integer m is called: *non-positive* if $m \leq 0$, *non-negative* if $m \geq 0$, *negative* if $m < 0$ and *positive* if $m > 0$.

From the definition, it follows that $m \leq n$ if and only if there exists $t \in \mathbb{N}$ such that $n = m + t$.

Theorem 6.3. *Let m and n be two integers. Then:*

(a) $m > 0$ if and only if $m \in \mathbb{Z}^+$.
(b) $m > 0$ if and only if $-m < 0$.
(c) If $mn > 0$, then either $m > 0$ and $n > 0$ or $m < 0$ and $n < 0$.

Proof.

(a) $m > 0 \Leftrightarrow m - 0 \in \mathbb{Z}^+$ (by definition) $\Leftrightarrow m \in \mathbb{Z}^+$ (since $m - 0 = m$).
(b) Assume that $m > 0$. By the Trichotomy law, exactly one of the following is true for the integer $-m$: $-m < 0$, $-m = 0$ or $-m > 0$. If $-m = 0$, then $m = -(-m) = -0 = 0$ (by Theorem 6.2) which is a contradiction. If $-m > 0$, then $m + (-m) > 0$ since $m > 0$ and $\mathbb{Z}^+$ is closed under addition. This also leads to a contradiction since $m + (-m) = 0$. This leaves us with the only possibility: $-m < 0$. The proof of the implication $-m < 0 \Rightarrow m > 0$ is done similarly.
(c) Left as an exercise (see Exercise (12) below).
$\square$

Theorem 6.4. *Let x, y and z be three integers.*

(a) If $x \leq y$, then $x + z \leq y + z$.
(b) If $x \leq y$ and $y \leq z$, then $x \leq z$.
(c) If $x < y$ and $z > 0$, then $xz < yz$.
(d) If $x < y$ and $z < 0$, then $xz > yz$.

Proof.   We only prove part (a). The other parts are left as exercises (see Exercises (18), (9) and (10) below). Assume $x \leq y$, then $y = x + t$ for some

$t \in \mathbb{N}$. We get:

$$\begin{aligned}
y + z &= (x + t) + z \\
&= x + (t + z) \text{ (Axiom A4)} \\
&= x + (z + t) \text{ (Axiom A3)} \\
&= (x + z) + t \text{ (Axiom A4)}.
\end{aligned}$$

This shows that $x + z \leq y + z$ and part (a) is proved. $\qquad\square$

The axioms we have seen so far are not proper to integers, they are also satisfied by real numbers. The next axiom, widely known in the literature as the *Well-ordering principal* provides an important property that distinguishes the natural numbers from other number systems. We will revisit this principle in a more general context in Chapter 7. We use the acronym WOP to refer to this axiom.

Axiom 1 (The well-ordering principle). *Every non-empty subset of* $\mathbb{N}$ *has a least element for the natural order on the integers.*

Clearly this is not satisfied by the set of real numbers as, for example, the open interval $]0, 1[$ has no smallest element.

Theorem 6.5 (The discreetness property of integers). *The following hold.*

(a) *There exists no integer between* 0 *and* 1.
(b) *For any integer n, there is no integer k with $n < k < n + 1$.*

Proof. Assume by contradiction that there exists an integer k with $0 < k < 1$. In particular $k \in \mathbb{N}$ and the subset $A = \{i \in \mathbb{N}; 0 < i < 1\}$ of $\mathbb{N}$ is non-empty. By the WOP, A has a smallest element e. In particular, $0 < e < 1$. Since $e > 0$, part (c) of Theorem 6.4 implies that $0 \cdot e < e \cdot e < 1 \cdot e$ or $0 < e^2 < e$ where $e^2 = e \cdot e$. Since $e < 1$, we get that $0 < e^2 < 1$ and consequently $e^2 \in A$. This is a contradiction to the fact that e is the least element of A (since $e^2 < e$). This proves part (a). For part (b), fix an integer n and assume that there exists an integer k with $n < k < n + 1$. Adding $-n$ from all members of the two inequalities, we get that $n + (-n) < k + (-n) < (n + 1) + (-n)$ or $0 < k - n < 1$. This contradicts part (a). The result follows. $\qquad\square$

A consequence of the above theorem is the following.

Corollary 6.1. 1 *is the smallest positive integer.*

The axiomatic approach to define properties of operations on a set we explored in this section is not unique to integers. It is standard to use such approach in defining discrete mathematical structures. The purpose of this section was, however, to give a formal definition of the set of integers and to get a flavor of how to build on axioms to prove some of their properties. Moving forward in this chapter, we will not bother much with proving properties using the above axioms. Basic properties of integers that you used before (like "m^2 is non-negative for any integer m" or "$(m-n)(m-n) = m^2 - n^2$") are assumed true without the need to provide a formal proof for them.

6.2.2 *Exercises*

(1) Explain why we require $x \neq 0$ for the cancellation property (Axiom A11 above)?

$\star$(2) Using the axioms of this section, prove that the identity element for the multiplication of integers is unique.

(3) Using the axioms of this section, prove that the additive inverse of an integer x is unique.

(4) Let m, n be two integers. Using the axioms of this section, prove that:

$\star$ (a) $-(-m) = m$ (c) $(-1)(-m) = m$
  (b) $(-1)m = -m$ $\star$ (d) $-(m+n) = -n - m$

(5) Let x, y be two integers. Use the axioms of this section to prove that:

$\star$ (a) $x(-y) = -(xy)$ (b) $(-x)y = -(xy)$ (c) $(-x)(-y) = xy$

(6) Using the axioms of this section, prove that $1 \in \mathbb{Z}^+$ (*Hint.* Proceed by contradiction. Use the Trichotomy law and part (c) of Exercise (15)).

(7) Let x, y and z be three integers. Prove the cancellation property for the addition: if $x + y = x + z$, then $y = z$.

(8) Using the axioms of this section, prove part (b) of Theorem 6.4.

$\star$(9) Using the axioms of this section, prove part (c) of Theorem 6.4.

(10) Using the axioms of this section, prove part (d) of Theorem 6.4.

(11) Let a, b, c and d be four integers. Provide a careful proof (using the axioms of this section) of each of the following identities.

$\star$ (a) $(a+b)(c+d) = ac + ad + bc + bd$
  (b) $(a-b)(a+b) = a^2 - b^2$

(12) Let m and n be two integers. Prove that if $mn > 0$, then either $m > 0$ and $n > 0$ or $m < 0$ and $n < 0$.

⋆(13) Fix an integer k and let $\Sigma_k = \{n \in \mathbb{Z}; n \geq k\}$. The purpose of this exercise is to prove that the well-ordering principle can be extended to the set Σ_k. Let T be a non-empty subset of Σ_k.

(a) Prove that $T' = \{n - k; n \in T\}$ is a non-empty subset of $\mathbb{N}$.
(b) Deduce that T' has a least element e.
(c) Prove that $e + k$ is the least element of T.

6.3   Division in $\mathbb{Z}$: Prime numbers

Unlike addition and multiplication, division of integers does not always result in an integer. In other words, the set $\mathbb{Z}$ is not closed under the division operation. For instance, $\frac{7}{3}$ is not an integer. However, there are instances where the division of two integers yields an integer like in the case of $\frac{21}{3}$. In this case, the result (7) is called the quotient and the remainder of the division is 0. The formal definition of the integer division is as follows.

Definition 6.4. Let m, n be two integers with $m \neq 0$. We say that m *divides* n (equivalently, m is a *divisor* or a *factor* of n) and we write $m \mid n$ if there exists an integer k such that $n = km$. If m does not divide n, we write $m \nmid n$. If m divides n, we also say that n is a *multiple* of m.

Divisibility by 2 partitions the integers into two classes. An integer is called *even* if it is divisible by 2 and *odd* otherwise.

We will always assume that $m \neq 0$ whenever we write $m \mid n$.

Example 6.1. Any non-zero integer n is a divisor of 0 since $0 = 0n$.

Example 6.2. Clearly, -2 is a divisor of each of the integers: $-6, -4, -2, 0, 2, 4, 6$ but is not a divisor of -3. It is also clear that 357 is a multiple of 3 and a multiple of 7 at the same time.

Example 6.3. How many integers between 80 and 234 are divisible by 7?

Solution. An integer divisible by 7 must have the form $7k$ for some $k \in \mathbb{Z}$. So the problem amounts to finding the number of integers k satisfying $80 \leq 7k \leq 234$. Dividing all sides with 7, we get that $\frac{80}{7} \leq k \leq \frac{234}{7}$. Since

k must be an integer, $12 \le k \le 33$ (sine $\frac{80}{7} \approx 11.4$ and $\frac{80}{7} \approx 33.4$). There are $33 - 12 + 1 = 22$ such integers. $\Diamond$

Theorem 6.6. *Let m, n, x be integers. Then:*

(a) $1 \mid x$, $-1 \mid x$ *and* $x \mid x$.
(b) *If* $m \mid n$ *and* $n \mid m$, *then* $m = n$ *or* $m = -n$.
(c) *If* $m \mid n$ *and* $n \mid x$, *then* $m \mid x$.
(d) *If* $x \mid m$, *then* $x \mid mn$.
(e) *Let* $m_1, \ldots, m_r$ *be integers such that* $x \mid m_i$ *for all* $i = 1, \ldots, r$. *Then* $x \mid (a_1 m_1 + \cdots + a_r m_r)$ *for any integers* $a_1, \ldots, a_r$.

Proof.

(a) The relations $x = x \cdot 1 = (-x)(-1)$ imply that $1 \mid x$ and $-1 \mid x$. The relation $x \mid x$ follows from the fact that $x = x \cdot 1$.
(b) Left as an exercise (see Exercise (4) below)
(c) Left as an exercise (see Exercise (5) below)
(d) If $x \mid m$, then $m = sx$ for some $s \in \mathbb{Z}$. So $mn = (sn)x$ and $x \mid mn$.
(e) For $i = 1, \ldots, r$, Write $m_i = s_i x$ for some integer s_i. For any integers $a_1, \ldots, a_r$, we have that $a_1 m_1 + \cdots + a_r m_r = a_1(s_1 x) + \cdots + a_r(s_r x) = (a_1 s_1 + \cdots + a_r s_r)x$. Since $a_1 s_1 + \cdots + a_r s_r$ is an integer, the result follows. $\square$

Example 6.4. Is it possible to find three integers a, b, c such that $3a + 6b + 12c = 76$?

Solution. We prove that no such integers exist. Suppose, by contradiction, that a, b, c are integers such that $3a + 6b + 12c = 76$. Since 3 is a common divisor of 3, 6 and 12, it is a divisor of $3a + 6b + 12c$ by part (d) of Theorem 6.6. But that means that 3 must divide 76 which is false. $\Diamond$

Theorem 6.7. *Let m, n be two integers.*

(a) *If* m, n *are positive and* $m \mid n$, *then* $m \le n$.
(b) *If* $m \mid 1$, *then* $m = -1$ *or* $m = 1$.

Proof.

(a) If $m \mid n$, then there exists $k \in \mathbb{Z}$ such that $n = km$. Since $n > 0$, Theorem 6.3 above implies that $k > 0$ (since $m > 0$). This implies that $1 \le k$ by Theorem 6.5. Multiplying both side by m, we get that

$m \leq mk$ (using Theorem 6.4 and the fact that $m > 0$). The result follows since $n = km$.

(b) Both -1 and 1 are divisors of 1 by Theorem 6.6. If $m \mid 1$, then there exists $k \in \mathbb{Z}$ such that $1 = km$. By Theorem 6.3, we know that k and m are either both positive or both negative. If k and m are both positive, then part (a) implies that $m \leq 1$. In this case, the discreetness property implies that $m = 1$. If k and m are both negative, then $-k > 0$ (Theorem 6.3) and the equation $(-k)(-m) = 1$ (equivalent to $km = 1$) implies that $-m = 1$ (using the same argument as before). This shows that $m = -1$ in this case. $\qquad\square$

Theorem 6.6 tells us in particular that every *positive* integer x is divisible by 1 and itself. If these are the only positive divisors of an integer $x \geq 2$, then x is called a prime number. Prime numbers play a central role in number theory and its applications.

Definition 6.5. An integer $p \geq 2$ is called a *prime number* (or simply *prime*) if the only positive divisors (or factors) of p are 1 and p itself. An integer $n \geq 2$ which is not prime is called a *composite number*. Note that 1 is neither a prime nor a composite number.

Example 6.5. 2, 3, 5, 7, 11, 13 are the first six prime numbers.

The following proposition characterizes composite integers.

Proposition 6.1. *An integer $n \geq 2$ is composite if and only if there exist integers a and b such that $n = ab$ and $1 < a < n$, $1 < b < n$.*

Proof. Follows easily from the definition of composite numbers. $\qquad\square$

The next result, known as the Fundamental Theorem of Arithmetic, illustrates the great importance of primes in number theory. It states that primes numbers are basically the "foundation" on which natural numbers are constructed. The proof of the Existence part of Theorem 6.17 was already established in Chapter 3 (see Example 3.36 of Section 3.6.2). The proof of the uniqueness part of the theorem is shown in Theorem 6.17 below.

Theorem 6.8 (The fundamental theorem of arithmetic). *Every integer $n \geq 2$ can be expressed in a unique way as $n = p_1^{t_1} p_2^{t_2} \cdots p_k^{t_k}$ with $p_1 < p_2 < \ldots < p_k$ are prime numbers, $t_1, t_2, \ldots, t_k$ are positive integers and $k \geq 1$.*

The expression $n = p_1^{t_1} p_2^{t_2} \cdots p_k^{t_k}$ of n as the product of powers of primes given in Theorem 6.8 is called the *prime factorization* of n.

Example 6.6. $1800 = 2 \cdot 2 \cdot 2 \cdot 3 \cdot 3 \cdot 5 \cdot 5 = 2^3 \cdot 3^2 \cdot 5^2$.

Remark 6.1. Let $n \geq 2$ be an integer.

(1) If n is prime, then its prime factorization is simply n.
(2) The Fundamental Theorem of Arithmetic says in particular that n must have at least one prime factor.

To find the prime factorization of a relatively small integer $n \geq 2$, we start by checking if n is even. If it is, then 2 is one prime factor and $\frac{n}{2}$ is an integer. If $\frac{n}{2}$ is also even, then $2 \cdot 2 = 2^2$ appears in the prime factorization of n. We continue this way until we reach a positive integer k such that $\frac{n}{k}$ is odd (if n itself is odd, then $k = 1$). In this case, we can test if the prime 3 is a divisor of $\frac{n}{k}$ following this simple rule: $3 \mid m$ if and only if 3 divides the sum of the digits of m. For example 3 divides the integer 213567 since it divides $2 + 1 + 3 + 5 + 6 + 7 = 24$. If 3 is not a divisor, we check if the prime 5 is a divisor of $\frac{n}{k}$. This is probably the easiest: $5 \mid m$ if and only if the last digit of m is either 0 or 5. If 5 is not a divisor, we check if the prime 7 is. There is no good strategy to determine if 7 divides an integer m, so we perform the long division by 7 in this case. Divisibility by the next prime, 11, can be determined by the following rule: $11 \mid m$ if and only if 11 divides the difference $a - b$ where a is the sum of the alternating digits of m starting at the first digit and b is the sum of the alternating digits of m starting at the second digit. For example, the integer $m = 43912011$ is divisible by 11 since in this case $a = 4+9+2+1 = 16$, $b = 3+1+0+1 = 5$ and 11 divides $a - b = 11$.

Example 6.7. We show the prime factorizations of $n = 60$ and $n = 3150$.

(1) 60 is even and $\frac{60}{2} = 30$ is also even. Now $\frac{30}{2} = 15$ is odd but divisible by 3 with quotient 5. We get that $60 = 2 \times 2 \times 3 \times 5 = 2^2 \times 3 \times 5$.
(2) $n = 3150$ is even and $\frac{n}{2} = 1575$ is odd. Note that $1 + 5 + 7 + 5 = 18$ is divisible by 3 which implies that 1575 is also divisible by 3: $\frac{1575}{3} = 525$. Note that $5 + 2 + 5 = 12$: divisible by 3, so 525 is divisible by 3 and $\frac{525}{3} = 175$. Since the sum of the digits of 175 is not divisible by 3, 175 is not divisible by 3. Clearly 175 is divisible by 5 and $\frac{175}{5} = 35$. In turns, 35 is divisible by 5 with quotient 7. This leads to the following prime factorization: $3150 = 2 \times 3^2 \times 5^2 \times 7$.

Determining if an integer is prime is an important question in many areas of mathematics and computer science. Testing the primality of integers has gained significant interest in recent years because of its close connection to public key cryptography and cyber security. At first glance, the problem seems to be easy to handel: given an integer $n \geq 2$, test if any of the integers between 2 and $\frac{n}{2}$ divides n evenly. If this is the case, n is composite by Proposition 6.1. This works fine with small integers, but clearly not practical for large ones. While there is no general effective algorithm to test primality of integers, there are some tests that can be effective in some particular cases. The following is well-known primality test.

Theorem 6.9. *If n is a composite integer, then there exists a prime divisor p of n such that $p \leq \sqrt{n}$.*

Proof. Assume n is a composite integer. By Proposition 6.1, there exist integers a and b such that $n = ab$ and $1 < a < n$, $1 < b < n$. Without loss of generality, we may assume that $a \leq b$. Then $a^2 \leq ab = n$ and so $a \leq \sqrt{n}$. Let p be a prime divisor of a (we know that such a divisor exists by remark 6.17 above). Then $p \leq a \leq \sqrt{n}$ and $p \mid n$ since $p \mid a$ and $a \mid n$ (by Theorem 6.6). $\square$

Example 6.8. Determine if the integer 107 is prime.

Solution. Assume that 107 is composite. Then by Theorem 6.9, there exists a prime divisor p of 107 such that $p \leq \sqrt{107}$. Since $100 < 107 < 121$, $10 < \sqrt{107} < 11$. The only primes less than 10 are $2, 3, 5$ and 7 and since none of them is a divisor of 107, our assumption that 107 is composite is false. We conclude that 107 is prime. $\Diamond$

6.3.1 Exercises

Recall that for integers a and b, the notation $a \mid b$ means that $a \neq 0$ and a divides b evenly.

(1) Let a, b, c and d be four integers. In each case, determine the truth value of the statement. Justify your answer.

   $\star$ (a) If a, b are two non-zero integers, then either $a \mid b$ or $b \mid a$.

     (b) If $a \mid b$ and $a \mid c$, then $a \mid bc$.

     (c) If $a \mid b$ and $a \mid c$, then $a^2 \mid bc$.

   $\star$ (d) If $a \mid b$ and $c \mid b$, then $ac \mid b$.

     (e) If $a \mid b$ and $c \mid b$, then $ac \mid b^2$.

(f) If $a \mid b$ and $c \mid b$, then $(a + c) \mid b$.

$\star$ (g) If $a \mid b$ and $c \mid d$, then $ac \mid bd$.

(h) If $a \mid b$ and $c \mid d$, then $(a + c) \mid (b + d)$.

$\star$ (i) If $a \mid bc$, then $a \mid b$ or $a \mid c$.

(2) Let m and n be two integers such that $m \mid n$. Prove that $(-m) \mid n$ and $m \mid (-n)$.

(3) Let m, n and k be three integers. Prove that if $m \mid n$ and $m \mid (n + k)$, then $m \mid k$.

$\star$(4) Prove part (b) of Theorem 6.6 above.

(5) Prove part (c) of Theorem 6.6 above.

(6) In each case, determine the truth value of the statement.

(a) If p and q are prime numbers, then $p + q$ is a prime number.

(b) If $p < q$ are prime numbers, then $q - p$ is a prime number.

$\star$ (c) If p is a prime number, then $p + 1$ is composite.

(d) If $p > 2$ is a prime number, then p must be odd.

$\star$ (e) If p and q are prime numbers, then $pq + 1$ is a prime number.

(f) If n is an odd integer, then $4n - 1$ is a prime number.

(7) How many integers between 807 and 2344 are divisible by 17?

(8) In each case, explain why the given integer is not prime.

(a) 121	(c) $2^{2023} + 4^{9087}$	(e) $3^{3333} - 3^{333}$
(b) 2022	$\star$ (d) $3^{2022} - 5^{4044}$	

(9) In each case, determine if the given integer is prime. Justify your answer.

(a) 11	(c) 1111	(e) 111111	(g) $\underbrace{111 \ldots 111}_{225 \text{ times}}$
(b) 111	(d) 11111	$\star$ (f) 111000111	

(10) If $n > 2$ is an even integer, prove that $2^n - 1$ is a composite number.

$\star$(11) Prove that $n^8 + 4$ is composite for any integer $n \geq 2$ (*Hint.* Expand $(n^4 + 2)^2 - 4n^4$)

$\star$(12) Two positive integers p and $p + 2$ are called *twin primes* if they are both prime numbers. Give the first eight twin primes couples $(p, p+2)$.

(13) Find a prime number p that can be expressed as the sum of:

(a) three distinct prime numbers

(b) four distinct prime numbers

(c) five distinct prime numbers

(14) Find the largest prime factor of 7040.

(15) In each case, find the prime factorization of the given integer.

(a) 120	(d) 315	(g) 675	(j) 1061775
(b) 128	(e) 528	$\star$ (h) 4851	
(c) 216	$\star$ (f) 544	(i) 13230	

(16) Let p be a prime number. If m, n are two positive integers such that $m \mid n$ and $m \mid n + p$, prove that $m = 1$ or $m = p$.

(17)(a) Give 3 prime numbers each of which is of the form of the form $2^n - 1$ for some integer n (*Hint.* See Exercise (10) above).

   (b) Show that there are no prime numbers of the form $3^n - 1$ for $n \geq 2$.

   (c) Show that there are no prime numbers of the form $4^n - 1$ for $n \geq 2$.

(18) An integer n is called a perfect cube if there exists an integer k such that $n = k^3$. Find all prime numbers p such that $p + 1$ is a perfect cube (*Hint.* $n^3 - 1 = (n - 1)(n^2 + n + 1)$)

$\star$(19) Let $n \geq 2$ be an integer. Prove that if $2^n - 1$ is a prime number, then n must be prime (*Hint.* Use the fact that for any real number x and any positive integer r, $(x^r - 1) = (x - 1)(x^{r-1} + x^{r-2} + \cdots + x + 1)$).

(20) A *Mersenne number* is a number of the form $M_p = 2^p - 1$ for some prime integer p. If M_p is prime, then we call it a *Mersenne prime*.

   (a) Give the first seven Mersenne numbers and indicate which one of them are Mersenne primes.

   (b) Give a counterexample to the statement: *Every Mersenne number is a Mersenne prime.*

   (c) Consider the sequence S_n, $n \geq 2$, of natural numbers defined recursively as follows: $S_2 = 4$ and for $S_n = S_{n-1}^2 - 2$ for $n \geq 3$. A well-known primality test for Mersenne numbers is Lucas test: M_p is prime if and only if M_p divides S_p for $p > 2$. Use Lucas test to check the primality of the Mersenne numbers M_7 and M_{11}.

(21) Use the primality test of Theorem 6.9 to prove that each of the following is a prime number.

(a) 91	(b) 119	$\star$ (c) 181	(d) 213

6.4   The division algorithm

So far we have dealt with "even" division of integers. That is division with no remainder. The division algorithm deals with the general division procedure of integers where the remainder is not necessarily zero.

Theorem 6.10 (The division algorithm). *Let m, n be two integers with $n > 0$. Then there exist unique integers q and r such that $0 \leq r < n$ and $m = qn + r$.*

Proof.   First, we prove the existence of integers q and r satisfying the relations $m = qn + r$ and $0 \leq r \leq n - 1$. Consider the set $S = \{m - \alpha n;\ \alpha \in \mathbb{Z}$ and $m - \alpha n \geq 0\}$. If $m > 0$ then $m \in S$ since $m = m - 0(n) > 0$ (choose $\alpha = 0 \in \mathbb{Z}$). If $m \leq 0$, take $\alpha = m - 1 \in \mathbb{Z}$. Then $m - \alpha n = m - (m - 1)n = m(1 - n) + n$. Since $1 - n \leq 0$ and $m \leq 0$, $m(1 - n) \geq 0$ and consequently $m(1 - n) + n > 0$. We conclude in this case that $m - (m - 1)n \in S$. This proves that S is a non-empty subset of $\mathbb{N}$. Using the well-ordering principal, S has a minimal element r. Write $r = m - qn \geq 0$ for some $q \in \mathbb{Z}$. If $r \geq n$, then $(m - qn) - n = m - (q + 1)n \geq 0$ which means that $m - (q + 1)n \in S$. Since $m - qn$ is the minimal element of S, we must have that $m - (q + 1)n \geq m - qn$. This implies that $n \leq 0$: a contradiction. This proves that $r < n$ and the existence of q and r with the required properties is established. For the uniqueness of r and q, assume that (q, r) and (q', r') are two pairs of integers satisfying $m = qn + r$ and $m = q'n + r'$ with $0 \leq r < n$ and $0 \leq r' < n$. Then, $(q' - q)n = (m - qn) - (m - q'n) = r - r' \leq r < n$. On the other hand, $(q' - q)n = q'n - qn = (m - r') - (m - r) = r - r' \geq -r' > -n$. We conclude that $-n < (q' - q)n < n$. Dividing all sides by n (this will preserve the inequalities since $n > 0$), we get that $-1 < q' - q < 1$. The only integer strictly between -1 and 1 is 0. This proves that $q' = q$. Finally, $r' = m - q'n = m - qn = r$. The uniqueness part is established. $\square$

Definition 6.6. With the notation of Theorem 6.10, We refer to m as the *dividend*, n as the *divisor*, q as the *quotient* and r as the *remainder* of the division of m by n.

An immediate consequence of the division algorithm is the following characterisation of even and odd integers.

Theorem 6.11. *An integer n is even if and only if $n = 2k$ for some $k \in \mathbb{Z}$ and n is odd if and only if $n = 2k + 1$ for some $k \in \mathbb{Z}$.*

Proof. Using the division algorithm of n by 2, we get that $n = 2q + r$ for some integers q and r with $0 \leq r < 2$. This means that r is either 0 or 1. If $r = 0$ then $n = 2k$ is even. If $r = 1$, then $n = 2k + 1$ is odd. The converses of these statements are also clearly true. $\qquad\square$

The name *Division Algorithm* attributed to Theorem 6.10 is a bit misleading. The theorem does not actually provide an algorithm to find the quotient and the remainder of the division. These can be found using the usual long division procedure. The theorem speaks only about the existence and the uniqueness of the quotient and the remainder, but it is an essential tool in proving results in number theory.

Example 6.9.

(a) Performing the long division of 123 by 17, we find that $123 = 4 \cdot 17 + 4$. Since $0 \leq 4 < 17$, the quotient and the remainder of division of 123 by 17 are 7 and 4, respectively.
(b) One has to be careful when the dividend is negative. For example, from the equation $-17 = -4(4) - 1$, one is tempted to say that the quotient and remainder of the division of -17 by 4 are -4 and -1, respectively. This is not true as the remainder in this case has to be in the set $\{0, 1, 2, 3\}$ (-1 is not). The equation $-17 = -5(4) + 3$ gives the correct information: the quotient is -5 and the remainder is 3.

6.4.1 *Representing integers in different bases*

Humans adopted base 10 as standard enumeration basis for the simple reason that we have 10 fingers (according to the belief of many historians). But different bases were certainly used by different civilizations throughout history. More recently, the simple base 2 became of great importance as it is used in computer languages (binary systems). First, let us think what does it mean to write an integer, say 423, in base 10? The number 423 has 4 in the hundredth place, 2 in the tenth place and 3 in the unit place. This means that we can express the integer as follows: $423 = 4 \cdot 100 + 2 \cdot 10 + 3 \cdot 1 = 4 \cdot 10^2 + 2 \cdot 10^1 + 3 \cdot 10^0$.

Given an integer $b \geq 2$, we can talk about the *base b* representation of integers in a similar way to our decimal system. In base b, every integer can be expressed using the digits from the set $S_b = \{0, 1, 2, \ldots, b-1\}$. So in base 10, there are ten digits: $0, 1, \ldots, 9$ and in base 2 there are only two: 0

and 1. More generally, if $a = a_{k-1}a_{k-2}\ldots a_1a_0$ (where each $a_i \in S_b$) is an integer written in base b, then a holds the value $a_{k-1}b^{k-1} + a_{k-2}b^{k-2} + \cdots + a_1b^1 + a_0b^0$ in decimal system. For the integer $a = a_{k-1}a_{k-2}\ldots a_1a_0$, the digits a_0 and a_{k-1} are called the least significant digit (LSD) and the most significant digit (MSD) respectively. If $c > b$, then an integer a written in base b can also be interpreted as a number in base c. This confusion can be avoided by specifying the base as a subscript and write $(a)_b$. Hence, $(123)_4 = (1 \times 4^2) + (2 \times 4^1) + (3 \times 4^0) = 27$ and $(123)_6 = (1 \times 6^2) + (2 \times 6^1) + (3 \times 6^0) = 51$.

Example 6.10. To write the integer 3451 in octal system (base $b = 8$), we look for integers $a_0, a_1, \ldots, a_k \in \{0, 1, \ldots, 7\}$ such that $a_k > 0$ and $3451 = a_k(8)^k + \cdots + a_1(8) + a_0$. Note that this representation can be written as $3451 = (a_k(8)^{k-1} + \cdots + a_1)(8) + a_0$. By the division algorithm, a_0 is the remainder upon division of 3459 by 8. Using long division, we know that $3451 = (431)(8) + 3$. So $431 = a_k(8)^{k-1} + \cdots + a_1$ and $a_0 = 3$. Similarly, a_1 is the remainder of the division of 431 by 8. We repeat the same process for the quotient 431 and so on until we get a zero quotient. The following table shows the quotient and the remainder (a_i) at every step:

Division by 8	Quotient	Remainder
3451	431	$3 = a_0$
431	53	$7 = a_1$
53	6	$5 = a_2$
6	0	$6 = a_3$

Reading the column of the remainders from bottom to top gives the following octal representation of 3451: $(3451)_8 = 6573$. We can check the answer as follows: $6(8)^3 + 5(8)^2 + 7(8) + 3 = 3451$.

Example 6.11. To write the decimal number 165 in the binary system, we apply the division algorithm with divisor 2 and we record the successive remainders (either 0 or 1) as shown in the following table.

Division by 2	Quotient	Remainder
165	82	1
82	41	0
41	20	1
20	10	0
10	5	0
5	2	1
2	1	0
1	0	1

Reading the column of remainders from bottom to top gives the following binary representation of 165: **10100101**.

Example 6.12. The decimal value of the binary number 1100101 is $1 \cdot 2^6 + 1 \cdot 2^5 + 0 \cdot 2^4 + 0 \cdot 2^3 + 1 \cdot 2^2 + 0 \cdot 2^1 + 1 \cdot 2^0 = 101$.

Example 6.13. If the octal representation (base 8) of an integer n is 102406, then its decimal value is $1 \cdot 8^5 + 0 \cdot 8^4 + 2 \cdot 8^3 + 4 \cdot 8^2 + 0 \cdot 8^1 + 6 \cdot 8^0 = 34054$.

6.4.2 *Exercises*

(1) The equation $57 = 13 \cdot 4 + 5$ is true. Does it result from the division algorithm of 57 by 4? Justify your answer.
(2) True or False: *"Every integer can be expressed in precisely one of the forms $4k$, $4k + 1$, $4k + 2$ or $4k + 3$ for some integer k."*
(3) For each of the pairs m, n of integers, find the quotient q and the remainder r of the division of m by n. Verify that the relation $m = qn + r$ is correct.

$\star$ (a) $m = 123$, $n = 17$ (d) $m = 435$, $n = 8$
    (b) $m = -321$, $n = 7$ $\star$ (e) $m = -462$, $n = 12$
    (c) $m = 0$, $n = 11$ (f) $m = 123$, $n = 321$

(4) In each case, give four integers with a remainder r upon division by $n = 11$.

(a) $r = 3$ (b) $r = 6$ (c) $r = 7$ (d) $r = 10$

(5) Using the division algorithm, write a general expression for an integer n upon its division by 7. Give three negative and three positive integers that have a remainder of 5 in the division by 11.
$\star$(6) Show that if n is an odd integer, then n^4 has a remainder of 1 upon division by 8.
(7) Convert each of the following binary numbers to decimal.

(a) 1011   (d) 11110111   (g) 111101010111101
(b) 10110 $\star$ (e) 1010101101
(c) 101101   (f) 111101101111

(8) Convert each of the following decimal numbers to binary.

    (a) 15   ⋆ (c) 225   (e) 987   (g) 1830   ⋆ (i) 12345
    (b) 185   (d) 456   ⋆ (f) 1598   (h) 4563

(9) Convert each of the following octal numbers (base 8) to decimal.

    (a) 67   (b) 123   ⋆ (c) 2345   (d) 4567

(10) Convert each of the decimal numbers to octal (base 8).

    (a) 67   (b) 123   ⋆ (c) 2389   (d) 14569

(11) Let $a, b, n \in \mathbb{Z}$ with $n > 0$. Use the division algorithm to prove the following: *a and b have the same remainder in the division by n if and only if $a - b$ is an integer multiple of n.*

(12) The division algorithm is not proper to integers. For example, there is a version of this algorithm for polynomials in one variable that goes as follows. If $f(x)$ and $g(x)$ are two polynomials (with real coefficients) such that $g(x)$ is not identically zero (not the zero polynomial), then there exist unique polynomials $q(x)$ (quotient) and $r(x)$ (remainder) such that $f(x) = q(x)g(x) + r(x)$ with either $r(x) = 0$ (the zero polynomial) or $\deg r(x) < \deg g(x)$ (deg stands for the degree of the polynomial). The long division of polynomials is used to find the quotient and the remainder. In each of the following cases, find the quotient $q(x)$ and the remainder $r(x)$ of the division of $f(x)$ by $g(x)$. Verify that the relation $f(x) = q(x)g(x) + r(x)$ is correct.

    (a) $f(x) = x^2 - 3x - 1$, $g(x) = x + 1$
    (b) $f(x) = -3x^3 + 2x^2 + 2x - 1$, $g(x) = x + 1$
 ⋆ (c) $f(x) = x^4 - 3x^3 + 2x^2 - 1$, $g(x) = x^2 + 1$
    (d) $f(x) = x^5 + x^3 + 2x^2$, $g(x) = x^3 + x$

⋆(13) Let $n \in \mathbb{Z}$. Prove that if the remainder of the division of n by 8 is 7, then the remainder of the division of n by 4 is 3.

(14) Let p be a prime number.

    (a) Prove that if p is odd, then $p = 4k + 1$ or $p = 4k + 3$ for some $k \in \mathbb{Z}$.
    (b) Prove that if $p \geq 5$, then $p = 6k + 1$ or $p = 6k + 5$ for some $k \in \mathbb{Z}$.

6.5 The Fundamental Theorem of Arithmetic, the gcd, the lcm and the Euclidean algorithm

In this section we look at the notions of greatest common divisor and the least common multiple of integers. In particular, the greatest common divisor plays a crucial role in number theory. We define it as an integer satisfying two properties (Properties (a) and (b) of Theorem 6.12 below) and we investigate two ways to computing it.

Definition 6.7. Let $x_1, \ldots, x_n$ be integers not all zero. A positive integer d is called a *common divisor* of $x_1, \ldots, x_n$ if $d \mid x_i$ for every $i \in \{1, \ldots, n\}$. A positive integer m is called a *common multiple* of $x_1, \ldots, x_n$ if m is an integer multiple of x_i for every $i \in \{1, \ldots, n\}$. Positive integers $x_1, \ldots, x_n$ are called *relatively prime* (or *coprime*) if the only common divisor of $x_1, \ldots, x_n$ is 1. They are called *pairwise relatively prime* if the only common divisor of any pair x_i and x_j $(i \neq j)$ of them is 1.

Example 6.14.

(a) 5 is a common divisor of 15, 25 and 125. In particular, 15, 25 and 125 are not relatively prime.
(b) The only common divisor of 3 and 7 is 1, so 3 and 7 are relatively prime.
(c) The integers 2, 3 and 7 are pairwise relatively prime.
(d) 108 is a common multiple of 2, 3, 6, 18, 27 and 54.

The next result paves the way to properly define the notion of the greatest common divisor of two integers.

Theorem 6.12. *If m, n are two integers not both zero, then there exists a unique positive integer g satisfying the two conditions:*

(a) g is a common divisor of m and n;
(b) If d is any common divisor of m and n then $d \mid g$.

Proof. Consider the set $S = \{mx + ny;\ x, y \in \mathbb{Z}, \text{ and } mx + ny > 0\}$. If both $m < 0$ and $n < 0$, then the integer $(-1)m + (-1)n \in S$ (choosing $x = y = -1 \in \mathbb{Z}$). If $m \leq 0$ and $n > 0$, then $(-1)m + (1)n \in S$ (choosing $x = -1$ and $y = 1$). Similarly $(1)m + (1)n \in S$ when $m > 0$ and $n > 0$. This shows that S is a non-empty subset of $\mathbb{N}$. By the well-ordering principal, S has a smallest element g that we can write as $g = \alpha m + \beta n$ for some

$\alpha, \beta \in \mathbb{Z}$. Using the division algorithm, write $m = ag + r$ for some $a, b \in \mathbb{Z}$ with $0 \leq r < g - 1$. Then

$$r = m - ag$$
$$= m - a(\alpha m + \beta n)$$
$$= \underbrace{(1 - a\alpha)}_{\in \mathbb{Z}} m + \underbrace{(-a\beta)}_{\in \mathbb{Z}} n.$$

If $r > 0$, then r would be an element of S as it is positive and a linear combination of m and n. But that means that $r \geq g$ since g is the smallest element of S. This is a contradiction to the fact that $r < g - 1$. We conclude that $r = 0$ and so $m = ag$ is divisible by g. Similarly, n is divisible by g. This proves that g satisfies condition (a). Next, assume that d is a common divisor of m and n. We need to prove that d divides g. Write $m = ad$ and $n = bd$ for some $a, b \in \mathbb{Z}$, then $g = \alpha m + \beta n = \alpha(ad) + \beta(bd) = (\alpha a + \beta b)d$ which means that d divides g. Finally, we prove the uniqueness of g. Assume that g' is another positive integer satisfying conditions (a) and (b) of the theorem. Since g is a common divisor of m and n and g' satisfies condition (b), g divides g'. Similarly, g' divides g. Since g and g' are both positive, we must have $g = g'$. $\qquad\qquad\square$

Definition 6.8. Let m and n be two integers not both zero. The unique integer satisfying the two properties of Theorem 6.12 is called the *greatest common divisor* of m and n, denoted by $\gcd(m, n)$.

An immediate consequence of Theorem 6.12 is the following.

Corollary 6.2. *If m and n are two integers not both zero, then $\gcd(m, n)$ is the largest common divisor of m and n.*

Remark 6.2. We could have defined condition (b) in Theorem 6.12 as follows: "If d is any common divisor of m and n, then $d \leq g$" but rather we chose the conclusion to be $d \mid g$. The reason behind this is simple: the notion of the greatest common divisor applies in general to domains where only binary operations are defined and where no natural order like the relation $\leq$ on the integers exists. So we want to keep the definition as general as possible.

Example 6.15.

(a) $\gcd(21, 1344) = 7$. To see this, note that the only positive divisors of 21 are 1, 3, 7 and 21. Of these, only 1 and 7 are also divisors of 1344.

(b) $\gcd(21, 64) = 1$ since 21 and 64 are relatively prime.

(c) $\gcd(-44, -16) = 4$

The next theorem explores the first properties of the gcd. Part (d) says in particular that we only need to deal with the gcd of non-negative integers. Each part of the theorem follows easily from the definition of the gcd. The proof is left to the reader.

Theorem 6.13. *Let a, b be integers not both zero.*

(a) $\gcd(a, b) = \gcd(b, a)$.
(b) If $a \neq 0$ and $b \neq 0$, then $\gcd(a, b) \leq \min(|a|, |b|)$.
(c) If $a \neq 0$, then $\gcd(a, 0) = \gcd(0, a) = |a|$.
(d) $\gcd(a, b) = \gcd(|a|, |b|)$.

Remark 6.3. The case $m = n = 0$ is excluded from Definition 6.8 since there exists no largest divisor of 0. We use the convention $\gcd(0, 0) = 0$. This way, $\gcd(n, 0) = \gcd(0, n) = |n|$ for every integer n.

The proof of Theorem 6.12 says more about the greatest common divisor than its existence and uniqueness. It reveals that $\gcd(m, n)$ is actually a linear combination of m and n with integer coefficients. As we will see below, this is very important feature of the gcd.

Theorem 6.14. *Let m, n be two integers not both zero, $g = \gcd(m, n)$.*

(a) (Bézout identity) $g = xm + yn$ for some integers x and y.
(b) m and n are relatively prime if and only if then there exist integers x and y such that $xm + yn = 1$.

Proof.

(a) This follows immediately from the proof of Theorem 6.12.
(b) If m and n are relatively prime, then $\gcd(m, n) = 1$ and the relation $xm + yn = 1$ for some $x, y \in \mathbb{Z}$ follows. Conversely, assume that $xm + yn = 1$ for some integers x and y. If d is a common divisor of m and n, then (by Theorem 6.6 above) d must divides $mx + ny$ and consequently, $d = \pm 1$ (the only divisors of 1). We conclude that $\gcd(m, n) = 1$ and m, n are relatively prime. $\qquad\square$

Example 6.16. Since $(3)5 + (-2)7 = 1$, we conclude that 5 and 7 are relatively prime. The same is true for the integers 33 and 430 since $(-13)(33) + (1)(430) = 1$.

Theorem 6.15. *Let n be a positive integer and let p be a prime number. Then $\gcd(n, p) = 1$ if and only if p does not divide n.*

Proof. Assume first that $\gcd(n, p) = 1$. Then p does not divide n since otherwise p would be a common divisor of n and p and consequently $p \mid \gcd(n, p) = 1$. This is a contradiction since p is a prime number. Conversely, assume that p does not divide n. If $\gcd(n, p) \neq 1$, then n and p have a common divisor d other than 1. But the only divisors of p are 1 and p, so $d = p$ a contradiction since $p \nmid n$. $\qquad\square$

Example 6.17. Since $11 \nmid 77778$ and $p = 11$ is prime, $\gcd(7778, 11) = 1$.

An important consequence of Theorem 6.14 is a result known as the Euclid's Lemma. The result is important on its own but it also serves as a tool to prove other major results.

Theorem 6.16 (Euclid's lemma). *Let m, n and x be three integers. If m and x are relatively prime and $m \mid nx$ then $m \mid n$.*

Proof. Since $m \mid nx$, there exists an integer t such that $nx = tm$. By Theorem 6.14, there exist integers α and β such that $\alpha m + \beta x = 1$. Then

$$
\begin{aligned}
n &= n \cdot 1 \\
&= n(\alpha m + \beta x) \text{ (since } \alpha m + \beta x = 1) \\
&= \alpha m n + \beta x n \\
&= \alpha m n + \beta t m \text{ (since } nx = tm) \\
&= (\alpha n + \beta t)m.
\end{aligned}
$$

This shows that $m \mid n$ as required. $\qquad\square$

Corollary 6.3. *Let m, n be two integers and let p be a prime number. If $p \mid mn$ then $p \mid m$ or $p \mid n$.*

Proof. Assume $p \mid mn$. If $p \nmid m$, then p and m are relatively prime by Theorem 6.15 and so $p \mid n$ by Euclid's lemma. $\qquad\square$

Example 6.18. Let $n \geq 1$ be an integer, p a prime number. If $p \mid n^2$, then $p \mid n$ by Corollary 6.3.

The result of Corollary 6.3 can be generalized to any number of integers as shown in the following.

Corollary 6.4. *Let $a_1, a_2, \ldots, a_n$ be positive integers and p a prime number. If $p \mid a_1 a_2 \cdots a_n$, then $p \mid a_i$ for some i.*

Proof. By induction on $n \geq 1$. The details are left to the reader as an exercise (see Exercise (26) below). $\qquad\square$

We are now ready to prove the Fundamental Theorem of Arithmetic briefly mentioned in Chapter 3. The existence part of the theorem was proved by induction in Chapter 3. In the proof below, we use the well-ordering principle of natural numbers for that end.

Theorem 6.17 (The Fundamental Theorem of Arithmetic). *Every integer $n \geq 2$ can be expressed as a product $n = p_1 p_2 \cdots p_k$ where $p_1 \leq p_2 \leq \cdots \leq p_k$ are prime numbers. Moreover, this expression is unique apart from the order of the p_i's. That is, if $n = q_1 q_2 \cdots q_l$ is another expression of n as a product of primes with $q_1 \leq q_2 \leq \cdots \leq q_l$, then $k = l$ and $p_i = q_i$ for every i.*

Proof. Let A be the set of all integers $n \geq 2$ that cannot be expressed as a product of primes. If $A \neq \emptyset$, then A must have a least element m by the well-ordering principle. In particular, m is not prime (since otherwise it would have the trivial prime factorization) and so $m = uv$ for some integers u, v with $2 \leq u, v \leq m - 1$. Since m is the smallest element of A, $u, v \notin A$ and so they can be expressed as product of prime numbers. But that means that $m = uv$ is also a product of prime numbers, a contradiction. So $A = \emptyset$ and every integer $n \geq 2$ can be expressed as a product of primes. For the uniqueness part, we use induction on k (the number of primes in one expression of n). For $k = 1$, we get that $p_1 = q_1 q_2 \cdots q_l$ and so p_1 must divide q_i for some i (by Corollary 6.4). Since p_1 and q_i are both prime numbers, $p_1 = q_i \leq q_l$. On the other hand $q_l \leq p_1$ (as a divisor of p_1) and so $p_1 = q_l$. Dividing the equation $p_1 = q_1 q_2 \cdots q_l$ by p_1, we get that $1 = q_1 q_2 \cdots q_{l-1}$. This is a contradiction unless $l = 1$ and $p_1 = q_1$. The result is true for $k = 1$. Let $k \geq 1$ and assume the result is true for k. If $p_1 p_2 \cdots p_k p_{k+1} = q_1 q_2 \cdots q_l$. Using the same argument of the base case, we get that $p_{k+1} \leq q_l$. Similarly, $q_l \leq p_{k+1}$ and so $p_{m+1} = q_l$. We conclude that $p_1 p_2 \cdots p_k = q_1 q_2 \cdots q_{l-1}$. By the induction hypothesis, $k = l - 1$ and $p_1 = q_1, p_2 = q_2, \ldots, p_k = q_{l-1}$. Therefore, $k + 1 = l$, $p_1 = q_1, p_2 = q_2, \ldots, p_{k+1} = q_l$. $\qquad\square$

Theorem 6.18. *Let m, n and x be three integers. Then $\gcd(m, nx) = 1$ if and only if $\gcd(m, n) = 1$ and $\gcd(m, x) = 1$.*

Proof. Assume first that $\gcd(m, nx) = 1$. If d is a common divisor of m and n, then d is a common divisor of m and nx and so $d \mid 1$. This shows that $\gcd(m, n) = 1$. Similarly, $\gcd(m, x) = 1$. Conversely, assume that $\gcd(m, n) = \gcd(m, x) = 1$. We need to prove that $\gcd(m, nx) = 1$. Assume by contradiction that $\gcd(m, nx) = g > 1$, then g must have a prime divisor p. In particular, $p \mid m$ and $p \mid nx$. By Corollary 6.3, $p \mid n$ or $p \mid x$. If $p \mid n$, then p is a common divisor of m and n. This is a contradiction since $\gcd(m, n) = 1$. We obtain a similar contradiction if we assume that $p \mid x$. We conclude that $\gcd(m, nx) = 1$. $\qquad\square$

Theorem 6.19. *If m, n are two integers not both zero and $g = \gcd(m, n)$, then* $\gcd\left(\frac{m}{g}, \frac{n}{g}\right) = 1$.

Proof. By Theorem 6.14, $g = \alpha m + \beta n$ for some integers α and β. Dividing both sides by g, we get that $\alpha\left(\frac{m}{g}\right) + \beta\left(\frac{m}{g}\right) = 1$. The result follows from part (b) of Theorem 6.14. $\qquad\square$

The following is another consequence of Euclid's lemma.

Theorem 6.20. *Let k be a positive integer and p a prime number. Then the only positive divisors of p^k are integers of the form p^s for some integer s with $0 \leq s \leq k$.*

Proof. Let d be an arbitrary positive divisor of p^k and let $d = q_1 q_2 \cdots q_r$ be the prime factorization of d. For each i, $q_i \mid p^k$ since $q_i \mid d$. By Corollary 6.4, q_i is a divisor of p for each i and since p is prime, $q_i = p$. This shows that d is of the form p^m for some non-negative integer m. Clearly, $m \leq k$ since otherwise p^m is not a divisor of p^k. $\qquad\square$

Example 6.19. Find all the positive divisors of 2401.

Solution. Notice that $2401 = 7^4$. Since 7 is prime, Theorem 6.20 tells us that the positive divisors of 7^4 are precisely $7^0 = 1$, $7^1 = 7$, $7^2 = 49$ and $7^4 = 2401$. $\qquad\Diamond$

We can extend the definition of the greatest common divisor to any finite number of integers, not all zero. First, we need the following.

Theorem 6.21. *Let x, y, z be three integers not all zero. Then* $\gcd(x, \gcd(y, z)) = \gcd(\gcd(x, y), z)$.

Proof. Let $g = \gcd(x, \gcd(y, z))$ and $g' = \gcd(\gcd(x, y), z)$. Then g divides x, y and z since it divides x and $\gcd(y, z)$. As a common divisor of x and y, g divides $\gcd(x, y)$ and so g is a common divisor of $\gcd(x, y)$ and z. This show that g divides $\gcd(\gcd(x, y), z) = g'$. Similarly, g' divides g. Since g and g' are positive integers, $g = g'$. $\qquad\square$

For three integers x, y, z not all zero, define their greatest common divisor, denoted by $\gcd(z, y, z)$, as being:

$$\gcd(z, y, z) = \gcd(x, \gcd(y, z)) = \gcd(\gcd(x, y), z), \qquad (6.1)$$

Theorem 6.21 tells us that relation (6.1) is well-defined. In general, given integers $a_1, a_2, \ldots, a_n$ not all zero, we define their greatest common divisor, denoted by $\gcd(a_1, a_2, \ldots, a_n)$, as follows. Let $d_1 = \gcd(a_1, a_2)$, $d_2 = \gcd(a_3, d_1)$, $\ldots$, $d_{n-1} = \gcd(a_n, d_{n-2})$. Then $\gcd(a_1, a_2, \ldots, a_n) = d_{n-1}$.

Theorem 6.22. *Let $a_1, a_2, \ldots, a_n$ be integers not all zero and let $g = \gcd(a_1, a_2, \ldots, a_n)$ as defined above. Then:*

(a) g is the unique positive integer satisfying the two conditions:

    (i) $g \mid a_i$ for all $i = 1, 2, \ldots, n$.
    (ii) If d is a common divisor of the a_i's, then $d \mid g$.

(b) There exist $\alpha_1, \alpha_2, \ldots, \alpha_n \in \mathbb{Z}$ such that $g = \alpha_1 a_1 + \alpha_2 a_2 + \cdots + \alpha_n a_n$.

The proof is by induction on n and the case $n = 2$ established in previous discussions. We leave the details of the proof as an exercise to the reader.

6.5.1 *The least common multiple*

We now turn our attention to the notion of the least common multiple. Similar to the greatest common divisor, we start by showing the existence and uniqueness.

Theorem 6.23. *Given non-zero integers $a_1, a_2, \ldots, a_n$, there exists a unique positive integer l satisfying the two conditions:*

(a) l is a common multiple of $a_1, a_2, \ldots, a_n$;
(b) If m is any non-zero common multiple of $a_1, a_2, \ldots, a_n$, then $l \mid m$.

Proof. Let A be the set of all common positive multiples of $a_1, a_2, \ldots, a_n$. Since $|a_1 a_2 \cdots a_n| \in A$, A is a non-empty subset of $\mathbb{N}$. By the well-ordering principle, A must have a least element l. Being an element of A, l is a positive common multiple of $a_1, a_2, \ldots, a_n$. If m is a non-zero common multiple of $a_1, a_2, \ldots, a_n$, the division algorithm allows us to write $m = \alpha l + r$ for some integers α and r with $0 \leq r < l$. Since m and l are both common multiples of $a_1, a_2, \ldots, a_n$, then so is $r = m - \alpha l$. If $r > 0$, then $r \in A$ (by definition of A) and so $l \leq r$ by the minimality of l in A: this is a contradiction. We conclude that $r = 0$ and so $l \mid m$. The uniqueness of l follows from the uniqueness of the smallest element of A. $\qquad\square$

Definition 6.9. Let $a_1, a_2, \ldots, a_n$ be non-zero integers. The unique positive integer satisfying the two conditions of Theorem 6.23 is called the *least common multiple* of $a_1, a_2, \ldots, a_n$ that we denote by $\mathrm{lcm}(a_1, a_2, \ldots, a_n)$.

Like the name suggests, $\mathrm{lcm}(a_1, a_2, \ldots, a_n)$ is the smallest among all positive common multiples of $a_1, a_2, \ldots, a_n$.

Example 6.20.

   (a) $\mathrm{lcm}(12, 45) = 180$.
   (b) $\mathrm{lcm}(12, 39, 108) = 1404$.
   (c) $\mathrm{lcm}(12, -24, -36, 108, -1512) = 1512$.

It should come as no surprise at this point that there is a relationship between the gcd and the lcm of a set of non-zero integers. If either one of gcd or the lcm is known, then the other can be computed easily.

Theorem 6.24. $\gcd(a, b) \cdot \mathrm{lcm}(a, b) = ab$ *for any positive integers a, b.*

Proof. Let $g = \gcd(a, b)$ and $l = \mathrm{lcm}(a, b)$. Let $\alpha, \beta, \gamma, \delta, \epsilon, \theta$ be integers satisfying: $g = \alpha a + \beta b$, $l = \gamma a = \delta b$, $a = \epsilon g$ and $b = \theta g$. Then:

$$gl = (\alpha a + \beta b)l$$
$$= \alpha a l + \beta b l$$
$$= \alpha a (\delta b) + \beta b (\gamma a)$$
$$= ab(\alpha \delta + \beta \gamma).$$

This shows in particular that $ab \mid gl$. On the other hand: $\frac{ab}{g} = \frac{\epsilon g b}{g} = \epsilon b$ and $\frac{ab}{g} = \frac{a\theta g}{g} = \theta a$. This shows that $\frac{ab}{g}$ is a common multiple of a and b and therefore $l \mid \frac{ab}{g}$ (by definition of the least common multiple) and so $gl \mid ab$. We conclude that $ab = gl$ since both ab, gl are positive. $\qquad\square$

Remark 6.4. Theorem 6.24 works for any non-zero integers (not necessarily positive). The general result states that $\gcd(a, b) \cdot \mathrm{lcm}(a, b) = |ab|$.

6.5.2 *Finding the* gcd *and the* lcm

For small integers m, n, the problem of computing $\gcd(m, n)$ and $\text{lcm}(m, n)$ can be solved relatively easily by finding the prime factorization of both integers. But for larger integers, the problem can be challenging.

6.5.2.1 *Using the Fundamental Theorem of Arithmetic to find the* gcd *and the* lcm

Let a, b be positive integers. Let $p_1 < p_2 < \ldots < p_n$ be the collection of all *distinct* primes appearing in the prime factorizations of a and b. Write:

$$a = p_1^{a_1} p_2^{a_2} \cdots p_n^{a_n}, \quad b = p_1^{b_1} p_2^{b_2} \cdots p_n^{b_n}$$

where the a_i's and the b_i's are non-negative integers with $a_i = 0$ if $p_i \nmid a$ and $b_i = 0$ if $q_i \nmid b$. For each $i \in \{1, 2, \ldots, n\}$, let $k_i = \min\{a_i, b_i\}$ and $s_i = \max\{a_i, b_i\}$. Let $g = p_1^{k_1} p_2^{k_2} \cdots p_n^{k_n}$ and $l = p_1^{s_1} p_2^{s_2} \cdots p_n^{s_n}$.

Theorem 6.25. *With the above notations, $g = \gcd(a, b)$ and $l = lcm(a, b)$.*

Proof. Since $k_i \leq a_i$, $p_i^{k_i} \mid p_i^{a_i}$ and so $g = p_1^{k_1} p_2^{k_2} \cdots p_n^{k_n} \mid p_1^{a_1} p_2^{a_2} \cdots p_n^{a_n} = a$. Similarly, $g \mid b$. If d is a common divisor of a and b, then each prime divisor of d is also a prime divisor of a and b. So d has a prime factorization of the form $p_1^{t_1} p_2^{t_2} \cdots p_n^{t_n}$ where the t_i's are non-negative integers with $t_i \leq k_i$ for each i. This shows that $d \mid g$. We conclude that $g = \gcd(a, b)$. This proves part (a). Part (b) is proved is left as an exercise. $\qquad\square$

Example 6.21. Using the prime factorization of each of the integers $a = 324$ and $b = 16632$, find $\gcd(a, b)$ and $\text{lcm}(a, b)$.

Solution. Using successive divisions by primes, we get that: $a = 2^2 \cdot 3^4$ and $b = 2^3 \cdot 3^3 \cdot 7 \cdot 11$. Then $\gcd(a, b) = 2^{\min\{2,3\}} \cdot 3^{\min\{4,3\}} \cdot 7^{\min\{0,1\}} \cdot 11^{\min\{0,1\}} = 2^2 \cdot 3^3 \cdot 7^0 \cdot 11^0 = 108$ and $\text{lcm}(a, b) = 2^{\max\{2,3\}} \cdot 3^{\max\{4,3\}} \cdot 7^{\max\{0,1\}} \cdot 11^{\max\{0,1\}} = 2^3 \cdot 3^4 \cdot 7^1 \cdot 11^1 = 49896$. $\qquad\Diamond$

We can extend this technique to find the gcd of three of more integers.

Example 6.22. Find the gcd and the lcm of the four integers $a_1 = 1848$, $a_2 = 6160$, $a_2 = 13475$ and $a_4 = 30492$.

Solution. The prime factorizations of each of the four integers are as follows: $a_1 = 2^2 \cdot 3 \cdot 7 \cdot 11$, $a_2 = 2^4 \cdot 5 \cdot 7 \cdot 11$, $a_3 = 5^2 \cdot 7^2 \cdot 11$ and $a_4 = 2^2 \cdot 3^2 \cdot 7 \cdot 11^2$. Therefore:

- $\gcd(a_1, a_2, a_3, a_4) = 2^{\min\{2,4,0,2\}} \cdot 3^{\min\{1,0,0,2\}} \cdot 5^{\min\{0,1,2,0\}} \cdot 7^{\min\{1,1,2,1\}} \cdot 11^{\min\{1,1,1,2\}} = 2^0 \cdot 3^0 \cdot 5^0 \cdot 7^1 \cdot 11^1 = 77$.
- $\operatorname{lcm}(a_1, a_2, a_3, a_4) = 2^{\max\{2,4,0,2\}} \cdot 3^{\max\{1,0,0,2\}} \cdot 5^{\max\{0,1,2,0\}} \cdot 7^{\max\{1,1,2,1\}} \cdot 11^{\max\{1,1,1,2\}} = 2^4 \cdot 3^2 \cdot 5^2 \cdot 7^2 \cdot 11^2 = 435600$. $\Diamond$

6.5.2.2   *The Euclidean algorithm*

There are two main issues with using the prime factorization method to find the gcd. First, it is usually hard to find the prime factors of large integers. Second, the method fails to express the $\gcd(a, b)$ as a linear combination (with integer coefficients) of a and b. This is an important feature of the gcd that we often want to exploit in applications. In this section we discuss another method, known as the Euclidean algorithm, to find the $\gcd(a, b)$ without finding the prime factorizations of a and b. It is based on successive use of the division algorithm of integers. Working the calculations backward, the Euclidean algorithm expresses the gcd as a linear combination of a and b in addition of computing its value. The algorithm dates back to the Greek mathematician Euclid (hence its name). It is based on the following result.

Theorem 6.26. *Let a, b be two positive integers.*

(1) For any integer q, $\gcd(a, b) = \gcd(b, a - qb)$.
(2) If r is the remainder of the division of a by b, then $\gcd(a, b) = \gcd(b, r)$.

Proof.

(1) Let $g = \gcd(a, b)$. Since g is a common divisor of a and b, it is a divisor of $a - qb$ and so it is a common divisor of b and $a - qb$. Moreover, if d is a common divisor of b and $a - qb$, then d is a divisor of $qb + (a - qb) = a$. So d is a common divisor of a and b and therefore d divides g. We conclude that $g = \gcd(b, a - qb)$.
(2) If r is the remainder of the division of a by b, then there exists an integer q such that $a = qb + r$. In particular, $\gcd(a, b) = \gcd(b, a - qb)$ (by the first part) $= \gcd(b, r)$. $\square$

Since $\gcd(a, b) = \gcd(|a|, |b|)$ for any integers a and b, we may assume that both a and b are positive integers for the purpose of computing their gcd. We may also assume that $a > b$ since if $a = b$, $\gcd(a, b) = a$. In a nutshell, the Euclidean algorithm uses the division algorithm and Theorem 6.26 above to reduce the problem of computing the $\gcd(a, b)$ to that of two smaller integers. The process continues until one of the two integers is zero. The following steps describes the algorithm to compute $g = \gcd(a, b)$ (with $a, b > 0$) and to express it as a linear combination of a and b.

(1) Apply the division algorithm to find integers q_1 and r_1 such that $a = q_1 b + r_1$ and $0 \leq r_1 < b$. Theorem 6.26 tells us that $\gcd(a, b) = \gcd(b, r_1)$.
(2) Apply the division algorithm to find integers q_2 and r_2 such that $0 \leq r_2 < r_1$ such that $b = q_2 r_1 + r_2$. By Theorem 6.26, we know that $\gcd(a, b) = \gcd(b, r_1) = \gcd(r_1, r_2)$.
(3) By repeating this process, we obtain a strictly decreasing sequence $r_1 > r_2 > r_3 > \ldots \geq 0$ where $r_1, r_2, \ldots$ are the remainders in the successive divisions. Since $r_i \geq 0$ for each i, this sequence of remainders must reach zero eventually.
(4) The successive applications of the division algorithm lead to the following equations:

$$\begin{aligned}
a &= q_1 b + r_1; & 0 &\leq r_1 < b \\
b &= q_2 r_1 + r_2; & 0 &\leq r_2 < r_1 \\
r_1 &= q_3 r_2 + r_3; & 0 &\leq r_3 < r_2 \\
&\;\;\vdots & &\;\;\vdots \\
r_{n-1} &= q_{n+1} r_n + r_{n+1}; & 0 &\leq r_{n+1} < r_n \\
r_n &= q_{n+2} r_{n+1} + 0.
\end{aligned}$$

(5) By Theorem 6.26 above, we have:

$$\gcd(a, b) = \gcd(b, r_1) = \ldots = \gcd(r_n, r_{n+1}) = \gcd(r_{n+1}, 0) = r_{n+1}.$$

We conclude that $\gcd(a, b)$ is the last non-zero remainder in the above sequence of remainders.
(6) Moreover, starting from the second to last of the equations in step (4) and working backwards, we get an expression of $\gcd(a, b)$ as a linear combination of a and b.

Example 6.23. In each case, find $\gcd(a, b)$ and express your answer as a linear combination of a and b.

(a) $a = 42$, $b = 351$.   (b) $a = 780$, $b = 2772$

Solution. (a) The following shows successive divisions with remainders underlined at every step:

$$351 = 8 \cdot 42 + \underline{15} \tag{6.2}$$

$$42 = 2 \cdot 15 + \underline{12} \tag{6.3}$$

$$15 = 1 \cdot 12 + \underline{3} \tag{6.4}$$

$$12 = 4 \cdot 3 + \underline{0}. \tag{6.5}$$

The last non-zero remainder is 3 and therefore, $\gcd(351, 42) = 3$. To write this gcd as a linear combination of 351 and 42, start at equation (6.4) and isolate 3 from that equation: $3 = 15 - (1 \cdot 12)$. Next, replace 12 by its value $42 - 2 \cdot 15$ from equation (6.3) and get

$$3 = 15 - (42 - 2 \cdot 15) = 3 \cdot 15 - 42. \tag{6.6}$$

From equation (6.2), $15 = 351 - 8 \cdot 42$. Replacing in (6.6), we get $3 = 3 \cdot 15 - 42 = 3 \cdot (351 - 8 \cdot 42) - 42 = 3 \cdot 351 + (-25) \cdot 42$.

(b) The following shows successive divisions with remainders underlined at every step:

$$2772 = 3 \cdot 780 + \underline{432} \tag{6.7}$$

$$780 = 1 \cdot 432 + \underline{348} \tag{6.8}$$

$$432 = 1 \cdot 348 + \underline{84} \tag{6.9}$$

$$348 = 4 \cdot 84 + \underline{12} \tag{6.10}$$

$$84 = 7 \cdot 12 + \underline{0} \tag{6.11}$$

The last non-zero remainder is 12 and therefore, $\gcd(2772, 780) = 12$. Working backwards from equation (6.10), we can express 12 as a linear combination of 2772 and 780 as shown in the following set of equations:

$$
\begin{aligned}
12 &= 384 - 4 \cdot 84 \quad \text{from (6.10)} \\
&= 384 - 4 \cdot (432 - 1 \cdot 348) \quad \text{from (6.9)} \\
&= 5 \cdot 348 - 4 \cdot 432 \\
&= 5 \cdot (780 - 1 \cdot 432) - 4 \cdot 432 \quad \text{from (6.8)} \\
&= 5 \cdot 780 - 9 \cdot 432 \\
&= 5 \cdot 780 - 9 \cdot (2772 - 3 \cdot 780) \quad \text{from (6.11)} \\
&= (-9) \cdot 2772 + 32 \cdot 780.
\end{aligned}
$$

$\Diamond$

6.5.3 *Exercises*

(1) Let $a \neq 0$ be an integer. Find each of the following:

(a) $\gcd(a, 0)$
(b) $\gcd(a, 1)$
(c) $\gcd(a, a)$

(d) $\gcd(-a, a)$
$\star$ (e) $\gcd(a, a^3)$
(f) $\gcd(a, a^3)$

(g) $\gcd(a^2, a^3)$
$\star$ (h) $\gcd(-2a, 3a)$

(2) Let $a_1, a_2, \ldots, a_n$ be positive integers. For each of the following statements, determine if it is true or false. Justify your answer.

$\star$ (a) If $a_1, a_2, \ldots, a_n$ are relatively prime, then $a_1, a_2, \ldots, a_n$ are pairwise relatively prime.

(b) If $a_1, a_2, \ldots, a_n$ are pairwise relatively prime, then $a_1, a_2, \ldots, a_n$ are relatively prime.

(c) If $a_1, a_2, \ldots, a_n$ are relatively prime, then any subset of these integers are also relatively prime.

$\star$ (d) If $a_1, a_2, \ldots, a_n$ are pairwise relatively prime, then any subset of these integers are also pairwise relatively prime.

(3) In each case, determine if the given integers are relatively prime, pairwise relatively prime or they have a common divisor $d \geq 2$.

(a) $21, 49, 63, 125, 360$
(b) $15, 16, 17, 23, 49, 121$

(c) $15, 21, 48, 51, 726$.

$\star$(4) Give all positive integers n such that $n \leq 35$ and $\gcd(n, 35) = 1$.

(5) Find the gcd and the lcm of each of the following pairs of integers using the method of prime factorization.

(a) $a = 6$, $b = 21$
(b) $a = 30$, $b = 66$
(c) $a = 56$, $b = 168$
(d) $a = 65$, $b = 105$
(e) $a = 126$, $b = 162$

$\star$ (f) $a = 285$, $b = 465$
(g) $a = 296$, $b = 666$
$\star$ (h) $a = 2010$, $b = 4095$
$\star$ (i) $a = 616$, $b = 255255$
(j) $a = 2133$, $b = 6320$

(6) In each case, use the Euclidean algorithm to find $\gcd(a, b)$ and express your answer as a linear combination of a and b.

(a) $a = 45$, $b = 60$
(b) $a = 60$, $b = 120$
$\star$ (c) $a = 84$, $b = 1320$

   (d) $a = 2010$, $b = 4095$

$\star$ (e) $a = 616$, $b = 255255$

   (f) $a = 2133$, $b = 6320$

(7) In each case, find the gcd and the lcm of the given set of integers.

  (a) 16, 28, 52 $\star$ (c) 56, 168, 256

  (b) 30, 66, 128   (d) 65, 90, 105

(8) In each case, three of the four integers a, b, $g = \gcd(a,b)$ and $l = \mathrm{lcm}(a,b)$ are given. Find the missing fourth integer.

  (a) $a = 54$, $g = 9$, $l = 702$ (c) $a = 572$, $b = 728$, $g = 52$

$\star$ (b) $b = 66$, $g = 22$, $l = 6468$ (d) $a = 105$, $b = 381$, $l = 13335$.

(9) In each case, use Theorem 6.20 to find all the divisors of the given integer.

  (a) 32   (d) 125   (g) 289 $\star$ (j) 2197

  (b) 81 $\star$ (e) 128   (h) 343   (k) 3125

  (c) 121   (f) 169   (i) 1331   (l) 6859

(10) Determine the truth value of the following statement: If a, b are positive integers such that $\gcd(a, b)$ is prime, then either a or b must be prime.

$\star$(11) Is it possible to find two positive integers a, b such that $a+b = 23456701$ and $\gcd(a, b) = 3$? Justify your answer.

(12) Prove that two consecutive integers are always relatively prime.

(13) If a, b are coprime positive integers, prove that $\mathrm{lcm}(x, y) = xy$.

(14) Given an integer n and a prime number p, what are the possible values of $\gcd(n, n + p)$?

(15) Let m, n and x be three positive integers. Prove that $\gcd(xm, xn) = x\gcd(m, n)$.

$\star$(16) Let m, n be two relatively prime integers. Prove that $\gcd(m+n, m-n)$ is either 1 or 2.

(17) Let m, n be two positive integers. Prove that $\gcd(m, n) = m$ if and only if $m \mid n$.

(18) Let m be a positive integer. Prove that the integers $2m+3$ and $3m+5$ are relatively prime.

(19) Give an example that shows that the result of Theorem 6.15 is not true if p is not prime.

(20) Let $m, n \in \mathbb{Z}$. Prove that $\gcd(a, b) = \gcd(a + nb, b)$ for any integer n.

$\star$(21) Let n be a positive integer and p a prime number. Prove that for any integer $k \geq 1$, $\gcd(n, p) = 1$ if and only if $\gcd(n, p^k) = 1$.

$\star$(22) Determine the truth value of the following statement: If a, b, c, x, y are integers satisfying the equation $ax + by = c$, then $c = \gcd(a, b)$.

(23) Let a, b and c be positive integers. Prove that if p is a prime number such that $p \mid a$ and $p \mid (a^c + b^c)$, then $p \mid b$.

(24) Let m, n and x be three integers. Prove that if $\gcd(m, n) = 1$, then $\gcd(mx, n) = \gcd(x, n)$.

(25) Let a, b and c be positive integers such that $\gcd(a, b) = \gcd(a, c)$. In each case, determine if the statement is necessarily true: (i) $\text{lcm}(a, b) = \text{lcm}(a, c)$; (ii) $\gcd(a, b) = \gcd(a, b, c)$.

$\star$(26) Prove Corollary 6.4.

(27) Let a, b be two coprime integers. Prove that for every integer d, the equation $ax + by = d$ has a integer solutions (i.e., there exist integers x, y satisfying $ax + by = d$).

6.6   Introduction to modular arithmetic

At the beginning of this chapter, we established the fact that every integer is either even or odd. This is equivalent to say that for every integer k, the remainder of the division of k by 2 is either 0 or 1. If the remainder is 0 (respectively 1), we say that k is *congruent to 0 modulo* 2 (respectively, *congruent to 1 modulo* 2) and we write $k \equiv 0 \,(\text{mod } 2)$ (respectively, $k \equiv 0 \,(\text{mod } 2)$). There is really nothing special about the integer 2 and the congruence notion can be generalized to any positive integer.

Definition 6.10. Let a, b, n be integers with $n \geq 1$. we say that a *is congruent to b modulo n*, and we write $a \equiv b \,(\text{mod } n)$ if $a - b$ is divisible by n. If a is not congruent to b modulo n, we write $a \not\equiv b \,(\text{mod } n)$.

Example 6.24.

(a) $12 \equiv -3 \,(\text{mod } 5)$ since $12 - (-3) = 15$ is divisible by 5.

(b) $15 \not\equiv 3 \,(\text{mod } 7)$ since $15 - 3 = 12$ is not divisible by 7.

(c) $-123 \equiv 64 \,(\text{mod } 17)$ since $-123 - 64 = -187$ is divisible by 17.

Immediate consequences of Definition 6.10 are given by the following theorem.

Theorem 6.27. *Let a, b, c, d and n be integers with $n \geq 1$.*

(a) $a \equiv a \,(mod\ n)$.
(b) If $a \equiv b \,(mod\ n)$, then $b \equiv a \,(mod\ n)$.
(c) If $a \equiv b \,(mod\ n)$ and $b \equiv c \,(mod\ n)$ then $a \equiv c \,(mod\ n)$.
(d) If $a \equiv b \,(mod\ n)$ and $c \equiv d \,(mod\ n)$ then $a + c \equiv b + d \,(mod\ n)$.
(e) If $a \equiv b \,(mod\ n)$ and $c \equiv d \,(mod\ n)$ then $ac \equiv bd \,(mod\ n)$.
(f) If $a \equiv b \,(mod\ n)$ and k is a non-negative integer, then $a^k \equiv b^k \,(mod\ n)$.

Proof. We only prove parts (d) and (e). The other parts are left as exercises. Assume that $a \equiv b \,(\text{mod } n)$ and $c \equiv d \,(\text{mod } n)$. Then there exist integers s and t such that $a - b = sn$ and $c - d = tn$. Adding the two equations, we get $(a + c) - (b + d) = (a - b) + (c - d) = sn + tn = (s + t)n$. This shows that $(a + c) - (b + d)$ is divisible by n and consequently $a + c \equiv b + d \,(\text{mod } n)$. This proves part (d). For part (e), notice that $ac - bd = ac - bd - ad + ad = a(c - d) + d(a - b) = atn + dsn = (at + ds)n$. This shows that $ac - bd$ is divisible by n and so $ac \equiv bd \,(\text{mod } n)$. $\qquad\square$

Example 6.25. Using the congruences: $11 \equiv 4 \,(\text{mod } 7)$, $19 \equiv 5 \,(\text{mod } 7)$, Theorem 6.27 tells us that:

(a) $(11 + 19) = 30 \equiv (4 + 5) = 9 \,(\text{mod } 7)$. But $9 \equiv 2 \,(\text{mod } 7)$ since $9 - 2 = 7$ is divisible by 7. So $30 \equiv 2 \,(\text{mod } 7)$.
(b) $(11 \cdot 19) = 209 \equiv (4 \cdot 5) = 20 \,(\text{mod } 7)$. But $20 \equiv 6 \text{ mod}(7)$ since $20 - 6 = 14$ is divisible by 7. So $209 \equiv 6 \,(\text{mod } 7)$.

Another characterization of the relation $a \equiv b \,(\text{mod } n)$ is given by the following theorem. Many textbooks adopt this characterization as the definition of the congruence relation.

Theorem 6.28. *Let a, b and n be integers with $n \geq 1$. Then $a \equiv b \,(mod\ n)$ if and only if a and b have the same remainder upon division by n.*

Proof. Assume first that $a \equiv b \,(\text{mod } n)$. Then there exists an integer t such that $a - b = tn$. Apply the division algorithm of a and b by n and write $a = q_1 n + r_1$ and $b = q_2 n + r_2$ with $q_1, q_2 \in \mathbb{Z}$ and $0 \leq r_1 < n$ and $0 \leq r_2 < n$. Therefore, $a - b = (q_1 - q_2)n + (r_1 - r_2)$. Since $a - b = tn$, $r_1 - r_2 = (t - q_1 + q_2)n$ is divisible by n. Since $-n < r_1 - r_2 < n$, the only possible value of $r_1 - r_2$ is 0. We conclude that $r_1 = r_2$ and a and b have the same remainder upon division by n. Conversely, assume that a and b have the same remainder upon division by n. Then we can write $a = q_1 n + r$ and

$b = q_2 n + r$ for some $q_1, q_2 \in \mathbb{Z}$ and $0 \leq r < n$. So $a - b = (q_1 - q_2)n$ is divisible by n and consequently $a \equiv b \,(\mathrm{mod}\ n)$. $\qquad\square$

Theorem 6.29. *Let a, b and n be integers with $n \geq 1$ such that $a \equiv b \,(mod\ n)$. Then $\gcd(a, n) = 1$ if and only if $\gcd(b, n) = 1$.*

Proof. Since $a \equiv b \,(\mathrm{mod}\ n)$, $a - b = \alpha n$ for some integer α. Assume that $\gcd(a, n) = 1$ and let $g = \gcd(b, n)$. Then $g \mid b$ and $g \mid n$ and therefore $g \mid b + \alpha n = a$. Since g is a common divisor of a and n and $\gcd(a, n) = 1$, $g = 1$. By symmetry, if $\gcd(b, n) = 1$ then $\gcd(a, n) = 1$. $\qquad\square$

If a, n are integers with $n \geq 1$, the notation $a \,(\mathrm{mod}\ n)$ stands for the remainder of the division of a by n. For instance, $23 \,(\mathrm{mod}\ 7) = 2$. Theorem 6.27 can be used effectively to compute $a \,(\mathrm{mod}\ n)$ for some large integers a as shown in the following example.

Example 6.26. Find $3^{12} \,(\mathrm{mod}\ 7)$.

Solution. At first glance, it appears that one has to compute $3^{12} = 531441$ and then apply the long division by 7 to figure out what the remainder is. This is one way but certainly not the most efficient one. Sine $3^2 = 9 \equiv 2 \,(\mathrm{mod}\ 7)$, $3^{12} = (3^2)^6 \equiv 2^6 \,(\mathrm{mod}\ 7)$ by the last part of Theorem 6.27. Now $2^3 = 8 \equiv 1 \,(\mathrm{mod}\ 7)$ which means that $2^6 = (2^3)^2 \equiv 1^2 = 1 \,(\mathrm{mod}\ 7)$ (again by the last part of Theorem 6.27). We conclude that $3^{12} \,(\mathrm{mod}\ 7) = 1$. $\quad\Diamond$

Example 6.27. Find $9 \cdot 12 \cdot 13 \,(\mathrm{mod}\ 5)$.

Solution. Since $9 \equiv 4 \,(\mathrm{mod}\ 5)$, $12 \equiv 2 \,(\mathrm{mod}\ 5)$ and $13 \equiv 3 \,(\mathrm{mod}\ 5)$, Theorem 6.27 tells us that $9 \cdot 12 \cdot 13 \equiv 4 \cdot 2 \cdot 3 = 24 \,(\mathrm{mod}\ 5)$. Now, $24 \equiv 4 \,(\mathrm{mod}\ 5)$ since $24 - 4 = 20$ is divisible by 5. We conclude that $9 \cdot 12 \cdot 13 \,(\mathrm{mod}\ 5) = 4$. $\quad\Diamond$

6.6.1 *Linear congruence equations*

The first equation you probably learnt to solve was a linear equation with one unknown: $ax = b$ where a and b are given constants with $a \neq 0$ and x is the unknown. Given a positive integer n, and integers a, b and n with $n \geq 1$, the goal of this section is to look at solutions of a similar type of equation but with the "equal" sign replaced with the "congruence $(\mathrm{mod}\ n)$" sign. That is, we would like to find the value(s) of the variable x such that:
$$ax \equiv b \,(\mathrm{mod}\ n).$$

Definition 6.11. A *linear congruence equation* is an equation of the form $ax \equiv b \,(\mathrm{mod}\ n)$ where a, b and n are integers with $n \geq 1$ and x is a variable ranging over the integers.

While the linear equation $ax = b$ with $a \neq 0$ has a unique solution (namely $x = \frac{b}{a}$), the congruence equation $ax \equiv b \,(\mathrm{mod}\ n)$ can have no solution or multiple solutions.

Example 6.28. Consider the linear congruence equation $2x \equiv 4 \,(\mathrm{mod}\ 6)$. Since $2 \cdot 2 = 4 \equiv 4 \,(\mathrm{mod}\ 6)$ and $2 \cdot 5 = 10 \equiv 4 \,(\mathrm{mod}\ 6)$, both $x = 2$ and $x = 5$ are solutions of the congruence equation.

Theorem 6.30. *Let a, b and n be three integers such that $n \geq 1$ and a, b not both zero, and let $g = \gcd(a, n)$.*

(a) The linear congruence equation

$$ax \equiv b \,(mod\ n) \tag{6.12}$$

    has solutions if and only if $g \mid b$.

(b) If $g \mid b$, and x_0 is a solution of (6.12), then any solution of (6.12) is of the form $x_0 + k\left(\frac{n}{g}\right)$ for some integer k. Moreover, the g solutions:

$$x_0,\ x_0 + \frac{n}{g},\ x_0 + 2\frac{n}{g}, \ldots, x_0 + (g-1)\frac{n}{g} \tag{6.13}$$

    are mutually non-congruent modulo n and every solution of (6.12) is congruent modulo n to one of them. In particular, there are exactly g distinct solutions of (6.12) modulo n.

Proof. Assume that x is a solution of (6.12), then there exists an integer m such that $b - ax = mn$ or $b = ax + mn$. Since $g \mid a$ and $g \mid n$, $g \mid (ax + mn) = b$. Conversely, if $g \mid b$, then $b = ug$ for some $u \in \mathbb{Z}$. By Bézout identity, $g = sa + tn$ for some integers s and t and so $b = ug = u(sa + tn) = a(su) + (ut)n$, or equivalently, $b - a(su) = (ut)n$. This shows that $a(su) \equiv b \,(\mathrm{mod}\ n)$ and hence su is a solution of (6.12). This proves part (a). For part (b), assume $g \mid b$ and x_0 is a fixed solution of the congruence (6.12). Choose an integer r such that $ax_0 - b = rn$. Let $s = \frac{a}{g}$ and $t = \frac{n}{g}$. We know, by Theorem 6.19, that $\gcd(s, t) = 1$. If x is any solution of (6.12), then there exists an integer α such that $ax - b = \alpha n$. Isolate b from the equation $ax_0 - b = rn$ and replace in $ax - b = \alpha n$, we get that $a(x - x_0) = (\alpha - r)n$. Divide both sides by g and obtain that $s(x - x_0) = (\alpha - r)t$. Since $\gcd(s, t) = 1$ and $t \mid s(x - x_0)$, Euclid's lemma tells us that $t \mid (x - x_0)$.

Choose $k \in \mathbb{Z}$ such that $x - x_0 = kt$, so $x = x_0 + kt = x_0 + k\frac{n}{g}$. Next, we prove that the solutions given in (6.13) are mutually non-congruent. Let $k, k' \in \{0, 1, \ldots, g - 1\}$, then $\left(x_0 + k\frac{n}{g}\right) - \left(x_0 + k'\frac{n}{g}\right) = (k - k')\frac{n}{g}$. Since $1 - g \le k - k' \le g - 1$, $(k - k')\frac{n}{g}$ is not an integer multiple of n and therefore the two solutions $x_0 + k\frac{n}{g}$ and $x_0 + k'\frac{n}{g}$ are not congruent modulo n. Finally, we prove that every solution of (6.12) is congruent modulo n to one in (6.13). If x is an arbitrary solution of (6.12), write $x = x_0 + k\frac{n}{g}$ for some integer k. If $0 \le k \le g - 1$, then x is one of the solutions in (6.13). If $k \ge g$, the we can find an integer α such that $0 \le k - \alpha g \le g - 1$. Let $m = k - \alpha g$ and $y = x_0 + m\frac{n}{g}$. Then y is one of the non-congruent solutions in (6.13) and $x - y = \alpha n$ which means that $x \equiv y \,(\mathrm{mod}\ n)$. We conclude that the g solutions in (6.13) are distinct modulo n and that every solution of (6.12) is congruent modulo n to one of them. $\qquad\square$

Theorem 6.30 suggests that if $g = \gcd(a, n) \mid b$, then we can find all solutions of the congruence $ax \equiv b \,(\mathrm{mod}\ n)$ by finding one solution x_0 first and then by adding to it the first $g - 1$ positive multiples of $\frac{n}{g}$.

Example 6.29. The congruence equation $4x \equiv 3 \,(\mathrm{mod}\ 6)$ has no solution since $\gcd(4, 6) = 2$ and 2 does not divide 3.

Example 6.30. Find all solutions to the congruence equation $8x \equiv 12 \,(\mathrm{mod}\ 28)$ such that $0 \le x < 28$.

Solution. Note first that $g = \gcd(8, 28) = 4$ and since $4 \mid 12$, we know that solutions exist and that there are exactly 4 non-congruent ones of them. By inspection, $x = 5$ is a solution since $8 \cdot 5 = 40 \equiv 12 \,(\mathrm{mod}\ 28)$. Since $\frac{n}{g} = \frac{28}{4} = 7$, the solutions of the congruence equation in the range from 0 to 27 are 5, $5 + 7 = 12$, $5 + 2 \cdot 7 = 19$ and $5 + 3 \cdot 7 = 26$. $\qquad\lozenge$

Example 6.31. Find all non-congruent solutions to $9x \equiv 12 \,(\mathrm{mod}\ 15)$.

Solution. Since $g = \gcd(9, 15) = 3 \mid 12$, we know that solutions exist and that there are exactly 3 non-congruent ones of them and any other solution is congruent to one of these three solutions. By inspection, $x = 3$ is a solution since $9 \cdot 3 = 27 \equiv 12 \,(\mathrm{mod}\ 15)$. Since $\frac{n}{g} = \frac{15}{3} = 5$, the non-congruent solutions of the congruence relation are 3, $3 + 5 = 8$ and $3 + 2 \cdot 5 = 13$. $\qquad\lozenge$

An important special case of Theorem 6.30 is when $\gcd(a, n) = 1$. The next theorem states that in this case, the set of non-congruent solutions of the congruence equation is a singleton. The proof of the theorem follows easily from Theorem 6.30 and is left as an exercise to the reader (see Exercise (13) below).

Theorem 6.31. *With the notations of Theorem 6.30, the linear congruence equation $ax \equiv b \, (mod \, n)$ has a unique solution modulo n if and only if* $\gcd(a, n) = 1$.

While Theorem 6.31 only talks about the existence of a unique solution, it does not tell us how to find it. It is however relatively easy to do that. If $\gcd(a, n) = 1$, then there exist integers s and t such that $sa + tn = 1$ (Bézout identity). The values of s and t can be computed using the Euclidean algorithm. Let $x = sb$, then:

$$
\begin{aligned}
ax - b &= a(sb) - b \\
&= b(1 - tn) - b \text{ (since } sa = 1 - tn) \\
&= (-bt)n.
\end{aligned}
$$

This proves that $ax \equiv b \, (\text{mod } n)$ and so $x = sb$ is a solution. The unique solution of the congruence $ax \equiv b \, (\text{mod } n)$ is then $sb \equiv \pmod{n}$ (the remainder of sb upon division by n).

Example 6.32. Find all non-congruent solutions to the congruence equation $3x \equiv 5 \, (\text{mod } 8)$.

Solution. Since $\gcd(3, 8) = 1$, the congruence relation has a unique solution modulo 8. The Euclidean algorithm gives the following expression of 1 as a linear combination of 3 and 8: $1 = 3 \cdot 3 + (-1) \cdot 8$. In particular, $s = 3$. By the above discussion, we know that $x = sb = (3)(5) = 15$ is a solution to the congruence relation. Since $15 \equiv 7 \, (\text{mod } 8)$, $x = 7$ is the unique solution to the congruence. $\qquad \Diamond$

Example 6.33. Find all non-congruent solutions to the congruence $5x \equiv 4 \, (\text{mod } 7)$.

Solution. Since 5 is prime, the congruence relation has a unique solution modulo 7 by Theorem 6.31. The Euclidean algorithm gives us that $1 = (-4) \cdot 5 + 3 \cdot 7$. In particular, $s = -4$ and $x = sb = (-4)(4) = -16$ is a solution to the congruence relation. Since $-16 \equiv 5 \, (\text{mod } 7)$, we conclude that $x = 5$ is the unique solution to the congruence. $\qquad \Diamond$

6.6.2 *The Chinese remainder theorem*

Consider the following scenario. A wedding planner is trying to organize seating of the invitees around tables. If the planner considers tables that sit 3 persons each, there is one invitee unseated. When tables that sit 5 persons each are considered, there are two invitees unseated. Finally, when tables that sit 7 persons each are considered, there are three invitees unseated. Can this information be used to know the smallest number of invitees to the wedding? If x is the number of invitees, then x must simultaneously satisfy the following three congruence equations: $x \equiv 1 \,(\mathrm{mod}\ 3)$, $x \equiv 2 \,(\mathrm{mod}\ 5)$ and $x \equiv 3 \,(\mathrm{mod}\ 7)$. So far, we have learnt how to find solutions (if any) of a single linear congruence equation. Of course, one could use an exhaustive approach by computing $x \,(\mathrm{mod}\ 3)$, $x \,(\mathrm{mod}\ 5)$ and $x \,(\mathrm{mod}\ 7)$ for several values of the positive integer x and see if we could reach a value for which all three congruences are satisfied. As you can imagine, this is could be a lengthy process, and we are not even sure if such a solution exists. The following theorem, know as the *Chinese remainder theorem*, gives us a tool to determine if a given system of linear congruence equations has a solution. In addition, the proof of the theorem presented below provides a formulation of the solution when it exists.

Theorem 6.32 (The Chinese remainder theorem). *Let* $n_1, n_2, \ldots,$ n_t *be pairwise relatively prime positive integers with* $n_i \geq 2$ *for each* i *and let* $n = n_1 \cdot n_2 \cdots n_t$. *Then for any integers* $a_1, a_2, \ldots, a_t$, *there exists a unique solution modulo* n *satisfying the following system of congruence relations:*

$$\begin{cases} x \equiv a_1 \,(mod\ n_1) \\ x \equiv a_2 \,(mod\ n_2) \\ \quad\vdots \\ x \equiv a_t \,(mod\ n_t) \end{cases}$$

Proof. For each i, let $m_i = \frac{n}{n_i}$. Since $\gcd(m_i, n_i) = 1$ for each i, the congruence $m_i x \equiv 1 \,(\mathrm{mod}\ n_i)$ has a unique solution x_i by Theorem 6.31. In particular, $m_i x_i = 1 + b_i n_i$ where $b_i \in \mathbb{Z}$ for each $i \in \{1, 2, \ldots, t\}$. Let $x = a_1 m_1 x_1 + a_2 m_2 x_2 + \cdots + a_t m_t x_t$. Then $x - a_i = a_1 m_1 x_1 + a_2 m_2 x_2 + \cdots + a_{i-1} m_{i-1} x_{i-1} + (a_i m_i x_i - a_i) + a_{i+1} m_{i+1} x_{i+1} + \cdots + a_t m_t x_t$. Since $n_i \mid m_j$ whenever $i \neq j$ and $(a_i m_i x_i - a_i) = a_i(1 + b_i n_i) - a_i = a_i b_i n_i$, $x - a_i$ is a multiple of n_i and therefore $x \equiv a_i \,(\mathrm{mod}\ n_i)$ for each i. This shows that x is a solution to each congruence relation in the above system. For the uniqueness of this solution modulo n, let y be any solution that satisfies each congruence relation in the above system. Since $x \equiv a_i \pmod{n_i}$ and $a_i \equiv y \pmod{n_i}$, we have that $x \equiv y \pmod{n_i}$ and therefore $n_i \mid x - y$ for each $i \in \{1, 2, \ldots, t\}$. Since the n_i's are pairwise relatively prime, $n = n_1 \cdot n_2 \cdots n_t \mid (x - y)$ and therefore $x \equiv y \pmod{n}$. $\qquad\square$

We can now solve the wedding planner problem presented at the beginning of this section.

Example 6.34. Find the smallest positive solution of the following system of three congruences: $x \equiv 1 \pmod 3$, $x \equiv 2 \pmod 5$ and $x \equiv 3 \pmod 7$.

Solution. Since 3, 5 and 7 are pairwise relatively prime, the Chinese remainder theorem tells us that the system has a unique solution modulo $n = 3 \cdot 5 \cdot 7 = 105$. Let $m_1 = \frac{105}{3} = 35$, $m_2 = \frac{105}{5} = 21$ and $m_3 = \frac{105}{7} = 15$. We start by solving each of the congruences $m_i x \equiv 1 \pmod{n_i}$ for each i. For the congruence $35x \equiv 1 \pmod 3$, $x_1 = 2$ is the unique solution modulo 3. For the congruence $21x \equiv 1 \pmod 5$, $x_2 = 1$ is the unique solution modulo 5. For the congruence $15x \equiv 1 \pmod 7$, $x_3 = 1$ is the unique solution modulo 7. From the proof of Theorem 6.32, we know that $x = m_1 a_1 x_1 + m_2 a_2 x_2 + m_3 a_3 x_3 = 157$ is a solution of the system of three congruence relations. Any other solution is congruent to 157 modulo $n = 3 \cdot 5 \cdot 7 = 105$. We conclude that the smallest positive solution to the system of congruences is $x = 157 - 105 = 52$. $\qquad\Diamond$

Example 6.35. Find the smallest positive solution of the following system of three congruences: $3x \equiv 1 \pmod 5$, $5x \equiv 2 \pmod 6$ and $4x \equiv 3 \pmod 7$.

Solution. The Chinese remainder theorem cannot be used directly as the congruences in the given system are not of the form $x \equiv a \pmod n$. But this is a problem we can work around. For the congruences $3x \equiv 1 \pmod 5$, note that $\gcd(3, 5) = 1$ and $2 \cdot 3 + (-1) \cdot 5 = 1$, so $2 \cdot 3 = 6 \equiv 1 \pmod 5$. Multiplying both sides of $3x \equiv 1 \pmod 5$ by 2, we then get that $x \equiv 2 \pmod 5$. Similar argument shows that $5x \equiv 2 \pmod 6$ is equivalent to $x \equiv 4 \pmod 6$ and $4x \equiv 3 \pmod 7$ is equivalent to $x \equiv 6 \pmod 7$. Since 5, 6 and 7 are pairwise relatively prime, the Chinese remainder theorem tells us that the system has a unique solution modulo $n = 5 \cdot 6 \cdot 7 = 210$. Let $m_1 = \frac{210}{5} = 42$, $m_2 = \frac{210}{6} = 35$ and $m_3 = \frac{210}{7} = 30$. For the congruence $42x \equiv 1 \pmod 5$ is $x_1 = 3$ is the unique solution modulo 5. For the congruence $35x \equiv 1 \pmod 6$ is $x_2 = 5$ is the unique solution modulo 6 and $x_3 = 4$ is the unique solution the congruence $30x \equiv 1 \, (\text{mod } 7)$ modulo 7. Therefore, $x = m_1 a_1 x_1 + m_2 a_2 x_2 + m_3 a_3 x_3 = 1672$ is a solution of the system of three congruence relations. Any other solution is congruent to 1672 modulo 210. We conclude that the smallest positive solution to the system of congruences is $x = 1672 - 7 \cdot 210 = 202$. $\qquad \Diamond$

It is important to point out here that not every system of congruences has solutions. For example, the system of the two congruences $x \equiv 5 \, (\text{mod } 6)$ and $x \equiv 7 \, (\text{mod } 9)$ has no integer solutions (see Exercise (15) below).

6.6.3 *Exercises*

(1) Prove part (a) of Theorem 6.27.

(2) Prove part (b) of Theorem 6.27.

(3) Prove part (c) of Theorem 6.27.

(4) Use induction on the non-negative integer k to prove part (f) of Theorem 6.27.

$\star$(5) Let a, b and n be integers with $n \geq 1$. Prove that if $a \equiv b \, (\text{mod } n)$ and d is a positive divisor of n, then $a \equiv b \, (\text{mod } d)$.

(6) Prove that for any integer n, $n^2 \equiv 0$ or $1 \, (\text{mod } 3)$.

(7) Prove that for any integer n, $n^4 \equiv 0$ or $1 \, (\text{mod } 3)$.

(8) Prove that for any integer n, $n^2 \equiv 0$ or $1 \, (\text{mod } 4)$.

$\star$(9) Prove that for any integer n, $n^4 \equiv 0$ or $1 \, (\text{mod } 5)$.

(10) Prove that for any integer n, $n^3 \equiv 0 \ 1$ or $6 \, (\text{mod } 7)$.

(11) Compute each of the following.

    (a) $2^{400} \, (\text{mod } 31)$ (c) $5^{206} \, (\text{mod } 17)$

   $\star$ (b) $3^{257} \, (\text{mod } 80)$ (d) $7^{22} \, (\text{mod } 16805)$

(12) In each case, a linear congruence equation $ax \equiv b \, (\text{mod } n)$ is given. Determine if the equation has solutions. If it does, give all non-congruent solutions.

    (a) $3x \equiv 1 \, (\text{mod } 5)$ (h) $24x \equiv 60 \, (\text{mod } 36)$

    (b) $3x \equiv 4 \, (\text{mod } 6)$ (i) $37x \equiv 38 \, (\text{mod } 148)$

    (c) $5x \equiv 20 \, (\text{mod } 8)$ $\star$ (j) $91x \equiv 78 \, (\text{mod } 143)$

  $\star$ (d) $8x \equiv 16 \, (\text{mod } 12)$ (k) $200x \equiv 15 \, (\text{mod } 30)$

    (e) $15x \equiv 39 \, (\text{mod } 42)$ (l) $225x \equiv 15 \, (\text{mod } 5)$

  $\star$ (f) $17x \equiv 3 \, (\text{mod } 7)$ (m) $90x \equiv 45 \, (\text{mod } 30)$

    (g) $21x \equiv 77 \, (\text{mod } 14)$ (n) $900x \equiv 11 \, (\text{mod } 4500)$

(13) Prove Theorem 6.31.

$\star$(14) Let a, b, c, n be integers with $n \geq 1$ such that $ac \equiv bc \,(\mathrm{mod}\ n)$. If $\gcd(c, n) = 1$, prove that $a \equiv b \,(\mathrm{mod}\ n)$.

(15) Prove that the system of the two congruences $x \equiv 5 \,(\mathrm{mod}\ 6)$ and $x \equiv 7 \,(\mathrm{mod}\ 9)$ has no integer solutions.

(16) Find the smallest positive solution of the following system of two congruences: $x \equiv 3 \,(\mathrm{mod}\ 5)$ and $x \equiv 6 \,(\mathrm{mod}\ 7)$.

(17) Find the smallest positive solution of the following system of three congruences: $x \equiv 2 \,(\mathrm{mod}\ 5)$, $x \equiv 3 \,(\mathrm{mod}\ 6)$ and $x \equiv 4 \,(\mathrm{mod}\ 7)$.

$\star$(18) Find the smallest positive solution of the following system of three congruences: $2x \equiv 1 \,(\mathrm{mod}\ 3)$, $3x \equiv 2 \,(\mathrm{mod}\ 5)$ and $4x \equiv 5 \,(\mathrm{mod}\ 11)$.

(19) A donut shop wants to donate n donuts to a local soccer tournament. The shop has three sizes of boxes: small, medium and large that can fit 5, 7 and 13 donuts each, respectively. If only small boxes are used for the donation, then two donuts are left. If only medium boxes are used, then three donuts are left. If only large boxes are used, then five donuts are left. What is the smallest possible value of n?

$\star$(20) A kid dropped a box full of glass marbles and lost most of them. When asked how many marbles were there in the box, the kid could not remember the exact number but he gave the following information. When the marbles are arranged in groups of five, there are three marbles left. When they are arranged in groups of eleven, there are six marbles left. When they are arranged in groups of thirteen, there are eight marbles left and when arranged in groups of seventeen, ten marbles are left. What is the smallest number of marbles the kid could have had in the box?

6.7   The Euler phi function

Definition 6.12. Given a positive integer n, we denote by $\phi(n)$ the number of integers k such that $1 \leq k \leq n$ and $\gcd(k, n) = 1$. The symbol ϕ is called the *Euler phi function*, or the *Euler totient function*. It is a function that assigns to every positive integer n the number of integers in the set $\{1, 2, \ldots, n\}$ that are relatively prime to n. We define $\phi(1) = 1$.

Example 6.36. The table below shows the values of $\phi(n)$ for $2 \leq n \leq 9$. The table shows in particulate the following values: $\phi(2) = 1$, $\phi(3) = 2$, $\phi(5) = 4$ and $\phi(7) = 6$. Note that each of the numbers $2, 3, 5$ and 7 is prime and that the value of the Euler phi function for each of them is one less than the value of the integer. This is no coincidence as shown in the next result.

n	$1 \le k \le n$	$1 \le k \le n$ and $\gcd(k, n) = 1$	$\phi(n)$
2	$1, 2$	1	1
3	$1, 2, 3$	$1, 2$	2
4	$1, 2, 3, 4$	$1, 3$	2
5	$1, 2, 3, 4, 5$	$1, 2, 3, 4$	4
6	$1, 2, 3, 4, 5, 6$	$1, 5$	2
7	$1, 2, 3, 4, 5, 6, 7$	$1, 2, 3, 4, 5.6$	6
8	$1, 2, 3, 4, 5, 6, 7, 8$	$1, 3, 5, 7$	4
9	$1, 2, 3, 4, 5, 6, 7, 8, 9$	$1, 2, 4, 5, 7, 8$	6

Theorem 6.33. *If p is a prime number then $\phi(p) = p - 1$.*

Proof. If p is a prime number, then the only divisors of p are 1 and p. Therefore each of the integers $1, 2, \ldots, p - 1$ is relatively prime with p and the result follows. $\qquad\square$

Theorem 6.34. *For any prime number p and any positive integer k, $\phi(p^k) = p^k - p^{k-1}$.*

Proof. Since p is prime, the only integers in the set $\{1, 2, \ldots, p^k\}$ that are not relatively prime with p^k are the multiples kp of p. If $1 \le kp \le p^k$ with k an integer, then $1 \le k \le p^{k-1}$. This shows that there are p^{k-1} integers between 1 and p^k which are *not relatively prime* with p^k and the result follows. $\qquad\square$

Example 6.37.

(a) $\phi(27) = \phi(3^3) = 3^3 - 3^2 = 18$
(b) $\phi(32) = \phi(2^5) = 2^5 - 2^4 = 16$
(c) $\phi(625) = \phi(5^4) = 5^4 - 5^3 = 500$
(d) $\phi(2401) = \phi(7^4) = 7^4 - 7^3 = 2058$

The next result asserts that the Euler phi function is *multiplicative* for coprime integers. That is, $\phi(mn) = \phi(m)\phi(n)$ when $\gcd(m, n) = 1$.

Theorem 6.35. *If m and n are relatively prime numbers, then $\phi(mn) = \phi(m)\phi(n)$.*

Proof. Arrange all the integers between 1 and mn in an $n \times m$ matrix as follows:

$$M = \begin{bmatrix} 1 & 2 & 3 & \cdots r & \cdots m \\ m+1 & m+2 & m+3 & \cdots m+r & \cdots 2m \\ 2m+1 & 2m+2 & 2m+3 & \cdots 2m+r & \cdots 3m \\ \vdots & \vdots & \vdots & \vdots \;\; \vdots & \vdots \;\; \vdots \\ (n-1)m+1 & (n-1)m+2 & (n-1)m+3 & \cdots (n-1)m+r & \cdots nm \end{bmatrix}$$

Since $\gcd(\alpha m + r, m) = \gcd(r, m)$ for any integers α and r, integers in column r of the matrix are relatively prime with m if and only if r is relatively prime with m. Therefore, there are $\phi(m)$ columns in the matrix such that for each one of these columns, each entry in the column is coprime to m and no entry in M outside of these $\phi(m)$ columns is coprime to n. Next, we prove that no two integers of the n entries $r, m+r, 2m+r, \ldots, (n-1)m+r$ of column r are congruent modulo n. Suppose to the contrary that $im + r \equiv jm + r \pmod{n}$ for some integers i, j with $0 \leq i < j \leq n - 1$. Then $(im + r) - (jm + r) = tn$ for some integer t or equivalently $(i - j)m = tn$. Since $n \mid (i - j)m$ and $\gcd(m, n) = 1$, $n \mid (i - j)$ by Euclid's lemma. This is a contradiction since $-n < i - j < 0$. So each integer in column r is congruent to exactly one element of the set $\{0, 1, 2, \ldots, n - 1\}$ modulo n. This implies that column r contains exactly $\phi(n)$ entries that are coprime to n. To summarize: there are exactly $\phi(m)$ columns in M all of which entries are coprime to m and each one of these columns contains exactly $\phi(n)$ integers relatively prime to n. This shows that the total number of integers in M which are coprime to both m and n is $\phi(m)\phi(n)$. By Theorem 6.18 above, we know that the number of integers which are coprime to both m and n is $\phi(mn)$. The result follows. $\qquad\square$

Example 6.38.

  (a) $\phi(14) = \phi(2 \cdot 7) = \phi(2)\phi(7) = 1 \cdot 6 = 6$ (since $\gcd(2, 7) = 1$).
  (b) $\phi(21) = \phi(3 \cdot 7) = \phi(3)\phi(7) = 2 \cdot 6 = 12$ (since $\gcd(3, 7) = 1$).
  (c) $\phi(104) = \phi(8 \cdot 13) = \phi(8)\phi(13) = 4 \cdot 12 = 48$ (since $\gcd(8, 13) = 1$).

Theorem 6.36. *Let $n \geq 2$ be an integer and suppose that $n = p_1^{t_1} p_2^{t_2} \cdots p_k^{t_k}$ is the prime factorization of n where $p_1, p_2, \ldots, p_k$ are all the distinct prime divisors of n and the t_i's are positive integers. Then*

$$\phi(n) = \left(p_1^{t_1} - p_1^{t_1-1}\right) \left(p_2^{t_2} - p_2^{t_2-1}\right) \cdots \left(p_k^{t_k} - p_k^{t_k-1}\right).$$

Proof. By induction on k, the number of prime factors of n. The case $k = 1$ is true by Theorem 6.34. Assume the result is true for $k \geq 1$ and let integer $n \geq 2$ be an integer having $k+1$ distinct primes factors $p_1, \ldots, p_{k+1}$ and write $n = p_1^{t_1} p_2^{t_2} \cdots p_k^{t_k} p_{k+1}^{t_{k+1}}$. Since the p_i's are distinct primes, the integers $p_1^{t_1} p_2^{t_2} \cdots p_k^{t_k}$ and $p_{k+1}^{t_{k+1}}$ are relatively prime. Theorem 6.35 tells us that $\phi \left(p_1^{t_1} p_2^{t_2} \cdots p_k^{t_k} p_{k+1}^{t_{k+1}} \right) = \phi \left(p_1^{t_1} p_2^{t_2} \cdots p_k^{t_k} \right) \phi \left(p_{k+1}^{t_{k+1}} \right)$. By the induction hypothesis and Theorem 6.34, $\phi \left(p_1^{t_1} p_2^{t_2} \cdots p_k^{t_k} \right) \phi \left(p_{k+1}^{t_{k+1}} \right) = \left(p_1^{t_1} - p_1^{t_1-1} \right) \left(p_2^{t_2} - p_2^{t_2-1} \right) \cdots \left(p_k^{t_k} - p_k^{t_k-1} \right) \left(p_{k+1}^{t_{k+1}} - p_{k+1}^{t_k} \right)$. The result follows by the principle of induction. $\square$

Example 6.39.

(a) $\phi(315) = \phi(3^2 \cdot 5 \cdot 7) = \left(3^2 - 3^1 \right) \left(5^1 - 5^0 \right) \left(7^1 - 7^0 \right) = 144.$
(b) $\phi(3600) = \phi(2^4 \cdot 3^2 \cdot 5^2) = \left(2^4 - 2^3 \right) \left(3^2 - 3^1 \right) \left(5^2 - 5^1 \right) = 960.$

Looking back at the examples in this section, one can notice that $\phi(n)$ is even for any integer $n \geq 3$. The following theorem confirms this observation.

Theorem 6.37. $\phi(n)$ *is even for any integer* $n \geq 3$.

Proof. Fix an integer $n \geq 3$. If $n = 2^k$ for some integer $k \geq 2$, then $\phi(n) = 2^k - 2^{k-1}$ (by Theorem 6.34 above), which is clearly even. If n is not a power of 2, then it must be divisible by some odd prime number p. In this case, write $n = mp^k$ for some positive integers k and m with $\gcd(p^k, m) = 1$. So;

$$\phi(n) = \phi\left(p^k \right) \phi(m) \text{ (Theorem 6.35)}$$
$$= \left(p^k - p^{k-1} \right) \phi(m) \text{ (Theorem 6.34)}$$
$$= p^{k-1}(p - 1)\phi(m).$$

Since $p - 1$ is even, $\phi(n)$ is also even in this case. $\square$

Example 6.40. There exists no positive integer k such that $\phi(k) = 3$. If such an integer exists, then $k \leq 2$ by Theorem 6.38. But $\phi(1) = \phi(2) = 1 \neq 3$. The result follows.

Example 6.41. Find all positive integers n such that $\phi(n) = 12$.

Solution. Let n be a positive integer such that $\phi(n) = 12$. Let $n = p_1^{t_1} p_2^{t_2} \ldots p_s^{t_s}$ be the prime factorization of n where the p_i's are all the distinct prime divisors of n and the t_i's are positive integers. Then $\phi(n) = 12 = $

$p_1^{t_1-1}(p_1-1)\,p_2^{t_2-1}(p_2-1)\cdots p_s^{t_s-1}(p_s-1)$. If $p_j > 13$ for some j, then $p_j - 1 > 12$ cannot be a divisor of 12. So $p_j \leq 13$ for every j. Note also that $11 - 1 = 10$ is not a divisor of 12, so 11 cannot be one of the prime divisors of n. That leaves us with the following possible prime divisors of n: 2, 3, 5, 7 and 13. Consequently $n = 2^a 3^b 5^c 7^d 13^e$ and thus

$$12 = (2^a - 2^{a-1})(3^b - 3^{b-1})(5^c - 5^{c-1})(7^d - 7^{d-1})(13^e - 13^{e-1}) \quad (6.14)$$

If $c \geq 2$, then 5 would be a divisor of 12 which is absurd. Thus $c \leq 1$. Similarly, $d \leq 1$ and $e \leq 1$ since 7 and 13 are not divisors of 12. If $c = 1$, then $n = 5k$ for some integer k relatively prime to 5. By Theorem 6.35, $\phi(n) = 12 = \phi(5)\phi(k) = 4\phi(k)$ and so $\phi(k) = 3$. This is not possible by Example 6.40. So $c = 0$ and $n = 2^a 3^b 7^d 13^e$. If $e = 1$, then $n = 13m$ for some integer m relatively prime to 13. By Theorem 6.35, $\phi(n) = 12 = 12\phi(m)$ and so $\phi(m) = 1$. Thus, $m = 1$ or $m = 2$ (the only two integers with Euler phi function equals to 1). This gives us the two values $n = 13$ and $n = 26$. If $e = 0$, then $n = 2^a 3^b 7^d$. If $d = 1$, then $n = 7k$ for some integer k coprime to 7. So $12 = 6\phi(k)$, or $\phi(k) = 2$. This shows that $k = 3, 4$ or 6 and thus $n = 21, 28$ or 42. If $d = 0$, then $n = 2^a 3^b$ and so $12 = (2^a - 2^{a-1})(3^b - 3^{b-1}) = 2^{a-1}(2-1)3^{b-1}(3-1) = 2^a 3^{b-1}$. Since $9 \nmid 12$, $b \leq 2$. Also $b \neq 1$ since otherwise $12 = 2^a$ for some integer a which is false. We conclude that $b = 2$ and $n = 9 \cdot 2^a$. In particular, $12 = \phi(9)\phi(2^a) = 6\phi(2^a)$ and $\phi(2^a) = 2$. This implies that $a = 2$. We conclude that $n = 9 \cdot 4 = 36$ in this case. In conclusion, the integers n such that $\phi(n) = 12$ are: 13, 21, 26, 28, 36 and 42. $\diamond$

Theorem 6.38. *Let m, n be two positive integers with $n \geq 2$ and $\gcd(m, n) = 1$. Let $b_1 < b_2 < \cdots < b_{\phi(n)}$ be the positive integers less than n and relatively prime to n. If $r_i = b_i m \pmod{n}$, then the r_i's are a permutation of the elements of set $U_n = \{b_1, b_2, \ldots, b_{\phi(n)}\}$.*

Proof.   Since $\gcd(m, n) = 1$ and $\gcd(b_i, n) = 1$, Theorem 6.18 tells us that $\gcd(b_i m, n) = 1$. Since $\gcd(b_i m, n) = \gcd(r_i, n)$, r_i is relatively prime to n and consequently $r_i \in U_n$ for each $i = 1, 2, \ldots, \phi(n)$. If $b_i m \equiv b_j m \pmod{n}$, then $(b_j - b_i)m$ is divisible by n. Since $\gcd(m, n) = 1$, $b_j - b_i$ must be divisible by n. But $b_j - b_i < n$, which means that $b_j - b_i = 0$ and consequently $mb_i = mb_j$. We conclude that $b_i m \not\equiv b_j m \pmod{n}$ for $i \neq j$ and therefore the r_i's are all distinct. $\square$

Example 6.42. Consider the integers $n = 9$ and $m = 11$. Clearly $\gcd(m, n) = 1$. The integers less than 9 and relatively prime to 9 are 1, 2, 4,

5, 7 and 8. Multiplying each of these integers my $m = 11$ gives the following list: 11, 22, 44, 55, 77 and 88. It is easy to verify that 11 (mod 9) = 2, 22 (mod 9) = 4, 44 (mod 9) = 8, 55 (mod 9) = 1, 77 (mod 9) = 5 and 88 (mod 9) = 7.

We now prove the main result of this section.

Theorem 6.39 (Euler's theorem). *If m and n are relatively prime positive integers with $n \geq 2$, then $m^{\phi(n)} \equiv 1 \pmod{n}$.*

Proof. Let $U_n = \{b_1, b_2, \ldots, b_{\phi(n)}\}$ be the set of all distinct positive integers less than n and relatively prime to n. By Theorem 6.38, $b_1 m, b_2 m, \ldots, b_{\phi(n)} m$ are just a permutation of the elements of U_n. Write

$$b_1 m \equiv b_1' \pmod{n}$$
$$b_2 m \equiv b_2' \pmod{n}$$
$$\vdots \quad \vdots \quad \vdots$$
$$b_{\phi(n)} m \equiv b_{\phi(n)}' \pmod{n}$$

where $\{b_1', b_2', \ldots, b_{\phi(n)}'\}$ is some permutation of $\{b_1, b_2, \ldots, b_{\phi(n)}\}$. Multiplying all n congruences together we get that

$$\left(b_1 b_2 \cdots b_{\phi(n)}\right) m^{\phi(n)} \equiv b_1' b_2' \cdots b_{\phi(n)}' \pmod{n} \qquad (6.15)$$

Let $b = b_1 b_2 \cdots b_{\phi(n)} = b_1' b_2' \cdots b_{\phi(n)}'$, then equation (1) becomes $bm^{\phi(n)} \equiv b \pmod{n}$ and so $n \mid b\left(m^{\phi(n)} - 1\right)$. Since $\gcd(b_i, n) = 1$ for each i, we have that $\gcd(b, n) = 1$ (Theorem 6.18) and so n divides $m^{\phi(n)} - 1$. The result follows. $\qquad \square$

Example 6.43. Since $\gcd(3, 8) = 1$, $3^{\phi(8)} \equiv 1 \pmod{8}$. Note that $3^{\phi(8)} = 3^4 = 81 \equiv 1 \pmod{8}$.

In what follows, we explore some important corollaries of Euler's theorem.

Corollary 6.5 (Fermat's little theorem). *Let $m \geq 1$ be an integer and let p be a prime number. If p is not a divisor of m, then $m^{p-1} \equiv 1 \pmod{p}$.*

Proof. Since p is a prime that does not divide m, $\gcd(p, m) = 1$. The result follows from Euler's theorem and the fact that $\phi(p) = p - 1$. $\qquad \square$

The next result is also referred to in the literature as Fermat's little theorem.

Corollary 6.6 (Fermat's little theorem). *For any positive integer m and any prime number p, $m^p \equiv m \pmod{p}$.*

Proof. Note first that since p is prime, $\gcd(m, p)$ is either 1 or p. If $\gcd(m, p) = 1$, then Theorem 6.5 implies that $m^{p-1} \equiv 1 \pmod{p}$. Multiplying both sides by m, we get that $m^p \equiv m \pmod{p}$. If $\gcd(m, p) = p$, then $m = kp$ for some integer k. Thus, $m^p - m = k^p p^p - kp = p(k^p p^{p-1} - k)$ is divisible by p and therefore $m^p \equiv m \pmod{p}$. $\qquad\square$

If the modulus n is prime, then Fermat's little theorem and its corollaries can be used efficiently to significantly reduce the calculations needed to compute the congruence $b^e \pmod{n}$.

Example 6.44. Since the prime $p = 7$ is not a divisor of 8, $8^6 = 262144 \equiv 1 \pmod{3}$. Similarly, $9^{10} = 3486784401 \equiv 1 \pmod{11}$.

Example 6.45. Find $3^{260} \pmod{7}$.

Solution. By Fermat's little theorem, we know that $3^6 \equiv 1 \pmod{7}$. Using the division algorithm, we have that $260 = 43 \cdot 6 + 2$, so $3^{260} = 3^{43 \cdot 6 + 2} = \left(3^6\right)^{43} \cdot 3^2 \equiv (1)^{43} \cdot 9 \pmod{7} \equiv 2 \pmod{7}$. $\qquad\lozenge$

Example 6.46. Compute $12^{12584} \pmod{7}$.

Solution. Since 7 is a prime number that does not divide 12, Fermat's little theorem tells that $12^6 \equiv 1 \pmod{7}$. Using the division algorithm, we have that $12584 = 6 \cdot 2097 + 2$. So $12^{12584} = 12^{(6 \cdot 2097 + 2)} = \left(12^6\right)^{2097} \cdot 12^2$. Now $\left(12^6\right)^{2097} \equiv 1^{2097} = 1 \pmod{7}$ and so $12^{12584} \equiv 12^2 = 144 \equiv 4 \pmod{7}$ $\quad\lozenge$

Example 6.47. Compute $1231^{71} \pmod{71}$.

Solution. By Corollary 6.6, $1231^{71} \equiv 1231 \pmod{71}$ since 71 is prime. Using the division algorithm, $1231 = 17 \cdot 71 + 24$. We conclude that $1231^{71} \pmod{71} = 24$. $\qquad\lozenge$

Corollary 6.7 (Euler's corollary). *Let p and q be two distinct prime numbers. If n is an integer relatively prime to both p and q, then $n^{(p-1)(q-1)} \equiv 1 \pmod{pq}$.*

Proof. Left as an exercise (see Exercise (7) below). $\qquad\square$

Example 6.48. $3^{24} = 3^{(5-1)(7-1)} \equiv 1 \pmod{35}$.

Ciphering and Deciphering digital messages often require computing large modular exponentiations, something like $12548^{858} \pmod{648}$. The bigger the integers in this modular expression $b^e \pmod{n}$, the more secure the transfer of data is. Needless to say that computing the value of 12548^{858} and then perform the long division to find its reminder upon division by 648 is not practical, even for powerful machine. But if the modulus n is not prime, Fermat's little theorem is not very helpful to compute the value of $b^e \pmod{n}$. If b is relatively small, it would be useful to express the exponent e in binary form; that is write e as a sum of powers of 2: $b = \sum_{i=0}^{n} a_i 2^i$ with $a_i \in \{0,1\}$ for each i. This will break the problem of computing $b^e \pmod{n}$ into a product of the form $\left(b^{2^{k_1}} \pmod{n}\right)\left(b^{2^{k_2}} \pmod{n}\right) \cdots \left(b^{2^{k_r}} \pmod{n}\right)$ with each $b^{2^{k_i}}$ $\pmod{n}$ is easier to compute than the original congruence.

Example 6.49. Compute $3^{2483} \pmod{14}$.

Solution. Here 14 is not prime but 3 is small. We start by writing 2483 as the sum of powers of 2:
$$2483 = 2048 + 256 + 128 + 32 + 16 + 2 + 1$$
$$= 2^{11} + 2^8 + 2^7 + 2^5 + 2^4 + 2^1 + 2^0.$$
Next, we compute $3^{2^k} \pmod{14}$ for $k = 0, 1, \ldots, 11$: $3^{2^0} = 3 \equiv 3 \pmod{14}$, $3^{2^1} = 9 \equiv 9 \pmod{14}$, $3^{2^2} = 81 \equiv 11 \pmod{14}$, $3^{2^3} = (81)^2 \equiv (11)^2 = 121 \equiv 9 \pmod{14}$, $3^{2^4} = \left(3^{2^3}\right)^2 \equiv 9^2 = 81 \equiv 11 \pmod{14}$. Similarly, $3^{2^5} = \left(3^{2^4}\right)^2 \equiv 9 \pmod{14}$. Continuing this way, we notice that for any positive integer k:
$$3^{2^k} \equiv \begin{cases} 9 \pmod{14} & \text{if } k \text{ is odd} \\ 11 \pmod{14} & \text{if } k \text{ is even} \end{cases}$$
We conclude that
$$3^{2483} = 3^{2^{11}+2^8+2^7+2^5+2^4+2^1+2^0}$$
$$= 3^{2^{11}} \cdot 3^{2^8} \cdot 3^{2^7} \cdot 3^{2^5} \cdot 3^{2^4} \cdot 3^{2^1} \cdot 3^{2^0}$$
$$\equiv 9 \cdot 11 \cdot 9 \cdot 9 \cdot 11 \cdot 9 \cdot 3 = 2381643 \pmod{14}$$
$$\equiv 5 \pmod{14}.$$
$\Diamond$

Example 6.50. Compute $8^{568} \pmod{12}$.

Solution. Like in the previous example, we start by writing 568 as the sum of powers of 2: $568 = 512 + 32 + 16 + 8 = 2^9 + 2^5 + 2^4 + 2^3$. Next we compute $8^{2^k} \pmod{12}$, $k = 1, \ldots, 9$: $8^{2^1} = 16 \equiv 4 \pmod{12}$, $8^{2^2} \equiv 4^2 = 16 \equiv 4 \pmod{12}$. This shows in particular that $8^{2^k} \equiv 4 \pmod{12}$ for any $k = 1, \ldots, 9$. We conclude that $8^{568} \equiv 4^4 = 256 \equiv 4 \pmod{12}$. $\Diamond$

6.7.1   *Exercises*

(1) Determine the truth value of the following statement: If p is a prime number, then $\phi(p^2) = (\phi(p))^2$.

(2) Prove that $\phi(2^a) = 2^{a-1}$ for any integer $a \geq 1$.

(3) Use techniques from this section to compute each of the following:

(a) $\phi(21)$	(g) $\phi(91)$	(m) $\phi(2020)$
(b) $\phi(26)$	(h) $\phi(105)$	(n) $\phi(19404)$
(c) $\phi(35)$	$\star$ (i) $\phi(143)$	$\star$ (o) $\phi(20200)$
(d) $\phi(55)$	(j) $\phi(187)$	(p) $\phi(189000)$
$\star$ (e) $\phi(65)$	(k) $\phi(756)$	q) $\phi(441000)$
(f) $\phi(77)$	(l) $\phi(1800)$	$\star$ (r) $\phi(58697100)$

(4) Prove that if $n \geq 3$ is an odd integer, then $\phi(2n) = \phi(n)$.

$\star$(5) Prove that if $n \geq 2$ is an even integer, then $\phi(2n) = 2\phi(n)$.

(6) Let $n \geq 2$ be an integer.

    (a) Give an example to show that the relation $\phi(3n) = 3\phi(n)$ is not true in general.

    (b) Give a necessary and sufficient condition on the integer n for the relation $\phi(3n) = 3\phi(n)$ to hold. Justify your answer.

(7) Give a proof of Euler's Corollary (Corollary 6.7 above).

(8) In each case, find (if any) all positive integers n such that $\phi(n) = a$. Justify your answer.

(a) $a = 2$	(c) $a = 4$	(e) $a = 6$	(g) $a = 8$
(b) $a = 3$	(d) $a = 5$	(f) $a = 7$	(h) $a = 14$

(9) Fermat's little theorem can be used in some cases to prove that a number n is not prime: pick an integer $a \geq 2$ and compute a^{n-1}, then n is not a prime if a^{n-1} is not congruent to 1 modulo n. Use this test to show that 889 is not prime.

(10) Use results from this section to compute each of the following.

(a) $4^5 \pmod 5$	(g) $3^{100} \pmod{16}$
(b) $5^{10} \pmod{11}$	$\star$ (h) $3^{1000} \pmod{17}$
(c) $12^{15638} \pmod{13}$	(i) $84^{60} \pmod{77}$
(d) $7^{16} \pmod{17}$	$\star$ (j) $128^{128} \pmod{703}$
$\star$ (e) $9^{1989} \pmod{19}$	(k) $13^{12769} \pmod{113}$
(f) $27^{22 \cdot 28} \pmod{(23 \cdot 29)}$	(l) $41^{72000000} \pmod{111}$

(11) Using the fact that 7541 is prime, compute 342378945^{7540} (mod 7541).

$\star$(12) Using the fact that 101 is prime, compute $7^{34567892123465412002}$ (mod 101).

(13)(a) Use induction to prove that $10^k \equiv 4$ (mod 6) for any integer $k \geq 1$.

$\star$ (b) Use part (a) to compute $10^{10^{100}}$ (mod 7).

(c) If today is a Monday, what day of the week is it in $10^{10^{100}}$ days from today?

(14) Compute $2^{22} + 3^{33} + 4^{44} + 5^{55} + 6^{66} + 7^{77} + 8^{88} + 9^{99} + 10^{1010}$ (mod 11).

(15) Find all values of the integer x in the range $1 \leq x \leq 10$ such that $x^{403} \equiv 4$ (mod 11).

(16) Let d, n be two positive integers such that $d \mid n$. Prove that $\phi(d) \mid \phi(n)$.

$\star$(17) Let m and n be two positive integers such that $\gcd(m, n) = 1$. Prove that $m^{\phi(n)} + n^{\phi(m)} \equiv 1$ (mod mn). Deduce that if p and q are distinct prime numbers, then $p^{q-1} + q^{p-1} \equiv 1$ (mod pq).

6.8 Caesar cipher: The RSA algorithm

In this section, we apply some of the theocratical ideas explored in this chapter to a public-key scheme known as the RSA Cryptosystem. Developed in the 1970s by Ron Rivest, Adi Shamir and Leonard Adleman (hence the acronym RSA). The RSA is an algorithm to encrypt and decrypt digital messages using two different keys, one public and another private. The original message is called the *plaintext*, and the encrypted one is called the *ciphertext*.

The idea of encrypting messages is not recent. The work *De vita Caesarum* (Latin for *About the Life of the Caesars*), written in 121 AD by the roman historian Suetonius talks about a deciphering practice used by Julius Caesar to send secret messages to his generals. Julius Caesar's scheme, often known as *Caesar cipher*, consists of a substitution technique in which each letter is shifted three positions forward in the alphabet with the letters x, y and z replaced by the letters a, b and c respectively. In general, we could use the same scheme by shifting the letters of the alphabet forward by a fixed number k with $0 \leq k \leq 25$, called the *key*. To represent the scheme mathematically, we start by assigning an integer to each letter. There are many ways to do that, but the *standard* way is $a \to 00$, $b \to 01$, ..., $z \to 25$ (using two-digits integers). If x is the integer assigned to a given letter α from the original plaintext message, then the encryption of α using a key k

is done using the rule $E_k(x) = (x + k) \pmod{26}$. Knowing the key k and the letter shifting scheme, Caesar's generals can decrypt the ciphertext by taking the "inverse" operation: $D_k(x) = (x - k) \pmod{26}$.

Example 6.51. Using Caesar cipher technique with key $k = 3$ and the standard numerical assignment of letters from the alphabet, encrypt the message DISCRETE MATH ROCKS.

Solution. Using the standard digital assignment of letters, the plaintext can be represented numerically as follows:

$$03\,08\,18\,02\,17\,04\,19\,04 \quad 12\,00\,19\,07 \quad 17\,14\,02\,10\,18. \tag{6.16}$$

Applying the encryption function $E(x) = (x + 3) \pmod{26}$, the encryption of message (6.16) is:

$$06\,11\,21\,05\,20\,07\,22\,07 \quad 18\,03\,22\,10 \quad 20\,17\,05\,13\,21.$$

which gives the encrypted message: GLVFUHWH SDWK URFNV. $\Diamond$

Example 6.52. You received the encrypted message EXB JROG QRZ from a friend using Caesar cipher technique with key $k = 3$ and the standard numerical assignment of letters from the alphabet. Decrypt the message and reconstruct the original plaintext message.

Solution. Using the standard digital assignment of letters, the encrypted message can be represented numerically as follows:

$$04\,23\,01 \quad 09\,17\,14\,06 \quad 16\,17\,25$$

Using the function $D_k(x) = (x - 3) \pmod{26}$ to decrypt the numerical message, we get:

$$01\,20\,24 \quad 06\,14\,11\,03 \quad 13\,14\,22$$

which corresponds to the plaintext message: BUY GOLD NOW. $\Diamond$

Remark 6.5. In the above examples, we avoided having messages with special symbols and punctuation. We also did not distinguish between upper and lower cases for the letters. All these constraints can be dealt with by expanding the set of integers used to represent the characters.

Caesar cipher technique is clearly not the most secure encryption scheme. With modern technology tools, the scheme can be easily hacked. There are, however, ways to improve its security. Exercise (4) below gives an improvement of Caesar's technique.

6.8.1 *The RSA scheme*

The main goal of the RSA algorithm is to securely transmit a message through public communication networks in a way that only intended receivers can decipher and read. The original message is first converted into a digital form as an integer m. There are many ways to do that, one of which is the Caesar cipher technique with key $k = 3$ that we adopt in this section. The effectiveness of the RSA algorithm is based on the fact that if n is a sufficiently large integer, then it is practically impossible (in a reasonable amount of time) to factor n into a product of two prime numbers. The RSA scheme to encrypt and decrypted the message m is described using the following steps.

(1) The receiver, say Bob, chooses two very large (but random) distinct prime numbers p and q (each having hundred digits or more). Many computer algorithms are designed to produced such numbers.

(2) Bob computes $n = pq$ and its Euler phi function $\phi(n) = (p-1)(q-1)$. It is important to realize here that without knowing the factorization $n = pq$, it would be practically impossible for anyone else to compute the value of $\phi(n)$. If p and q are large enough, then $1 \leq m < n$ (if $m \geq n$, the message m can be divided into smaller pieces each smaller than n with an even number of digits). The primes p and q are kept secret but their product n is made public. Notice that $\gcd(m, n) = 1$ unless m is a multiple of either p or q. There are $pq - \phi(pq) = pq - (p-1)(q-1) = p+q-1$ integers in the set $\{1, 2, \ldots, n\}$ which are *not* relatively prime to n and the probability that $\gcd(m, n) \neq 1$ is therefore $\frac{p+q-1}{pq} = \frac{1}{q} + \frac{1}{p} - \frac{1}{pq}$. Since p and q are fairly large with hundreds of digits each, the chance that $\gcd(m, n) \neq 1$ is almost zero and we can safely work with the assumption that m and n are indeed relatively prime (even in the highly unlikely event that $\gcd(m, n) \neq 1$, there are ways to alter the message m and force $\gcd(m, n)$ to be 1).

(3) Bob chooses an integer e such that $1 < e < \phi(n)$ and $\gcd(e, \phi(n)) = 1$. Many choices are possible for the integer e, but a choice widely used in the scheme is the prime number $e = 2^{16} - 1 = 65537$. The pair (n, e) is known as Bob's *public key*.

(4) Since $\gcd(e, \phi(n)) = 1$, there exist integers c, d (that we can compute using the Euclidean algorithm) such that $c\phi(n) + de = 1$. This means that $ed \equiv 1 \pmod{\phi(n)}$. Since it is impossible to anyone other than Bob to compute $\phi(n)$, the value of the integer d is not known to the

public. The pair (n, d) is known as Bob's *private key*. It should be noted here that some modular arithmetic techniques might be needed to make sure that $1 \leq d < \phi(n)$.

(5) If someone, say Alice, wants to send the message m to Bob, she would compute $c = m^e \pmod{n}$ using the public key (n, e) and sends the encrypted message c to Bob.

(6) To decipher c and get the original message m, Bob uses his private key (n, d) and computes $c^d \pmod{n}$. Since $ed \equiv 1 \pmod{\phi(n)}$, we have $ed = 1 + k\phi(n)$ for some integer k. Therefore, $c^d \pmod{n} = (m^e)^d \pmod{n} = m^{ed} \pmod{n} \equiv m^{1+k\phi(n)} \equiv m \left(m^{\phi(n)}\right)^k \equiv m \pmod{n}$ since $m^{\phi(n)} \equiv 1 \pmod{n}$ by Euler's theorem.

We illustrate the RSA algorithm with an example using small prime numbers for simplicity.

Example 6.53. Consider the plaintext message STOP. We use the RSA algorithm with $p = 31$, $q = 41$ and $e = 11$ to encrypt the message. Then we check our answer by decrypting our encrypted message. With the choices of p and q , $n = pq = 1271$, $\phi(n) = (p-1)(q-1) = 1200$. Notice that the choice of e is valid since $\gcd(11, 1200) = 1$. Using the standard numerical assignment of letters, the plaintext message STOP is converted into the integer $m = 18191415$. Since $m > n$, we divide m into four blocks: 18, 19 14 and 15 each smaller than n. We need to compute the congruences $18^{11} \pmod{1271}$, $19^{11} \pmod{1271}$, $14^{11} \pmod{1271}$ and $15^{11} \pmod{1271}$. For the congruence $18^{11} \pmod{1271}$, notice that $18^5 = 1889568 \equiv 862 \pmod{1271}$, $18^{10} \equiv 862^2 = 743044 \equiv 780 \pmod{1271}$, $18^{11} = 18^{10} \cdot 18 \equiv 780 \cdot 18 = 14040 \equiv 59 \pmod{1271}$. For the congruence $19^{11} \pmod{1271}$: $19^5 = 2476099 \equiv 191 \pmod{1271}$, $19^{10} \equiv 191^2 = 36481 \equiv 893 \pmod{1271}$, so $19^{11} = 19^{10} \cdot 19 \equiv 893 \cdot 19 = 16967 \equiv 444 \pmod{1271}$. Similar calculations show that $14^{11} \equiv 1063 \pmod{1271}$ and $15^{11} \equiv 480 \pmod{1271}$. The message is then encrypted as $59\,444\,1063\,480$. We now check that $59\,444\,1063\,480$ decrypts to m. First, we need to find the private key d such that $ed \equiv 1 \pmod{\phi(n)}$. For this, the Euclidean algorithm gives us that $1 \cdot 1200 + (-109) \cdot 11 = 1$. Therefore, $d = -109 \equiv -109 + 1200 = 1091 \pmod{1200}$. To decrypt the message $59\,444\,1063\,480$, we need compute the congruences: $59^{1091} \pmod{1271}$, $444^{1091} \pmod{1271}$, $1063^{1091} \pmod{1271}$ and $480^{1091} \pmod{1271}$. For $59^{1091} \pmod{1271}$, notice that $59^5 = 714924299 \equiv 780 \pmod{1271}$, $59^{10} \equiv 780^2 = 608400 \equiv 862 \pmod{1271}$, $59^{20} \equiv 862^2 = 743044 \equiv$

780 (mod 1271), $59^{40} \equiv 780^2 \equiv 862$ (mod 1271), $59^{80} \equiv 862^2 \equiv 780$ (mod 1271). Continuing this way, we see that $59^{160} \equiv 862$ (mod 1271), $59^{320} \equiv 780$ (mod 1271) and $59^{640} \equiv 862$ (mod 1271). So, $59^{1091} = 59^{640} \cdot 59^{320} \cdot 59^{80} \cdot 59^{40} \cdot 59^{10} \cdot 59 \equiv 862 \cdot 780 \cdot 780 \cdot 862 \cdot 862 \cdot 59$ (mod 1271) $= 780^2 \cdot 862^3 \cdot 59 \equiv 862 \cdot 780 \cdot 862 \cdot 59 \equiv 862 \cdot 59 = 50858 \equiv 18$ (mod 1271). Similar calculations show that $444^{1091} \equiv 19$ (mod 1271), $1063^{1091} \equiv 14$ (mod 1271) and $480^{1091} \equiv 15$ (mod 1271). Therefore, the message $59\,444\,1063\,480$ decrypts to $m = 18191415$.

6.8.2 *Exercises*

(1) Encrypt each of the following using Caesar cipher technique with key $k = 3$ and the standard numerical assignment of letters from the alphabet.

⋆ (a) MOON	(c) TRAPEZOID
(b) CLEAR	⋆ (d) OXYGEN

⋆(2) The following is a numerical message encrypted using Caesar cipher technique with key $k = 3$ and the standard numerical assignment of letters from the alphabet:

$$21\,07\,14\,14 \quad 01\,17\,23\,20 \quad 10\,17\,23\,21\,07.$$

What was the original plaintext?

(3) Decrypt the following message using the Caesar cipher technique with key $k = 3$ and the standard numerical assignment of letters from the alphabet: PDNH BRXU PRYH QRZ.

⋆(4) One way to improve the security of Caesar's cipher is to use the function $f_{a,b}(x) = ax + b$ (mod 26) where $a, b \in \{0, 1, \ldots, 25\}$. We say that $f_{a,b}(x)$ represents a valid encryption function, known as the *affine cipher*, if the congruence $y \equiv f_{a,b}(x)$ (mod 26) has a unique solution in the set $\{0, 1, \ldots, 25\}$ for any integer y.

(a) Prove that $f(x) = ax + b$ (mod 26) is a valid encryption function if and only if $\gcd(a, 26) = 1$.

(b) How many different affine ciphers are there?

(c) Consider the affine cipher $f_{5,7}(x) = 5x + 7$ (mod 26). Starting with the standard numerical assignment of letters from the alphabet, find the letter corresponding to each of the following:

$$\text{(i)} \ A \qquad \text{(ii)} \ F \qquad \text{(iii)} \ J \qquad \text{(iv)} \ M \qquad \text{(v)} \ Z$$

    (d) Give the decryption function $g_{5,7}(x)$ of the affine cipher $f_{5,7}(x) = 5x + 7 \pmod{26}$ (this is the function that expresses x in terms of y in $y \equiv 5x + 7 \pmod{26}$).

    (e) Using the decryption function $g_{5,7}(x)$ of the affine cipher $f_{5,7}(x) = 5x + 7 \pmod{26}$, decrypt the message HMHUWZU.

(5) What security issue could arise if $p = q$ in the RSA scheme?

(6) In each case, the pair (p, q) of prime numbers and the integer e are given of a certain RSA cryptosystem. Give the private key (n, d) of the cryptosystem.

    (a) $(p, q) = (13, 17)$, $e = 31$ (c) $(p, q) = (53, 73)$, $e = 79$

   $\star$ (b) $(p, q) = (23, 37)$, $e = 47$ $\star$ (d) $(p, q) = (179, 211)$, $e = 139$

(7) In each case, an RSA encryption scheme is given in the form $c = m^e \pmod{n}$ with relatively small integer n. Find the decryption scheme $m = c^d \pmod{n}$.

    (a) $c = m^{13} \pmod{35}$ (c) $c = m^{31} \pmod{91}$

   $\star$ (b) $c = m^7 \pmod{55}$ $\star$ (d) $c = m^{17} \pmod{221}$

(8) An RSA system is designed using the prime numbers $p = 101$, $q = 127$ and a public key $(12600, 373)$. Using this system, encrypt the message MOVE. Start by representing the message with an integer using Caesar cipher technique with key $k = 3$.

(9) Using the RSA system of Exercise (8), decrypt each of the following numerical messages.

    (a) 17   (b) 27   (c) 35   (d) 43

$\star$(10) Show that knowing n and $\phi(n)$ in a RSA cryptosystem is equivalent to factoring n as a product of two primes p and q.

Chapter 7

Binary Relations

7.1 Introduction

The notion of binary relations is one of great importance in mathematics. The Oxford English dictionary gives the following definition of the word "relation": *The way in which two or more people or things are connected; a thing's effect on or relevance to another.* On a human level, the word could have many interpretations. Some classical examples include a relation between two friends, a parent and a child, two siblings, ... In particular, we talk about a "couple" to describe two people in a romantic relationship. However, in order to have a couple, the two people involved must "recognize" the relationship. It could very well happen that person a is in love with person b but not the other way around. In a more general context, a binary relation from a set A to a set B (hence the name "binary") is a rule to determine if a given element a of A is in relation with a given element b of B. In this chapter, we develop the general notion of binary relations, study their properties and the various ways to represent them. One case of particular interest for us in this chapter is when $A = B$. In this case we talk about a binary relation *on* the set A. Two types of binary relations on a given set are of particular importance: the equivalence relation and the partial order. We explore the properties of these two types in details as well as some of their applications.

7.2 Basic definitions and terminology

Consider the two sets $A = \{2, 3, 4, 5, 6, 7, 8\}$ and $B = \{0, 9, 15, 25\}$. Define a relation R from A to B as follows: the element $a \in A$ is said to be in relation with the element $b \in B$ if and only if a divides b. For example, 3 is

309

in relation with both 9 and 15 but it is not in relation with 25. The relation can be fully described by listing all couples (a, b) with $a \in A$ and $b \in B$ such that a divides b. In other words, the relation R is completely determined by the subset $\{(2,0), (3,0), (3,9), (3,15), (4,0), (5,0), (5,15), (5,25)\}$ of the Cartesian product $A \times B$. This motivates the following definition.

Definition 7.1. Given two sets A and B, a *binary relation from A to B* is a subset R of the Cartesian product $A \times B$. If $A = B$, then R is called a relation *on the set A*. If $R \subseteq A \times B$ is a binary relation and $(a, b) \in R$, we say that the element a of A is in *relation* with the element b of B (or simply, a is *related* to b) and we write $a \, R \, b$. If $a \in A$ is not in relation with $b \in B$, we write $a \, \not\!R \, b$.

Before we look at some examples, we make some remarks.

(1) We can extend the notion of a binary relation given in Definition 7.1 to an *n-ary relation* defined as a subset of the Cartesian product $A_1 \times A_2 \times \cdots \times A_n$ of n sets. However, our main focus in this book is on binary relations. In what follows, the word *relation* means *binary relation* unless otherwise specified.

(2) Given any two sets A and B, the empty set $\emptyset$ and the set $A \times B$ are relations from A to B as subsets of $A \times B$. The relation $\emptyset$ is called the *empty relation* and $A \times B$ is called the *universal* relation. In the empty relation, no element of A is in relation with any of the elements of B. In the universal relation every element of A is in relation with every element of B.

(3) If R is a relation on a set A, it could very well happens that $(a, b) \in R$ but $(b, a) \notin R$ (or equivalently, $a \, R \, b$ but $b \, \not\!R \, a$.)

(4) If $A = \emptyset$ or $B = \emptyset$, then the only relation from A to B or from B to A is the empty relation.

Example 7.1. Consider the sets $A = \{a, b, c\}$ and $B = \{\{1\}, 2, 3\}$.

- $R_1 = \{(a, \{1\}), (c, 3), (b, 3)\}$ is a relation from A to B.
- $R_2 = \{(b, a), (c, b), (c, c)\}$ is a relation on A.
- $R_3 = \{(\{1\}, \{1\}), (3, 2), (2, 3)\}$ is a relation on B.
- $R_4 = \{(\{1\}, a), (2, b), (3, a)\}$ is a relation from B to A.

The above example shows relations defined by explicit listing of their elements as subsets of the Cartesian product. Very often, relations are defined in terms of a description of the properties that make two elements to be in relation. This is in particular useful when we are dealing with infinite relations (as sets).

Example 7.2. Recall that if $m, n \in \mathbb{Z}$, the notation $m \mid n$ stands for "*m divides n evenly*" (so the remainder of the division of n by m is zero). This defines a binary relation R on the set $\mathbb{Z}$. The couples $(2, 2)$, $(2, 4)$, $(3, 18)$ and $(11, 121)$ are elements of R while none of $(4, 2)$, $(3, 17)$ and $(11, 120)$ is. Note that $(0, n) \notin R$ for any integer n (as 0 does not divide any number) but $(n, 0) \in R$ for any non-zero integer n.

Example 7.3. On the set A of all binary strings of length 4 (there are $2^4 = 16$ such strings), we define a relation R as follows. If $x, y \in A$, then $x \, R \, y$ if and only if x and y have the same last digit. So, $1101 \, R \, 0001$, $1010 \, R \, 0000$ but $1101 \, \not\!R \, 1010$ and $1111 \, \not\!R \, 1010$.

Example 7.4. Define a relation R on the set $\mathbb{R}^2 = \mathbb{R} \times \mathbb{R}$ as follows. If (x, y), (a, b) are in $\mathbb{R}^2$, then $(x, y) R(a, b)$ if and only if $x - y = a - b$. So, $(1, 2)R(-4, -3)$, $\left(\frac{1}{2}, \frac{1}{3}\right) R \left(0, -\frac{1}{6}\right)$ but $(4, 2) \, \not\!R \, (5, 2)$ and $(0, 1) \, \not\!R \, (1, 1)$.

Example 7.5. Define a relation R on the set $A = \{1, 3, 5, 7, 9\}$ as follows: $x \, R \, y \Leftrightarrow x \leq y$. In this case we can list all the elements of R: $(1, 1)$, $(1, 3)$, $(1, 5)$, $(1, 7)$, $(1, 9)$, $(3, 3)$, $(3, 5)$, $(3, 7)$, $(3, 9)$, $(5, 5)$, $(5, 7)$, $(5, 9)$, $(7, 7)$, $(7, 9)$ and $(9, 9)$.

Example 7.6. Given a function $f : \mathbb{R} \to \mathbb{R}$, define a relation R on $\mathbb{R}$ as follows. If $x, y \in \mathbb{R}$, then $x \, R \, y$ if and only if $y = f(x)$. A point (x, y) of the Cartesian plane $\mathbb{R} \times \mathbb{R}$ is an element of R if an only if $y = f(x)$. In other words, R is the graph of the function f in the usual sense of modern Calculus. For instance, the following graph (subset of $\mathbb{R} \times \mathbb{R}$) represents the relation on $\mathbb{R}$ defined by the function $f(x) = x^2$.

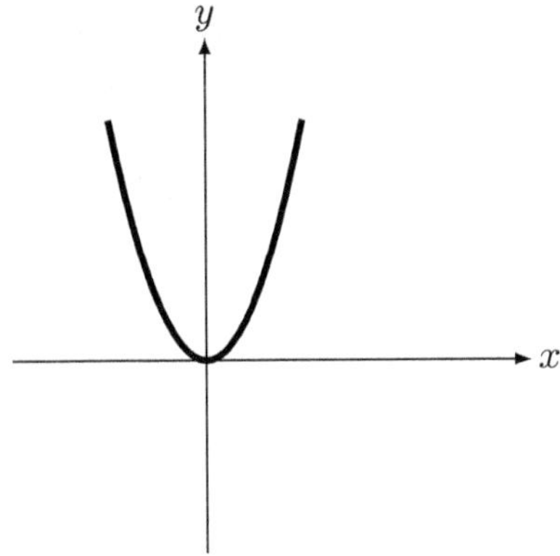

The following definition starts to draw a certain link between relations and functions. Later in this chapter, we will establish a more rigourous link.

Definition 7.2. Let $R \subseteq A \times B$ be a relation from A to B. The *domain* of R, denoted by $\mathrm{Dom}(R)$, is defined to be the following subset of A: $\mathrm{Dom}(R) = \{a \in A;\, (a,b) \in R \text{ for some } b \in B\}$. The *range* (also called the *image*) of R, denoted by $\mathrm{Ran}(R)$, is the following subset of B: $\mathrm{Ran}(R) = \{b \in B;\, (a,b) \in R \text{ for some } a \in A\}$.

Example 7.7. Let us look back at the relations given in Example 7.1. The following table gives the domain and the range of each one of them.

	Domain	Range
R_1	A	$\{\{1\},3\}$
R_2	$\{b,c\}$	A
R_3	B	B
R_4	B	$\{a,b\}$

Example 7.8. For the relation "m *divides* n" on $\mathbb{Z}$ defined in Example 7.2, $\mathrm{Dom}(R) = \mathbb{Z}\setminus\{0\}$ (since any non-zero integer divides (at least) itself) and $\mathrm{Ran}(R) = \mathbb{Z}$ (note that $0 \in \mathrm{Ran}(R)$ since n divides 0 for any $n \in \mathbb{Z}$).

7.2.1   *Exercises*

(1) In each case, list all possible binary relations from A to B.

  (a) $A = \emptyset$, $B = \{a,b,c\}$ (c) $A = \{1\}$, $B = \{a,b,c\}$

  (b) $A = \{1\}$, $B = \{a,b\}$ $\star$ (d) $A = \{1,2\}$, $B = \{a,b\}$

(2) Consider the relation $R = \{(m,n) \in \mathbb{Z} \times \mathbb{Z};\, 3m - 2n = -1\}$ on $\mathbb{Z}$. In each case, determine the truth value of the given statement.

  $\star$ (a) $0\,R\,1$ (c) $-1\,R\,0$ $\star$ (e) $1\,R\,2$ (g) $-2\,R\,2$

   (b) $-1\,R-1$ (d) $0\,R-1$ (f) $2\,R\,1$ $\star$ (h) $3\,R\,5$

(3) In each case, a relation R on $\mathbb{R}$ is given. Determine $\mathrm{Dom}(R)$ and $\mathrm{Ran}(R)$.

  (a) for all $x,y \in \mathbb{R}$: $x\,R\,y \Leftrightarrow 2x - 3y = 1$.

  (b) for all $x,y \in \mathbb{R}$: $x\,R\,y \Leftrightarrow x^2 + y^2 = 9$.

(4) Consider the sets $A = \{\{1\}, 2, \{3,4\}\}$ and $B = \{1, \{2\}, 3\}$. In each case, determine if the given set represents a binary relation from A to B, from B to A, on the set A or on the set B. Determine the domain and the range of each relation.

  (a) $R_1 = \{(1, \{1\}), (3, \{3,4\}), (\{2\}, 2)\}$.

  $\star$ (b) $R_2 = \{(\{1\}, 2), (\{1\}, \{3,4\}), (2, \{3,4\}), (2, 2)\}$.

  (c) $R_3 = \{(\{1\}, 1), (2, \{2\}), (2, 3), (\{3,4\}, \{2\}), (2, 1)\}$.

  (d) $R_4 = \{(\{3,4\}, \{3,4\})\}$.

  $\star$ (e) $R_5 = \{(\{2\}, \{2\}), (1, 1), (3, 3)\}$.

(5) C_1, C_2, C_3 and C_4 are four north American cities. The following table gives the price p_{ij} (in dollars per minute) of making a phone call from C_i to C_j. For example, a call from C_2 to C_4 costs $p_{24} = 2.1$ dollars per minute.

	C_1	C_2	C_3	C_4
C_1	0.1	0.5	0.3	1.2
C_2	0.6	0.2	1.0	2.1
C_3	0.3	0.85	0.6	1.8
C_4	1.2	2.0	1.8	0.4

Note that due to various tax schemes, calls from C_i to C_j and from C_j to C_i are not necessarily the same. Note also that the diagonal values in the table represent costs of calling within the same city. Let $A = \{C_1, C_2, C_3, C_4\}$. For each of the given binary relations on A, list all the elements of the relation.

  $\star$ (a) $C_i \, R \, C_j \Leftrightarrow p_{ij} < p_{ji}$

  (b) $C_i \, S \, C_j \Leftrightarrow p_{ij} < 1$

  (c) $C_i \, T \, C_j \Leftrightarrow p_{ij} = p_{ji}$

  (d) $C_i \, U \, C_j \Leftrightarrow p_{ij} + p_{ji} \leq 1.5$

(6) The shaded area in the following diagram represents the region of the Cartesian plane in the first quadrant bounded by the graph of the parabola $f(x) = x^2$ and the line $x = 2$ as shown. Give subsets A and B of $\mathbb{R}$ and a relation R from A to B represented by this region.

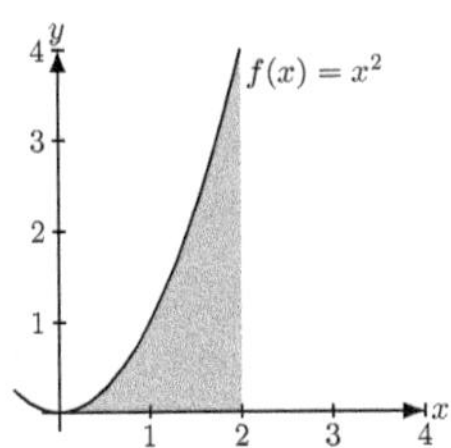

(7) In each case, list all ordered pairs in the relation R from the set $A = \{-1, 0, 1, 2, 3, 4\}$ to the set $B = \{1, 2, 3, 6, 12\}$.

  (a) $a\,R\,b \Leftrightarrow a$ is a divisor of b.

$\star$ (b) $a\,R\,b \Leftrightarrow a + b \leq 6$.

  (c) $a\,R\,b \Leftrightarrow \gcd(a, b) = 1$.

$\star$ (d) $a\,R\,b \Leftrightarrow \operatorname{lcm}(a, b) = 12$.

(8) Define a relation R on the set $\mathbb{R}^2 = \mathbb{R} \times \mathbb{R}$ as follows: $(x, y)\,R\,(a, b) \Leftrightarrow xb = ya$. In each case, determine if the given pair is an element of R.

  (a) $((1, 2), (-4, -8))$

$\star$ (b) $((-4, -8), (1, 2))$

  (c) $((4, 2), (5, 2))$

  (d) $\left(\left(\frac{1}{2}, \frac{1}{3}\right), \left(1, \frac{2}{3}\right)\right)$

$\star$ (e) $\left((\sqrt{2}, -1), \left(1, \frac{1}{\sqrt{2}}\right)\right)$

  (f) $\left(\left(\frac{1}{1-\sqrt{2}}, 1\right), (1 + \sqrt{2}, -1)\right)$

(9) Let A, B be two sets, R a relation from A to B. For a subset A' of A, define $R(A')$ as being the following subset of B:

$$R(A') = \{b \in B;\ (a, b) \in R \text{ for some } a \in A'\}.$$

If $A' = \{a\}$, then $R(A')$ is simply denoted by $R(a)$. On the set $A = \{1, 2, 3, 4\}$, a relation R is given by the list of its elements: $(1, 1)$, $(1, 3)$, $(1, 4)$, $(2, 1)$, $(2, 2)$, $(2, 3)$, $(3, 2)$, $(3, 4)$, $(4, 1)$, $(4, 3)$, $(4, 4)$. Find each of the following.

$\star$ (a) $R(1)$ (b) $R(4)$ $\star$ (c) $R(\{1, 2\})$ (d) $R(\{3, 4\})$

(10) Refer to the notations of Exercise (9). Consider the relation R defined on $\mathbb{R}$ as follows: $x\,R\,y \Leftrightarrow 4x^2 + 9y^2 = 1$. Find each of the following.

$\star$ (a) $R(-1)$ (c) $R\left(\frac{1}{2}\right)$ $\star$ (e) $R\left(\frac{1}{8}\right)$ $\star$ (g) $R(\{1, 2\})$

  (b) $R(1)$ (d) $R\left(\frac{1}{3}\right)$ (f) $R\left(-\frac{1}{9}\right)$

(11) Refer to the notations of Exercise (9). Let R be a relation from a set A to a set B and let A_1, A_2 be two subsets of A.

  (a) Prove that $R(A_1 \cup A_2) = R(A_1) \cup R(A_2)$.

$\star$ (b) Prove that $R(A_1 \cap A_2) \subseteq R(A_1) \cap R(A_2)$.

  (c) Give an example of a set A, two subsets A_1, A_2 of A and a relation R on A such that $R(A_1 \cap A_2) \neq R(A_1) \cap R(A_2)$.

(12) Refer to the notations of Exercise (9). Let R and S be two relations on a set A. Prove that $R = S$ if and only if $R(x) = S(x)$ for every x in A.

(13) If A, B are two finite sets such that $|A| = m$ and $|B| = n$. How many binary operations from A to B are there?

7.3 Representations of binary relations: Boolean matrices

Consider two finite sets $A = \{a_1, a_2, \ldots, a_n\}$ and $B = \{b_1, b_2, \ldots, b_m\}$ and let R be a relation from A to B. Aside from listing all its elements, R can be represented using a tabular form (a matrix) or a pictorial form (a digraph). Each of these forms of representation has its advantages depending on the given setting. A tabular representation of R consists of a binary (also called *Boolean*) $n \times m$ matrix (n rows and m columns) where each entry is either 0 or 1 (hence the name). We denote the matrix by $M_R = [m_{ij}]$ where the entry m_{ij} at the ith row and the jth column is defined as follows:

$$m_{ij} = \begin{cases} 0 & \text{if } a_i \not{R} \, b_j \\ 1 & \text{if } a_i R b_j \end{cases}$$

This matrix is called the *adjacency matrix* of R. It is important to keep in mind that the definition of the adjacency matrix of the relation R is relative to the orders $a_1, a_2, \ldots, a_n$ and $b_1, b_2, \ldots, b_m$ of the elements of A and B. In a pictorial representation of R, elements of A and B are represented by points (or by small circles) in the plane, called *vertices*. If $a_i R b_j$ then a directed *edge* (a straight or a curved line), from the vertex representing a_i to that representing b_j is drawn. The resulting picture is called a *directed graph* or a *digraph* for short.

Example 7.9. Consider the two sets $A = \{a_1, a_2, a_3\}$ and $B = \{b_1, b_2, b_3\}$. In each case, represent the binary relation from A to B using a binary matrix and a digraph.

(a) $R_1 = \{(a_1, b_1), (a_1, b_2), (a_1, b_3), (a_2, b_2), (a_3, b_1), (a_3, b_3)\}$
(b) $R_2 = \{(a_1, b_1), (a_2, b_2), (a_2, b_3), (a_3, b_1)\}$

Solution. For R_1, the matrix representation is:

$$M_{R_1} = \begin{array}{c} \\ a_1 \\ a_2 \\ a_3 \end{array} \begin{array}{ccc} b_1 & b_2 & b_3 \\ \left[\begin{array}{ccc} 1 & 1 & 1 \\ 0 & 1 & 0 \\ 1 & 0 & 1 \end{array}\right] \end{array}$$

For the digraph representation of R_1, we start by drawing the six vertices (elements of A and B). If $a_i R b_j$, we draw an oriented edge from vertex a_i to vertex b_j:

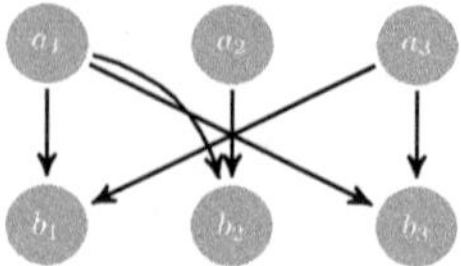

For R_2, the adjacency matrix is:

$$M_{R_2} = \begin{array}{c@{}c} & \begin{array}{ccc} b_1 & b_2 & b_3 \end{array} \\ \begin{array}{c} a_1 \\ a_2 \\ a_3 \end{array} & \left[\begin{array}{ccc} 1 & 0 & 0 \\ 0 & 1 & 1 \\ 1 & 0 & 0 \end{array}\right] \end{array}$$

and the digraph is:

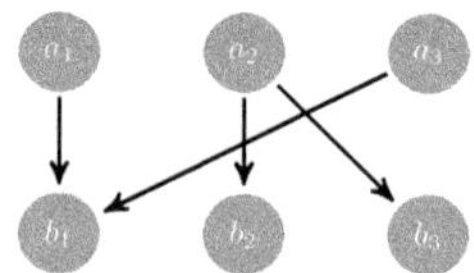

$$\Diamond.$$

If A is a finite set of cardinality n and R is a binary relation R on A, then the adjacency matrix of R is square (same number of rows as the number of columns) of size $n \times n$. For the graph representation of R, we draw a loop around the vertex representing the element $a \in A$ if $a\,R\,a$.

Example 7.10. On the set $A = \{a, b, c, d\}$, a relation R is defined as follows: $R = \{(a, b), (a, c), (b, a), (b, b), (b, c), (c, d), (d, a), (d, d)\}$. The adjacency matrix of R is the following 4×4 matrix:

$$M_R = \begin{array}{c@{}c} & \begin{array}{cccc} a & b & c & d \end{array} \\ \begin{array}{c} a \\ b \\ c \\ d \end{array} & \left[\begin{array}{cccc} 0 & 1 & 1 & 0 \\ 1 & 1 & 1 & 0 \\ 0 & 0 & 0 & 1 \\ 1 & 0 & 0 & 1 \end{array}\right] \end{array}$$

and the digraph of R is:

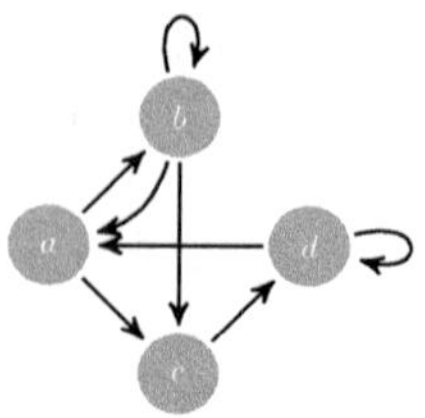

Example 7.11. On the set $A = \{1, 2, 3, 4, 9, 12\}$, consider the relation R defined as follows: $x\,R\,y \Leftrightarrow \gcd(x, y) = 1$. Note that R is "symmetric" in the sense that if $(a, b) \in R$, then $(b, a) \in R$. We have: $\gcd(2, 3) = \gcd(2, 9) = 1$, $\gcd(3, 2) = \gcd(3, 4) = 1$, $\gcd(4, 3) = \gcd(4, 9) = 1$, $\gcd(9, 2) = \gcd(9, 4) = 1$. So R consists of the following couples: $(1, 1)$, $(1, 2)$, $(1, 3)$, $(1, 4)$, $(1, 9)$, $(1, 12)$, $(2, 1)$, $(2, 3)$, $(2, 9)$, $(3, 1)$, $(3, 2)$, $(3, 4)$, $(4, 1)$, $(4, 3)$, $(4, 9)$, $(9, 1)$, $(9, 2)$, $(9, 4)$ and $(12, 1)$. The adjacency matrix of R is:

$$
M_R = \begin{array}{c}
\\ 1 \\ 2 \\ 3 \\ 4 \\ 9 \\ 12
\end{array}
\begin{array}{c}
\begin{array}{cccccc} 1 & 2 & 3 & 4 & 9 & 12 \end{array} \\
\left[\begin{array}{cccccc}
1 & 1 & 1 & 1 & 1 & 1 \\
1 & 0 & 1 & 0 & 1 & 0 \\
1 & 1 & 0 & 1 & 0 & 0 \\
1 & 0 & 1 & 0 & 1 & 0 \\
1 & 1 & 0 & 1 & 0 & 0 \\
1 & 0 & 0 & 0 & 0 & 0
\end{array}\right]
\end{array}
$$

and the digraph of R is:

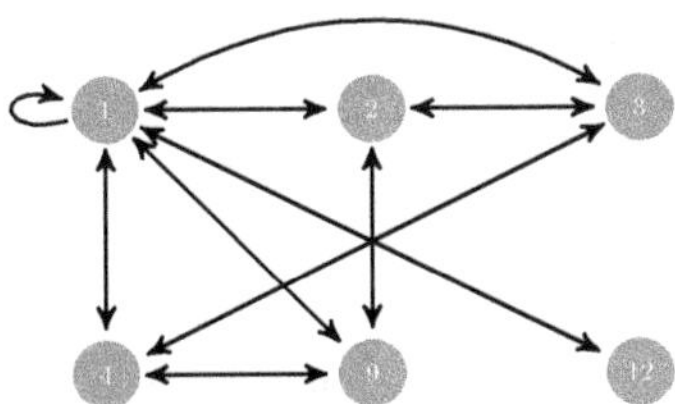

7.3.1 *Properties of Boolean matrices*

Having only two possibilities (0 or 1) for each entry in a Boolean matrix allows us to use the rules of propositional logic (or Boolean algebra) with "True" replaced by 1 and "False" replaced with 0. Namely: $1 \wedge 1 = 1$, $1 \wedge 0 = 0 \wedge 1 = 0$, $0 \wedge 0 = 0$, $1 \vee 1 = 1$, $1 \vee 0 = 0 \vee 1 = 1$ and $0 \vee 0 = 0$. In this section, we define new operations on Boolean matrices that we use later in studying properties of binary relations.

Definition 7.3. Let $A = [a_{ij}]$ and $B = [b_{ij}]$ be two Boolean matrices of the same size $m \times n$. The *join* of A and B, denoted by $A \vee B$, is the $m \times n$ matrix $C = [c_{ij}]$ where $c_{ij} = a_{ij} \vee b_{ij}$. In other words,

$$
c_{ij} = \begin{cases} 1 & \text{iff } a_{ij} = 1 \text{ or } b_{ij} = 1 \\ 0 & \text{iff } a_{ij} = b_{ij} = 0 \end{cases}
$$

The *meet* of A and B, denoted by $A \wedge B$, is the $m \times n$ matrix $D = [d_{ij}]$ where $d_{ij} = a_{ij} \wedge b_{ij}$. In other words,

$$d_{ij} = \begin{cases} 1 & \text{iff } a_{ij} = b_{ij} = 1 \\ 0 & \text{iff } a_{ij} = 0 \text{ or } b_{ij} = 0 \end{cases}$$

Example 7.12. For the Boolean matrices

$$A = \begin{bmatrix} 1\,0\,0\,1 \\ 0\,0\,1\,1 \\ 1\,1\,1\,0 \end{bmatrix}, \quad B = \begin{bmatrix} 0\,0\,0\,1 \\ 1\,0\,1\,1 \\ 0\,1\,1\,0 \end{bmatrix} :$$

$$A \vee B = \begin{bmatrix} 1\,0\,0\,1 \\ 1\,0\,1\,1 \\ 1\,1\,1\,0 \end{bmatrix}, \quad A \wedge B = \begin{bmatrix} 0\,0\,0\,1 \\ 0\,0\,1\,1 \\ 0\,1\,1\,0 \end{bmatrix}.$$

Properties of the conjunction and the disjunction connectives we saw in Chapter 1 can be extended to the join and meet operations on Boolean matrices.

Theorem 7.1. *Let A, B and C be three Boolean matrices of the same size $m \times n$. Let $\mathbf{0}$ (respectively $\mathbf{1}$) be the matrix of size $m \times n$ with each entry equals to 0 (respectively 1). The following hold.*

(a) $A \vee B = B \vee A$

(b) $A \wedge B = B \wedge A$

(c) $(A \vee B) \vee C = A \vee (B \vee C)$

(d) $(A \wedge B) \wedge C = A \wedge (B \wedge C)$

(e) $A \vee (B \wedge C) = (A \vee B) \wedge (A \vee C)$

(f) $A \wedge (B \vee C) = (A \wedge B) \vee (A \wedge C)$

(g) $A \wedge \mathbf{0} = \mathbf{0}$, $A \vee \mathbf{0} = A$

(h) $A \wedge \mathbf{1} = A$, $A \vee \mathbf{1} = \mathbf{1}$

(i) $A \wedge A$, $A \vee A = A$

Proof. Each of the properties in the theorem follows from a similar property in Propositional Logic. Details are left as exercises. $\square$

The join and meet operations on Boolean matrices are performed "component wise" just like the addition of matrices. We now introduce the notion of Boolean product which is similar to standard matrix multiplication. Unlike the join and the meet operations, the Boolean product, $A \odot B$, of two Boolean matrices A and B requires that the number of columns of A is the same as the number of rows of B.

Definition 7.4. Let $A = [a_{ij}]$ and $B = [b_{ij}]$ be two Boolean matrices such that A is of size $m \times n$ and B is of size $n \times p$. We define the *Boolean product* of A and B, denoted by $A \odot B$, as being the $m \times p$ matrix $C = [c_{ij}]$ with:

$$c_{ij} = (a_{i1} \wedge b_{1j}) \vee (a_{i2} \wedge b_{2j}) \vee \cdots \vee (a_{in} \wedge b_{nj}).$$

Example 7.13. Consider the following two Boolean matrices:

$$A = \begin{bmatrix} 1\;0\;0\;1 \\ 0\;0\;1\;1 \\ 1\;1\;1\;0 \end{bmatrix}, \quad B = \begin{bmatrix} 0\;0 \\ 1\;1 \\ 1\;1 \\ 0\;1 \end{bmatrix}.$$

Since A is of size 3×4 and B is of size 4×2, $A \odot B$ exists but $B \odot A$ does not. Moreover, $A \odot B$ is a 3×2 given as follows:

$A \odot B$

$$= \begin{bmatrix} (1 \wedge 0) \vee (0 \wedge 1) \vee (0 \wedge 1) \vee (1 \wedge 0) & (1 \wedge 0) \vee (0 \wedge 1) \vee (0 \wedge 1) \vee (1 \wedge 1) \\ (0 \wedge 0) \vee (0 \wedge 1) \vee (1 \wedge 1) \vee (1 \wedge 0) & (0 \wedge 0) \vee (0 \wedge 1) \vee (1 \wedge 1) \vee (1 \wedge 1) \\ (1 \wedge 0) \vee (1 \wedge 1) \vee (1 \wedge 1) \vee (0 \wedge 0) & (1 \wedge 0) \vee (1 \wedge 1) \vee (1 \wedge 1) \vee (0 \wedge 1) \end{bmatrix}$$

$$= \begin{bmatrix} 0\;1 \\ 1\;1 \\ 1\;1 \end{bmatrix}$$

$\diamond$

A quick way to compute the entry c_{ij} (at the ith row and the jth column) of the Boolean product $A \odot B$ consists of arranging row i of A and column j of B side by side as two vectors with n components each and to compare their corresponding entries. If there exists at least one pair (a_{ik}, b_{kj}) of corresponding entries with $a_{ik} = b_{kj} = 1$, then $c_{ij} = 1$. Otherwise, $c_{ij} = 0$:

$$\begin{bmatrix} a_{i1} \\ a_{i2} \\ \vdots \\ a_{in} \end{bmatrix} \begin{matrix} \leftrightarrow \\ \leftrightarrow \\ \vdots \\ \leftrightarrow \end{matrix} \begin{bmatrix} b_{1j} \\ a_{2j} \\ \vdots \\ a_{nj} \end{bmatrix}$$

The *identity matrix* of size n is the square matrix $I_n = [\alpha_{ij}]$ of size $n \times n$ such that $\alpha_{ij} = 1$ if $i = j$ and $\alpha_{ij} = 0$ if $i \neq j$. So all entries of I_n are 0 except the ones on the diagonal which are equal to 1 each. For example

$$I_2 = \begin{bmatrix} 1\;0 \\ 0\;1 \end{bmatrix}, \quad I_3 = \begin{bmatrix} 1\;0\;0 \\ 0\;1\;0 \\ 0\;0\;1 \end{bmatrix}, \quad I_4 = \begin{bmatrix} 1\;0\;0\;0 \\ 0\;1\;0\;0 \\ 0\;0\;1\;0 \\ 0\;0\;0\;1 \end{bmatrix}.$$

The following theorem summarizes some important properties of the Boolean product. The proof of the theorem is omitted.

Theorem 7.2. *Let A, B and C be three Boolean matrices such that each of the operations listed below is well-defined.*

(a) $A \odot (B \odot C) = (A \odot B) \odot C$ *(the Boolean product is associative).*
(b) If A is an $m \times n$ Boolean matrix, then $A \odot I_n = A$ and $I_m \odot A = A$.

Definition 7.5. Let A be a square $n \times n$ Boolean matrix. If k is a non-negative integer, we define the *Boolean power* of A as being the $n \times n$ matrix $A^{[k]}$ defined as follows:

$$A^{[k]} = \begin{cases} I_n & \text{if } k = 0 \\ \underbrace{A \odot A \odot \cdots \odot A}_{k \text{ times}} & \text{if } k \neq 0 \end{cases}$$

Remark 7.1.

(a) The definition of $A^{[r]}$ makes sense by the associativity of the Boolean product of matrices.
(b) The Boolean power of a square (Boolean) matrix A of size $n \times n$ can be defined recursively as follows:

$$A^{[0]} = I_n, \quad A^{[n]} = A \odot A^{[n-1]} \quad \text{for } n \geq 1,$$

Example 7.14. Consider the Boolean matrix:

$$A = \begin{bmatrix} 1 & 0 & 1 \\ 0 & 1 & 0 \\ 1 & 1 & 1 \end{bmatrix}$$

(a) Compute $A^{[2]}$, $A^{[3]}$ and $A^{[4]}$.
(b) Can you guess what $A^{[n]}$ is in general for any integer $n \geq 2$? Justify your answer using induction.

Solution. For part (a):

$$A^{[2]} = A \odot A = \begin{bmatrix} 1 & 0 & 1 \\ 0 & 1 & 0 \\ 1 & 1 & 1 \end{bmatrix} \odot \begin{bmatrix} 1 & 0 & 1 \\ 0 & 1 & 0 \\ 1 & 1 & 1 \end{bmatrix} = \begin{bmatrix} 1 & 1 & 1 \\ 0 & 1 & 0 \\ 1 & 1 & 1 \end{bmatrix}$$

$$A^{[3]} = A \odot A^{[2]} = \begin{bmatrix} 1 & 0 & 1 \\ 0 & 1 & 0 \\ 1 & 1 & 1 \end{bmatrix} \odot \begin{bmatrix} 1 & 1 & 1 \\ 0 & 1 & 0 \\ 1 & 1 & 1 \end{bmatrix} = \begin{bmatrix} 1 & 1 & 1 \\ 0 & 1 & 0 \\ 1 & 1 & 1 \end{bmatrix}.$$

For part (b), notice that $A^{[4]} = A \odot A^{[3]} = A \odot A^{[2]} = A^{[3]} = A^{[2]}$ since $A^{[3]} = A^{[2]}$. We show, using induction on n, that

$$A^{[n]} = A^{[2]} = \begin{bmatrix} 1 & 1 & 1 \\ 0 & 1 & 0 \\ 1 & 1 & 1 \end{bmatrix}$$

for any integer $n \geq 2$. The property is clearly true for $n = 2$. Assume that $A^{[n]} = A^{[2]}$ for some $n \geq 2$, then $A^{[n+1]} = A \odot A^{[n]} = A \odot A^{[2]}$ (by the induction hypothesis) $= A^{[3]} = A^{[2]}$ (by part (a)). We conclude that $A^{[n]} = A^{[2]}$ for any $n \geq 2$. $\Diamond$

7.3.2 *Exercises*

$\star$(1) Let $A = \{1, 2, 3, 4, 9, 12\}$. We define a binary relation R on A as follows: $x \, R \, y \Leftrightarrow x \mid y$ where the notation $x \mid y$ means "x *divides* y". List all the elements and give the adjacency matrix and the digraph representation of R.

(2) Consider the two sets $A = \{a, b, c, d\}$, $B = \{x, y, z\}$. In each case, give the adjacency matrix and a digraph representation of the binary relation R from A to B.

   (a) $R = \{(a, x), (a, z), (c, y), (c, z)\}$
   (b) $R = \{(a, y), (b, y), (c, y), (d, x), (d, y), (d, z)\}$
   (c) $R = \{(a, x), (a, y), (a, z), (b, x), (b, y), (b, z), (c, x), (c, y), (c, z)\}$

(3) Consider the two sets $A = \{a, b, c, d\}$, $B = \{x, y, z, w\}$.

   (a) The adjacency matrix of a binary relation R from A to B is given below. List all the elements of R.

$$M = \begin{array}{c} \\ a \\ b \\ c \\ d \end{array} \overset{\displaystyle x \; y \; z \; w}{\begin{bmatrix} 1 & 1 & 1 & 0 \\ 0 & 0 & 1 & 0 \\ 0 & 0 & 0 & 1 \\ 1 & 0 & 0 & 1 \end{bmatrix}}$$

   $\star$ (b) The digraph of a binary relation S from A to B is given below. List all the elements of S.

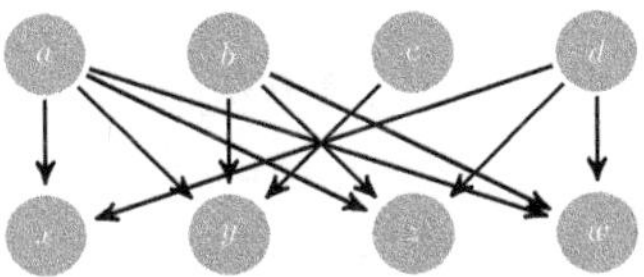

(4) In each case, represent the relation R on the set A with a Boolean matrix and with a digraph.

   (a) $A = \left\{-2, -1, -\frac{1}{\sqrt{3}}, -\frac{1}{\sqrt{2}}, 0, 1, 2, 3, \frac{1}{\sqrt{3}}, \frac{1}{\sqrt{2}}\right\}$, and for $a, b \in A$: $a \, R \, b$ if and only if $a^2 + b^2 = 1$.

   (b) $A = \{-3, -2, -1, 0, 1, 4, 6, 9, 12\}$, and for $a, b \in A$: $a \, R \, b$ if and only if a is a divisor of b.

$\star$ (c) $A = \{-3, -2, -1, 0, 1, 4, 6, 9, 12\}$, and for $a, b \in A$: $a\,R\,b$ if and only if $|a - b| \le 3$.

(5) Give an example of two square Boolean matrices A and B of the same size such that $A \odot B \ne B \odot A$.

$\star$(6) If A is an $m \times n$ Boolean matrix, prove that $A \odot I_n = A$ and $I_m \odot A = A$.

(7) Let A be a Boolean matrix. Prove that $A \wedge A = A$ and $A \vee A = A$.

(8) Find $A^{[k]}$ for $k \in \{2, 3, 4, 5\}$ where A is the Boolean matrix:

$$A = \begin{bmatrix} 1 & 0 & 1 \\ 0 & 1 & 0 \\ 1 & 1 & 0 \end{bmatrix}.$$

(9) Consider the following three Boolean matrices:

$$A = \begin{bmatrix} 1 & 0 \\ 1 & 0 \end{bmatrix}, \quad B = \begin{bmatrix} 1 & 1 \\ 1 & 0 \end{bmatrix}, \quad C = \begin{bmatrix} 0 & 1 & 0 \\ 1 & 0 & 1 \end{bmatrix}.$$

Find each of the following matrices, if it is defined.

(a) $A \wedge B$	(d) $A \odot C$	(g) $A \odot (B \vee C)$
$\star$ (b) $A \vee B$	$\star$ (e) $A \wedge (B \odot C)$	$\star$ (h) $(A \odot B) \odot (A \odot C)$
(c) $A \odot B$	(f) $(A \odot B) \odot C$	(i) $\left(A^{[2]} \odot B^{[3]}\right) \vee C$

(10) In each case, find $A \odot B$ and $B \odot A$, if defined.

(a) $A = \begin{bmatrix} 1 & 0 & 0 & 1 \\ 0 & 1 & 0 & 1 \\ 0 & 1 & 1 & 1 \\ 0 & 1 & 0 & 1 \end{bmatrix}$, $B = \begin{bmatrix} 0 & 0 \\ 1 & 1 \\ 1 & 0 \\ 1 & 1 \end{bmatrix}$

$\star$ (b) $A = \begin{bmatrix} 1 & 0 & 0 & 1 \\ 0 & 1 & 0 & 1 \\ 0 & 1 & 1 & 1 \\ 0 & 1 & 0 & 1 \end{bmatrix}$, $B = \begin{bmatrix} 1 & 1 & 0 & 0 \\ 0 & 0 & 1 & 1 \\ 1 & 0 & 1 & 1 \\ 1 & 1 & 0 & 0 \end{bmatrix}$

(11) The *complement* of an $m \times n$ Boolean matrix A is the $m \times n$ matrix A' obtained from A by replacing every 0 with 1 and every 1 with 0. For example, if $A = \begin{bmatrix} 1 & 1 & 0 \\ 0 & 0 & 0 \\ 1 & 0 & 1 \end{bmatrix}$, then $A' = \begin{bmatrix} 0 & 0 & 1 \\ 1 & 1 & 1 \\ 0 & 1 & 0 \end{bmatrix}$. Given two $m \times n$ Boolean matrices A and B, find $A \wedge A'$ and $A \vee A'$.

(12) Given two square Boolean matrices A and B of the same size $n \times n$, what is the number of operations required to compute $A \odot B$? (for example, two operations are required to compute the value of $a \wedge (b \vee c)$ where $a, b, c \in \{0, 1\}$).

7.4 Functions as relations

In Chapter 5, we defined a function from a set A to a set B as being a "rule of correspondence" that assigns to *every* element of A a *unique* element of B. Although this definition captures the essence of what we expect from the concept of a function to accomplish, it has some ambiguity as the term "rule of correspondence" is not very precise mathematically. The purpose of this section is to introduce a more formal definition of a function using the notion of binary relations.

Definition 7.6. Let A, B be two sets. A binary relation R from A to B is called a *function* (or a *mapping*) if it satisfies the following two conditions:

(1) (*Existence*) For every $x \in A$, there exists $y \in B$ such that $(x, y) \in R$.
(2) (*Uniqueness*) If $x \in A$ and $y, z \in B$ are such that (x, y) and (x, z) are in R, then $y = z$.

In other words, a function from A to B is a subset R of $A \times B$ such that for every $x \in A$, there *exists* a *unique* $y \in B$ with $(x, y) \in R$. Using the terminology of functions used in Chapter 5, the first condition in Definition 7.6 states that every element of A has an image in B and the second states that every image is unique. If the relation $R \subseteq A \times B$ is a function, we usually use the *functional notation* $R : A \to B$ to refer to R. Moreover, if R is a function from A to B and $(x, y) \in R$, we usually write $y = R(x)$ and say that y is the *image* of x.

Remark 7.2. Some authors distinguish between two types of functions: *partial* and *total*. A relation from A to B is called a *partial function* if the uniqueness condition in Definition 7.6 above holds but not necessarily the existence condition. If both conditions in Definition 7.6 are satisfied, then R is called a total function. The word *function* in this book means a total function, unless otherwise specified.

If the binary relation $R \subseteq A \times B$ is a function, then the notion of injection and surjection can be stated as follows:

(1) R is injective if and only if the following condition holds:
$$(\forall a \in A)(\forall a' \in A)(\forall b \in B)\,((a, b) \in R \wedge (a', b) \in R \Leftrightarrow a = a').$$

(2) R is surjective if and only if the following condition holds:
$$(\forall b \in B)\,(\exists a \in A)\,((a, b) \in R).$$

(3) As usual, R is called bijective if and only if it is injective and surjective at the same time.

Example 7.15. Let $A = \{1, 2, 3, 4\}$, $B = \{a, b, c, d\}$. In each case, determine if the given binary relation from A to B is a function. If the relation is a function, determine if it is injective or/and surjective.

(1) $R_1 = \{(1, b), (3, b), (2, a), (4, c), (1, a)\}$
(2) $R_2 = \{(1, a), (2, b), (3, a)\}$
(3) $R_3 = \{(2, d), (1, b), (4, a), (3, c)\}$
(4) $R_4 = \{(1, b), (2, b), (3, b), (4, b)\}$

Solution. The relation R_1 is not a function since $(1, b)$ and $(1, a)$ are both elements of R_1 which contradicts the uniqueness condition of Definition 7.6 (the image of $1 \in A$ is not unique). The relation R_2 is not a function since $(4, y)$ is not in R for any $y \in B$ which contradicts the existence condition of Definition 7.6 (the element $4 \in A$ has no image). Relations R_3 and R_4 are functions as they satisfy both conditions of Definition 7.6. Relation R_3 is bijective since it is injective and surjective. Relation R_4, on the other hand, is neither injective nor surjective.

7.4.1 *Exercises*

(1) Let $A = \{\alpha, \beta, \gamma, \eta\}$, $B = \{\oslash, \uplus, \bullet, \star\}$. In each case, determine if the given binary relation from A to B is a function. If the relation is a function, determine if it is injective, surjective.

    (a) $R_1 = \{(\beta, \star), (\gamma, \uplus), (\alpha, \oslash), (\eta, \bullet)\}$
    (b) $R_2 = \{(\beta, \star), (\alpha, \uplus), (\eta, \star), (\gamma, \bullet)\}$
    (c) $R_3 = \{(\beta, \star), (\gamma, \uplus), (\alpha, \oslash), (\eta, \bullet), (\beta, \uplus)\}$
    (d) $R_4 = \{(\alpha, \uplus), (\gamma, \bullet), (\eta, \star)\}$

(2) Let $A = \{-1, 0, 1, 2, \{1, 2\}\}$, $B = \{a, b, c, d\}$. In each case, determine if the given binary relation from A to B is a function. If you say it is, determine if it is injective, surjective.

    (a) $R_1 = \{(1, c), (\{1, 2\}, a), (1, a), (-1, d)\}$
  $\star$ (b) $R_2 = \{(-1, c), (\{1, 2\}, a), (0, c), (2, d), (1, b)\}$
    (c) $R_3 = \{(-1, c), (\{1, 2\}, c), (0, c), (1, c), (2, c)\}$
    (d) $R_4 = \{(-1, a), (\{1, 2\}, b), (1, d), (1, a), (2, d), (1, b)\}$

(3) In each case, determine if the given binary relation on the set A is a function from A to A. If the relation is a function, determine if it is injective or surjective.

    (a) $R_1 = \{(x, y) \in \mathbb{Z} \times \mathbb{Z};\ x + y = 1\}$, $A = \mathbb{Z}$.
  $\star$ (b) $R_2 = \{(x, y) \in \mathbb{R}^* \times \mathbb{R}^*;\ xy = 1\}$, $A = \mathbb{R}^* = \mathbb{R} \setminus \{0\}$.

   (c) $R_3 = \{(x,y) \in \mathbb{R} \times \mathbb{R};\ x^2 + y^2 = 1\}$, $A = \mathbb{R}$.
$\star$ (d) $R_4 = \{(x,y) \in \mathbb{Q} \times \mathbb{Q};\ y = |x|\}$, $A = \mathbb{Q}$.
   (e) $R_5 = \{(x,y) \in \mathbb{R} \times \mathbb{R};\ x + y^2 = 1\}$, $A = \mathbb{R}$.

(4) In each case, the digraph representing a binary relation R on the set $A = \{1, 2, 3, 4, 5\}$ is given. Determine if R is a function. If you say it is, determine if it is injective, surjective.

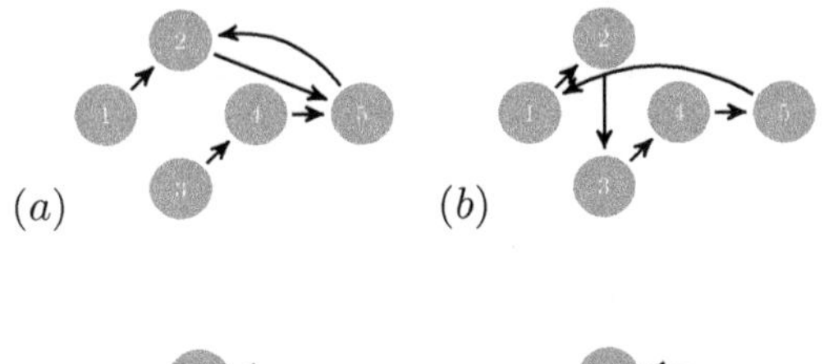

        (a)   (b)

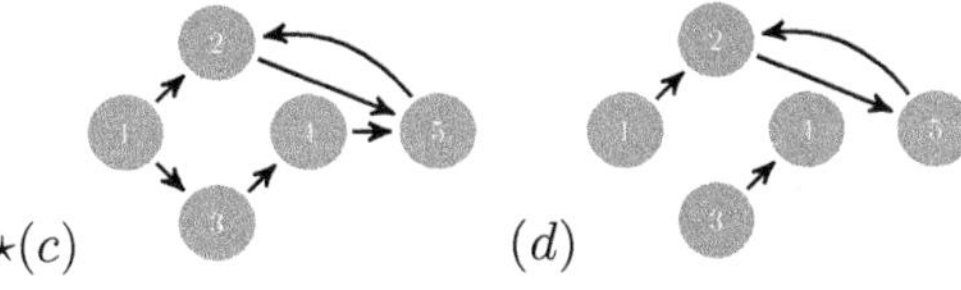

        $\star(c)$ (d)

(5) Consider the two sets $A = \{a, b, c, d\}$ and $B = \{x, y, z, w\}$. In each case, the adjacency matrix M_R of a binary relation R from A to B is given. Determine if R is a function. If you say it is, determine if it is injective, surjective.

(a) $M_R = \begin{array}{c} \\ a \\ b \\ c \\ d \end{array}\begin{array}{c} x\ y\ z\ w \\ \left[\begin{array}{cccc} 1 & 1 & 1 & 0 \\ 0 & 0 & 1 & 0 \\ 0 & 0 & 0 & 1 \\ 1 & 0 & 0 & 1 \end{array}\right] \end{array}$

(c) $M_R = \begin{array}{c} \\ a \\ b \\ c \\ d \end{array}\begin{array}{c} x\ y\ z\ w \\ \left[\begin{array}{cccc} 0 & 0 & 0 & 0 \\ 0 & 0 & 0 & 0 \\ 1 & 0 & 0 & 0 \\ 1 & 0 & 0 & 0 \end{array}\right] \end{array}$

$\star$ (b) $M_R = \begin{array}{c} \\ a \\ b \\ c \\ d \end{array}\begin{array}{c} x\ y\ z\ w \\ \left[\begin{array}{cccc} 0 & 0 & 0 & 1 \\ 0 & 0 & 1 & 0 \\ 1 & 0 & 0 & 0 \\ 1 & 0 & 0 & 0 \end{array}\right] \end{array}$

(d) $M_R = \begin{array}{c} \\ a \\ b \\ c \\ d \end{array}\begin{array}{c} x\ y\ z\ w \\ \left[\begin{array}{cccc} 1 & 0 & 0 & 0 \\ 0 & 1 & 0 & 0 \\ 0 & 0 & 1 & 0 \\ 0 & 0 & 0 & 1 \end{array}\right] \end{array}$

7.5   New relations from old

Since binary relations from a set A to a set B are just subsets of the Cartesian product $A \times B$, we can talk about their intersection, union, complement and difference exactly the same way these operations are defined on sets. There is, however, a benefit of looking at these operations in terms of the underlined relations defined on the elements of A and B as in the following definition.

Definition 7.7. Let A, B be two sets, R and S two binary relations from A to B and let $a \in A$, $b \in B$ be arbitrary elements.

(a) The *intersection* of R and S is the relation from A to B, denoted by $R \cap S$, defined as follows: $a\,(R \cap S)\,b \Leftrightarrow aRb$ and aSb.

(b) The *union* of R and S is the relation from A to B, denoted by $R \cup S$, defined as follows: $a\,(R \cup S)\,b \Leftrightarrow aRb$ or aSb.

(c) The *difference* of R and S is the relation from A to B, denoted by $R\backslash S$, defined as follows: $a\,(R\backslash S)\,b \Leftrightarrow aRb$ and $a\,\not{S}\,b$.

(d) The *negation* of R is the relation from A to B, denoted by $\neg R$, defined as follows: $a\,(\neg R)\,b \Leftrightarrow a\,\not{R}\,b$.

Remark 7.3. If we look at the operations in Definition 7.7 as operations on sets, then the underlying universal set is $A \times B$. For instance, $\neg R$ is the same (as sets) as $(A \times B)\backslash R$.

Example 7.16. Let $A = \{1, 2, 3, 4\}$, $B = \{a, b, c\}$. Consider the following binary relations from A to B: $R = \{(1, b), (2, a), (3, a), (3, b), (4, c)\}$, $S = \{(1, a), (1, b), (2, b), (2, c), (3, a)\}$. Determine the relations $R \cap S$, $R \cup S$, $R\backslash S$, $S\backslash R$, $\neg R$ and $\neg S$ in terms of their elements.

Solution. (1) $R \cap S = \{(1, b), (3, a)\}$
(2) $R \cup S = \{(1, a), (1, b), (2, a), (2, b), (2, c), (3, a), (3, b), (4, c)\}$
(3) $R\backslash S = \{(2, a), (3, b), (4, c)\}$
(4) $S\backslash R = \{(1, a), (2, b), (2, c)\}$
(5) $\neg R = \{(1, a), (1, c), (2, b), (2, c), (3, c), (4, a), (4, b)\}$
(6) $\neg S = \{(1, c), (2, a), (3, b), (3, c), (4, a), (4, b), (4, c)\}$ $\Diamond$

Definition 7.8. Let $R \subseteq A \times B$ be a binary relation from A to B. The *inverse relation* of R, denoted by R^{-1}, is the relation from B to A defined as follows: $R^{-1} = \{(b, a) \in B \times A;\ (a, b) \in A \times B\}$. Alternatively, for any $y \in B$ and $x \in A$: $y\,R^{-1}\,x \Leftrightarrow x\,R\,y$.

Example 7.17. For the relations R and S given in Example 7.16, the inverses R^{-1} and S^{-1} are the following relations from B to A:

- $R^{-1} = \{(a, 2), (a, 3), (b, 1), (b, 3), (c, 4)\}$
- $S^{-1} = \{(a, 1), (a, 3), (b, 1), (b, 2), (c, 2)\}$.

Theorem 7.3. *Let R and S be two relations from A to B. Then:*

(a) $Dom\left(R^{-1}\right) = Ran\left(R\right)$

(b) $Ran\left(R^{-1}\right) = Dom\left(R\right)$

(c) $(R \cap S)^{-1} = R^{-1} \cap S^{-1}$

(d) $(R \cup S)^{-1} = R^{-1} \cup S^{-1}$

(e) $(\neg R)^{-1} = \neg\left(R^{-1}\right)$

(f) $\left(R^{-1}\right)^{-1} = R$

Proof. We prove parts (b) and (d). The other parts are left as exercises. For part (b), let $x \in A$ be an arbitrary element. Then

$$x \in \mathrm{Ran}\left(R^{-1}\right) \Leftrightarrow \exists y \in B;\ (y, x) \in R^{-1}$$
$$\Leftrightarrow \exists y \in B;\ (x, y) \in R$$
$$\Leftrightarrow x \in \mathrm{Dom}(R).$$

For part (d), consider $y \in B$ and $x \in A$. Then

$$(y, x) \in (R \cup S)^{-1} \Leftrightarrow (x, y) \in R \cup S$$
$$\Leftrightarrow (x, y) \in R \,\mathrm{or}\, (x, y) \in S$$
$$\Leftrightarrow (y, x) \in R^{-1} \,\mathrm{or}\, (y, x) \in S^{-1}$$
$$\Leftrightarrow (y, x) \in R^{-1} \cup S^{-1}. \qquad \square$$

In some applications, it is convenient and practical to represent relations between finite sets by their adjacency matrices. The following theorem gives a quick way to determine the adjacency matrix of each of the relations $R \cap S$, $R \cup S$, R^{-1} and $\neg R$ in terms of the adjacency matrices of R and S. Part (c) of the theorem uses the notation M^T to denote the *transpose* of an $m \times n$ matrix M: this is defined to be the $n \times m$ matrix whose rows are the same as the columns of M.

Theorem 7.4. *Let R and S be relations from A to B with adjacency matrices M_R and M_S respectively. Then:*

(a) $M_{R \cap S} = M_R \wedge M_S$ *(the meet of M_R and M_S).*

(b) $M_{R \cup S} = M_R \vee M_S$ *(the join of M_R and M_S).*

(c) $M_{R^{-1}} = (M_R)^T$.

(d) $M_{\neg R} = (M_R)'$ *(obtained from R by changing every 0 to 1 and every 1 to 0).*

Proof. We prove only parts (b) and (c), the other two parts are left as exercises. Write $M_R = [r_{ij}]$, $M_S = [s_{ij}]$ and $M_{R \cup S} = [t_{ij}]$. For $a_i \in A$ and $b_j \in B$, $a_i\,(R \cup S)\,b_j$ if and only if $a_i\,R\,b_j$ or $a_i\,S\,b_j$. Then $t_{ij} = 1$ if and only if $r_{ij} = 1$ or $s_{ij} = 1$. In others words $t_{ij} = r_{ij} \vee s_{ij}$. This proves part (b).

For part (c), write $M_{R^{-1}} = [v_{ij}]$. For $a_i \in A$ and $b_j \in B$, $a_i \, R \, b_j$ if and only if $b_j \left(R^{-1}\right) a_i$. In other words, $r_{ij} = 1$ if and only if $v_{ji} = 1$. This shows that $M_{R^{-1}} = (M_R)^T$. $\qquad\square$

Example 7.18. Two relations R and S from a set A containing four elements to a set B containing five elements are given in terms of their adjacency matrices:

$$
M_R = \begin{bmatrix} 1 & 0 & 1 & 0 & 0 \\ 0 & 1 & 0 & 1 & 1 \\ 0 & 0 & 0 & 1 & 0 \\ 1 & 0 & 0 & 0 & 0 \end{bmatrix}, \quad
M_S = \begin{bmatrix} 0 & 0 & 1 & 0 & 0 \\ 0 & 1 & 0 & 0 & 0 \\ 0 & 0 & 0 & 0 & 0 \\ 1 & 0 & 0 & 0 & 1 \end{bmatrix}.
$$

Find the adjacency matrix of each of the relations $R \cap S$, $R \cup S$, R^{-1}, $\neg R$, S^{-1} and $\neg S$.

Solution. We use the above theorem.

(a)

$$
M_{R \cap S} = M_R \wedge M_S = \begin{bmatrix} 1 & 0 & 1 & 0 & 0 \\ 0 & 1 & 0 & 1 & 1 \\ 0 & 0 & 0 & 1 & 0 \\ 1 & 0 & 0 & 0 & 0 \end{bmatrix} \wedge \begin{bmatrix} 0 & 0 & 1 & 0 & 0 \\ 0 & 1 & 0 & 0 & 0 \\ 0 & 0 & 0 & 0 & 0 \\ 1 & 0 & 0 & 0 & 1 \end{bmatrix} = \begin{bmatrix} 0 & 0 & 1 & 0 & 0 \\ 0 & 1 & 0 & 0 & 0 \\ 0 & 0 & 0 & 0 & 0 \\ 1 & 0 & 0 & 0 & 0 \end{bmatrix}.
$$

(b)

$$
M_{R \cup S} = M_R \vee M_S = \begin{bmatrix} 1 & 0 & 1 & 0 & 0 \\ 0 & 1 & 0 & 1 & 1 \\ 0 & 0 & 0 & 1 & 0 \\ 1 & 0 & 0 & 0 & 0 \end{bmatrix} \vee \begin{bmatrix} 0 & 0 & 1 & 0 & 0 \\ 0 & 1 & 0 & 0 & 0 \\ 0 & 0 & 0 & 0 & 0 \\ 1 & 0 & 0 & 0 & 1 \end{bmatrix} = \begin{bmatrix} 1 & 0 & 1 & 0 & 0 \\ 0 & 1 & 0 & 1 & 1 \\ 0 & 0 & 0 & 1 & 0 \\ 1 & 0 & 0 & 0 & 1 \end{bmatrix}.
$$

(c)

$$
M_{R^{-1}} = (M_R)^T = \begin{bmatrix} 1 & 0 & 0 & 1 \\ 0 & 1 & 0 & 0 \\ 1 & 0 & 0 & 0 \\ 0 & 1 & 1 & 0 \\ 0 & 1 & 0 & 0 \end{bmatrix}, \quad
M_{\neg R} = (M_R)' = \begin{bmatrix} 0 & 1 & 0 & 1 & 1 \\ 1 & 0 & 1 & 0 & 0 \\ 1 & 1 & 1 & 0 & 1 \\ 0 & 1 & 1 & 1 & 1 \end{bmatrix}.
$$

(d)

$$
M_{S^{-1}} = (M_S)^T = \begin{bmatrix} 0 & 0 & 0 & 1 \\ 0 & 1 & 0 & 0 \\ 1 & 0 & 0 & 0 \\ 0 & 0 & 0 & 0 \\ 0 & 0 & 0 & 1 \end{bmatrix}, \quad
M_{\neg S} = (M_S)' = \begin{bmatrix} 1 & 1 & 0 & 1 & 1 \\ 1 & 0 & 1 & 1 & 0 \\ 1 & 1 & 1 & 1 & 1 \\ 0 & 1 & 1 & 1 & 0 \end{bmatrix}.
$$

$$\diamond$$

7.5.1 *Composition of relations*

Like the case of functions, the composition of relations is at the heart of many important applications.

Definition 7.9. Let A, B and C be three sets, R a relation from A to B and S a relation from B to C. The *composition* of R and S, denoted by $S \circ R$, is the relation from A to C defined by:

$$S \circ R = \{(a,c) \in A \times C;\ (a,b) \in R \text{ and } (b,c) \in S \text{ for some } b \in B\}.$$

Some remarks are in order before we look at some examples.

(1) The reader might find it more natural to use the notation $R \circ S$ rather than $S \circ R$ in Definition 7.9. The order $S \circ R$ we use in this book is consistent with the notations in Chapter 5 about the composition of functions.
(2) When the sets A, B and C are finite, the elements of $S \circ R$ can be determined systematically as follows: for each pair $(a,b) \in R$, look for pairs in S of the form (b,c). Then pair (a,c) is an element of $S \circ R$.
(3) Unlike the composition of functions, we still can define $S \circ R$ even if the domain of S is not the same as the range of R. More specifically, if $R \subseteq A \times B$ and $S \subseteq C \times D$ are relations, then we can look at R and S as subsets of $A \times (B \cup C)$ and $(B \cup C) \times D$, respectively. Hence, we can talk about $S \circ R$ as the subset of $A \times C$ as defined above. But in all what follows, we always assume the setting of Definition 7.9.

Example 7.19. Let $A = \{1,2,3\}$, $B = \{a,b,c\}$ and $C = \{\gamma, \uplus, \bullet\}$. Relations R and S from A to B and from B to C (respectively) are given as follows:

- $R = \{(1,a),(1,b),(2,c),(3,a),(3,b)\}$.
- $S = \{(a,\uplus),(a,\bullet),(b,\gamma),(b,\bullet),(c,\gamma),(c,\uplus)\}$.

To find the elements of the relation $S \circ R$ from A to C, we look for pairs $(a,b) \in R$ and $(b,c) \in S$ and then we include (a,c) in $S \circ R$. The elements of $S \circ R$ are then: $(1,\uplus)$, $(1,\bullet)$, $(1,\gamma)$, $(2,\gamma)$, $(2,\uplus)$, $(3,\uplus)$, $(3,\bullet)$ and $(3,\gamma)$.

Example 7.20. Two relations R and S on $\mathbb{R}$ are defined as follows: $a\,R\,b \Leftrightarrow b = a^2$ and $a\,S\,b \Leftrightarrow b = 2^a$. In this example, we find expressions for $S \circ R$ and $R \circ S$ and we give two elements in each of the two relations. Let $a, c \in \mathbb{R}$.

Then:

$$a\,(S \circ R)\,c \Leftrightarrow a\,R\,b \wedge b\,S\,c \text{ for some } b \in \mathbb{R}$$
$$\Leftrightarrow b = a^2 \text{ and } c = 2^b \text{ for some } b \in \mathbb{R}$$
$$\Leftrightarrow c = 2^{a^2}$$

So $a\,(S \circ R)\,c \Leftrightarrow c = 2^{a^2}$. On the other hand:

$$a\,(R \circ S)\,c \Leftrightarrow a\,S\,b \wedge b\,R\,c \text{ for some } b \in \mathbb{R}$$
$$\Leftrightarrow b = 2^a \wedge c = b^2 \text{ for some } b \in \mathbb{R}$$
$$\Leftrightarrow c = (2^a)^2$$
$$\Leftrightarrow c = 2^{2a}$$

So $a\,(R \circ S)\,c \Leftrightarrow c = 2^{2a}$. For $a = -1$, $2^{a^2} = 2$ and $2^{2a} = \frac{1}{4}$. For $a = 1$, $2^{a^2} = 2$ and $2^{2a} = 4$. So $(-1, 2)$ and $(1, 2)$ are elements of $S \circ R$ and $\left(-1, \frac{1}{4}\right)$ and $(1, 4)$ are elements of $R \circ S$.

Remark 7.4. Like in the case of functions, composition of relations is not a commutative operation. For instance, $R \circ S \neq S \circ R$ in Example 7.20 since $\left(-1, \frac{1}{4}\right)$ is an element of $R \circ S$ but not an element of $S \circ R$. The associativity of relations composition, on the other hand, holds as stated in the next theorem. As a consequence, one can write $T \circ S \circ R$ with no need of parentheses.

Theorem 7.5. *The composition of relations is an associative operation. In other words, if $R \subseteq A \times B$, $S \subseteq B \times C$ and $T \subseteq C \times D$ are three relations then $T \circ (S \circ R) = (T \circ S) \circ R$.*

Proof. Since we are dealing with equality of sets, it is wise to start by verifying that $T \circ (S \circ R)$ and $(T \circ S) \circ R$ are indeed subsets of the same set. Since $S \circ R$ is a subset of $A \times C$ and T is a subset of $C \times D$, $T \circ (S \circ R)$ is a subset of $A \times D$. Now, R is a subset of $A \times B$ and $T \circ S$ is a subset of $B \times D$ and, so $(T \circ S) \circ R$ is a subset of $A \times D$. We now proceed to prove the equality of the two sets. Let $(a, d) \in A \times D$:

$$(a, d) \in T \circ (S \circ R) \Leftrightarrow (\exists c \in C)\,((a, c) \in S \circ R \wedge (c, d) \in T)$$
$$\Leftrightarrow (\exists c \in C)\,((\exists b \in B)\,((a, b) \in R \wedge (b, c) \in S) \wedge (c, d) \in T)$$
$$\Leftrightarrow (\exists b \in B)(\exists c \in C)\,((a, b) \in R \wedge (b, c) \in S \wedge (c, d)).\ (*)$$

A similar argument shows that the statement $(a, d) \in (T \circ S) \circ R$ is also equivalent to the expression $(*)$ above. The result follows. $\square$

Let A, B and C be finite sets. Given two relations $R \subseteq A \times B$, $S \subseteq B \times C$ defined by their adjacency matrices M_R and M_S respectively, the following theorem determines how to find the adjacency matrix of the composition relation $S \circ R$ in terms of M_R and M_S.

Theorem 7.6. *Let A, B and C be finite sets, R a relation from A to B and S a relation from B to C. If M_R and M_S are the respective adjacency matrices of R and S, then the adjacency matrix of the composite relation $S \circ R$ is the Boolean product $M_R \odot M_S$ of M_R and M_S: $M_{S \circ R} = M_R \odot M_S$.*

Proof. Write $A = \{a_1, \ldots, a_m\}$, $B = \{b_1, \ldots, b_p\}$ and $C = \{c_1, \ldots, c_n\}$. Then M_R is of size $m \times p$ and M_S is of size $p \times n$. By definition of the composition of relations, the ordered pair (a_i, c_j) of $A \times C$ is in $S \circ R$ if and only if there exists an element $b_k \in B$ such that $(a_i, b_k) \in R$ and $(b_k, c_j) \in S$. This means that the entry at the (i, j) position in the matrix $M_{S \circ R}$ is equal to 1 if and only if there exists $k \in \{1, \ldots, p\}$ such that the entry at the (i, k) position in M_R is equal to 1 and the same is true for the entry at the (k, j) position in M_S. This is exactly the definition of the (i, j) entry in the matrix $M_R \odot M_S$. $\qquad\square$

Example 7.21. Let $A = \{a_1, a_2, a_3\}$, $B = \{b_1, b_2, b_3, b_4\}$ and $C = \{c_1, c_2\}$. Binary relations R and S from A to B and from B to C (respectively) are defined as follows: $R = \{(a_1, b_1), (a_1, b_4), (a_2, b_3), (a_3, b_1), (a_3, b_2)\}$ and $S = \{(b_1, c_1), (b_1, c_2), (b_4, c_2)\}$. Find the adjacency matrix of $S \circ R$ using the adjacency matrices of R and S. List all elements of $S \circ R$.

Solution. We start by writing the matrices M_R and M_S:

$$
M_R = \begin{array}{c} \\ a_1 \\ a_2 \\ a_3 \end{array}
\begin{array}{c} b_1\ b_2\ b_3\ b_4 \\ \left[\begin{array}{cccc} 1 & 0 & 0 & 1 \\ 0 & 0 & 1 & 0 \\ 1 & 1 & 0 & 0 \end{array}\right] \end{array},
\quad
M_S = \begin{array}{c} \\ b_1 \\ b_2 \\ b_3 \\ b_4 \end{array}
\begin{array}{c} c_1\ c_2 \\ \left[\begin{array}{cc} 1 & 1 \\ 0 & 0 \\ 0 & 0 \\ 0 & 1 \end{array}\right] \end{array}.
$$

So

$$
M_{S \circ R} = M_R \odot M_S =
\begin{bmatrix} 1 & 0 & 0 & 1 \\ 0 & 0 & 1 & 0 \\ 1 & 1 & 0 & 0 \end{bmatrix}
\odot
\begin{bmatrix} 1 & 1 \\ 0 & 0 \\ 0 & 0 \\ 0 & 1 \end{bmatrix}
=
\begin{bmatrix} 1 & 1 \\ 0 & 0 \\ 1 & 1 \end{bmatrix}.
$$

From $M_{S \circ R}$, we get all the elements of $S \circ R$:

$$
S \circ R = \{(a_1, c_1), (a_1, c_2), (a_3, c_1), (a_3, c_2)\}. \qquad \Diamond
$$

Theorem 7.7. *If $R \subseteq A \times B$ and $S \subseteq B \times C$ are two relations then $(S \circ R)^{-1} = R^{-1} \circ S^{-1}$.*

Proof. Note first that $S \circ R$ is a subset of $A \times C$, so $(S \circ R)^{-1}$ is a subset of $C \times A$. Same is true for the composite $R^{-1} \circ S^{-1}$. Let $(c, a) \in C \times A$.

$$
\begin{aligned}
(c, a) \in (S \circ R)^{-1} &\Leftrightarrow (a, c) \in S \circ R \\
&\Leftrightarrow (\exists b \in B)((a, b) \in R \wedge (b, c) \in S) \\
&\Leftrightarrow (\exists b \in B)((b, a) \in R^{-1} \wedge (c, b) \in S^{-1}) \\
&\Leftrightarrow (c, a) \in R^{-1} \circ S^{-1}.
\end{aligned}
$$

$\square$

Given a binary relation R on a set A and $n \in \mathbb{N}$, then $R^{[n]} = R \circ R \circ \cdots \circ R$ is the composition of R with itself n times. Here we use the convention that $R^{[0]} = I_A = \{(x, x); \, x \in A\}$ that we call the *identity relation* on A. Note that $R^{[n]}$ can be defined recursively as follows: $R^{[0]} = I_A$ and $R^{[n+1]} = R^{[n]} \circ R$ for $n \geq 0$.

Example 7.22. On the set $A = \{a, b, c, d\}$, a relation R is defined as follows: $R = \{(a, a), (b, a), (c, b), (d, c)\}$. Find $R^{[2]}$, $R^{[3]}$ and $R^{[4]}$. What is $R^{[n]}$ for $n \geq 3$.

Solution. The adjacency matrix of R is

$$
M_R = \begin{array}{c} \\ a \\ b \\ c \\ d \end{array}
\begin{array}{c} a\ b\ c\ d \\
\begin{bmatrix}
1 & 0 & 0 & 0 \\
1 & 0 & 0 & 0 \\
0 & 1 & 0 & 0 \\
0 & 0 & 1 & 0
\end{bmatrix}
\end{array}
$$

The adjacency matrix of $R^{[2]}$ is $M_R \odot M_R$:

$$
M_R \odot M_R =
\begin{bmatrix}
1 & 0 & 0 & 0 \\
1 & 0 & 0 & 0 \\
0 & 1 & 0 & 0 \\
0 & 0 & 1 & 0
\end{bmatrix}
\odot
\begin{bmatrix}
1 & 0 & 0 & 0 \\
1 & 0 & 0 & 0 \\
0 & 1 & 0 & 0 \\
0 & 0 & 1 & 0
\end{bmatrix}
=
\begin{bmatrix}
1 & 0 & 0 & 0 \\
1 & 0 & 0 & 0 \\
1 & 0 & 0 & 0 \\
0 & 1 & 0 & 0
\end{bmatrix}.
$$

This gives us $R^{[2]} = \{(a, a), (b, a), (c, a), (d, b)\}$. Since $R^{[3]} = R^{[2]} \circ R$, the adjacency matrix of $R^{[3]}$ is $M_R \odot M_{R^{[2]}}$:

$$
M_R \odot M_{R^{[2]}} =
\begin{bmatrix}
1 & 0 & 0 & 0 \\
1 & 0 & 0 & 0 \\
0 & 1 & 0 & 0 \\
0 & 0 & 1 & 0
\end{bmatrix}
\odot
\begin{bmatrix}
1 & 0 & 0 & 0 \\
1 & 0 & 0 & 0 \\
1 & 0 & 0 & 0 \\
0 & 1 & 0 & 0
\end{bmatrix}
=
\begin{bmatrix}
1 & 0 & 0 & 0 \\
1 & 0 & 0 & 0 \\
1 & 0 & 0 & 0 \\
1 & 0 & 0 & 0
\end{bmatrix}.
$$

So $R^{[3]} = \{(a,a), (b,a), (c,a), (d,a)\}$. A quick calculation shows that $M_{R^{[4]}} = M_{R^{[3]}}$ and so $R^{[4]} = R^{[3]}$. It follows that $R^{[5]} = R^{[4]} \odot R = R^{[3]} \odot R = R^{[4]} = R^{[3]}$. A simple proof by induction shows that $R^{[n]} = R^{[3]}$ for any $n \geq 3$. $\diamond$

Theorem 7.8. *Given a binary relation R on a set A and any non-negative integers $m, n \in$, then $R^{[m+n]} = R^{[m]} \circ R^{[n]}$ and $\left(R^{[m]}\right)^{[n]} = R^{[mn]}$.*

Proof. Left as an exercise for the reader (see Exercise (34) below). $\square$

Theorem 7.8 shows in particular that the composition of powers of a relation R is commutative: $R^{[m]} \circ R^{[n]} = R^{[n]} \circ R^{[m]} = R^{[m+n]}$ for any non-negative integers m, n.

7.5.2 *Exercises*

$\star$(1) Let $A = \{1, 2, 3\}$, $B = \{a, b, c, d\}$. Two relations from A to B are given as follows: $R = \{(1, a), (1, b), (1, c), (2, b), (3, c), (3, d)\}$ and $S = \{(1, a), (1, c), (1, d), (2, a), (2, d), (3, a), (3, d)\}$. Give the elements of $R \cap S$, $R \cup S$, $R \backslash S$, $S \backslash R$, $\neg R$, $\neg S$, R^{-1} and S^{-1}.

(2) Let A be a group of people. Relations R and S on A are defined as follows: $a\,R\,b$ if and only if a is a sibling of b; $a\,S\,b$ if and only if a is younger than b. Describe in words the relations $R \cap S$ and $R \cup S$ on A.

$\star$(3) Let A be the set of all humans. A binary relation R on A is defined as follows: $a\,R\,b$ if and only if a is a parent of b. Describe in words the meaning of $a\,R^2\,b$ and $a\,R^{-1}\,b$ for $a, b \in A$.

$\star$(4) Let A be the set of all commercial airlines operating in North America and let B the set of all North American cities with a commercial airport. We define two relations R and S from A to B as follows: $a\,R\,b$ if and only if airline a operates a direct flight from Ottawa to b; $a\,S\,b$ if and only if airline a has only indirect flights from Ottawa to b (with at least one stop in a city other than b). Describe in words the meaning of each of the following relations: $R \cap S$, $R \cup S$, $R \backslash S$ and $S \backslash R$.

(5) Consider the standard relations $>$ (strictly greater than) and $<$ (strictly smaller than) on the set $\mathbb{R}$ of real numbers. Give a description of each of the relations $\neg >$, $\neg <$, $> \cap <$, $> \cup <$, $>^{-1}$ and $<^{-1}$.

(6) Given two relations R and S from A to B, we can talk about their *symmetric difference* as subsets of $A \times B$: $R \oplus S = (R \cup S) \backslash (R \cap S)$. Find $R \oplus S$ and $S \oplus R$ for the relations R and S in Exercise (1).

$\star$(7) A relation R on $\mathbb{Z}$ is defined by: $R = \{(a,b) \in \mathbb{Z} \times \mathbb{Z};\ a - b \text{ is even}\}$. Describe the relations R^{-1} and $\neg R$.

(8) Two relations R and S are defined on the set $\mathbb{R}$ as follows: $a\,R\,b \Leftrightarrow b = 2^k a$ for some $k \in \mathbb{Z}$ and $a\,S\,b \Leftrightarrow b - a$ is a rational number. Determine the truth value of each of the following statements.

$\star$ (a) $1\,R^{-1}\,3$

(b) $1\,S^{-1}\,3$

$\star$ (c) $1\,(R \cap S)\,3$

(d) $1\,(R \cup S)\,3$

(e) $\frac{1}{2}\,R^{-1}\,\frac{1}{16}$

(f) $\frac{1}{2}\,S^{-1}\,\frac{1}{16}$

(g) $\frac{1}{2}\,(R \cap S)\,\frac{1}{16}$

$\star$ (h) $\frac{1}{2}\,(R \cup S)\,\frac{1}{16}$

(9) On the set $A = \{a, b, c, d\}$, two relations R and S are given by their adjacency matrices:

$$
M_R = \begin{array}{c@{}c} & \begin{array}{cccc} a & b & c & d \end{array} \\ \begin{array}{c} a \\ b \\ c \\ d \end{array} & \left[\begin{array}{cccc} 1 & 1 & 1 & 0 \\ 0 & 0 & 1 & 0 \\ 0 & 0 & 0 & 1 \\ 1 & 0 & 0 & 1 \end{array}\right] \end{array},\quad
M_S = \begin{array}{c@{}c} & \begin{array}{cccc} a & b & c & d \end{array} \\ \begin{array}{c} a \\ b \\ c \\ d \end{array} & \left[\begin{array}{cccc} 0 & 0 & 1 & 1 \\ 1 & 0 & 1 & 0 \\ 0 & 1 & 0 & 1 \\ 1 & 0 & 1 & 1 \end{array}\right] \end{array}
$$

Determine the truth value of each of the following statements.

(a) $a\,R^{-1}\,d$

$\star$ (b) $c\,S^{-1}\,a$

$\star$ (c) $a\,(R \cap S)\,b$

(d) $a\,(R \cup S)\,b$

(e) $b\,(R^{-1} \cap S^{-1})\,d$

(f) $c\,(S^{-1} \cup R)\,c$

$\star$ (g) $d\,(R \cap S)^{-1}\,b$

(h) $a\,(R \backslash S)\,b$

(10) On the set $A = \{1, 2, 3, 4, 5\}$, relations R and S are given by their digraph representations. In each case, find the digraph representation of each of the following relations: R^{-1}, S^{-1}, $R \cap S$, $R \cup S$, $R \backslash S$ and $S \backslash R$.

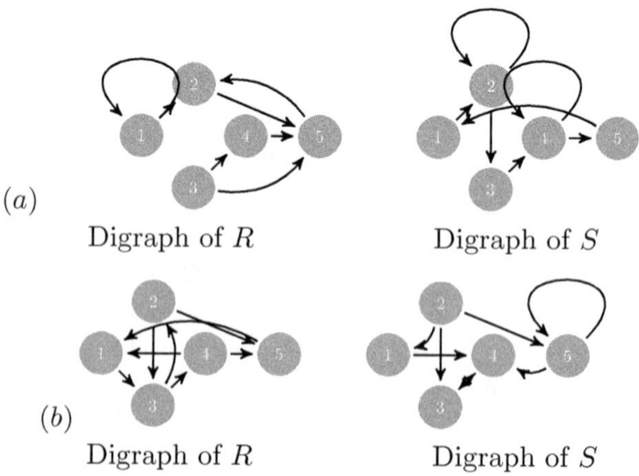

(a)   Digraph of R Digraph of S

(b)   Digraph of R Digraph of S

(11) Prove part (a) of Theorem 7.3.

(12) Let R and S be two relations from A to B. Prove:

   $\star$ (a) $\mathrm{Dom}\,(R \cap S) \subseteq \mathrm{Dom}\,(R) \cap \mathrm{Dom}\,(S)$.

     (b) $\mathrm{Dom}\,(R \cup S) = \mathrm{Dom}\,(R) \cup \mathrm{Dom}\,(S)$.

     (c) $\mathrm{Ran}\,(R \cap S) \subseteq \mathrm{Ran}\,(R) \cap \mathrm{Ran}\,(S)$.

     (d) $\mathrm{Ran}\,(R \cup S) = \mathrm{Ran}\,(R) \cup \mathrm{Ran}\,(S)$.

(13) Give an example of two sets A and B, and two relations R and S from A to B such that $\mathrm{Dom}\,(R \cap S) \neq \mathrm{Dom}\,(R) \cap \mathrm{Dom}\,(S)$.

$\star$(14) Give an example of two sets A and B, and two relations R and S from A to B such that $\mathrm{Ran}\,(R \cap S) \neq \mathrm{Ran}\,(R) \cap \mathrm{Ran}\,(S)$.

(15) R and S are two relations from A to B. Prove:

     (a) $\left(R^{-1}\right)^{-1} = R$

     (b) $(\neg R)^{-1} = \neg\left(R^{-1}\right)$

   $\star$ (c) $(R \cap S)^{-1} = R^{-1} \cap S^{-1}$

(16) Let $A = \{a, b, c, d\}$ and $B = \{x, y, z, w\}$. Two relations R and S from A to B are given by their adjacency matrices M_R and M_S respectively. Find the adjacency matrix of each of the following relations: R^{-1}, S^{-1}, $R \cap S$, $R \cup S$, $R \backslash S$ and $S \backslash R$.

$$
\star(a)\quad M_R = \begin{array}{c} \\ a \\ b \\ c \\ d \end{array}\begin{bmatrix} x & y & z & w \\ 1 & 1 & 1 & 0 \\ 0 & 0 & 1 & 0 \\ 0 & 0 & 0 & 1 \\ 1 & 0 & 0 & 1 \end{bmatrix}, \qquad
M_S = \begin{array}{c} \\ a \\ b \\ c \\ d \end{array}\begin{bmatrix} x & y & z & w \\ 0 & 0 & 0 & 1 \\ 0 & 0 & 1 & 0 \\ 1 & 0 & 0 & 0 \\ 1 & 0 & 0 & 0 \end{bmatrix}
$$

$$
(b)\quad M_R = \begin{array}{c} \\ a \\ b \\ c \\ d \end{array}\begin{bmatrix} x & y & z & w \\ 0 & 0 & 1 & 0 \\ 1 & 0 & 1 & 0 \\ 1 & 0 & 0 & 1 \\ 1 & 0 & 1 & 1 \end{bmatrix}, \qquad
M_S = \begin{array}{c} \\ a \\ b \\ c \\ d \end{array}\begin{bmatrix} x & y & z & w \\ 1 & 0 & 1 & 1 \\ 1 & 0 & 1 & 0 \\ 0 & 0 & 0 & 0 \\ 1 & 1 & 1 & 1 \end{bmatrix}
$$

(17) Let R be a binary relation on a set A. Describe the relations $R \circ \emptyset$ and $\emptyset \circ R$ on A.

(18) Let $>$ and $=$ denote the standard "strictly greater than" and "equal" on the set $\mathbb{R}$ of all real numbers. Describe in words the meaning of the relation $> \circ = $ (the composition of $>$ and $=$).

$\star$(19) On the set A of all humans, we define two binary relations R and S as follows: $a\,R\,b$ if and only if a is a daughter of b and $a\,S\,b$ if and only if a is a sibling of b. Describe in words the meaning of the relations $S \circ R$ and $R \circ S$.

(20) On the set $A = \{a, b, c, d\}$, relations R and S are given as follows:

- $R = \{(a, a), (a, b), (a, c), (b, b), (c, c), (c, d), (d, b)\}$.
- $S = \{(a, a), (a, c), (a, d), (b, a), (b, d), (c, a), (d, a), (d, d)\}$.

Find the relations $R \circ R$, $S \circ S$, $R \circ S$ and $S \circ R$ in terms of their elements.

(21) Let R be a relation on a non-empty set A and let $I_A = \{(x, x);\ x \in A\}$ be the identity relation on A.

   (a) Prove that $I_A \circ R = R \circ I_A = R$.

   (b) Is it true that either $R \circ R^{-1}$ or $R^{-1} \circ R$ must be equal to I_A? Justify your answer.

(22) On the set $\mathbb{R}$, relations R and S are defined as follows: $m\,R\,n \Leftrightarrow n = m^2$ and $m\,S\,n \Leftrightarrow n = m^8$. In each case, determine the truth value of the statement. Justify your answer.

$\star$ (a) $1\,(R \circ S) - 1$	(e) $1\,(S \circ R) - 1$	$\star$ (i) $-1\,(R \circ R)\,1$
(b) $-1\,(R \circ S)\,1$	(f) $1\,(S \circ R)\,1$	(j) $\sqrt{3}\,(R \circ R)\,9$
(c) $-1\,(R \circ S) - 1$	(g) $-1\,(S \circ R) - 1$	(k) $\sqrt{2}\,(S \circ R)\,256$
(d) $0\,(R \circ S)\,0$	$\star$ (h) $0\,(S \circ R)\,0$	$\star$ (l) $-\sqrt[16]{2}\,(S \circ S)\,16$

(23) With the same notations as in Exercise (22), prove that $R \circ S = S \circ R$.

(24) Let A, B and C be three sets, $R \subseteq A \times B$, $S \subseteq B \times C$ and $T \subseteq B \times C$ three binary relations. Prove:

   $\star$ (a) $(S \cup T) \circ R = (S \circ R) \cup (T \circ R)$

   (b) $(S \cap T) \circ R \subseteq (S \circ R) \cap (T \circ R)$

(25) With the notations of Exercise (24), give an example to show that

$$(S \cap T) \circ R \neq (S \circ R) \cap (T \circ R).$$

$\star$(26) Let R be a binary relation on a set A. Given $n \in \mathbb{N}$, define a new relation $R^{[-n]}$ on A as follows: $R^{[-n]} = \left(R^{-1}\right)^{[n]}$. For the relation $R = \{(a, b), (a, d), (b, a), (d, b)\}$ on $A = \{a, b, c, d\}$, find $R^{[-2]}$ and $R^{[-3]}$.

(27) Let $A = \{a, b, c, d\}$, $B = \{1, 2, 3\}$ and $C = \{x, y, z, w\}$. Two relations R and S from A to B and from B to C respectively are given by their digraphs as shown below. In each case, give the elements of $S \circ R$.

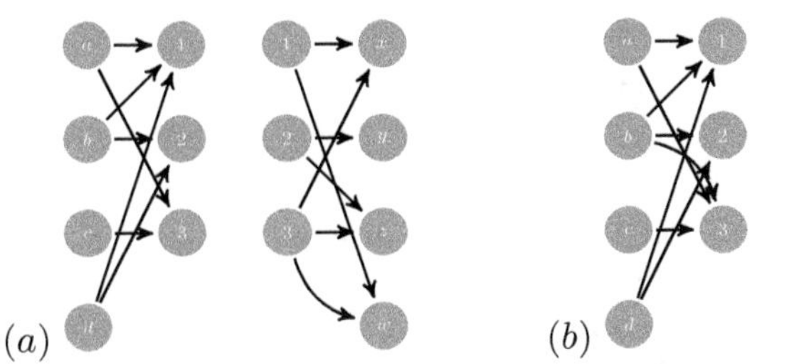

(a) Digraph of R Digraph of S (b) Digraph of R Digraph of S

⋆(28) Let $A = \{a_1, a_2, a_3, a_4\}$, $B = \{b_1, b_2, b_3\}$ and $C = \{c_1, c_2, c_3\}$. Binary relations R from A to B and S from B to C are defined as follows:
$R = \{(a_1, b_2), (a_1, b_3), (a_2, b_1), (a_2, b_3), (a_3, b_1), (a_3, b_2), (a_4, b_3)\}$ and
$S = \{(b_1, c_1), (b_1, c_3), (b_2, c_1), (b_2, c_2), (b_3, c_2)\}$. Using the adjacency matrices of R and S, find the adjacency matrix of the relation $S \circ R$ and list all elements of this relation.

(29) A binary relation R on the set $A = \{a, b, c, d\}$ is given by its adjacency matrix M_R:

$$M_R = \begin{array}{c} \\ a \\ b \\ c \\ d \end{array} \begin{array}{c} a\ b\ c\ d \\ \left[\begin{array}{cccc} 1 & 1 & 1 & 0 \\ 0 & 0 & 1 & 0 \\ 0 & 0 & 0 & 1 \\ 1 & 0 & 0 & 1 \end{array}\right] \end{array}.$$

Determine the truth value of each of the following.

(a) aRa (d) dRc (g) $bR^{[2]}c$ (j) $bR^{[3]}b$
(b) aRc (e) $aR^{[2]}a$ (h) $dR^{[2]}a$ (k) $cR^{[3]}d$
(c) bRa (f) $aR^{[2]}d$ (i) $aR^{[3]}d$ (l) $bR^{[3]}c$

(30) On the set $A = \{a, b, c, d\}$, two relations R and S are given by their adjacency matrices:

$$M_R = \begin{array}{c} \\ a \\ b \\ c \\ d \end{array} \begin{array}{c} a\ b\ c\ d \\ \left[\begin{array}{cccc} 1 & 1 & 1 & 0 \\ 0 & 0 & 1 & 0 \\ 0 & 0 & 0 & 1 \\ 1 & 0 & 0 & 1 \end{array}\right] \end{array}, \quad M_S = \begin{array}{c} \\ a \\ b \\ c \\ d \end{array} \begin{array}{c} a\ b\ c\ d \\ \left[\begin{array}{cccc} 0 & 0 & 1 & 1 \\ 1 & 0 & 1 & 0 \\ 0 & 1 & 0 & 1 \\ 1 & 0 & 1 & 1 \end{array}\right] \end{array}.$$

Determine the truth value of each of the following.

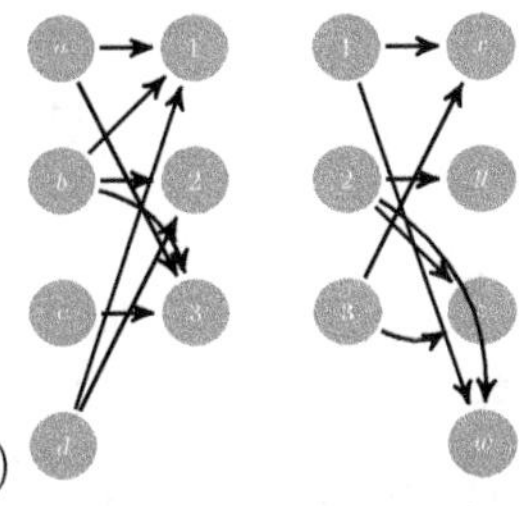

$\star$ (a) $a\,(R \circ S)\,d$ $\star$ (d) $b\,(R \circ S)^{-1}\,a$ (g) $d\,(R^{-1} \circ S)^{-1}\,b$

(b) $c\,(S^{-1} \circ R^{-1})\,a$ $\star$ (e) $b\,(R^{-1} \circ S^{-1})\,d$ (h) $a\,(R \circ S^{-1})\,b$

(c) $a\,(S \circ S)\,b$ (f) $c\,(S^{-1} \circ R)\,c$

(31) $R = \{(a,a),(a,b),(b,c),(c,c),(c,d),(d,d)\}$ is a binary relation defined on the set $A = \{a,b,c,d\}$. Draw a directed graph for each of the relations R, $R^{[2]}$ and $R^{[3]}$.

(32) Consider the three sets $A = \{a,b,c,d\}$, $B = \{x,y,z,w\}$ and $C = \{1,2,3\}$. Relations R from A to B and S from B to C are given by their adjacency matrices M_R and M_S respectively. In each case, find the adjacency matrix of the relation $S \circ R$. Give all the elements of $S \circ R$.

$$\star\text{(a)}\quad M_R = \begin{array}{c} \\ a \\ b \\ c \\ d \end{array}\begin{array}{c} x\ y\ z\ w \\ \left[\begin{array}{cccc} 1 & 1 & 1 & 0 \\ 0 & 0 & 1 & 0 \\ 0 & 0 & 0 & 1 \\ 1 & 0 & 0 & 1 \end{array}\right] \end{array},\quad M_S = \begin{array}{c} \\ x \\ y \\ z \\ w \end{array}\begin{array}{c} 1\ 2\ 3 \\ \left[\begin{array}{ccc} 0 & 0 & 0 \\ 0 & 0 & 1 \\ 1 & 0 & 0 \\ 1 & 0 & 0 \end{array}\right] \end{array}$$

$$\text{(b)}\quad M_R = \begin{array}{c} \\ a \\ b \\ c \\ d \end{array}\begin{array}{c} x\ y\ z\ w \\ \left[\begin{array}{cccc} 0 & 0 & 1 & 0 \\ 0 & 0 & 1 & 0 \\ 1 & 0 & 0 & 1 \\ 1 & 0 & 0 & 1 \end{array}\right] \end{array},\quad M_S = \begin{array}{c} \\ x \\ y \\ z \\ w \end{array}\begin{array}{c} 1\ 2\ 3 \\ \left[\begin{array}{ccc} 0 & 0 & 1 \\ 0 & 0 & 0 \\ 0 & 0 & 1 \\ 1 & 0 & 1 \end{array}\right] \end{array}$$

(33) Let $A = \{a,b,c,d\}$. Two relations R and S on A are given by their adjacency matrices M_R and M_S respectively as follows:

$$M_R = \begin{array}{c} \\ a \\ b \\ c \\ d \end{array}\begin{array}{c} a\ b\ c\ d \\ \left[\begin{array}{cccc} 1 & 1 & 1 & 0 \\ 0 & 0 & 1 & 0 \\ 0 & 0 & 0 & 1 \\ 1 & 0 & 0 & 1 \end{array}\right] \end{array},\quad M_S = \begin{array}{c} \\ a \\ b \\ c \\ d \end{array}\begin{array}{c} a\ b\ c\ d \\ \left[\begin{array}{cccc} 0 & 0 & 0 & 1 \\ 0 & 0 & 1 & 0 \\ 1 & 0 & 0 & 0 \\ 1 & 0 & 0 & 0 \end{array}\right] \end{array}$$

Find the adjacency matrix of each of the relations $S^{-1} \circ R$, $R^{-1} \circ S$ and $(S \circ R)^{-1}$.

(34) Prove Theorem 7.8 using induction.

$\star$(35) On the set $\mathbb{Z}$, we define a relation R as follows: $m\,R\,n \Leftrightarrow m+n$ is odd. Find $R^{[2]}$, $R^{[3]}$ and $R^{[4]}$. In general, what is $R^{[n]}$ for $n \geq 1$?

(36) On the set $\mathbb{R}$, we define a relation R as follows: $x\,R\,y \Leftrightarrow y = x+1$. Find $R^{[2]}$ and $R^{[3]}$. What is $R^{[n]}$ for $n \geq 0$? Prove your answer using induction on n.

7.6 Paths and connectivity

We start by defining the notion of a path in a relation R on a set A. We will revisit this notion in the context of graph theory in Chapter 9.

Definition 7.10. Let R be a binary relation on a set A, $a, b \in A$ and $n \in \mathbb{N}$. A *path* of length n from a to b in R is a *finite* sequence: (x_0, x_1), (x_1, x_2), ..., (x_{k-1}, x_k), ..., (x_{n-1}, x_n) of elements of R with $x_0 = a$ and $x_n = b$. A path of length n from a to b can be represented by a sequence $a, x_1, \ldots, x_{n-1}, b$ of $n + 1$ elements in A such that $a \, R \, x_1$, $x_1 \, R \, x_2$, ..., $x_{n-1} \, R \, b$. If $a = b$, then the path is called a *cycle*. Elements a, b are called the initial and the terminal points (respectively) of the path.

Definition 7.10 gives a rather abstract way of defining a path in a relation. Given a finite set A and a binary relation R on A, the digraph G_R representing R gives a more visual way of a path in R that actually justifies the name. In G_R, a path from vertex a to vertex b is just a sequence of directed edges $v_0 = a \to v_1 \to v_2 \to \cdots \to v_{n-1} \to v_n = b$ such that the first edge starts at a, the last edge ends at b and for each edge, the terminal vertex is the same as the initial vertex of the following edge. With this interpretation of a path, the length is just the number of edges used in the path. Clearly, a path of length 1 is just one edge in the graph.

Example 7.23. On the set $A = \{a, b, c, d, e\}$, define a relation R with elements: (a, b), (a, c), (b, a), (b, c), (b, e), (c, d), (c, e), (d, a), (d, e), (e, b) and (e, d). The digraph representation of R is as follows:

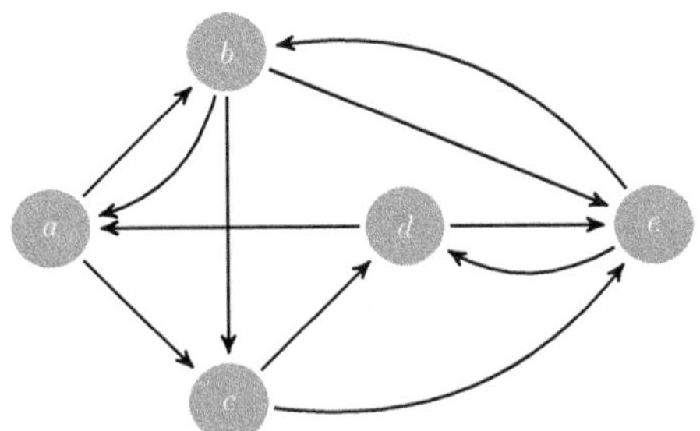

The following diagram shows four paths in R:

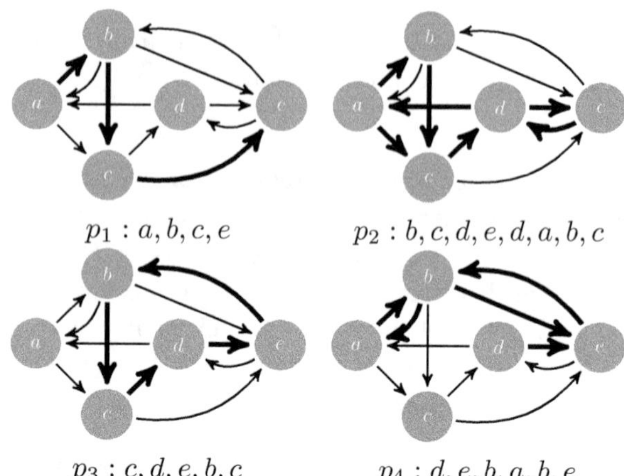

$$p_1 : a, b, c, e \qquad\qquad p_2 : b, c, d, e, d, a, b, c$$

$$p_3 : c, d, e, b, c \qquad\qquad p_4 : d, e, b, a, b, e$$

Path p_1 is from a to e of length 3, path p_2 is from b to c of length 7, p_3 is a cycle at c of length 4 and path p_4 is from d to e of length 5.

Definition 7.11. Given a binary relation R on a set A and an integer $n \geq 0$, we define the relation $R^{(n)}$ on A as follows: $R^{(0)} = R \cap I_A$ (where I_A is the identity relation on A) and for $n \geq 1$, $R^{(n)}$ is the set of pairs $(a, b) \in A \times A$ such that there exists a path of length n from a to b in R.

Note that $R^{(0)}$ is the set of all pairs (a, a) of $A \times A$ such that $a \, R \, a$ and $R^{(1)}$ is simply R. Using the digraph G_R of R, the relation $R^{(n)}$ can be interpreted as follows: $x \, R^{(n)} \, y$ if and only if it is possible to reach vertex y in G_R by a sequence of n edges starting at vertex x and following the directions of the edges.

Example 7.24. For the relation R of Example 7.23, the elements (a, e), (d, b) and (c, a) are in the relation $R^{(2)}$. The elements (b, a), (c, b) and (b, b) belong to the relation $R^{(3)}$.

Example 7.25. Let C be the set of all north American cities with a commercial airport. On C, we consider the binary relation R defined as follows: $x \, R \, y$ if and only if there is a direct flight from x to y. The relation $R^{(2)}$ is then defined as follows: $x \, R^{(2)} \, y$ if and only if there exists $z \in C$ such that one can fly directly from x to z and from z to y. In general, given a positive integer n, the relation $R^{(n)}$ is defined as follows: $x \, R^{(n)} \, y$ if and only if one can get from x to y with stops in exactly $n - 1$ north American cities (not necessarily different).

Given a binary relation R on a set A and $n \in \mathbb{N}$, recall that $R^{[n]} = R \circ R \circ \cdots \circ R$ is the composition of R with itself n times (with $R^{[0]} = I_A$, the identity relation on A). The next result tells us that $R^{[n]}$ and $R^{(n)}$ are actually the same relation on A for $n \geq 1$.

Theorem 7.9. *If R is a binary relation on a set A, then $R^{[n]} = R^{(n)}$ for any integer $n \geq 1$.*

Proof. We proceed by induction on n. The theorem is true for $n = 1$ as $R^{(1)} = R^{[1]} = R$. Let $n \geq 1$ be an integer and assume that $R^{[n]} = R^{(n)}$. If $(a, b) \in R^{[n+1]} = R^{[n]} \circ R$, then there exists $c \in A$ such that $(a, c) \in R$ and $(c, b) \in R^{[n]}$. By the induction hypothesis, $(c, b) \in R^{(n)}$ which means there exists a path p of length n from c to b in R. Adding (a, c) to path p produces a path of length $n + 1$ from a to b in R. Hence $(a, b) \in R^{(n+1)}$. This proves that $R^{[n+1]} \subseteq R^{(n+1)}$. Conversely, let $(a, b) \in R^{(n+1)}$ then there exits a path $(a, x_1), (x_1, x_2), \ldots, (x_n, b)$ in R of length $n + 1$ from a to b. In particular, there exists a path of length n from x_1 to b and so $(x_1, b) \in R^{(n)}$. By the induction hypothesis, $(x_1, b) \in R^{[n]}$. By definition of the composition of relations, $(a, b) \in R^{[n]} \circ R = R^{[n+1]}$. We conclude that $R^{[n]} = R^{(n)}$ for any positive integer n. $\qquad\square$

Given a binary relation R on the set A and an integer $n \geq 1$, Theorem 7.9 allows us to eliminate distinction between $R^{[n]}$ and $R^{(n)}$. From this point on, we will simply write R^n to denote either one of $R^{[n]}$ and $R^{(n)}$ for $n \geq 1$.

Theorem 7.10. *Let R be a relation on a set A. If M_R is the adjacency matrix of R, then the adjacency matrix of the relation R^n is the Boolean power $M_R^{[n]}$ of M_R for any integer $n \geq 1$.*

Proof. This is a direct consequence of Theorems 7.6 and 7.9 above. $\quad\square$

Given a binary relation R on a set $A = \{a_1, a_2, \ldots, a_n\}$ and an integer $n \geq 1$, Theorem 7.10 tells us that the entry (i, j) of $(M_R)^{[n]}$ is equal to 1 if and only if there exists a path of length n joining a_i and a_j in the digraph.

Example 7.26. On the set $A = \{a, b, c, d, e\}$, consider the binary relation R represented by the following digraph:

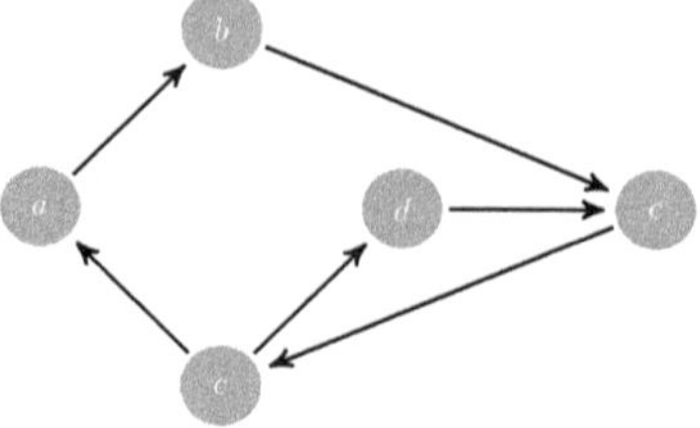

Find all vertices of the digraph that can be joined by a path of length 3.

Solution. The adjacency matrix of R is

$$
M = \begin{array}{c} \\ a \\ b \\ c \\ d \\ e \end{array}
\begin{array}{c} a\ b\ c\ d\ e \\
\begin{bmatrix}
0 & 1 & 0 & 0 & 0 \\
0 & 0 & 0 & 0 & 1 \\
1 & 0 & 0 & 1 & 0 \\
0 & 0 & 0 & 0 & 1 \\
0 & 0 & 1 & 0 & 0
\end{bmatrix}
\end{array}
$$

The adjacency matrix of R^2 is $M^{[2]} = M \odot M$:

$$
\begin{array}{c} a \\ b \\ c \\ d \\ e \end{array}
\begin{bmatrix}
0 & 1 & 0 & 0 & 0 \\
0 & 0 & 0 & 0 & 1 \\
1 & 0 & 0 & 1 & 0 \\
0 & 0 & 0 & 0 & 1 \\
0 & 0 & 1 & 0 & 0
\end{bmatrix}
\odot
\begin{bmatrix}
0 & 1 & 0 & 0 & 0 \\
0 & 0 & 0 & 0 & 1 \\
1 & 0 & 0 & 1 & 0 \\
0 & 0 & 0 & 0 & 1 \\
0 & 0 & 1 & 0 & 0
\end{bmatrix}
=
\begin{bmatrix}
0 & 0 & 0 & 0 & 1 \\
0 & 0 & 1 & 0 & 0 \\
0 & 1 & 0 & 0 & 1 \\
0 & 0 & 1 & 0 & 0 \\
1 & 0 & 0 & 1 & 0
\end{bmatrix}
$$

and the adjacency matrix of R^3 is $M^{[3]} = M^{[2]} \odot M$:

$$
\begin{array}{c} a \\ b \\ c \\ d \\ e \end{array}
\begin{bmatrix}
0 & 0 & 0 & 0 & 1 \\
0 & 0 & 1 & 0 & 0 \\
0 & 1 & 0 & 0 & 1 \\
0 & 0 & 1 & 0 & 0 \\
1 & 0 & 0 & 1 & 0
\end{bmatrix}
\odot
\begin{bmatrix}
0 & 1 & 0 & 0 & 0 \\
0 & 0 & 0 & 0 & 1 \\
1 & 0 & 0 & 1 & 0 \\
0 & 0 & 0 & 0 & 1 \\
0 & 0 & 1 & 0 & 0
\end{bmatrix}
=
\begin{bmatrix}
0 & 0 & 1 & 0 & 0 \\
1 & 0 & 0 & 1 & 0 \\
0 & 0 & 1 & 0 & 1 \\
1 & 0 & 0 & 1 & 0 \\
0 & 1 & 0 & 0 & 1
\end{bmatrix}
$$

So R^3 consists of the following couples: (a, c), (b, a), (b, d), (c, c), (c, e), (d, a), (d, d), (e, b) and (e, e). For each one of these couples, there exists a path of length 3 starting at the first component and ending at the second one. For example, we can go from d to a following the three edges: (d, e), (e, c) and (c, a). Note also that the matrix of R^3 tells us that there is no way to get from a to d following three edges in the digraph.   ◊

The powers R^n, $n \geq 1$ of a relation R on a set A can be used to create a new relation on A known as the connectivity relation.

Definition 7.12. Let R be a relation on a set A. The *connectivity relation* of R, denoted by R^∞, is defined as follows: $a\,R^\infty\,b$ if and only if there exists a path of some positive length from a to b in R.

The connection between R^∞ and the relations R^n on a set A is described in the following theorem.

Theorem 7.11. *If R is a binary relation on set A, then $R^\infty = \cup_{n=1}^{\infty} R^n$.*

Proof. If $(x, y) \in R^\infty$, then there exists a path p in R of a certain length $k \geq 1$ joining x to y in R. In particular, $(x, y) \in R^k \subseteq \cup_{n=1}^{\infty} R^n$. Conversely, if $(x, y) \in \cup_{n=1}^{\infty} R^n$ then $(x, y) \in R^k$ for some integer $k \geq 1$. This means that there exists a path p of length k from x to y in R. By definition of the connectivity relation, $(x, y) \in R^\infty$. $\qquad\square$

Theorem 7.11 tells us that we have to find the union of an infinite number of sets in order to find R^∞, something not usually practical to do. But when the set A is finite, we can restrict to a finite union to compute R^∞ as shown in the following theorem.

Theorem 7.12. *If R is a relation on a finite set A containing n elements, then $R^\infty = \cup_{i=1}^{n} R^i$.*

Proof. Let $(x, y) \in R^\infty$, then there exists a path of positive length from x to y in R. Let m be the length of the *shortest path* (i.e., the path using the least number of edges) from x to y in R. If $m > n$, then a path p of length m would have a cycle γ in it since A has only n elements:

$$(x, x_1), (x_1, x_2), \ldots, \underbrace{(x_i, x_{i+1}), \ldots, (x_k, x_i)}_{\gamma}, \ldots, (x_{m-1}, y).$$

Removing the edges of γ from the path p yields a path p' from x to y of length strictly less than m. This is a contradiction. We conclude that $(x, y) \in R^m$ with $m \leq n$. This proves that $R^\infty \subseteq \cup_{i=1}^{n} R^i$. The other inclusion is immediate since $R^i \subseteq R^\infty$ for $i = 1, \ldots, n$. $\qquad\square$

Remark 7.5. With the notations of Theorem 7.12, if $k > n$ then the relation R^k does not contain any new element which is not already in $\cup_{i=1}^{n} R^i$. Equivalently, if $k > n$ then the digraph of R^k does not add any new edge not existing already in one of the digraphs of the relations R^i, $i = 1, \ldots, n$.

Theorem 7.13. *Let R be a relation on a set A containing n elements and let M_R be its adjacency matrix. Then the adjacency matrix of the connectivity relation R^∞ is given by: $M_{R^\infty} = M_R \vee M_R^{[2]} \vee \cdots \vee M_R^{[n]}$ where, as usual, $M_R^{[i]}$ is the ith Boolean power of M_R.*

Proof.   This is a direct consequence of Theorems 7.4, 7.10 and 7.12. $\square$

Example 7.27. Let $A = \{a, b, c, d\}$ and let R be the relation on A given by the digraph below. Determine the adjacency matrix, the digraph and the elements of the connectivity relation R^∞ of R.

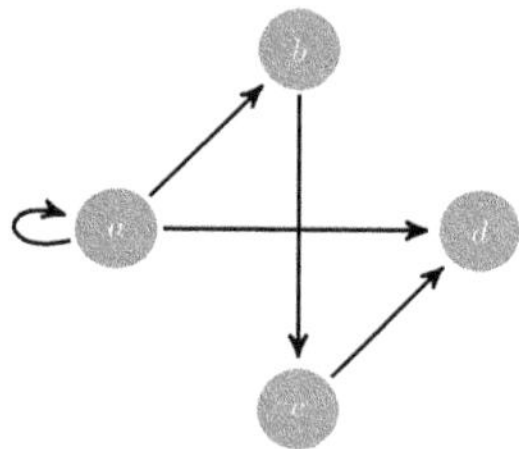

Solution. The adjacency matrix of R is

$$
M_R = \begin{array}{c@{}c} & \begin{array}{cccc} a & b & c & d \end{array} \\ \begin{array}{c} a \\ b \\ c \\ d \end{array} & \left[\begin{array}{cccc} 1 & 1 & 0 & 1 \\ 0 & 0 & 1 & 0 \\ 0 & 0 & 0 & 1 \\ 0 & 0 & 0 & 0 \end{array}\right] \end{array}
$$

The adjacency matrix of R^2 is $M_R^{[2]} = M_R \odot M_R$:

$$
\begin{array}{c@{}c} & \begin{array}{cccc} a & b & c & d \end{array} \\ \begin{array}{c} a \\ b \\ c \\ d \end{array} & \left[\begin{array}{cccc} 1 & 1 & 0 & 1 \\ 0 & 0 & 1 & 0 \\ 0 & 0 & 0 & 1 \\ 0 & 0 & 0 & 0 \end{array}\right] \end{array}
\odot
\begin{array}{c@{}c} & \begin{array}{cccc} a & b & c & d \end{array} \\ \begin{array}{c} a \\ b \\ c \\ d \end{array} & \left[\begin{array}{cccc} 1 & 1 & 0 & 1 \\ 0 & 0 & 1 & 0 \\ 0 & 0 & 0 & 1 \\ 0 & 0 & 0 & 0 \end{array}\right] \end{array}
=
\begin{array}{c@{}c} & \begin{array}{cccc} a & b & c & d \end{array} \\ \begin{array}{c} a \\ b \\ c \\ d \end{array} & \left[\begin{array}{cccc} 1 & 1 & 1 & 1 \\ 0 & 0 & 0 & 1 \\ 0 & 0 & 0 & 0 \\ 0 & 0 & 0 & 0 \end{array}\right] \end{array}
$$

and the adjacency matrix of R^3 is $M_R^{[3]} = M_R^{[2]} \odot M_R$:

$$
\begin{array}{c@{}c} & \begin{array}{cccc} a & b & c & d \end{array} \\ \begin{array}{c} a \\ b \\ c \\ d \end{array} & \left[\begin{array}{cccc} 1 & 1 & 1 & 1 \\ 0 & 0 & 0 & 1 \\ 0 & 0 & 0 & 0 \\ 0 & 0 & 0 & 0 \end{array}\right] \end{array}
\odot
\begin{array}{c@{}c} & \begin{array}{cccc} a & b & c & d \end{array} \\ \begin{array}{c} a \\ b \\ c \\ d \end{array} & \left[\begin{array}{cccc} 1 & 1 & 0 & 1 \\ 0 & 0 & 1 & 0 \\ 0 & 0 & 0 & 1 \\ 0 & 0 & 0 & 0 \end{array}\right] \end{array}
=
\begin{array}{c@{}c} & \begin{array}{cccc} a & b & c & d \end{array} \\ \begin{array}{c} a \\ b \\ c \\ d \end{array} & \left[\begin{array}{cccc} 1 & 1 & 1 & 1 \\ 0 & 0 & 0 & 0 \\ 0 & 0 & 0 & 0 \\ 0 & 0 & 0 & 0 \end{array}\right] \end{array} .
$$

Finally, the adjacency matrix of R^4 is $M_R^{[4]} = M_R^{[3]} \odot M_R$:

$$
\begin{array}{c}
\begin{array}{cccc} a & b & c & d \end{array} \\
\begin{array}{c} a \\ b \\ c \\ d \end{array}
\begin{bmatrix} 1 & 1 & 1 & 1 \\ 0 & 0 & 0 & 0 \\ 0 & 0 & 0 & 0 \\ 0 & 0 & 0 & 0 \end{bmatrix}
\end{array}
\odot
\begin{array}{c}
\begin{array}{cccc} a & b & c & d \end{array} \\
\begin{array}{c} a \\ b \\ c \\ d \end{array}
\begin{bmatrix} 1 & 1 & 0 & 1 \\ 0 & 0 & 1 & 0 \\ 0 & 0 & 0 & 1 \\ 0 & 0 & 0 & 0 \end{bmatrix}
\end{array}
=
\begin{array}{c}
\begin{array}{cccc} a & b & c & d \end{array} \\
\begin{array}{c} a \\ b \\ c \\ d \end{array}
\begin{bmatrix} 1 & 1 & 1 & 1 \\ 0 & 0 & 0 & 0 \\ 0 & 0 & 0 & 0 \\ 0 & 0 & 0 & 0 \end{bmatrix}
\end{array}
= M_R^{[3]}.
$$

By Theorem 7.13, the adjacency matrix of R^∞ is $M_{R^\infty} = M_R \vee M_R^{[2]} \vee M_R^{[3]} \vee M_R^{[4]} = M_R \vee M_R^{[2]} \vee M_R^{[3]}$ since $M_R^{[4]} = M_R^{[3]}$ and $M_R^{[3]} \vee M_R^{[3]} = M_R^{[3]}$:

$$
M_{R^\infty} =
\begin{array}{c}
\begin{array}{cccc} a & b & c & d \end{array} \\
\begin{array}{c} a \\ b \\ c \\ d \end{array}
\begin{bmatrix} 1 & 1 & 0 & 1 \\ 0 & 0 & 1 & 0 \\ 0 & 0 & 0 & 1 \\ 0 & 0 & 0 & 0 \end{bmatrix}
\end{array}
\vee
\begin{array}{c}
\begin{array}{cccc} a & b & c & d \end{array} \\
\begin{array}{c} a \\ b \\ c \\ d \end{array}
\begin{bmatrix} 1 & 1 & 1 & 1 \\ 0 & 0 & 0 & 1 \\ 0 & 0 & 0 & 0 \\ 0 & 0 & 0 & 0 \end{bmatrix}
\end{array}
\vee
\begin{array}{c}
\begin{array}{cccc} a & b & c & d \end{array} \\
\begin{array}{c} a \\ b \\ c \\ d \end{array}
\begin{bmatrix} 1 & 1 & 1 & 1 \\ 0 & 0 & 0 & 0 \\ 0 & 0 & 0 & 0 \\ 0 & 0 & 0 & 0 \end{bmatrix}
\end{array}
=
\begin{array}{c}
\begin{array}{cccc} a & b & c & d \end{array} \\
\begin{array}{c} a \\ b \\ c \\ d \end{array}
\begin{bmatrix} 1 & 1 & 1 & 1 \\ 0 & 0 & 1 & 1 \\ 0 & 0 & 0 & 1 \\ 0 & 0 & 0 & 0 \end{bmatrix}
\end{array}.
$$

We conclude that $R^\infty = \{(a, a), (a, b), (a, c), (a, d), (b, c), (b, d), (c, d)\}$ and the digraph of R^∞ is the following:

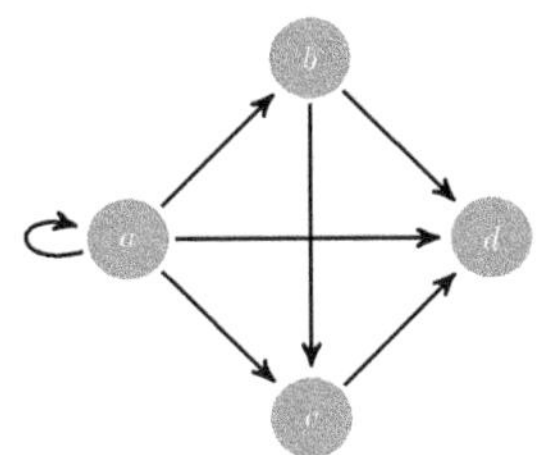

$\Diamond$

7.6.1 *Exercises*

(1) Consider the following digraph:

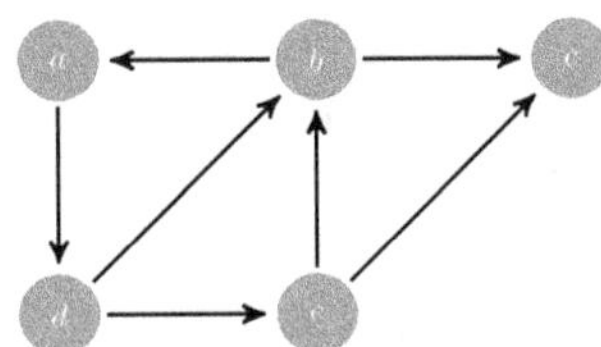

Find:

(a) two paths of length 3 each from a to c;

(b) A cycle of length 3 at a;

(c) A path of length 4 from a to c;

(d) A cycle of length 4 at b.

(2) The digraph in Exercise (1) represents a relation R on the set $A = \{a, b, c, d, e\}$. Draw the digraphs of R^2 and R^3.

(3) Let R be a relation on a set A and let $x \in A$. For any integer $n \geq 1$, we define $R^n(x)$ as the following subset of A: $R^n(x) = \{y \in A;\ x\,R^n\,y\}$. We also define $R^\infty(x) = \{y \in A;\ x\,R^n\,y$ for some $n \geq 1\}$. A relation R on the set $A = \{a, b, c, d, e\}$ given by the following digraph:

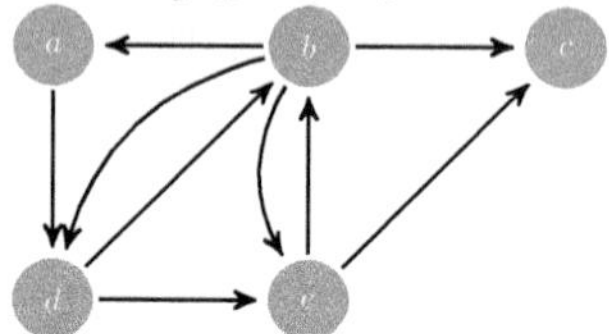

Find each of the following as a subset of A:

$\star$ (a) $R^2(a)$	(d) $R^2(c)$	$\star$ (g) $R^4(e)$	(j) $R^\infty(b)$
$\star$ (b) $R^3(a)$	(e) $R^2(b)$	(h) $R^2(d)$	$\star$ (k) $R^\infty(c)$
(c) $R^4(a)$	$\star$ (f) $R^3(e)$	(i) $R^\infty(a)$	(l) $R^\infty(e)$

(4) Let $A = \{1, 2, 3\}$. In each case, find the connectivity relation of the given relation on A.

(a) $R = \{(2, 2), (2, 3)\}$

(b) $R = \{(1, 1), (2, 2)\}$

(c) $R = \{(1, 2), (1, 3)\}$

(d) $R = \{(1, 2), (1, 3), (2, 3), (3, 1)\}$

(5) Let $A = \{a, b, c\}$. In each case, a relation R on A is given by its adjacency matrix M_R. Find the adjacency matrix of the connectivity relation R^∞ of R.

$$(a)\ M_R = \begin{array}{c} \\ a \\ b \\ c \end{array}\!\!\begin{array}{ccc} a\ b\ c \\ \begin{bmatrix} 1 & 0 & 1 \\ 0 & 0 & 0 \\ 0 & 1 & 1 \end{bmatrix} \end{array} \quad (b)\ M_R = \begin{array}{c} \\ a \\ b \\ c \end{array}\!\!\begin{array}{ccc} a\ b\ c \\ \begin{bmatrix} 1 & 0 & 0 \\ 1 & 0 & 1 \\ 0 & 1 & 1 \end{bmatrix} \end{array} \quad (c)\ M_R = \begin{array}{c} \\ a \\ b \\ c \end{array}\!\!\begin{array}{ccc} a\ b\ c \\ \begin{bmatrix} 1 & 1 & 1 \\ 1 & 0 & 1 \\ 0 & 1 & 1 \end{bmatrix} \end{array}$$

(6) Let $A = \{a, b, c, d\}$. In each case, a relation R on A is given by its adjacency matrix M_R. Find the adjacency matrix of the connectivity relation R^∞ of R.

$$\star(a)\ M_R = \begin{array}{c} \\ a \\ b \\ c \\ d \end{array}\!\!\begin{array}{cccc} a\ b\ c\ d \\ \begin{bmatrix} 0 & 0 & 0 & 1 \\ 1 & 0 & 1 & 0 \\ 0 & 1 & 0 & 1 \\ 0 & 0 & 0 & 0 \end{bmatrix} \end{array} \quad (b)\ M_R = \begin{array}{c} \\ a \\ b \\ c \\ d \end{array}\!\!\begin{array}{cccc} a\ b\ c\ d \\ \begin{bmatrix} 1 & 0 & 1 & 0 \\ 1 & 0 & 1 & 0 \\ 0 & 0 & 0 & 1 \\ 0 & 0 & 0 & 0 \end{bmatrix} \end{array} \quad (c)\ M_R = \begin{array}{c} \\ a \\ b \\ c \\ d \end{array}\!\!\begin{array}{cccc} a\ b\ c\ d \\ \begin{bmatrix} 1 & 0 & 0 & 1 \\ 1 & 0 & 1 & 0 \\ 1 & 0 & 0 & 1 \\ 0 & 0 & 1 & 0 \end{bmatrix} \end{array}$$

(7) Let $A = \{a, b, c, d\}$. A relation R on A is defined as follows:
$$R = \{(a, a), (a, b), (b, a), (b, c), (c, a), (c, c), (d, c)\}.$$

    (a) Find the adjacency matrix of the connectivity relation R^∞ of R.

    (b) Define the relation R^* on A as follows: $a\, R^*\, b$ if and only if $a = b$ or $a\, R^\infty\, b$. Find the adjacency matrix of R^*.

(8) In each case, find the connectivity relation of the relation R on the set $A = \{a, b, c, d\}$ given by its digraph.

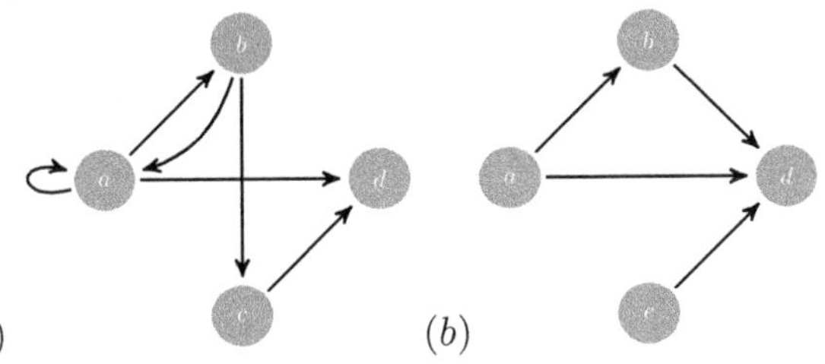

        (a) (b)

(9) If R and S are two binary relations on a set A, prove that $(R \cap S)^n \subseteq R^n \cap S^n$ for any integer $n \geq 1$.

$\star$(10) Consider the relation $R = \{(k, k+1);\ k \in \mathbb{Z}\}$ on the set $\mathbb{Z}$.

    (a) Use induction to prove that $a\, R^n\, b$ if and only if $a = b - n$ for any integer $n \geq 1$.

    (b) For $a, b \in \mathbb{Z}$, give an expression describing $a\, R^\infty\, b$.

7.7 Properties of binary relations

In many applications of mathematics and computer science, binary relations satisfying certain properties are of particular importance. In this section, we discuss some of these properties in preparation for the two most important classes of binary relations: the equivalence relations and the partial orders.

Definition 7.13. A relation R on a set A is called *reflexive* if $(x, x) \in R$ for any $x \in A$. Equivalently, R is reflexive if and only if $x\, R\, x$ for any $x \in A$.

Example 7.28. On the set $\mathbb{Z}^*$ of non-zero integers, the relation "divides" is reflexive since any non-zero integer divides itself.

Example 7.29. On the set of all humans, the relation "has the same age as" is reflexive.

Example 7.30. On the set $A = \{a, b, c, d\}$, consider the following four relations:

- $R_1 = \{(a,a),(a,b),(c,a),(b,b),(b,a),(d,d),(d,b),(a,c),(b,d)\}$.
- $R_2 = \{(a,b),(a,a),(c,c),(b,b),(b,a),(d,d),(d,c)\}$.
- $R_3 = \{(a,a),(b,b),(c,c),(d,d)\}$.
- $R_4 = \emptyset$.

Relation R_1 is not reflexive since it does not contain the element (c,c). Relation R_2 is reflexive since it contains all couples (x,x) for $x \in A$. Same is true for relation R_3. Relation R_4 is not reflexive: for instance $(a,a) \notin R_4$.

Example 7.31. The relation $\{(m,n) \in \mathbb{Z} \times \mathbb{Z};\ m^2 + n^2 \geq 16\}$ on $\mathbb{Z}$ is not reflexive. For example, the relation does not contain (among others) the couple $(1,1)$.

It is easy to determine if a relation is reflexive from its digraph representation: every vertex in the graph must have a loop.

Definition 7.14. A relation R on a set A is called *symmetric* if for all $x, y \in A$, $(y,x) \in R$ whenever $(x,y) \in R$. In other words, R is symmetric if and only if the implication $(x\,R\,y \Rightarrow y\,R\,x)$ is true for any $x, y \in A$.

A word of warning: the symmetry of a relation R does not mean that xRy for any elements x and y. Definition 7.14 simply says that in order for the relation R to be symmetric, the condition $x\,R\,y$ must imply that $y\,R\,x$ for any elements x, y of A. In particular, R is not symmetric only when we can find a pair $(x,y) \in R$ with $(y,x) \notin R$.

Example 7.32. On the set $\mathbb{Z}^*$ of non-zero integers, the relation "divides" is not symmetric since, for example, 2 divides 4 but 4 does not divide 2.

Example 7.33. On the set of all humans, the relation "has the same age as" is symmetric: If x has the same age as y, then clearly y has the same age as x.

Example 7.34. With the notations of Example 7.30:

- Relation R_1 is symmetric: every time $(x,y) \in R_1$, (y,x) is also in R_1.
- Relation R_2 is not symmetric: $(d,c) \in R_2$ but $(c,d) \notin R_2$.
- Relation R_3 is symmetric as it only consists of pairs (x,x) with $x \in A$.
- The empty relation R_4 is symmetric (see Exercise (1) below).

Example 7.35. The relation $R = \{(x,y) \in \mathbb{R} \times \mathbb{R};\ x < y^2\}$ on $\mathbb{R}$ is not symmetric since, for example, $-1\,R\,2$ but $2\,\not{R}\,-1$.

In the diagraph of a symmetric relation, every time there is an edge from a to b then there must be an edge from b to a in the graph.

Definition 7.15. A relation R on a set A is called *transitive* if for any $x, y, z \in A$, the conditions $(x, y) \in R$ and $(y, z) \in R$ imply that $(x, z) \in R$. Alternatively, R is transitive if and only if for any $x, y, z \in A$:

$$x \, R \, y \text{ and } y \, R \, z \Rightarrow x \, R \, z.$$

Example 7.36. On the set $\mathbb{Z}^*$ of non-zero integers, the relation "divides" is transitive: if a, b, c are non-zero integers such that a divides b and b divides c, then clearly a divides c.

Example 7.37. On the set of all humans, the relation "has the same age as" is transitive: If x has the same age as y and y has the same age as z then x has the same age as z.

Example 7.38. For the relations in Example 7.30:

- R_1 is not transitive: $(a, b) \in R_1$ and $(b, d) \in R_1$ but $(a, d) \notin R_1$.
- R_2 and R_3 are transitive since every time (x, y) and (y, z) are in these relations, the same is true for (x, z).
- $R_4 = \emptyset$ is transitive (see Exercise (1) below).

Example 7.39. The relation $R = \{(x, y) \in \mathbb{Z} \times \mathbb{Z}; \, x - y \text{ is even}\}$ on $\mathbb{Z}$ is transitive: if $(x, y) \in R$ and $(y, z) \in R$ then both $x - y$ and $y - z$ are even. So $x - y = 2s$ and $y - z = 2t$ for some $s, t \in \mathbb{Z}$. We conclude that $x - z = (x - y) + (y - z) = 2s + 2t = 2(s + t)$ is also even and so $(x, z) \in R$.

Theorem 7.14. *Let R be a relation on a set A. If R is transitive, then $R^n \subseteq R$ for any integer $n \geq 1$.*

Proof. We proceed by induction on n. Assume that R is a transitive relation on A. The property is true for $n = 1$ since $R^n = R$ in this case. Let $n \geq 1$ and assume that $R^n \subseteq R$. If $x, y \in A$ are such that $x \, R^{n+1} \, y$ then there exists $z \in A$ such that $x \, R \, z$ and $z \, R^n \, y$ (by definition of the composition of relations). By the induction hypothesis, $z \, R \, y$. Since $x \, R \, z$ and $z \, R \, y$, the transitivity of R implies that $x \, R \, y$. This prove that $R^{n+1} \subseteq R$ and the result is true for any $n \geq 1$ by the principle of induction. $\qquad\square$

Theorem 7.15. *Let R is a relation on a set A and let $n \geq 1$ be an integer. Then:*

(a) If R is reflexive, then R^n is also reflexive.
(b) If R is symmetric, then R^n is also symmetric.
(c) If R is transitive, then R^n is also transitive.

Proof.   We prove parts (a) and (c) using induction on n. Part (b) is left as an exercise (see Exercise (20) below). For part (a), the property is true for $n = 1$ since $R^1 = R$. Let $n \geq 1$ and assume that R^n is reflexive. For any $x \in A$, we have that $x\,R\,x$ (since R is reflexive) and $x\,R^n\,x$ (by the induction hypothesis). Therefore $x\,(R^n \circ R)\,x$ by definition of the composition of relations. This shows that $x\,R^{n+1}\,x$ for any $x \in A$ and the result follows. For part (c), the result is true for $n = 1$ since $R^1 = R$. Let $n \geq 1$ and assume that R^n is transitive. Let $x, y, z \in A$ such that $x\,R^{n+1}\,y$ and $y\,R^{n+1}\,z$. Since $R^{n+1} = R^n \circ R$, then exist $a, b \in A$ such that $x\,R\,a$, $a\,R^n\,y$, $y\,R\,b$ and $b\,R^n\,z$. By Theorem 7.14, $a\,R\,y$ and so $a\,R\,b$ since $y\,R\,b$ and R is transitive. The conditions $x\,R\,a$ and $a\,R\,b$ imply (by the transitivity of R) that $x\,R\,b$. Now $x\,R\,b$ and $b\,R^n\,z$ imply that $x\,R^{n+1}\,z$. This shows that R^{n+1} is transitive and the result follows by the principle of induction. $\square$

Other types of binary relations are given in the following definition.

Definition 7.16. Let R be a binary relation on a set A.

(a) We say that R is *irreflexive* if $(x, x) \notin R$ for any $x \in A$. Alternatively, R is irreflexive if for any $x \in A$, $x\,\not\!R\,x$.

(b) We say that R is *antisymmetric* if for any $x, y \in A$, the conditions $(x, y) \in R$ and $(y, x) \in R$ imply that $x = y$. Alternatively, R is antisymmetric if for any $x, y \in A$, we have: $x\,R\,y$ and $y\,R\,x$ imply that $x = y$.

(c) We say that R is *asymmetric* if for all $x, y \in A$, we have: $(x, y) \in R \Rightarrow (y, x) \notin R$. Alternatively, R is asymmetric if for any $x, y \in A$, the condition $x\,R\,y$ implies that $y\,\not\!R\,x$.

Example 7.40. On the set $\mathbb{Z}$, the relation $R = \{(m, n) \in \mathbb{Z} \times \mathbb{Z};\ m < n\}$ is irreflexive since no integer is strictly less than itself.

Example 7.41. On the set of all positive integers, the relation "divides" is antisymmetric since if m, n are two positive integers such that m divides n and n divides m, then $m = n$.

Example 7.42. The relation $R = \{(1, 2), (1, 3), (2, 3)\}$ on the set $A = \{1, 2, 3\}$ is asymmetric since if $(x, y) \in R$, then $(y, x) \notin R$.

7.7.1 *Exercises*

(1) Prove that the empty relation on a set A is symmetric and transitive.

(2) In each case, a relation R on the set $A = \{a, b, c, d\}$ is given by its digraph. Determine if R is reflexive, symmetric transitive, irreflexive, asymmetric or antisymmetric.

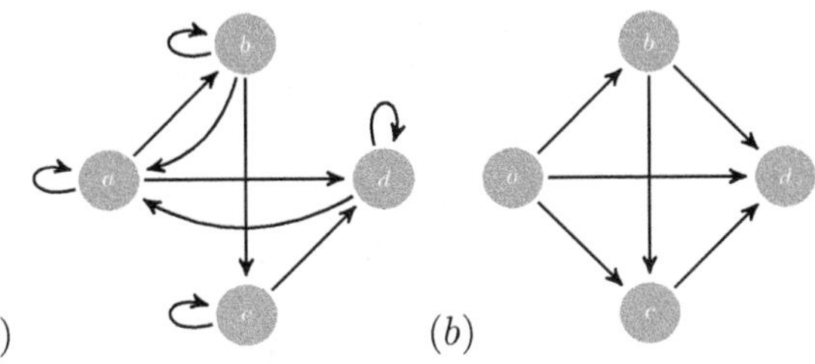

$\star(a)$ (b)

(3) Consider the following relations on the set $A = \{\alpha, \beta, \gamma, \tau\}$:

- $R_1 = \{(\alpha, \alpha), (\beta, \beta), (\gamma, \gamma), (\alpha, \beta), (\tau, \tau), (\beta, \alpha), (\tau, \alpha), (\alpha, \gamma), (\alpha, \tau)\}$
- $R_2 = \{(\alpha, \tau), (\tau, \beta), (\beta, \gamma), (\gamma, \alpha)\}$
- $R_3 = \{(\alpha, \alpha), (\beta, \beta), (\gamma, \gamma), (\tau, \tau)\}$
- $R_4 = \{(\alpha, \beta), (\beta, \gamma), (\gamma, \gamma), (\tau, \alpha)\}$

Complete the following table:

	Reflexive	Symmetric	Transitive	Irreflexive	Antisymmetric	Asymmetric
R_1	Yes					
R_2						
R_3						
R_4		No				

$\star$(4) Consider the following relations on the set $A = \{1, 2, 3, 4, 5\}$:

- $R_1 = \{(1, 1), (2, 3), (3, 4), (2, 2), (3, 3), (3, 2), (3, 3), (4, 4), (5, 5)\}$
- $R_2 = \{(1, 2), (2, 3), (3, 4), (4, 5)\}$
- $R_3 = \{(1, 3), (3, 4), (4, 1), (1, 4), (4, 3)\}$
- $R_4 = \{(1, 1), (2, 2), (3, 3), (4, 4), (5, 5)\}$

Complete the following table:

	Reflexive	Symmetric	Transitive	Irreflexive	Antisymmetric	Asymmetric
R_1		No				
R_2						
R_3						
R_4	Yes					

(5) A certain committee of t members $m_1, m_2, \ldots, m_t$ met on the side of a conference. For each of the relations given below, determine if it is reflexive, symmetric or transitive.

⋆ (a) $m_i\,R_1\,m_j$ if and only if m_i shook hand with m_j at the conference.

(b) $m_i\,R_2\,m_j$ if and only if m_i spoke to m_j at the conference.

(c) $m_i\,R_3\,m_j$ if and only if m_i works at the same place as m_j.

(d) $m_i\,R_4\,m_j$ if and only if m_i is older than m_j.

⋆ (e) $m_i\,R_5\,m_j$ if and only if m_i met m_j for the first time at the conference.

(6) Let A be the set of all humans. For each of the following binary relations on A, determine if it is reflexive, symmetric or transitive. Justify your answer.

(a) $x\,R_1\,y$ if and only if x and y speak a common language.

(b) $x\,R_2\,y$ if and only if x and y share a common parent.

⋆ (c) $x\,R_3\,y$ if and only if x is a cousin of y.

⋆ (d) $x\,R_4\,y$ if and only if x and y have either the same height or the same eye color.

(7) In each case, the adjacency matrix of a binary relation on the set $A = \{a, b, c, d, e\}$ is given. Determine if the relation is reflexive, irreflexive, symmetric, antisymmetric, asymmetric or transitive. Justify your answer.

$$
\begin{array}{c}
\begin{array}{c}
\begin{matrix} a & b & c & d & e \end{matrix} \\
\begin{matrix} a \\ b \\ c \\ d \\ e \end{matrix}
\begin{bmatrix}
1 & 1 & 0 & 0 & 1 \\
1 & 1 & 0 & 0 & 0 \\
0 & 0 & 1 & 0 & 0 \\
0 & 0 & 0 & 1 & 0 \\
1 & 0 & 0 & 0 & 1
\end{bmatrix} \\
\star(a)
\end{array}
\quad
\begin{array}{c}
\begin{matrix} a & b & c & d & e \end{matrix} \\
\begin{matrix} a \\ b \\ c \\ d \\ e \end{matrix}
\begin{bmatrix}
1 & 0 & 1 & 0 & 1 \\
1 & 0 & 1 & 0 & 0 \\
0 & 0 & 0 & 1 & 0 \\
0 & 0 & 0 & 1 & 0 \\
0 & 0 & 0 & 1 & 0
\end{bmatrix} \\
(b)
\end{array}
\quad
\begin{array}{c}
\begin{matrix} a & b & c & d & e \end{matrix} \\
\begin{matrix} a \\ b \\ c \\ d \\ e \end{matrix}
\begin{bmatrix}
0 & 0 & 1 & 1 & 1 \\
1 & 0 & 0 & 0 & 1 \\
0 & 0 & 0 & 1 & 0 \\
0 & 0 & 0 & 0 & 0 \\
0 & 0 & 0 & 1 & 0
\end{bmatrix} \\
(c)
\end{array}
\quad
\begin{array}{c}
\begin{matrix} a & b & c & d & e \end{matrix} \\
\begin{matrix} a \\ b \\ c \\ d \\ e \end{matrix}
\begin{bmatrix}
0 & 0 & 0 & 0 & 1 \\
0 & 0 & 1 & 1 & 0 \\
0 & 1 & 0 & 0 & 0 \\
0 & 1 & 0 & 1 & 0 \\
1 & 0 & 0 & 0 & 0
\end{bmatrix} \\
\star(d)
\end{array}
$$

(8) Let $\mathcal{U}$ be a non-empty set and let X be a fixed subset of $\mathcal{U}$. For each of the relations on the power set $\wp(\mathcal{U})$ of $\mathcal{U}$ given below, determine if it is reflexive, symmetric, or transitive.

(a) $A\,R\,B \Leftrightarrow A \cap X = B \cap X$.

⋆ (b) $A\,R\,B \Leftrightarrow A \cup X = B \cup X$.

(c) $A\,R\,B \Leftrightarrow A \backslash X = B \backslash X$.

⋆ (d) $A\,R\,B \Leftrightarrow A \backslash B = \emptyset$.

(9) In each case, determine if the given relation R on the set A is reflexive, symmetric, or transitive.

⋆ (a) $A = \mathbb{R}$, $x\,R\,y \Leftrightarrow xy = 0$.

(b) $A = \mathbb{R}$, $x\,R\,y \Leftrightarrow x^2 - y = 0$.

⋆ (c) $A = \mathbb{Z}$, $m\,R\,n \Leftrightarrow |m - n| \leq 3$.

(d) $A = \{f : \mathbb{R} \to \mathbb{R};\ f\ f$ is a function$\}$, $f\,R\,g \Leftrightarrow f(0) = g(0)$.
(e) $A = \mathbb{N}$, $m\,R\,n \Leftrightarrow m - n > 1$.
$\star$ (f) $A = \mathbb{N}\backslash\{0\}$, $m\,R\,n \Leftrightarrow \gcd(m,n) = 1$.

(10) Let $\mathcal{L}$ be the set of all well-formed formulas of propositional logic. A relation R on $\mathcal{L}$ is defined as follows: $p\,R\,q$ if and only if the implication $(p \to q)$ is a tautology. Is R reflexive? Symmetric? Transitive? Justify your answer.

(11) Point to the error in the proof of the following (false) theorem.
Theorem. *Let R be a binary relation on a set A. If R is symmetric and transitive, then it is also reflexive.*
Proof. Let $a, b \in A$. If $a\,R\,b$, then $b\,R\,a$ since R is symmetric. Now $a\,R\,b$ and $b\,R\,a$ imply that $a\,R\,a$ since R is transitive.

(12) Let R be a relation on a set A.

(a) Show that R is reflexive if and only if its complement $\neg R$ is irreflexive.
(b) Show that R is reflexive if and only if its inverse R^{-1} is reflexive.

(13) Give an example of a non-empty relation on $\mathbb{R}$ which is:

(a) Reflexive but neither symmetric nor transitive
$\star$ (b) symmetric but neither reflexive nor transitive
(c) Transitive but neither reflexive nor symmetric

(14) Let R and S be two binary relations on a set A. Prove:

(a) If R and S are reflexive, then so is $R \cap S$.
(b) If R and S are reflexive, then so is $R \cup S$.
(c) If R and S are symmetric, then so is $R \cap S$.
$\star$ (d) If R and S are symmetric, then so $R \cup S$.
$\star$ (e) If R and S are transitive, then so is $R \cap S$.

(15) Give an example of two transitive relations R and S on a set A such that the relation $R \cup S$ is not transitive.

(16) Give an example of two symmetric relations R and S on a set A such that $S \circ R$ is not symmetric.

(17) Let R and S be two binary relations on a set A. Prove that if R and S are reflexive, then $S \circ R$ and $R \circ S$ are also reflexive.

$\star$(18) Give an example of two transitive relations R and S on a set A such that $S \circ R$ is not transitive.

(19) Let R be a symmetric and transitive relation on a set A such that for every $a \in A$, there exists $b \in A$ with $a\,R\,b$. Prove that R is reflexive.

$\star$(20) Let R be a symmetric relation on a set A. Use induction to prove that R^n is also symmetric for any integer $n \geq 1$.

(21) Prove that if R and S are symmetric relations on a set A, then $S \circ R$ is symmetric if and only if $S \circ R = R \circ S$.

$\star$(22) Let R and S be two transitive relations on a set A such that $S \circ R \subseteq R \circ S$. Prove that $R \circ S$ is transitive.

(23) Let R be a relation on a set A. Prove:

    (a) R is symmetric if and only if $R^{-1} = R$.

    (b) R is asymmetric if and only if $R \cap R^{-1} = \emptyset$.

  $\star$ (c) R is antisymmetric if and only if $R \cap R^{-1} \subseteq I_A$ where $I_A = \{(a, a); a \in A\}$ is the identity relation on A.

7.8   Closures of a relation

When a relation R on a set A fails to satisfy a certain property P, it is possible to add pairs from $A \times A$ to R in order to form a new relation on A that contains R and satisfies P. In general, this can be achieved in more than one way, but naturally we wish to add the smallest number of pairs possible to achieve that. In other words, we wish to find the *smallest* relation on A that contains R and that satisfies P. In this section, the property P we consider is either reflexibility, symmetry or transitivity.

Definition 7.17. For a binary relation R on a set A, the *reflexive closure* of R, is the **smallest** relation on A, denoted by $r(R)$, which is reflexive and which contains R. In other words, the reflexive closure of R is the relation $r(R)$ on A that satisfies the following conditions:

  (i) $r(R)$ is reflexive;

 (ii) $R \subseteq r(R)$;

(iii) If R' is a reflexive relation on A such that $R \subseteq R'$, then $r(R) \subseteq R'$.

If the reflexive closure of R exists, then it is unique. To see this, assume S and T are two reflexive closures of R. Then S and T are both reflexive and $R \subseteq S$, $R \subseteq T$. By definition of the reflexive closure, we must have that $S \subseteq T$ and $T \subseteq S$ which imply that $S = T$.

Theorem 7.16. *Let R be a relation on a set A. Then:*

  (a) $r(R) = R \cup I_A$ *where* $I_A = \{(x, x); x \in A\}$ *(identity relation on A).*

  (b) R *is reflexive if and only if* $r(R) = R$.

Proof. The relation $R \cup I_A$ is clearly reflexive and contains R. Moreover, if R' is a reflexive relation containing R, then R' must contain I_A (since it is reflexive) and so it contains $R \cup I_A$. This proves part (a). Part (b) is left as an exercise (see Exercise (4) below). $\qquad\square$

Example 7.43. The relation $R = \{(1,1), (1,2), (2,3), (3,3), (4,1)\}$ on the set $A = \{1,2,3,4\}$ is clearly not reflexive. Its reflexive closure is $r(R) = R \cup I_A = \{(1,1), (1,2), (2,3), (3,3), (4,1), (2,2), (4,4)\}$ obtained from R by adding pairs (x,x) which are not already in R.

Example 7.44. Consider the relation $R = \{(x,y) \in \mathbb{R} \times \mathbb{R}; \ x < y\}$ on the set $\mathbb{R}$. Clearly, R is not reflexive. Its reflexive closure is $r(R) = R \cup I_{\mathbb{R}} = \{(x,y) \in \mathbb{R} \times \mathbb{R}; \ x \leq y\}$.

Definition 7.18. For a binary relation R on a set A, the *symmetric closure* of R, is the **smallest** relation on A, denoted by $s(R)$, which is symmetric and which contains R. In other words, $s(R)$ is the relation on A that satisfies the following conditions:

(i) $s(R)$ is symmetric;
(ii) $R \subseteq s(R)$;
(iii) If R' is a symmetric relation on A such that $R \subseteq R'$, then $s(R) \subseteq R'$.

Like the reflexive closure, the symmetric closure is unique when it exists. The following theorem gives an explicit expression for $s(R)$.

Theorem 7.17. *Let R be a relation on a set A. Then:*

(a) $s(R) = R \cup R^{-1}$.
(b) R *is symmetric if and only if* $s(R) = R$.

Proof. We only prove part (a), part (b) is left as an exercise (see Exercise (4) below). Let $S = R \cup R^{-1}$, then $R \subseteq S$. If $(x,y) \in S$, then either $(x,y) \in R$ or $(x,y) \in R^{-1}$. If $(x,y) \in R$, then $(y,x) \in R^{-1} \subseteq S$. If $(x,y) \in R^{-1}$ then $(y,x) \in \left(R^{-1}\right)^{-1} = R \subseteq S$. We conclude that S is symmetric. If T is a symmetric relation on A containing R, then $R^{-1} \subseteq T$ (by the symmetry of T) and so $S = R \cup R^{-1} \subseteq T$. $\qquad\square$

Example 7.45. The relation R of Example 7.43 above is not symmetric. For instance, $(1,2) \in R$ but $(2,1) \notin R$. Note that $R^{-1} = \{(1,1), (2,1), (3,2), (3,3), (1,4)\}$ so $s(R) = R \cup R^{-1} = \{(1,1), (1,2), (2,1), (2,3), (3,2), (3,3), (4,1), (1,4)\}$.

Example 7.46. Consider the relation $R = \{(x, y) \in \mathbb{R} \times \mathbb{R}; x < y\}$ on the set $\mathbb{R}$. The symmetric closure of R is $s(R) = R \cup R^{-1} = \{(x, y) \in \mathbb{R} \times \mathbb{R}; x < y \text{ or } y < x\} = \mathbb{R}^2 \setminus I_{\mathbb{R}}$. Notice that this is the relation "$\neq$" on $\mathbb{R}$. Geometrically, $s(R)$ represents the whole plane $\mathbb{R}^2$ from which we remove the line $y = x$.

While reflexive and symmetric closures are easy to compute, the transitive closure requires more work and it is not as intuitive as the other two closures. The construction of the transitive closure will pave the way to some interesting applications. First, we give the definition.

Definition 7.19. For a binary relation R on a set A, the *transitive closure* of R, denoted by $t(R)$, is the **smallest** relation on A which is transitive and which contains R. In other words, $t(R)$ is the relation on A that satisfies the following conditions:

  (i) $t(R)$ is transitive;
  (ii) $R \subseteq t(R)$;
 (iii) If R' is a transitive relation on A such that $R \subseteq R'$, then $t(R) \subseteq R'$.

As you can probably guess by now that the transitive closure is unique and if a relation R on A is transitive then $t(R) = R$. If R is not transitive, then there must exist three elements x, y and z of A such that $(x, y) \in R$, $(y, z) \in R$ but $(x, z) \notin R$. So one must add the pair (x, z) to $t(R)$. This strategy of keep adding pairs is not very effective as it becomes harder to keep track of the elements added and to compare to already existing ones in the relation. For example, consider the relation $R = \{(1, 2), (2, 3), (3, 4)\}$ on the set $A = \{1, 2, 3, 4\}$. Since $(1, 2) \in R$ and $(2, 3) \in R$, we must add the pair $(1, 3)$. Since $(2, 3) \in R$ and $(3, 4) \in R$, we must add the pair $(2, 4)$. We then get the relation $R^* = \{(1, 2), (2, 3), (3, 4), (1, 3), (2, 4)\}$. Unfortunately, R^* is still not transitive since $(1, 2) \in R^*$ and $(2, 4) \in R^*$ but $(1, 4) \notin R^*$) and so it cannot be the transitive closure of R. Obviously, we need a more constructive approach to build $t(R)$.

Definition 9.18 above introduced the connectivity relation R^∞ of R as being the set of pairs connected with a path of positive length in R. The next result shows that R^∞ is actually equal to the transitive closure of R. In the case of a finite set, this result introduces an algorithm to effectively compute the transitive closure.

Theorem 7.18. *If R is a binary relation on a set A, then $t(R) = R^\infty$.*

Proof. We need to prove that R^∞ is transitive, that it contains R and that any transitive relation containing R must also contain R^∞. Clearly, $R \subseteq R^\infty$. If (x, y) and (y, z) are in R^∞, then there exists a path of length $i \geq 1$ from x to y and a path of length $j \geq 1$ from y to z. This implies that there exists a path of length $i + j \geq 2$ from x to z and so $(x, z) \in R^\infty$. This proves that R^∞ is transitive. If S is a transitive relation on A containing R, then clearly $R^n \subseteq S^n$ for any positive integer n. Since S is transitive, $S^n \subseteq S$ for any positive integer n by Theorem 7.14 above. We conclude that $R^n \subseteq S$ for any positive integer n. By Theorem 7.11 above, $R^\infty = \cup_{n \geq 1} R^n \subseteq S$. This proves that $t(R) = R^\infty$. $\qquad\square$

7.8.1   *Warshall's algorithm*

Given a relation R on a set A, here is what we know so far from the previous results:

- $t(R) = R^\infty = \cup_{k \geq 1} R^k$.
- If A is a finite set with n elements, then $t(R) = \cup_{i=1}^{n} R^i$.
- If A is a finite set with n elements, then the adjacency matrix of $t(R) = R^\infty$ is $M_R \vee M_R^{[2]} \vee \cdots \vee M_R^{[n]}$ where M_R is the adjacency matrix of R and $M_R^{[k]}$ is the kth Boolean power of M_R.

Definition 7.20. If $(a, x_1), (x_1, x_2), \ldots, (x_k, b)$ is a path from a to b in R, then $x_1, x_2, \ldots, x_k$ are called the *interior vertices* (or *interior points*) of the path.

Example 7.47. For the relation R represented by the digraph below, the vertices c, d, e, d, a, b are interior vertices of path $p : b, c, d, e, d, a, b, e$.

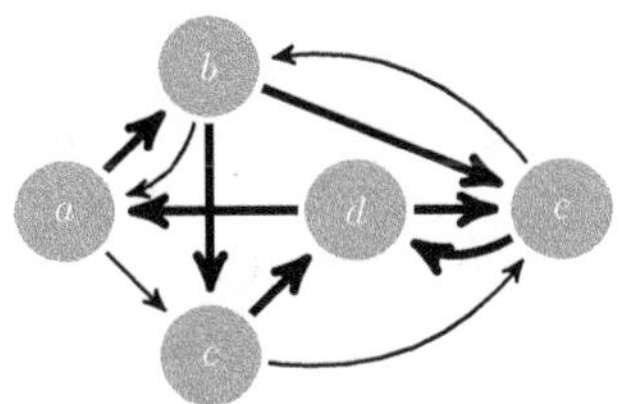

Given a relation R on the set $A = \{a_1, a_2, \ldots, a_n\}$, we introduce a new class of Boolean matrices W_k related to R as follows. First assume that the order $a_1, a_2, \ldots, a_n$ of the elements of A is arbitrary but fixed. All matrices that follow are computed with respect to this order. For $k = 0$, define W_0 to be the adjacency matrix M_R of R. For $k \geq 1$, the matrix W_k is an

$n \times n$ Boolean matrix (each entry is either 0 or 1) with the (i, j)-entry equals to 1 if and only if there exists a path from a_i to a_j with interior vertices belonging to the set $\{a_1, a_2, \ldots, a_k\}$ (first k elements of A). Note that W_n is the adjacency matrix of the connectivity relation R^∞ of R since the (i, j)-entry of W_n is equal to 1 if and only if there exists a path from a_i to a_j with interior vertices belonging to the set $\{a_1, a_2, \ldots, a_n\} = A$. This is simply a path of positive length in R from a_i to a_j. Using Theorem 7.18 above, the adjacency matrix of the transitive closure $t(R)$ of R is equal to W_n. This suggests that we turn our attention to the computation of the matrices W_k.

A key question to answer is the following: If the matrix W_{k-1} is constructed for $k \geq 1$, how can we construct the next matrix W_k? The answer is in the following observations. Write $W_{k-1} = [u_{ij}]$ and $W_k = [v_{ij}]$ where the notation α_{ij} stands for the entry at the ith row and the jth column of the matrix $\mathcal{M} = [\alpha_{ij}]$. If $u_{ij} = 1$, then there exists a path p from a_i to a_j with interior vertices in $\{a_1, \ldots, a_{k-1}\}$. In particular, path p has its interior vertices in $\{a_1, \ldots, a_k\}$, which means that $v_{ij} = 1$. Now assume that $u_{ik} = u_{kj} = 1$, then there exists path p_1 from a_i to a_k and path p_2 from a_k to a_j both having interior vertices in the set $\{a_1, \ldots, a_{k-1}\}$:

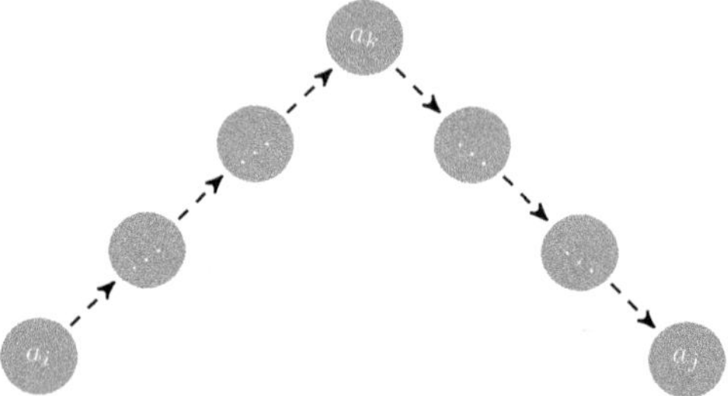

This gives us a path from a_i to a_j having interior vertices in the set $\{a_1, \ldots, a_k\}$. In other words, $v_{ij} = 1$. In summary: if $u_{ij} = 1$ or $u_{ik} = u_{kj} = 1$, then $v_{ij} = 1$. We prove that the converse is also true. Assume $v_{ij} = 1$, then there exists a path p from a_i to a_j with interior vertices in the set $\{a_1, a_2, \ldots, a_k\}$. Consider the following two cases.

Case 1. *The element a_k is not an interior vertex of p.* In this case, the path p can be thought of as having its interior vertices in the set $\{a_1, \ldots, a_{k-1}\}$, which implies that $u_{ij} = 1$.

Case 2. *a_k is an interior vertex of p.* In this case, we can assume that a_k appears exactly once in p by leaving out all cycles (if any) at a_k

(hence creating a shorter path). This way, we have a path from a_i to a_k and another from a_k to a_j both of which have interior vertices in the set $\{a_1, \ldots, a_{k-1}\}$. This means that $u_{ik} = u_{kj} = 1$.

This proves that $v_{ij} = 1 \Leftrightarrow u_{ij} = 1$ or $u_{ik} = u_{kj} = 1$, which translates into the following Boolean relation:

$$v_{ij} = u_{ij} \vee (u_{ik} \wedge u_{kj}). \tag{7.1}$$

Here is a description of Warshall's algorithm to compute the adjacency matrix of $t(R)$ using the above discussion: start with the matrix $W_0 = M_R$, construct W_k from W_{k-1} for $k = 1, \ldots, n$ based on relation (7.1). By the above discussions, W_n is the adjacency matrix of the transitive closure $t(R)$ of R.

When constructing W_k from W_{k-1} for $k \geq 1$, it helps to keep the following facts in mind.

- All the 1's in W_{k-1} are transferred to the same positions in W_k.
- An entry of 0 in the (i, j) position of W_{k-1} is changed to 1 in the same position in W_k if and only if the elements in the (i, k) and the (k, j) positions in W_{k-1} are both 1.
- If all entries of row i of W_{k-1} are 0's, then all entries in row i of W_k are also 0's.

Example 7.48. Consider the relation $R = \{(a, b), (b, a), (b, d), (d, a), (d, c)\}$ on the set $A = \{a, b, c, d\}$. Use Warshall's algorithm to find the transitive closure of R. Draw a digraph for $t(R)$.

Solution. First note that the relation R is not transitive. For example, $(a, b), (b, a) \in R$ but $(a, a) \notin R$. The adjacency matrix of R (relative to the order a, b, c, d of the elements of A) is:

$$\begin{bmatrix} 0 & 1 & 0 & 0 \\ 1 & 0 & 0 & 1 \\ 0 & 0 & 0 & 0 \\ 1 & 0 & 1 & 0 \end{bmatrix}.$$

This is the matrix W_0 in Warshall's algorithm. To construct W_1 from W_0, all 1's in W_0 are kept in the same positions in W_1 and an entry of 0 in the (i, j)-position of W_0 is changed to 1 in the same position of W_1 if the $(i, 1)$ and the $(1, j)$ positions in W_0 are both 1. For instance, the $(2, 2)$-position in W_1 is 1 since each of the $(2, 1)$ and the $(1, 2)$ positions in W_0 is 1. On the other hand, the $(2,3)$-position in W_1 is 0 since the $(2,1)$-position in W_0 is 1 but the $(1,3)$-position is 0. Continuing to reason this way, we get the Warshall matrices W_1, W_2, W_3 and W_4 as follows:

$$W_1 = \begin{bmatrix} 0 & 1 & 0 & 0 \\ 1 & 1 & 0 & 1 \\ 0 & 0 & 0 & 0 \\ 1 & 1 & 1 & 0 \end{bmatrix}, \ W_2 = \begin{bmatrix} 1 & 1 & 0 & 1 \\ 1 & 1 & 0 & 1 \\ 0 & 0 & 0 & 0 \\ 1 & 1 & 1 & 1 \end{bmatrix}, \ W_3 = \begin{bmatrix} 1 & 1 & 0 & 1 \\ 1 & 1 & 0 & 1 \\ 0 & 0 & 0 & 0 \\ 1 & 1 & 1 & 1 \end{bmatrix}, \ W_4 = \begin{bmatrix} 1 & 1 & 1 & 1 \\ 1 & 1 & 1 & 1 \\ 0 & 0 & 0 & 0 \\ 1 & 1 & 1 & 1 \end{bmatrix}.$$

We conclude that the transitive closure of R consists of the following pairs: (a,a), (a,b), (a,c), (a,d), (b,a), (b,b), (b,c), (b,d), (d,a), (d,b), (d,c), (d,d). The following is a digraph representation of both R and $t(R)$:

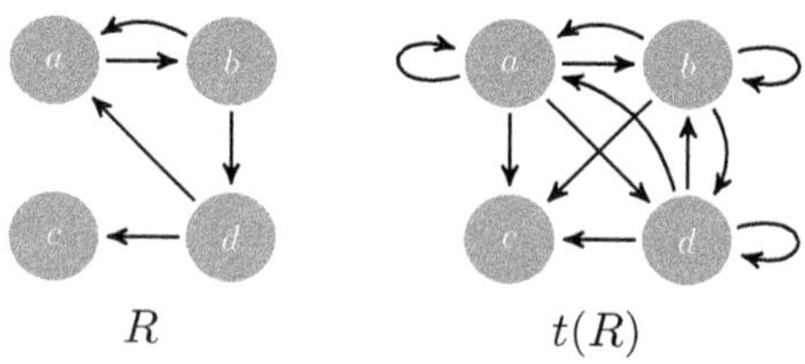

$$R \qquad\qquad\qquad t(R)$$

$\Diamond$

7.8.2   *Exercises*

(1) A relation R on $A = \{a,b,c,d\}$ is given by its elements: (a,a), (a,b), (a,d), (b,d), (c,a), (c,c), (b,a). Find the reflexive closure $r(R)$ of R.

(2) A relation R on $A = \{a,b,c,d\}$ is given by its elements: (a,d), (a,b), (a,a), (b,d), (c,a), (c,b). Find the symmetric closure $s(R)$ of R.

(3) Let A be a non-empty set. Find the reflexive, symmetric and transitive closures for each of the following relations on A:

    (a) $\emptyset$ (c) $I_A = \{(x,x); \ x \in A\}$

    (b) $A \times A$

(4) Let R be a relation on a set A. Prove that:

    (a) R is reflexive if and only if $r(R) = R$.

    (b) R is symmetric if and only if $s(R) = R$.

    (c) R is transitive if and only if $t(R) = R$.

(5) Let $A = \{a,b,c,d,e\}$. In each case find the reflexive, symmetric, and transitive closures of the given relation R on A.

(a) $R = \{(a,a), (a,e), (b,b), (c,a)\}$.
(b) $R = \{(a,b), (a,c), (a,e), (b,a), (b,c), (c,a), (c,b), (d,a), (e,d)\}$.
(c) $R = \{(a,e), (b,a), (b,d), (c,d), (d,a), (d,c), (e,a), (e,b), (e,c), (e,e)\}$.

(6) In each case, a relation R is given on a set A. Describe in words the relations $r(R)$, $s(R)$ and $t(R)$.

    (a) $A = \mathbb{Z}$, R is the relation "greater than".

$\star$ (b) $A = \mathbb{Z}\backslash\{0\}$, $a\,R\,b$ if and only if a divides b.

    (c) A is the set of all people, $a\,R\,b$ if and only if a is a parent of b.

    (d) $A = \mathbb{N}$, $a\,R\,b$ if and only if $a - b = 1$.

$\star$ (e) $A = \mathbb{Z}$, $m\,R\,n$ if and only if $m + n$ is even.

(7) In each case, the adjacency matrix of a binary relation R on the set $A = \{a, b, c\}$ is given. Give the adjacency matrix of $r(R)$, $s(R)$ and $t(R)$ (use Warshall's algorithm to compute the adjacency matrix of $t(R)$).

$$(a)\ \begin{array}{c} \\ a \\ b \\ c \end{array}\begin{array}{ccc} a & b & c \\ \end{array}\begin{bmatrix} 1 & 1 & 0 \\ 1 & 1 & 0 \\ 0 & 0 & 1 \end{bmatrix} \quad \star(b)\ \begin{bmatrix} 0 & 1 & 0 \\ 1 & 0 & 0 \\ 1 & 0 & 1 \end{bmatrix} \quad (c)\ \begin{bmatrix} 1 & 1 & 0 \\ 1 & 1 & 0 \\ 1 & 1 & 1 \end{bmatrix} \quad (d)\ \begin{bmatrix} 0 & 1 & 0 \\ 0 & 1 & 0 \\ 0 & 0 & 1 \end{bmatrix}$$

(8) In each case, the adjacency matrix of a binary relation R on the set $A = \{a, b, c, d\}$ is given. Give the adjacency matrix of $r(R)$, $s(R)$ and $t(R)$ (use Warshall's algorithm to compute the adjacency matrix of $t(R)$).

$$(a)\ \begin{bmatrix} 1 & 0 & 0 & 0 \\ 1 & 1 & 1 & 0 \\ 0 & 0 & 1 & 0 \\ 0 & 0 & 0 & 1 \end{bmatrix} \quad (b)\ \begin{bmatrix} 1 & 1 & 0 & 0 \\ 0 & 1 & 1 & 0 \\ 1 & 0 & 1 & 1 \\ 0 & 0 & 0 & 1 \end{bmatrix} \quad \star(c)\ \begin{bmatrix} 0 & 0 & 1 & 0 \\ 1 & 0 & 1 & 0 \\ 1 & 0 & 1 & 0 \\ 1 & 0 & 0 & 1 \end{bmatrix} \quad (d)\ \begin{bmatrix} 1 & 0 & 0 & 1 \\ 1 & 0 & 1 & 0 \\ 1 & 0 & 1 & 0 \\ 1 & 1 & 1 & 1 \end{bmatrix}$$

(9) In each case, the adjacency matrix of a binary relation R on the set $A = \{a, b, c, d, e\}$ is given. Give the adjacency matrix of $r(R)$, $s(R)$ and $t(R)$ (use Warshall's algorithm to compute the adjacency matrix of $t(R)$).

$$(a)\ \begin{bmatrix} 1 & 1 & 0 & 0 & 1 \\ 1 & 1 & 0 & 0 & 0 \\ 0 & 0 & 1 & 0 & 0 \\ 0 & 0 & 0 & 1 & 0 \\ 1 & 0 & 0 & 0 & 1 \end{bmatrix} \quad \star(b)\ \begin{bmatrix} 1 & 0 & 1 & 0 & 1 \\ 1 & 0 & 1 & 0 & 0 \\ 0 & 0 & 0 & 1 & 0 \\ 0 & 0 & 0 & 1 & 0 \\ 0 & 0 & 0 & 1 & 0 \end{bmatrix} \quad (c)\ \begin{bmatrix} 0 & 0 & 1 & 1 & 1 \\ 1 & 0 & 0 & 0 & 1 \\ 1 & 1 & 0 & 1 & 0 \\ 0 & 1 & 0 & 1 & 0 \\ 1 & 0 & 1 & 1 & 0 \end{bmatrix}$$

(10) In each case, the digraph representing a binary relation R on the set $A = \{1, 2, 3, 4, 5\}$ is given. Draw the digraphs of the relations $r(R)$, $s(R)$ and $t(R)$.

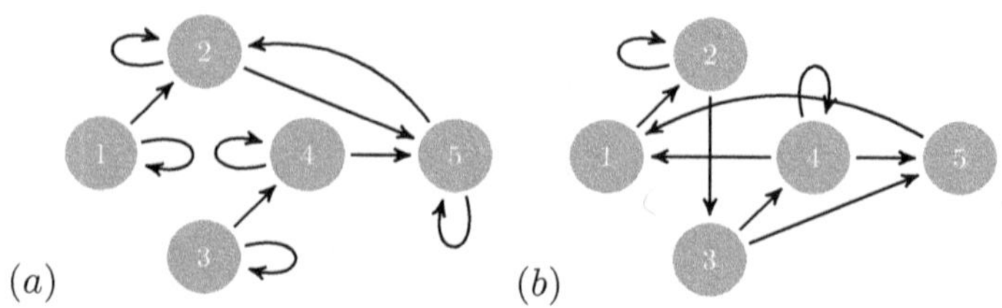

(a) $\qquad\qquad\qquad\qquad\qquad\qquad\qquad$ (b)

(11) Let R be a binary relation on a set A.

    (a) Prove that if R is reflexive, then so are $s(R)$ and $t(R)$.

  $\star$ (b) Prove that if R is symmetric then so are $r(R)$ and $t(R)$.

    (c) Prove that if R is transitive then so is $r(R)$.

$\star$(12) Give an example of a transitive relation R on a set A such that $s(R)$ is not transitive.

(13) For a relation R on a set A, define the *reflexive symmetric closure* of R as being the reflexive closure of $s(R)$ that we denote by $rs(R)$. In other words, $rs(R) = r\,(s(R))$. Similarly, we can define other double closures of R: $sr(R) = s\,(r(R))$, $rt(R) = r\,(t(R))$, $tr(R) = t\,(r(R))$, $st(R) = s\,(t(R))$ and $ts(R) = t\,(s(R))$. Prove:

    (a) $rs(R) = sr(R)$ $\quad\star$ (b) $rt(R) = tr(R)$ $\qquad$ (c) $st(R) \subseteq ts(R)$

(*Hint.* You might find the results of Exercise (11) helpful).

(14) Give an example of a relation R on a set A such that $st(R) \neq ts(R)$.

7.9 Equivalence relations

A relation on a set A which is reflexive, symmetric and transitive can be used to partition A into classes of "alike" elements. This is very useful in mathematics as it allows us to deal with one representative of a class rather than each element individually.

Definition 7.21. A relation R on a set A is called an *equivalence relation* on A if it is reflexive, symmetric and transitive at the same time. If R is an equivalence relation on A and $a, b \in A$ are such that $a\,R\,b$, we say that a and b are *similar* or *equivalent* and we sometimes write $a \sim b$.

Example 7.49. For any set A, the complete relation $A \times A$ is clearly an equivalence relation on A.

Example 7.50. Define a relation R on $\mathbb{Z}$ as follows: $m \, R \, n \Leftrightarrow m - n$ is even. Prove that R is an equivalence relation on $\mathbb{Z}$.

Solution. If $m \in \mathbb{Z}$, then $m - m = 0$ is even and so $m \, R \, m$. The relation is reflexive. If $m, n \in \mathbb{Z}$ are such that $m \, R \, n$, then $m - n$ is even which implies that $n - m = -(m - n)$ is also even and so $n \, R \, m$. The relation is symmetric. Finally, if $m, n, t \in \mathbb{Z}$ are such that $m \, R \, n$ and $n \, R \, t$ then $m - n$ and $n - t$ are even and so $m - t = (m - n) + (n - t)$ is even (as the sum of two even integers). We conclude that $m \, R \, t$ and the relation is transitive. Hence, R is an equivalence relation on $\mathbb{Z}$. $\Diamond$

Example 7.51. Example 7.50 is a particular case of the following (more general) equivalence relation on $\mathbb{Z}$, called the *congruence modulo n* relation. It is widely used in many area of mathematics and computer science. Fix an integer $n \geq 2$. We say that integers $a, b \in \mathbb{Z}$ are *congruent modulo n*, and we write $a \equiv b \, (\text{mod } n)$, if and only if $a - b$ is divisible by n. The proof that this defines an equivalence relation on $\mathbb{Z}$ is left to the reader as an exercise (see Exercise (11) below).

Example 7.52. On the set $\mathbb{R}^* = \mathbb{R} \backslash \{0\}$ of all non-zero real numbers, consider the relation R defined as follows: $x \, R \, y \Leftrightarrow \frac{x}{y} = 2^k$ for some $k \in \mathbb{Z}$. For example, $\frac{1}{8} \, R \, 4$ since $\frac{\frac{1}{8}}{4} = \frac{1}{32} = 2^{-5}$ (with $-5 \in \mathbb{Z}$) and $32 \, R \, 4$ since $\frac{32}{4} = 8 = 2^3$ (with $3 \in \mathbb{Z}$). Prove that R is an equivalence relation on $\mathbb{R}^*$.

Solution. If $x \in \mathbb{R}^*$, then $x \, R \, x$ since $\frac{x}{x} = 1 = 2^0$ and $0 \in \mathbb{Z}$. The relation is reflexive. If $x, y \in \mathbb{R}^*$ are such that $x \, R \, y$, then $\frac{x}{y} = 2^k$ for some $k \in \mathbb{Z}$ which implies that $\frac{y}{x} = 2^{-k}$ with $-k \in \mathbb{Z}$. This shows that $y \, R \, x$ and R is symmetric. Finally, if $x, y, z \in \mathbb{R}^*$ are such that $x \, R \, y$ and $y \, R \, z$, then $\frac{x}{y} = 2^k$ and $\frac{y}{z} = 2^l$ for some integers k and l. Then $\frac{x}{z} = \left(\frac{x}{y}\right)\left(\frac{y}{z}\right) = 2^k 2^l = 2^{k+l}$ with $k + l \in \mathbb{Z}$. This shows that $x \, R \, z$. The relation is transitive. We conclude that R is an equivalence relation on $\mathbb{R}^*$. $\Diamond$

Example 7.53. Let $\mathbb{Z}^* = \mathbb{Z} \backslash \{0\}$ be the set of all non-zero integers. On the set $A = \mathbb{Z} \times \mathbb{Z}^*$ we define a relation R as follows: $(a, b) \, R \, (m, n) \Leftrightarrow an = bm$. Show that R is an equivalence relation on A.

Solution. For any $(a, b) \in A$, we have that $ab = ba$ which means that $(a, b) \, R \, (a, b)$. The relation is reflexive. For $(a, b), (x, y) \in A$, if $(a, b) \, R \, (x, y)$

then $ay = bx$ which can be written as $xb = ya$. This last equation means that $(x, y) R (a, b)$ and R is symmetric. For the transitivity, assume $(a, b) R (x, y)$ and $(x, y) R (s, t)$ for some $(a, b), (x, y)$ and $(s, t) \in A$. Then $ay = bx$ and $xt = ys$. From the first equation, we get that $x = \frac{ay}{b}$ (this is possible since $b \neq 0$). Replacing in the second equation, we get that $\left(\frac{ay}{b}\right) t = ys$ which simplifies to $\frac{a}{b} t = s$ (again, this is possible since $y \neq 0$) or $at = bs$. This is precisely the definition of $(a, b) R (s, t)$. We conclude that R is transitive. Thus, R is an equivalence relation on A. $\Diamond$

Example 7.54. On the set Σ_n of all binary strings of length $n \geq 1$, we consider the relation: $x R y$ if and only if x and y have the same number of 0's. Prove that R is an equivalence relation on Σ_n.

Solution. Clearly, any binary string has the same number of 0's as itself, so R is reflexive. If $x, y \in \Sigma_n$ are such that x has the same number of 0's as y, then y has the same number of 0's as x. The relation R is then symmetric. For the transitivity, assume $x, y, z \in \Sigma_n$ are such that x and y have the same number of 0's and y and z have the same number of 0's. Then x and z have the same number of 0's. The relation is transitive and hence it is an equivalence relation on A. $\Diamond$

Example 7.55. On the set Σ_n of all binary strings of length $n \geq 1$, we consider the relation R defined as follows: $x R y \Leftrightarrow x$ and y end with same digit. Prove that R is an equivalence relation on Σ_n.

Solution. Left as an exercise (see Exercise (12) below). $\Diamond$

In some cases, we need to look at a certain relation R on A in terms of its effect on the elements of a subset B of A.

Definition 7.22. Given a relation R on a set A and a subset B of A, the *restriction* of R to B, denoted by R/B, is the relation $R \cap (B \times B)$ on B.

Theorem 7.19. *If R is an equivalence relation on a non-empty set A, then for any non-empty subset B of A the restriction R/B of R to B is an equivalence relation on B.*

Proof. Left as an exercise (see Exercise (21) below). $\square$

Example 7.56. On the set $\mathbb{Z}^* = \mathbb{Z} \backslash \{0\}$ of all non-zero integers, the relation R defined by: $x R y \Leftrightarrow \frac{x}{y} = 2^k$ for some $k \in \mathbb{Z}$ is an equivalence relation on $\mathbb{Z}^*$. This is true since the relation in this case is the restriction of the equivalence relation on $\mathbb{R}^*$ given in Example 7.52 above to the subset $\mathbb{Z}^*$ of $\mathbb{R}^*$.

Theorem 7.20. *If R and S are two equivalence relations on a set A, then $R \cap S$ is also an equivalence relation on A.*

Proof. For $x \in A$, we know that $x\,R\,x$ and $x\,S\,x$ since both R and S are reflexive. This implies that $x\,(R \cap S)\,x$ and so $R \cap S$ is reflexive. If $x, y \in A$ are such that $x\,(R \cap S)\,y$, then $x\,R\,y$ and $x\,S\,y$. Since R and S are both symmetric, $y\,R\,x$ and $y\,S\,x$ and so $y\,(R \cap S)\,x$. The relation $R \cap S$ is symmetric. For the transitivity, let $x, y, z \in A$ such that $x\,(R \cap S)\,y$ and $y\,(R \cap S)\,z$. Then $x\,R\,y$, $x\,S\,y$, $y\,R\,z$ and $y\,S\,z$. By the transitivity of R and S, we get that $x\,R\,z$ and $x\,S\,z$ which implies that $x\,(R \cap S)\,z$. We conclude that $R \cap S$ is an equivalence relation on A. $\qquad\square$

Note that Theorem 7.20 is not true in general for the union of equivalence relations (see Exercise (34 below)).

Example 7.57. On the set Σ_n of all binary strings of length $n \geq 1$, we consider the relation R defined as follows: $x\,R\,y$ if and only if x and y have the same number of 0's and x and y end with same digit. Then R is an equivalence relation on Σ_n. This is a direct consequence of Theorem 7.20 and Examples 7.54 and 7.55 above.

The next result may come as a surprise as it shows that the composition of equivalence relations on a set is not always an equivalence relation.

Theorem 7.21. *Let R and S be two equivalence relations on a set A. Then $S \circ R$ is an equivalence relation on A if and only if $S \circ R = R \circ S$.*

Proof. Assume first that $S \circ R = R \circ S$. We need to prove that $S \circ R$ is an equivalence relation on A. If $x \in A$, then $(x, x) \in R$ and $(x, x) \in S$ since both R and S are reflexive. By the definition of $S \circ R$, $(x, x) \in S \circ R$ and so $S \circ R$ is reflexive. Assume next that $(x, y) \in S \circ R$. Then there exists $z \in A$ such that $(x, z) \in R$ and $(z, y) \in S$. Since R and S are symmetric, $(z, x) \in R$ and $(y, z) \in S$. Conditions $(y, z) \in S$ and $(z, x) \in R$ imply that $(y, x) \in R \circ S = S \circ R$. The relation $S \circ R$ is symmetric. For the transitivity of $S \circ R$, assume that $x, y, z \in A$ are such that $(x, y) \in S \circ R$ and $(y, z) \in S \circ R$. By our assumption, $(y, z) \in R \circ S$. Let $s, t \in A$ be such that $(x, s) \in R$, $(s, y) \in S$, $(y, t) \in S$ and $(t, z) \in R$. By the transitivity of S, $(s, t) \in S$. So we have that $(x, s) \in R$, $(s, t) \in S$ and $(t, z) \in R$. The last two statements imply that $(s, z) \in R \circ S$. Since $R \circ S = S \circ R$, there exists $p \in A$ such that $(s, p) \in R$ and $(p, z) \in S$. The conditions $(x, s) \in R$ and $(s, p) \in R$ imply that $(x, p) \in R$ by the transitivity of R. Now we have that $(x, p) \in R$ and $(p, z) \in S$ which means that $(x, z) \in S \circ R$. This shows that $S \circ R$ is indeed transitive, and hence it is an equivalence relation on A. For the other direction, assume that $S \circ R$ is an equivalence relation on A, we need to prove that $S \circ R = R \circ S$. If $(x, y) \in S \circ R$, then $(y, x) \in S \circ R$ by the symmetry of $S \circ R$. So there exists $z \in A$ such that $(y, z) \in R$ and $(z, x) \in S$. By the symmetry of R and S, we get that $(z, y) \in R$ and $(x, z) \in S$ and consequently, $(x, y) \in R \circ S$. This shows that $S \circ R \subseteq R \circ S$. The inclusion $R \circ S \subseteq S \circ R$ is proven similarly. $\qquad \square$

Theorem 7.22. *If R is an equivalence relation on a set A, then R^n is also an equivalence relation on A for any integer $n \geq 0$.*

Proof. The result is true for $n = 0$ as $R^0 = \{(x, x); x \in A\}$ is an equivalence relation on A. Let $n \geq 0$ and assume that R^n is an equivalent relation on A. Then $R^{n+1} = R^n \circ R$ is also an equivalence relation on A by Theorem 7.21 since in this case, $R^n \circ R = R \circ R^n$. The result follows by the principle of induction. $\qquad \square$

7.9.1 *Equivalence classes and partition of a set*

An equivalence relation on a set A breaks A into subsets of "similar" elements that we call equivalence classes. Once the classes are identified, each class can be represented by a single element. This is a very common procedure in dealing with mathematical structures.

Definition 7.23. Given an equivalence relation R on a set A and an element a of A, the *equivalence class* of a with respect to R is the subset of A, denoted by $[a]_R$ (or sometimes $\overline{a}^R$), of all elements of A which are in relation with a: $[a]_R := \{x \in A; x R a\}$. When there is no ambiguity regarding the relation R, we write $[a]$ (or $\overline{a}$) for the equivalence class of a. The set of all equivalence classes of R is called the *quotient set* of A that we denote by A/R. So, $A/R := \{[a]; a \in A]\}$.

Example 7.58. In this example, we give a description of the equivalence class $[1]$ of the element $1 \in \mathbb{Z}$ for the equivalence relation given in Example 7.52 above and we give three elements of $[1]$ other than 1. By definition:

$$[1] = \{x \in \mathbb{R}^*; \, x \, R \, 1\}$$
$$= \left\{x \in \mathbb{R}^*; \, \frac{x}{1} = 2^k \text{ for some } k \in \mathbb{Z}\right\}$$
$$= \left\{2^k \text{ for some } k \in \mathbb{Z}\right\}.$$

So $[1]$ is the collection of all integer powers of 2. Hence $2^{-3} = \frac{1}{8}$, $2^{-1} = \frac{1}{2}$, $2^2 = 4$ and $2^6 = 64$ are elements of $[1]$.

Example 7.59. Describe the equivalence class $[(1,2)]$ of the element $(1,2) \in \mathbb{Z} \times \mathbb{Z}^*$ for the equivalence relation given in Example 7.53 above. Give three elements of $(1,2)$ other than $(1,2)$.

Solution. Using the definition of an equivalence class:

$$[(1,2)] = \{(a,b) \in A; \, (a,b) \, R \, (1,2)\} = \{(a,b) \in A; 2a = b\} = \{(a,2a); a \in \mathbb{Z}^*\}.$$

In particular $(-1,-2)$, $(2,4)$ and $(3,6)$ are elements of $[(1,2)]$. $\quad\Diamond$

Example 7.60. Find the equivalence class of the binary string $x = 10111$ of the set Σ_5 for the equivalence relation on Σ_5 defined in Example 7.54 above.

Solution. A binary string of Σ_5 is related to $x = 10111$ if and only of it has the same number of 0's, namely one 0. So $[10111]$ consists of all binary strings of length 5 that have exactly one 0: 01111, 10111, 11011, 11101, 11110. $\quad\Diamond$

Example 7.61. Let $n \geq 2$ be an integer. Consider the congruence modulo n relation $\equiv \pmod{n}$ on $\mathbb{Z}$ defined in Example 7.51 above. Give a description of the equivalence classes of this relation.

Solution. Note first that there are exactly n possible remainders in the division by n, namely $0, 1, \ldots, n-1$. Given $k \in \mathbb{Z}$, write $k = \alpha n + r$ for some $\alpha, r \in \mathbb{Z}$ with $0 \leq r < n$ (using the division algorithm). In particular, $k - r = \alpha n$ which means that $k \equiv r \pmod{n}$. This shows that $k \in [r]$ for some $r \in \{0, 1, \ldots, n-1\}$. The distinct equivalence classes of the congruence modulo n relation are then $[0], [1], \ldots, [n-1]$. For instance, there are only three distinct equivalence classes of the congruence modulo 3, namely $[0]$, $[1]$ and $[2]$. $\quad\Diamond$

Given an equivalence relation R on a set A, the next theorem paves the way for using the equivalence classes of R as a tool to partition A into classes of "similar" elements.

Theorem 7.23. *Let R be an equivalence relation on a non-empty set A and let $a, b \in A$. The following conditions are equivalent.*

(i) $a\,R\,b$ $\qquad\qquad$ *(ii)* $[a] = [b]$ $\qquad\qquad$ *(iii)* $[a] \cap [b] \neq \emptyset$

Proof. We prove the cycle of implications: $(i) \Rightarrow (ii) \Rightarrow (iii) \Rightarrow (i)$. For the first implication, assume that $a\,R\,b$. If $x \in [a]$, then $x\,R\,a$ and so $x\,R\,b$ since $a\,R\,b$ and R is transitive. This shows that $x \in [b]$, and consequently $[a] \subseteq [b]$. The other inclusion is proven similarly. For the implication $(ii) \Rightarrow (iii)$, assume that $[a] = [b]$. Since $a \in [a]$ and $[a] = [b]$, $a \in [a] \cap [b]$ and therefore $[a] \cap [b] \neq \emptyset$. This proves (iii). For the implication $(iii) \Rightarrow (i)$, assume $[a] \cap [b] \neq \emptyset$, and choose an element $x \in [a] \cap [b]$. In particular $a\,R\,x$ and $x\,R\,b$ by definition of an equivalence class. Since R is transitive, $a\,R\,b$ and (i) is true. $\qquad\qquad\qquad\qquad\qquad\qquad\qquad\qquad\qquad\qquad\square$

Theorem 7.23 says in particular that two equivalence classes cannot overlap at just a proper (non-empty) subset of A. They are either disjoint (their intersection is empty) or they coincide (exactly the same subset of A). In other words, as soon as two equivalence classes have one element in common, they are equal.

Theorem 7.24. *Let R be an equivalence relation on a non-empty set A.*

(a) The quotient set A/R defines a partition on A.
(b) Conversely, any partition $\mathcal{F} = \{A_i;\ i \in I\}$ on A defines an equivalence relation $R_{\mathcal{F}}$ on A with $A/R_{\mathcal{F}} = \mathcal{F}$.

Proof.

(a) Since R is reflexive, $a \in [a]$ for any element a of A. In particular, $[a] \neq \emptyset$ for any $[a] \in A/R$. If $[a], [b] \in A/R$ are such that $[a] \neq [b]$, then $[a] \cap [b] = \emptyset$ by Theorem 7.23. This shows that distinct elements of A/R are pairwise disjoint. Finally, note that $[a] \subseteq A$ for any $a \in A$ which means that $\cup_{a \in A}[a] \subseteq A$. The other inclusion is also true since if $x \in A$, then $x \in [x] \subseteq \cup_{a \in A}[a]$. Thus $A = \cup_{a \in A}[a]$ and the quotient set A/R forms a partition on A.
(b) Assume that $\mathcal{F} = \{A_i;\ i \in I\}$ is a partition of A. Define a relation $R_{\mathcal{F}}$ on A as follows: $x\,R_{\mathcal{F}}\,y$ if and only if $x, y \in A_i$ for some $i \in I$. We prove

that $R_{\mathcal{F}}$ is an equivalence relation on A. Given $x \in A$, there exists $i \in I$ such that $x \in A_i$ since $A = \cup_{i \in I} A_i$. So $x\,R_{\mathcal{F}}\,x$ and $R_{\mathcal{F}}$ is reflexive. If $x, y \in A$ are such that $x\,R_{\mathcal{F}}\,y$, then $x, y \in A_i$ for some $i \in I$ and so $y, x \in A_i$ for some $i \in I$. This proves that $y\,R_{\mathcal{F}}\,x$ and $R_{\mathcal{F}}$ is symmetric. For the transitivity of $R_{\mathcal{F}}$, let $x, y, z \in A$ such that $x\,R_{\mathcal{F}}\,y$ and $y\,R\,z$. Then there exist $i, j \in I$ such that $x, y \in A_i$ and $y, z \in A_j$. Since $y \in A_i \cap A_j$, we must have that $A_i = A_j$ (by definition of a partition) and so $x, z \in A_i = A_j$. This implies that $x\,R_{\mathcal{F}}\,z$. We conclude that $R_{\mathcal{F}}$ is an equivalence relation on A. Finally, if $[x] \in A/R_{\mathcal{F}}$, then there exists a *unique* $i \in I$ such that $x \in A_i$. This shows that $[x] = A_i$, and therefore $A/R_{\mathcal{F}} \subseteq \mathcal{F}$. Conversely, if $A_i \in \mathcal{F}$ then $A_i = [x]_{\mathcal{F}}$ for any $x \in A_i$ and so $A_i \in A/R_{\mathcal{F}}$. This proves that $A/R_{\mathcal{F}} = \mathcal{F}$.

$\square$

Example 7.62. Let $n \geq 2$ be an integer. For the congruence modulo n relation on $\mathbb{Z}$, Example 7.61 tells us that there are exactly n distinct equivalence classes, namely $[0], [1], \ldots, [n-1]$. The quotient set in this case is $\{[0], [1], \ldots, [n-1]\}$ and this is a partition of the set $\mathbb{Z}$. The notation $\mathbb{Z}_n$ is widely used for the quotient set in this case.

Remark 7.6. Given an equivalence relation R on a finite set $A = \{a_1, \ldots, a_n\}$, the partition of A defined by R is determined as follows. Start by finding the elements of $[a_1]$ and cross out all elements of $[a_1]$ from the set A. In particular, a_1 is crossed out since $a_1 \in [a_1]$. Move to the next uncrossed element of A, find its equivalence class and cross out all its elements from A. Repeat the process until all elements of A are crossed out.

Example 7.63. On the subset $A = \{-6, -5, -3, -2, -1, 0, 3, 4, 5, 6, 10, 12\}$ of $\mathbb{Z}$, the congruence relation modulo 3 on $\mathbb{Z}$ induces an equivalence relation R on A by Theorem 7.19 above. The partition of A defined by R is determined by the following classes: $[-6] = \{-6, -3, 0, 3, 6, 12\}$, $[-5] = \{-5, -2, 4, 10\}$, $[-2] = \{-2\}$.

Example 7.64. Consider the equivalence relation R given in Example 7.54 above on the set $\Sigma_3 = \{111, 110, 101, 100, 011, 010, 001, 000\}$ of all binary strings of length 3. The partition of Σ_3 defined by R is determined by the following classes: $[111] = \{111\}$, $[110] = \{110, 101, 011\}$, $[100] = \{100, 010, 001\}$, $[000] = \{000\}$.

Example 7.65. The relation R on $\mathbb{R}^* = \mathbb{R}\backslash\{0\}$ defined by $x\,R\,y \Leftrightarrow \frac{x}{y} = 2^k$ for some $k \in \mathbb{Z}$ is an equivalence relation (Example 7.52 above). The partition of the subset $A = \{-32, -14, -7, -5, -2, -1, 1, 2, 3, 6, 64\}$ of $\mathbb{R}^*$ induced by R is determined by the following classes: $[-32] = \{-32, -2, -1\}$, $[-14] = \{-14, -7\}$, $[-5] = \{-5\}$, $[1] = \{1, 2, 64\}$, $[3] = \{3, 6\}$.

Example 7.66. Let $A = \{1, 2, 3, 4\}$. Determine the equivalence relation R on A corresponding to the partition $\wp = \{\{1, 2\}, \{3, 4\}\}$ of A.

Solution. Since R is an equivalence relation on A, it must contain $(1, 1)$, $(2, 2)$, $(3, 3)$ and $(4, 4)$. Moreover, since the subset $\{1, 2\}$ belongs to the partition we must have $(1, 2) \in R$ and $(2, 1) \in R$. Similarly, $(3, 4) \in R$ and $(4, 3) \in R$. There are no other pairs in R. Therefore: $R = \{(1, 1), (2, 2), (3, 3), (4, 4), (1, 2), (2, 1), (3, 4), (4, 3)\}$. $\Diamond$

7.9.2   *Well-defined operations on the quotient set*

Consider the rule $f : \mathbb{Q} \to \mathbb{Z}$ defined by $f\left(\frac{a}{b}\right) = a$. We have: $f\left(\frac{1}{2}\right) = 1$ and $f\left(\frac{2}{4}\right) = 2$. But $\frac{1}{2} = \frac{2}{4}$, so the rule assigns two different images to the same element. This is what we referred to as a *not well-defined function* in Chapter 5. Note that every element of $\mathbb{Q}$ has infinitely many representations, like $\frac{1}{2} = \frac{2}{4} = \frac{3}{6} = \dots$. In fact, $\mathbb{Q}$ can be thought of as the quotient set of the equivalence relation on $\mathbb{Z} \times \mathbb{Z}^*$ given in Example 7.53 above and as such, every element of $\mathbb{Q}$ is the representative of its equivalence class. So $\frac{1}{2}$ is just one representative of all the elements $\frac{1}{2}, \frac{2}{4}, \frac{3}{6}, \frac{4}{8}, \dots$. This suggests that one has to be careful when defining binary operations on the elements of the quotient set A/R or functions having A/R as domain.

Definition 7.24. Let R be an equivalence relation on a set A. If $\times$ is a binary operation on the elements of A, we can define a binary operation $\otimes$ on the elements of A/R as follows: $[x] \otimes [y] = [x \times y]$. We say that the operation $\otimes$ is *well-defined* if the following condition holds:

$$[x] = [a] \text{ and } [y] = [b] \Rightarrow [x \otimes y] = [a \otimes b].$$

In other words, $\otimes$ is well-defined if it is independent of the choice of the representatives of the equivalence classes.

Example 7.67. Let $n \geq 2$ be an integer. Consider the congruence modulo n equivalence relation on the set $\mathbb{Z}$: $a \equiv b \,(\mathrm{mod}\, n) \Leftrightarrow a - b$ is a multiple of n. We saw earlier that the quotient set in this case is

$\mathbb{Z}_n = \{[0], [1], [2], \ldots, [n-1]\}$. The standard addition of integers can be extended to an addition on $\mathbb{Z}_n$ as follows: $[x] \oplus [y] = [x+y]$. Prove that $\oplus$ is a well-defined operation on $\mathbb{Z}_n$.

Solution. Let $[x], [y], [a], [b] \in \mathbb{Z}_n$ be such that $[x] = [a]$ and $[y] = [b]$. Then $x - a = \alpha n$ and $y - b = \beta n$ for some $\alpha, \beta \in \mathbb{Z}$. Adding the two equations together, we get that $(x-a) + (y-b) = \alpha n + \beta n$ which can be written as $(x+y) - (a+b) = (\alpha + \beta)n$. This means that $(x+y) \equiv (a+b) \pmod{n}$ and so $[x+y] = [a+b]$. We conclude that $[x] \oplus [y] = [a] \oplus [b]$ and thus the operation $\oplus$ on $\mathbb{Z}_n$ is well-defined. $\diamond$

Example 7.68. Recall that if $A = \begin{bmatrix} a & b \\ c & d \end{bmatrix}$ is a 2×2 matrix, then the determinant of A, denoted by $\det(A)$, is defined to be the real number $\det(A) = ad - bc$. On the set $\mathbb{M}_{22}$ of all 2×2 matrices with real entries, we define the binary relation R as follows: $A \, R \, B$ if and only if $\det(A) = \det(B)$.

(a) Prove that R is an equivalence relation on $\mathbb{M}_{22}$.
(b) Define an operation $\oplus$ on $\mathbb{M}_{22}/R$ as follows: $[A] \oplus [B] = [A+B]$ where $A + B$ stands for the standard matrix addition (componentwise). Determine if $\oplus$ is well-defined.

Solution. The fact that R is an equivalence relation on $\mathbb{M}_{22}$ is straightforward and left to the reader. Consider the following matrices:

$$A = \begin{bmatrix} 1 & 0 \\ 2 & 4 \end{bmatrix}, B = \begin{bmatrix} 2 & -1 \\ 0 & 2 \end{bmatrix}, C = \begin{bmatrix} 1 & 0 \\ 0 & 0 \end{bmatrix}, D = \begin{bmatrix} 1 & -1 \\ -1 & 1 \end{bmatrix}.$$

We have that $\det(A) = \det(B) = 4$ and $\det(C) = \det(D) = 0$. This means that $[A] = [B]$ and that $[C] = [D]$. Note that

$$A + C = \begin{bmatrix} 2 & 0 \\ 2 & 4 \end{bmatrix}, B + D = \begin{bmatrix} 3 & -2 \\ -1 & 3 \end{bmatrix}.$$

Since $\det(A + C) = 8$ and $\det(C + D) = 7$, $(A+C) \, \not\!\!R \, (B+D)$ and so $[A+C] \neq [B+D]$. This means that $[A] \oplus [C] \neq [B] \oplus [D]$ and the operation $\oplus$ on $\mathbb{M}_{22}/R$ is not well-defined. $\diamond$

Definition 7.25. Let R be an equivalence relation on a set A. A function $f : A/R \to B$ is *well-defined* if and only if the following condition holds: $(\forall x \in A)(\forall y \in A)\,([x] = [y] \Rightarrow f([x]) = f([y]))$. In other words, the function $f : A/R \to B$ is well defined if and only if the condition $x \, R \, y$ implies that $f([x]) = f([y])$.

Example 7.69. Recall the equivalence relation R on $\mathbb{Z} \times \mathbb{Z}^*$ given in Example 7.53: $(a, b)\, R\, (x, y)$ if and only if $ay = bx$. The function $f : \mathbb{Z} \times \mathbb{Z}^*/R \to \mathbb{Z}$ defined by $f\left(\left[(a, b)\right]\right) = a + b$ is not well-defined since $[(1, 2)] = [(2, 4)]$ but $f\left(\left[(1, 2)\right]\right) = 3 \neq f\left(\left[(2, 4)\right]\right) = 6$.

Example 7.70. Consider the function $f : \mathbb{Z}_{12} \to \mathbb{Z}_4$ defined by $f\left([x]_{12}\right) = [x]_4$, where $[x]_{12}$ and $[x]_4$ are the equivalence classes of the integer x modulo 12 and 4 respectively. If $x, y \in \mathbb{Z}$ are such that $[x]_{12} = [y]_{12}$, then $x - y = 12n$ for some $n \in \mathbb{Z}$. In particular, $x - y = 4(3n)$ and so $x \equiv y \pmod 4$. This means that $[x]_4 = [y]_4$ and so $f\left([x]_{12}\right) = f\left([y]_{12}\right)$. The function f is then well-defined.

7.9.3   *Equivalence relation generated by an arbitrary relation*

Given a relation R on a set A, we want to add just enough pairs from $A \times A$ to R to make it an equivalence relation on A. Like we did in Section 7.8 on closures of relations, we would like to find the *smallest* equivalence relation on A that contains R.

Definition 7.26. Given a relation R on a set A, we define the *equivalence relation on A generated by R* as being the smallest equivalence relation on A that contains R as a subset. In other words, the relation S is the equivalence relation on A generated by R if the following conditions hold:

(a) S is an equivalence relation on A;
(b) $R \subseteq S$;
(c) If T is an equivalence relation on A such that $R \subseteq T$, then $S \subseteq T$.

Theorem 7.25. *For a relation R on a set A, the equivalence relation on A generated by R is equal to $tsr\,(R)$ (the transitive closure of the symmetric closure of the reflexive closure of R).*

Proof.   The fact that $tsr(R)$ is an equivalence relation on R is left as an exercise (see Exercise (39) below). Clearly, $R \subseteq tsr(R)$ by definition of the closures of a relation. Remains to show that $tsr(R)$ is the smallest equivalence relation on A containing R. To see this, assume that T is an equivalence relation on A such that $R \subseteq T$. Since T is reflexive, $r(R) \subseteq T$. Since T is symmetric and $r(R) \subseteq T$, $sr(R) \subseteq T$. Lastly, since T is transitive and $sr(R) \subseteq T$, $tsr(R) \subseteq T$. $\square$

Remark 7.7. The order in $tsr\,(R)$ is important. For example, it could very well happens that $str(R)$ is not an equivalence relation on A (see Exercise (8.51) below).

Example 7.71. Find the equivalence relation on the set $A = \{a, b, c, d\}$ generated by the relation $R = \{(a, b), (b, c)\}$.

Solution. We compute $tsr(R)$. First, $r(R) = R \cup I_A$ consists of the elements: (a, a), (b, b), (c, c), (d, d), (a, b), (b, c). The symmetric closure of $r(R)$ is $sr(R) = r(R) \cup (r(R))^{-1}$. It consists of the elements (a, a), (b, b), (c, c), (d, d), (a, b), (b, c), (b, a), (c, b). For the transitive closure of $sr(R)$, we start by writing the adjacency matrix of $sr(R)$:

$$
\begin{array}{c}
\begin{array}{cccc} a & b & c & d \end{array} \\
\begin{array}{c} a \\ b \\ c \\ d \end{array}
\begin{bmatrix}
1 & 1 & 0 & 0 \\
1 & 1 & 1 & 0 \\
0 & 1 & 1 & 0 \\
0 & 0 & 0 & 1
\end{bmatrix}
\end{array}
$$

This is the matrix W_0 in Warshall's algorithm. The matrices W_1, W_2, W_3 and W_4 are as follows:

$$
W_1 = \begin{array}{c}
\begin{array}{cccc} a & b & c & d \end{array} \\
\begin{array}{c} a \\ b \\ c \\ d \end{array}
\begin{bmatrix}
1 & 1 & 0 & 0 \\
1 & 1 & 1 & 0 \\
0 & 1 & 1 & 0 \\
0 & 0 & 0 & 1
\end{bmatrix}
\end{array}
\quad
W_2 = \begin{array}{c}
\begin{array}{cccc} a & b & c & d \end{array} \\
\begin{array}{c} a \\ b \\ c \\ d \end{array}
\begin{bmatrix}
1 & 1 & 1 & 0 \\
1 & 1 & 1 & 0 \\
1 & 1 & 1 & 0 \\
0 & 0 & 0 & 1
\end{bmatrix}
\end{array}
$$

$$
W_3 = \begin{array}{c}
\begin{array}{cccc} a & b & c & d \end{array} \\
\begin{array}{c} a \\ b \\ c \\ d \end{array}
\begin{bmatrix}
1 & 1 & 1 & 0 \\
1 & 1 & 1 & 0 \\
1 & 1 & 1 & 0 \\
0 & 0 & 0 & 1
\end{bmatrix}
\end{array}
\quad
W_4 = \begin{array}{c}
\begin{array}{cccc} a & b & c & d \end{array} \\
\begin{array}{c} a \\ b \\ c \\ d \end{array}
\begin{bmatrix}
1 & 1 & 1 & 0 \\
1 & 1 & 1 & 0 \\
1 & 1 & 1 & 0 \\
0 & 0 & 0 & 1
\end{bmatrix}
\end{array}
$$

The equivalence relation generated by R consists of the following pairs: (a, a), (b, b), (c, c), (d, d), (a, b), (a, c), (b, a), (b, c), (c, a), (c, b).

7.9.4 *Exercises*

(1) Prove that the empty relation is an equivalence relation on $\emptyset$.
(2) On the set $A = \{1, 2, 3, 4, 5\}$, consider the relation R consisting of the pairs $(1, 1)$, $(2, 2)$, $(1, 2)$, $(2, 1)$, $(2, 5)$, $(5, 4)$, $(3, 3)$, $(2, 4)$, $(4, 3)$ and $(4, 4)$. Explain why R does not satisfy *any* of the properties of an equivalence relation.
(3) In each case, the digraph representing a binary relation R on the set $A = \{1, 2, 3, 4\}$ is given. Determine if the relation is an equivalence relation on A. If you say it is not, specify the properties that R fails to satisfy.

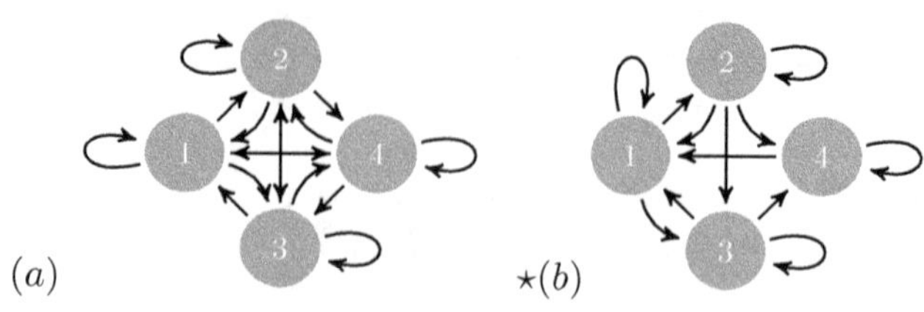

(a) $\star(b)$

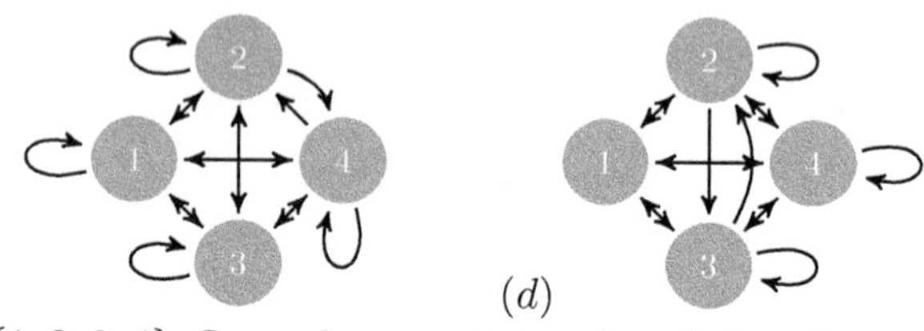

(c) (d)

(4) Let $A = \{1, 2, 3, 4\}$. In each case, determine if the given relation is an equivalence relation on A. If you say that the relation is not an equivalence relation, determine which of the properties (reflexivity, symmetry, transitivity) the relation does not satisfy.

   (a) $R_1 = \{(1,1), (2,2), (2,3), (3,3), (3,2), (4,4)\}$
   (b) $R_2 = \{(1,1), (2,2), (3,1), (3,3), (4,4), (1,3), (3,4), (1,4), (4,1)\}$
   (c) $R_3 = \{(1,1), (1,2), (1,3), (2,1), (2,2), (2,3), (3,1), (4,4)\}$

(5) In each case, determine if the relation R on the set A is an equivalence relation. If you say it is not, specify which properties R fails to satisfy.

   (a) $A = \mathbb{R}$, $R = \{(x, y) \in A \times A;\ x \leq y\}$.
$\star$ (b) A is the set of all humans, $x\,R\,y$ if and only if x is a cousin of y.
   (c) $A = \mathbb{Z}$, $R = \{(a, b) \in A \times A;\ b = ma$ for some integer $m\}$.
   (d) A is the set of all binary strings, $s\,R\,t$ if and only if s and t have the same first bit.
$\star$ (e) A is the set of all functions from $\mathbb{R}$ to $\mathbb{R}$, $f\,R\,g \Leftrightarrow f(-1) = g(-1)$ or $f(1) = g(1)$.

(6) In each case, the adjacency matrix of a binary relation R on the set $A = \{a, b, c\}$ is given. Determine if the relation is an equivalence relation on A. If you say it is not, specify the properties that R fails to satisfy.

$$(a)\ \begin{array}{c}a\\b\\c\end{array}\begin{bmatrix} 1 & 1 & 0 \\ 1 & 1 & 0 \\ 0 & 0 & 1 \end{bmatrix} \quad (b)\ \begin{array}{c}a\\b\\c\end{array}\begin{bmatrix} 0 & 1 & 0 \\ 1 & 0 & 0 \\ 1 & 0 & 1 \end{bmatrix} \quad (c)\ \begin{array}{c}a\\b\\c\end{array}\begin{bmatrix} 1 & 1 & 1 \\ 1 & 1 & 0 \\ 1 & 1 & 1 \end{bmatrix} \quad (d)\ \begin{array}{c}a\\b\\c\end{array}\begin{bmatrix} 1 & 1 & 0 \\ 1 & 1 & 0 \\ 0 & 0 & 1 \end{bmatrix}$$

(7) In each case, the adjacency matrix of a binary relation R on the set $A = \{a, b, c, d\}$ is given. Determine if the relation is an equivalence relation on A. If you say it is not, specify the properties that R fails to satisfy.

$$\star(a)\quad \begin{array}{c} \\ a \\ b \\ c \\ d \end{array}\begin{array}{cccc} a & b & c & d \\ \end{array}\left[\begin{array}{cccc} 1 & 1 & 1 & 0 \\ 1 & 1 & 1 & 0 \\ 1 & 1 & 1 & 0 \\ 0 & 0 & 0 & 1 \end{array}\right] \qquad (b)\quad \begin{array}{c} \\ a \\ b \\ c \\ d \end{array}\begin{array}{cccc} a & b & c & d \\ \end{array}\left[\begin{array}{cccc} 1 & 1 & 0 & 0 \\ 1 & 1 & 1 & 0 \\ 0 & 1 & 1 & 1 \\ 0 & 0 & 1 & 1 \end{array}\right]$$

$$(c)\quad \begin{array}{c} \\ a \\ b \\ c \\ d \end{array}\begin{array}{cccc} a & b & c & d \\ \end{array}\left[\begin{array}{cccc} 1 & 0 & 1 & 0 \\ 0 & 1 & 0 & 0 \\ 1 & 0 & 1 & 1 \\ 0 & 0 & 1 & 1 \end{array}\right] \qquad (d)\quad \begin{array}{c} \\ a \\ b \\ c \\ d \end{array}\begin{array}{cccc} a & b & c & d \\ \end{array}\left[\begin{array}{cccc} 1 & 1 & 0 & 1 \\ 1 & 1 & 1 & 1 \\ 0 & 1 & 1 & 1 \\ 1 & 1 & 1 & 1 \end{array}\right]$$

(8) Consider the subset $A = \{-8, -7, -4, -2, -1, 0, 1, 3, 4, 5, 8, 12, 24\}$ of $\mathbb{Z}$. The congruence relation modulo 4 on $\mathbb{Z}$ induces an equivalence relation R on A by Theorem 7.19 above. Give the partition of A defined by R.

(9) Let A be the subset of $\mathbb{Z} \times \mathbb{Z}^*$ consisting of the elements $(2, 3)$, $(-2, -3)$, $(1, 2)$, $(0, 3)$, $(3, 6)$, $(4, 8)$, $(-1, 3)$, $(0, -2)$, $(1, -3)$, $(-1, -4)$, $(6, 9)$ and $(4, 16)$. Give the partition of A defined by the equivalence relation on A induced by the relation R on $\mathbb{Z} \times \mathbb{Z}^*$ defined in Example 7.53 above.

$\star$(10) R is an equivalence relation on a finite set A containing 12 elements. Assume that $a, b, c \in A$ are such that $[a]$ contains 6 elements, $[b]$ contains a unique element and $[c]$ contains 4 elements. How many elements are there in the quotient set A/R of A?

(11) For an integer $n \geq 2$, let $\equiv_n$ denote the congruence modulo n relation on the set $\mathbb{Z}$: $a \equiv_n b \,(\mathrm{mod}\ n)$ if and only if $a - b$ is divisible by n.

(a) Prove that $\equiv_n$ is an equivalence relation on $\mathbb{Z}$.
(b) Describe the equivalence class of 0 for the relation $\equiv_n$.

(12) On the set Σ_n of all binary strings of length $n \geq 1$, we consider the following relation: $x\,R\,y$ if and only if x and y end with same digit. Prove that R is an equivalence relation on Σ_n. How many equivalence classes of R are there?

$\star$(13) On the set Σ_4 of all binary strings of length 4, define a relation R as follows: $w\,R\,z$ if and only if w and t have the same first digit and the same last digit.

(a) Prove that R is an equivalence relation on Σ_4.
(b) Give the partition on Σ_4 determined by R.
(c) Is the relation: "$w\,R\,z$ if and only if w and t have the same first digit **or** the same last digit" an equivalence relation on Σ_4? Justify your answer.

(14) On the set $\mathbb{Q}$ of all rational numbers, define a relation R as follows: $x\,R\,y \Leftrightarrow a - b \in \mathbb{Z}$.

    (a) Prove that R is an equivalence relation on $\mathbb{Q}$.

    (b) Describe the equivalence class of 0.

(15) If $n \geq 2$ is an integer, we denote by $\kappa(n)$ the smallest **prime** divisor of n (remember that 1 is not considered a prime number). For example, $\kappa(6) = 2$ and $\kappa(9) = 3$. On the set $\mathbb{P} = \{n \in \mathbb{N};\ n \geq 2\}$, define a relation R as follows: $a\,R\,b \Leftrightarrow \kappa(a) = \kappa(b)$. For example, $6\,R\,12$ but $2\,\not{R}\,3$.

    (a) Show that R is an equivalence relation on $\mathbb{P}$.

    (b) Describe the equivalence class of 2.

(16) On the set $\mathbb{R}$, define a binary relation R as follows: $x\,R\,y \Leftrightarrow x - y = k\sqrt{2}$ for some integer $k \in \mathbb{Z}$.

    (a) Prove that R is an equivalence relation on $\mathbb{R}$.

    (b) Describe the equivalence class of 0.

$\star$(17) Consider the relation R on the set $\mathbb{Z}$ defined as follows: $a\,R\,b \Leftrightarrow 3a + 5b$ is even.

    (a) Prove that R is an equivalence relation on $\mathbb{Z}$.

    (b) How many elements in the quotient set $\mathbb{Z}/R$ are there? Justify your answer.

(18) On the set $\Gamma = \mathbb{N} \times \mathbb{N}$, we define a relation R as follows: $(a, b)\,R\,(c, d)$ if and only if $ab = cd$.

    (a) Prove that R is an equivalence relation on Γ.

    (b) Give the elements of the equivalence class of each of the following elements of Γ.

        (i) $(1, 2)$ (ii) $(3, 4)$ (iii) $(9, 2)$

    (c) Give an equivalence class of R containing exactly three elements.

(19) On the set $\mathbb{N}$, we define a relation R as follows: $a\,R\,b$ if and only if $a^2 + b^2$ is even.

    (a) Show that R is an equivalence relation on $\mathbb{N}$.

    (b) Determine the partition of $\mathbb{N}$ defined by R.

$\star$(20) A relation R on the set $\mathbb{R}^* = \mathbb{R}\backslash\{0\}$ is defined as follows: $a\,R\,b \Leftrightarrow |a|b = a|b|$. As usual, $|x|$ stands for the absolute value of x.

(a) Show that R is an equivalence relation on $\mathbb{R}^*$.

(b) Describe the partition of $\mathbb{R}^*$ induced by R: how many equivalence classes? What subset of $\mathbb{R}^*$ does each class represent?

(21) Prove Theorem 7.19 above.

(22) In each case a partition $\wp$ of the set $A = \{1, 2, 3, 4\}$ is given. List the elements of the corresponding equivalence relation R on A.

(a) $\wp = \{\{1\}, \{2, 3\}, \{4\}\}$
(b) $\wp = \{\{1, 2, 3\}, \{4\}\}$
(c) $\wp = \{\{1, 3\}, \{2, 4\}\}$

(23) In each case a partition $\wp$ of the set $A = \{1, 2, 3, 4, 5, 6\}$ is given. List the elements of the corresponding equivalence relation R on A.

(a) $\wp = \{\{1, 3\}, \{2, 4\}, \{5\}, \{6\}\}$
$\star$ (b) $\wp = \{\{1, 2, 3\}, \{4, 5, 6\}\}$
(c) $\wp = \{\{1, 3\}, \{2, 4\}, \{5, 6\}\}$

(24) In each case, determine if the given rule represents a well-defined function.

$\star$ (a) $f : \mathbb{Z}_3 \to \mathbb{Z}_4$ defined by $f([x]_3) = [x]_4$
(b) $f : \mathbb{Z}_3 \to \mathbb{Z}_{12}$ defined by $f([x]_3) = [x]_{12}$
$\star$ (c) $f : \mathbb{Z}_{18} \to \mathbb{Z}_6$ defined by $f([x]_6) = [x]_6$
(d) $f : \mathbb{Z}_3 \to \mathbb{Z}$ defined by $f([x]_3) = (-1)^x$
(e) $f : \mathbb{Z}_4 \to \mathbb{Z}$ defined by $f([x]_4) = (-1)^x$
$\star$ (f) $f : \mathbb{Z}_4 \to \mathbb{Z}$ defined by $f([x]_4) = 2^x \pmod 3$

(25) For each of the following binary relations on the set $\mathbb{R}^2 = \mathbb{R} \times \mathbb{R}$, prove that the relation is an equivalence relation on $\mathbb{R}^2$ and give a complete geometric description of its equivalence classes (are they circles, lines, ellipses, ...?)

(a) $(x, y) R (a, b) \Leftrightarrow x + y = a + b$.
$\star$ (b) $(x, y) S (a, b) \Leftrightarrow x^2 + y = a^2 + b$
(c) $(x, y) T (a, b) \Leftrightarrow x^2 + y^2 = a^2 + b^2$.

(26) If $\vec{u} = (x, y)$ and $\vec{v} = (x', y')$ are two vectors (elements) in $\mathbb{R}^2$, the **dot product** of $\vec{u}$ and $\vec{v}$, denoted by $\vec{u}.\vec{v}$, is the real number $\vec{u}.\vec{v} = xx' + yy'$. Fix the vector $\vec{t} = (1, -1)$ of $\mathbb{R}^2$ and define a relation R on $\mathbb{R}^2$ as follows: $\vec{u} R \vec{v} \Leftrightarrow \vec{u}.\vec{t} = \vec{v}.\vec{t}$.

(a) Prove that R is an equivalence relation on $\mathbb{R}^2$.

(b) Give a complete geometric description of the equivalence class of the vector $\vec{u} = (1, 1) \in \mathbb{R}^2$.

(27) A relation R on $\mathbb{Z}$ is defined as follows: $a\,R\,B \Leftrightarrow a^2 - b^2$ is a multiple of 3.

  (a) Prove that R is an equivalence relation on $\mathbb{Z}$.

  (b) Let $A = \{-5, -4, -3, -2, -1, 0, 1, 2, 3, 4, 5\}$. Give the partition of A induced by the relation R.

(28) Let $A = \{a, b, c, d\}$. On the powerset $\wp(A)$ of A, we consider the following relation: $X\,R\,Y \Leftrightarrow |X| = |Y|$, where $|E|$ stands for the cardinality of the set E.

  (a) Prove that R is an equivalence relation on $\wp(A)$.

  (b) Give the partition of $\wp(A)$ determined by R.

(29) (Calculus required) Consider the set $\mathcal{C}[0, 1]$ of all continuous functions on the interval $[0, 1]$ with values in $\mathbb{R}$. Define a relation R on $\mathcal{C}[0, 1]$ as follows: $f\,R\,g \Leftrightarrow \int_0^1 f(x)dx = \int_0^1 g(x)dx$. Show that R is an equivalence relation on $\mathcal{C}[0, 1]$. Give two elements of the equivalence class of the constant function 1 ($f(x) = 1$ for any $x \in [0, 1]$).

$\star$(30) Let A be the set of all non-zero real numbers of the form $a + b\sqrt{2}$ where $a, b \in \mathbb{Q}$. For example, $\frac{1}{2} - \frac{2}{3}\sqrt{2}$, $2 + 3\sqrt{2}$, $-1 + \frac{5}{3}\sqrt{2}$ are elements of A. Note that $\mathbb{Q}^* = \mathbb{Q}\setminus\{0\} \subset A$. Define a relation R on A as follows: $x\,R\,y \Leftrightarrow \frac{x}{y} \in \mathbb{Q}$.

  (a) Prove that R is an equivalence relation on A.

  (b) What is the equivalence class of 1?

  (c) If $x = a + b\sqrt{2} \in A$ with $b \neq 0$, show that $[x] = [q + \sqrt{2}]$ for some $q \in \mathbb{Q}$.

(31) On the set $\mathbb{M}_{22}$ of all 2×2 matrices with entries in $\mathbb{R}$, we define the following relation: $A\,R\,B \Leftrightarrow A = \lambda B$ for some $\lambda \in \mathbb{R}$ with $\lambda \neq 0$.

Recall that if $\lambda \in \mathbb{R}$ and $A = \begin{bmatrix} a_{11} & a_{12} \\ a_{21} & a_{22} \end{bmatrix} \in \mathbb{M}_{22}$, then λA is the matrix:

$$\begin{bmatrix} \lambda a_{11} & \lambda a_{12} \\ \lambda a_{21} & \lambda a_{22} \end{bmatrix} \in \mathbb{M}_{22}.$$

  (a) Show that R is an equivalence relation on $\mathbb{M}_{22}$.

  (b) Consider the subset $S = \{A_1, A_2, \ldots, A_8\}$ of $\mathbb{M}_{22}$ where

$$A_1 = \begin{bmatrix} 1 & -1 \\ 2 & 3 \end{bmatrix}, \quad A_2 = \begin{bmatrix} 1 & 3 \\ -1 & 0 \end{bmatrix}, \quad A_3 = \begin{bmatrix} 2 & 0 \\ 0 & 2 \end{bmatrix}, \quad A_4 = \begin{bmatrix} 1 & 0 \\ 0 & 0 \end{bmatrix}$$

$$A_5 = \begin{bmatrix} -3 & 3 \\ -6 & -9 \end{bmatrix}, \quad A_6 = \begin{bmatrix} -1 & -3 \\ 1 & 0 \end{bmatrix}, \quad A_7 = \begin{bmatrix} 1 & 0 \\ 0 & 1 \end{bmatrix}, \quad A_8 = \begin{bmatrix} 2 & -2 \\ 4 & 6 \end{bmatrix}.$$

Give the partition of S defined by the restriction of R to S.

$\star$(32) Assume that R is a symmetric and transitive relation on a non-empty set A. Prove that R is an equivalence relation on A if and only the following condition holds: $(\forall x \in A)(\exists y \in A)\,((x, y) \in R)$.

(33) The purpose of this exercise is to generalize the result of Theorem 7.20 above. Assume that $\{R_i;\ i \in I\}$ is a collection of equivalence relations on a non-empty set A. Prove that the relation $R = \cap_{i \in I}$ is also an equivalence relation on A.

$\star$(34) Give an example of a set A and two equivalence relations R and S on A such that $R \cup S$ is not an equivalence relation on A.

(35) Let R be an equivalence relation on a set A and let $g : A \to B$ be a function satisfying $g(x) = g(y)$ whenever $x, y \in A$ are such that $x\,R\,y$. Prove that the function $f : A/R \to B$ defined by $f([x]) = g(x)$ is well-defined.

$\star$(36) Let $n \geq 2$ be an integer. On the set $\mathbb{Z}_n$ of all equivalence classes of the congruence relation modulo n on $\mathbb{Z}$, we define the operation $\otimes$ as follows: $[x] \otimes [y] = [xy]$ where xy stands for the standard multiplication of integers. Prove that $\otimes$ is a well-defined operation on $\mathbb{Z}_n$.

(37) On the set $\mathbb{R}$ of all real numbers, define a relation R as follows: $x\,R\,y \Leftrightarrow |x| = |y|$ where the symbol $|x|$ stands for the absolute value of x.

   (a) Prove that R is an equivalence relation on $\mathbb{R}$.

   (b) If $x \in \mathbb{R}$, what is the equivalence class $[x]$ of x?

   (c) On the quotient set $\mathbb{R}/R$, we define two operations $\oplus$ and $\otimes$ as follows: $[x] \oplus [y] = [x + y]$ and $[x] \otimes [y] = [xy]$. Show that $\otimes$ is well-defined but $\oplus$ is not.

(38) If R is a relation on a non-empty set A, let Σ_R be the set of all equivalence relations on A containing R.

   (a) Prove that Σ_R in not empty.

   (b) Prove that the equivalence relation on A generated by R is equal to $\cap_{S \in \Sigma_R} S$. (*Hint.* Use Exercise (33) above)

(39) If R is a relation on a set A, prove that $tsr(R)$ is an equivalence relation on A (*Hint.* Use Exercise (11) of the previous section).

$\star$(40) Consider the relation $R = \{(1, 1), (2, 1), (2, 3)\}$ on the set $A = \{1, 2, 3\}$. Find $str(R)$ and prove that it is not an equivalence relation on A.

(41) In each case, find the equivalence relation on the set $A = \{a, b, c, d\}$ generated by the relation R on A.

   (a) $R = \{(a, b), (a, c)\}$.

   (b) $R = \{(a, b), (b, c), (d, a)\}$.

$\star$ (c) $R = \{(a, a), (a, b), (b, c), (c, c), (c, d)\}$.

   (d) $R = \{(a, a), (b, b), (c, c), (d, d), (a, d)\}$.

7.10   Order relation

While an equivalence relation on a set A partitions A into classes of "similar" elements, an order relation on A compares its elements and sets an environment of order on the set. Using order relations, we can rank the elements of the set and have a framework in which some elements of the set are "larger" than others.

Definition 7.27. A binary relation R on a set A is called a *partial order* if it is reflexive, antisymmetric and transitive at the same time. If R is a partial order on A, we say that A is a *partially ordered set*, or a *poset* (for short) that we usually denoted by the pair (A, R).

Some authors refer to the relation in Definition 7.27 as *weak partial order*. In the literature, there are also references to the notion of *strict partial order* which is defined as being an asymmetric, transitive and irreflexive relation. The following result tells us that a partial order on a set A gives rise to a strict partial order on A and that the converse is also true.

Theorem 7.26. *If R is a strict partial order on a set A, then $R \cup I_A$ is a partial order on A. If R is a partial order on A then $R \backslash I_A$ is a strict partial order on A. As usual, $I_A = \{(x, x);\ x \in A\}$.*

Proof.   Let R be a strict partial order on a set A, and let $T = R \cup I_A$. Then T is reflexive since it contains I_A. Let $x, y \in A$ such that $(x, y) \in T$ and $(y, x) \in T$. If $x \neq y$, then $(x, y) \notin I_A$ and therefore $(x, y) \in R$. By the asymmetry of R, $(y, x) \notin R$ and so $(y, x) \in I_A$, a contradiction. We conclude that $x = y$ and T is antisymmetric. For the transitivity of T, let $x, y, z \in A$ such that $x\,T\,y$ and $y\,T\,z$ then ($x\,R\,y$ or $x = y$) and ($y\,R\,z$ or $y = z$). So ($x\,R\,y$ and $y\,R\,z$) or ($x\,R\,y$ and $y = z$) or ($x = y$ and $y\,R\,z$) or ($x = y$ and $y = z$). These cases lead to $x\,R\,z$ or $x = z$ which means that $x\,T\,z$. We conclude that T is a partial order. The second result of the theorem is left as an exercise (see Exercise (4) below). $\square$

Example 7.72. The standard order $\leq$ (less than or equal) on $\mathbb{R}$ is clearly a partial order relation on $\mathbb{R}$. The relation $<$ (strictly less than), on the other hand, is a strict partial order on $\mathbb{R}$.

Similar to the notation of the standard order $\leq$ on $\mathbb{R}$, the notation $(A, \preccurlyeq)$ is widely used in the literature to denote an arbitrary poset. The reader should be careful not to confuse the symbol $\preccurlyeq$ with the standard order on $\mathbb{R}$. In some cases, we talk about two posets $(A, \preccurlyeq)$ and $(B, \preccurlyeq)$ but we really mean two different sets with two different partial order relations. If $(A, \preccurlyeq)$ is a poset, then $\prec$ is the corresponding strict partial order $\preccurlyeq \backslash I_A$ on A. So for $a, b \in A$, the notation $a \prec b$ means $a \preccurlyeq b$ and $a \neq b$.

Example 7.73. Let A be a set, $\wp(A)$ its powerset. Consider the inclusion relation $\subseteq$ on $\wp(A)$. Since $X \subseteq X$ for any $X \in \wp(A)$, $\subseteq$ is reflexive. If $X, Y \in \wp(A)$ are such that $X \subseteq Y$ and $Y \subseteq X$, then $X = Y$ which implies that $\subseteq$ is antisymmetric. Finally, if $X \subseteq Y$ and $Y \subseteq Z$ for $X, Y, Z \in \wp(A)$, then clearly $X \subseteq Z$. This shows that $\subseteq$ is transitive. Hence, $\subseteq$ is a partial order on $\wp(A)$.

Example 7.74. On the set $\mathbb{N}^*$ of all positive integers, consider the relation R defined by $x \, R \, a$ if and only if x divides a. The relation is reflexive since any positive integer divides itself. It is antisymmetric since if m divides n and n divides m then $m = n$. It is transitive since if m divides n and n divides p then m divides p. We conclude that R is a partial order on $\mathbb{N}^*$. The same relation is not a partial order on the set $\mathbb{Z}^*$ of all non-zero integers since it is not antisymmetric on $\mathbb{Z}^*$. For example, 2 divides -2 and -2 divides 2, but $2 \neq -2$.

Example 7.75. On the set $\mathbb{Z} \times \mathbb{Z}$, define a relation R as follows: $(x, y) \, R \, (a, b)$ if and only if $x \leq a$ and $y \leq b$. For any $(x, y) \in \mathbb{Z} \times \mathbb{Z}$, $(x, y) \, R \, (x, y)$ since $x \leq x$ and $y \leq y$. The relation R is reflexive. If $(x, y), (a, b) \in \mathbb{Z} \times \mathbb{Z}$ are such that $(x, y) \, R \, (a, b)$ and $(a, b) \, R \, (x, y)$ then $x \leq a$, $y \leq b$, $a \leq x$ and $b \leq y$. This shows that $x = a$ and $y = b$ and so $(x, y) = (a, b)$. The relation R is antisymmetric. For (x, y), (a, b), $(c, d) \in \mathbb{Z} \times \mathbb{Z}$, if $(x, y) \, R \, (a, b)$ and $(a, b) \, R \, (c, d)$ then $x \leq a$, $y \leq b$, $a \leq c$ and $b \leq d$. This shows that $x \leq c$ and $y \leq d$ by the transitivity of $\leq$ on $\mathbb{Z}$. So $(x, y) \, R \, (c, d)$ and R is transitive. We conclude that R is a partial order on $\mathbb{Z} \times \mathbb{Z}$.

Example 7.76. Let $(A, \preccurlyeq)$ and $(B, \preccurlyeq)$ be two posets. An important strict order on $A \times B$ is the *lexicographic order* (also known as the dictionary order) $\prec_{lex}$ defined as follows:

$$(a, b) \prec_{lex} (x, y) \Leftrightarrow ((a \prec x) \text{ or } (a = x \text{ and } b \prec y)). \tag{7.2}$$

To compare two pairs (a, b) and (x, y) of $A \times B$, we start by comparing the first components a and x of the two pairs. If $a \prec x$ as elements of A, then $(a, b) \prec_{lex} (x, y)$. If $a = x$, we compare the second components: if $b \prec y$ as elements of B, then we also have $(a, b) \prec_{lex} (x, y)$. In other words, the first component dominates and we only look at the second component if we have a "tie" in comparing the first components. The proof that $\prec_{lex}$ is a strict partial order on $A \times B$ is left as an exercise (see Exercise (5) below).

Example 7.77. Consider the lexicographic order $\prec_{lex}$ on $\mathbb{N} \times \mathbb{N}$ induced by the natural order $\leq$ on $\mathbb{N}$. Arrange the following pairs in increasing order: $(3, 2)$, $(3, 1)$, $(5, 2)$, $(2, 4)$, $(2, 6)$, $(4, 7)$, $(4, 2)$, $(1, 5)$, $(4, 3)$, $(1, 8)$ and $(5, 9)$.

Solution. For the lexicographic order, the first component dominates. The largest first component among the given pairs is 5 and the smallest is 1. If two pairs have the same first component, then the one with the largest second component dominates. The arrangement of the pairs in increasing order is: $(1, 5)$, $(1, 8)$, $(2, 4)$, $(2, 6)$, $(3, 1)$, $(3, 2)$, $(4, 2)$, $(4, 3)$, $(4, 7)$, $(5, 2)$, $(5, 9)$. $\Diamond$

Given n posets $(A_1, \preccurlyeq_1)$, $(A_2, \preccurlyeq_2)$, ..., $(A_n, \preccurlyeq_n)$, we can generalize the notion of the lexicographic order to $A_1 \times A_2 \times \cdots \times A_n$ similar to the formulation (7.2) above: $(a_1, a_2, \ldots, a_n) \prec_{lex} (b_1, b_2, \ldots, b_n)$ if and only if $a_1 \prec b_1$ or there exists $k \geq 1$ such that $a_1 = b_1$, ..., $a_k = b_k$ and $a_{k+1} \prec b_{k+1}$.

Example 7.78. On the set $\mathbb{N}^3 = \mathbb{N} \times \mathbb{N} \times \mathbb{N}$, consider the lexicographic order $\prec_{lex}$ induced by the natural order $\leq$ on $\mathbb{N}$. Arrange the elements $(3, 2, 2)$, $(5, 3, 1)$, $(3, 2, 5)$, $(2, 4, 5)$, $(2, 6, 9)$, $(1, 9, 9)$, $(5, 4, 5)$, $(5, 4, 3)$ and $(1, 6, 1)$ of $\mathbb{N}^3$ in increasing order with respect to the order $\prec_{lex}$.

Solution. Like before, the first component dominates. For the three pairs with first component 5, we have $(5, 3, 1) \prec_{lex} (5, 4, 3) \prec_{lex} (5, 4, 5)$. Note that the first two components in $(5, 4, 3)$, $(5, 4, 5)$ are equal so we compare the third components in each triplet and we get $(5, 4, 3) \prec_{lex} (5, 4, 5)$. Continuing this way, we get the following ordering of the elements in increasing order: $(1, 6, 1)$, $(1, 9, 9)$, $(2, 4, 5)$, $(2, 6, 9)$, $(3, 2, 2)$, $(3, 2, 5)$, $(5, 3, 1)$, $(5, 4, 3)$ and $(5, 4, 5)$. $\Diamond$

If you are wondering why the lexicographic order is also referred to as the dictionary order, the following might shed a bit of light on that. Consider an alphabet $\mathcal{A} = \{a_1, a_2, \ldots, a_n\}$ on which we have the order $a_1 \prec a_2 \cdots \prec a_n$

on the characters. Recall that a *word* on $\mathcal{A}$ of *length* t is an ordered string $x_1 x_2 \cdots x_t$ of t characters from $\mathcal{A}$. There is a unique word of length zero called the empty word and denoted by λ. Notice that a word $x_1 x_2 \cdots x_s$ of length s on $\mathcal{A}$ can be identified with the s-tuple $(x_1, x_2, \ldots, x_s)$ of $\mathcal{A}^s$. Let $\mathcal{A}^*$ be the set of all words (of any length) on the alphabet $\mathcal{A}$. The lexicographic order $\prec_{lex}$ on $\mathcal{A}^*$ is defined as follows: $\lambda \prec_{lex} x$ for any word x of positive length, and for words $x = x_1 x_2 \cdots x_s$, $y = y_1 y_2 \cdots y_t$ with $s \geq 1$ and $s \geq 1$, $x \prec_{lex} y$ if and only if either one of the following conditions holds:

(L1) $s < t$ and $x_i = y_i$ for $i = 1, \ldots, s$.
(L2) There exist $k < \min(s, t)$ such that $x_i = y_i$ for $i = 1, \ldots, k$ and
$x_{k+1} \prec y_{k+1}$.

Example 7.79. Consider the English alphabet $\mathcal{A} = \{a, b, \ldots, y, z\}$ with characters ordered in the usual way: $a \prec b \prec \ldots \prec y \prec z$. The corresponding lexicographic order on the set $\mathcal{A}^*$ of all English words is the same as the order of words in the dictionary (with the difference that $\mathcal{A}^*$ contains infinitely many elements as a word could be something like *aabbgg* with no actual meaning in English). For example:

$$\text{math} \prec \text{mathematical} \prec \text{mathematician} \prec \text{mathematics} \prec \text{maths}.$$

Example 7.80. Words with alphabet $\mathcal{A} = \{0, 1\}$ are just binary strings. Consider the order $0 \prec 1$ on $\mathcal{A}$. Arrange the following stings in increasing order with respect to the lexicographic order on $\mathcal{A}^*$: 11, 0110000, 10011, 110101, 000101, 10101, 100101, 0101010.

Solution. A word starting with a 0 is smaller than any word starting with a 1. In our list, there are three words starting with a 0, so these would be at the beginning of our chain. Among these three words, 000101 is the smallest, followed by 0101010 and then 0110000. Arranging words starting with a 1 using a similar argument gives us the following chain:

$$000101 \prec 0101010 \prec 0110000 \prec 100101 \prec 10011 \prec 10101 \prec 11 \prec 110101.$$

Example 7.81. (Monomial ordering) When dealing with polynomials, we often need to arrange the terms in a certain order. The problem is easy for polynomials in one variables as the terms in this case are arranged according to the exponents of the variable in the polynomial. So we write, for instance, $p = 3x^4 - 2x^3 + 3x + 1$ to indicate that the leading term of

the polynomial is the monomial $3x^4$ as it has the highest degree. Things are not that obvious when dealing with polynomials in several variables. For example, what is the leading term in the following polynomial in three variables x, y and z?

$$f = 3x^2y^4z^2 - 2x^5y^2z^8 + 5xy^4z^3 + 2y^4z^5 \qquad (7.3)$$

The answer clearly depends on the choice of the order on the monomials of f and on the order on the variables x, y and z. Choosing the order $z \prec y \prec x$ on the variables, then any monomial $x^ay^bz^c$ can be identified with the triplet $(a, b, c) \in \mathbb{N}^3$. For instance $(2, 1, 3)$ represents the monomial x^2yz^3. This allows us to order monomials in three variables using the lexicographic order on $\mathbb{N}^3$ corresponding to the order $z \prec y \prec x$ on the variables. Hence the leading term of the polynomial f in (7.3) is $-2x^5y^2z^8$ and the terms in f can be arranged in increasing order as follows:

$$f = -2x^5y^2z^8 + 3x^2y^4z^2 + 5xy^4z^3 + 2y^4z^5.$$

Theorem 7.27. *Let $(A, \preccurlyeq)$ be a poset and $B \subseteq A$. The following hold:*

(a) *$(A, \succcurlyeq)$ is also a poset (called the* dual poset*) where $\succcurlyeq$ is the inverse relation $\preccurlyeq^{-1}$ of $\preccurlyeq$.*

(b) *The restriction of $\preccurlyeq$ to B is a partial order on B. We say in this case that $(B, \preccurlyeq)$ is a subposet of $(A, \preccurlyeq)$.*

Proof. The proof of each part is straightforward and left as an exercise to the reader (see Exercise (14) below). $\qquad\square$

7.10.1 *Digraph of a partial order: The Hasse diagram*

Given a partial order $\preccurlyeq$ on a finite set $A = \{a_1, \ldots, a_n\}$, what does the digraph representation of $\preccurlyeq$ look like? First we know that a loop must exist at every edge of the digraph since $\preccurlyeq$ is reflexive. Second, if there is an edge from a_i to a_j (with $a_i \neq a_j$), there cannot be one in the other direction since the relation is antisymmetric. Finally, if there is an edge between a_i and a_j and another between a_j and a_k, then there must exist an edge between a_i and a_k by the transitivity of $\preccurlyeq$. This suggests that a "cleanup" can be made on the digraph G representing $\preccurlyeq$ to make it look simpler and easier to read. Start by removing all loops at the vertices of G (we know that they exist, but no point of drawing them). Next, we can remove any edge (a_i, a_k) whenever $a_i \preccurlyeq a_j$ and $a_j \preccurlyeq a_k$ for some $a_j \in A$ (we know such an edge must exist by the transitivity of $\preccurlyeq$ but again, no need to draw it). Finally, in order to eliminate the need to draw arrows on the edges (to indicate directions), we can arrange each edge (a_i, a_j) in a way that vertex a_i is below vertex a_j. In other words, if $a_i \preccurlyeq a_j$, the initial vertex of edge (a_i, a_j) is drawn below to the terminal vertex of the edge. The "clean" diagram we obtain this way for the relation $\preccurlyeq$ is called the **Hasse diagram** of $\preccurlyeq$.

Example 7.82. Let $A = \{1, 2, 4, 3, 6\}$. In this example, we draw the Hasse diagram for the poset $(A, \,|\,)$ where $a \mid b$ stands for "a divides b". We start by drawing a digraph of the division relation. We do that while keeping in mind that vertex a_i is drawn below vertex a_j whenever $a_i \mid a_j$.

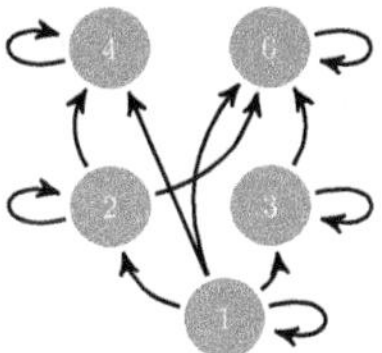

Start by removing loops from the graph, then remove the edges $(1, 4)$ and $(1, 6)$ resulting from the transitivity and finally replace the oriented edges by straight lines.

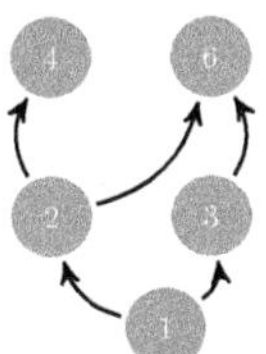

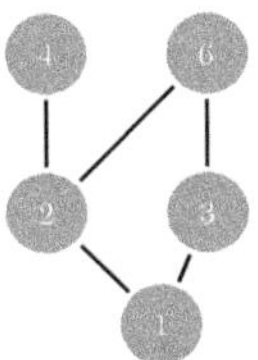

Remove loops and transitivity edges   Replace arrows with straight lines

The graph on the right is the Hasse diagram of the partial order.

Example 7.83. Let $A = \{1, 2, 3\}$, $\wp(A)$ its powerset. In this example, we draw the Hasse diagram of the poset $(\wp(A), \subseteq)$ where $\subseteq$ is the inclusion relation on $\wp(A)$. Note first that the elements of $\wp(A)$ are $\emptyset$, $\{1\}$, $\{2\}$, $\{3\}$, $\{1, 2\}$, $\{1, 3\}$, $\{2, 3\}$ and A. Note also that $\emptyset$ is in relation with every other element of $\wp(A)$ and every element of $\wp(A)$ is in relation with A. This means that the vertex corresponding to $\emptyset$ should be at the bottom of the Hasse diagram and that corresponding to A should be at the top:

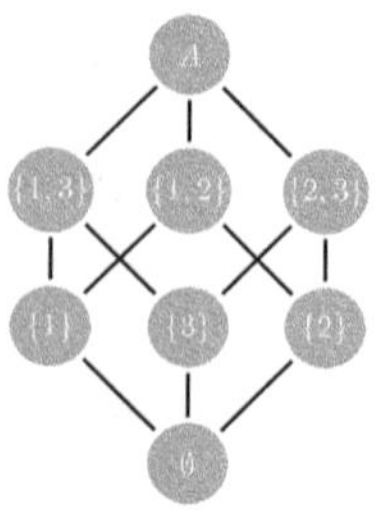

7.10.2   *Total order*

The term "partial" in a *partial order relation* defined above refers to the fact that there could exist two elements of the poset that are not comparable.

Definition 7.28. Let $(A, \preccurlyeq)$ be a poset. We say that the elements $a, b \in A$ are *comparable* if either $a \preccurlyeq b$ or $b \preccurlyeq a$. The elements are called *incomparable* otherwise.

Example 7.84. In the poset $(\mathbb{N}, |)$, the elements 2 and 3 are incomparable since $2 \nmid 3$ and $3 \nmid 2$ but 3 and 9 are comparable since 3 divides 9.

Example 7.85. In the post $(\mathbb{R}, \leq)$, every pair of real numbers are comparable since $x \leq y$ or $y \leq x$ for any $x, y \in \mathbb{R}$.

The above two examples motivate the following definition.

Definition 7.29. A partial order $\preccurlyeq$ on a set A is called a *total order* if every pair of elements of A are comparable. In other words, $\preccurlyeq$ is a total order on A if and only if $x \preccurlyeq y$ or $y \preccurlyeq x$ for any $x, y \in A$. If $\preccurlyeq$ is a total order on A, we say that the poset $(A, \preccurlyeq)$ is a *totally ordered set*. Elements of a totally ordered set can be arranged in linear sequence or a chain. For this reason, a total order R on a set A is sometimes referred to as a *linear order*) and the poset (A, R) as a *chain*.

Example 7.86. The partial order given in Example 7.73 above is not total. Given two subsets X and Y of a set A, it could very well happen that neither $X \subseteq Y$ nor $Y \subseteq X$ is true.

Example 7.87. The natural order $\leq$ on $\mathbb{R}$ is clearly a total order.

7.10.3 *Exercises*

(1) Give all possible partial orders on a set A of cardinality 2.

(2) On the set H of all humans who ever lived, define a relation R as follows: $x\,R\,y$ if and only if x and y are the same person or x is an ancestor of y. Is R a partial order? Justify your answer.

(3) In each case, a relation R on the set $\mathbb{Z}$ is given. Determine if R is a partial order on $\mathbb{Z}$. If you say that the relation is not a partial order, indicate which properties it fails to satisfy.

$\star$ (a) $m\,R\,n \Leftrightarrow |m - n| \leq 1$.

$\star$ (b) $m\,R\,n \Leftrightarrow m = n$ or $|m| < |n|$.

   (c) $m\,R\,n \Leftrightarrow m + n$ is even.

   (d) $m\,R\,n \Leftrightarrow m^2 \leq n^2$.

$\star$ (e) $m\,R\,n \Leftrightarrow m^3 \leq n^3$.

   (f) $m\,R\,n \Leftrightarrow m$ and n have the same set of prime divisors.

   (g) $m\,R\,n \Leftrightarrow m = n$ or $r_m < r_n$ where r_k is the remainder of the division of the integer k by 5.

(4) Let R be a partial order on a set A. Prove that $R \backslash I_A$ is a strict partial order on A.

(5) Prove that the lexicographic order defined in Example 7.76 above is indeed a strict partial order.

(6) Let $A = \{a, b, c, d\}$. In each case, the adjacency matrix of a binary relation R on A is given. Determine if the relation is a partial order on A. If you say it is not, specify the properties that R fails to satisfy.

$$\star(a)\quad \begin{array}{c} \\ a \\ b \\ c \\ d \end{array}\begin{array}{cccc} a\ b\ c\ d \\ \left[\begin{array}{cccc} 1 & 1 & 1 & 1 \\ 0 & 1 & 1 & 1 \\ 0 & 0 & 1 & 1 \\ 0 & 0 & 0 & 1 \end{array}\right]\end{array} \qquad (b)\quad \begin{array}{c} \\ a \\ b \\ c \\ d \end{array}\begin{array}{cccc} a\ b\ c\ d \\ \left[\begin{array}{cccc} 1 & 0 & 0 & 0 \\ 1 & 1 & 0 & 0 \\ 0 & 1 & 1 & 1 \\ 0 & 0 & 1 & 1 \end{array}\right]\end{array}$$

$$(c)\quad \begin{array}{c} \\ a \\ b \\ c \\ d \end{array}\begin{array}{cccc} a\ b\ c\ d \\ \left[\begin{array}{cccc} 1 & 1 & 1 & 1 \\ 0 & 1 & 0 & 0 \\ 0 & 0 & 1 & 0 \\ 0 & 1 & 1 & 1 \end{array}\right]\end{array} \qquad (d)\quad \begin{array}{c} \\ a \\ b \\ c \\ d \end{array}\begin{array}{cccc} a\ b\ c\ d \\ \left[\begin{array}{cccc} 1 & 0 & 0 & 1 \\ 1 & 1 & 1 & 1 \\ 0 & 1 & 1 & 1 \\ 1 & 1 & 1 & 1 \end{array}\right]\end{array}$$

(7) Using the lexicographic order on the set of all binary strings with $0 \prec 1$ (like in Example 7.80 above), arrange in each case the given collection strings in increasing order.

  $\star$ (a) 111, 1110000, 10011, 0110101, 000101, 10101, 100101, 1110101010.

   (b) 011011, 111011, 11110000, 011111, 111, 010011, 0110101, 0001111, 1010110, 100101, 0101010.

(8) Using the lexicographic ordering on the monomials in three variables x, y, z with $z \prec y \prec x$, arrange the following sets of monomials in increasing order.

(a) y^3z^2, xy^2z^3, xy^2z^2, x^3, z^7, x^3yz.

$\star$ (b) x^5, x^4y, y^3z^3, yz^5, x^3y^2, x^3z^2, x^3yz, xy^5, z^7, x^2yz^3.

(9) Repeat the previous exercise using the order $y \prec x \prec z$ on the variables.

$\star$(10) Using the lexicographic ordering on the monomials in four variables x, y, z, w with $z \prec y \prec x \prec w$ (as in Example 7.81) to arrange the terms in the following polynomial in increasing order: $f = 3w^3xyz^2 + xy^3z^2 - 5w^3xy^2z^2 + xyz^2 - xy + z^2w^2 + 2xyz^2w^2$.

(11) Using the lexicographic order $\prec$ on $\mathbb{N} \times \mathbb{N}$ induced by the natural order $\leq$ on $\mathbb{N}$, arrange each of the following sets of pairs in $\mathbb{N} \times \mathbb{N}$ in increasing order:

(a) $(4,0)$, $(4,1)$, $(5,6)$, $(2,0)$, $(1,8)$, $(3,4)$, $(3,9)$.

(b) $(5,2)$, $(4,1)$, $(5,7)$, $(2,3)$, $(2,1)$, $(1,7)$, $(4,2)$, $(5,5)$, $(4,3)$, $(1,0)$, $(3,9)$.

(12) Using the lexicographic order $\prec$ on $\mathbb{N} \times \mathbb{N} \times \mathbb{N}$ induced by the natural order $\leq$ on $\mathbb{N}$, arrange each of the following sets of triplets in $\mathbb{N} \times \mathbb{N} \times \mathbb{N}$ in increasing order:

$\star$ (a) $(1,4,0)$, $(2,4,1)$, $(1,0,6)$, $(0,2,6)$, $(1,8,0)$, $(3,4,7)$, $(2,3,9)$.

(b) $(5,2,2)$, $(5,2,1)$, $(5,0,7)$, $(2,3,0)$, $(2,1,2)$, $(1,0,7)$, $(4,2,3)$, $(5,5,5)$, $(2,4,3)$, $(1,0,4)$, $(3,9,9)$.

$\star$(13) Let (A, R) and (B, S) be two posets. Define a relation T on $A \times B$ as follows: $(a, b)\, T\, (x, y)$ if and only if $a\,R\,x$ and $b\,S\,y$. Prove that $(A \times B, T)$ is also a poset.

(14) Prove Theorem 7.27.

(15) In each case, draw a Hasse diagram for the poset $(A, \preccurlyeq)$.

(a) $A = \{2, 4, 5, 7, 9\}$, $\preccurlyeq$ is the relation "less than or equal to".

$\star$ (b) $A = \{2, 3, 4, 8, 9, 12, 15, 36\}$, $\preccurlyeq$ is the divisibility relation.

(c) $A = \{\{1\}, \{2\}, \{4\}, \{1,2\}, \{1,4\}\}$ and $\preccurlyeq$ stands for the inclusion relation.

(16) Let $A = \{1, 2, 3\}$ and consider the poset $(\wp(\wp(A)), \subseteq)$ consisting of the powerset of the power set of A with the inclusion relation. In each case, determine if the elements a and b of the poset are comparable.

(a) $a = \emptyset$, $b = \{\emptyset\}$.

(b) $a = \{\{1\}\}$, $b = \{\{2\}, \{1,2\}\}$.

(c) $a = \{\{1,2\}, \{3\}\}$, $b = \{\{1\}, \{2\}, \{1,2\}, \{1,3\}\}$.

(d) $a = \{\{1,2,3\}\}$, $b = \{\{2\}, \{1,2,3\}, \{3\}\}$.

(e) $a = \{\emptyset\}$, $b = \{\{1, \emptyset\}, \{2, \emptyset\}\}$.

(17) Prove that the lexicographic order $\preccurlyeq_{lex}$ on $\mathbb{N} \times \mathbb{N}$ induced by the natural order $\leq$ on $\mathbb{N}$ is a *total* order on $\mathbb{N} \times \mathbb{N}$.

(18) Let P the set of all pairs (m, n) of positive integers such that $\gcd(m, n) = 1$ (as usual the term gcd stands for the *greatest common divisor*). On the set $P \times P$, we define the relation $\preccurlyeq$ as follows: $(a, b) \preccurlyeq (m, n) \Leftrightarrow an \leq bm$. Prove that $\preccurlyeq$ is a total order on $P \times P$.

$\star$(19) Give a subset A of $\mathbb{N}$ of cardinality 7 such that the subposet $(A, |)$ of $(\mathbb{N}, |)$ is totally ordered.

(20) Consider the set $\mathbb{R}^{\circ} = \mathbb{R} \cup \{\omega, \kappa\}$ where ω, κ are just two symbols that do not represent real numbers. On the set R°, consider the relation $\preccurlyeq$ defined by: $\preccurlyeq = \{(x, y) \in \mathbb{R} \times \mathbb{R}; \ x \leq y\} \cup \{(\omega, x); \ x \in \mathbb{R}\} \cup \{(x, \kappa); \ x \in \mathbb{R}\}$. Prove that $\preccurlyeq$ is a total order on $\mathbb{R}^{\circ}$. (The poset $(\mathbb{R}^{\circ}, \preccurlyeq)$ is called the *extended real line*. In this poset, ω is smaller than any other real number and κ is greater than any real number. Traditionally, ω and κ are denoted by the symbols $-\infty$ and ∞, respectively).

$\star$(21) The divisibility relation on $\mathbb{N}$ is not a total order, but if restricted to a suitable subset of $\mathbb{N}$, it could become one. Prove that the divisibility relation is a total order on the subset $\mathcal{G}_k = \{k^n; \ n \in \mathbb{N}\}$ of $\mathbb{N}$ where $k \geq 2$ is a fixed integer.

7.11 Special elements in a poset, lattices, well-ordering principle and topological sorting

When dealing with order relations on a given set, some elements are of particular interest. One would normally search for elements of the poset which are larger or smaller than any other element of the set. The existence of such elements is not guaranteed, even for finite posets. We also make the distinction between a *minimal* element and the *least* element (similarly *maximal* element and *greatest* element) in a poset.

Definition 7.30. Let $(A, \preccurlyeq)$ be a poset and let $a \in A$.

(a) We say that a is a *maximal* element of A if there exists no element x of A such that $a \prec x$. Equivalently, a is maximal if the implication $(a \preccurlyeq x \rightarrow x = a)$ is true for any $x \in A$.

(b) We say that a is a *minimal* element of A if there exists no element x of A such that $x \prec a$. Equivalently, a is minimal if the implication $(x \preccurlyeq a \rightarrow x = a)$ is true for any $x \in A$.

(c) We say that a is a *greatest* element of A if $x \preccurlyeq a$ for any $x \in A$.

(d) We say that a is a *least* element of A if $a \preccurlyeq x$ for any $x \in A$.

Least and greatest elements are unique if they exist in a poset as shown in the following theorem.

Theorem 7.28. *If a poset has a least (respectively greatest) element, then it is unique.*

Proof. We prove the uniqueness of the least element. The proof of that of the greatest element is similar and left to the reader as an exercise (see Exercise (7) below). Assume $a, a' \in A$ are two least elements of the poset $(A, \preccurlyeq)$. Then $a \preccurlyeq a'$ and $a' \preccurlyeq a$ by definition of a least element. The antisymmetry of $\preccurlyeq$ implies that $a = a'$. $\qquad\qquad\square$

In the Hasse diagram of a poset $(A, \preccurlyeq)$, maximal elements (when they exist) can be found on top with no connections leading up to other elements. Similarly, minimal elements can be found at the bottom of the diagram with no connections leading down to other elements. The greatest and the least elements (if they exist) are connected to every other element in the diagram by a path leading downward for the greatest element and upward for the least element. Notice that the least element is a minimal element and the greatest element is a maximal element, when they exist. The other way around is not true in general.

Example 7.88. Consider the poset $(\mathbb{Z}_+, \leq)$ where $\mathbb{Z}_+$ is the set of positive integers and $\leq$ is the standard order on integers. The element 1 is the least element (hence a minimal element) but the poset has no maximal element (hence no greatest element).

Example 7.89. The poset $(\mathbb{Z}, \leq)$ has no maximal and no minimal elements, hence no least or greatest elements.

Example 7.90. Let A be a non-empty set. The poset $(\wp(A), \subseteq)$ has A as the greatest element and $\emptyset$ as the least element.

Example 7.91. Consider the poset $(S, |)$ where $S = \{1, 2, 3, 4, 5, 6, 7\}$ and $|$ stands for the divisibility relation. Clearly, 1 is the least element of the poset. Each of the elements 4, 5, 6 and 7 is maximal (as none of them divides another element in S) but none of them is the greatest element of the poset.

Example 7.92. Consider the poset $(S, |)$ where $S = \{2, 3, 4, 5, 7\}$ and $|$ stands for the divisibility relation. The poset has no least element and no greatest element. The element 2 is minimal and each of the elements $3, 5, 7$ is maximal and minimal at the same time. The element 4 is maximal but not minimal.

Theorem 7.29. *Every finite non-empty poset $(A, \preccurlyeq)$ has at least one minimal element and at least one maximal element.*

Proof. We only prove the existence of a minimal element. The proof for a maximal element is done similarly and is left as an exercise (see Exercise (15) below). Fix $a_1 \in A$. If a_1 is a minimal element, we are done. If a_1 is not a minimal element, then there exists $a_2 \in A$ such that $a_2 \prec a_1$. If a_2 is minimal, we are done. If not, there exists $a_3 \in A$ such that $a_3 \prec a_2$. Note that $a_3 \neq a_1$. Since A is finite, this process will eventually end with a chain $a_k \prec a_{k-1} \prec \cdots \prec a_2 \prec a_1$ of elements of A with the property that no element $a \in A$ exists such that $a \prec a_k$. This means that the element a_k is minimal. $\square$

Definition 7.31. Let $(A, \preccurlyeq)$ be a poset, $B \subseteq A$ and $a \in A$. We say that a is a *lower bound of B* (respectively, an *upper bound of B*) if $a \preccurlyeq b$ for any $b \in B$ (respectively, $b \preccurlyeq a$ for any $b \in B$).

Example 7.93. For the subset $B = [0, 1]$ (the subset of all real numbers between 0 and 1 inclusively) of the poset $(\mathbb{R}, \leq)$, -1 and 0 are lower bounds of B and 1 and $\frac{3}{2}$ are upper bounds. In fact, any real number $x \leq 0$ is a lower bound and any $y \geq 1$ is an upper bound for B.

Example 7.94. For the subset $B = \{4, 6, 8, 12, 24, 28\}$ of the poset $(\mathbb{N}, |)$, the elements 1 and 2 are lower bounds and 168 and 336 are upper bounds.

Example 7.95. Let $A = \{a, b, c, d\}$. An order relation on A is given by the following Hasse diagram:

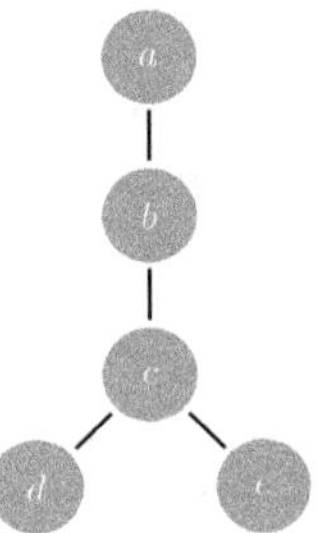

Clearly, a is the greatest element of the poset, but the poset has no least element. Elements d and e are minimal. For the subset $B = \{b, c\}$ of A, elements d and e are lower bounds and b and a are upper bounds.

Definition 7.32. Let $(A, \preccurlyeq)$ be a poset and $B \subseteq A$.

(a) An element $a \in A$ is called a *greatest lower bound of B* or *infimum* of B, denoted by $\text{glb}(B)$ or $\inf(B)$, if a is a lower bound of B and $a' \preccurlyeq a$ for any lower bound a' of B.

(b) An element $a \in A$ is called a *least upper bound of B* or *supremum* of B, denoted by $\text{lub}(B)$ or $\sup(B)$, if a is an upper bound of B and $a \preccurlyeq a'$ for any upper bound a' of B.

Like the least and the greatest elements, we can easily prove that $\inf(B)$ and $\sup(B)$ are unique when they exist (See Exercise (8) below).

Example 7.96. For the subset $B = [0, 1]$ of the poset $(\mathbb{R}, \leq)$, $\inf(B) = 0$ and $\sup(B) = 1$.

Example 7.97. Consider the poset $A = \{a, b, c, d, e, x, y\}$ given by the following Hasse diagram:

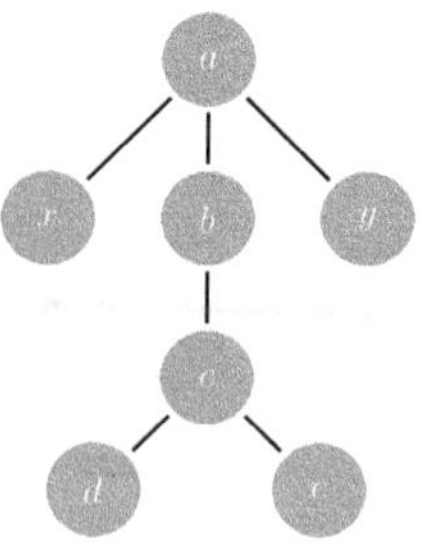

The subset $B = \{a, x, y\}$ has no lower bound since no element of A is a smaller than any element of B. This means in particular that $\inf(B)$ does not exist. On the other hand, it is easy to see that $\sup(B) = a$. For the subset $C = \{a, b, c, d\}$, $\inf(C) = d$ and $\sup(C) = a$.

Example 7.98. Consider the poset $(\mathbb{N}^*, |)$ where $\mathbb{N}^* = \mathbb{N} \setminus \{0\}$ and $|$ stands for the division of integers. For $m, n \in \mathbb{N}^*$, $\inf(\{m, n\}) = \gcd(m, n)$ (the *greatest common divisor* of m and n) and $\sup(\{m, n\}) = \text{lcm}(m, n)$ (the *least common multiple* of m and n).

Example 7.98 can be generalized to any finite subset of $\mathbb{N}^*$ as follows: If $B = \{a_1, \ldots, a_n\} \subseteq \mathbb{N}^*$, then $\inf(B) = \gcd(a_1, \ldots, a_n)$ and $\sup(B) = \text{lcm}(a_1, \ldots, a_n)$.

Example 7.99. For the subset $B = \{4, 6, 8, 12, 24, 28\}$ of the poset $(\mathbb{N}, |)$, $\inf(B) = 2$ and $\sup(B) = 168$.

7.11.1 *Lattices*

In this section, we define a special type of posets which have some resemblance to the structure of Boolean algebras we introduced in Chapter 2. First, we need some definitions.

Definition 7.33. Let $(A, \preceq)$ be a poset and let x and y be two elements of A. If $\inf \{x, y\}$ exists, we call it the *meet* of x and y and we denote it by $x \wedge y$. If $\sup \{x, y\}$ exists, we call it the *join* of x and y and we denote it by $x \vee y$.

Example 7.100. In the poset $(\mathbb{N}^*, |)$, $m \wedge n$ is the greatest common divisor of m and n, and $m \vee n$ is the least common multiple of m and n. For instance, $4 \wedge 6 = 2$ and $4 \vee 6 = 12$.

Example 7.101. Let A be any set. In the poset $(\wp(A), \subseteq)$, $X \wedge Y = X \cap Y$ and $X \vee Y = X \cup Y$ for any $X, Y \in \wp(A)$. The proof of this is left as an exercise (see Exercise (19) below).

Definition 7.34. A poset $(A, \preceq)$ is called a *lattice* if both $x \wedge y$ and $x \vee y$ exist for every pair x, y of elements of A.

Example 7.102. The set $(\mathbb{R}, \leq)$ is a lattice since for every $x, y \in \mathbb{R}$, we have $x \wedge y = \min(x, y)$ and $x \vee y = \max(x, y)$.

Example 7.103. The poset $(\mathbb{N}^*, |)$ (where $|$ stands for the divisibility relation) is a lattice since for every $x, y \in \mathbb{N}^*$, both $x \wedge y$ and $x \vee y$ exist as shown in Example 7.100 above.

Example 7.104. Given any set A, Example 7.101 shows that the poset $(\wp(A), \subseteq)$ is a lattice.

Example 7.105. For a non-empty set A with at least two elements, the poset of all non-empty subsets of A with the inclusion relation is not a lattice since the meet of two disjoint subsets of A (namely, the empty set) is not in the poset.

Example 7.106. Let $\mathcal{D}_{12} = \{1, 2, 3, 4, 6, 12\}$ be the set of all divisors of 12 in $\mathbb{N}$. Then $(\mathcal{D}_{12}, |)$ is clearly a lattice (verify this) and its Hasse diagram is the following:

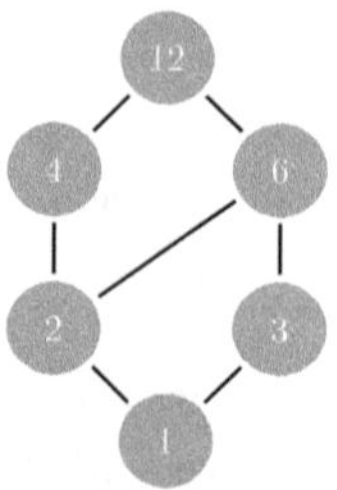

In general, if n is a positive integer, then the set D_n of all positive integer divisors of n is a lattice for the relation of divisibility. If $x, y \in D_n$ then $x \wedge y = \gcd(x, y) \in D_n$ and $x \vee y = \mathrm{lcm}(x, y) \in D_n$ since both x, y are divisors of n.

Example 7.107. Let $A = \{2, 3, 5, 6, 7, 11, 12, 35\}$ equipped with the division relation $|$. The poset $(A, |)$ is not a lattice. For instance, $\inf \{2, 3\}$ does not exist in A.

Basic properties of the meet and the join are given in the following.

Theorem 7.30. *Let $(A, \preccurlyeq)$ be a lattice and let x, y, z and w be arbitrary elements of A. The following hold:*

(a) $x \wedge y \preccurlyeq x \preccurlyeq x \vee y$ and $x \wedge y \preccurlyeq y \preccurlyeq x \vee y$.
(b) $x \wedge y = x \Leftrightarrow x \preccurlyeq y$.
(c) $x \vee y = y \Leftrightarrow x \preccurlyeq y$.
(d) $x \wedge y = x \Leftrightarrow x \vee y = y$.
(e) If $x \preccurlyeq y$, then $x \wedge z \preccurlyeq y \wedge z$.
(f) If $x \preccurlyeq y$, then $x \vee z \preccurlyeq y \vee z$.
(g) If $x \preccurlyeq y$ and $z \preccurlyeq w$, then $x \wedge z \preccurlyeq y \wedge w$.
(h) If $x \preccurlyeq y$ and $z \preccurlyeq w$, then $x \vee z \preccurlyeq y \vee w$.

Proof. We prove parts $(a), (b), (d), (e)$ and (g). The other parts are left as exercises. Part (a) follows directly from the definitions of the meet and the join of two elements. For part (b), assume that $x \wedge y = x$ then $x \preccurlyeq y$ by definition of the meet. Conversely, if $x \preccurlyeq y$ then x is a lower bound for $\{x, y\}$ (since $x \preccurlyeq x$ and $x \preccurlyeq y$) and so $x \preccurlyeq x \wedge y$ by definition of $x \wedge y$. But $x \wedge y \preccurlyeq x$ by part (a), so $x \wedge y = x$ by the antisymmetry of $\preccurlyeq$. For part (d), let $x, y \in A$ then $x \wedge y = x$ if and only if $x \preccurlyeq y$ (by part (b)) if and only if $x \vee y = y$ (by part (c)). For part (e), assume that $x \preccurlyeq y$. Since $x \wedge z \preccurlyeq x$ (by definition of $x \wedge z$), the transitivity of $\preccurlyeq$ implies that $x \wedge z \preccurlyeq y$. By definition, we also have that $x \wedge z \preccurlyeq z$. We conclude that $x \wedge z$ is a lower bound for $\{y, z\}$ which implies that $x \wedge z \preccurlyeq y \wedge z$. For part (g), assume that $x \preccurlyeq y$ and $z \preccurlyeq w$. Then $x \wedge z \preccurlyeq y \wedge z$ and $y \wedge z \preccurlyeq y \wedge w$ by part (e). By the transitivity of $\preccurlyeq$, we get that $x \wedge z \preccurlyeq y \wedge w$. $\qquad\square$

If x and y are two comparable elements of a poset $(A, \preccurlyeq)$, say $x \preccurlyeq y$, then both their meet and their join exist since $x \wedge y = x$, $x \vee y = y$. A consequence of this is the following.

Theorem 7.31. *Every totally ordered poset is a lattice.*

The converse of Theorem 7.31 is not true in general (see Example 7.103 above).

Theorem 7.32. *Let $(A, \preccurlyeq)$ be a lattice. For any $a, b, c \in A$, the following hold.*

(a) $a \wedge a = a$ *and* $a \vee a = a$ *(Indempotence)*
(b) $a \wedge b = b \wedge a$ *and* $a \vee b = b \vee a$ *(Commutativity)*
(c) $(a \wedge b) \wedge c = a \wedge (b \wedge c)$ *and* $(a \vee b) \vee c = a \vee (b \vee c)$ *(Associativity)*
(d) $a \wedge (a \vee b) = a$ *and* $a \vee (a \wedge b) = a$ *(Absorption)*

Proof. We only prove part (c), the other parts are left as exercises. Equality between two elements in a poset is usually shown using the antisymmetry property of the partial order. So we need to prove that $(a \wedge b) \wedge c \preccurlyeq a \wedge (b \wedge c)$ and $a \wedge (b \wedge c) \preccurlyeq (a \wedge b) \wedge c$. By definition of the greatest lower bound, we have: $(a \wedge b) \wedge c \preccurlyeq a \wedge b$ and $a \wedge b \preccurlyeq a$. So $(a \wedge b) \wedge c \preccurlyeq a$ by transitivity. Also, $a \wedge b \preccurlyeq b$, so $(a \wedge b) \wedge c \preccurlyeq b \wedge c$ by part (e) of Theorem 7.30. We conclude that $(a \wedge b) \wedge c$ is a lower bound for the set $\{a, b \wedge c\}$ which implies that $(a \wedge b) \wedge c \preccurlyeq a \wedge (b \wedge c)$. The relation $a \wedge (b \wedge c) \preccurlyeq (a \wedge b) \wedge c$ is proven similarly. $\qquad\square$

Theorem 7.32 shows in particular that we don't need parentheses when we write $(a \wedge b) \wedge c$ or $(a \vee b) \vee c$ in a lattice. We simply write $a \wedge b \wedge c$ and $a \vee b \vee c$ to denote the greatest lower bound and the least upper bound (respectively) of the set $\{a, b, c\}$. More generally, if $a_1, \ldots, a_n$ are n elements of a lattice, we write $a_1 \wedge \cdots \wedge a_n$ for $\inf \{a_1, \ldots, a_n\}$ and $a_1 \vee \cdots \vee a_n$ for $\sup \{a_1, \ldots, a_n\}$.

7.11.2 *The well-ordering principle and the well-ordered sets*

The poset $(\mathbb{N}, |)$ has 1 as the least element. Interestingly enough, Example 7.92 above shows a non-empty subset of $(\mathbb{N}, |)$ with no least element. If the divisibility relation on $\mathbb{N}$ is replaced with the natural order $\leq$, then every non-empty subset S of $\mathbb{N}$ would have a least element. This is a well-known axiom in the literature known as the *well-ordering principle* (WOP for short). It is important to note that the WOP is false if $\mathbb{N}$ is replaced with $\mathbb{Z}$. For example, the subset of $\mathbb{Z}$ of all negative integers has no least element.

Axiom (The well-ordering principle). Every non-empty subset of $\mathbb{N}$ has a least element for the standard order $\leq$.

At first glance, the WOP seems to be a simple fact that is completely detached and independent from any other major mathematical results. But remember that this principle is equivalent to the principle of mathematical induction we saw in Chapter 3. Every result provable with (simple or strong) induction is also provable with the well-ordering principle, and vice versa.

We generalize the well-ordering principle to any poset.

Definition 7.35. A poset $(A, \preccurlyeq)$ is called *well-ordered* and the partial order $\preccurlyeq$ on A is called a *well-order* if every *non-empty* subset of A has a least element.

Example 7.108. If A is a non-empty set, the poset $(\wp(A), \subseteq)$ is not-well ordered in general (see Exercise (14) below).

Theorem 7.33. *Every well-ordered set is totally ordered.*

Proof. Assume $(A, \preccurlyeq)$ is a well-ordered set. If $x, y \in A$, then the subset $\{x, y\}$ must then have a least element. The least element is either x or y, which means that either $x \preccurlyeq y$ or $y \preccurlyeq x$. Therefore any two elements of A are comparable and $\preccurlyeq$ is a total order. $\qquad\square$

The converse of the Theorem 7.33 is not true in general (see Exercise (14) below), but it is when the poset is finite.

Theorem 7.34. *Every totally ordered finite poset is well-ordered.*

Proof. Left as an Exercise (see Exercise (16) below). $\qquad\square$

7.11.3 *Application: The topological sorting*

If you are a contractor in charge of building a house, then a critical part of your job is to organize the tasks so no trade steps on the foot of another in the process. If A is the set of all tasks required to build a house, then one needs to create an order on A that establishes the priority of tasks and dependency of one task on others (for instance, the cement foundation cannot be poured before the digging is done). Some tasks can be performed simultaneously (for example, exterior landscaping can be done before, after or at the same time, as priming interior walls). Assume for simplicity that there are six tasks T_1, ..., T_6 involved in building a house and that these tasks must be performed respecting the order given by the Hasse diagram below. As a contractor, you have a clear idea on the order in which tasks should be performed. The challenge is to be able to write them using a logical order for all the trades to follow. In other words, you need to find a total order R on the set $A = \{T_1, \ldots, T_6\}$ that is *compatible* with the partial order $\preccurlyeq$ on A in the following sense: if $T_i \preccurlyeq T_j$, then $T_i \, R \, T_j$. Such a compatible total order is called a *topological sorting*.

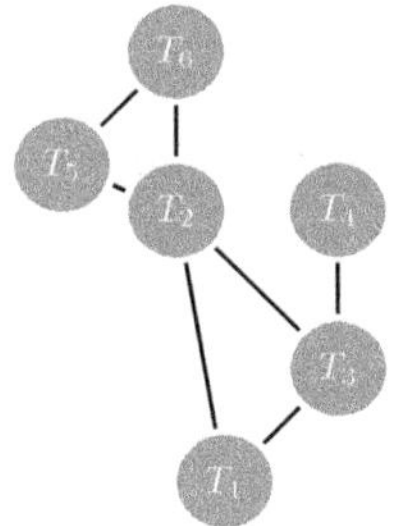

Fig. 7.1   Tasks sorting

Definition 7.36. Let $(A, \preccurlyeq)$ be a finite poset. A *topological sorting* of A is a total order R on A such that for every elements $x, y \in A$: if $x \preccurlyeq y$ then $x \, R \, y$.

But how do we order incomparable elements? The proof of the next theorem illustrates one way to achieve that.

Theorem 7.35. *Any finite poset has a topological sorting.*

Proof.   Let $(A, \preccurlyeq)$ be a finite poset of cardinality n. Start by choosing an arbitrary minimal element a_1 of A and remove it from of the poset

(Theorem 7.29 guarantees the existence of a such an element). In the Hasse diagram representing the poset, this step consists of removing the vertex corresponding to a_1 together with all edges pointing upwards from a_1. If $A_1 = A \backslash \{a_1\} \neq \emptyset$, then A_1 is also a finite poset with the same inherited partial order $\preccurlyeq$ from A and so A_1 must have a minimal element a_2. Let $A_2 = A \backslash \{a_1, a_2\}$ and repeat the same process for A_2 as long as it is non-empty. Since A is finite, the process must terminate with the empty set. Consider the total order R on A induced by the order in which the elements are removed in the above process:

$$a_1 \, R \, a_2 \, R \cdots R \, a_{n-1} \, R \, a_n \qquad (7.4)$$

To prove that R is compatible with $\preccurlyeq$, assume that $x, y \in A$ are such that $x \preccurlyeq y$. Then x must be removed before y in the sequence (7.4) since otherwise y is not a minimal element of a subset of A as described above. In other words, $x \, R \, y$ and R is compatible with $\preccurlyeq$. $\qquad \square$

The proof of Theorem 7.35 gives a clear algorithm to build a topological sorting on a finite poset $(A, \preccurlyeq)$. However, the topological sorting described in the proof is not unique as it depends on the choice of the minimal element we decide to remove at every step.

Example 7.109. Find a topological sorting on the poset $A = \{T_1, \ldots, T_6\}$ of house building tasks given by the Hasse diagram in Figure 7.1 above.

Solution. From the Hasse diagram of Figure 7.1, we see that T_1 is a minimal element of A. Removing T_1 along with the edges moving upwards from it produces the Hasse diagram of the poset $A_1 = (A \backslash \{T1\}, \preccurlyeq)$ shown as diagram (2) of Figure 7.2 below. In A_1, T_3 is a minimal element and removing it along with the edges pointing upwards results in diagram (3). Element T_4 is minimal in $A_2 = A \backslash \{T_1, T_3\}$, so we remove it from A_2 and get diagram (4). Next, T_2 is minimal in Hasse diagram (4), so we remove it together with the corresponding edges pointing and get diagram (5). Finally, removing the minimal element T_5 and the corresponding edge leaves us with only the one-vertex diagram (6) of Figure 7.2. Arranging the elements of A by the order they were successively removed yields the following topological order on A: T_1, T_3, T_4, T_2, T_5, T_6.

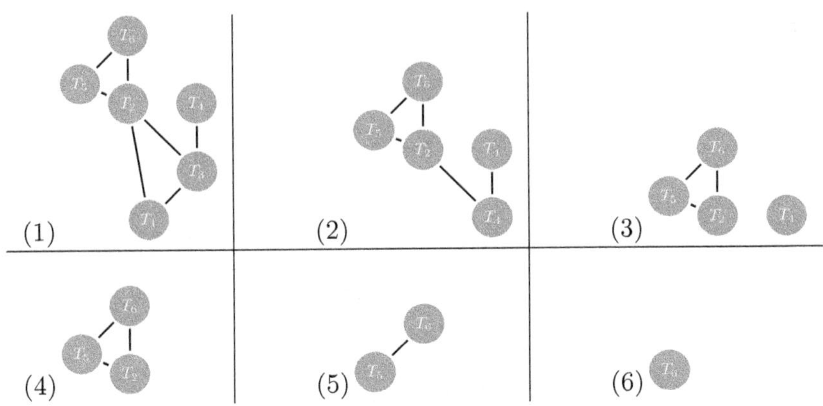

Fig. 7.2   Topological sorting of the poset of Figure 7.1

7.11.4   *Exercises*

⋆(1) On the $A = \{a, b, c, d, e, f, g, h, i\}$, a partial order is given by its Hasse diagram:

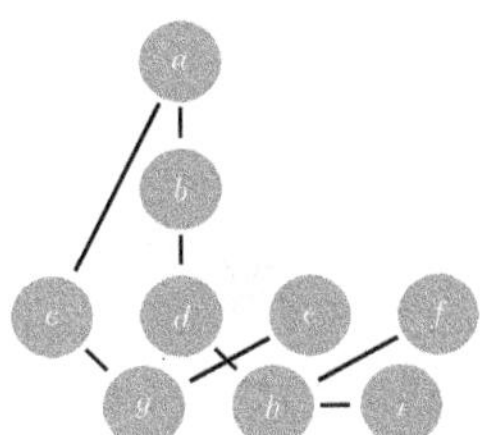

Find each of the following (if they exist):

(a) all minimal elements   (c) the least element
(b) all maximal elements   (d) the greatest element

(2) For the poset $(A, |)$ where $A = \{2, 4, 5, 6, 9, 11, 12, 18, 36, 48, 60, 72\}$, find each of the following (if they exist):

(a) all minimal elements   (c) the least element
⋆ (b) all maximal elements   ⋆ (d) the greatest element

(3) Let

- $A = \{1, 2, 3, 4, 5\}$;
- $X = \{\{2\}, \{3\}, \{4\}, \{5\}, \{1, 2\}, \{3, 4\}, \{2, 4, 5\}\} \subset \wp(A)$.

For the subposet $(X, \subseteq)$ of $(\wp(A), \subseteq)$, find each of the following (if they exist):

$\star$ (a) all minimal elements $\star$ (c) the least element
    (b) all maximal elements   (d) the greatest element

(4) For the subset $B = \{3, 4, 6, 8, 12, 24, 35, 96\}$ of the poset $(\mathbb{N}^*, \mid)$, find $\inf(B)$ and $\sup(B)$.

(5) Let $A = \{a, b, c, d, e, f, g, h, i, j\}$. The Hasse diagram of a partial order relation $\preceq$ on A is given.

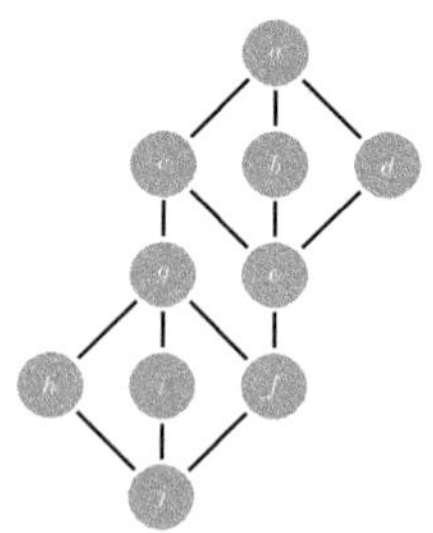

Find each of the following.

    (a) $\sup\{j, e\}$ (e) $\inf\{g, d\}$ $\star$ (i) $\sup\{c, e, f, g, i, j, h\}$
$\star$ (b) $\sup\{h, d\}$ $\star$ (f) $\inf\{i, h\}$ $\star$ (j) $\inf\{c, e, f, g, i, j, h\}$
    (c) $\sup\{i, b\}$ (g) $\sup\{a, c, e, b\}$
    (d) $\inf\{i, b\}$ (h) $\inf\{a, c, e, b\}$

(6) Let $A = \{a, b, c, d\}$. In each case, a subset X of $\wp(A)$ is given. Determine $\inf(X)$ and $\sup(X)$ (if they exist) in the poset $(\wp(A), \subseteq)$.

$\star$ (a) $X = \{\{a\}, \{b\}\}$
    (b) $X = \{\{a\}, \{b\}, \{a, b, c\}\}$
    (c) $X = \{\{a\}, \{a, b\}, \{a, c\}, \{a, b, c\}\}$
$\star$ (d) $X = \{\{a\}, \{b\}, \{d\}, \{a, b\}, \{b, d\}, \{a, b, d\}\}$
    (e) $X = \{\emptyset, \{a\}, \{b\}, \{c\}, \{a, b, c\}\}$
$\star$ (f) $X = \{\{a, b, d\}, \{b, d\}, \{c, b, d\}, \{a, b, c, d\}\}$

(7) Prove that if a poset has a greatest element, then it must be unique.

(8) Let $(A, \preceq)$ be a poset, and $B \subseteq A$.

    (a) Prove that if $\inf(B)$ exists, it is unique.
$\star$ (b) Prove that if $\sup(B)$ exists, it is unique.

(9) In each case, give a possible poset that has a:

    ★ (a) minimal element but no maximal element
       (b) maximal element but no minimal element
    ★ (c) minimal element but no least element
       (d) maximal element but no greatest element

★(10) Consider the subposet $(S, \leq)$ of $(\mathbb{Q}, \leq)$ where $S = \left\{ \frac{1}{n}; n \in \mathbb{N}\backslash\{0\} \right\}$ and $\leq$ is the standard order on $\mathbb{Q}$. Prove that the poset $(S, \leq)$ does not have a least element but has a greatest element.

(11) Assume that $(A, \preccurlyeq)$ is a poset that has a greatest element z. Prove that z is the only maximal element of the poset.

(12) Assume that $(A, \preccurlyeq)$ is a poset that has a least element t. Prove that t is the only minimal element of the poset.

(13) Let $A = \{1, 2, 3\}$. Prove that $(\wp(A), \subseteq)$ is not well-ordered by giving a non-empty subset of $\wp(A)$ that has no least element.

★(14) Give an example of a totally ordered poset which is not well-ordered.

★(15) Prove that a non-empty finite poset must have a maximal element and a minimal element.

(16) Prove that every totally ordered *finite* poset is well-ordered (*Hint.* Use Exercise (15)).

(17) Hasse diagrams representing some posets are given below. In each case determine if the poset is a lattice. Justify your answer.

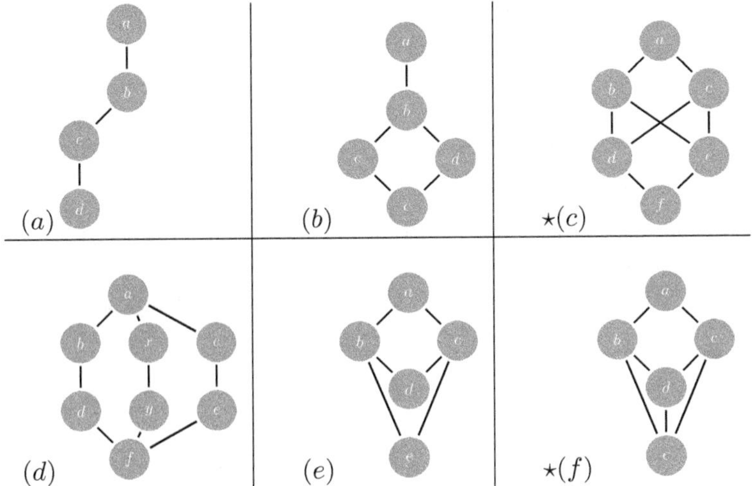

★(18) Draw a possible Hasse diagram of a finite poset which has a greatest element and a least element but which is not a lattice.

(19) Let A be a non-empty set, $\wp(A)$ its powerset. Given two elements X and Y of the poset $(\wp(A), \subseteq)$, prove that $\inf(\{X, Y\}) = X \cap Y$ and $\sup(\{X, Y\}) = X \cup Y$. Conclude that $(\wp(A), \subseteq)$ is a lattice.

(20) In each case, show that the subposet $(A, |)$ of $(\mathbb{N}, |)$ is not a lattice:
(a) $A = \{2, 3, 4, 5, 6\}$; (b) $A = \{2, 4, 6, 8, 16, 24\}$

(21) Let $(A, \preccurlyeq)$ be a lattice, $a, b, c \in A$. Prove the following.

    (a) $a \wedge a = a$ and $a \vee a = a$
    (b) $a \wedge b = b \wedge a$ and $a \vee b = b \vee a$
 $\star$ (c) $a \wedge (a \vee b) = a$
    (d) $a \vee (a \wedge b) = a$

(22) Let $(A, \preccurlyeq)$ be a well-ordered poset and let ∞ be any element not in A. On the set $Z = A \cup \{\infty\}$, we define a relation R as follows: $x \, R \, y$ if and only if either $x, y \in A$ and $x \preccurlyeq y$ or $y = \infty$.

    (a) Prove that R is a partial order relation on Z.
    (b) Prove that (Z, R) is well-ordered.

$\star$(23) Let $(A, \preccurlyeq)$ be a lattice. Use mathematical induction to prove that for any non-empty finite subset B of A, $\inf(B)$ and $\sup(B)$ exist.

(24) A lattice $(A, \preccurlyeq)$ is called *bounded* if it has a least element (usually denoted by 0) and a greatest element (usually denoted by 1). In each case, determine if the lattice is bounded. Clearly specify the elements 0 and 1 in the lattice if they exist.

  $\star$ (a) $(\mathbb{N} \backslash \{0\}, |)$ $\star$ (c) $(D_{12}, |)$ (e) $([0, 1], \leq)$ $\star$ (g) $([0, 1[, \leq)$
     (b) $(\mathbb{R}, \leq)$ $\star$ (d) $(\wp(\mathbb{N}), \subseteq)$ (f) $(]0, 1], \leq)$

(25) Prove that every finite lattice is bounded (see Exercise (24)).

(26) Let $(A, \preccurlyeq)$ be a bounded lattice (see Exercise (24)) with least element 0 and greatest element 1. Prove that:

    (a) For any $a \in A$, $a \vee 0 = a$ and $a \wedge 0 = 0$.
    (b) For any $a \in A$, $a \vee 1 = 1$ and $a \wedge 1 = a$.

(27) We say that the lattice $(A, \preccurlyeq)$ is *distributive* if for every $a, b, c \in A$:

    • $a \vee (b \wedge c) = (a \vee b) \wedge (a \vee c)$
    • $a \wedge (b \vee c) = (a \wedge b) \vee (a \wedge c)$.

    (a) In each case, prove that the given lattice is distributive.

      $\star$ (i) $(\mathbb{N} \backslash \{0\}, \leq)$
      (ii) $(\wp(S), \subseteq)$ for any set S

(iii) $(D_6, |)$

$\star$ (b) Give an example of a non-distributive lattice.

(28) In each case, show that the lattice given by its Hasse diagram is not distributive (see Exercise (27) for the definition of a distributive lattice).

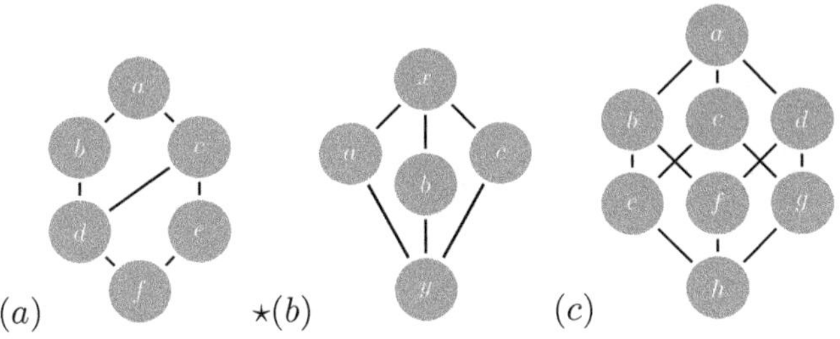

(a) $\star(b)$ (c)

(29) A lattice $(A, \preccurlyeq)$ is called *modular* if the following condition holds for any $a, b, c \in A$: If $a \preccurlyeq c$ then $a \vee (b \wedge c) = (a \vee b) \wedge c$. Prove that a distributive lattice must be modular (see Exercise (27)).

$\star$(30) Let $(A, \preccurlyeq)$ be a bounded distributive lattice (see Exercises (24) and (27)) with least element 0 and a greatest element 1. For an element $x \in A$, we say that the element $y \in A$ is a *complement* of x if $x \wedge y = y \wedge x = 0$ and $x \vee y = y \vee x = 1$.

(a) Prove that a complement of an element x, if it exists, it must be unique. We denote the complement of x by x'.

(b) Prove that $0' = 1$ and $1' = 0$.

(31) In each case, a bounded distributive lattice $(A, \preccurlyeq)$ is given (see Exercises (24) and (27)). Give the least element and the greatest element of the lattice. For each $x \in A$, give the complement x' if it exists.

(a) $A = \wp(S)$ for some non-empty set S with the inclusion relation.

(b) $A = D_{105}$ (the set of all positive integer divisors of 105) with the divisibility relation.

$\star$(32) Give an example of a bounded lattice A and an element x of A that has no complement (see Exercises (24) and (27)).

(33) Let $(L_1, \preccurlyeq_1)$ and $(L_2, \preccurlyeq_2)$ be two lattices with meet operations $\wedge_1$ and $\wedge_2$ and join operations $\vee_1$ and $\vee_2$, respectively. A function $f : L_1 \to L_2$ is called a *lattice homomorphism* if $f(x \wedge_1 y) = f(x) \wedge_2 f(y)$ and $f(x \vee_1 y) = f(x) \vee_2 f(y)$ hold for any $x, y \in L_1$. We say that f is an order preserving map if the following condition hold: if $x, y \in L_1$ are

such that $x \preceq_1 y$ then $f(x) \preceq_2 f(y)$. Prove that if $f : L_1 \to L_2$ is a lattice homomorphism, then it is also an order preserving map.

$\star(34)$ In each case, a finite poset $(A, \preceq)$ is given by its Hasse diagram. Give a topological sorting of the set A.

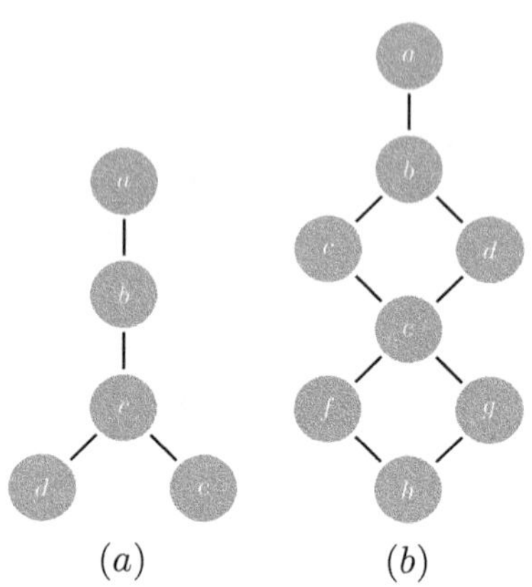

(a) (b)

Chapter 8

Basic Combinatorics: The Art of Counting Without Counting

8.1 Introduction

In a department of Mathematics at a certain university, there are 32 professors: 18 men and 14 women. In how many ways the chair of the department can form a committee of 8 professors such that the committee has at least 4 women but Joseph and Monica (two professors) are not on the committee at the same time? Of course, one can approach this problem by considering all possibilities for such a committee. But it does not take much effort to realize that exhausting all possible cases is not the optimal approach to solve this type of problems.

Combinatorics is the area of Mathematics that deals with counting problems without actually exhausting all possible cases. It is a very broad area of study with many techniques and tools and it has a wide range of applications in areas like set theory, probability, statistics and others.

8.2 Basic counting principles

In this section, we explore the basic counting principles that are the foundation of every counting technique. These principles are then combined with other techniques like permutations and combinations and the pigeonhole principle to produce more powerful counting tools.

8.2.1 *The sum principle*

The first counting principle, known as the *sum principle*, is based on the following simple set theory result. Recall that if E is a finite set, then the notation $|E|$ stands for the number of elements (cardinality) of E.

405

Theorem 8.1 (The sum principle). *Let $n \geq 2$ be an integer. If $A_1, A_2, \ldots, A_n$ are finite and pairwise disjoint subsets (i.e., $A_i \cap A_j = \emptyset$ for $i \neq j$) of a universal set $\mathcal{U}$, then $|A_1 \cup A_2 \cup \ldots A_n| = |A_1| + |A_2| + \cdots + |A_n|$.*

Proof. We prove the result for $n = 2$, the proof of the general case follows by induction on n and left as an exercise (See Exercise (1) below). Let $A = \{a_1, a_2, \ldots, a_n\}$ and $B = \{b_1, b_2, \ldots, b_m\}$ be two finite and disjoint sets. Then $A \cup B = \{a_1, a_2, \ldots, a_n, b_1, b_2, \ldots, b_m\}$ and so $|A \cup B| = n + m = |A| + |B|$. $\qquad\square$

In the context of counting, the sum principle can be stated as follows: if $E_1, E_2, \ldots, E_n$ are pairwise mutually exclusive events such that E_i can occur in k_i ways for $i = 1, 2, \ldots, n$, then the event "E_1 or E_2 or ... or E_n" can occur in $k_1 + k_2 + \cdots + k_n$ ways.

Example 8.1. Students in a certain course must write a project at the end of the course. The professor provides three lists of projects: list A has 6 projects, list B has 8 projects and list C has 10 projects. If all projects are different, a student can choose a project in $6 + 8 + 10 = 24$ different ways.

Example 8.2. A regular deck of 52 cards consist of 4 suits: hearts, diamonds, spades and clubs. Two suits (hearts and diamonds) are colored red and the other two (spades and clubs) are colored black. Each suit consists of 13 cards: 10 value cards (Ace to 10) and 3 picture cards (Jack, Queen, and King). In how many ways can we draw from the deck a diamond or a spade card? An Ace or a 2?

Solution. There are 13 ways to draw a diamond and 13 ways to draw a spade. By the sum principle, there are $13 + 13 = 26$ ways to draw a diamond or a spade card. There are 4 ways to draw an Ace and 4 ways to draw a 2. By the sum principle, there are $4 + 4 = 8$ ways to draw an Ace or a 2. $\quad\lozenge$

Example 8.3. Two regular dice are rolled simultaneously. In how many ways can we get an odd sum of the values on the faces of the dice?

Solution. An odd sum means either a sum of 3, 5, 7, 9 or 11. There are two ways to obtain a sum of 3: $(1, 2)$ and $(2, 1)$, four ways to obtain a sum of 5: $(1, 4)$, $(4, 1)$, $(2, 3)$ and $(3, 2)$, six ways to get a sum of 7: $(1, 6)$, $(6, 1)$, $(2, 5)$, $(5, 2)$, $(3, 4)$ and $(4, 3)$, four ways to get a sum of 9: $(3, 6)$, $(6, 3)$, $(4, 5)$ and $(5, 4)$. Finally, there are two ways to get a sum of 11: $(5, 6)$ and $(6, 5)$.

Hence there are $2 + 4 + 6 + 4 + 2 = 18$ ways we can get an odd sum of the values on the faces of the dice. $\Diamond$

8.2.2 The product principle

We look next at the product principle of counting. This principle counts the numbers of n-tuples in the Cartesian product of n finite sets. First, we announce and prove the principle for two sets. Theorem 8.1 generalizes the result for n sets.

Theorem 8.2. *If A, B are finite sets with $|A| = m$ and $|B| = n$, then $|A \times B| = mn$.*

Proof. The result is true if $A = \emptyset$ or $B = \emptyset$ since $\emptyset \times U = \emptyset$ for any set U. So we may assume that A and B are non-empty. Assume the set A is fixed with m elements $x_1, \ldots, x_m$. We prove the result by induction on $n = |B|$. If $n = 1$, then $B = \{y\}$ and $A \times B = \{(x_1, y), \ldots, (x_m, y)\}$. The result is true in this case since $A \times B$ has $m = m \times 1 = |A| \cdot |B|$ elements. Let $n \geq 1$ and assume that $|A \times U| = |A| \cdot |U|$ for any set U of cardinality $n - 1$ or less. Let B be a set of cardinality n and write $B = U \cup \{y\}$ where $y \in B$ and $U = B \backslash \{y\}$. It is easy to verify that $A \times B = (A \times U) \cup (A \times \{y\})$ and that the sets $A \times U$ and $A \times \{y\}$ are disjoint. By the sum principle and the induction hypothesis, $|A \times B| = |A \times U| + |A \times \{y\}| = m(n - 1) + m(1) = mn$. The result follows by the principle of induction. $\square$

We now prove the generalized version of the product principle.

Theorem 8.3 (The product principle). *If $A_1, A_2, \ldots, A_n$ are finite sets then $|A_1 \times A_2 \times \cdots \times A_n| = |A_1| \cdot |A_2| \cdots |A_n|$.*

Proof. We use induction on n. The result is trivially true for $n = 1$. Let $n \geq 1$, and assume that the result is true for $n - 1$ sets. Then

$$
\begin{aligned}
|A_1 \times A_2 \times \cdots \times A_n| &= |(A_1 \times A_2 \times \cdots \times A_{n-1}) \times A_n| \\
&= |A_1 \times A_2 \times \cdots \times A_{n-1}| \cdot |A_n| \ \text{(By Theorem 8.2)} \\
&= |A_1| \cdot |A_2| \cdots |A_n| \ \text{(By the induction hypothesis)}.
\end{aligned}
$$

Note that in the above proof, we identified an element $((a_1, \ldots, a_{n-1}), a_n)$ of $(A_1 \times A_2 \times \cdots \times A_{n-1}) \times A_n$ with the element $(a_1, \ldots, a_{n-1}, a_n)$ of $A_1 \times A_2 \times \cdots \times A_{n-1} \times A_n$. The result is true for any $n \geq 1$ by the induction principle. $\square$

In the context of counting, the product principle counts the number of ordered lists, or equivalently n-tuples or strings, of the form $x_1x_2\cdots x_n$ where x_1 can take t_1 values (from a set A_1), and for each of these values, x_2 can take t_2 values (from a set A_1), ect... Then the product principle says that we can form the list $x_1x_2\cdots x_n$ (or the n-tuple $(x_1, x_2, \ldots, x_n)$) in $t_1 \cdot t_2 \cdots \cdot t_n$ ways.

Example 8.4. A test consists of 9 multiple-choice questions. Each question has 6 possible answers (choices). How many answer keys are possible for this test?

Solution. We can think of an answer key as an ordered list $a_1a_2\ldots a_9$ where a_i is a possible answer to question i. There are 6 possible values for each a_i and each value is independent of the others. By the product principal, there are $6^9 = 10077696$ possible answer keys. $\qquad\qquad \Diamond$

Example 8.5. A meal in a certain restaurant consists of a choice of a main dish, a drink and a dessert. There are 13 choices for the main dish, 8 choices for a drink and 10 choices for a dessert. In how many ways can a meal be ordered?

Solution. Since a meal is a triplet (main dish, drink, dessert), there are $13 \cdot 8 \cdot 10 = 1040$ ways to order a meal by the product principle. $\qquad\qquad \Diamond$

Let Σ be a non-empty set of k symbols, and let $n \geq 0$ be an integer. A *string of length n* is an ordered sequence of n elements of Σ. If $n = 0$, the string is called the *empty string* and usually denoted by λ. If $k = 2$ (respectively $k = 3$), the string is called *binary* (respectively *ternary*). In this context, Σ is called an *alphabet* and its symbols are called *letters*. The study of strings is important in computer science. Binary strings are of particular importance because of their "two- states nature" (on and off) that can be used to physically implement procedures in electrical devices. Binary strings are usually represented by sequences of 0's and 1's that we call "bits". The number of bits in a binary is the length of the string. For example, 1000010110 is a binary string of length 10.

Example 8.6 (Number of binary strings of length n). *Let $n \geq 0$ be an integer. The product principle tells us that the number of binary string $x_1x_2\cdots x_n$ of length n is $\underbrace{2 \cdot 2 \cdots \cdot 2}_{n\ times} = 2^n$. This is true since each x_i is* *chosen from the set $\{0, 1\}$ and the choice of a bit in the string is independent from the choice of any other bit in the string.*

Example 8.7. How many binary strings:

(a) of length 10?
(b) of length 10 that start with 010?
(c) of length 10 that end with 11?
(d) of length 10 that start with 010 **and** end with 11?
(e) of length 10 that start with 10 **or** with 011?

Solution. (a) The number of binary strings of length 10 is $2^{10} = 1024$.

(b) The number of binary strings of length 10 that start with 010 is equivalent to the number of binary strings of length 7 since the first three bits are fixed. There are $2^7 = 128$ such strings.

(c) Similar to the previous part, the number of binary strings of length 10 that end with 11 is $2^8 = 256$.

(d) In this case, the first three and the last two bits of the string are already fixed. We are then counting the number of binary strings of length 5: $2^5 = 32$.

(e) Let A (respectively B) be the set of all binary strings of length 10 that start with 10 (respectively with 011). We need to compute $|A \cup B|$. Since $A \cap B = \emptyset$ (no binary string starts with 10 and 011 at the same time), the sum principle tells us that $|A \cup B| = |A| + |B| = 2^8 + 2^7 = 384$. $\diamond$

Example 8.8. How many four-digit positive integers with at least one occurrence of the digit 4 are there?

Solution. A four-digit positive integers has the form $x_1 x_2 x_3 x_4$ where $x_1 \geq 1$ and $x_2, x_3, x_4 \geq 0$. There are 9 choices for x_1 and for each of these choices, there are 10 choices for each of x_2, x_3 and x_4. By the product principle, there is a total of $9 \cdot 10^3 = 9000$ four-digit positive integers. Among these, there are $8 \cdot 9^3 = 5832$ four-digit positive integers with no occurrence of the digit 4. Therefore, there are $9000 - 5832 = 3168$ four-digit positive integers with at least one occurrence of the digit 4. $\diamond$

Example 8.9 (Number of functions). Let A and B be two finite sets of cardinalities m and n, respectively. How many functions $f : A \longrightarrow B$ from A to B are there? At first glance, this does not seem like a problem we can solve using either one of the previous two principles. In order to use a counting principle, we have to rethink the way we represent a function between two finite sets. Write $A = \{x_1, \ldots, x_m\}$ and $B = \{y_1, \ldots, y_n\}$ and assume that the order $x_1, \ldots, x_m$ is fixed on the elements of A. Then a function from A to B can be completely identified by the list of images $(f(x_1), f(x_2), \ldots, f(x_m))$ of the elements of A. So an m-tuple $(i_1, \ldots, i_m)$ where $i_k \in B$ for each k represents the function $f : A \longrightarrow B$ defined by $f(x_1) = i_1, \ldots, f(x_m) = i_m$. Counting the number of functions from A to B is then equivalent to counting the number of m-tuples $(i_1, \ldots, i_m)$ where each component is an element of B. By the product principle, there are n^m such m-tuples and so is a total of n^m functions from A to B.

Example 8.10 (Number of injective functions). With the notations of the previous example, how many *injective* functions from A to B are there? Note first that if $m > n$, then no injective function from A to B exists. Assume that $m \leq n$. Like in the previous example, we represent a function with an m-tuple $(i_1, \ldots, i_m)$ of elements of B. If we require that the function is injective, then the components must be pairwise distinct. That is, $i_s \neq i_t$ for $s \neq t$. There are n possible values for i_1, and for each these values, there are $n - 1$ possible values for i_2 and so on. By the product principle, there is a total of $n(n-1)(n-2)\cdots(n-m+1)$ m-tuples $(i_1, \ldots, i_m)$ such that $i_s \neq i_t$ for $s \neq t$. Note that $\frac{n!}{(n-m)!} = n(n-1)(n-2)\cdots(n-m+1)$. We conclude that the number of injective functions $f : A \longrightarrow B$ is given by:

$$\begin{cases} 0 & \text{if } m > n \\ \frac{n!}{(n-m)!} & \text{if } m \leq n \end{cases}$$

Example 8.11. Let A, B be two sets with $|A| = 5$ and $|B| = 8$. Then, there are $8^5 = 32768$ functions from A to B. Among these functions, $\frac{8!}{(8-5)!} = 6720$ are injective. On the other hand, there are $5^8 = 390625$ functions from B to A, none of which is injective.

8.2.3   *The principle of inclusion–exclusion*

The sum principle counts the number of elements in the union of pairwise disjoint sets. Very often, the sets we deal with in applications are not disjoint. This section deals with counting the number of elements in the union of any number of finite sets. As a consequence we will be able to count the number of surjective functions between finite sets. We start by stating and proving the principle for two sets. Unlike the principle of sum, the generalization of the inclusion–exclusion principle requires a bit more effort as we will discover below.

Theorem 8.4. *Let A and B be two finite subsets of a universal set $\mathcal{U}$. Then* $|A \cup B| = |A| + |B| - |A \cap B|$.

Proof. Clearly, the sets A and $B \backslash A$ are disjoint. Moreover, $A \cup (B \backslash A) = A \cup B$. By the sum principle,

$$|A \cup B| = |A \cup (B \backslash A)| = |A| + |B \backslash A| . \tag{8.1}$$

On the other hand, the sets $B \backslash A$ and $A \cap B$ are also disjoint with union equals to B. The sum principle gives us that

$$|B| = |(B \backslash A) \cup (A \cap B)| = |B \backslash A| + |A \cap B| \tag{8.2}$$

Subtracting Equation (8.2) from (8.1), we get that $|A \cup B| - |B| = |A| - |A \cap B|$ which leads to desired result. $\qquad\square$

Example 8.12. 1000 residents of a certain city were surveyed about their mean of transportation around the city. The survey revealed that 233 residents use the bus system, 346 use the subway system and 487 don't use any public transportation at all. How many, of the surveyed residents, use both public transportation systems? How many use the bus system only? How many use the subway system only?

Solution. Let B and S be the subsets of the surveyed residents who use the bus and the subway systems, respectively. The number of surveyed residents who use either one of the two public transportation systems is $|B \cup S| = 1000 - 487 = 513$. By the inclusion–exclusion principle, $|B \cap S| = |B| + |S| - |B \cup S| = 233 + 346 - 513 = 66$. So, there are 66 persons who use both transportation systems. The number of people who use the bus system only is $|B| - |B \cap S| = 233 - 66 = 167$ and the number of people who use the subway system only is $|S| - |B \cap S| = 346 - 66 = 280$. $\qquad\Diamond$

Example 8.13. We draw one card from a regular deck of 52 cards. In how many ways can we get a Queen or a red card?

Solution. There are 4 ways to draw a Queen, 26 ways to draw a red card and 2 ways to draw a red Queen. By the principle of inclusion–exclusion, there are $4 + 26 - 2 = 28$ ways to draw a Queen or a red card. $\qquad\Diamond$

Example 8.14. How many integers between 25 and 1256 are divisible by 3 or by 5?

Solution. Let A and B be the sets of all integers between 25 and 1256 that are divisible by 3 and by 5, respectively. We need to compute $|A \cup B|$. To find $|A|$, we need to find how many integers k satisfy $25 \le 3k \le 1256$. Dividing both inequalities by 3, we get that $\frac{25}{3} \le k \le \frac{1256}{3}$. Since $\frac{25}{3} \approx 8.33$ and $\frac{1256}{3} \approx 418.66$ and k is an integer, we get that $9 \le k \le 418$. There are $418 - 9 + 1 = 410$ such integers and so $|A| = 410$. A similar argument shows that $|B| = 247$. Since 3 and 5 are relatively prime, the set of integers between 25 and 1256 which are divisible by 3 and by 5, namely $A \cap B$, is the same as the set of integers which are divisible by 15. A similar argument as above shows that $|A \cap B| = 82$. By the principle of inclusion–exclusion, $|A \cup B| = |A| + |B| - |A \cap B| = 410 + 247 - 82 = 575$. $\diamond$

Example 8.15. How many binary strings of length 10 either begin with 11 **or** end with 101?

Solution. Let A be the set of binary strings of length 10 starting with 11 and B the set of binary strings of length 10 ending with 101. We need to compute $|A \cup B|$. There are 2^8 binary strings starting with 11, 2^7 binary strings ending with 101 and 2^5 binary strings starting with 11 and ending with 101. So, $|A| = 2^8$, $|B| = 2^7$ and $|A \cap B| = 2^5$. By the principle of inclusion–exclusion, $|A \cup B| = |A| + |B| - |A \cap B| = 2^8 + 2^7 - 2^5 = 352$. $\diamond$

We now generalize Theorem 8.4 to any finite number of finite sets. Recall first that if $J = \{1, \ldots, k\}$ and $A_1, A_2, \ldots, A_n$ are finite subsets of a universal set $\mathcal{U}$, then the notations $\cup_{j \in J} A_j$ and $\cap_{j \in J} A_j$ stand for $A_1 \cup A_2 \cup \cdots \cup A_k$ and $A_1 \cap A_2 \cap \cdots \cap A_k$ respectively.

Theorem 8.5 (Principle of inclusion–exclusion). *Let $A_1, A_2, \ldots, A_n$ be finite subsets of a universal set $\mathcal{U}$ with $n \ge 2$. Then:*

$$|A_1 \cup A_2 \cup \cdots \cup A_n| = \sum_{J \subseteq \{1,2,\ldots,n\}, J \ne \emptyset} (-1)^{|J|-1} |\cap_{j \in J} A_j|$$

$$= \sum_{i=1}^{n} |A_i| - \sum_{1 \le i < j \le n} |A_i \cap A_j| + \sum_{1 \le i < j < k \le n} |A_i \cap A_j \cap A_k|$$

$$- \cdots + (-1)^{n-1} |A_1 \cap A_2 \cap \cdots \cap A_n|.$$

Before we proceed with the proof of the theorem, let us examine what the result says for $n = 3$ and $n = 4$. In the case of three sets, the principle of inclusion–exclusion states that:

$$|A_1 \cup A_2 \cup A_3| = |A_1| + |A_2| + |A_3| - |A_1 \cap A_2|$$
$$- |A_1 \cap A_3| - |A_2 \cap A_3| + |A_1 \cap A_2 \cap A_3|$$

and in the case of four sets:

$$|A_1 \cup A_2 \cup A_3 \cup A_4| = |A_1| + |A_2| + |A_3| + |A_4|$$
$$- |A_1 \cap A_2| - |A_1 \cap A_3| - |A_1 \cap A_4| - |A_2 \cap A_3| - |A_2 \cap A_4| - |A_3 \cap A_4|$$
$$+ |A_1 \cap A_2 \cap A_3| + |A_1 \cap A_2 \cap A_4| + |A_1 \cup A_3 \cup A_4| + |A_2 \cap A_3 \cap A_4|$$
$$- |A_1 \cap A_2 \cap A_3 \cap A_4|.$$

We now present the proof of Theorem 8.5 using induction on n.

Proof. The result is true for $n = 2$ by Theorem 8.4. Let $n \geq 2$ and assume that the result the result is true for any $n - 1$ finite subsets of $\mathcal{U}$. Then:

$$\left| \bigcup_{k=1}^{n} A_k \right| = \left| \left(\bigcup_{k=1}^{n-1} A_k \right) \cup A_n \right| = \left| \bigcup_{k=1}^{n-1} A_k \right| + |A_n| - \left| \left(\bigcup_{k=1}^{n-1} A_k \right) \cap A_n \right|$$

(Theorem 8.4)

$$= \left| \bigcup_{k=1}^{n-1} A_k \right| + |A_n| - \left| \bigcup_{k=1}^{n-1} (A_k \cap A_n) \right| \quad \text{(Distributivity)}$$

$$= \sum_{I \subseteq \{1,2,\ldots,n-1\}, I \neq \emptyset} (-1)^{|I|-1} |\cap_{i \in I} A_j| + |A_n|$$

$$- \sum_{I \subseteq \{1,2,\ldots,n-1\}, I \neq \emptyset} (-1)^{|I|-1} |\cap_{i \in I} (A_i \cap A_n)|$$

(induction hypothesis)

Notice that:

$$|A_n| - \sum_{I \subseteq \{1,2,\ldots,n-1\}, I \neq \emptyset} (-1)^{|I|-1} |\cap_{i \in I} (A_i \cap A_n)|$$

$$= |A_n| + \sum_{I \subseteq \{1,2,\ldots,n-1\}, I \neq \emptyset} (-1)^{|I|} |(\cap_{i \in I} A_i) \cap A_n|$$

$$= |A_n| + \sum_{I \subseteq \{1,2,\ldots,n-1\}, I \neq \emptyset} (-1)^{|I|} |\cap_{i \in I \cup \{n\}} A_i|$$

$$= \sum_{I \subseteq \{1,2,\ldots,n-1\}} (-1)^{|I \cup \{n\}|-1} |\cap_{i \in I \cup \{n\}} A_i|$$

Consequently,

$$
\left| \bigcup_{k=1}^{n} A_k \right| = \sum_{I \subseteq \{1,2,\ldots,n-1\}, I \neq \emptyset} (-1)^{|I|-1} \left| \cap_{i \in I} A_j \right|
$$

$$
+ \sum_{I \subseteq \{1,2,\ldots,n-1\}} (-1)^{|I \cup \{n\}|-1} \left| \cap_{i \in I \cup \{n\}} A_i \right|
$$

$$
= \sum_{J \subseteq \{1,2,\ldots,n\}, J \neq \emptyset, n \notin J} (-1)^{|J|-1} \left| \cap_{j \in J} A_j \right|
$$

$$
+ \sum_{J \subseteq \{1,2,\ldots,n\}, n \in J} (-1)^{|J|-1} \left| \cap_{j \in J} A_j \right|
$$

$$
= \sum_{J \subseteq \{1,2,\ldots,n\}, J \neq \emptyset} (-1)^{|J|-1} \left| \cap_{j \in J} A_j \right|.
$$

The result is then true by the principle of mathematical induction. $\qquad \square$

Example 8.16. How many integers in the set $S = \{10, 11, 12, \ldots, 200\}$ are relatively prime with 105 ?

Solution. Note that $105 = 3 \cdot 5 \cdot 7$ and since $3, 5$ and 7 are pairwise relatively prime, an integer is not relatively prime with 105 if and only if it is divisible by at least one of the integers 3, 5 or 7. Let A, B and C be the subsets of S of integers divisible by 3, 5 and 7 respectively. We need to compute $|S| - |A \cup B \cup C|$. Using an argument similar to the one used in Example 8.14, we get that $|A| = 64$, $|B| = 37$ and $|C| = 27$. Now, $A \cap B$, $A \cap C$ and $B \cap C$ are the subsets of S of all integers divisible by 15, 21 and 35, respectively. The same argument as before shows that $|A \cap B| = 13$, $|A \cap C| = 9$ and $|B \cap C| = 5$. Finally, $A \cap B \cap C$ is the subset of S of all integers divisible by $3 \cdot 5 \cdot 7 = 105$ and $|A \cap B \cap C| = 1$. By the inclusion–exclusion principle, $|A \cup B \cup C| = |A| + |B| + |C| - |A \cap B| - |A \cap C| - |B \cap C| + |A \cap B \cap C| = 64 + 37 + 27 - 13 - 9 - 5 + 1 = 102$ and so $|S| - |A \cup B \cup C| = 89$. $\qquad \Diamond$

Example 8.17. How many positive integers less than or equal to 500 are not divisible by 6, 7, or 8?

Solution. Let A_i denote the set of integers between 1 and 500 that are divisible by i, where $i \in \{6, 7, 8\}$. Using an argument like in previous examples, we get: $|A_6| = 83$, $|A_7| = 71$ and $|A_8| = 62$. Any integer divisible by 6 and 7 is divisible by 42 since 6 and 7 are relatively prime, and so $|A_6 \cap A_7| = 11$. Similarly, $A_7 \cap A_8$ is the set of positive integers less than or equal to 500 which are not divisible 56 and so $|A_7 \cap A_8| = 8$. For $|A_6 \cap A_8|$, notice that 6 and 8 are not relatively prime, but an integer is divisible by 6 and 8 at the same if and only if it is divisible by their least common multiple, namely 24. We then get that $|A_6 \cap A_8| = 20$. Since the least common multiple of 6, 7 and 8 is 168, we find that $|A_6 \cap A_7 \cap A_8| = 2$. By the inclusion–exclusion principle, $|A_6 \cup A_7 \cup A_8| = |A_6| + |A_7| + |A_8| - |A_6 \cap A_7| - |A_6 \cap A_8| - |A_7 \cap A_8| + |A_6 \cap A_7 \cap A_8| = 83 + 71 + 62 - 11 - 8 - 20 + 2 = 179$. We conclude that the number of positive integers less than or equal to 500 which are not divisible by 6, 7, or 8 is equal to $500 - |A_6 \cup A_7 \cup A_8| = 500 - 179 = 321$. $\Diamond$

The following example shows how the inclusion–exclusion principle can be used to count surjective functions. Later on, we will see a general expression for the number of surjective functions between two finite sets using binomial coefficients (see Theorem 8.14 below).

Example 8.18. Let A be a finite set of cardinality $n \geq 1$ and let B be a set of cardinality 3. Find the number of surjective functions from A to B.

Solution. Write $A = \{x_1, \ldots, x_n\}$, $B = \{y_1, y_2, y_3\}$. If $n \leq 2$, we know that there are no surjective functions from A to B. Assume that $n \geq 3$. Let $\Gamma = \{f : A \to B; f \text{ is a function}\}$, $\Gamma_i = \{f \in \Gamma; y_i \notin \text{Im}(f)\}$ for $i = 1, 2, 3$. A function $f : A \to B$ is non-surjective if and only if $f \in \Gamma_1 \cup \Gamma_2 \cup \Gamma_3$ (either y_1, y_2 or y_3 is not in the image of f). By the inclusion–exclusion principle, $|\Gamma_1 \cup \Gamma_2 \cup \Gamma_3| = |\Gamma_1| + |\Gamma_2| + |\Gamma_3| - |\Gamma_1 \cap \Gamma_2| - |\Gamma_1 \cap \Gamma_3| - |\Gamma_2 \cap \Gamma_3| + |\Gamma_1 \cap \Gamma_2 \cap \Gamma_3|$. An element of Γ_1 is a function $f : A \to B$ that does not have the element y_1 as an image, so it is a function from A to $\{y_2, y_3\}$. We know that the number of such functions is 2^n, and so $|\Gamma_1| = 2^n$. Similarly, $|\Gamma_2| = |\Gamma_3| = 2^n$. A function in $\Gamma_1 \cap \Gamma_2$ is a function from A to $\{y_3\}$. Consequently, $|\Gamma_1 \cap \Gamma_2| = 1^n = 1$. Similarly, $|\Gamma_1 \cap \Gamma_3| = |\Gamma_2 \cap \Gamma_3| = 1$. Finally, $|\Gamma_1 \cap \Gamma_2 \cap \Gamma_3| = 0$ since $\Gamma_1 \cap \Gamma_2 \cap \Gamma_3 = \emptyset$. So, $|\Gamma_1 \cup \Gamma_2 \cup \Gamma_3| = 3 \cdot 2^n - 3$ and the number of surjective functions from A to B is $|\Gamma| - |\Gamma_1 \cup \Gamma_2 \cup \Gamma_3| = 3^n - (3 \cdot 2^n - 3) = 3^n - 3 \cdot 2^n + 3$. $\Diamond$

8.2.4 *Exercises*

$\star$(1) Use induction to prove Theorem 8.1 above.

(2) Find the number of ways to draw from a regular deck of 52 cards:

    (a) A king or a queen   ⋆ (c) A value card or a Jack

    (b) A value card (Ace to 10)   (d) A red card or a King

(3) Two regular dice are rolled simultaneously. In how many ways can we get an even sum on the faces of the dice?

⋆(4) Two regular dice are rolled simultaneously. In how many ways can we get a sum of 6 or 9 on the faces of the dice?

(5) In a certain ice cream shop, you have the choice between 18 flavors, 6 toppings, 3 cone sizes (small, medium and large) and 4 cone types. You can pick only one flavor and one toping when you buy an ice cream cone.

    ⋆ (a) In how many ways can you get an ice cream cone?

      (b) In how many ways can you get a chocolate ice cream cone? (chocolate is one of the flavors)

      (c) In how many ways can you get a medium chocolate ice cream cone?

(6) How many non-empty binary strings of length *at most* 6 are there?

(7) By a *word*, we simply mean a succession of letters of the English alphabet. A word does not necessarily have a meaning. For example *exrvtyuio* is a word.

      (a) How many words with five letters?

    ⋆ (b) How many words with five letters but no letter is repeated?

      (c) How many words with five letters starting with "A" and ending with "Z"?

    ⋆ (d) How many words with five letters containing *at most* one occurrence of the letter "A"?

(8) In Mathland, a common car licence plate consists of three letters followed by three digits.

      (a) How many licence plates can be found in Mathland?

    ⋆ (b) How many licence plates with distinct letters?

    ⋆ (c) How many licence plates with none of the letters and none of the digits are repeated?

      (d) How many licence plates use only vowels (A, E, I, O, U) and odd digits?

(9) In Canada, every postal code is of the form $\square\,\triangle\,\square\,\triangle\,\square\,\triangle$, where $\square$ is a letter and $\triangle$ is a digit.

      (a) How many postal codes are there?

    ⋆ (b) How many postal codes if the first letter is either K or H?

(c) How many postal codes if the last digit is either 2 or 3?

(d) How many postal codes are there if the first letter is either K or H and the last digit is either 2 or 3?

(10) Let A, B be two finite sets such that $|A| = 7$ and $|B| = 5$.

(a) How many functions from A to B?

(b) How many injective functions from A to B?

(c) How many functions from B to A?

(d) How many injective functions from B to A?

(e) How many functions from $\wp(A)$ to $\wp(B)$?

(f) How many injective functions from $\wp(A)$ to $\wp(B)$?

(g) How many functions from $\wp(B)$ to $\wp(A)$?

(h) How many injective functions from $\wp(B)$ to $\wp(B)$?

(11) Write the principle of inclusion–exclusion for five sets.

(12) Let A_1, A_2 and A_3 be three subsets of a universal set $\mathcal{U}$. If $A_1 \cap A_2 \cap A_3 = \emptyset$, can we confirm that $|A_1 \cup A_2 \cup A_3| = |A_1| + |A_2| + |A_3|$? Justify your answer.

$\star$(13) In how many ways can a photographer align 7 persons to take a photo?

(14) We draw one card from a standard deck of 52 cards. In how many ways can we draw a value card or a black card?

(15) How many binary strings of length 13 that start with 0101 or end with 00?

(16) A computer password must have 5, 6 or 7 characters. Each character is either a capital letter, a digit (between 0 and 9), or one of the five special symbols: $, &, #, @ and !.

$\star$ (a) How many possible passwords are there?

(b) How many passwords containing none of the special symbols?

(c) How many passwords containing *at least one* of the five special symbols?

$\star$(17) JoeMart is a department store which also offers an online shopping option. A survey asked 1000 people about their shopping habits at JoeMardt. Out of the 1000 persons surveyed, 550 said that they shop online, 580 said that they shop in store and 200 said that they shop both online and in store. How many, of the surveyed people, do not shop at JoeMart at all?

(18) How many integers between 353 and 34565 which are:

(a) divisible by 5?

(b) divisible by 11?

(c) divisible by 5 and by 11?

(d) divisible by 5 or by 11?

(e) divisible by 5, not by 11?

(f) divisible by 11, not by 5?

(g) not divisible by 5 and not divisible by 11?

(19) How many integers between 500 and 2500 inclusive are not divisible by neither one of the integers 2, 3 and 5?

(20) How many integers between 250 and 7500 inclusive are not divisible by neither one of the integers 5, 6, or 9?

(21) A study done on a 100 people in a small town shows that 65 people have curly hair, 45 have green eyes, 42 have long legs, 15 have green eyes and long legs, 25 have curly hair and long legs and 20 have curly hair and green eyes. Find the number of people who have:

(a) Curly hair and green eyes but not long legs.

(b) Green hair and long legs but not curly hair.

(c) Green eyes but not curly hair nor long legs.

$\star$(22) A survey of n university students shows the following information: 80 students took linear algebra, 32 students took calculus II, 63 students took discrete math, 17 students took both linear algebra and calculus II, 25 students took both linear algebra and discrete math, 16 students took both calculus II and discrete math, 9 students took all three classes and 21 students did not take any of the three courses. What is the value of n?

(23) A and B are two finite sets with $|A| = 6$ and $|B| = 4$. How many surjective functions from A to B? How many surjective functions from B to A?

(24) In how many ways a committee can be formed in a class of 18 students if at least one of three students Alex, Bob and Carl must be on the committee? Use two different ways to solve this problem, one that uses the principle of inclusion–exclusion and another that does not.

$\star$(25) How many passwords of length 9 can be formed using only the four letters A, B, C and D? How many of these passwords are such that each of the four letters appears at least once in the password?

8.3   The pigeonhole principle

The simple version of the pigeonhole principle says the following: *"If more than n pigeons occupy n pigeonholes, then at least one pigeonhole must be occupied by two or more pigeons."* At first glance, this obvious principle seems to be out of context in this chapter and almost has nothing to do with counting. But as we will see through various examples, this simple idea could be a key to unlock the solutions to some complex and difficult problems. Let us start with a simple example.

Example 8.19. What is the smallest number n of persons in a group in order to ensure that at least two persons in the group: (a) are born on the same day of the week? (b) are born the same month?

Solution. There are 7 days in a week. Each one of these days can be thought of as a pigeonhole, and each person as a "pigeon". If $n = 8$, then at least one pigeonhole (day of the week) contains at least two "pigeons" (persons). If $n \leq 7$, there is no guarantee that at least two pigeons will share the same pigeonhole. So the minimal value of n is 8. A similar argument shows we need at least $n = 13$ persons in the group to ensure that at least two are born the same month. $\diamondsuit$

We know give the general version the pigeonhole principle.

Theorem 8.6 (The pigeonhole principle). *Let k and n be two positive integers. If n objects are distributed in k boxes, then there exists at least one box containing at least $\lceil \frac{n}{k} \rceil$ objects.*

Proof. Assume by contradiction that the number of objects in every box is less than or equal to $\lceil \frac{n}{k} \rceil - 1$. Then the total number of objects satisfies: $n \leq k \left(\lceil \frac{n}{k} \rceil - 1 \right)$. Notice that $\lceil x \rceil < x + 1$ for any real number x as shown in the following graph (with the line $y = x + 1$):

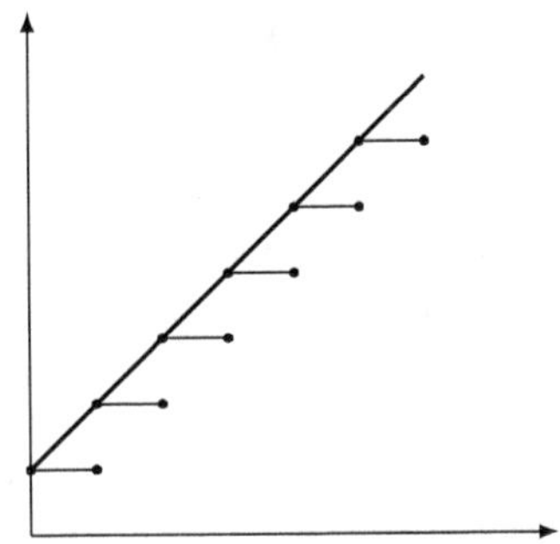

We conclude that $n \leq k \left(\lceil \frac{n}{k} \rceil - 1 \right) < k \left(\frac{n}{k} + 1 - 1 \right) = n$: This is a contradiction. The result follows. $\square$

When using the pigeonhole principle, it is important to specify three things: the objects, the boxes and in what manner the objects are placed in the boxes.

Example 8.20. There are 235 students in a discrete math class. Show that the professor can choose at least 20 students from the class who are born the same month.

Solution. Here, the objects are the 235 students and the boxes are the 12 months of the year. A student is "placed" in the box corresponding to his/her month of birth. By the pigeonhole principle, there exists at least one box (a month) containing at least $\lceil \frac{235}{12} \rceil = 20$ objects (students). $\Diamond$

Example 8.21. Prove that among 12 non-negative integers chosen at random, we can always choose at least 3 integers that have the same remainder in the division by 5.

Solution. There are 5 remainders in the division by 5, namely $0, 1, 2, 3$ and 4. We use the pigeonhole principle with the objects being the 12 integers, the boxes being the 5 remainders in the division by 5. Each of the 12 integers is placed in the box corresponding to its remainder in the division by 5. By the pigeonhole principle, there exists at least one box (remainder) containing at least $\lceil \frac{12}{5} \rceil = 3$ integers. These 3 integers have the same remainder in the division by 5. $\Diamond$

Example 8.22. A certain paragraph contains n random words from the English alphabet (here a word does not need to have a meaning, just a string of letters) with each word consisting of at least two letters. What is the minimal value of n in order to ensure that at least four words in the paragraph start with the same substring of length 2 (i.e., the same first two letters in the same order)?

Solution. The number of strings consisting of two letters from the English alphabet is $26^2 = 676$. We use the pigeonhole principle with the objects being the n words in the paragraph, the boxes are the 676 distinct strings of length 2. A word is placed in the box corresponding to its initial substring of length 2. For example, the word *drive* is placed in the box labeled "dr". By the pigeonhole principle, there exits at least one box containing at least $\lceil \frac{n}{676} \rceil$ objects (words). In order to have at least four words starting with the same first two letters, we need that $3 < \frac{n}{676} \leq 4$. In other words, the minimum value of n is $3 \cdot 676 + 1 = 2029$ words. $\Diamond$

The following example shows an application of the pigeonhole principle in geometry.

Example 8.23. 11 points are place on the perimeter of a circle of radius 2 units. Prove that there exist at least two of the 11 points that are less than 1.5 units apart from each other (*Hint.* the side length of a regular polygon with n sides inscribed in a circle of radius R is $2R\sin\left(\frac{\pi}{n}\right)$).

Solution. Inscribe a regular polygon with 10 sides inside the circle. This subdivide the circumference of the circle into 10 arcs of the same length each. Since the 11 points lie on the circumference, the pigeonhole principle tells us that at least one arc must have at least $\lceil\frac{11}{10}\rceil = 2$ points. The largest distance separating two points on the same arc is equal to the side length of the regular polygon, namely $4\sin\left(\frac{\pi}{10}\right) \approx 1.24 < 1.5$ units. $\qquad\qquad\diamond$

The next example shows a more theocratical application of the pigeonhole principle. In the proof of the result of the example, we use the notation $\overline{x} = x - \lfloor x \rfloor$ for any real number x where $\lfloor x \rfloor$ is the largest integer smaller than or equal to x. Clearly $0 \leq \overline{x} < 1$.

Example 8.24 (Dirichlet's theorem). *Let x be a real number. Prove that for any integer $n \geq 1$, there exists a rational number $\frac{a}{b}$ such that $1 \leq b \leq n$ and $\left|x - \frac{a}{b}\right| < \frac{1}{nb}$.*

Solution. Let x be a real number and $n \geq 1$ an integer. For $i \in \{1, 2, \ldots, n+1\}$, let $a_i = \overline{ix}$. Since each of these $n+1$ elements is in $[0, 1[$ and $[0, 1[$ is the union of the n subintervals $[0, \frac{1}{n}[$, $[\frac{1}{n}, \frac{2}{n}[$, $\ldots$, $[\frac{n-1}{n}, 1[$ (of length $\frac{1}{n}$ each), the pigeonhole principle tells us that there exists at least one subinterval containing at least two elements a_i and a_j with $1 \leq i < j \leq n$. In particular, $|jx - \lfloor jx \rfloor - (ix - \lfloor ix \rfloor)| < \frac{1}{n}$. Rearranging, we get that $|bx - a| < \frac{1}{n}$ where $a = \lfloor jx \rfloor - \lfloor ix \rfloor$ and $b = j - i$. Dividing the last inequality by b we get that $\left|x - \frac{a}{b}\right| < \frac{1}{bn}$ as required. $\qquad\diamond$

8.3.1 *Exercises*

(1) Prove that in a town of 5750 people, the mayor can choose at least 16 people who have the same birthday.

⋆(2) Prove that among 132 non-negative integers chosen at random, we can choose at least 19 integers that have the same remainder in the division by 7.

(3) Show that in any set of 15 non-negative integers, we can choose at least two integers with difference divisible by 14.

(4) Every item produced by a certain company comes with a label of type "digit-letter-letter-digit-letter". The company allows two items to have the same label if they are produced at least two years apart. To date, the company has produced 60 millions items. What can be said about the number of items having the same label?

$\star$(5) Let S be a non-empty set of n integers, $n \geq 2$. Prove that there exists a non-empty subset T of S such that the sum of the elements of T is a multiple of n.

(6) For a social event, a company ordered 90 sandwiches of each of the following kinds: ham, tuna, turkey, veggie and cheese for its employees. Unfortunately the sandwiches arrived rapped but unlabeled. A group of 15 employ want to have the same kind of sandwich, regardless of that kind. What is the minimum number of sandwiches the waiter has to bring to their table in order to guarantee that each of the 15 individuals will have the same sandwich?

(7) A certain paragraph contains n random words (strings of letters, not necessarily meaningful) from the English alphabet with each word containing at least three letters. What is the minimum value of n to make sure that at least three words in the paragraph start with the same substring of length 3?

(8) A box contains a large number of colored balls: red, green, yellow, blue and purple. What is the minimum number of balls that must be drawn from the box (without looking) in order to make sure that at least 10 balls are of the same color?

$\star$(9) At a certain University, students can obtain one of the following grades on any course they take: A^+, A, A^-, B^+, B, C^+, C, D, E and F. What is the minimum number of students who need to be registered in a class to make sure that at least 9 students receive the same grade?

(10) 500 points are placed inside a $1\,m \times 2\,m$ rectangle such that not all the points are on the same line. Show that at least three of these points form a triangle of area less than or equal to $50\,\mathrm{cm}^2$.

(11) Prove that among 11 distinct numbers selected at random from the set $\{1, 2, \ldots, 20\}$, we can choose at least two of these numbers with difference equal to 10 (*Hint.* Consider the boxes labeled as $\{1, 11\}$, $\{2, 12\}$, $\{3, 13\}$, $\ldots$).

(12) In the space of three dimensions, consider nine points (x_i, y_i, z_i) with integer coordinates. Show that we can choose at least two points with

the property that the midpoint of the line segment joining the two points has integer coordinates.

(13) Let $n \geq 1$ be an integer. Show that there is a positive multiple of n whose digits are all 0's and 1's (*Hint.* Consider the integers 1, 11, 111,...)

(14) What is the minimal number of times a pair of dice must be rolled (simultaneously) to guarantee 7 occurrences of the same value on both dice?

$\star$(15) Consider the set $A = \{2, 3, 4, \ldots, 49, 50\}$. Prove that if we choose n distinct integers in A with $n \geq 16$, then there are at least two of the chosen integers that are not relatively prime (*Hint.* How many prime numbers are there in A?). Is the result true when $n = 15$?

(16) Let n be an even positive integer. Show that if $\frac{n}{2} + 1$ integers are chosen from the set $\{1, 2 \ldots, n\}$ then one of the integers must divide another. (*Hint.* Every positive integer can be written in the form $2^a \cdot b$ for some non-negative integers a_j and b with $1 \leq b \leq n$ odd).

8.4   Permutation

In a meeting of 20 persons, in how many ways can we choose 10 persons and arrange them in a row to take a photo? Every photo consists of two steps: first *choosing* 10 persons out of 20 and second *arranging* the 10 chosen persons in a row to take a photo. The main difference between *choosing* and *arranging* in this context is the order of elements. When choosing 10 persons out of 20, the order in which these persons are selected is not important. However, arranging the 10 person in a row to take a photo takes the order into consideration. In this section, we tackle the arrangement part of the above problem. We learn how to count the number of ways to arrange objects in a row, around a circle, with and without repeating the objects.

8.4.1   *Permutation without repetition*

Definition 8.1. Let n, r be two integers such that $0 \leq r \leq n$. An *r-permutation* of n objects $a_1, \ldots, a_n$ is an *ordered* linear arrangement of r objects chosen from the set $\{a_1, a_2, \ldots, a_n\}$ *without repetition* of elements. A *permutation* of the objects $a_1, a_2, \ldots, a_n$ is just an n-permutation of these objects.

In what follows, the word permutation means "permutation without repetition" unless otherwise indicated. Later on, we investigate permutation with repetitions. Note that in some textbooks, a permutation of n objects is defined simply as a bijection from the set $\{1, 2, \ldots, n\}$ to itself.

Example 8.25. There are six 2-permutations of the objects a, b, c, namely: ab, ba, ac, ca, bc, cb. There are also six permutations of these three objects, namely: *abc, acb, bac, bca, cab* and *cba*.

For $0 \leq r \leq n$, the number of r-permutations without repetition of n objects is denoted by $P(n, r)$.

Theorem 8.7. *Let n, r be two integers satisfying $0 \leq r \leq n$. Then:*

(a) $P(n, r) = \frac{n!}{(n-r)!}$.
(b) *The number of permutations of n objects is $n!$.*

Proof. An r-permutation of n objects $a_1, \ldots, a_n$ can be thought of as an ordered list $a_{i_1} \ldots a_{i_r}$ of r elements of the set $\{a_1, \ldots, a_n\}$. By the product rule, there are $n(n-1) \cdots (n-r+1)$ such ordered list. Notice that $\frac{n!}{(n-r)!} = n(n-1) \cdots (n-r+1)$ which proves part (a). Part (b) is an immediate consequence of part (a) using $r = n$ and the fact that $0! = 1$. $\square$

Example 8.26. You have a box containing 15 books that you want to organize on bookshelves. The order in which the books appear on a bookshelf is important to you.

(a) In how many ways can you place the 15 books on a bookshelf if it can fit all the books?
(b) If the bookshelf can fit only 7 books, in how many ways can you place 7 books on it?

Solution. (a) Since the order in which the books appear on a bookshelf is important, there are $15! = 1307674368000$ ways of arranging the 15 books on the bookshelf.

(b) Arranging 7 books on the bookshelf amounts to a 7-permutations of 15 objects: $P(15, 7) = \frac{15!}{(15-7)!} = 32432400$ ways. $\Diamond$

Example 8.27. After a series of interviews, the list of candidates for three positions: Dean, Vice-Dean research and Vice-Dean academic is narrowed to 7 people. Each of the 7 final candidates can fill any of the three positions. In how many ways these positions can be filled?

Solution. Filling the three positions corresponds to a 3-permutation of 7 objects: there are $P(7,3) = \frac{7!}{(7-3)!} = 210$ such permutations. $\Diamond$

Example 8.28. How many different words can we obtain by permuting letters of the word *EQUATIONS* if:

(a) all letters are used;
(b) only 5 letters are used;
(c) all letters are used but the word starts with a vowel or ends with a vowel;
(d) all letters are used but all vowels are side by side.

Solution. Notice first that the 9 letters of the word *EQUATIONS* are pairwise distinct. This is a key observation for the solution.

(a) This corresponds to the number of permutations of 9 objects. There are $9! = 362880$ such words.
(b) This corresponds to the number of 5-permutations of 9 objects. There are $P(9,5) = \frac{9!}{4!} = 15120$ such words.
(c) There are five vowels in the word: A, E, U, I and O. There are 5 ways to choose the first letter and for each of these ways, there are $8! = 40320$ ways to permute the other letters. By the product principle, there are $5 \cdot 40320 = 201600$ words starting with a vowel. Similarly, there are 201600 words ending with a vowel. The number of words starting and ending with a vowel is $5 \cdot 7! \cdot 4 = 100800$. By the principle of inclusion–exclusion, there are $201600 + 201600 - 100800 = 302400$ words starting or ending with a vowel.
(d) We treat the vowels A, E, U, I and O appearing in the word as one object. There are $5! = 120$ ways to permute the 5 objects $AEUIO$, Q, T, N and S. For each of these ways, there are $5! = 120$ ways of permuting the letters in $AEUIO$. By the product principle, there are $120 \cdot 120 = 14400$ words using all letters and such that all vowels are side by side. $\Diamond$

Example 8.29. How many different numbers can we form using the digits $3, 4, 7, 8$ and 9 if no digit occurs more than once in the number?

Solution. There are $P(5,1)$ numbers with one digit, $P(5,2)$ numbers with two digits, $P(5,3)$ numbers with three digits, $P(5,4)$ numbers with four digits and $P(5,5)$ numbers with five digits. By the sum principle, there are

$P(5,1) + P(5,2) + P(5,3) + P(5,4) + P(5,5) = 5 + 20 + 60 + 1 = 86$ such numbers. $\diamond$

8.4.2   Circular permutation

The permutations we have considered so far count how many ways we can arrange objects in a linear manner. When objects are arranged on a circle, the permutation is called circular. For circular permutations, it is understood that two arrangements of objects are the same if one can be obtained from the other by a rotation. This means that the "physical" locations of the objects on the circle are irrelevant in counting the number of circular arrangements, only the "relative" positions count. For instance, the following two arrangements are considered the same as each object has the same left and right neighbors:

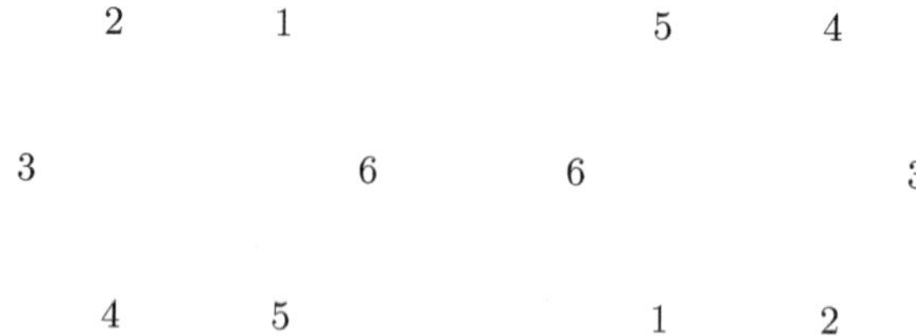

Theorem 8.8. *Let* $n \geq 2$ *be an integer. The number of circular permutations of* n *objects with no repetition is* $(n-1)!$.

Proof. Imagine n chairs around a table and n persons to sit on the chairs. Label the chairs using the numbers $1, 2, \ldots, n$. Label the persons using the symbols $p_1, p_2, \ldots, p_n$. Start by seating person p_1 on chair 1, person p_2 on chair 2, ..., person p_n on chair n. Then move p_1 to chair 2, person p_2 on chair 3, ..., person p_n on chair 1. After repeating these moves n times, we get the original seating arrangement. Since there are $n!$ ways of assigning seats to the n persons and each assignment of seats corresponds to n equivalent circular permutations of the n persons, we have $\frac{n!}{n} = (n-1)!$ different circular permutations. $\square$

Example 8.30. There are $6! = 720$ ways to arrange 7 people around a table.

Example 8.31. In how many ways can six women and six men be seated around a table such that they occupy alternate seats? (so no two men and no two women are side by side).

Solution. The six women can be seated in $5! = 120$ ways around the table. For the men to occupy alternate positions, they need to be seated in the empty seats between the women. There are $6! = 720$ ways to seat the men this way (since each seating of the men corresponds to a linear permutation). By the product rule, there are $5! \cdot 6! = 86400$ ways to arrange the men and the women persons around the table. $\Diamond$

Example 8.32. Joe invited 15 guests to his house. Among the guests, Mike and Bob, Joe's best friends. In how many ways can everyone at the party (including the host) be seated around a table in such a way that Mike and Bob are on either side of Joe?

Solution. We treat the trio Joe, Mike and Bob as one "person". So, we have to arrange 14 persons on a table. There are $13!$ ways to do that. Since Mike and Bob can be on either side of Joe, there are $2 \cdot 13! = 12454041600$ ways of arranging people around the table at the party. $\Diamond$

Example 8.33. When dealing with necklaces formed using n beads of different colors, the situation is a bit different from arranging n people around a table. If the necklace can be worn on both sides, then the number of ways we can make a necklace using n beads of different colors is $\frac{(n-1)!}{2}$ sine each arrangement of the beads on the necklace is actually two arrangements at the same time. Since the two belong to the same necklace, they become indistinguishable.

8.4.3 *Permutations of objects not all distinct*

Interesting counting problems often involve permutations of objects some of which are indistinguishable. For instance, how many different words can be formed by permuting all letters of the word *POPCORNE*? Unlike what we did in Example 8.28 above, the answer is not $8!$ since the 8 letters of the word are not pairwise distinct.

Theorem 8.9. *Assume that we have n objects formed by n_1 indistinguishable objects of type 1, n_2 indistinguishable objects of type 2, ..., and n_k indistinguishable objects of type k where $n_1 + n_2 + \cdots + n_k = n$. Then the number of different linear permutations of the n objects, denoted by $P(n, n_1, \ldots, n_k)$, is $P(n, n_1, \ldots, n_k) = \frac{n!}{n_1! n_2! \cdots n_k!}$.*

Proof. Represent each of the n_1 indistinguishable objects of type 1 with the letter A. These objects can made distinguishable by adding subscripts to A, say $A_1, A_2, \ldots, A_{n_1}$. For each arrangement of the n objects in which the n_1 indistinguishable A's appear, there are $n_1!$ ways to arrange the distinguishable A_i's. In general, for each arrangement of the n objects in which the n_i indistinguishable objects of type i appear, there are $n_i!$ ways to arrange the corresponding indexed distinguishable objects. By the product principle, there are $n_1!n_2!\cdots n_k!P(n, n_1, \ldots, n_k)$ possible permutations (without repetition) of the n objects when made distinguishable. Since the number of permutations of n distinguishable objects is $n!$, we get that $n_1!n_2!\cdots n_k!P(n, n_1, \ldots, n_k) = n!$ and the result follows. $\square$

Example 8.34. The word *POPCORNE* has $n = 8$ letters: 1 of "type" C, 1 of "type" E, 1 of "type" N, 2 of "type" O, 2 of "type" P and 1 of "type" R. The number of different words obtained by permuting all letters of the word *POPCORNE* is then $\frac{8!}{1!1!1!2!2!1!} = 10080$.

Example 8.35. MISSISSAUGA is a city in Ontario, Canada.

(a) How many different words (not necessarily with a meaning) do we obtain by permuting all the letter of MISSISSAUGA?

(b) How many of these words have the four S together?

(c) How many of these words have the two I together?

(d) How many of these words have the two I together *or* the four S together?

(e) How many of these words are such that the four S are not all together *and* the two I are not together?

Solution. Note first that there are 11 letters in the word MISSISSAUGA: 2 A, 2 I, 1 G, 1 M, 4 S and 1 U.

(a) There are $\frac{11!}{2!2!1!1!4!1!} = 415800$ different words obtained by permuting all the letter of MISSISSAUGA.

(b) We treat the four S as one object: $\frac{8!}{1!2!1!2!1!1!} = 10080$.

(c) We treat the two I as one object: $\frac{10!}{1!1!4!2!1!1!} = 75600$.

(d) Let Σ be the set of all words having the two I together, and let Γ be the set of all words having all the four S together. We need to compute $|\Sigma \cup \Gamma|$. From parts (b) and (c), we know that $|\Sigma| = 201600$ and $|\Gamma| = 75600$. The number of words with the four S together and the two I together is $|\Sigma \cap \Gamma| = \frac{7!}{1!1!1!2!1!1!} = 2520$. By the inclusion–exclusion principle, $|\Sigma \cup \Gamma| = 10080 + 75600 - 2520 = 83160$.

(e) This is $415800 - |\Sigma \cup \Gamma| = 415800 - 83160 = 332640$. $\Diamond$

Example 8.36. In how many different ways can we arrange 3 blue balls, 6 red balls and 5 green balls in row?

Solution. Each arrangement of the balls in a row corresponds to a permutation of the word $BBBRRRRRRGGGGG$. There are $\frac{14!}{3!6!5!} = 168168$ such arrangements. $\Diamond$

8.4.4 *Permutations with repetitions*

Consider three objects labeled a, b and c. If repetition is allowed, how many 2-permutations are possible of these three objects? In this case, it is easy to list all the possibilities: aa, ab, ac, ba, bb, bc, ca, cb and cc: 9 such permutations. We can think of each of these permutations as being an ordered pair (x, y) where $x, y \in \{a, b, c\}$. By the product principle, there are $3 \cdot 3 = 9$ such pairs. In general, an r-permutation with repetition of n distinct objects $a_1, \ldots, a_n$ is an arrangement of r elements of the set $\{a_1, \ldots, a_n\}$ with repetitions allowed.

Theorem 8.10. *The number of r-permutations of n elements with repetitions allowed is n^r.*

Proof. Left as an easy exercise (see Exercise (13) below). $\square$

Example 8.37. A stair case has 6 steps. In how many ways can the steps be colored using 8 different colors?

Solution. Each coloring of the six steps corresponds to a 6-permutation with repetitions of 8 objects (8 colors). Theorem 8.10 tells us that there are $8^6 = 262144$ ways to color the steps. $\Diamond$

Example 8.38. You have a rectangular area of size 1×7 (square units) that you want to tile. There are three sizes of tiles you can use: 1×1, 1×2 and 1×3 (in square units). In how many ways can you do the job assuming that tiles of the same size are indistinguishable?

Solution. Let t_i, $i = 1, 2, 3$ denotes a tile of size $1 \times i$. The following diagram represents (in different shades) the two tilings $t_1 t_2 t_3 t_1$ and $t_3 t_1 t_2 t_1$ of the rectangle:

Notice that each tiling $t_{i_1} t_{i_2} \cdots t_{i_k}$ of the rectangle must satisfy $i_1 + i_2 + \cdots + i_k = 7$. The tiling of the rectangle can be done using tiles of type: (i) t_1 only; (ii) t_1 and t_2 only; (iii) t_1 and t_3 only; (iv) t_2 and t_3 only; (v) t_1, t_2 and t_3. In case (i), there is one single tiling with six 1×1 tiles. In case (ii), every tiling is a permutation of either one of $t_1 t_2 t_2 t_2$, $t_1 t_1 t_1 t_2 t_2$ or $t_1 t_1 t_1 t_1 t_1 t_2$. There are $\frac{4!}{1!3!} = 4$ permutations of $t_1 t_2 t_2 t_2$, $\frac{5!}{3!2!} = 10$ permutations of $t_1 t_1 t_1 t_2 t_2$ and $\frac{6!}{5!1!} = 6$ permutations of $t_1 t_1 t_1 t_1 t_1 t_2$. In case (iii), a tiling of the rectangle is a permutation of $t_1 t_3 t_3$ or $t_1 t_1 t_1 t_1 t_3$. There are $\frac{3!}{1!2!} = 3$ permutations of $t_1 t_3 t_3$ and $\frac{5!}{4!1!} = 5$ permutations of $t_1 t_1 t_1 t_1 t_3$. In case (iv), a tiling is a permutation of $t_2 t_2 t_3$ with $\frac{3!}{2!1!} = 3$ such permutations. Finally, in case (v) a tiling is a permutation of $t_1 t_1 t_2 t_3$ with $\frac{4!}{2!1!1!} = 12$ such permeations. By the sum principle, there are $1 + 4 + 10 + 6 + 3 + 5 + 3 + 12 = 44$ ways to tile the rectangle. $\Diamond$

Given r labeled objects $1, 2, \ldots, r$ and n labeled boxes $1, 2, \ldots, n$, every distribution of the objects in the boxes corresponds to an r-permutation of n objects. There are n^r such permutations by Theorem 8.10 above. For example, if we have six objects labeled a, b, c, d, e, f and four boxes labeled $1, 2, 3$ and 4, then 122342 corresponds to placing a in box 1, b, c and f in box 2, d in box 3 and e in box 4. Theorem 8.10 above can then be interpreted as the number of ways of distributing r labeled objects in n labeled boxes as follows.

Theorem 8.11. *The number of ways to distribute r labeled objects in n labeled boxes is n^r.*

Example 8.39. There are $4^7 = 16384$ ways to distribute 7 balls of different colors in 4 different boxes.

8.4.5 *Exercises*

(1) How many 2-permeations of the objects a, b, c, d? List them all.
$\star$(2) How many 3-permeations of the objects a, b, c, d? List them all.
(3) There are 15 athletes competing in a sport competition.

    (a) If at the end of the competition, athletes are ranked from 1 to 15. How many possible rankings if no ties can occur?

    $\star$ (b) If at the end of the competition, only the top five athletes are ranked. How many possible outcomes if no ties can occur?

(4) In how many ways can we arrange 12 persons: (a) in a row? (b) around a table?

$\star$(5) In how many ways can we make a necklace with 13 beads of different colors if the necklace can be worn: (a) on one side only? (b) on both sides?

(6) Solve each of the following equations for n.

    (a) $P(n, 2) = 20$
    (b) $P(2n, 3) = 12P(n, 3)$
  $\star$ (c) $P(n, 5) = 32P(n - 1, 3)$

(7) Two women, three men and a child sit around a table.

    (a) How many seating arrangements are possible?
    (b) How many seating arrangements are possible if the child must sit directly between the two women?

$\star$(8) Five women and seven men are to be seated in a row. In each case, find in how many ways this can be done.

    (a) There are no restrictions on seating arrangements.
    (b) The women are sitting together (no man is between two women).
    (c) The men are sitting together and the women are sitting together.

(9) A locker code consists of 4 digits arranged in a specific linear order. How many possible codes are there? How many possible codes if no digit is repeated in the code?

$\star$(10) How many different words (with possibly no meaning in English) can we obtain by permuting all the letters of the word *"COVERING"*? How many of these words are such that the letter "I" immediately precedes the letter "N"?

(11) How many different six-digit integers can be formed using every digit of the integer 223334 exactly once?

(12) Consider the word *"SAVOURY"*. In each case, find the number of different words (just a string of letters with possibly no meaning) we obtain using the letters of the word, assuming no repetition.

    (a) All the letters are used.
  $\star$ (b) Four letters are used at a time.
    (c) All the letters are used and the word starts with a vowel.
    (d) All the letters are used and the word starts with a vowel or ends with a vowel.
  $\star$ (e) All the letters are used and vowels are together (no consonant between two vowels).

(13) Prove Theorem 8.10.

(14) We want to form a 3-digit or a 4-digit integer using digits from the set $\{2, 3, 4, 5, 6, 7, 8, 9\}$.

   (a) How many different integers can we form if the digits are pairwise distinct?

   (b) How many different integers can we form if repetition of digits is allowed?

(15) In how many ways can six persons $P_1, \ldots, P_6$ be seated around a circular table if:

   (a) P_1 and P_2 are side by side?

   (b) P_1 and P_2 are not side by side?

$\star$(16) A series of five projectors placed in a row at a certain airport is used to send signals to airplanes near by. Each of the projectors can be lit in five different colors. A "signal" consists of lighting one or more of the projectors at the same time but with different colors. For signals of two or more colors, changing the order in which the colors appear changes the interpretation of the signal. For example "Green-Blue-Red" and "Red-Green-Blue" are two different signals. Also, the positions of lit projectors are important. How many different signals are there?

(17) Find the number of binary strings of length 11 with six 0's and five 1's.

(18) Find the number of ternary strings of length 14 with six 0's, four 1's and four 2's.

$\star$(19) You have 22 marbles that you want to arrange in a row. In how many ways this can be done if five of the marbles are blue, four are green, six are red and 7 are yellow? Marbles of the same color are indistinguishable.

(20) You have a rectangular area of size 1m by 8m that you want to tile. There are three colors of tiles you can use, each is of size 1×1: blue, green and red tiles. In how many ways can you do the job assuming that tiles of the same color are indistinguishable?

(21) You have a rectangular area of size 1m by 8m that you want to tile. There are three sizes of tiles you can use: 1×1, 1×2, 1×3 and 1×4. In how many ways can you do the job assuming that tiles of the same size are indistinguishable?

(22) Let n be a positive integer. Show that $\frac{(2n)!}{2^n}$ is an integer using:

   (a) induction on n;

   (b) a combinatorial proof (a proof using one of the counting techniques of this section).

(23) Interpret the result of Theorem 8.11 as the number of functions between two sets.

$\star$(24) In how many ways can r labeled objects be distributed in n labeled boxes if every box must have at most one object?

8.5   Combinations

A lottery ticket consists of choosing k numbers from a set of n numbers $(1 \leq k < n)$ and the order we select these k numbers is not important. The ticket with numbers $2, 5, 9, 11, 34, 45$ is the same as the ticket with numbers $34, 45, 9, 11, 2, 5$ (they both have the exact chance of winning). In this section, we look at unordered arrangements of objects that we call combinations.

Definition 8.2. Let $0 \leq r \leq n$ be two integers. An r-combination without repetition of n distinct objects $a_1, \ldots, a_n$ is a subset of $\{a_1, \ldots, a_n\}$ consisting of r elements.

Example 8.40. The 2-combinations of the objects a, b, c, d are: $\{a, b\}$, $\{a, c\}$, $\{a, d\}$, $\{b, c\}$, $\{b, d\}$ and $\{c, d\}$. The 3-combinations of a, b, c, d are: $\{a, b, c\}$, $\{a, b, d\}$, $\{a, c, d\}$ and $\{b, c, d\}$.

The number of r-combinations of n elements is denoted by $C(n, r)$ or, more universally, by $\binom{n}{r}$. This number is referred to as the *binomial coefficient* because of its use in the binomial theorem (see Theorem 8.16 below).

Theorem 8.12. *Let $0 \leq r \leq n$ be two integers. Then $\binom{n}{r} = \frac{n!}{r!(n-r)!}$.*

Proof. We know, from the previous section, that the number of r-permutations of n objects is $P(n, r) = \frac{n!}{(n-r)!}$. Let us think how we can form an r-permutation. First, we choose r objects from n: there are $\binom{n}{r}$ ways to do that. Second, we consider all possible permutations of these r objects: there are $r!$ ways to do that. By the product principle, there are $r!\binom{n}{r}$ possible r-permutations of n elements. The result follows from the equality $\frac{n!}{(n-r)!} = r!\binom{n}{r}$. $\qquad\square$

Some immediate consequences of Definition 8.12 are given in the following remark.

Remark 8.1. Let $0 \leq r \leq n$ be integers. Then:

(a) $\binom{n}{0} = \binom{n}{n} = 1$ (b) $\binom{n}{r} = \binom{n}{n-r}$ (c) $\binom{n}{r} = 0$ for $r > n$

Equalities involving binomial coefficients like part (b) of the above remark are called *combinatorial identities* since they can be established using a *combinatorial proof* (or combinatorial argument). Such a proof consists of realizing both sides of the identity as the cardinality of the same finite set counted in two different ways.

Example 8.41. You have five friends and you want to invite one or more of them to a golf trip. In how many ways can you do that?

Solution. You can invite either one, two, three, four or all five of your friends at a time. By the sum principle, there are $\binom{5}{1} + \binom{5}{2} + \binom{5}{3} + \binom{5}{4} + \binom{5}{5} = 5 + 10 + 10 + 5 + 1 = 31$ ways to invite your friends for the trip. $\Diamond$

Example 8.42. Find the number of binary strings of length 10 containing: (a) exactly three 0's; (b) at most three 0's; (c) at least three 0's; (d) the same number of 0's as the number of 1's.

Solution. Let A_i be the set of binary string of length 10 containing exactly i 0's, $0 \leq i \leq 10$. Then the number of strings in A_i is the number of ways of choosing the i positions for the 0's among the 10 possible positions, namely $|A_i| = \binom{10}{i} = \frac{10!}{i!(10-i)!}$. For part (a), we use $i = 3$ and get $|A_3| = \frac{10!}{3!7!} = 120$ binary strings containing exactly three 0's. For part (b), we compute $|A_0 \cup A_1 \cup A_2 \cup A_3| = \frac{10!}{0!(10)!} + \frac{10!}{1!9!} + \frac{10!}{2!8!} + \frac{10!}{3!7!} = 1 + 10 + 45 + 120 = 176$. For part (c), we compute $|A_3 \cup A_4 \cup \cdots \cup A_{10}| = 120 + 210 + 252 + 210 + 120 + 45 + 10 + 1 = 968$. For (d), we compute the number of different strings obtained by permuting all the bits in the string 1111100000: there are $\frac{10!}{5!5!} = 252$ such permutations. $\Diamond$

Example 8.43. In how many ways can the letters of the word MAKARIA be arranged if no two A's are adjacent?

Solution. Note first that aside from the three A's in the word, each of the other four letters appears exactly once. There are 4! ways of arranging the letters M, K, R and I. For each arrangement of M, K, R and I, there are five spaces where the three A's can be placed so not two of them are adjacent. For example, the spaces determined by the arrangement MKRI are shown as small boxes as follows:

Therefore, there are $\binom{5}{3} = 10$ possible choices for the positions of the three A's. By the product principle, there are $4! \cdot 10 = 240$ arrangements of the letters of the word MAKARIA such that no two A's are adjacent. $\diamond$

Theorem 8.13. *(Pascal's identity) Let r, n be two integers such that $1 \leq r \leq n$. Then:*

$$\binom{n}{r-1} + \binom{n}{r} = \binom{n+1}{r}.$$

Proof. We present a combinatorial proof of the result. An algebraic proof is left to the reader as an exercise (see Exercise (4) below). Let $A = \{a_1, \ldots, a_{n+1}\}$ be a set of cardinality $n + 1$ and let $B = A \backslash \{a_1\}$. For a subset X of A of cardinality r, either $a_1 \in X$ or $a_1 \notin X$. In the first case, $X = Y \cup \{a_1\}$ for some subset Y of B containing $r - 1$ elements. There are $\binom{n}{r-1}$ such subsets Y. In the second case, X is a subset of B containing r elements. There are $\binom{n}{r}$ such subsets. By the sum principle, there are $\binom{n}{r-1} + \binom{n}{r}$ subsets of A containing r elements. We know that there are $\binom{n+1}{r}$ subsets of A containing r elements and the result follows. $\square$

The binomial coefficients $\binom{n}{k}$, $0 \leq k \leq n$ are arranged in a triangle form with the coefficients $\binom{i-1}{0}, \binom{i-1}{1}, \ldots, \binom{i-1}{i-1}$ at the ith row ($i \geq 1$). The triangle is known as *Pascal's triangle* (after the mathematician and philosopher Blaise Pascal (1623-1662)). The following diagram shows the first eight rows of the Pascal's triangle. On the left, the triangle is given in terms of the binomial coefficients and on the right, it is given in terms of the numerical values of these coefficients.

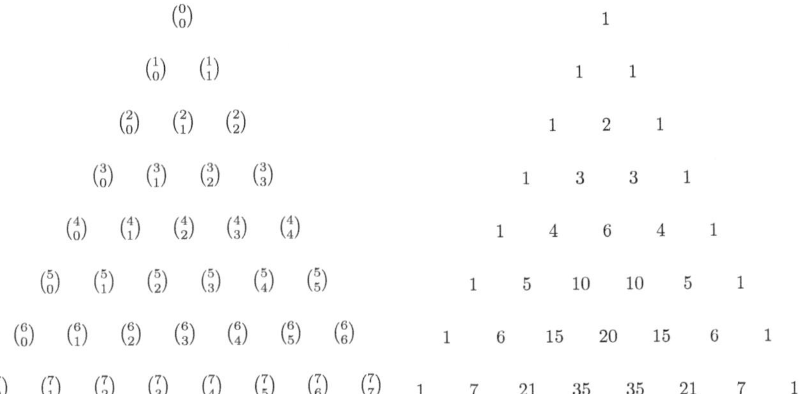

The following are important observations about Pascal's triangle:

- Every row starts and ends with 1.
- Row i of the triangle contains i coefficients.
- In every row, the sum of every pair of adjacent coefficients is equal to the coefficient between the pair in the following row. This is a consequence of Pascal's identity. For example, in the seventh row of Pascal triangle, adding the adjacent elements 6 and 15 results in the element 21 (in the middle of 6 and 15) in the eighth row.

8.5.1 *Number of surjective functions: Derangements*

Using binomial coefficients, we can give a formula that counts the number of surjective functions between two finite sets A and B with $|A| \geq |B|$.

Theorem 8.14. *Let A, B be two finite sets with $|A| = m$, $|B| = n$ and $m \geq n$. Then the number of surjective functions from A to B is given by* $n^m - \binom{n}{1}(n-1)^m + \binom{n}{2}(n-2)^m - \cdots + (-1)^{n-1}\binom{n}{n-1}1^m$.

Proof. Write $B = \{b_1, b_2, \ldots, b_n\}$. Let γ be the set of all functions from A to B and let γ_i be the subset of γ of all functions $f : A \to B$ such that the element b_i of B is not in the image of f. Then $\gamma_1 \cup \gamma_2 \cup \ldots \cup \gamma_n$ represents the subset of γ of all non-surjective functions. A function $f \in \gamma_i$ is a function from A to the set $B_i = B\backslash\{b_i\}$. Since there are $(n-1)^m$ functions from A to B_i, $|\gamma_i| = (n-1)^m$ for each $i \in \{1, 2, \ldots, m\}$. For $i \neq j$, an element in $\gamma_i \cap \gamma_j$ is a function from A to $B_{i,j} = B\backslash\{b_i, b_j\}$. Since there are $(n-2)^m$ such functions, $|\gamma_i \cap \gamma_j| = (n-2)^m$. Similarly, if $i_1, i_2, \ldots, i_k$ are pairwise distinct integers in the set $\{1, 2, \ldots, n\}$, then $|\gamma_{i_1} \cap \gamma_{i_2} \cap \cdots \cap \gamma_{i_k}| = (n-k)^m$. By the generalized inclusion–exclusion principle, $|\gamma_1 \cup \gamma_2 \cup \cdots \cup \gamma_n| = \sum_{i=1}^{n} |\gamma_i| - \sum_{1 \leq i < j \leq n} |\gamma_i \cap \gamma_j| + \cdots + |\gamma_1 \cap \gamma_2 \cap \cdots \cap \gamma_n|$. For each $k \in \{1, 2, \ldots, n\}$, there are $\binom{n}{k}$ sets of the form $\gamma_{i_1} \cap \gamma_{i_2} \cap \cdots \cap \gamma_{i_k}$, each of cardinality $(n-k)^m$. Therefore, $|\gamma_1 \cup \gamma_2 \cup \cdots \cup \gamma_n| = \binom{n}{1}(n-1)^m - \binom{n}{2}(n-2)^m + \binom{n}{3}(n-3)^m + \cdots + (-1)^n\binom{n}{n-1}(1)^m$. The result follows from the fact the set of surjective functions from A to B is $\gamma\backslash\cup_{i=1}^{n}\gamma_i$ and $|\gamma| = n^m$. $\qquad\square$

Example 8.44. There are $4^6 - \binom{4}{1}(4-1)^6 + \binom{4}{2}(4-2)^6 - \binom{4}{3}(4-3)^6 = 1304$ surjective functions from a set of cardinality 6 to a set of cardinality 4.

Definition 8.3. A *derangement* of some objects placed in a certain original configuration is a permutation of the objects that keeps no object in its original position.

Example 8.45. Consider six objects given with initial configuration $a_1a_2a_3a_4a_5a_6$. Then $a_2a_3a_4a_5a_6a_1$ and $a_3a_4a_5a_6a_1a_2$ are derangements of the six objects but $a_3a_2a_1a_5a_4a_6$ is not as it keeps a_2 (and a_6) in its original position.

Theorem 8.15. *The number D_n of derangements of n objects is given by:*

$$D_n = n! \left(\frac{1}{2!} - \frac{1}{3!} + \cdots + (-1)^n \frac{1}{n!} \right).$$

Proof. Let $a_1a_2 \ldots a_n$ be the original configuration of the n objects. For $1 \le i \le n$, let A_i be the set of all permutations of the n objects that keep the element a_i in its original position. Then each element in A_i is just a permutation of $a_1, a_2, \ldots, a_{i-1}, a_{i+1}, \ldots, a_n$ and $|A_i| = (n-1)!$. If $i \ne j$, then $A_i \cap A_j$ is the set of all permutations of the n objects that keep both a_i and a_j in their original positions and so $|A_i \cap A_j| = (n-2)!$. In general, if $1 \le i_1 < i_2 < \ldots < i_k \le n$ are integers, then $|A_{i_1} \cap A_{i_2} \cap \ldots \cap A_{i_k}| = (n-k)!$. For $1 \le k \le n$, there are $\binom{n}{k}$ intersections $A_{i_1} \cap A_{i_2} \cap \ldots \cap A_{i_k}$ each of cardinality $(n-k)!$. By the generalized inclusion–exclusion principle, $|A_1 \cup A_2 \cup \cdots \cup A_n| = \sum_{i=1}^{n} |A_i| - \sum_{1 \le <j \le n} |A_i \cap A_j| + \cdots + (-1)^{n-1}|A_1 \cap A_2 \cap \cdots \cap A_n| = \binom{n}{1}(n-1)! - \binom{n}{2}(n-2)! + \cdots + (-1)^{n-1}\binom{n}{n-1}(1)!$. The total number of permutations of the n object is $n!$, which implies that the number of derangements of these objects is $n! - |A_1 \cup A_2 \cup \cdots \cup A_n| = n! - \binom{n}{1}(n-1)! + \binom{n}{2}(n-2)! - \cdots + (-1)^n \binom{n}{n-1}(1)! = n! - \frac{n!}{1!} + \frac{n!}{2!} - \frac{n!}{3!} + \cdots + (-1)^n \frac{n!}{n!} = n! \left(\frac{1}{2!} - \frac{1}{3!} + \cdots + (-1)^n \frac{1}{n!} \right)$. $\square$

Example 8.46. The number of derangements of the objects *abcd* is $4! \left(\frac{1}{2!} - \frac{1}{3!} + \frac{1}{4!} \right) = 9$. The following is the list of these derangements: *badc, bcda, bdac, cadb, cdab, cdba, dabc, dcab, dcba*.

Example 8.47. Six students wrote a final exam. After marking the exam, the professor gave back the marked papers to the students without looking at the names. What is the chance that none of the students got his/her exam paper?

Solution. There are 6! ways of giving back the exam to the students (permutations of 6 objects). Among theses, there are D_6 ways of giving back the exam such that none of the students receives his/her exam (derangements of six objects). Therefore, the chance that none of the students received his/her exam is $\frac{D_6}{6!} = \frac{6!\left(\frac{1}{2!} - \frac{1}{3!} + \frac{1}{4!} - \frac{1}{5!} + \frac{1}{6!} \right)}{6!} = \frac{53}{144}$. $\diamond$

8.5.2   *Exercises*

(1) Compute each of the following:

    (a) $\binom{15}{12}$ (b) $\binom{20}{15}$ $\star$ (c) $\binom{50}{42}$ (d) $\binom{100}{93}$

(2) In each case, find the value(s) of the non-negative integer n (if any) satisfying the given equation:

    $\star$ (a) $\binom{n+2}{2} = 15$ $\star$ (c) $\binom{16}{n+2} = \binom{16}{n}$

    (b) $\binom{n}{6} = \binom{n}{8}$ (d) $P(n,2) + \binom{n}{2} = 30$

(3) Prove parts (a) and (b) of Remark 8.1 using: (i) an algebraic approach; (ii) a combinatorial approach.

(4) Prove Pascal's identity using an algebraic approach (using the formula $\binom{n}{k} = \frac{n!}{k!(n-k)!}$).

(5) In each case, determine the number of binary strings of length 8 satisfying the given property.

    (a) The binary string has exactly five 1's.
    (b) The binary string has at most five 1's.
    (c) The binary string has at least five 1's.
    (d) The binary string has the same number of 0's as the number of 1's.

$\star$(6) There are 10 students in a class. In how many ways can the teacher choose 5 students and place them around a circular table?

(7) Let $A = \{1, 2, \ldots, 20\}$ be the set of the first 20 positive integers and let B denote the subset $\{1, 2, \ldots, 9\}$ of A. In each case, find the number of subsets X of A satisfying the given property.

    (a) $|X| = 10$ (c) $|X| = 10$ and $|X \cap B| = 4$
    $\star$ (b) $|X \cap B| = 4$ (d) $|X| = 5$ and $X \cap B = \emptyset$

(8) In a certain university class, there are 15 engineering students and 10 science students. The professor has to form a committee of seven students in the class. The committee must have a president, a vice-president, and five other members. In how many ways can the professor form such a committee if the president must be an engineering student and the vice-president a science student and no restrictions on the other members?

(9) A certain exam consists of 40 questions. Students must answer 35 questions of their choosing. In how many ways a student can choose the questions on the exam if:

  (a) There are no restrictions?
  (b) Students must answer 18 questions from the first 20 questions and 17 from the last 20 questions?
  ⋆ (c) Student must choose at least 18 questions from the first 20 questions?

(10) In a certain department, there are 5 men and 4 women. In how many ways can the chair of the department form a committee of 6 persons of the department if the number of men on the committee must be greater than or equal to the number of women?

(11) In a certain department, there are 6 men and 7 women. In how many ways can the chair of the department form a committee of 5 persons of the department if either Joe (a man in the department) or Jessica (a woman in the department) must be on the committee but not both at the same time?

⋆(12) Seven new teaching positions are announced in a school board: three positions in mathematics, two in computer science, one in English and one in geography. Eight men and seven women (all qualified to teach any of the four subjects) applied for these positions. In how many different ways the school board can fill theses positions if:

  (a) There are no restrictions?
  (b) Exactly three women must be hired?

(13) A task force committee of 6 people must be formed from a larger group of 10 chemists, 5 politicians, 8 economists and 15 biologists. In how many ways the task force can be formed if:

  (a) There are no restrictions?
  (b) It must have exactly three economists?
  (c) It must have at least three chemists?
  (d) It must have exactly four chemists but no economist?
  ⋆ (e) It must have exactly two politicians and exactly two biologists but Ron (a politician) and Rebecca (a biologist) cannot be on the task force at the same time?

(14) Let k be a positive integer. Show that the product of any consecutive k integers must be divisible by $k!$.

$\star$(15) Let n, r be a positive integers, $n_1, n_2, \ldots, n_r$ be r be non-negative integers such that $n_1 + n_2 + \cdots + n_r = n$. Prove that:

$$\binom{n}{n_1}\binom{n-n_1}{n_2}\binom{n-n_1-n_2}{n_3}\cdots\binom{n-n_1-n_2-\cdots-n_{r-1}}{n_r}$$
$$= \frac{n!}{n_1!n_2!n_3!\cdots n_r!}.$$

$\star$(16) Let n, r, k be three integers satisfying $0 \le r \le k \le n$. Show that

$$\binom{n}{k}\binom{k}{r} = \binom{n}{r}\binom{n-r}{k-r}$$

using: (a) an algebraic approach; (b) a combinatorial approach.

(17) Let $n \ge 0$ be an integer. Prove the following identity: $\sum_{k=0}^{n}\binom{n}{k} = 2^n$ using: (a) an algebraic proof; (b) a combinatorial proof.

(18) Give the integers on the 9th row of Pascal's triangle.

(19) Let m, n, r be three non-negative integers with $r \le \min\{m, n\}$.

   (a) Give a combinatorial proof of the following identity (known as the Vandermond's Identity): $\binom{m+n}{r} = \sum_{k=0}^{n}\binom{m}{k}\binom{n}{r-k}$ (*Hint.* Think of two ways to compute the number of committees containing r members from two groups of people, one of size m and a second of size n).

   (b) Deduce that $\sum_{k=0}^{n}\binom{n}{k}^2 = \binom{2n}{n}$.

(20) Let A and B be two finite sets with $|A| = m$ and $|B| = n$. In each case, find the number of surjective functions from A to B.

   (a) $m = 6$, $n = 3$ (c) $m = 8$, $n = 4$ (e) $m = 10$, $n = 7$
   (b) $m = 7$, $n = 4$ (d) $m = 9$, $n = 6$ (f) $m = 9$, $n = 16$

$\star$(21) Eight guests checked their coats at a party.

   (a) In how many ways none of the eight guests gets the right coat at the end of the party?

   (b) In how many ways at least three guests get the wrong coat at the end of the party?

(22) In how many ways can the letters of the word VISITING be arranged if no two I's are adjacent?

(23) Show that the number of derangements D_n of n objects satisfies the following recursive definition: $D_1 = 0$, $D_2 = 1$ and $D_n = (n-1)(D_{n-1} + D_{n-2})$ for $n \ge 3$.

(24) Let $n \geq 1$ be an integer. Prove that D_n is odd if and only if D_{n+1} is even.

8.6 The binomial theorem

The following are two well-known expansions that you have seen and used before: $(x + y)^2 = x^2 + 2xy + y^2$, $(x + y)^3 = x^3 + 3x^2y + 3xy^2 + y^3$. The purpose of this section is to generalize these formulas to find an expression for the expansion of $(x + y)^n$ for arbitrary non-negative integer n. Let us start by noticing that $(x + y)^2 = (x + y)(x + y) = xx + xy + yx + yy = x^2 + 2xy + y^2$ and $(x + y)^3 = (x + y)(x + y)(x + y) = xxx + xxy + xyx + xyy + yxx + yxy + yyx + yyy = x^3 + 3x^2y + 3xy^2 + y^3$. In general, expanding $(x + y)^n = (x + y)(x + y) \cdots (x + y)$ yields a sum of $n + 1$ terms, each of which is of the form $a_k x^k y^{n-k}$ for some non-negative integers $k \geq 0$ and a_k. The coefficient a_k of $x^k y^{n-k}$ represents the number of "words" of length n of the alphabet $\{x, y\}$ containing exactly k x's. There are $\binom{n}{k}$ such words. We conclude that $a_k = \binom{n}{k}$. This proves the following theorem.

Theorem 8.16 (The binomial theorem). *Let x, y be two variables, $n \geq 1$ an integer. Then*

$$(x + y)^n = \sum_{k=0}^{n} \binom{n}{k} x^{n-k} y^k. \tag{8.3}$$

Note that the binomial theorem can be stated as $(x + y)^n = \sum_{k=0}^{n} \binom{n}{k} x^k y^{n-k}$ by the symmetry of the variables x and y.

Using the binomial theorem, the coefficients appearing in the expansion $(x + y)^n = \sum_{k=0}^{n} \binom{n}{k} x^{n-k} y^k$ are just the entries on the $n + 1$ row of the Pascal triangle.

Example 8.48. To write the expansion of $(x + y)^7$, we look at the coefficients in the eighth row of the Pascal triangle: $1, 7, 21, 35, 35, 21, 7, 1$. So, $(x + y)^7 = x^7 + 7x^6y + 21x^5y^2 + 35x^4y^3 + 35x^3y^4 + 21x^2y^5 + 7xy^6 + y^7$.

Example 8.49. Find the coefficient of x^3y^7 in the expansion of $(2x - 3y)^{10}$.

Solution. Of course one can multiply $(2x - 3y)$ by itself 10 times, regroup the alike terms and find the required coefficient. But as you can imagine, that would not be practical considering the time this will take and the errors that could occur along the way. This is where the binomial theorem comes in handy. Using the binomial theorem with x replaced by $2x$ and y by $-3y$, we get that:

$$(2x - 3y)^{10} = \sum_{k=0}^{10} \binom{10}{k} (2x)^{10-k}(-3y)^k = \sum_{k=0}^{10} 2^{10-k}(-1)^k 3^k \binom{10}{k} x^{10-k} y^k.$$

The coefficient of $x^3 y^7$ is obtained by setting $k = 7$ in $2^{10-k}(-1)^k 3^k \binom{10}{k}$:
$2^{10-7}(-1)^7 3^7 \cdot \binom{10}{7} = -2099520.$ $\Diamond$

Example 8.50. Find the coefficients of x^4 and x^6 in the expansion of $\left(2x^{-2} - x^3\right)^8$.

Solution. Using the binomial theorem:

$$\left(2x^{-2} - x^3\right)^8 = \sum_{k=0}^{8} \binom{8}{k}(2x^{-2})^{8-k}\left(-x^3\right)^k$$

$$= \sum_{k=0}^{8} (-1)^k 2^{8-k} \binom{8}{k} x^{-2(8-k)} \left(x^3\right)^k$$

$$= \sum_{k=0}^{8} (-1)^k 2^{8-k} \binom{8}{k} x^{5k-16}.$$

To find the coefficient of x^4, we set $5k - 16 = 4$ and get $k = 4$. The coefficient of x^4 is then $(-1)^4 2^{8-4} \binom{8}{4} = 1120$. For the coefficient of x^6, we need to find k such that $5k - 16 = 6$. But since this equation has no integer solution, the coefficient of x^6 in the expansion is 0. $\Diamond$

Example 8.51. Let $n \geq 0$ be an integer. Use the binomial theorem to prove that $\binom{n}{0} + \binom{n}{1} + \cdots + \binom{n}{n} = 2^n$.

Solution. The result follows immediately by setting $x = y = 1$ in the binomial theorem. $\Diamond$

Example 8.52. Use the binomial theorem to compute the value of 1.1^5.

Solution. Use $x = 1$, $y = 0.1$ in the binomial theorem:

$1.1^5 = (1 + 0.1)^5$

$$= \binom{5}{0} 1^{5-0}(0.1)^0 \binom{5}{1} 1^{5-1}(0.1)^1 + \binom{5}{2} 1^{5-2}(0.1)^2 + \binom{5}{3} 1^{5-3}(0.1)^3$$

$$+ \binom{5}{4} 1^{5-4}(0.1)^4 + \binom{5}{5} 1^{5-5}(0.1)^5$$

$$= 1 + 5(0.1) + 10(0.1)^2 + 10(0.1)^3 + 5(0.1)^4 + (0.1)^5 = 1.61051. \quad \Diamond$$

The adjective *binomial* in the binomial theorem refers to the fact that there are two variables in $(x + y)^n$. We now give a generalization of the binomial theorem, known as the *multinomial theorem*, that uses any (finite) number of variables.

Theorem 8.17. *Let $x_1, \ldots, x_s$ be s variables and let $n \geq 1$ be an integer. Then*

$$(x_1 + x_2 + \cdots + x_s)^n = \sum_{n_1 + n_2 + \cdots + n_s = n} \frac{n!}{n_1! n_2! \cdots n_s!} x_1^{n_1} x_2^{n_2} \cdots x_s^{n_s}.$$

In the above sum, each n_i is an integer satisfying $0 \leq n_i \leq n$.

Proof. Like in the proof of the binomial theorem, the expansion of $(x_1 + x_2 + \cdots + x_s)^n$ amounts to a sum of terms of the form $a_{n_1, n_2, \ldots, n_s} x_1^{n_1} x_2^{n_2} \cdots x_s^{n_s}$ for some non-negative integers $a_{n_1, n_2, \ldots, n_s}$ and $n_1, n_2, \ldots, n_s$ with $n_1 + n_2 + \cdots + n_s = n$. Regrouping "alike terms" in the expansion amounts to finding the number of ways to rearranging the words of length n from the alphabet $\{x_1, x_2, \cdots, x_s\}$ containing exactly n_i times the symbol x_i, for $i = 1, \ldots, s$. We know that there are $\frac{n!}{n_1! n_2! \ldots n_s!}$ such words and the result follows. $\square$

The term $\frac{n!}{n_1! n_2! \ldots n_s!}$ with $n_1 + n_2 + \cdots + n_s = n$ is called a *multinomial coefficient* that we denote by $\binom{n}{n_1, n_2, \ldots, n_s}$.

Example 8.53. In the expansion of $(x + y + 2z + 3t)^9$, the coefficient of $x^3 y^2 z^4$ is $2^4 \cdot \frac{9!}{3! 2! 4!} = 20160$ by the multinomial theorem.

8.6.1 *Exercises*

$\star$(1) Find the coefficient of $x^2 y^9$ in the expansion of $(x + 2y)^{11}$.

(2) Find the coefficient of x^{10} and x^{11} in the expansion of $\left(2x - \frac{1}{x^2}\right)^{16}$.

(3) Let x, y be two real numbers with $y \geq 0$. Expand $\left(3x^2 + 2\sqrt{y}\right)^4$.

$\star$(4) Let n be a positive integer. Use the binomial theorem to prove that:

$$\binom{n}{0} - \binom{n}{1} + \binom{n}{2} - \cdots + (-1)^n \binom{n}{n} = 0.$$

$\star$(5) Let n be a positive integer. What is the value of the sum:

$$\binom{n}{0} + 2\binom{n}{1} + 4\binom{n}{2} + \cdots + 2^n \binom{n}{n}.$$

(6) Compute each of the following multinomial coefficients.

  (a) $\binom{8}{2,2,4}$ $\star$ (b) $\binom{12}{2,3,3,4}$ (c) $\binom{14}{3,3,3,5}$ (d) $\binom{20}{2,3,4,5,6}$

(7) Write the complete expansion of each of the following expressions.

  (a) $(x+y+z)^3$ $\star$ (b) $(x+y+z)^4$ (c) $(x+y+z)^5$

(8) Find the coefficient of xy^2z in the expansion of $(x+2y-3z+2)^5$.

$\star$(9) Find the coefficient of x^6y^2 in the expansion of $\left(3x^2+2y-\frac{1}{y^2}\right)^{14}$.

(10) Prove the binomial theorem (Theorem 8.16) using mathematical induction on $n \geq 1$.

(11) Find the coefficient of $x^2yz^2w^3$ in the expansion of:

  (a) $(x+y+z+w)^8$ (c) $(2x+2y-z+3w)^8$
  $\star$ (b) $(-2x+3y+2z-w)^8$ (d) $(-x+3y-2z+w)^8$

$\star$(12) Find the integer n satisfying the equation $\sum_{k=0}^{40}\binom{40}{k}15^k = n^{160}$.

(13) In each case, determine the sum of the coefficients in the expansion of the given expression.

  (a) $(x+y)^5$ (b) $(x+y+z)^8$ $\star$ (c) $(3x-2y-4z+6w)^9$

8.7   Combinations with repetition

A certain pastry shop offers four kinds of pies: Apple (A), Blueberry (B), Lemon (L) and Pecan (P). For instance, 3 apples pies, 1 Blueberry pie, 2 lemon pies and 2 Pecan pie is one possible order. This corresponds to the following "multiset" $\{A, A, A, B, L, L, P, P\}$ which represents a *combination with repetition* of 8 elements of the set $\{A, B, L, P\}$. In how many ways can one place an order of eight pies? It does take much to realize that it is quite tedious to count all possible such orders. One way to avoid counting all possible orders and to use some results previously established is to associate a binary string to each order of eight pies. To understand the connection between pie orders and binary strings, consider the following table showing three different orders:

A	B	L	P
○ ○	○ ○	○ ○	○ ○
○ ○○ ○		○ ○	○ ○
	○ ○ ○	○ ○	○ ○ ○

The first row represents an order of two pies of each sort. The second row represents the order of 4 apple pies, no Blueberry pies, 2 Lemon pies and 2 Pecan pies. The third row represents the order of no Apple pies, 3 Blueberry pies, 2 Lemon pies and 3 Pecan pies. If we treat each vertical separator in a row as 1 and each $\circ$ as a 0, then each order (a row in the table) corresponds to a binary strings with three 1's and eight 0's. This way, the first order in the table corresponds to the binary string 00100100100, the second row corresponds to 00001100100 and the third row corresponds to 10001001000. So the number of possible orders of pies corresponds to the number of binary strings of length 11 having exactly three 1's: there are $\binom{11}{3} = 165$ such strings.

There is really nothing special about the above pie ordering example. Once the connection between combinations with repetition and binary strings with fixed length and fixed number of 1's is understood, the following theorem follows easily.

Theorem 8.18. *The number of r-combinations of n objects with repetitions allowed is $\binom{n+r-1}{r}$. This is also the number of ways of distributing r indistinguishable objects in n distinct containers.*

Proof. Each r-combinations with repetitions of n objects can be considered as a binary strings of length $n + r - 1$ having exactly $(n-1)$ 1's (or exactly r 0's) as explained above. The number of such strings is $\binom{n+r-1}{r}$ as required. $\qquad\square$

Example 8.54. An ice cream shop has 13 different flavors. An ice cream cone comes with 4 flavors. In how many ways can one buy an ice cream cone?

Solution. Here $n = 13$ and $r = 4$, so the number of different ice cream cones that can be bought is $\binom{13+4-1}{4} = 1820$. $\qquad\Diamond$

Example 8.55. How many non-negative *integer* solutions to the equation $x_1 + x_2 + x_3 + x_4 + x_5 = 12$?

Solution. Here, the problem is equivalent to distributing $r = 12$ indistinguishable objects in $n = 5$ different containers (the five variables). The required number is then $\binom{12+5-1}{12} = \binom{16}{12} = 1820$. $\qquad\Diamond$

Example 8.56. How many non-negative integer solutions to the inequality $x_1 + x_2 + x_3 + x_4 + x_5 < 12$?

Solution. Introduce a new variable $x_6 = 12 - (x_1 + x_2 + x_3 + x_4 + x_5)$. Then $x_6 \geq 1$ and the problem is now reduced to finding the number of integer solutions to the equation $x_1 + x_2 + x_3 + x_4 + x_5 + x_6 = 12$ with $x_i \geq 0$ for $i = 1, \ldots, 5$ and $x_6 \geq 1$. This is equivalent to finding non-negative integer solutions to the equation $y_1 + y_2 + y_3 + y_4 + y_5 + y_6 = 11$ where $y_i = x_i \geq 0$ for $i = 1, \ldots, 5$ and $y_6 = x_6 - 1 \geq 0$. This is the same as the number of ways of distributing 11 indistinguishable objects in 6 different containers. The required number is then $\binom{11+6-1}{11} = \binom{16}{11} = 4368$. $\qquad\qquad \diamond$

Example 8.57. You have 10 identical chocolate bars that you want to distribute among 4 kids: Andrea, Bill, Céline and Doug. In how many ways this can be done if:

(a) There are no restrictions?
(b) No kid is left with no chocolate bar?
(c) Andrea gets at least two chocolate bars and each of the other three kids gets at least one bar?

Solution.   (a) If there are no restrictions, this amounts to distributing $r = 10$ indistinguishable objects (chocolate bars) in $n = 4$ different containers (the four kids). The required number is then $\binom{10+4-1}{10} = \binom{13}{10} = 286$.

(b) For this part, we start by making sure every kid gets one chocolate bar. That leaves $10 - 4 = 6$ bars to distribute among the 4 kids. There are $\binom{6+4-1}{6} = \binom{9}{6} = 84$ ways to do that.

(c) Starting with two bars to Andrea and one bar to each of the other three kids leaves 5 chocolate bars to be distributed among the four kids. There are $\binom{5+4-1}{5} = 56$ ways to do that. $\qquad \diamond$

Example 8.58. In how many ways can we distribute 9 chocolate chip cookies and 10 oatmeal cookies to 7 kids if:

(a) There are no restrictions?
(b) Each kid must have at least one chocolate chip cookie?
(c) Each kid must have at least one cookie from each kind?

Solution.   (a) There are $\binom{9+7-1}{9} = \binom{15}{9} = 5005$ ways of distributing the 9 chocolate chip cookies to the 7 kids. For each one of these ways, there are $\binom{10+7-1}{10} = \binom{16}{10} = 8008$ ways of distributing the 10 oatmeal cookies to the 7 kids. By the product rule, there are $5005 \times 8008 = 40080040$ ways of distributing all the cookies to the 7 kids.

(b) If each kid must have at least one chocolate chip cookie, then there are $\binom{2+7-1}{2} = \binom{8}{2} = 28$ ways to distribute the remaining two chocolate chip cookies. For each one of these ways, there are there are $\binom{10+7-1}{10} = \binom{16}{10} = 8008$ ways of distributing the 10 oatmeal cookies to the 7 kids. By the product rule, there are $28 \times 8008 = 224224$ ways of distributing all the cookies to the 7 kids such that each kid must have at least one chocolate chip cookie.

(c) Start by giving out one chocolate chip and one oatmeal cookie for each kid. Like before, there are $\binom{2+7-1}{2} = \binom{8}{2} = 28$ ways to distribute the remaining two chocolate chip cookies. For each one of these ways, there are there are $\binom{3+7-1}{3} = \binom{9}{3} = 84$ ways of distributing the remaining 3 oatmeal cookies to the 7 kids. By the product rule, there are $28 \times 84 = 2352$ ways of distributing all the cookies to the 7 kids such that each kid must have at least one cookie from each kind. $\diamond$

Example 8.59. Let n, r be positive integers. What is the number of terms in the expansion of $(x_1 + x_2 + \cdots + x_n)^r$? Deduce the number of terms in the expansion of $(x + y + z + w + t)^{12}$.

Solution. By the multinomial theorem, we know that

$$(x_1 + x_2 + \cdots + x_n)^r = \sum_{a_1+a_2+\cdots+a_n=r} \frac{r!}{a_1!a_2!\cdots a_n!} x_1^{a_1} x_2^{a_2} \cdots x_n^{a_n}.$$

The number of terms in the sum on the right is equal to the number of non-negative integer solutions to the equation $a_1 + a_2 + \cdots + a_n = r$. We have seen that this number is equal to $\binom{n+r-1}{r} = \frac{(n+r-1)!}{r!(n-1)!}$. For the expansion of $(x + y + z + w + t)^{12}$, we use $r = 12$ and $n = 5$ and find that the required number of terms is $\frac{(5+12-1)!}{12!(5-1)!} = 1820$. $\diamond$

8.7.1 *Exercises*

$\star$(1) A chocolate shop offers 10 sorts of chocolates squares.

    (a) In how many ways can you order two dozens of chocolate squares?

    (b) In how many ways can you order three dozens of chocolate squares assuming that you want at least two squares of each sort?

(2) Consider the equation $x_1 + x_2 + x_3 + x_4 + x_5 + x_6 = 18$. In each case, find the number of integer solutions to the equation that satisfy the given conditions for each $i = 1, \ldots, 6$.

  (a) $x_i \geq 0$ $\star$ (b) $x_i \geq 1$ (c) $x_i \geq 2$

(3) Joe has a large collection of marbles in n different colors with $n \leq 10$. Counting the number of ways he can select 12 marbles, Joe finds out that he has 50388 ways to do that. What is the value of n?

(4) How many positive integer solutions to the inequality $x_1 + x_2 + x_3 + x_4 + x_5 < 14$?

(5) In how many ways can we distribute 7 apples, 9 bananas and 8 oranges to 6 kids if:

  (a) There are no restrictions?

  (b) Each kid must have at least one banana?

  $\star$ (c) Each kid must have at least one of the three fruits?

  (d) Each kid must have at least one banana or at least one orange?

$\star$(6) Find the number of integer solutions of the equation $x_1 + x_2 + x_3 + x_4 = 37$ such that $2 \leq x_i \leq 10$ for each $i = 1, 2, 3, 4$.

$\star$(7) How many terms in the expansion of $(x_1 + \cdots + x_7)^{21}$?

(8) Find the number of ways to color 13 identical white balls using three different colors: red, blue and green, such that at least 3 balls are colored red and at least 4 balls are colored blue.

(9) In each case, find the number of ways to put 26 balloons of the same size in 6 labeled boxes.

  (a) The balloons are all of the same color.

  $\star$ (b) Half of the balloons are blue and the other half are red.

  (c) The 26 balloons are of different colors.

  $\star$ (d) The 26 balloons are of different colors and the boxes are labeled $1, \ldots, 6$ with Box 1 containing 2 balloons, Box 2 containing 4 balloons and each one of Boxes 3, 4, 5 and 6 contains 5 balloons.

(10) You have 9 types of postcards with 4 cards of each type. In how many ways can you send *all* the cards to 7 friends if:

  (a) There are no restrictions?

  (b) One of the friends must receive at least one card of each type?

$\star$(11) Let n be a positive integer. A *composition* of n is an ordered sequence $\alpha = (\alpha_1, \alpha_2, \ldots, \alpha_r)$ where $\alpha_i \geq 1$ is an integer and $\alpha_1 + \alpha_2 + \cdots + \alpha_r = n$. Each α_i is called a *summand* of the composition α. For example, there are 8 possible compositions of 4, namely: (4), $(1, 3)$, $(2, 2)$, $(3, 1)$, $(1, 1, 2)$, $(1, 2, 1)$, $(2, 1, 1)$ and $(1, 1, 1, 1)$. For a positive integer n, prove that:

(a) The number of compositions of n into k summands ($k \geq 1$ is an integer) is $\binom{n-1}{k-1}$.

(b) The number of all compositions of n is 2^{n-1} (*Hint.* Use Example (8.51) above).

(12) Let $n \geq 1$ be an integer. Find the number of triplets (i, j, k) of integers such that $1 \leq i \leq j \leq k \leq n$. Find that number for $n = 5$.

$\star$(13) In how many ways can 12 men and 4 women be arranged in a row such that the end positions of the row are occupied by men and no two women are side by side?

Chapter 9

Basics of Graph Theory

9.1 Introduction and a bit of history

Throughout history, many towns and cities around the world became famous as the birth places of influential individuals whose discoveries revolutionized science, art and literature. Graph theory contributed to put one city on the map, not because it was the birth place of the scientist who first worked on it, but strangely enough because of a puzzle created by its inhabitants and was the origin of that important branch of mathematics.

Our story begins in the 18th century, in the vibrant city of Königsberg in German Prussia at that time (the city is now known as Kaliningrad and it is part of Russia). The city was a thriving cultural and economical center on the banks of the Pregel river. Geographically, the city is formed by two large islands connected to each others and to the main land (North and South shores of Pregel) by a total of seven bridges. The following is an illustration of the city and its seven bridges that can be found in *Topographia Prussiae et Pomerelliae* by Zeiller M. (published in Frankfurt in 1650).

Started as an entertainment activity, it is said the inhabitants of Königsberg used to try to trace a path around the four regions of the city in which they seek to cross each of the seven bridges exactly once. The puzzle spread

451

quickly and became well known in the region. But it wasn't until the mid 1730s that the Swiss mathematician Leonhard Euler (1707–1783) published his solution to the puzzle in which he confirmed mathematically what many inhabitants had already guessed: such a tour in Königsberg was not possible. Of course, a brilliant mind like Euler did not seek to solve a silly puzzle for just the fun of it but rather because he saw in its solution a model to represent and solve more complex problems. The main idea in Euler's solution is that the geographical map of the city, the sizes of the four regions and the dimension of every bridge are all irrelevant. All what really matters is how each region is connected to the other regions of the city. For his solution, Euler represented the city with a simple diagram in which each of the four (land) regions is illustrated by a point in the plane and each bridge connecting two regions is illustrated by an edge (a straight line or a curve) between the corresponding points. Figure 9.1 illustrates the diagram drawn by Euler in his solution and it is the first illustration of what is now known as a *graph*. In the diagram, the four land regions are labeled by the capital letters A, B, C and D (with A and D being the two islands, B and C the two shores) and bridges are labeled by the letters a, b, c, d, e, f and g. Using this graph, Euler answered the puzzle of Könesberg bridges easily by considering the "number of connections" from each point in the graph.

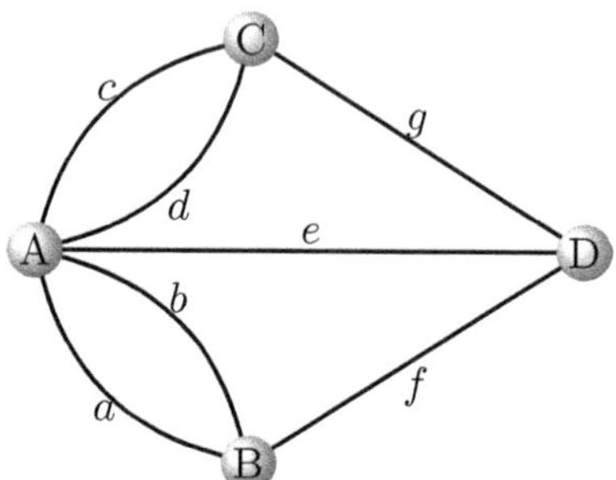

Fig. 9.1   The bridges of Könesberg

This original approach of Euler to solve the Könesberg puzzle paved the way for a new and exciting area of mathematics, now known as *graph theory*. This theory found its way to most modern day applications like finding the shortest path in a network, ranking hyperlinks, flow of computations in computer science, organizing sport tournaments and many others. Research in this area helped to better understand and to improve the structure of the world wide web, seen as a giant graph with the webpages as "points" and the hyperlinks connecting them as "edges".

9.2 Basic definitions and terminology

You are probably familiar with the term *graph* from your Calculus classes as a visual way to represent a function. Namely, if f is a real-valued function with domain $D \subseteq \mathbb{R}$, then the graph of f is the subset $\{(x, f(x)); x \in D\}$ of the Cartesian plane. In all what follows , the term *graph* will serve a completely different purpose and should not be confused with that used in Calculus.

Definition 9.1. An **undirected graph** is a triplet $G = (V, E, \phi_G)$, where:

- V is a non-empty finite set with elements called the *vertices* (the singular term is a *vertex*) of the graph.
- E is a subset of the set $\{\{x, y\}; x, y \in V\})$. An element $e = \{x, y\}$ of E is called an *edge* of the graph with *endpoints* x and y.
- $\phi_G : E \to \{\{x, y\}; x, y \in V\}$ is the function that assigns to each edge of G its end points. The function ϕ_G is called the *incidence function* of G.

The number of vertices of G is called the *order* of the graph and the number of edges in G is called the *size* of the graph. A graph of order 1 (one vertex) and size 0 (no edges) is called the *trivial graph*. A graph of order $n \geq 2$ and size 0 (no edges) is called an *empty graph*.

You might be wondering at this point why do we need the notion of the incidence function in the above definition. The reason is purely technical and has to do with labeling of edges. If we are only given the set $V = \{v_1, \ldots, v_n\}$ of vertices and the set $E = \{e_1, \ldots, e_r\}$ of edges of a graph G, then there is no way for us to draw these edges unless the endpoints are specified. The incidence function ϕ_G can be omitted (often the case in this book) provided that each edge of G is specified by the subset $\{u, v\}$ of its endpoints.

Example 9.1. The Könesberg graph of Figure 9.1 is an example of an undirected graph. The edges are labeled a, b, c, d, e, f and g and the incidence function is given as follows: $\phi(a) = \{A, B\}$, $\phi(b) = \{A, B\}$, $\phi(c) = \{A, C\}$, $\phi(d) = \{A, C\}$, $\phi(e) = \{A, D\}$, $\phi(f) = \{B, D\}$, $\phi(g) = \{C, D\}$.

As you can imagine, representing an undirected graph with a topological figure (a drawing in the plane) can be done in many ways. An important question we will consider later is wether two figures actually represent the same graph. For example, the following diagram shows three different representations of the same graph.

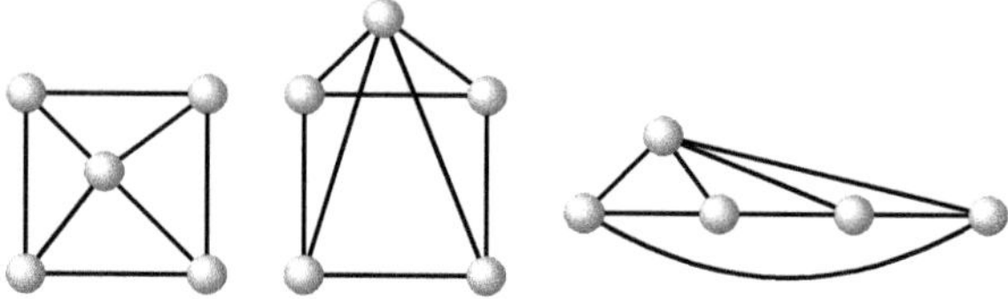

The term "undirected" in Definition 9.1 refers to the fact that there is no orientation on the edges of the graph. This is illustrated by labeling each edge with the set $\{u, v\}$ of its endpoints (where the order of the vertices is irrelevant). There are however some situations where an orientation is needed on the edges. For instance, if the graph represents a network of streets then each edge (street) must indicate a direction. In this case, an edge e in the graph is represented by an *ordered* pair (u, v) of elements of vertices where u is called the origin (or the *source*) of e and v is called the *endpoint* (or the *target*) of e. A graph with directed edges is called *directed graph* or simply a *digraph*. More formally, we have the following definition.

Definition 9.2. A *directed graph* , also known as a *digraph*, is a quadruple $G = (V, E, s, t)$ where V is a finite set (the set of vertices), $E \subseteq V \times V$ (the set of edges), and s, t are two functions from E to V (known as the source function and the target function, respectively) such that for every $e \in E$, $s(e)$ is the origin and $t(e)$ is the endpoint of e. The order and size of a directed graph are defined the same way as in the case of undirected graph.

Like in the case of an undirected graph, the source and the target functions can be omitted if the origin and the endpoint of each edge are given. When representing a digraph with a drawing, the orientation of an edge is indicated by an arrow head pointing from the origin of the edge to its endpoint.

Example 9.2. The following is a representation of the digraph G with vertices v_1, v_2, v_3, v_4 and edges (v_1, v_2), (v_1, v_4), (v_2, v_1), (v_2, v_4), (v_3, v_1),

(v_3, v_2), (v_3, v_4), (v_4, v_2).

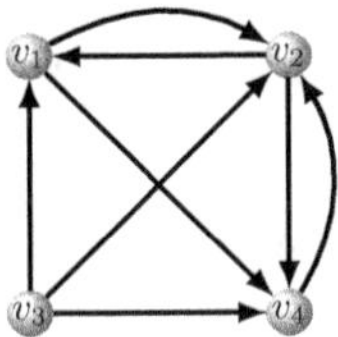

Although most of the terminologies and results that follow apply to both undirected and directed graphs (with few exceptions), our main focus in this book is on undirected graphs. **From this point on, and unless otherwise mentioned, the word *graph* means an undirected graph of finite order.** There will be topics that require using directed graphs, in which case we use the term *digraph*. Typically, a graph G will be denoted by the pair $G = (V, E)$ where it is understood that V is the set of vertices and E is the set of edges. The incidence function is omitted except in the case where there an ambiguity about their endpoints.

Definition 9.3. Let $G = (V, E)$ be a graph, $v \in V$ a vertex of G. An edge of G of the form $\{v, v\}$ is called a *loop* at v (so the endpoints of a loop are the same vertex). Two edges e and e' of G are called *parallel* or *multiple* if they have the same endpoints. A graph with parallel edges is called a *multigraph*. A graph without loops and without multiple edges is called a *simple graph*.

Note that the incidence function of a simple graph is injective.

Example 9.3. Consider the following three graphs.

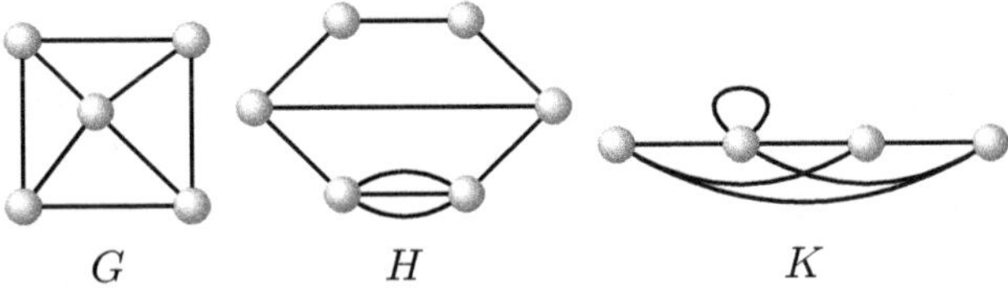

$$G \qquad\qquad H \qquad\qquad K$$

Graph G is simple as it has no multiple edges or loops. Graph H is not simple as it has multiple edges and so is graph K as it has a loop.

For digraphs, we assume that there exists at most one edge going from vertex v_i to vertex v_j. We don't consider digraphs with multiples edges going in the same direction from v_i to v_j.

Definition 9.4. Let G be a graph. Two vertices u and v of G are called *adjacent* (or sometimes *neighbors*) if $\{u, v\}$ is an edge of G (i.e., u, v are connected by an edge of G). An edge e of G is said to be *incident* to a vertex v of G if v is one of the endpoints of e.

Example 9.4. For the graph given in the diagram below: vertices v_1 and v_5 are adjacent while vertices v_1 and v_3 are not. Edge a is incident to both vertices v_1 and v_2 and edge c is incident to both vertices v_3 and v_6.

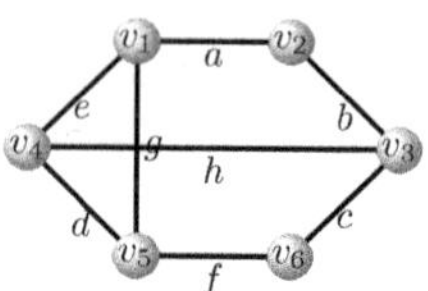

9.2.1   *Some common simple graphs*

Some simple graphs appear frequently in the theory and applications of graph theory. The following is a list of some of these well-known graphs.

- **The complete graph.** This is a simple graph of order $n \geq 2$, denoted by K_n. In the complete graph, every pair of distinct vertices are adjacent. More formally, $K_n = (V, E)$ with $V = \{v_1, \ldots, v_n\}$ and $E = \{\{v_i, v_j\}; 1 \leq i < j \leq n\}$. The following diagram shows the first four complete graphs:

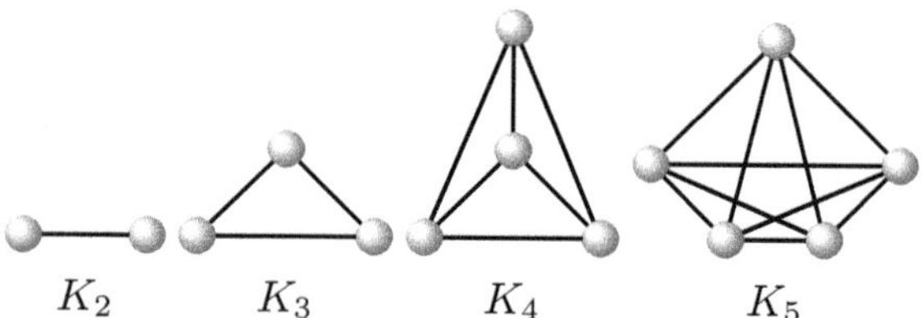

- **The cycle.** This is a simple graph of order $n \geq 3$, denoted by C_n, with vertices linked in a circular way. More formally, C_n is the graph with vertices v_1, v_2, $\ldots$, v_n and edges $\{v_1, v_2\}$, $\{v_2, v_3\}$, $\{v_3, v_4\}$, $\ldots$, $\{v_{n-1}, v_n\}$, $\{v_n, v_1\}$. The following diagram shows the first four cycles.

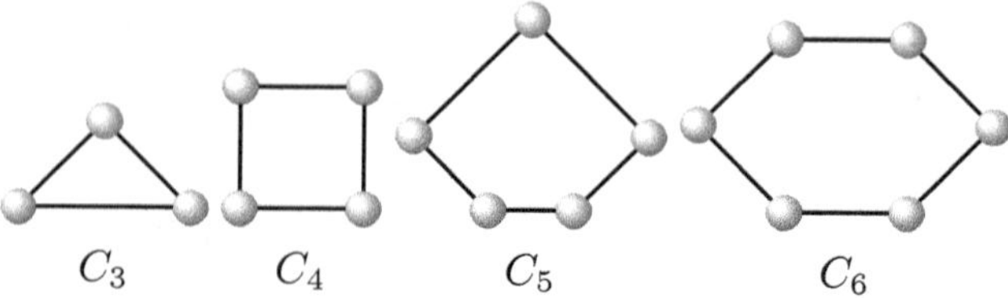

- **The wheel.** This is a simple graph obtained from C_n by adding a vertex v in the center of C_n and connecting it with one edge to every other vertex of C_n. So the wheel, denoted by W_n, has $n+1$ vertices $v_1, v_2, \ldots, v_n, v$ and the set of edges is $E = E_1 \cup E_2$ where:

  - $E_1 = \{\{v_1, v_2\}, \{v_2, v_3\}, \{v_3, v_4\}, \ldots, \{v_{n-1}, v_n\}, \{v_n, v_1\}\}$
  - $E_2 = \{\{v_i, v\}; i = 1, \ldots, n\}$

The following are the first four wheels.

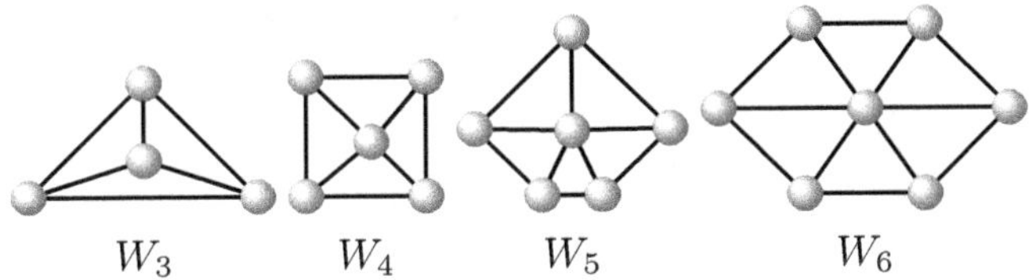

It should be noted here that in some textbooks, the wheel W_n is constructed from cycle C_{n-1} rather than C_n. In this case, W_n would be of order n rather than $n+1$.

- **The path.** This simple graph of order $n \geq 1$, denoted by P_n, has its vertices linked in a linear way, creating $n-1$ edges. More precisely, P_n is a simple graph with n vertices $v_1, v_2, \ldots, v_n$ and $n-1$ edges: $\{v_1, v_2\}$, $\{v_2, v_3\}$, $\ldots$, $\{v_{n-1}, v_n\}$. The following are the first four paths.

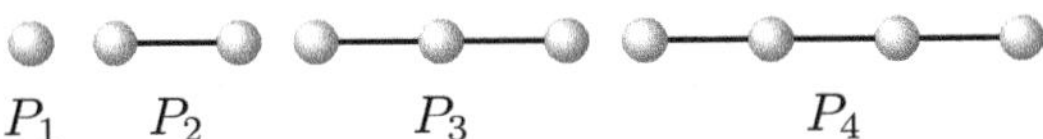

- **The complete bipartite graph** Let m and n be two positive integers. The complete bipartite graph of order $m + n$ is a simple graph, denoted by $K_{m,n}$, with set of vertices V partitioned into two subsets $V_1 = \{x_1, \ldots, x_m\}$ and $V_2 = \{y_1, \ldots, y_n\}$ (so $V = V_1 \cup V_2$ and $V_1 \cap V_2 = \emptyset$) of cardinalities m and n respectively. Every vertex in V_1 is adjacent to every vertex in V_2 and no two vertices belonging to the same subset (V_1 or V_2) are adjacent. So $K_{m,n} = (V, E)$ with:

$$V = \{x_1, x_2, \ldots, x_m\} \cup \{y_1, y_2, \ldots, y_n\}$$
$$E = \{\{x_i, y_j\}; 1 \leq i \leq m, \, 1 \leq j \leq n\}.$$

The following shows the complete bipartite graphs $K_{1,1}$, $K_{2,2}$, $K_{3,2}$, $K_{4,3}$ and $K_{3,3}$.

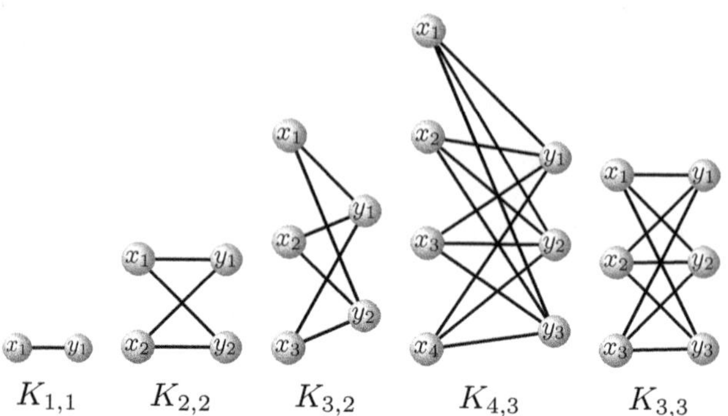

9.2.2 *Adjacency and incidence matrices*

Representing a graph with a topological figure is useful to visually detect some properties of the graph, but not so much to store it in a computer. Luckily, there are ways to turn a pictorial representation of a graph into a two-dimensional array of numbers (a matrix) that can be stored digitally. Matrix representation also allows us to apply classic results from linear algebra to graph theory. In this section, we look at two matrices associated with a given graph. The reader is assumed to be familiar with basic operations on matrices.

Definition 9.5. Let $G = (V, E)$ be a graph with $V = \{v_1, v_2, \ldots, v_n\}$, $E = \{e_1, e_2, \ldots, e_m\}$, and incidence function ϕ_G.

(a) The *adjacency matrix* of G relative to the ordering $v_1, v_2, \ldots, v_n$ of its vertices, is the square $n \times n$ matrix $A = [a_{ij}]$ with a_{ij} (the entry on the ith row and the jth column of A) is equal to the cardinality of the subset $\{e_k \in E;\ \phi_G(e_k) = \{v_i, v_j\}\}$. In other words, the entry a_{ij} corresponds to the number of edges having the vertices v_i and v_j as their endpoints. In particular, the adjacency matrix $A = [a_{ij}]$ of a *simple* graph can be defined as follows:

$$a_{ij} = \begin{cases} 1 & \text{if } v_i \text{ and } v_j \text{ are adjacent in } G \\ 0 & \text{otherwise} \end{cases}$$

(b) The *incidence matrix* of G relative to the ordering $v_1, v_2, \ldots, v_n$ of the vertices and the ordering $e_1, e_2, \ldots, e_m$ of the edges is the $n \times m$ matrix $M = [m_{ij}]$ where the entry m_{ij} is equal to the number of times edge e_j is incident to vertex v_i. More formally:

$$m_{ij} = \begin{cases} 2 & \text{if } \phi_G(e_j) = \{v_i, v_i\} \text{ (edge } e_j \text{ is a loop on vertex } v_i) \\ 1 & \text{if } \phi_G(e_j) = \{v_i, v_k\} \text{ for some } k \neq i \\ 0 & \text{otherwise} \end{cases}$$

Rows and columns of the adjacency matrix are labeled by the vertices of the graph. For the incidence matrix, rows are labeled by the vertices while columns are labeled by the edges.

Remark 9.1. It is important to remember that the adjacency and incidence matrices of a graph are defined relative to a given ordering on the vertices and edges of the graph. So technically, it is incorrect to talk about *the* adjacency matrix and *the* incidence matrix of a graph since these are not unique. However, once an ordering is specified and fixed, both matrices are unique. It is also important to keep in mind that for a simple graph, both the adjacency and the incidence matrices are boolean (entries consisting of 0's and 1's only).

Example 9.5. In each case, write the adjacency matrix of the Könesberg graph relative to the given ordering of its vertices.

(a) A, B, C, D (b) C, D, A, B (c) D, A, C, B.

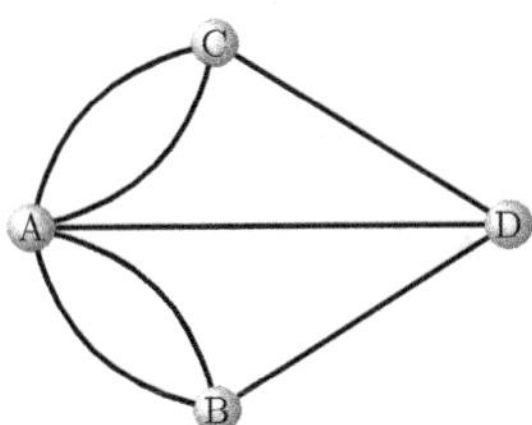

Solution. Since the graph has 4 vertices, its adjacency matrix is of size 4×4. In each case, we label both the rows and the columns of the adjacency matrix using the given ordering on the vertices.

$$\begin{array}{c} \begin{array}{cccc} A & B & C & D \end{array} \\ \begin{array}{c} A \\ B \\ C \\ D \end{array} \begin{bmatrix} 0 & 2 & 2 & 1 \\ 2 & 0 & 0 & 1 \\ 2 & 0 & 0 & 1 \\ 1 & 1 & 1 & 0 \end{bmatrix} \\ (a) \end{array} \qquad \begin{array}{c} \begin{array}{cccc} C & D & A & B \end{array} \\ \begin{array}{c} C \\ D \\ A \\ B \end{array} \begin{bmatrix} 0 & 1 & 2 & 0 \\ 1 & 0 & 1 & 1 \\ 2 & 1 & 0 & 2 \\ 0 & 1 & 2 & 0 \end{bmatrix} \\ (b) \end{array} \qquad \begin{array}{c} \begin{array}{cccc} D & A & C & B \end{array} \\ \begin{array}{c} D \\ A \\ C \\ B \end{array} \begin{bmatrix} 0 & 1 & 1 & 1 \\ 1 & 0 & 2 & 2 \\ 1 & 2 & 0 & 0 \\ 1 & 2 & 0 & 0 \end{bmatrix} \\ (c) \end{array} \qquad \Diamond$$

Example 9.6. Find the incidence matrix of the Könesberg graph relative to the ordering A, B, C, D of its vertices and the ordering a, b, c, d, e, f, g of its edges.

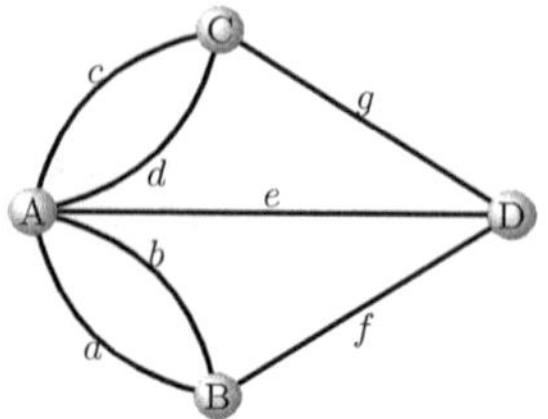

Solution. Since the graph has 4 vertices and 7 edges, its incidence matrix is of size 4×7. We label the rows of the matrix with A, B, C and D, the columns with a, b, c, d, e, f and g.

$$
\begin{array}{c}
\begin{array}{ccccccc} a & b & c & d & e & f & g \end{array} \\
\begin{array}{c} A \\ B \\ C \\ D \end{array}
\left[\begin{array}{ccccccc}
1 & 1 & 1 & 1 & 1 & 0 & 0 \\
1 & 1 & 0 & 0 & 0 & 1 & 0 \\
0 & 0 & 1 & 1 & 0 & 0 & 1 \\
0 & 0 & 0 & 0 & 1 & 1 & 1
\end{array}\right]
\end{array}
$$

$\Diamond$

Remark 9.2. Let $G = (V, E)$ be a graph with adjacency matrix $A = [a_{ij}]$.

- If G is simple, then every diagonal entry of A is 0 since diagonal entries correspond to loops in the graph.
- A is *symmetric*, meaning that $a_{ij} = a_{ji}$ for every i, j (the entry at the ith row and the jth column is the same as the entry at the jth row and ith column). This means in particular that it suffices to form one half (upper or lower) of the matrix, the other half is just the reflection in the main diagonal of the matrix.

The definition of the adjacency matrix can be extended to a digraph. This should come as no surprise as it was already used in the context of binary relations in Chapter 7. Remember first our assumption on digraphs: no multiple edges going from vertex v_i to vertex v_j.

Definition 9.6. Let $G = (V, E)$ be a digraph with $V = \{v_1, v_2, \ldots, v_n\}$. The adjacency matrix of G with respect to the ordering $v_1, v_2, \ldots, v_n$ of the vertices is the $n \times n$ matrix $A = [a_{ij}]$ with entry a_{ij} on the ith row and the jth column of A equals to 1 if there is an edge from v_i to v_j and 0 otherwise.

Unlike undirected graphs, the adjacency matrix of a digraph needs not to be symmetric.

Example 9.7. Find the adjacency matrix of the following digraph.

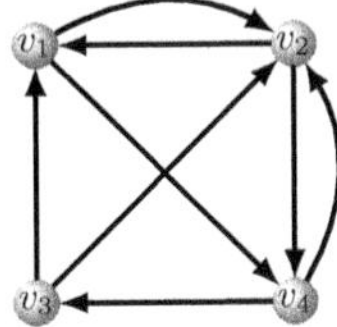

Solution. Since the graph has 4 vertices, its adjacency matrix is 4×4. Like in the case of undirected graphs, we label the rows and the columns with the vertices of the graph.

$$
\begin{array}{c}
\begin{array}{cccc} v_1 & v_2 & v_3 & v_4 \end{array} \\
\begin{array}{c} v_1 \\ v_2 \\ v_3 \\ v_4 \end{array}
\left[
\begin{array}{cccc}
0 & 1 & 0 & 1 \\
1 & 0 & 0 & 1 \\
1 & 1 & 0 & 0 \\
0 & 1 & 1 & 0
\end{array}
\right]
\end{array}
$$

Notice that the matrix is not symmetric.

9.2.3 *Exercises*

(1) In each case, the graph $G = (V, E)$ represents a certain real life situation. The vertices are given and an adjacency rule between vertices is specified. Determine if the model requires a directed or an undirected graph.

★ (a) V is the set of all towns in eastern Ontario. Towns u and v are adjacent if and only if there exists at least one direct road that goes from u to v without crossing another town.

(b) V is the set of all species in a certain ecosystem. Species u and v are adjacent if and only if u and v compete for food resources.

★ (c) V is the set of species in an ecosystem. Species u and v are adjacent if and only if u preys on v.

(d) V is a group of mathematicians. Vertices u and v are adjacent if and only if mathematicians u and v co-authored a research article.

(e) V is the collection of all web pages. Pages u and v are adjacent if and only if page u has a link to page v.

(2) In each case, describe a possible scenario that can be represented by the given graph. Make sure you clearly specify what the vertices represent and the rule of adjacency between two vertices.

* (a) The complete graph K_6
  (b) The Cycle C_6
  (c) the wheel W_6

(d) The complete bipartite graph $K_{3,3}$

(3) The following is the map of a part of the western region of the United States showing seven states.

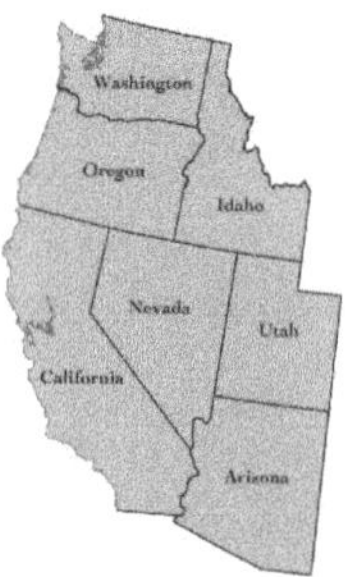

Represent this region with a graph $G = (V, E)$ where V is the set of states in that region and such that two vertices are adjacent if and only if the corresponding states share a common land border.

*(4) The following diagram represents an office space. Letters $A, B, \ldots, L$ represent the offices and the letter O represents the outside of the space. Small thick rectangles represent doors between different spaces. Draw a graph $G = (V, E)$ where $V = \{A, B, \ldots, L, O\}$ and where two vertices are adjacent if and only if one can get from one vertex (an office or the outside) to the other using one of the doors.

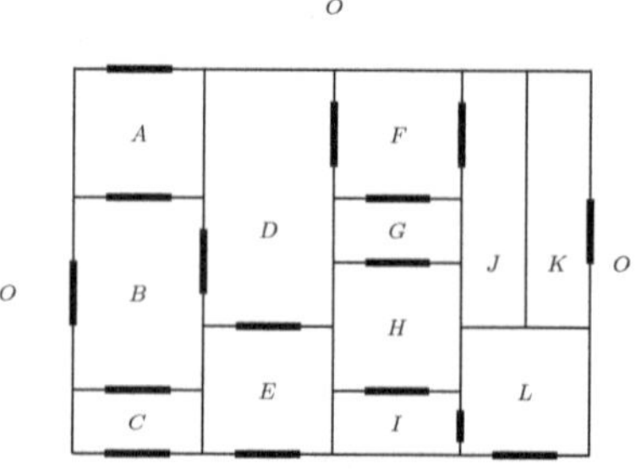

(5) Given a positive integer n, the n-**cube** is the graph $Q_n = (V, E)$ defined as follows: V is the set of all binary strings of length n and two vertices are adjacent if and only if the corresponding binary strings differ by exactly one bit. Draw the cubes Q_1, Q_2 and Q_3.

(6) Given a non-empty set A, every binary relation R on A can be represented by a digraph G where each element of A is a vertex and (a, b) is a directed edge of G if and only if $a\,R\,b$. If the relation is symmetric, then it can be represented by an undirected graph since in this case $a\,R\,b$ implies that $b\,R\,a$. In each case, determine if the binary relation R on the set A can be represented by an undirected graph or a digraph and draw the diagram representing the relation.

  (a) $A = \{1, 2, 3, 4\}$, $a\,R\,b$ if and only if $xy \geq 4$.
  (b) $A = \{1, 2, 3, 4, 5, 12, 15\}$, $a\,R\,b$ if and only if b is a multiple of a.
  (c) $A = \{1, 2, 3, 4, 5, 6, 7, 8, 9\}$, $a\,R\,b$ if and only if $x + y < 10$.
  $\star$ (d) $A = \{1, 2, 3, 4, 5, 6, 7, 8, 9\}$, $a\,R\,b$ if and only if $2a - 3b = 0$.

(7) Given a graph G, the *underlying simple graph* of G is the simple graph obtained by removing all loops and all but one edge from each collection of parallel edges. For the graph given below, draw the underlying simple graph.

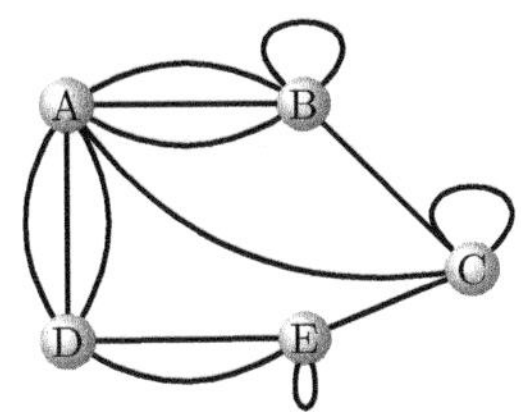

(8) In each case, write the adjacency matrix of the graph in Exercise (7) relative to the given ordering of the vertices.

  (a) A, B, C, D, E $\star$ (b) E, C, D, A, B (c) D, A, C, E, B

(9) Write the incidence matrix of the Könesberg graph relative to the ordering B, D, A, C of its vertices and the ordering d, a, g, f, c, e, b of its edges.

$\star$(10) Find the incidence matrix of the given graph below relative to the ordering A, B, C, D, E, F of its vertices and the ordering

$a_1, a_2, a_3, a_4, a_5, a_6, a_7, a_8$ of its edges.

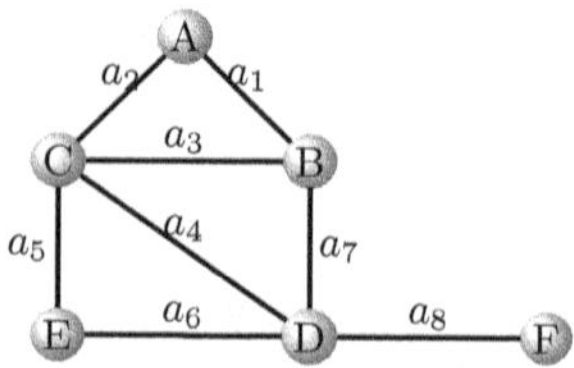

(11) In each case, the adjacency matrix of a graph G is given. What is the order of G? Is G a simple graph? Give a possible drawing of G.

$$(a) \begin{bmatrix} 0 & 1 & 1 \\ 1 & 0 & 1 \\ 1 & 1 & 1 \end{bmatrix} \quad (b) \begin{bmatrix} 0 & 0 & 1 & 1 \\ 0 & 0 & 0 & 1 \\ 1 & 0 & 0 & 1 \\ 1 & 1 & 1 & 0 \end{bmatrix} \quad \star(c) \begin{bmatrix} 0 & 2 & 1 & 1 \\ 2 & 0 & 3 & 1 \\ 1 & 3 & 0 & 3 \\ 1 & 1 & 3 & 0 \end{bmatrix} \quad (d) \begin{bmatrix} 1 & 2 & 1 & 2 & 0 \\ 2 & 1 & 1 & 1 & 0 \\ 1 & 1 & 0 & 2 & 0 \\ 2 & 1 & 2 & 0 & 1 \\ 0 & 0 & 0 & 1 & 0 \end{bmatrix}$$

(12) Let $G = (V, E)$ be a simple graph with $V = \{v_1, \ldots, v_n\}$. Let $A(G)$ be the adjacency matrix of G relative to the ordering $v_1, \ldots, v_n$ of its vertices. What does the sum of the entries of the ith row of $A(G)$ represent?

(13) Let $G = (V, E)$ be a simple graph with $V = \{v_1, \ldots, v_n\}$ and $E = \{e_1, e_2, \ldots, e_m\}$. Let $I(G)$ be the incidence matrix of G relative to the ordering $v_1, \ldots, v_n$ of its vertices and the ordering $e_1, e_2, \ldots, e_m$ of its edges. What does the sum of the entries of the ith row of $I(G)$ represent? What does the sum of the entries of the jth column of $I(G)$ represent?

$\star$(14) Given a set V containing n elements, how many *different* simple graphs with vertex set V are there?

9.3   Degree of a vertex, the Handshaking lemma and graphical sequences

Now that we have explored the basic terminology of graph theory, we start to look at the first important results. One of the oldest and simplest results known in graph theory is widely known as the handshaking lemma. The name comes from its connection with the number of handshakes in a gathering of a given number of people. In that context, the result states that the number of persons who shook hands with an odd number of persons must be even. The result uses the notion of the *degree* of a vertex in a graph. We also look in this section at the question of determining if a given sequence of natural numbers is the degree sequence of vertices of a certain graph.

Definition 9.7. Let $G = (V, E)$ be a graph of order n and let v be a vertex of G. The *degree* of v, denoted by $\deg_G(v)$ or simply $\deg(v)$ (if there is no ambiguity), is defined as the number of edges of G incident to v, with *each loop on v counted twice*. A vertex of degree 0 is called *isolated* and a vertex of degree 1 is called *pendant* (or a *leaf* in the contexts of a *tree* that we will study later). A sequence $d = (d_1, d_2, \ldots, d_n)$ of non-negative integers is called *a degree sequence* of G if there exists a certain ordering $v_1, v_2, \ldots, v_n$ of the vertices of G such that $\deg(v_i) = d_i$ for $i = 1, 2, \ldots, n$. *The degree sequence* of G is the arrangement of the degrees of *all* vertices of G in a non-increasing fashion (from the largest to the smallest).

A degree sequence of a graph $G = (V, E)$ is just an arrangement of *all* members of the multiset $\{\deg(v); v \in V\}$ (with possible repetitions). For example, if $d_1 = (3, 3, 2, 1, 1, 1)$ is a degree sequence of a graph G, then $d_2 = (1, 3, 2, 3, 1, 1)$ and $d_3 = (1, 3, 1, 2, 1, 3)$ can also be realized as degree sequences of G by relabeling the vertices. Among all the degree sequences the graph G can have, only one is non-increasing, namely *the degree sequence* of G.

Example 9.8. For the Könesberg graph:

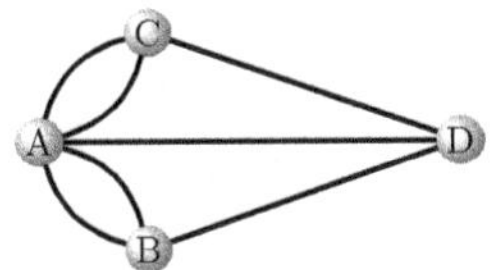

$\deg(A) = 5$, $\deg(B) = \deg(C) = \deg(D) = 3$. The degree sequence of the graph is then $(5, 3, 3, 3)$.

Remark 9.3. If $G = (V, E)$ is a simple graph of order n, then $\deg(v) \leq n-1$ for any vertex v. This is due to the fact that multiple edges and loops are not allowed in simple graphs.

Example 9.9. There is no simple graph with degree sequence $(5, 4, 2, 1)$ since the graph is of order 4 and has one vertex of degree 5.

Example 9.10. Construct, if possible, a graph with $(5, 3, 3, 1)$ as the degree sequence.

Solution. The graph must have four vertices, one pendent, two of degree 3 and one of degree 5. Remember that loops count for two in the degree of a vertex. Since there are no restrictions on the graph beside the order and the degree sequence, there are many ways to construct it. The following is one possibility:

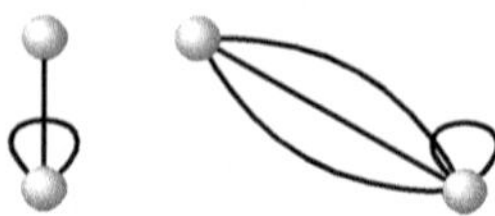

$\Diamond$

The last example gives the false impression that any finite non-increasing list d of natural numbers can be realized as the degree sequence of a certain graph. The following example shows that this is not true.

Example 9.11. Construct, if possible, a graph with $(1, 1, 1)$ as the degree sequence.

Solution. The graph, if it exists, must have three vertices that we label by u, v and w of degree 1 each. This means in particular that no loop exists at any of the vertices. Start by drawing an edge between u and v, giving already degree 1 to each of these two vertices. But w is also of degree 1 which means that it has to be connected to either u or v. This is not possible as one of u or v would have a degree greater than or equal to 2. $\Diamond$

Theorem 9.1 gives a much quicker justification of the result of Example 9.11 (see Example 9.12 below).

Theorem 9.1 (The Handshaking lemma). *In any graph $G = (V, E)$, the sum of the degrees of all vertices is twice the number of edges:*

$$\sum_{v \in V} \deg(v) = 2|E|.$$

Proof. Every edge $\{u, v\}$ (vertices u and v are not necessarily different) contributes with 1 toward the value of $\deg(u)$ and with 1 toward the value of $\deg(v)$. So the sum $\sum_{v \in V} \deg(v)$ of degrees of all vertices counts every edge twice. $\square$

Corollary 9.1. *In any graph:*

(a) the sum of degrees of all vertices is even;
(b) the number of vertices with odd degrees is even.

Proof. The first part is a direct consequence of the handshaking lemma. The second part is left as an exercise (see Exercise (20) below). $\square$

Example 9.12. The sequence $(1, 1, 1)$ cannot be the degree sequence of a graph as the sum of the elements of the list is odd. The same is true for the sequence $(7, 7, 5, 5, 4, 4, 4, 4, 3, 3, 2, 1)$.

The following example justifies the *handshaking lemma* name associated with Theorem 9.1.

Example 9.13. There are n people present in a gathering, $n \geq 2$. Everyone shook hands with everyone else at the gathering. If there were 78 handshakes in total, what is the value of n?

Solution. The situation can be represented with a graph G of order n. The vertices of G are the n persons at the gathering and the edges of G are the handshakes that took place. Therefore, $\deg(v) = n - 1$ for every vertex v of G. By the handshaking lemma, $n(n - 1) = 2 \times 78$ or $n^2 - n - 156 = 0$. Solving for n gives $n = -12$ or $n = 13$. We conclude that here 13 people were present at the gathering. $\Diamond$

Example 9.14. How many edges in a graph with degree sequence $(5, 4, 4, 3, 2)$?

Solution. Let e be the number of edges in the graph. By the handshaking lemma, $5 + 4 + 4 + 3 + 2 = 2e$. We conclude that $e = 9$. $\Diamond$

Example 9.15. Every vertex in a graph G has degree 3, 4, or 5. If G has 12 vertices and 27 edges, and has as many vertices of degree 3 as vertices of degree 4, how many vertices of each degree G has?

Solution. Let x, y and z be the number of vertices of degree 3, 4 and 5, respectively. By the handshaking lemma, $3x + 4y + 5z = 2|E| = 58$. In addition, we know that $x + y + z = 12$ and $x = y$. This gives the following system of linear equations

$$3x + 4y + 5z = 54$$
$$x + y + z = 12$$
$$x - y = 0$$

It is easy to verify that $x = y = 2$, $z = 8$ is the unique solution of this system. Hence G has two vertices of degree 3, two vertices of degree 4, and eight vertices of degree 5. $\Diamond$

9.3.1 *Graphical sequences*

Examples 9.10 and 9.11 above raise an important question: Given a finite sequence $d = (d_1, d_2, \ldots, d_n)$ of non-negative integers, how can we decide if d is a degree sequence for some graph G? The handshaking lemma gives a necessary condition: $\sum_{i=1}^{n} d_i$ must be even. It is also a sufficient condition if we don't require the graph to be simple as shown in the following theorem.

Theorem 9.2. *Let* $d = (d_1, d_2, \ldots, d_n)$ *be a sequence of non-negative integers such that* $\sum_{i=1}^{n} d_i$ *is even. Then there exists a graph* G *having* d *as a degree sequence.*

Proof. First assume that d_i is odd for every i. Corollary 9.1 above tells us that n must be even in this case. Consider the graph G with n vertices $v_1, \ldots, v_n$ and the following edges: $\{v_1, v_2\}$, $\{v_3, v_4\}$, $\ldots$, $\{v_{n-1}, v_n\}$ (this is possible since n is even) together with $\frac{d_i - 1}{2}$ loops around each vertex v_i. For each i, $\deg(v_i) = 2\left(\frac{d_i - 1}{2}\right) + 1 = d_i$. So the graph G has $d = (d_1, d_2, \ldots, d_n)$ as a degree sequence. Assume next that at least one of the d_i's is even. Relabel the sequence $(d_1, d_2, \ldots, d_n)$ if necessary so that $d_1, \ldots, d_k$ are all odd and $d_{k+1}, \ldots, d_n$ are all even for some index $k \geq 1$. By the previous discussion, we can form a graph $G_1 = (V_1, E_1)$ with $(d_1, \ldots, d_k)$ as a degree sequence. Consider the graph G_2 with set of vertices $V_2 = \{v_{k+1}, \ldots, v_n\}$ and set of edges E_2 formed by drawing $\frac{d_i}{2}$ loops around each vertex v_i for $i = k+1, \ldots, n$. In G_2, $\deg(v_i) = 2\frac{d_i}{2} = d_i$ and so $(d_{k+1}, \ldots, d_n)$ is a degree sequence of G_2. The sequence $d = (d_1, d_2, \ldots, d_n)$ is clearly a degree sequence of the graph $G = (V, E)$ with $V = V_1 \cup V_2$ and $E = E_1 \cup E_2$. $\square$

If we require that the graph is *simple*, then the result of Theorem 9.2 is no longer true. Example 9.9 above gives an example of a sequence of non-negative integers with even sum but which cannot be realized as a degree sequence of a simple graph.

Definition 9.8. A non-increasing sequence $d = (d_1, d_2, \ldots, d_n)$ of natural numbers is called *graphical* if it is the degree sequence of some *simple* graph.

Given a non-increasing sequence $d = (d_1, d_2, \ldots, d_n)$ of natural numbers, we know so far two necessary conditions for the sequence to be graphical: $\sum_{i=1}^{n} d_i$ is even (the handshaking lemma) and $d_i \leq n-1$ for any $i = 1, \ldots, n$ (Remark 9.3 above). For example, the sequence $(4, 4, 3, 3, 1)$ is not graphical since the sum of the entries is not even. Similarly, the sequence $(5, 4, 4, 3)$ is not graphical as it has only four terms with one term (the first) greater than 4. These two necessary conditions are, however, not sufficient for the sequence to be graphical. For instance, the sequence $(3, 3, 3, 1)$ satisfies both conditions but it is not graphical (Exercise (29) below). Two main characterizations of graphical sequences are well known in the literature. The first is due to Havel, Hakimi and the second to Erdös and Gallai. In this book, we look at the Havel–Hakimi characterization as it offers a more algorithmic approach to the graphical sequence problem. It also allows the construction of the simple graph with the given sequence as the degree sequence.

Theorem 9.3 (Havel–Hakimi). *Let* $n \geq 2$ *be an integer. A non-increasing sequence* $l = (d_1, d_2, \ldots, d_{n-1}, d_n)$ *of natural numbers with* $d_1 \geq 1$ *is graphical if and only if the sequence*

$$l_1 = (\underbrace{d_2 - 1, d_3 - 1, \ldots, d_{d_1+1} - 1}_{d_1 \ terms}, d_{d_1+2}, \ldots, d_n)$$

is graphical.

Before the proof, note first how we obtain the sequence l_1 from the graphical sequence l: remove d_1 (the largest term) from l, then subtract 1 from each of the next d_1 terms in l. This is possible since $d_1 \leq n - 1$ so when d_1 is removed from l, the number of terms left in the sequence is $n - 1 \geq d_1$.

Proof. Assume first that the sequence l_1 is graphical. By definition, there exists a simple graph $G_1 = (V_1, E_1)$ of order $n - 1$ having l_1 as the degree sequence. Label the vertices of G_1 as $v_2, v_3, \ldots, v_{d_1+1}, v_{d_1+2}, \ldots, v_n$ with $\deg(v_i) = d_i - 1$ for $2 \leq i \leq d_1 + 1$ and $\deg(v_i) = d_i$ for $d_1 + 2 \leq i \leq n$. Let G be the graph obtained from G_1 by adding a vertex v_1 to V_1 and connecting it to each of the vertices $v_2, v_3, \ldots, v_{d_1+1}$ by one edge. The graph G is clearly simple with $\deg(v_i) = d_i$ for each $i = 1, 2, \ldots, n$. We conclude that $l = (d_1, d_2, \ldots, d_{n-1}, d_n)$ is graphical. Conversely, assume that l is graphical and let G be a simple graph of order n with degree sequence l and vertices $v_1, v_2, \ldots, v_n$ labeled so that $\deg(v_i) = d_i$ for $i = 1, 2, \ldots, n$. If vertex v_1 is adjacent to each of the vertices $v_2, v_3, \ldots, v_{d_1+1}$, then we are done since the graph G_1 obtained from G by removing vertex v_1 and all its incident edges is clearly simple and has $l_1 = (d_2 - 1, d_3 - 1, \ldots, d_{d_1+1} - 1, d_{d_1+2}, \ldots, d_n)$ as the degree sequence. So assume there exists $i \in \{2, \ldots, d_1 + 1\}$ such that v_1 is not adjacent to v_i. Since $\deg(v_1) = d_1$, there exists an index $k \geq d_1 + 2$ such that v_k is adjacent to v_1. We also have that $\deg(v_k) \leq \deg(v_i)$ as the sequence l is non-increasing. If $\deg(v_k) = \deg(v_i)$, then exchanging vertices v_i and v_k does not change the degree sequence of G but increases by one the number of neighbors of v_1 which belong to the set $\{v_2, v_3, \ldots, v_{d_1+1}\}$.

If $\deg(v_k) < \deg(v_i)$, vertex v_i has more neighbors than vertex v_k and consequently there exists an index $m \notin \{i, k\}$ such that v_m is adjacent to v_i but not to v_k. In particular, $v_m \neq v_1$ as v_m is a neighbor of v_i but v_1 is not. By removing edges $\{v_1, v_k\}$ and $\{v_i, v_m\}$ and adding edges $\{v_1, v_i\}$ and $\{v_k, v_m\}$, we don't actually change the degree of any vertex in G but once again we increase by one the number of neighbors of v_1 which belong to the set $\{v_2, v_3, \ldots, v_{d_1+1}\}$. We repeat this process, each time increasing by one the number of neighbors of v_1 in the set $\{v_2, v_3, \ldots, v_{d_1+1}\}$. Since the number of vertices is finite, the above procedure must terminate after a finite number of steps and when it does, it produces a graph with the same degree list as G and such that the neighbors of vertex v_1 are precisely $v_2, \ldots, v_{d_1+1}$. The proof is complete. $\qquad\square$

Given a non-increasing sequence $d_1 \geq d_2 \geq \ldots, d_{n-1} \geq d_n$ (with $n \geq 2$ and $d_1 \geq 1$) of non-negative integers, Theorem 9.3 gives an algorithm to determine if the sequence is graphical. Keep in mind that if either one of the necessary conditions $\sum_{i=1}^{n} d_i$ is even and $d_i \leq n-1$ for any $i = 1, \ldots, n$ is not satisfied, the sequence is not graphical. We describe the algorithm using the following steps.

- **Step 1.** If $d_i = 0$ for each i, the sequence is graphical (it is the degree sequence of the edgeless graph of order n). If $d_i \geq 1$ for at least one index i, proceed to step 2.
- **Step 2.** Remove d_1 from the sequence and then subtract 1 from each of the next d_1 remaining terms.
- **Step 3.** If necessary, rearrange the sequence obtained in step 2 in a non-increasing order and return to step 1 .
- **Step 4.** The process stops when we obtain either:

  (i) a sequence consisting of 0's only or
  (ii) a sequence containing a negative integer.

  The original sequence is graphical if and only if we obtain a sequence as in (i).

Remark 9.4. Note that we can stop the algorithm as soon as we reach a new sequence that we can recognize as being graphical or not graphical. For instance, if the algorithm produces the sequence $(2, 1, 1)$ at one stage, we can conclude that the original sequence is graphical since $(2, 1, 1)$ is. Similarly, if the algorithm produces the sequence $(2, 2, 1)$ at one stage, we conclude that the original sequence is not graphical since $(2, 2, 1)$ is not.

If l is a graphical sequence, the Havel–Hakimi algorithm can be "reversed" to actually construct a graph with l as the degree sequence. This can be done in a recursive fashion as follows. Let $l_0 = l, l_1, l_2, \ldots, l_k$ be the series of non-increasing sequences obtained from the Havel–Hakimi algorithm starting with the original sequence l and ending with the zero sequence l_k. Start with the edgeless graph G_k whose degree sequence is l_k. If graph G_i with degree sequence l_i ($i \geq 1$) is constructed, graph G_{i-1} can be constructed by adding a vertex whose degree is the largest term in l_{i-1} and which is adjacent to those vertices in G_i whose degrees were reduced (by 1) in l_i. Continue this way until graph G_0 is obtained. It is worth mentioning here that in some cases it is hard to keep track of the vertices whose degrees were reduced at every stage. It is usually easier to add one vertex and necessary edges to construct a new graph having the required degree sequence at a given stage.

Example 9.16. Determine if the sequence $l_0 = (5, 3, 3, 3, 2, 2, 1, 1)$ is graphical. If you say it is, construct a graph G with l as the degree sequence.

Solution. Using Havel–Hakimi algorithm:

$$l_0 = (5, 3, 3, 3, 2, 2, 1, 1) \text{ is graphical} \Leftrightarrow$$
$$l_1 = (2, 2, 2, 1, 1, 1, 1) \text{ is graphical} \Leftrightarrow$$
$$l_2 = (1, 1, 1, 1, 1, 1) \text{ is graphical} \Leftrightarrow$$
$$l_3 = (0, 1, 1, 1, 1) \leftrightarrow (1, 1, 1, 1, 0) \text{ is graphical (rearranged)} \Leftrightarrow$$
$$l_4 = (0, 1, 1, 0) \leftrightarrow (1, 1, 0, 0) \text{ is graphical (rearranged)} \Leftrightarrow$$
$$l_5 = (0, 0, 0) \text{ is graphical}$$

Since the algorithm ends with a zero sequence, the original sequence is graphical. The following diagram shows the sequence of graphs obtained by repeated applications of the above algorithm together with their degree sequences. The starting graph is the edgeless graph and the last graph has the original sequence l as the degree sequence.

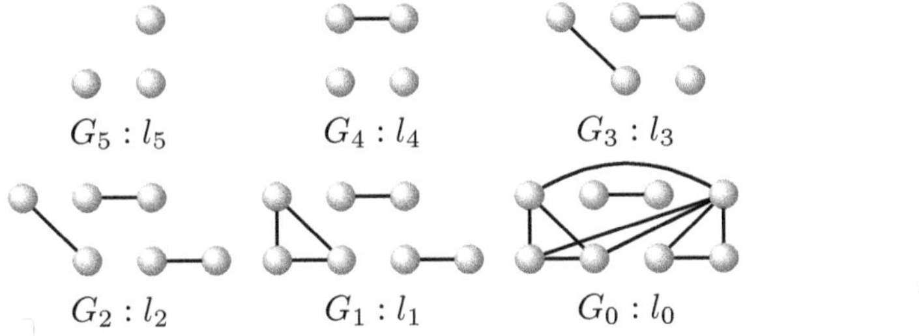

Example 9.17. We know that the sequence $l = (6, 5, 4, 3, 2, 2, 1)$ is not graphical because the sum of its entries is odd (23). Prove this fact using the Havel–Hakimi algorithm.

Solution. Using the algorithm of Havel–Hakimi: $(6, 5, 4, 3, 2, 2, 1)$ is graphical $\Leftrightarrow$ $(4, 3, 2, 1, 1, 0)$ is graphical $\Leftrightarrow$ $(2, 1, 0, 0, 0)$ $\Leftrightarrow$ $(0, -1, 0, 0)$ is graphical. Since the last sequence contains a negative integer, it is not graphical and so the original sequence is not graphical. $\Diamond$

Example 9.18. Determine if the sequence $l_0 = (6, 4, 2, 2, 2, 1, 1)$ is graphical. If you say it is, draw a graph G with l as the degree sequence.

Solution. Using Havel–Hakimi algorithm: $l_0 = (6, 4, 2, 2, 2, 1, 1)$ is graphical $\Leftrightarrow$ $l_1 = (3, 1, 1, 1, 0, 0)$ is graphical $\Leftrightarrow$ $l_2 = (0, 0, 0, 0, 0)$ is graphical. We conclude that the sequence $(6, 4, 2, 2, 2, 1, 1)$ is indeed graphical. The following diagram shows the sequence of graphs obtained by reversing the Havel–Hakimi algorithm. Graph G_0 below is the simple graph with l as degree sequence.

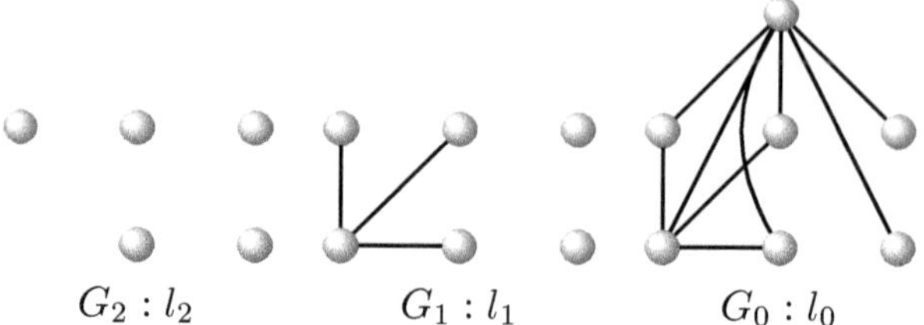

$G_2 : l_2$ $G_1 : l_1$ $G_0 : l_0$

$\Diamond$

9.3.2   *Exercises*

(1) Give the degree sequence of each of the following three graphs:

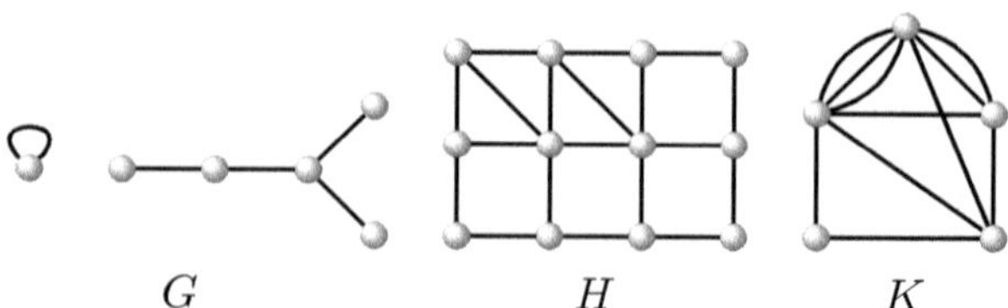

G H K

(2) Give the degree sequence of each of the following graphs: K_9, C_9, W_9, P_9 and $K_{6,3}$.

$\star$(3) The following graph is widely known in the literature as the *Petersen graph*. It appears in many areas of graph theory. Verify the result of the handshaking lemma for the Petersen graph.

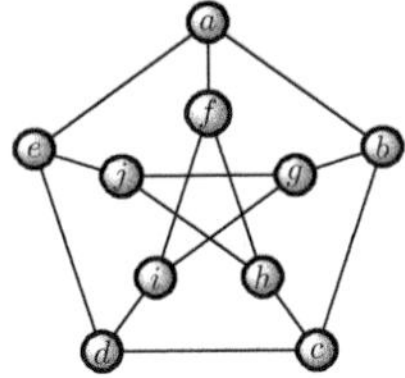

(4) Prove that the number of edges in the wheel graph W_n is $2n$ using the handshaking lemma.

$\star$(5) Prove that the number of edges in the complete bipartite graph $K_{m,n}$ is mn using the handshaking lemma.

(6) The following is the adjacency matrix of a graph G:

$$A = \begin{bmatrix} 0 & 1 & 1 & 0 & 1 & 0 & 0 \\ 1 & 0 & 0 & 1 & 1 & 1 & 0 \\ 1 & 0 & 0 & 0 & 1 & 0 & 1 \\ 0 & 1 & 0 & 0 & 1 & 1 & 0 \\ 1 & 1 & 1 & 1 & 0 & 1 & 0 \\ 0 & 1 & 0 & 1 & 1 & 0 & 0 \\ 0 & 0 & 1 & 0 & 0 & 0 & 0 \end{bmatrix}.$$

   (a) What is the order of G?

   (b) What is the size of G?

   (c) Is G simple?

  $\star$ (d) Write the degree sequence of G.

(7) What is the size of a graph with degree sequence $(7, 6, 6, 4, 3, 2)$?

$\star$(8) Let G be a graph with degree sequence $(7, 5, 4, 4, 3, 2, 1)$. What is the size of G? Can G be simple?

(9) The degree sequence of a graph G is $(4, 4, 2, 2, 2, 2)$. What is the size of G? Give a possible drawing for G.

$\star$(10) Every vertex in a graph G is of degree 4. What is the order of G if its size is 16? Give a possible drawing for G.

(11) A graph G has three vertices of degree 4, two vertices of degree 3 and the rest of the vertices are of degree 2 each. If the graph has 14 edges, what is the number of vertices of degree 2 in G? Give a possible drawing for G.

$\star$(12) A graph G has seven vertices of degree 1, three vertices of degree 2, seven vertices of degree 3, and the rest of the vertices are of degree 4 each. What is the order of G if its size is 21?

(13) 12 people were present at a meeting. Each person shook hand once with everyone else at the meeting. How many handshakes took place in total at the meeting?

(14) At a party, everyone shook hands once with everyone else. There were 105 handshakes in total. How many people were at the party?

$\star$(15) In a certain office, there are n computers with $n \geq 2$. Each computer is connected with a single cable to every other computer in the office. If there are 66 connection cables in total, how many computers are there in the office?

(16) Prove that it is impossible to have a meeting of 11 persons in which every person shook hands with *exactly* 5 other persons in the meeting.

(17) Every vertex in a graph G is of degree 3 or 5. If G has 12 vertices and 27 edges, how many vertices of each degree does G have?

$\star$(18) Every vertex in a graph G is of degree 2, 3, or 4. The number of vertices of degree 3 is double the number of vertices of degree 4. If G is of order 12 and size 16, how many vertices of each degree in G?

(19) Every vertex in a graph G has degree 1, 3 or 5 and the number of vertices of degree 3 is one more than the number of vertices of degree 5. If G has a total of 6 vertices and 10 edges, how many vertices of each degree does G have?

(20) Prove that in any graph, the number of vertices with odd degrees is even.

(21) Given an integer $d \geq 0$, we say that the graph G is *d-regular* if all vertices of G have the same degree d.

   (a) Each of the following is d-regular for some integer $d \geq 0$. Find d in each case: C_n, K_n, $K_{n,n}$ and Q_3 (see Exercise (5) of the previous section for the definition of Q_n).

   (b) If d is odd, prove that a d-regular graph must have an even number of vertices.

  $\star$ (c) If d is odd, prove that in a d-regular graph the number of edges must be a multiple of d.

  $\star$ (d) Prove that in a 6-regular graph, the number of edges is three times the number of vertices.

(22) Prove that the number of edges in the cycle graph C_n for $n \geq 3$ is equal to n using: (a) an induction argument on n; (b) the handshaking lemma.

$\star$(23) Prove that the number of edges in the complete graph K_n for $n \geq 1$ is equal to $\frac{n(n-1)}{2}$ using: (a) a combinatorial approach; (b) the handshaking lemma.

(24) What is the largest size a simple graph of order $n \geq 1$ can have?

(25) Prove the handshaking lemma by induction on the order of the graph.

(26) Prove the handshaking lemma by induction on the size of the graph.

(27) Let G be a simple graph of order $n \geq 2$. In this problem we prove that G has at least two vertices of the same degree.

  (a) Prove that G cannot have simultaneously a node of degree 0 and another of degree $n - 1$ (*Hint.* Use a proof by contradiction).

  (b) Use part (a) and the pigeonhole principal to prove that there must be at least one pair of different vertices of G with the same degree.

  (c) As an application, prove that if some handshakes took place in a meeting of two or more persons, then there are at least two persons in the meeting who had the same number of handshakes.

(28) Let $G = (V, E)$ be a graph of order $n \geq 1$. If G has $n - 1$ edges, prove that G has at least one vertex v with $\deg(v) < 2$ (*Hint.* Use a proof by contradiction).

(29) Without using the Havel–Hakimi algorithm, determine if the given sequence is graphical. If you say it is, draw a graph realizing the sequence and if you say it is not, explain why.

(a) $(2,2,2)$	(d) $(5,5,4,4,3,2,1,1)$	$\star$ (g) $(3,3,3,1)$
(b) $(2,1,1)$	$\star$ (e) $(4,1,1,1,1)$	
(c) $(3,2,1,0)$	$\star$ (f) $(6,4,4,3,3)$	

(30) In each case determine if the sequence is graphical using the Havel–Hakimi algorithm. If you say it is, draw a graph realizing the sequence.

$\star$ (a) $(4,4,3,2,2,1)$	(f) $(6,6,6,4,4,3,3)$
(b) $(6,6,5,4,3,3,1)$	(g) $(9,7,7,7,3,3,3,2,2,2)$
(c) $(6,5,5,5,4,4,2,1)$	(h) $(9,9,9,9,8,7,5,4,3,2,1)$
$\star$ (d) $(8,7,6,6,5,3,2,2,2,1)$	(i) $(7,6,6,6,5,5,4,1)$
(e) $(6,6,6,4,4,2,2)$	

9.4   Subgraphs and new graphs from old

Consider the following two graphs $G = (V, E)$ and $G' = (V', E')$.

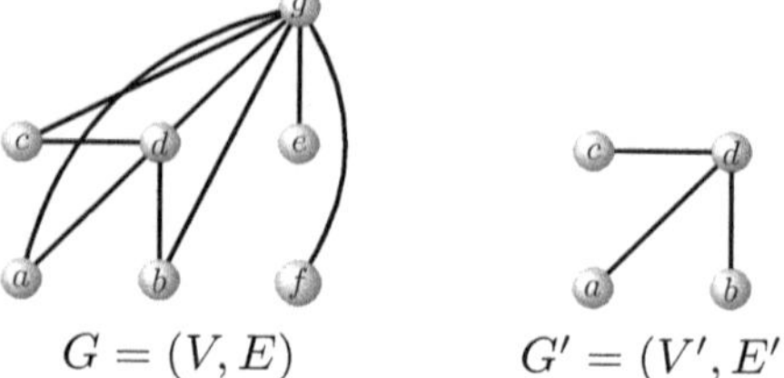

$$G = (V, E) \qquad G' = (V', E')$$

Observe that $V' \subseteq V$ (every vertex of G' is a vertex of G) and $E' \subseteq E$ (every edge of G' is an edge of G). We can see that G' is a graph "embedded" within the graph G. We say in this case that G' is a *subgraph* of G. The formal definition of a subgraph is as follows.

Definition 9.9. Let $G = (V, E)$ and $G' = (V', E')$ be two graphs. We say that G' is a **subgraph** of G, and we write $G' \leq G$, if $\emptyset \neq V' \subseteq V$, $E' \subseteq E$ and V' contains the endpoints of any edge in E'.

The last condition in Definition 9.9 states that each edge in G' has the same end points in G' as in G. In terms of incidence functions, the condition states that the incidence function of G' is the *restriction* of the incidence function of G to the set E'. If H is a subgraph of G which is not the same as G, we say that H is a *proper subgraph* of G. If we need to stress on the fact that H is a proper subgraph of G, we write $H < G$.

The following facts follow immediately from the definition:

(1) The trivial graph (a single vertex with no edges) is a subgraph of any non-empty graph.

(2) If $e = \{a, b\}$ is an edge of a graph G, then $G' = (\{a, b\}, E = \{e\})$ is a subgraph of G.

(3) Any graph is a subgraph of itself.

(4) If G' is a subgraph of G and G'' is a subgraph of G', then G'' is a subgraph of G.

Example 9.19. For the graph G given below, give all subgraphs of order 3 and size 2.

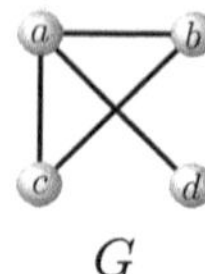

$$G$$

Solution. There are $\binom{4}{3} = 4$ subsets of V of cardinality 3, namely $V_1 = \{a, b, c\}$, $V_2 = \{a, b, d\}$, $V_3 = \{a, c, d\}$ and $V_4 = \{b, c, d\}$. There are three subgraphs of G with vertex set V_1 and size 2, namely $G_1 = (V_1, \{\{a, b\}, \{a, c\}\})$, $G_2 = (V_1, \{\{a, b\}, \{b, c\}\})$ and $G_3 = (V_1, \{\{a, c\}, \{b, c\}\})$. There exists one subgraph of G with vertex set V_2 and size 2, namely $G_4 = (V_2, \{\{a, b\}, \{a, d\}\})$. Similarly, there exists one subgraph of G with vertex set V_3 and size 2, namely $G_5 = (V_3, \{\{a, c\}, \{a, d\}\})$. Finally, there is no subgraph of G with vertex set V_4 and size 2. Subgraphs $G_1, \ldots, G_5$ are drawn in the following diagram.

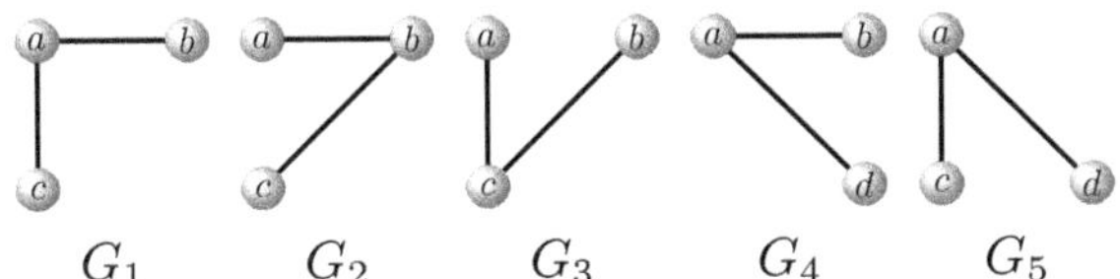

$\Diamond$

Given a graph $G = (V, E)$ with $|V| \geq 2$, there are two ways of forming a subgraph of G:

- Remove a vertex v of G together with all the edges of G incident to v. The resulting subgraph is denoted by $G - v$ and often referred to as a *vertex-deleted subgraph* of G. More generally, if $W \subseteq V$ then $G - W$ is the subgraph of G obtained by deleting all vertices of G that are in W together with all edges of G incident to any of the vertices in W.

- Delete an edge e of G but not its end points. The resulting subgraph is denoted by $G - e$ and often referred to as an *edge-deleted subgraph* of G. So $G - e$ has the same vertex set as G but its set of edges is $E \backslash \{e\}$. More generally, if F is any subset of E, then $G - F$ is the subgraph obtained from G by removing all edges in F but not their end points.

Example 9.20. In the following diagram, a graph G and four of its subgraphs are given. Graph G_1 is the subgraph $G - g$, G_2 is the subgraph $G - \{g, b\}$, G_3 is the subgraph $G - \{a, g\}$ and $G_4 = G - F$ where $F = \{\{a, g\}, \{b, d\}, \{b, g\}, \{c, g\}, \{f, g\}\}$.

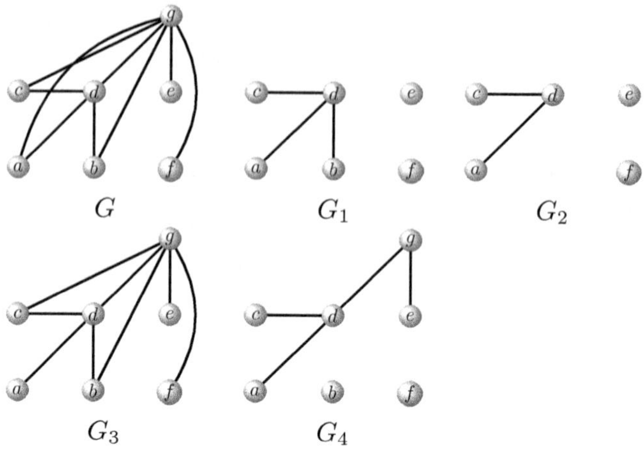

9.4.1   *Induced and spanning subgraphs*

Definition 9.10. Let $G = (V, E)$ be a graph, W a subset of V and F a subset of E.

(a) The *subgraph of G induced* by W (*vertex-induced subgraph*) is the subgraph $G[W] = (W, E')$ where $E' = \{\{x, y\} \in E;\ x, y \in W\}$. So the set of vertices of $G[W]$ is W and its set of edges consists of *all* edges of G with both endpoints in W.

(b) The *subgraph of G induced* by F (*edge-induced subgraph*) is the subgraph $G[F] = (V', F)$ where V' is the set of vertices in V which are endpoints of an edge in F. So the vertices of $E[F]$ are all endpoints of edges in F and the set of edges of $E[F]$ is F. An *induced subgraph* of G is either a vertex-induced or an edge-induced subgraph of G.

(c) If H is a subgraph of G with the same set of vertices as G, then we say that H *spans* G or that H is a *spanning subgraph* of G.

Example 9.21. For the graph $G = (V, E)$ given below, consider the subsets $W = \{e, f, g, h\}$, $W' = \{a, b, c, e, f, g\}$ of V and the subset $F = \{\{a, c\}, \{a, e\}, \{c, g\}, \{e, g\}, \{e, f\}\}$ of E. Draw the subgraphs $G[W]$, $G[W']$ and $G[F]$. Draw one spanning subgraph H of G.

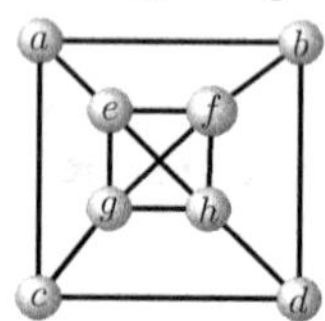

Solution. The subgraphs are given in the following diagram.

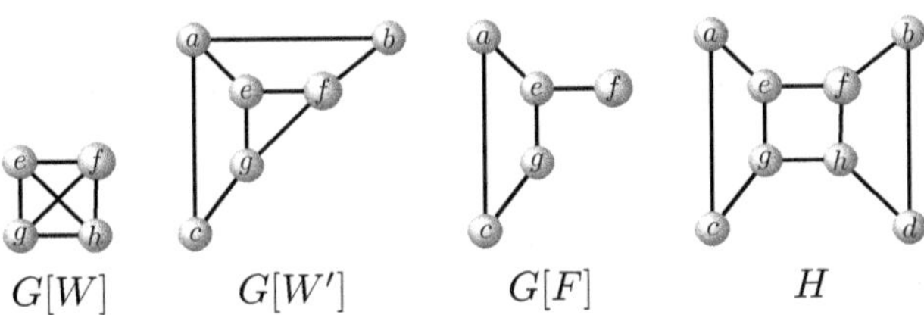

$$G[W] \qquad G[W'] \qquad G[F] \qquad H$$

$\Diamond$

9.4.2 *New subgraphs from old*

Several operations can be used to generate new graphs from given ones. We introduce some of these operations using the language of subgraphs.

Definition 9.11. Let $G_1 = (V_1, E_1)$, $G_2 = (V_2, E_2)$ be two subgraphs of a graph $G = (V, E)$.

(a) The *union* of G_1 and G_2, denoted by $G_1 \cup G_2$, is the subgraph (V_3, E_3) where $V_3 = V_1 \cup V_2$, and $E_3 = E_1 \cup E_2$. A common form of graph union is the *disjoint union* where $V_1 \cap V_2 = \emptyset$.
(b) The *intersection* of G_1 and G_2, denoted by $G_1 \cap G_2$, is the subgraph (V_3, E_3) where $V_3 = V_1 \cap V_2$, and $E_3 = E_1 \cap E_2$. If $V_1 \cap V_2 = \emptyset$, then $G_1 \cap G_2$ is the empty graph.
(c) The *ringsum* of G_1 and G_2, denoted by $G_1 \oplus G_2$, is the subgraph (V_3, E_3) where $V_3 = V_1 \cup V_2$ and $E_3 = E_1 \oplus E_2$. So, the set of vertices of $G_1 \oplus G_2$ is the same as that of $G_1 \cup G_2$ and its edges are those edges of G that are either in G_1 or in G_2 but not in both.

Example 9.22. The following diagrams shows a graph G and two of its subgraphs, G_1 and G_2.

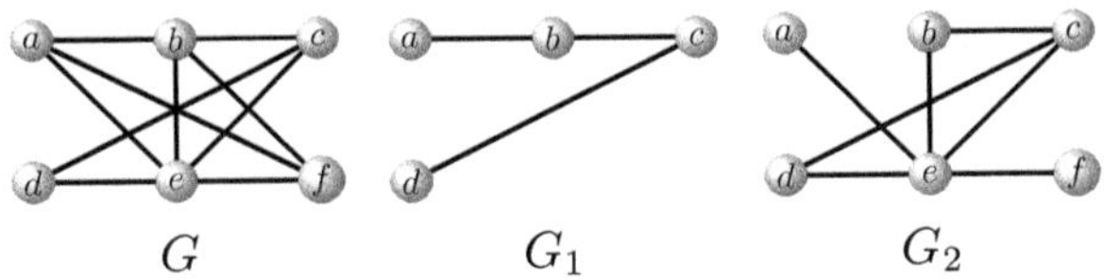

$$G \qquad G_1 \qquad G_2$$

The subgraphs $G_1 \cup G_2$, $G_1 \cap G_2$ and $G_1 \oplus G_2$ of G are as follows.

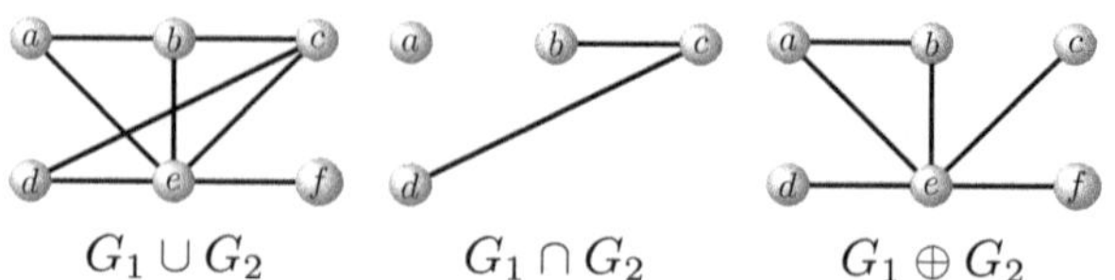

$$G_1 \cup G_2 \qquad\qquad G_1 \cap G_2 \qquad\qquad G_1 \oplus G_2$$

The following theorem lists some properties of the above graph operations. The proof of each of these properties follows directly from the definition and it is left to the reader to verify.

Theorem 9.4. *Let G_1, G_2 be two subgraphs of a graph G.*

(1) *The union, intersection and ringsum of graphs are commutative operations.*

(2) $G_1 \cup G_1 = G_1$, $G_1 \cap G_1 = G_1$.

(3) $G_1 \oplus G_1$ *is an edgeless graph with the same order as G_1.*

(4) *If G_1 and G_2 are edge-disjoint (no edge in common) with at least one vertex in common, then $G_1 \cap G_2$ is an edgeless graph and $G_1 \cup G_2 = G_1 \oplus G_2$.*

(5) *If G_1 and G_2 are vertex-disjoint (no vertex in common), then $G_1 \cap G_2$ is the empty graph (no vertices and no edges).*

9.4.3 *Complement of a graph*

Definition 9.12. Let $G = (V, E)$ be a graph, $H = (W, F)$ a subgraph of G. The *complement* of H in G is the subgraph $\overline{H} = (V', E')$ such that $E' = E \backslash F$ and V' consists of all vertices of G which are incident to edges in $E \backslash F$ or those vertices isolated in G but not contained in W.

Example 9.23. The following diagram shows a graph G, a subgraph H of G and the complement $\overline{H}$ of H in G.

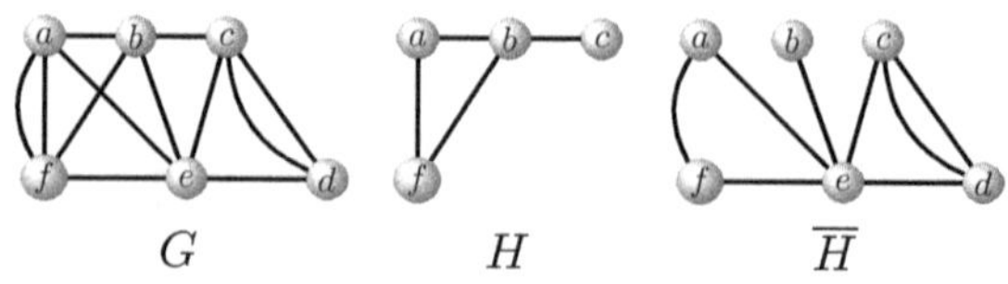

$$G \qquad\qquad H \qquad\qquad \overline{H}$$

If G is a *simple* graph of order n, then G can be considered as subgraph of the complete graph K_n and we can then talk about the complement of G in K_n.

Definition 9.13. Let $G = (V, E)$ be a *simple* graph. The *complement* of G, denoted by $\overline{G}$, is the complement of G as a subgraph of K_n. In other words, $\overline{G} = (V, \overline{E})$ where $\overline{E} = \{\{u, v\}; u, v \in V, \{u, v\} \notin E\}$. So $\overline{G}$ has the same set of vertices as G and two vertices u and v are adjacent in $\overline{G}$ if and only if they are *not* adjacent in G.

Example 9.24. The following diagram shows a simple graph G together with its complement $\overline{G}$:

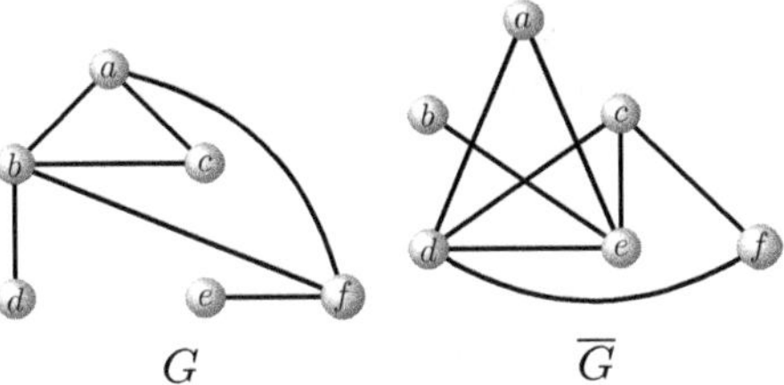

$$G \qquad \overline{G}$$

Note that if G is a simple graph of order $n \geq 1$ and size m, then $\overline{G}$ is of order n and size $\binom{n}{2} - m = \frac{n(n-1)}{2} - m$

9.4.4 *Line graph, Hamiltonian closure of a graph*

Definition 9.14. Let $G = (V, E)$ be a loopless graph. The *line graph* of G, denoted by $L(G)$, is the graph with E as the set of vertices (so $L(G)$ has a vertex for every edge in G) such that two vertices in $L(G)$ are adjacent if and only the corresponding edges in G have a common endpoint.

Example 9.25. The diagram below shows a graph G together with its line graph $L(G)$.

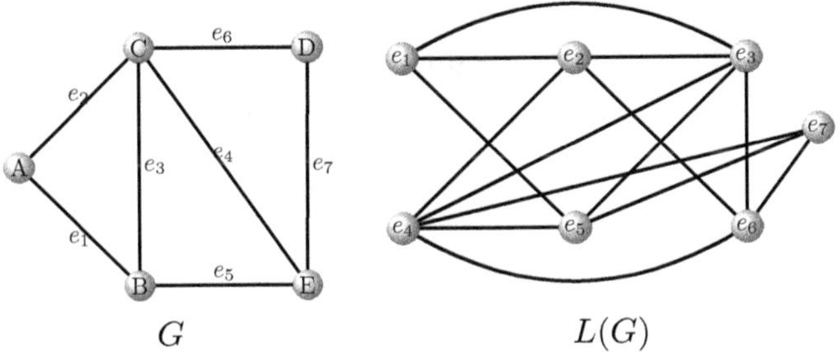

$$G \qquad\qquad L(G)$$

Definition 9.15. Let G be a simple graph of order n. *The Hamiltonian closure* of G is the graph, denoted by $cl(G)$, obtained from G by successively adding edges between non-adjacent vertices u, v of G satisfying $\deg(u) + \deg(v) \geq n$ until this can no longer be done.

A natural question that arises from Definition 9.15 is whether $cl(G)$ depends on the order in which we choose to add edges to G. It could happen that different orders in which we choose to add edges yield different closures. In other words, is the notion of Hamiltonian closure of a graph well-defined? The answer is yes as shown in the next result.

Theorem 9.5. *For a simple graph G of order n, the Hamiltonian closure $cl(G)$ does not depend on the order in which we choose to add edges following the terminology of Definition 9.15.*

Proof. Let G_1 and G_2 be two graphs obtained from G by adding the sequences of edges $S_1 = (a_1, \ldots, a_r)$ and $S_2 = (b_1, \ldots, b_s)$ respectively to non-adjacent vertices whose degree sum is at least n until no more such additions are possible. Our goal is to show that $G_1 = G_2$ by showing that each of the two sequences S_1 and S_2 contains all edges of the other. Assume to the contrary that for some $k \in \{1, \ldots, r-1\}$, $a_1, \ldots, a_k$ are all edges in G_2 but a_{k+1} is not. Let u, v be the end points of a_{k+1} and let H be the graph obtained from G by adding edges $a_1, \ldots, a_k$ to G. Then $\deg_H(u) \geq \deg_G(u)$ and $\deg_H(v) \geq \deg_G(v)$. So $\deg_H(u) + \deg_H(v) \geq \deg_G(u) + \deg_G(v) \geq n$. But since H is a subgraph of G_2 (all edges of H are also edges of G_2), $\deg_{G_2}(u) + \deg_{G_2}(v) \geq n$ and so edge $a_{k+1} = \{u, v\}$ must be in G_2. This is a contradiction. We conclude that $E_1 = E_2$. $\qquad\square$

Example 9.26. The following figure shows a graph G and successive additions of edges (shown in bold) until the closure is reached.

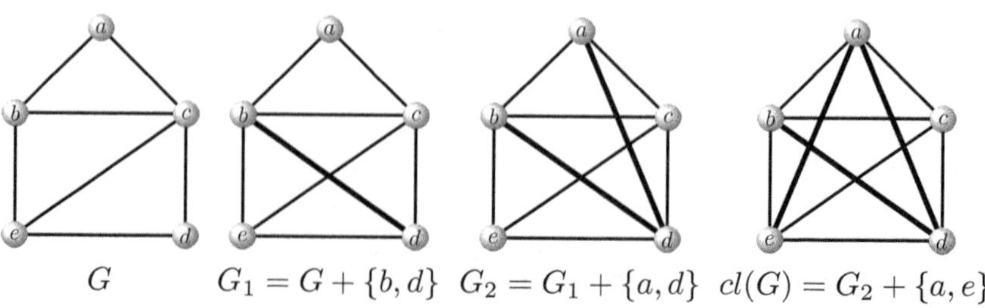

$$G \qquad\qquad G_1 = G + \{b, d\} \quad G_2 = G_1 + \{a, d\} \quad cl(G) = G_2 + \{a, e\}$$

9.4.5 *Exercises*

⋆(1) Consider the graph G:

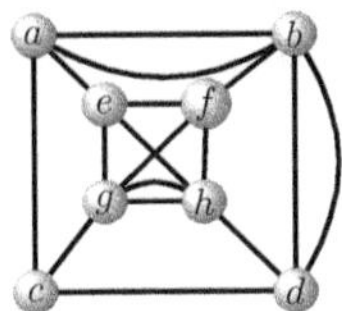

For each of the following graphs, determine if it is a subgraph of G, Justify your answer.

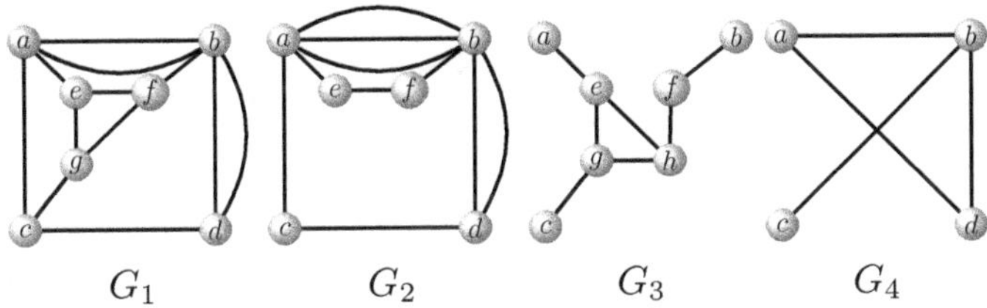

G_1 G_2 G_3 G_4

(2) For the graph G given below, what is the number of subgraphs of G of order 3 and size 2 ?

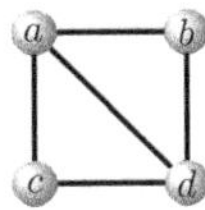

(3) Let $V = \{a, b, c\}$ and $E = \{\{a, b\}, \{a, c\}\}$. Draw the graph $G = (V, E)$ and all its possible subgraphs.

⋆(4) For the graph $G = (V, E)$ given below, let $U = \{a, b, d\} \subseteq V$ and $F = \{\{a, b\}, \{a, c\}, \{b, d\}\} \subseteq E$. Draw the induced subgraphs $G[U]$ and $G[F]$ of G.

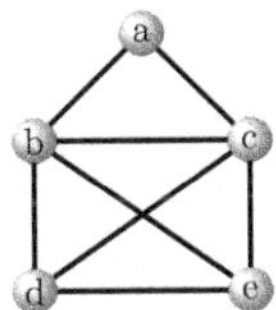

(5) Consider the graph G:

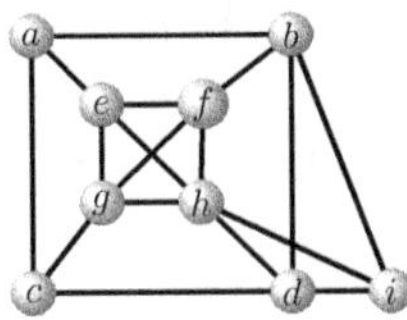

Draw each of the following subgraphs of G.

(a) $G - i$.

(b) $G - e'$ where e' is the edge $\{a, b\}$.

(c) $G - V'$ where V' is the subset $\{h, i\}$ of vertices of G.

(d) $G - F'$ where F' is the subset $\{\{b, i\}, \{h, d\}, \{f, g\}, \{d, i\}\}$ of edges of G.

(e) The subgraph of G induced by the subset $V' = \{a, b, c, d, e\}$ of vertices of G.

(6) Consider the graph G:

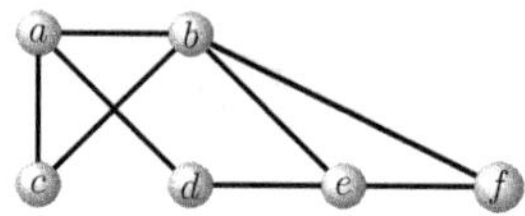

(a) Give a subgraph G_1 of G which is not a spanning subgraph.

(b) Give a subgraph G_2 of G which is induced but not spanning.

(c) Give a subgraph G_3 of G which is not an induced subgraph.

(d) Give a subgraph G_4 of G which spanning but not induced.

(e) Give a subgraph G_5 of G which is neither spanning nor induced.

$\star$(7) Let $G = (V, E)$ be a graph and let $\mathcal{S}(G)$ be the collection of all subgraphs of G. On the set $\mathcal{S}(G)$, define a binary relation R as follows: if $H_1, H_2 \in \mathcal{S}(G)$ then $H_1 \, R \, H_2 \Leftrightarrow H_1$ is a subgraph of H_2. Is R reflexive? Symmetric? Antisymmetric? Transitive? Justify your answer.

(8) Let G be a simple graph of order n and size m.

$\star$ (a) How many *spanning* subgraphs G has?

(b) How many *vertex-induced* subgraphs G has?

(c) How many spanning subgraphs K_n has? (Use part (a))

(d) How many vertex-induced subgraphs K_n has? (Use part (b))

$\star$ (e) How many spanning subgraphs with exactly k edges K_n has?

(9) Let $G = (V, E)$ be a simple graph. A *clique* in G is any non-empty subset U of V such that the subgraph $G[U]$ of G induced by U is *complete*. In other words, $U \subseteq V$ is a clique if any two distinct vertices in U are neighbors. The size of a clique U is the cardinality of U. The *clique number* of G, denoted by $w(G)$, is defined to be the largest possible size of a clique in G. In each case, find the clique number of the given graph.

(a) $K_n,\ n \geq 1$

$\star$ (b) $K_{m,n},\ m, n \geq 1$

(c) $C_n,\ n \geq 3$

(d) $W_n,\ n \geq 3$

(10) (See Exercise (9) for the definition of a clique) For the graph G given below:

(a) Give three cliques of size 3 and draw their corresponding induced subgraphs in G.

$\star$ (b) Give three cliques of size 3 and draw their corresponding induced subgraphs in G.

(c) Give the clique number of G.

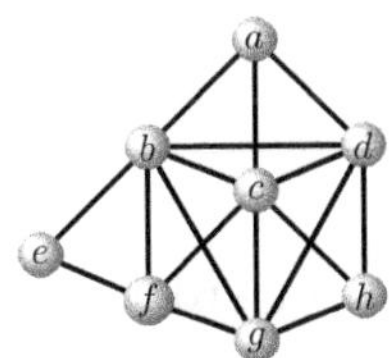

(11) For a graph $G = (V, E)$, a subset U of V is called *independent* if the subgraph $G[U]$ of G induced by U is edgeless. In other words, $U \subseteq V$ is independent if no two vertices in U are adjacent in G. The *size* of an independent set U is the cardinality of U. The *independence number* of G is the largest size of an independent set in G. For the graph in Exercise (9):

(a) Give three independent sets of size 2.

(b) Give an independent of size 3.

(c) Give the independence number of G.

$\star$(12) What is the independence number (see Exercise (11)) of the complete graph K_n?

(13) Consider the following graph G:

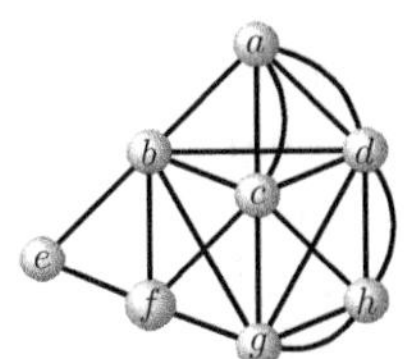

In each case, draw the complement of the given subgraph in G.

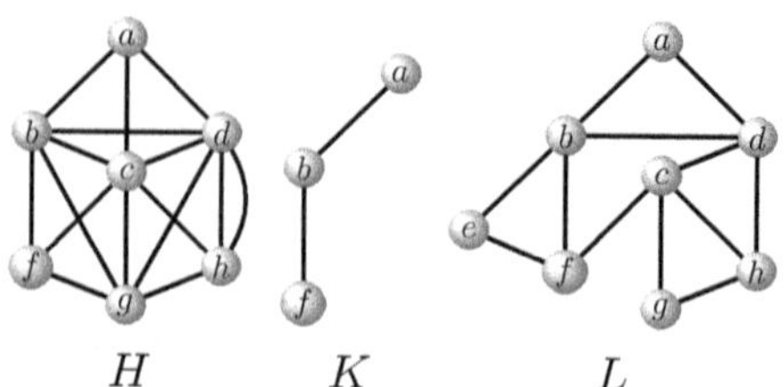

$$H \qquad K \qquad L$$

$\star$(14) In each case, draw the complement of the given simple graph.

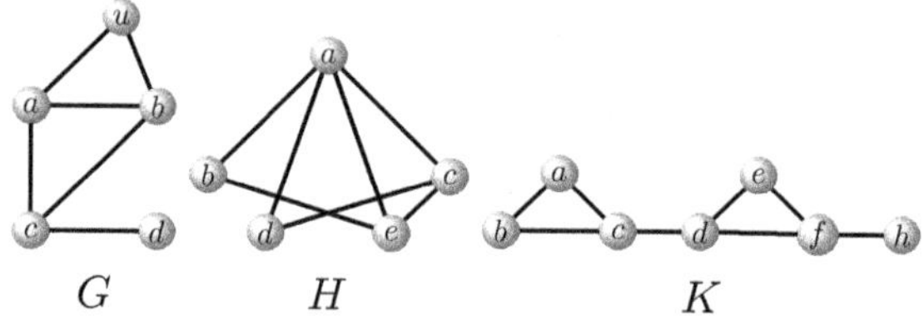

$$G \qquad H \qquad K$$

(15) Let G be a simple graph of order $n \geq 1$ and size k. Let $\overline{G}$ be the complement of G and let $v \in V$.

   (a) Prove that $\deg_G(v) + \deg_{\overline{G}}(v)$.

   (b) what is the size of $\overline{G}$ of G?

$\star$(16) Let $G_1 = (V_1, E_1)$ and $G_2 = (V_2, E_2)$ be two graphs. Prove that $G_1 \oplus G_2$ is of size 0 if and only if $E_1 = E_2$.

$\star$(17) Let $G_1 = (V_1, E_1)$ and $G_2 = (V_2, E_2)$ be two graphs with no vertices in common. The *join* of G_1 and G_2, denoted by $G_1 + G_2$, is the graph (V_3, E_3) where $V_3 = V_1 \cup V_2$ and $E_3 = E_1 \cup E_2 \cup F$ with $F = \{\{u, v\}; u \in V_1, v \in V_2\}$. So an edge in $G_1 + G_2$ is either an edge in G_1 or an edge in G_2 or an edge joining a vertex in G_1 with another in G_2.

   (a) Illustrate the graph $C_3 + C_4$ with a drawing.

   (b) What is the join of the complete graphs K_m and K_n?

   (c) Write the wheel W_n as the join of two simple graphs.

(18) Prove or disprove: The line graph of any loopless graph is always simple.

(19)(a) Let G be a simple graph and let $L(G)$ be its line graph. If a, b are two vertices of G with $\deg(a) = m$ and $\deg(b) = n$, what is the degree of the edge $e = \{a, b\}$ as a vertex of $L(G)$?

   (b) If G is a k-regular graph with $k \geq 1$, prove that the line graph of G is a $2(k-1)$-regular graph.

$\star$(20) Let G be a simple graph and let $L(G)$ be its line graph. If v is a vertex in G such that $\deg(v) = d \geq 1$, prove that v contributes with $\frac{d(d-1)}{2}$ edges to the set of edges of $L(G)$.

(21) Let $(d_1, d_2, \ldots, d_n)$ be the degree sequence of a simple graph G. Find:

    (a) The number of vertices of $L(G)$ in terms of $d_1, d_2, \ldots, d_n$;

    (b) The number of edges of $L(G)$ in terms of $d_1, d_2, \ldots, d_n$ (*Hint.* Use Exercise (20)).

(22) Give an example of a simple graph G of order 4 with: (a) $cl(G) \neq G$; (b) $cl(G) = G$.

$\star$(23) Draw the Hamiltonian closure of the following graph:

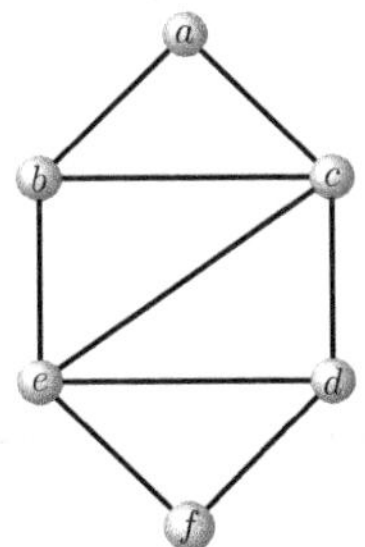

9.5   Walks, trails, paths, cycles and graph connectivity

Our first illustration of a graph in this chapter was about a network of regions connected by bridges. As graphs are often used to represent a network of some sort, it is natural to investigate ways to move from one vertex to another in the graph.

Definition 9.16. Let $G = (V, E)$ be a graph with incidence function ϕ. For vertices $v_0, v_k \in V$ (not necessarily distinct) and $k \in \mathbb{N}$, a $v_0 v_k$-*walk* (or a *walk* between v_0 and v_k) in G is a finite alternating sequence of the form: $W = v_0 e_1 v_1 e_2 v_2 \ldots e_{k-1} v_{k-1} e_k v_k$ where each v_i is a vertex, each e_i is an edge and $\phi(e_i) = v_{i-1} v_i$ for all $i = 1, 2, \ldots, k$ (so e_i is an edge with endpoints v_{i-1} and v_i). Vertices v_0 and v_k are called the *initial* and the *terminal* points of walk W, respectively. Vertices $v_1, \ldots, v_{k-1}$ are called the *internal* vertices of the walk. The *length* of a walk is the number of edges used in the walk, including multiplicities. A walk of length 0 consists of just one vertex and no edges and it is called *trivial*. Unless otherwise specified, all walks considered are assumed non-trivial.

Example 9.27. In the graph G given in the following diagram:

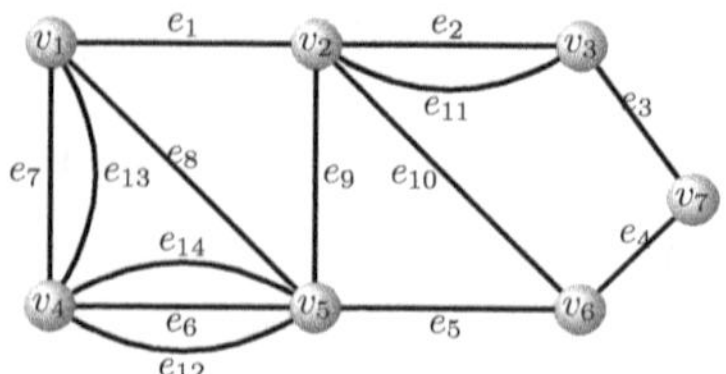

- $W_1 : v_1 e_7 v_4 e_6 v_5 e_{12} v_4 e_{14} v_5 e_9 v_2$ is a $v_1 v_2$-walk of length 5:

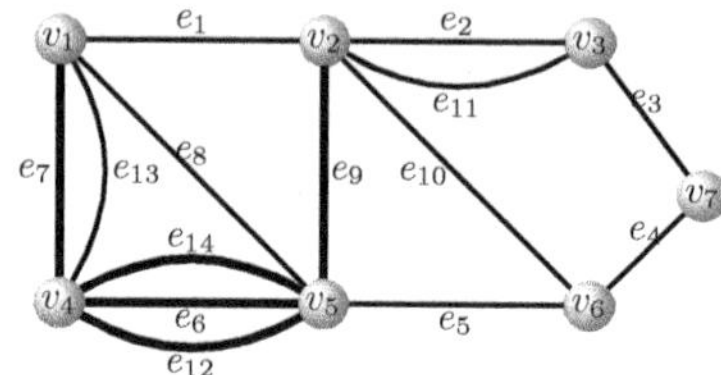

- $W_2 : v_2 e_9 v_5 e_5 v_6 e_4 v_7 e_3 v_3 e_2 v_2$ is a $v_2 v_2$-walk of length 5:

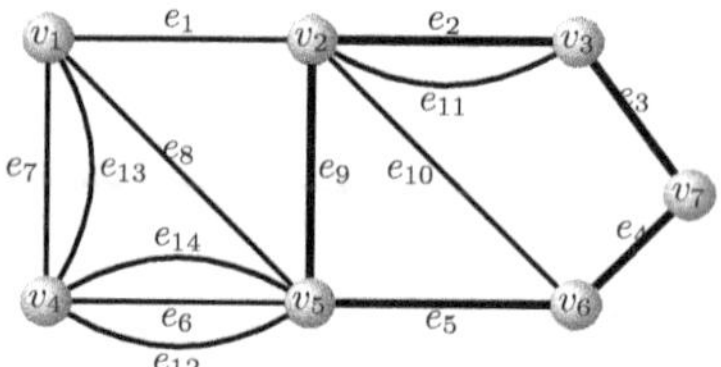

- $W_3 : v_5 e_8 v_1 e_8 v_5 e_{12} v_4 e_{13} v_1 e_7 v_4 e_6 v_5 e_5 v_6 e_4 v_7$ is a $v_1 v_7$-walk of length 8:

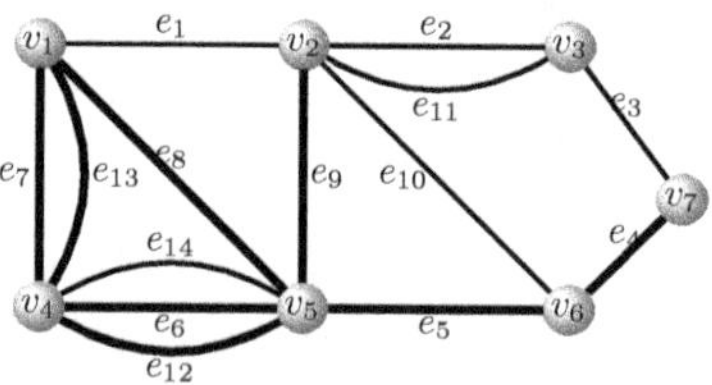

Remark 9.5. In a simple graph, a walk $W = v_0 e_1 v_1 e_2 v_2 \ldots e_{k-1} v_{k-1} e_k v_k$ is completely determined by its set of vertices and can be written as $W = v_0 v_1 v_2 \ldots v_{k-1} v_k$. The reason is obvious: since the graph is simple, it does not have parallel edges and so there is no ambiguity of which edge to use when a pair of consecutive vertices appears in the walk.

Definition 9.17. A $v_0 v_k$-walk $W = v_0 e_1 v_1 e_2 v_2 \ldots e_{k-1} v_{k-1} e_k v_k$ in a graph G is called:

- *closed* if $v_0 = v_k$, and *open* otherwise;
- a *trail* if no edge is repeated in G (so the edges are pairwise distinct);
- a *circuit* if it is a closed trail;
- a *path* if it has no repeated vertices (so the vertices are pairwise distinct). A trivial path consists of just one vertex;
- a *cycle* if it is a closed path (no repeated internal vertices).

Note that in a path, no vertices are repeated and consequently no edges are repeated. In particular, every path is a trail. It is also clear from the definition that every cycle is a circuit.

Example 9.28. For each of the three walks W_1, W_2 and W_3 of Example 9.27 above, the following table indicates whether the walk is closed, a trail, a circuit, a path or a cycle. We use the symbol $\checkmark$ to indicate that the property is satisfied and $\times$ if it is not.

	Closed	Trail	Circuit	Path	Cycle
W_1	$\times$	$\checkmark$	$\times$	$\times$	$\times$
W_2	$\checkmark$	$\checkmark$	$\checkmark$	$\times$	$\checkmark$
W_3	$\times$	$\times$	$\times$	$\times$	$\times$

9.5.1 *Graph connectivity*

Definition 9.18. Let $G = (V, E)$ be a graph and let $u, v \in V$. We say that vertex u is connected to vertex v if there exists a walk between u and v. A vertex v is assumed to be connected to itself by the trivial walk. The graph G is called *connected* if any two of its vertices are connected. A graph that is not connected is called *disconnected*.

The condition of having a *walk* between two distinct vertices in the above definition can be replaced with the more "suitable" condition of having a *path* between the two vertices as shown by the following theorem.

Theorem 9.6. *Let x and y be two distinct vertices in a graph G. Then there exists a walk between x and y in G if and only if there exists a path between x and y in G.*

Proof. One direction is easy: any path between x and y in G is in particular a walk between x and y in G. Conversely, assume that there exists a walk W between x and y given by the sequence $v_0 e_1 v_1 e_2 v_2 \ldots e_{k-1} v_{k-1} e_k v_k$ with $x = v_0$ and $y = v_k$. If none of the vertices

$v_0, v_1, \ldots, v_{k-1}, v_k$ appears more than once in W, then W is already a path and the proof is done. If one of the vertices appears more than once, then we can choose indices i and j such that $i < j$ and $v_i = v_j$. Remove all vertices $v_i, v_{i+1}, \ldots, v_{j-1}$ together with the edges between these vertices in W. We obtain this way an xy-walk W' having fewer vertices and edges than W. If W' has no repeated vertices then it is a path between x and y and we are done. If W' has repeated vertices, we repeat the same procedure as above. Since the number of vertices is finite, this process has to end and when it does we end up with a path between x and y in G. $\square$

In light of Theorem 9.6, Definition 9.18 can be stated as follows: two vertices u and v of a graph G are connected if there exists a uv-path in G and the graph is called *connected* if there exists a path between any two of its vertices.

Example 9.29. In the diagram below, the graph G is disconnected. For instance, there is no path between vertices v_1 and v_2 in G. Graph H, on the other hand, is connected as any two vertices are connected by a path.

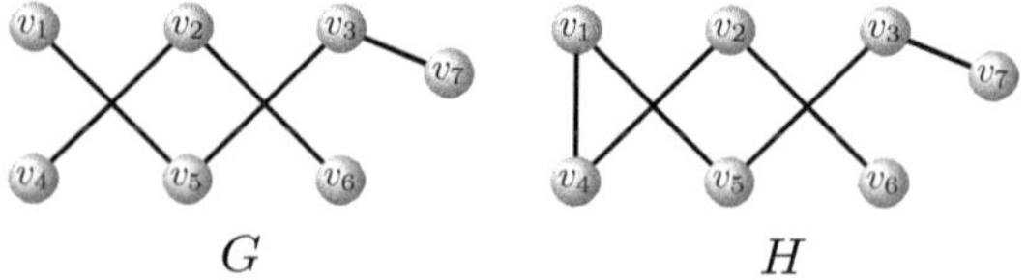

Definition 9.19. Let $G = (V, E)$ be a graph. For any vertex $v \in V$, let $C(v)$ be the set of all vertices of G connected to v by a path in G. The subgraph $G[C(v)]$ of G induced by $C(v)$ is called the connected component of G containing v.

Since we assumed that every vertex u is connected to itself via the trivial path, $u \in C(u)$ and consequently u belongs to the connected component $G[C(u)]$.

Theorem 9.7. *Let $G = (V, E)$ be a graph and let $u, v \in V$.*

(a) There exists a uv-path in G if and only if $G[C(u)] = G[C(v)]$.

(b) A connected component of G is a maximal connected subgraph of G. That is, any connected component is a connected subgraph of G which is not a proper subgraph of any other connected subgraph of G.

Proof. Part (a) is left as an exercise (see Exercise (9) below). For part (b), let γ be a connected component, and write $\gamma = G[C(w)]$ for some $w \in V$. If

u, v are two vertices in γ, then u, w are connected and v, w are connected. Therefore, u, v are connected. This proves that γ is a connected graph. Let κ be a connected subgraph of G containing γ as a subgraph. Then every vertex of κ belongs to $C(w)$ (and hence to γ) since κ is connected. This shows that γ and κ have the same set of vertices. On the other hand, every edge $e = \{u, v\}$ of κ connects two vertices of $C(w)$ and therefore must be an edge of $G[C(w)] = \gamma$ by definition of an induced subgraph. We conclude that $\gamma = \kappa$ as they have the same set of vertices and the same set of edges. $\qquad\square$

The number of connected components of a graph G is denoted by $\omega(G)$. For example, for the graph G of Example 9.29, $\omega(G) = 2$ as G has two connected subgraphs, one with vertex set $\{v_1, v_3, v_5, v_7\}$ and another with vertex set $\{v_2, v_4, v_6\}$ as illustrated in the following diagram.

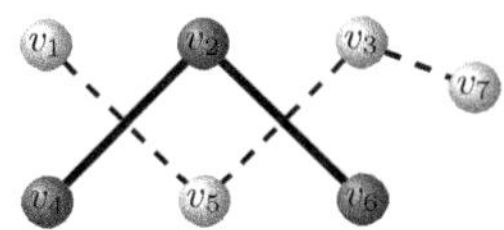

Example 9.30. The following diagram shows a graph G and its four connected components.

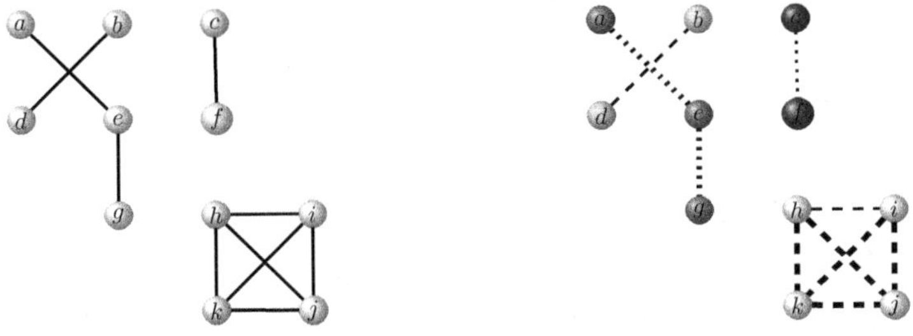

Graph G The four connected components of G

In what follows we give some useful remarks.

(1) A graph G is connected if and only if $\omega(G) = 1$. A connected graph has a unique connected component, namely itself.

(2) If γ is a connected component of a graph G, then adding a new vertex or a new edge of G to γ would result in a disconnected subgraph of G. That follows from the maximality of a connected component.

(3) Any graph G can be written as $G = \gamma_1 \cup \cdots \cup \gamma_n$ where $\gamma_1, \ldots, \gamma_n$ are the connected components of G and such that no vertex in γ_i is connected to any vertex in γ_j by an edge in G for $i \neq j$.

Definition 9.20. Let $G = (V, E)$ be a graph and let $u, v \in V$. The *distance* between u and v, denoted by $\mathrm{dis}(u, v)$, is defined as the smallest length of a path between u and v. If there is no path between the two vertices, then the distance is defined (conventionally) as being ∞. If vertices u, v are adjacent, then clearly $\mathrm{dis}(u, v) = 1$. The *eccentricity* of a vertex v, denoted by $\epsilon(v)$, is the largest distance between v and another vertex of G. The *diameter* of G, denoted by $\mathrm{diam}(G)$, is the maximum eccentricity of any vertex in the graph. The diameter of G is defined to be ∞ if there exists a pair of vertices (u, v) in G with $\mathrm{dis}(u, v) = \infty$.

Example 9.31. For the following graph G:

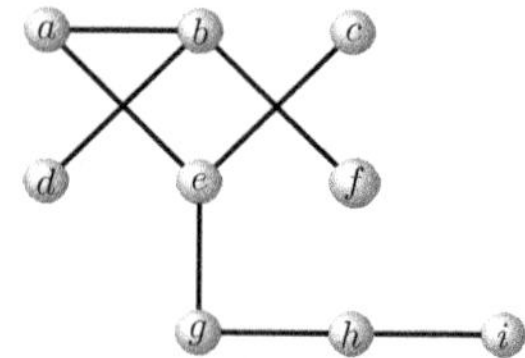

$\mathrm{dis}(a, d) = 2$, $\mathrm{dis}(a, i) = 4$, $\mathrm{dis}(a, e) = 1$, $\mathrm{dis}(c, i) = 4$, $\mathrm{dis}(b, f) = 1$, $\mathrm{dis}(f, i) = 6$, $\epsilon(a) = 4$, $\epsilon(f) = 6$ and the diameter of the graph is 6.

9.5.2   *Vertex-connectivity, edge-connectivity*

Graph connectivity comes comes in different forms. Some graphs can loose their connectivity by simply removing a vertex or an edge while others are more "rigid" when it comes to preserving their connectivity.

Example 9.32. Consider the following three connected graphs:

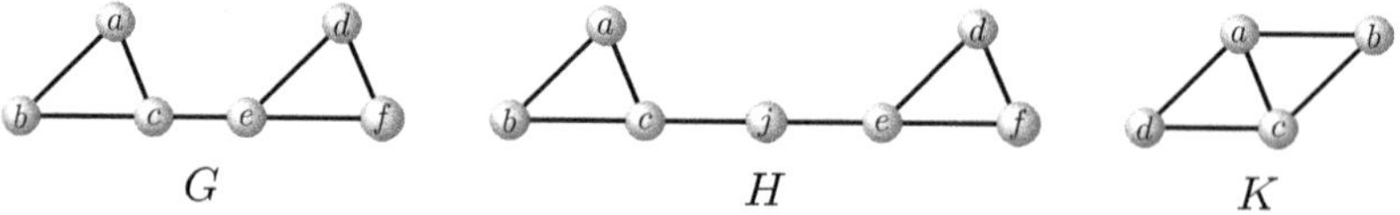

Removing edge $\{c, e\}$ from graph G would result in a disconnected graph. The same is true if vertex j is removed from graph H together with its incident edges. Removing a single edge or a single vertex from graph K does not affect the connectivity of the graph.

Definition 9.21. Let $G = (V, E)$ be a non-trivial graph.

(a) A proper subset U of V is called a *vertex-cut* of G if $G - U$ is either disconnected or equal to the trivial graph. If $U = \{v\}$ is a vertex-cut of G then the vertex v is called a *cut vertex* (or *articulation point*).

(b) A subset F of edges in G is called an *edge-cut* if $G - F$ is either disconnected or equal to the trivial graph. If $\{e\}$ is an edge-cut, then the edge e is called a *cut edge* (or a *bridge*).

Example 9.33. Consider the three graphs of Example 9.32. For graph G, vertex c is a cut vertex and edge $\{c, e\}$ is a bridge. For graph H, vertices c, j and e are cut vertices and edges $\{c, j\}$ and $\{j, e\}$ are bridges. Graph K has $U = \{a, c\}$ as a vertex-cut and $F = \{\{a, b\}, \{a, c\}, \{a, d\}\}$ is an edge-cut but has neither cut vertices nor cut edges.

Remark 9.6. One can define the notion of a cut vertex and a cut edge as follows: a vertex v (respectively an edge e) of a graph G is called a cut vertex (respectively a cut edge) if $G - v$ (respectively, $G - e$) has more connected components than G. In general, a vertex v (respectively, an edge e) of G is a cut vertex (respectively a cut edge) if its removal disconnects a connected component of G or reduces G to a trivial graph. Thus, we can restrict ourselves to connected graphs when we study vertex or edge cuts.

Theorem 9.8. *Let $G = (V, E)$ be a connected graph of order $n \geq 2$.*

(a) G has a vertex-cut if and only if G is not the complete graph K_n.
(b) G has an edge-cut.

Proof.

(a) Assume that G has a vertex-cut U. If G is the complete graph K_n, then $G - U$ is another complete graph (hence connected): a contradiction. So $G \neq K_n$. Conversely, if $G \neq K_n$, then there exists a pair (u, v) of non-adjacent vertices in G. The set $U = V \backslash \{u, v\}$ is a vertex-cut of G since $G - U$ is disconnected.
(b) Since G is connected with at least two vertices, we can choose a vertex v in G of positive degree. The set of all edges incident to v is clearly an edge-cut of the graph. $\qquad\square$

Definition 9.22. Let G be a simple connected graph of order $n \geq 2$.

(a) If $G \neq K_n$, the *vertex-connectivity* of G, denoted by $\kappa(G)$, is defined as the minimal cardinality of a vertex-cut of G (we know that a vertex-cut exists in this case by Theorem 9.8). If $G = K_n$, then $\kappa(G)$ is defined to be $n - 1$. In other words, $\kappa(G)$ is the *smallest* number of vertices of G whose removal (together with their incident edges) results in a disconnected graph or the trivial graph K_1. For an integer $m \geq 0$, we say that G is m-vertex-connected if $\kappa(G) \geq m$.

(b) The *edge-connectivity* of G, denoted by $\kappa'(G)$, is the minimal cardinality of an edge-cut of G. So, $\kappa'(G)$ is the smallest number of edges of G whose removal results in a disconnected graph or the trivial graph. For a an integer $m \geq 0$, we say that G is m-edge-connected if $\kappa'(G) \geq m$.

We define $\kappa(K_1) = \kappa'(K_1) = 0$.

Example 9.34. Find the vertex-connectivity and the edge-connectivity of each of the following three graphs.

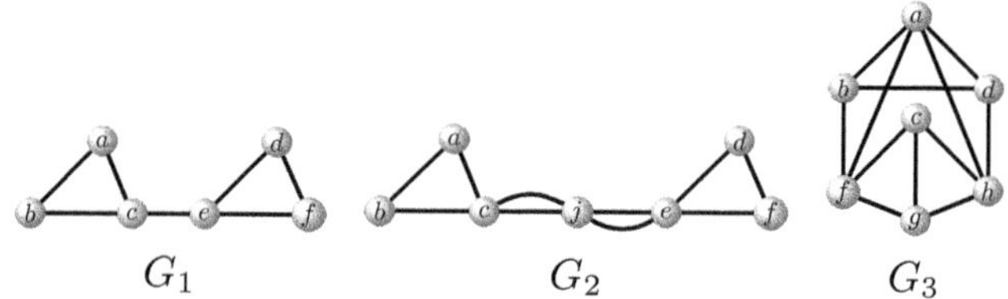

$$G_1 \qquad\qquad\qquad G_2 \qquad\qquad\qquad G_3$$

Solution. Removing vertex c (or e) from G_1 disconnects the graph. Also, removing edge $\{c, e\}$ from G_1 disconnects G_1. We conclude that $\kappa(G_1) = \kappa'(G_1) = 1$. Removing vertex j from G_2 disconnects the graph, so $\kappa(G_2) = 1$. Removing a single edge from G_2 does not change its connectivity, but removing two edges (for instance the parallel edges between vertices c and j) disconnects the graph. So $\kappa'(G_2) = 2$. For G_3, it is clear that removing a single vertex or a single edge will not disconnect the graph, so $\kappa(G_3) \geq 2$ and $\kappa'(G_3) \geq 2$. Removing vertices f and h from G_3 results in a disconnected graph. So $\kappa(G_3) = 2$. A quick inspection shows that it is impossible to disconnect G_3 by removing only two edges. It takes removing at least three edges to disconnect G_3. For instance, removing all edges incident to vertex b will do the job as this leaves an isolated vertex. We conclude that $\kappa'(G_3) = 3$. $\diamond$

Theorem 9.9. *For every integer $n \geq 1$, $\kappa'(K_n) = n - 1$.*

Proof. We may assume that $n \geq 2$ since the result is true for $n = 1$ ($\kappa'(K_1) = 0$ by definition). Let F be a minimal edge-cut of K_n. Then $|F| = \kappa'(K_n)$ and $K_n - F$ consists of two connected components C_1 and C_2 (if we "return" an edge from F back in $K_n - F$, the new graph becomes connected and a single edge can only connect two components). Let k be the order of C_1, then C_2 has order $n - k$. The edges in F are precisely those edges $\{u, v\}$ of K_n with u in C_1 and v in C_2. There are $k(n-k)$ such edges. Now, $(k-1)(n-k-1) = k(n-k) - n + 1$ and since $k - 1 \geq 0$ and $n - k - 1 \geq 0$, $k(n-k) - n + 1 \geq 0$ and so $|F| = \kappa'(K_n) = k(n-k) \geq n - 1$. On the other hand, removing the $n - 1$ edges incident to a vertex v of K_n would isolate v and disconnect the graph, so $\kappa'(K_n) \leq n - 1$. We conclude that $\kappa'(K_n) = n - 1$. $\qquad\square$

The following is a characterization of a cut vertex in a connected graph.

Theorem 9.10. *Let $G = (V, E)$ be a connected graph of order $n \geq 3$. A vertex v in G is a cut vertex if and only if there exist two vertices u, w such that $u \neq v$, $w \neq v$ and v belongs to every uw-path in G.*

Proof. If v is cut vertex of G then $G - v$ is disconnected. Let u, w be vertices in two different connected components C_1 and C_2 of $G - v$. Any uw-path in G must go through vertex v since otherwise C_1 and C_2 would be connected. Conversely, assume that u, w are two vertices of G with $u, w \neq v$ such that v belongs to every uw-path in G. Then $G - v$ is disconnected since there is no uw-path in $G - v$ and so v is a cut vertex of G. $\qquad\square$

There is a similar characterization of a bridge (cut edge) in a connected graph.

Theorem 9.11. *Let $G = (V, E)$ be a connected graph. An edge e is a bridge of G if and only if it is not part of a cycle (closed path) in the graph.*

Proof. Suppose that $e = \{u, v\}$ is a bridge of G. Then $G - e$ is disconnected and the endpoints u and v of e are on two different components of $G - e$. Since G is connected, there exists a path p joining u and v in G and p must contain edge e. If e lies on a cycle γ in G then we can delete e from p and replace it with a part of cycle γ to obtain a path joining u and v in $G - e$: a contradiction. Conversely, assume that edge $e = \{u, v\}$ does not belong to any cycle in G. If e is not a bridge, then $G - e$ is connected and there must be a path p between u and v in $G - e$. Since e is not an edge of $G - e$, it is not part of the path p. Adding edge e to p creates a cycle in G containing e: a contradiction. So e is a bridge. $\qquad\square$

For a graph $G = (V, E)$, the set $\{\deg(v); v \in V\}$ is a non-empty subset of $\mathbb{N}$ and so it has a least element that we denote by $\delta(G)$. The following theorem gives the relationships between $\kappa(G)$, $\kappa'(G)$ and $\delta(G)$.

Theorem 9.12. *For a graph $G = (V, E)$, $\kappa(G) \leq \kappa'(G) \leq \delta(G)$.*

Proof.   Let $v \in V$ be such that $\deg(v) = \delta$. Removing v together with all δ edges incident to v would clearly isolate v and disconnect G. This proves that $\kappa'(G) \leq \delta(G)$. Let F be an edge-cut of G of minimal cardinality. Then $G - F$ is disconnected and the vertex set V of G is partitioned into two subsets V_1 and V_2 such that every edge of F is of the form $\{u, v\}$ with $u \in V_1$ and $v \in V_2$. For $i = 1, 2$, let U_i be the set of all vertices in V_i which are incident with some edge in F. Removing the vertices of either U_1 or U_2 disconnects G. We conclude that $\kappa(G) \leq |F| = \kappa'(G)$. $\qquad\qquad\square$

9.5.3 *Exercises*

(1) Let G be a connected graph. In each case, determine if the statement is true or false. Justify your answer.

   (a) Every trail is a walk.
   (b) Every closed walk is a cycle.
$\star$ (c) Every cycle is a path.
   (d) Every cycle is a trail.
$\star$ (e) There are trails that are not paths.
   (f) A walk with no repeated vertices is a path.
$\star$ (g) If G has no bridges, then G has no articulation points.

(2) For the Petersen graph:

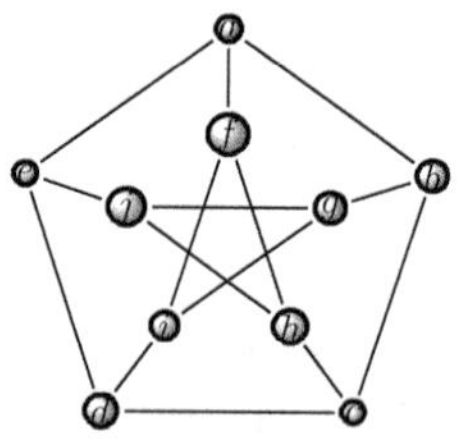

$\star$ (a) Find a path of length k for $k = 5, 6, 8$ and 9.
   (b) Find a cycle of length k for $k = 5, 6, 8$ and 9
$\star$ (c) Find each of the following: $\mathrm{dis}(a, f)$, $\mathrm{dis}(a, g)$, $\mathrm{dis}(b, i)$, $\mathrm{dis}(c, e)$, $\mathrm{dis}(h, i)$, $\mathrm{dis}(d, j)$ and $\mathrm{dis}(a, g)$.
   (d) What is the diameter of the graph?

(3) Determine if the graph G given below is connected. If you say it is not,

determine the number of its connected components.

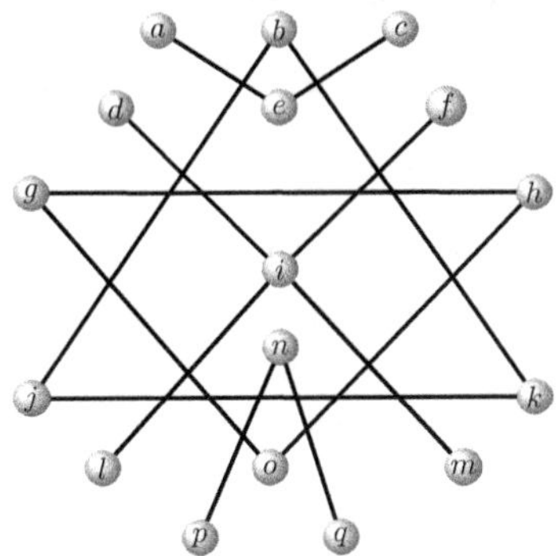

(4) In each case, draw a connected graph G different from the complete graph K_n and of order $n \geq 4$ satisfying the given conditions.

   (a) $\kappa(G) = \kappa'(G) = \delta(G)$.

$\star$ (b) $\kappa(G) = \kappa'(G) < \delta(G)$.

   (c) $\kappa(G) < \kappa'(G) = \delta(G)$.

$\star$ (d) $\kappa(G) < \kappa'(G) < \delta(G)$.

(5) Draw a graph of order 9 such that $\kappa(G) = \kappa'(G) = \delta(G) = 1$.

(6) Let G be a d-regular simple graph for some non-negative integer d (so $\deg(v) = d$ for every vertex v of G). Prove that if $d \leq 2$, then $\kappa(G) = \kappa'(G)$.

(7) Find the vertex-connectivity and the edge-connectivity of each of the following graphs.

$\star$ (a) P_n $(n \geq 2)$

$\star$ (b) C_n $(n \geq 4)$

$\star$ (c) W_n $(n \geq 4)$

   (d) $K_{3,3}$

   (e) Q_3 (see Exercise (5) of Section 9.2.3 for the definition of Q_n).

(8) Determine the vertex-connectivity and the edge-connectivity of each of the following graphs.

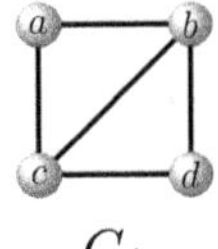
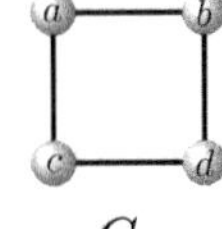
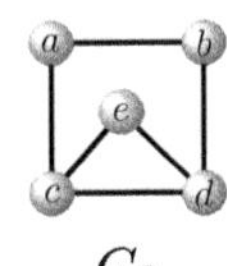
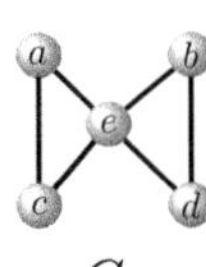

G_1 G_2 G_3 G_4

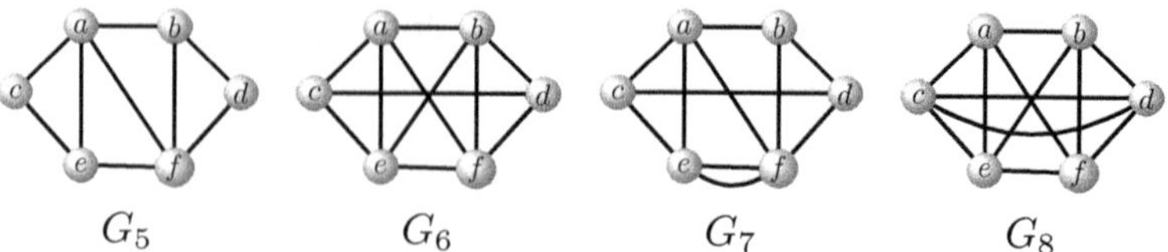

$$G_5 \qquad\qquad G_6 \qquad\qquad G_7 \qquad\qquad G_8$$

(9) Let $G = (V, E)$ be a graph and let $u, v \in V$. Prove that there exists a uv-path in G if and only if $G[C(u)] = G[C(v)]$.

$\star$(10) Let $G = (V, E)$ be a graph. Consider the binary relation R on V defined by $u\,R\,v$ if and only there exists a uv-path in G.

    (a) Prove that R is an equivalence relation on V.

    (b) Let X be an be an equivalence class of R. Prove that the subgraph $G[X]$ induced by X is a connected component of G.

$\star$(11) Let $G = (V, E)$ be a connected graph. Prove that if $\deg(v)$ is even for every $v \in V$, then G has no bridge.

(12) Prove that if G is a non-trivial connected graph, then its line graph $L(G)$ is also connected.

(13)(a) Let G be a connected graph of order $n \geq 1$. Use a proof by induction on n to show that G must have at least $n - 1$ edges.

    (b) Let G be a graph of order 9 with 5 vertices of degree 2 and 4 of degree 1. Is G connected?

(14) Prove that a graph $G = (V, E)$ is disconnected if and only if V can be written as the union of two non-empty subsets V_1 and V_2 such that no edge in G has one endpoint in V_1 and another in V_2.

$\star$(15) Let $G = (V, E)$ be a connected graph in which there is a unique path between any pair of vertices. If $v \in V$ is such that $\deg(v) \geq 2$, prove that $G - v$ is disconnected.

(16) Given a connected graph G and two paths p_1 and p_2 of G of maximal length l (so any path of G has length less than or equal to l), prove that p_1 and p_2 must have at least one vertex in common.

$\star$(17) Let G be a simple graph. Show that either G or its complement $\overline{G}$ is connected.

(18) Let $G = (V, E)$ be a simple graph of order $n \geq 1$.

    (a) What is the maximum number of edges in G?

    (b) If G is not connected, what is the maximum number of edges in G? (*Hint.* One way to solve this problem is to consider the complement $\overline{G}$ of G and use Exercises (13) and (17).)

9.6 Bipartite graphs

The complete bipartite graph $K_{m,n}$ we encountered earlier is a particular member of a larger family of graphs that we study in this section.

Definition 9.23. A graph $G = (V, E)$ is called **bipartite** if the vertex set V can be partitioned into two non-empty subsets V_1 and V_2 (so $V_1 \cap V_2 = \emptyset$ and $V_1 \cup V_2 = V$) such that every edge of G joins a vertex of V_1 and a vertex of V_2 (so no two edges of either V_1 and V_2 are adjacent). The set $\{V_1, V_2\}$ is called a **bipartition** of G, and the subsets V_1 and V_2 are called the **partite sets** of G.

An immediate consequence of Definition 9.23 is that there are no loops in a bipartite graph.

Example 9.35. The complete bipartite graph $K_{m,n}$ is bipartite in the sense of Definition 9.23. Subsets $\{x_1, x_2, \ldots, x_m\}$ and $\{y_1, y_2, \ldots, y_n\}$ (see the definition of $K_{m,n}$ given in Section 9.2.1 above) are the two partite sets of $K_{m,n}$.

Example 9.36. Consider the graph G given in the diagram below. Let $V_1 = \{v_1, v_3\}$ and $V_2 = \{v_2, v_4\}$. Every edge of G joins a vertex of V_1 to a vertex of V_2. Moreover, $\{V_1, V_2\}$ is clearly a partition of $V = \{v_1, v_2, v_3, v_4\}$. So G is bipartite and $\{V_1, V_2\}$ is a bipartition of G.

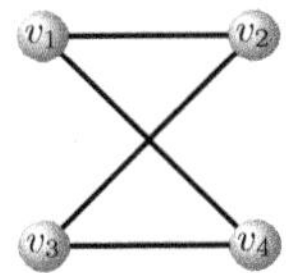

Example 9.37. The cycle C_3 is not bipartite. To see this, let $V = \{v_1, v_2, v_3\}$ be the set of vertices of C_3 as shown in the diagram.

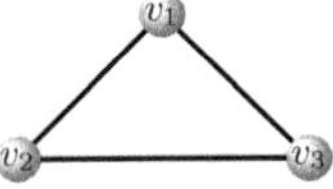

Assume that C_3 is bipartite and let $\{V_1, V_2\}$ be a bipartition of G. Without loss of generality, we may assume that $V_1 = \{v_1\}$ and $V_2 = \{v_2, v_3\}$. This contradicts the definition of bipartite graph since the edge $\{v_2, v_3\}$ joins two vertices of V_2. We conclude that C_3 is not bipartite.

Example 9.38. The cycle graph C_6 is bipartite. To see this, let $V = \{v_1, \ldots, v_6\}$ be the vertex set of C_6 as shown in the following diagram.

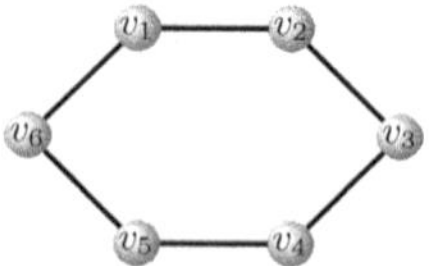

Then the subsets $V_1 = \{v_1, v_3, v_5\}$ and $V_2 = \{v_2, v_4, v_6\}$ form a partition of V such that every edge of G joins one vertex of V_1 to one vertex of V_2. We conclude that G is bipartite and that $\{V_1, V_2\}$ is a bipartition of G.

Example 9.39. We prove that the cycle C_7 is not bipartite. Let $\{v_1, \ldots, v_7\}$ be the vertex set of C_7. Assume by contradiction that C_7 is bipartite with bipartition $\{V_1, V_2\}$. Without loss of generality, we can assume that vertex v_1 belongs to V_1. Then vertex v_2 has to be in V_2 (since $\{v_1, v_2\}$ is an edge of C_7), vertex v_3 into V_1 again, and so on. Vertex v_7 ends up in subset V_1. This is a contradiction since v_1 and v_2 are adjacent. We conclude that C_7 is not bipartite.

There is nothing particular about C_3 and C_7 that prevents them from being bipartite except the fact that 3 and 7 are odd integers. The proof that C_7 is not bipartite given in the previous example can be generalized to any cycle C_n where $n \geq 3$ is odd (see Exercise (4) below). As we will see in the next theorem, the existence of an odd cycle in a graph is exactly what prevents the graph from being bipartite.

Theorem 9.13. *A non-trivial graph of order at least 2 is bipartite if and only if it contains no cycle of odd length.*

Proof.   Assume that G is a bipartite graph with bipartition $\{V_1, V_2\}$ and let $\gamma = v_1 v_2 \ldots v_n v_1$ be a cycle of G of length $n \geq 1$. Without loss of generality, we may assume that $v_1 \in V_1$. So, $v_2 \in V_2$, $v_3 \in V_1$, and so on. In general, $v_i \in V_1$ if i is odd and $v_i \in V_2$ if i is even. If n is odd, then $v_n \in V_1$ and so v_1 is in V_2, which is a contradiction. We conclude that every cycle of G must have an even length. Conversely, assume that every cycle of G has an even length. We need to show that G is bipartite. Assume first that G is connected, and fix a vertex v of G. We show that $\{V_1, V_2\}$ is a bipartition of G where:

- $V_1 = \{u \in V; \text{the shortest path from } v \text{ to } u \text{ has an even length}\}$;
- $V_2 = \{w \in V; \text{the shortest path from } v \text{ to } w \text{ has an odd length}\}$.

Since $v \in V_1$, $V_1 \neq \emptyset$. Since G is assumed to be connected with at least two vertices, there exists an edge $\{v, w\}$ in G for some $w \in V$ different from v. In particular, $w \in V_2$ and $V_2 \neq \emptyset$. If $t \in V$ is arbitrary, then there exists a path connecting v and t (because G is connected) and the shortest path between v and t is either of even or odd length. This shows that $V = V_1 \cup V_2$. Clearly, $V_1 \cap V_2 = \emptyset$. Next, let $u, u' \in V_1$ and assume by contradiction that they are adjacent. Let $\gamma_1 = u_1 u_2 \ldots u_{2k+1}$ and $\gamma_2 = w_1 w_2 \ldots w_{2l+1}$ be two paths of shortest length from v to u and from v to u' respectively (so $u_1 = w_1 = v$, $u_{2k+1} = u$, $w_{2l+1} = u'$ and both γ_1, γ_2 have even lengths). Let z be the last (i.e., the farthest from v) vertex that γ_1 and γ_2 have in common. The path $\gamma_{12} = u_1 u_2 \ldots z$ is one of shortest length from v to z and the same is true for the path $\gamma_{22} = w_1 w_2 \ldots z$. We conclude that γ_{12} and γ_{22} have the same length and so there exists i such that $u_i = w_i = z$. If i is even, then it easy to check that the cycle $\epsilon = u_i u_{i+1} \ldots u_{2k+1} w_{2l+1} w_{2l} \ldots w_i$ has an odd length (its length is of the form "odd $+1+$odd"). We arrive at the same conclusion if i is odd. We have then constructed a cycle of odd length in G, a contradiction to our assumption. This contradiction is a result of our assumption that two vertices in V_1 are adjacent. We conclude that there is no edge of G connecting two vertices in V_1. A similar argument shows that the same is true for vertices in V_2. This shows that $\{V_1, V_2\}$ is bipartition of G and the graph is bipartite if it is connected. Now let G be any graph (not necessarily connected) with no cycles of odd length. Let $G_1, G_2, \ldots, G_m$ be the connected non-trivial components of G and let V_0 be the set of all isolated vertices of G. For each i, G_i is a non-trivial connected graph with no cycles of odd length (otherwise G would have one). By the previous discussion, G_i is bipartite for each $i = 1, \ldots, m$. For each i, let $\{V_{i1}, V_{i2}\}$ be a bipartition of G_i. It is not hard to see that $\{V_1, V_2\}$ is a bipartition of G where $V_1 = V_{11} \cup V_{21} \cup \cdots \cup V_{m1} \cup V_0$ and $V_2 = V_{12} \cup V_{22} \cup \cdots \cup V_{m2}$ (the details are left to the reader as an easy exercise). The graph is then bipartite. $\qquad\square$

Corollary 9.2. *Every non-empty subgraph of a bipartite graph is bipartite.*

Proof. Left as an exercise (see Exercise (2) below). $\qquad\square$

Example 9.40. The following graph is not bipartite. It contains a cycle of length 5 traced in bold in the diagram.

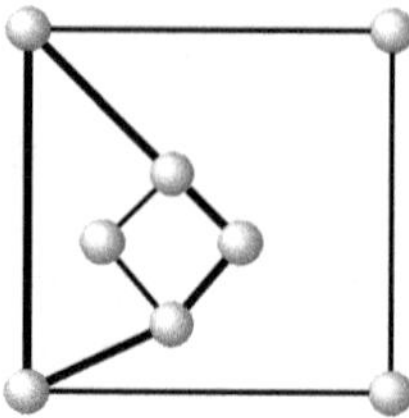

Example 9.41. The following graph is bipartite since it is not possible to trace a cycle of odd length in it (of course, this is not a formal proof that the graph is bipartite, it is just an observation).

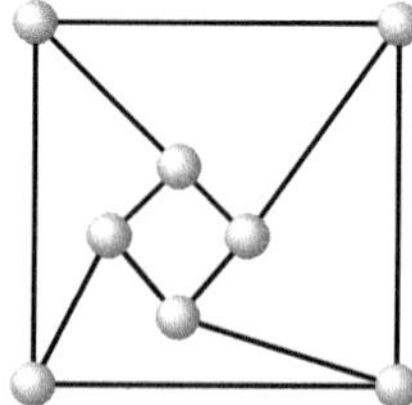

Example 9.42. The Petersen graph is not bipartite as it clearly contains cycles of odd length.

Theorem 9.13 is useful to determine if a graph is not bipartite by locating a cycle of odd length. It gives a characterization of being bipartite in terms of a structural property of the graph. However, the theorem is not practical to actually prove that a graph is bipartite as it requires checking the length of every cycle in the graph. There is another algorithmic approach to check if a graph is bipartite and if it is, to produce a bipartition. First a definition.

Definition 9.24. We say that a graph G has a *proper 2-vertex-coloring* if the vertices of G can be colored using only 2 colors such that the endpoints of each edge of G receive different colors.

Theorem 9.14. *A graph $G = (V, E)$ is bipartite if and only if it has a proper 2-vertex-coloring.*

Proof.   Assume G is bipartite with bipartition $\{V_1, V_2\}$. Color the vertices in V_1 using one color and the vertices in V_2 using a second color. Since each edge has one endpoint in V_1 and another in V_2, the endpoints of each edge receive different colors. The graph has a proper 2-vertex-coloring.

Conversely, assume that G has a proper 2-vertex-coloring and suppose the two colors used are red and blue. Let V_1, V_2 be the subsets of V colored red and blue, respectively. Since the endpoints of each edge receive different colors, each edge must have one endpoint in V_1 and another in V_2. We conclude that G is bipartite with bipartition $\{V_1, V_2\}$. $\square$

This suggests the following algorithm, known as the *two-colors algorithm*, to determine if a graph $G = (V, E)$ is bipartite. The proof of the correctness of the algorithm is omitted but it naturally follows from Theorem 9.14. We can assume that the graph is connected and non-trivial since the algorithm can be applied to each connected component of G.

Step 1. Choose an arbitrary vertex v of G and assign color 1 to it.

Step 2. Assign color 2 to all neighbors of v.

Step 3. Choose an arbitrary vertex of color 2 and assign color 1 to all its uncolored neighbors (if such neighbors exist).

Step 4. Repeat with a vertex of color 1, and so on.

Step 5. The process terminates if we encounter a situation where we have to overlap colors on a given vertex. In this case the graph is not bipartite.

Step 6. If all vertices of the graph are colored and no two adjacent vertices are assigned the same color, then G has a proper 2-vertex-coloring and it is therefore bipartite. Moreover, if V_1 and V_2 represent the subsets of V of colors 1 and 2 respectively, then $\{V_1, V_2\}$ is a bipartition of the graph.

Example 9.43. Apply the two-colors algorithm to determine if the following graph is bipartite. If it is, give a bipartition of the graph.

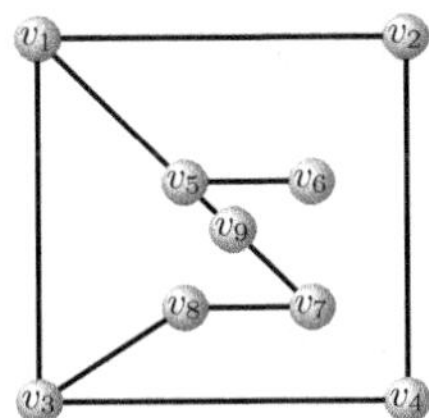

Solution. Start at vertex v_1, assign color 1 to it (light color). Color its neighbors v_2, v_3 and v_5 with color 2 (darker color). Next, assign color 1 to the only uncolored neighbor v_4 of v_2. Next, assign color 1 to uncolored neighbors v_6 and v_9 of vertex v_5. Assign color 2 to the only uncolored neighbor v_7 of v_9. Finally, assign color 2 to the only uncolored neighbor

v_8 of v_7 . All vertices are now colored and no two adjacent vertices are assigned the same color. We conclude that the graph is bipartite. Moreover, $V_1 = \{v_1, v_4, v_6, v_8, v_9\}$ and $V_2 = \{v_2, v_3, v_5, v_7\}$ form a bipartition of G.

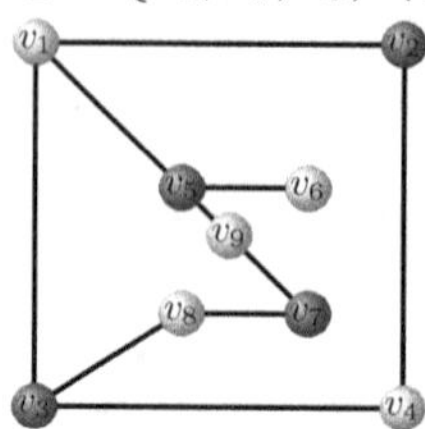

$\Diamond$

9.6.1　*Exercises*

(1) True or False: a non-empty edgeless graph is bipartite.

(2) Prove Corollary 9.2.

$\star$(3) The *star graph* S_n of order $n + 1$ is defined as the graph with $V = \{0, 1, \ldots, n\}$ as the set of vertices and $E = \{\{0, i\}; \ i = 1, \ldots, n\}\}$ as the set of edges.

 (a) Draw S_n for $n = 2, 3, 4, 5, 6$.

 (b) Prove that S_n is bipartite and give a bipartition of it.

(4) In each case, determine for what value(s) of the integer n (if any), the graph is bipartite. Justify your answer.

 $\star$ (a) K_n for $n \geq 1$ (c) P_n for $n \geq 2$

 (b) C_n for $n \geq 3$ $\star$ (d) W_n for $n \geq 3$

$\star$(5) Prove that the cube graph Q_n is bipartite for $n \geq 1$ (see Exercise (5) of Section 9.2.3 for the definition of Q_n).

(6) Let $G = (V, E)$ be a bipartite graph of positive size and let (V_1, V_2) be a bipartition of G.

 (a) Prove that $\sum_{v \in V_1} \deg(v) = \sum_{v \in V_2} \deg(v) = |E|$ (*Hint.* Use induction on the size of G)

 (b) If G is d-regular $(d > 0)$, prove that $|V_1| = |V_2|$.

(7) In each case, determine if the graph is bipartite. If it is, show a bipartition. If you say it is not, show a cycle of odd length.

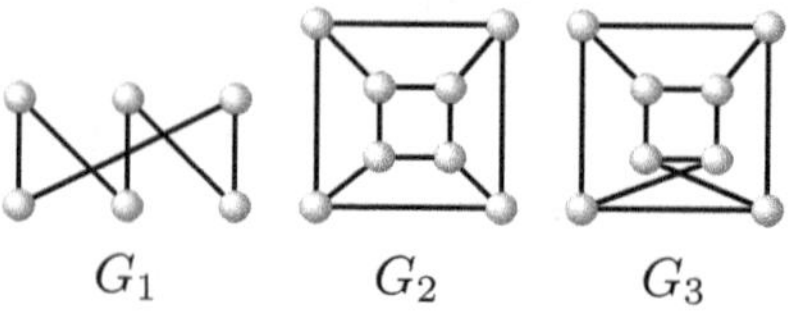

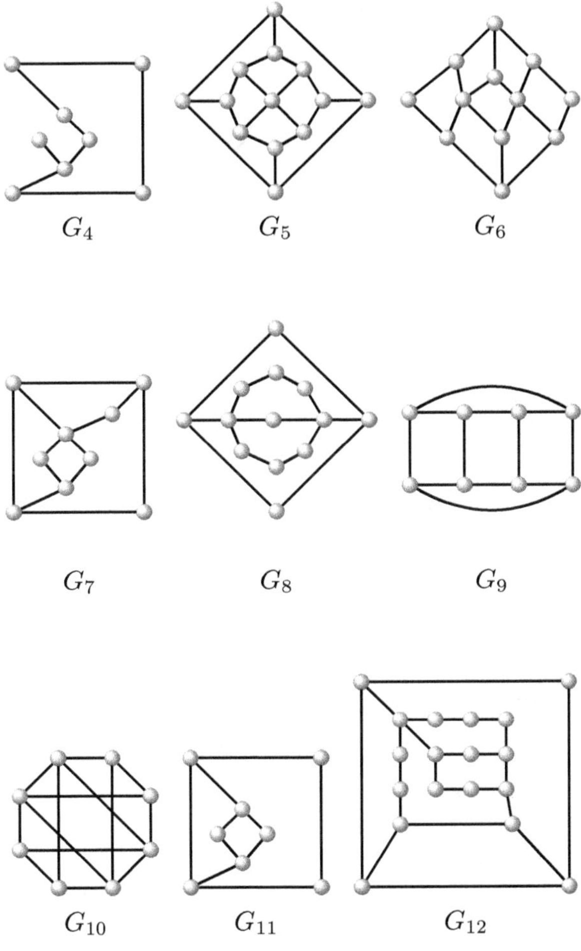

(8) Consider the following graph G.

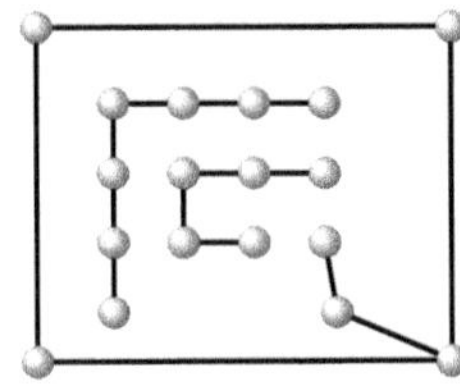

(a) How many connected components does G have?

(b) Determine if G is bipartite.

$\star$(9) Consider the following graph G:

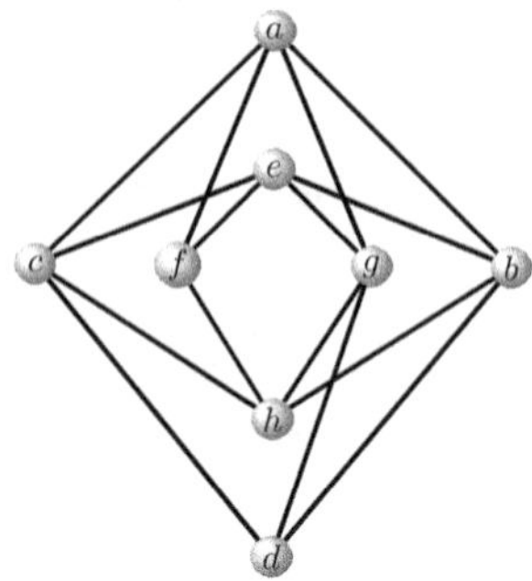

(a) Show that G is bipartite by exhibiting a bipartition.

(b) Write the adjacency matrix of G relative to the ordering a, b, c, d, e, f, g, h of its vertices.

(c) Find an ordering of the vertices of G with respect to which the adjacency matrix of G has the form

$$\begin{bmatrix} \mathbf{0} & B \\ B^T & \mathbf{0} \end{bmatrix}$$

where B is a square matrix, B^T is its transpose and $\mathbf{0}$ is a square matrix consisting of 0's.

$\star$(10) Let $G = (V, E)$ be a bipartite simple graph with $|V| = n$ and $|E| = e$.

(a) If (V_1, V_2) is a bipartition of the graph with $|V_1| = k$ and $|V_2| = n-k$, prove that $e \leq k(n - k)$.

(b) Use your answer to part (a) to prove that $e \leq \frac{n^2}{4}$ (*Hint.* What is the maximum value of the function $f(k) = k(n - k)$?)

(c) Give an example of a simple graph G of order n and size e such that $e > \frac{n^2}{4}$. Show that G is not bipartite by exhibiting a cycle of odd length in it.

(11) The notion of bipartite graphs can be generalized as follows. Given an integer $n \geq 2$, we say that a graph $G = (V, E)$ is *n-partite* if the set V can be portioned into subsets $V_1, \ldots, V_n$ (so $V_i \neq \emptyset$ for any i, $V = V_1 \cup \cdots \cup V_n$ and $V_i \cap V_j = \emptyset$ for $i \neq j$) such that every edge of G joins a vertex in V_i to a vertex in V_j where $i \neq j$. In this case, $\{V_1, \ldots, V_n\}$ is called *n-partition* of G. If $n = 2$, this coincides with the definition of bipartite graphs.

(a) Draw a possible simple non-bipartite, 3-partite graph of order 4. Label the vertices and give a 3-partition of G.

(b) Draw a possible simple non-bipartite, 3-partite graph of order 9. Label the vertices and give a 3-partition of G.

$\star$ (c) Draw a possible simple non-bipartite, 4-partite graph of order 12. Label the vertices and give a 4-partition of G.

9.7 Matching

A certain company bought 5 different new machines $M_1, \ldots, M_5$ to be added to its line of production. Each machine requires a certified technician to operate on it. After going through the list of candidates, the company decided to hire 5 technicians $T_1, \ldots, T_5$ to operate on the new machines. In the following table, a "$\checkmark$" in the (T_i, M_j) cell indicates that technician T_i is certified to operate on machine M_j and a "$\times$" in that cell indicates that T_i cannot operate on M_i:

	M_1	M_2	M_3	M_4	M_5
T_1	$\times$	$\checkmark$	$\times$	$\times$	$\checkmark$
T_2	$\times$	$\times$	$\checkmark$	$\times$	$\checkmark$
T_3	$\times$	$\times$	$\checkmark$	$\checkmark$	$\checkmark$
T_4	$\checkmark$	$\times$	$\times$	$\times$	$\times$
T_5	$\checkmark$	$\times$	$\times$	$\times$	$\checkmark$

The company has now to find a "matching" of technicians and machines such that each machine is operated on by one specialist. This can be modeled by the graph G on the left in the diagram below where each technician and each machine is represented by a vertex and each edge connects a machine with the technicians who are certified to operate on it. A good matching corresponds then to a set of edges in the graph that are pairwise disjoint (no common endpoints between any pair of edges) as illustrated with the edges in bold in the graph M on the right in the diagram.

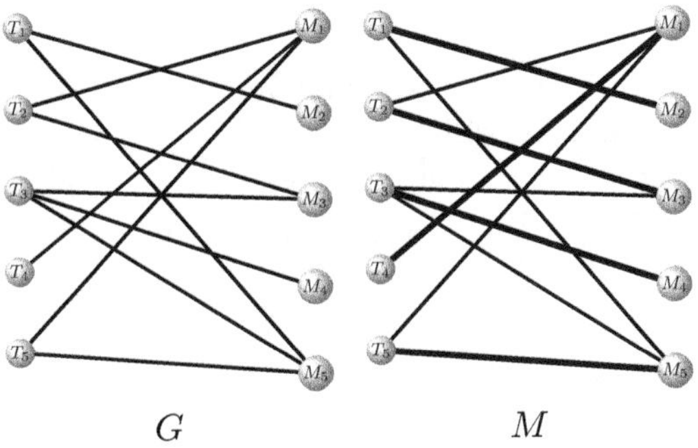

Matching in a graph is an interesting topic on its own, but our main focus will be on matching in bipartite graphs.

9.7.1   *Matching in general graphs*

Definition 9.25. Let $G = (V, E)$ be a graph. A *matching* in G is a subset M of E such that no two edges in M have an endpoint in common. The *size* of a matching M is defined as being $|M|$, the number of edges in M. A matching of G is called *maximal* if it is not a subset of any other matching of G. A matching of G is called *maximum* if it has a maximum size among all matchings of G. A vertex v of G is said to be *M-covered* (or that M covers v) by a matching M if v is incident to some edge in M. Otherwise v is said to be *M-exposed*. A matching M is called *perfect* (or *complete*) if every vertex of G is *M*-covered.

It is clear from the definition that a graph with a perfect matching must have an even number of vertices. It is also clear that every maximum matching is maximal and every perfect matching is maximum.

Example 9.44. In each of the following four graphs, a subset M of edges is drawn in bold. Determine if M is a matching in the graph. If you say it is, determine if it is maximal, maximum, perfect or none of these types.

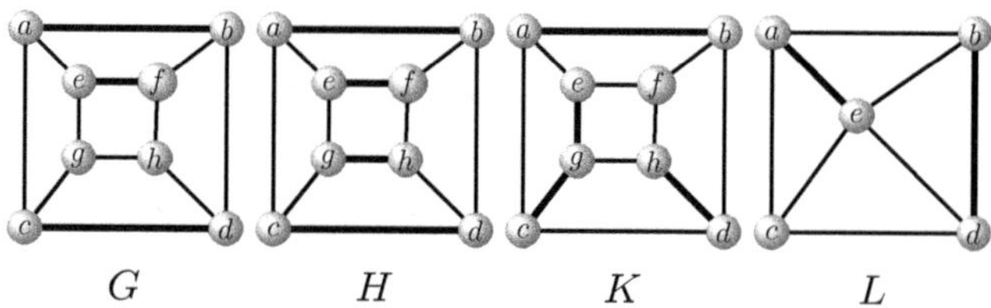

Solution. In graph G, the set is a matching but is not maximal as we can add edge $\{g, h\}$ and we still have a matching. It is not maximum, nor perfect. In graph H the given set is a perfect matching (hence maximum and maximal) as it covers every vertex. The set of bold edges in graph K is not a matching as edges $\{c, g\}$ and $\{e, g\}$ have an end point in common. In graph L, the set of bold edges is a maximum matching (hence maximal) since no matching with three edges can exist in the graph, but it is not perfect. $\Diamond$

9.7.2 *Exercises*

$\star$(1) Given a matching M of size $r \geq 1$, what is the number of vertices covered by M?

(2) In each of the following graphs, a subset M of edges is drawn in bold. Determine if M is a matching of the graph. If you say it is, determine if it maximal, maximum, perfect or none of these properties.

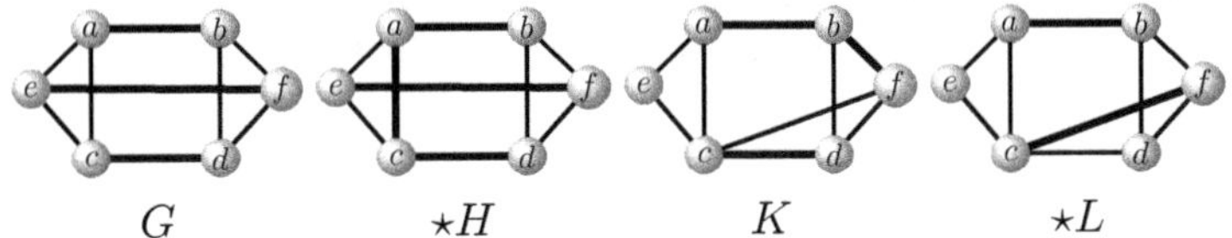

G $\qquad$ $\star H$ $\qquad$ K $\qquad$ $\star L$

(3) Draw three different perfect matching in the following graph.

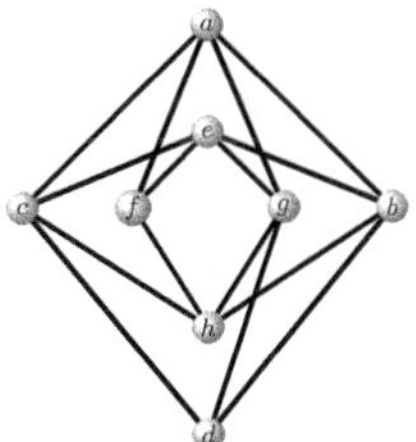

(4) For each of the graphs given below, determine if a perfect matching exists. If you say that a perfect matching does not exist, explain why and give a maximum matching of the graph.

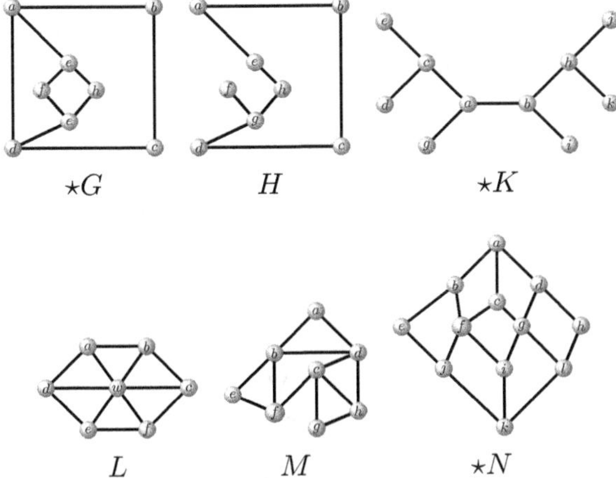

$\star G$ $\qquad$ H $\qquad$ $\star K$

L $\qquad$ M $\qquad$ $\star N$

$\star$(5) Let G be a graph. Prove that every maximum matching in G is maximal. Prove that the converse is not true by providing an example of a graph G with a maximal matching which is not maximum.

(6) Prove that any perfect matching in a graph is also maximum.

(7) For a graph $G = (V, E)$, a set $W \subseteq V$ is called a *vertex cover* if every edge $e \in E$ has at least one endpoint in W. If M is a matching of G and W is a vertex cover of G, find a relation between $|M|$ and $|W|$.

(8) Prove that a matching M in a graph G is maximal if every edge in G has a common endpoint with at least one edge in M.

$\star$(9) How many perfect matchings are there in the cycle graph C_n for $n \geq 1$?

(10) How many perfect matchings are there in the compete graph K_n for $n \geq 1$?

9.8   Matching in bipartite graphs and Hall's Theorem

Matching in bipartite graphs, also known as *bipartite matching*, arises naturally in a wide range of applications. Problems like optimizing matching on a dating site or maximizing the use of classrooms when scheduling classes at a certain university can be reduced to finding a perfect or maximum matching in a bipartite graph.

Example 9.45. Joe is renovating his house. He needs to hire trade people to do the following five tasks: electrical (e), flooring (f), landscaping (l), plumping (p) and roofing (r). His friend suggested five contractors: Adam (A), Brenda (B), Carl (C), Dina (D) and Ethane (E). Adam is a licensed electrician and a plumper, Brenda can do plumping and flooring, Carl can do roofing, flooring and landscaping, Dina can only do landscaping and Ethane is expert in all the above trades except electrical work. Assuming Joe wants to hire all five contractors, can he assign each contractor a renovation task he or she is qualified to do?

Solution. Joe's dilemma can be illustrated using the following bipartite graph $G = (V, E)$:

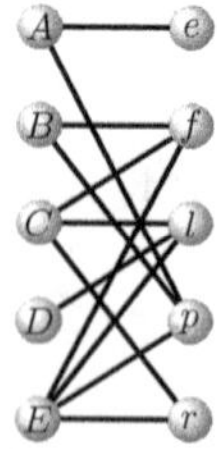

with bipartition sets $\{X, Y\}$ where X is the set of five contractors and Y is the set of five tasks to be accomplished. An edge joining a vertex x in X

to a vertex y in Y represents a task y that contractor x is qualified to do. What Joe is looking for is a perfect matching in the graph G. The following diagram shows two such matchings (in thicker lines).

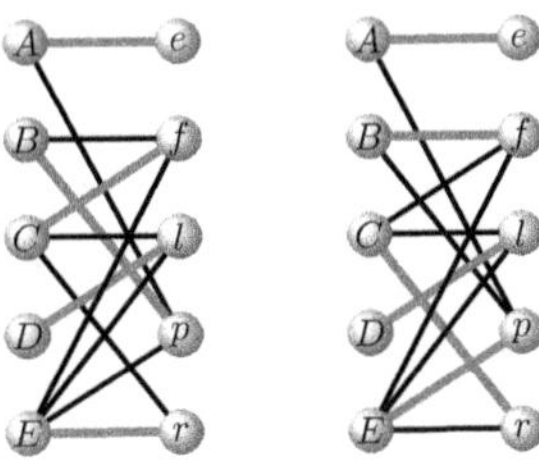

$\Diamond$

Matching in a bipartite graph has the same meaning as given in Definition 9.25 above, but one can exploit the fact that there are no edges between vertices belonging to the same partition set to be a little more precise.

Definition 9.26. Let $G = (V, E)$ be a *bipartite* graph with partition sets $X = \{x_1, \ldots, x_n\}$ and $Y = \{y_1, \ldots, y_m\}$. A *matching* in G is a subset M of edges of G such that if $\{x_i, y_j\} \in M$ and $\{x_r, y_s\} \in M$, then $i \neq r$ and $j \neq s$. The notions of maximal and maximum matchings, exposed and matched vertices are defined the same way as in Definition 9.25 above. A *perfect matching of X onto Y* (respectively a *perfect matching of Y onto X*) is a matching M in G such that no vertex in X (respectively in Y) is M-exposed. A matching of G is called *perfect* if it is at the same time a perfect matching of X onto Y and of Y onto X.

Remark 9.7. It is not hard to see that the notion of perfect matching in a bipartite graph as given in Definition 9.26 coincides with the notion of a perfect matching in general as given in Definition 9.25 above. It is also not hard to see that if M is a perfect matching in a bipartite graph G with bipartition sets X and Y, then $|M| = |X| = |Y|$.

Example 9.46. In the diagram below, three bipartite graphs G, H and K are given. In each graph, the partition sets are X and Y with X is the collection of vertices on the left and Y is that of the vertices on the right. It is easily seen that the collection of the thick edges in each graph represents a matching in the graph. The matching in G is perfect (hence maximum) as it covers every vertex. Note that $|X| = |Y| = |M|$ in G. The matching in H could not be perfect as $|X| \neq |Y|$. It is however a perfect matching of Y onto X. The matching in K is not maximum (hence not perfect). For instance, one can add the edge $\{E, p\}$ to the set of thick edges and still have a matching.

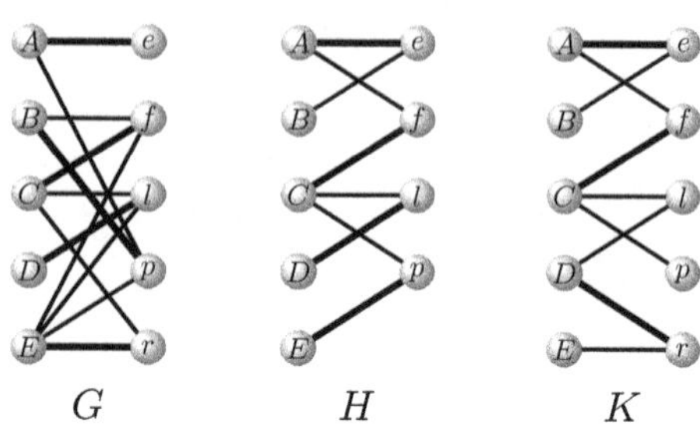

$$G \qquad\qquad H \qquad\qquad K$$

Example 9.46 begs the questions: Given a bipartite graph G with partition sets X and Y, how can we determine if G has a perfect matching of X onto Y? How can we determine if G has a perfect matching? Philip Hall (an algebraist) gave answers to these questions in the 30's of the twentieth century. Hall came up with a necessary and sufficient condition for a matching in a bipartite graph to be perfect from one partition set onto the other. We start with a definition.

Definition 9.27. Let $G = (V, E)$ be a bipartite graph with bipartition sets X and Y. For any subset U of X, define the *neighborhood* of U in Y, that we denote by $N(U)$, to be the subset of Y consisting of all those vertices in Y adjacent to at least one vertex in U:

$$N(U) = \{v \in Y; \ \exists u \in U \text{ with } \{u, v\} \in E\}.$$

Example 9.47. In graph G of Example 9.46 above with bipartition sets $X = \{A, B, C, D, E\}$ and $Y = \{e, f, l, p, r\}$, consider the two subsets $U_1 = \{A, B\}$ and $U_2 = \{A, C\}$ of X. Then $N(U_1) = \{e, f, p\}$ and $N(U_2) = \{e, f, l, p, r\} = Y$.

Hall's condition relates the existence of a perfect matching to this idea of neighborhood. Namely, we have the following.

Theorem 9.15. *(Hall's Theorem) Let $G = (V, E)$ be a bipartite graph with bipartition sets X and Y. Then G has a perfect matching of X onto Y if and only $|U| \leq |N(U)|$ for any subset U of X.*

The condition

$$|U| \leq |N(U)| \text{ for every subset } U \text{ of } X \qquad\qquad (9.1)$$

is known in the literature as *Hall's condition*.

Proof. First assume that a perfect matching M of X onto Y exists and let U be a subset of X. Let $M(U)$ be the subset of Y defined by:

$$M(U) = \{v \in Y;\ \{u,v\} \in M \text{ for some } u \in U\}.$$

Since $N(U)$ contains all neighbors of vertices of U in Y and not only those covered by M, $M(U) \subseteq N(U)$ and consequently $|M(U)| \leq |N(U)|$. By definition of a perfect matching of X onto Y, we have that $|U| = |M(U)| \leq |N(U)|$. Conversely, assume that G satisfies Hall's condition. We need to prove that G has a perfect matching of X onto Y. We proceed by induction on the cardinality of X. If $X = \{u\}$ is a singleton, then the condition $|X| \leq |N(X)|$ implies that there exists at least one vertex v in Y adjacent to u. The edge $\{u,v\}$ is a perfect matching of X onto Y in this case and the result is true. Let $n \geq 1$ and assume that a perfect matching of X onto Y exists for any bipartite graph with bipartition (X,Y) satisfying Hall's condition and such that $|X| \leq n$. Let G be a bipartite graph with bipartition (X,Y) satisfying Hall's condition and such that $|X| = n+1$. There are two possibilities: (1) for every proper subset U of X, $|U| < |N(U)|$; or (2) there exists a proper subset U of X such that $|U| = |N(U)|$.

(1) $|U| < |N(U)|$ **for any proper subset** U **of** X**.** Choose a vertex $u \in X$. Since $|\{u\}| < |N(\{u\})|$, we can choose a vertex v in Y adjacent to u. Let G' be the graph obtained from G by removing vertices u, v and deleting all edges of G incident to these two vertices. Clearly G' remains a bipartite graph with bipartition sets $X' = X\backslash\{u\}$ and $Y' = Y\backslash\{v\}$. We claim that G' satisfies Hall's condition. To see this, let U be a subset of X' and let $N'(U)$ be the subset of Y' of all neighbors of vertices in U. Since either $N(U) = N'(U)$ or $N(U) = N'(U) \cup \{v\}$, $|N(U)| \leq |N'(U)|+1$. We conclude that $|U| < |N(U)| \leq |N'(U)|+1$ or equivalently, $|U| \leq |N'(U)|$. This proves the claim. By the induction hypothesis, there exists a perfect matching M' of X' onto Y' in the graph G'. Adding the edge $\{u,v\}$ to M' creates a perfect matching of X onto Y in G.

(2) **There exists a proper subset** U **of** X **with** $|N(U)| = |U|$**.** Consider the subgraphs $G' = G[U \cup N(U)]$ and $G'' = G[(X\backslash U) \cup (Y\backslash N(U))]$ induced by $U \cup N(U)$ and $(X\backslash U) \cup (Y\backslash N(U))$ respectively. So G' is the graph with vertex set $U \cup N(U)$ and the edges of G' are those edges of G with both end points in $U \cup N(U)$. The graph G'' is defined similarly. Clearly, G' is bipartite with bipartition sets U and $N(U)$ and G'' is bipartite with bipartition sets $X\backslash U$ and $Y\backslash N(U)$. We show next that each of G' and G'' satisfies Hall's condition. For G', Consider

a subset S of U and let $N'(S)$ be the set of vertices in $N(U)$ which are adjacent to at least one vertex in S. By construction, $N'(S) = N(S)$ and since $|S| \leq |N(S)|$, Hall's condition holds for G'. For the graph G'', let A be a subset of $X \backslash U$ and let $N''(A)$ be the set of vertices in $Y \backslash N(U)$ which are adjacent to at least one vertex in A. Note that any vertex in Y adjacent to one vertex in A is either in $N''(A)$ or in $N(U)$ which means that $N(A) \subseteq N''(A) \cup N(U)$. Moreover, it is not hard to see that $N(A \cup U) = N''(A) \cup N(U)$. Assume by contradiction that $|A| > |N''(A)|$, then

$$\begin{aligned}
|N(A \cup U)| &= |N''(A) \cup N(U)| \\
&= |N''(A)| + |N(U)| \, (N''(A) \text{ and } N(U) \text{ are disjoint}) \\
&< |A| + |U| \, (\text{By assumption and since } |N(U)| = |U|) \\
&= |A \cup U| \, (A \text{ and } U \text{ are disjoint}).
\end{aligned}$$

This is a contradiction to the fact that $|Z| \leq |N(Z)|$ for any subset Z of X. We conclude that $|A| \leq |N''(A)|$ and so G'' satisfies Hall's condition. Since $|U| < n+1$ and $|X \backslash U| < n+1$ (because U is a proper subset of X), the induction hypothesis tells that G' has a perfect matching M' of U onto $N(U)$ and G'' has a perfect matching of $X \backslash U$ onto $Y \backslash N(U)$. Clearly $M = M' \cup M''$ is a perfect matching in G of X onto Y. The proof in complete.

$\square$

The following is a direct consequence of Hall's Theorem.

Corollary 9.3. *Let $G = (V, E)$ be a bipartite graph with bipartition sets X and Y. Then G has a perfect matching if and only $|X| = |Y|$ and $|U| \leq |N(U)|$ for any subset U of X.*

Example 9.48. Consider the following bipartite graph G with bipartition sets $X = \{A, B, C, D\}$ and $Y = \{e, f, l, p, r\}$:

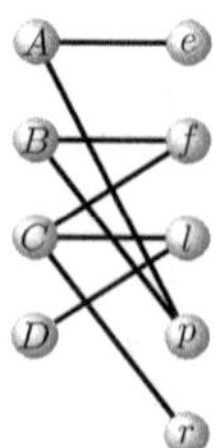

Use Hall's Theorem to determine if G has a perfect matching of X onto Y.

Solution. We verify that Hall's condition is satisfied for the empty subset of X. Each entry in the first column of the table below is a non-empty subset U of X. The corresponding entry in the second column is the set $N(U)$ of all neighbors of vertices in U.

U	$N(U)$	$|U| \leq |N(U)|$
$\{A\}$	$\{e,p\}$	True
$\{B\}$	$\{f,p\}$	True
$\{C\}$	$\{f,l,r\}$	True
$\{D\}$	$\{l\}$	True
$\{A,B\}$	$\{e,f,p\}$	True
$\{A,C\}$	$\{e,f,l,p,r\}$	True
$\{A,D\}$	$\{e,l,p\}$	True
$\{B,C\}$	$\{f,l,p,r\}$	True
$\{B,D\}$	$\{f,l,p\}$	True
$\{C,D\}$	$\{f,l,r\}$	True
$\{A,B,C\}$	$\{e,f,l,p,r\}$	True
$\{A,B,D\}$	$\{e,f,l,p\}$	True
$\{A,C,D\}$	$\{e,f,l,p,r\}$	True
$\{B,C,D\}$	$\{f,l,p,r\}$	True
$\{A,B,C,D\}$	$\{e,f,l,p,r\}$	True

The table shows that the inequality $|U| \leq |N(U)|$ holds for any subset U of vertices of G. By Hall's Theorem, the graph has a perfect matching of X onto Y. $\Diamond$

Example 9.49. The following diagram shows a bipartite graph with bipartition sets $X = \{A, B, C, D\}$ and $Y = \{e, f, l, p, r\}$.

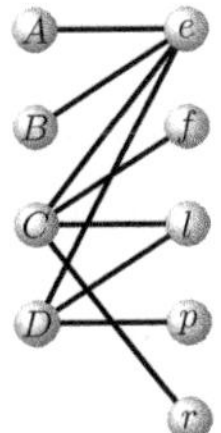

The graph does not have a perfect matching of X onto Y since Hall's condition is not satisfied for the subset $U = \{A, B\}$ of X as $N(U) = \{e\}$ (so $|N(U)| < |U|$).

Example 9.50. Consider the following bipartite graph.

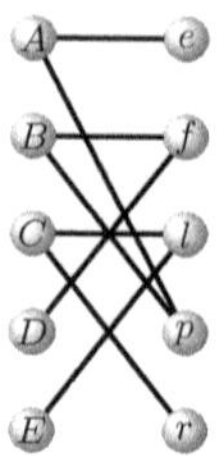

The table below shows subsets U and $N(U)$ for every non-empty subset U of X. From the table, one can see that Hall's condition is satisfied.

U	$N(U)$
$\{A\}$	$\{e,p\}$
$\{B\}$	$\{f,p\}$
$\{C\}$	$\{l,r\}$
$\{D\}$	$\{f\}$
$\{E\}$	$\{l\}$
$\{A,B\}$	$\{e,f,p\}$
$\{A,C\}$	$\{e,p,l,r\}$
$\{A,D\}$	$\{e,f,p\}$
$\{A,E\}$	$\{e,l,p\}$
$\{B,C\}$	$\{f,l,p,r\}$
$\{B,D\}$	$\{f,p\}$
$\{B,E\}$	$\{f,l,p\}$
$\{C,D\}$	$\{f,l,r\}$
$\{C,E\}$	$\{l,r\}$
$\{D,E\}$	$\{f,l\}$
$\{A,B,C\}$	$\{f,e,l,p,r\}$

U	$N(U)$
$\{A,B,D\}$	$\{f,e,p\}$
$\{A,B,E\}$	$\{f,e,l,p\}$
$\{A,C,D\}$	$\{f,e,l,p,r\}$
$\{A,C,E\}$	$\{e,l,p,r\}$
$\{A,D,E\}$	$\{f,e,l,p\}$
$\{B,C,D\}$	$\{f,l,p,r\}$
$\{B,C,E\}$	$\{f,l,p,r\}$
$\{B,D,E\}$	$\{f,l,p\}$
$\{C,D,E\}$	$\{f,l,r\}$
$\{A,B,C,D\}$	$\{f,e,l,p,r\}$
$\{A,B,C,E\}$	$\{f,e,l,p,r\}$
$\{A,B,D,E\}$	$\{f,e,l,p,r\}$
$\{A,C,D,E\}$	$\{f,e,l,p,r\}$
$\{B,C,D,E\}$	$\{f,l,p,r\}$
$\{A,B,C,D,E\}$	$\{f,e,l,p,r\}$

Since $|X| = |Y|$, the graph has a perfect matching. The thick edges in the following diagram show a perfect matching in G.

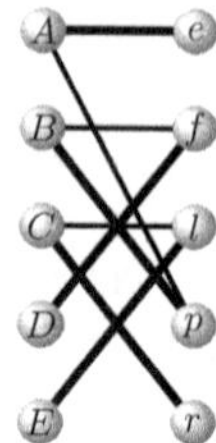

9.8.1　*Exercises*

⋆(1) Let $G = (V, G)$ be a bipartite graph with bipartition sets X and Y. We have seen that the condition $|X| = |Y|$ is necessary to have a perfect matching in G. Is the condition sufficient? Justify your answer

(2) Prove that the definition of a perfect matching in a bipartite graph given in Definition 9.26 above coincides with the definition of a perfect matching in a general graph as given in Definition 9.25.

(3) Let $d \geq 1$ be an integer and let G be a bipartite d-regular graph with bipartition sets X and Y. Prove that $|X| = |Y|$ and use Hall's Theorem to prove that G has a perfect matching.

⋆(4) From an 8×8 chessboard formed by 64 unit squares (1×1 each), we remove the top left and bottom right corners. Using results from this section, determine if the resulting chessboard can be covered by 2×1 rectangular tiles such that each square on the truncated chessboard is covered by exactly one tile (no two tiles overlap).

(5) The diagram below shows a bipartite Graph G with bipartition sets X (vertices on the left) and Y (vertices on the right). Using Hall's condition, prove that G has a perfect matching of X onto Y.

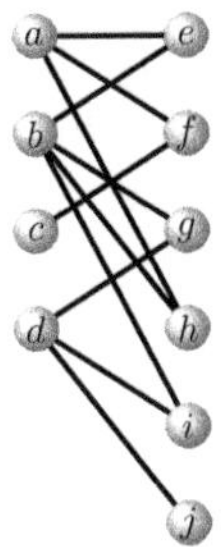

(6) Three bipartite graphs are given.

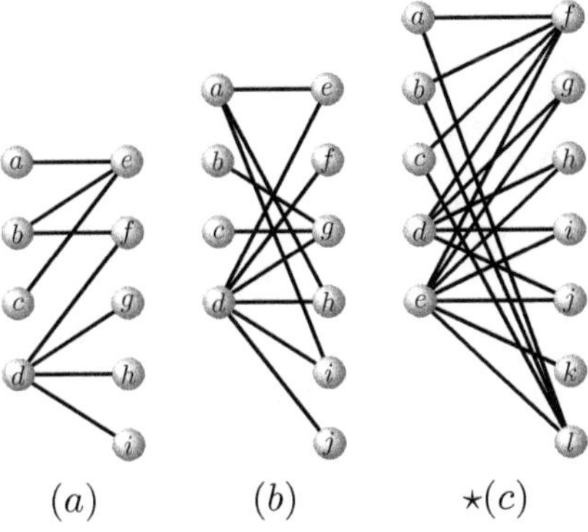

(a)　　　　　(b)　　　　⋆(c)

In each case, prove that the graph does not have a perfect matching of the set of vertices on the left onto the set of vertices on the right.

$\star$(7) A standard deck of cards consists of 52 cards divided into 4 suits of 13 cards each: hearts, diamonds, spades and clubs. There are 13 values in each suit: Ace, 2, 3, ..., 9, 10, J, Q and K. A deck is shuffled and the 52 cards are placed in an arrangement of 13 rows and 4 columns. Show that it is always possible to choose one card from each of the 13 rows and end up with 13 cards of pairwise different values.

(8) Ada, Bea, Cam, Dee and Joe are applying for teaching at a certain school board. Positions available are in Math, Physics, History, Chemistry and Geography. Ada is qualified to teach Math, Physics and History. Bea is qualified to teach Physics, Chemistry and Geography. Cam is qualified to teach Physics and History. Dee is qualified to teach History and Chemistry. Finally, Joe is qualified to teach physics only. Each candidate is assigned to teach only one subject in various schools. The school board decided to hire all five applicants. Can the board do the hiring in such a way that each applicant teaches exactly one subject he or she is qualified to teach? justify your answer using matching in bipartite graphs.

(9) Let G be a bipartite graph with bipartition sets X and Y and without isolated vertices such that for any edge $\{x, y\}$ of G with $x \in X$ and $y \in Y$, $\deg(x) \geq \deg(y)$. Show that G has a perfect matching from X onto Y.

9.9   Isomorphism of simple graphs

Consider the two graphs:

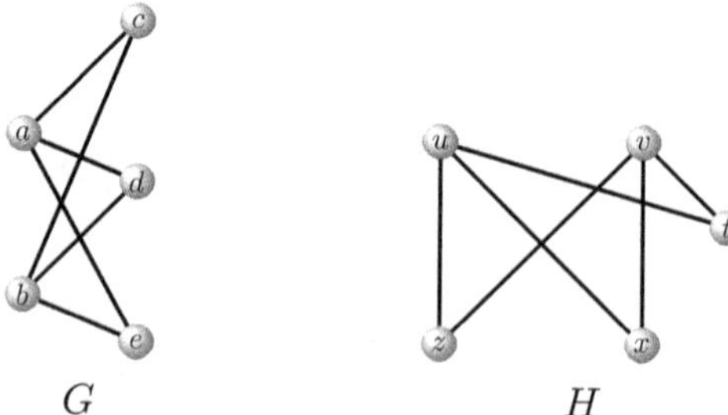

$$G \qquad\qquad\qquad\qquad H$$

At first glance, the two graphs look different. But a closer look shows that the graphs differ only in the way they are drawn. It is just a matter of relabeling the vertices of one and redrawing it to get the other. In other words, G and H are two topological representations of the same graph. In the language of graphs, we say that G and H are *isomorphic* and the relabeling function of vertices is called a *graph isomorphism* . For the sake of simplicity, we only discuss the notion of isomorphism for simple graphs, so *all graphs in this section are assumed to be simple.*

Definition 9.28. We say that the two graphs $G = (V, E)$ and $G' = (V', E')$ are *isomorphic*, and we write $G \cong G'$, if there exists a bijection $f : V \to V'$ that preserves adjacency of vertices. That is to say, for any vertices $u, v \in V$: $\{u, v\} \in E$ if and only if $\{f(u), f(v)\} \in E'$. Such a bijection is called a *graph isomorphism.*

First properties of isomorphic graphs are given in the following theorem.

Theorem 9.16. *Let $G = (V, E)$ and $G' = (V', E')$ be two isomorphic graphs and $f : V \to V'$ be a graph isomorphism.*

(a) $|V| = |V'|$ (isomorphic graphs have the same order).
(b) For any $v \in V$, $\deg(v) = \deg(f(v))$.
(c) $|E| = |E'|$ (isomorphic graphs have the same size).
(d) G and G' have the same degree sequence.

Proof. Part (a) Follows from the fact that f is a bijection between V and V'. For part (b), assume that $\deg(v) = k$ and let $v_1, \ldots, v_k$ be all the neighbors of v and $w_1, \ldots, w_t$ the vertices in G not adjacent to v. By definition of a graph isomorphism, $f(v)$ is adjacent to $f(v_1), \ldots, f(v_k)$ and not adjacent to $f(w_1), \ldots, f(w_l)$ in G'. In particular, $\deg(f(v)) = k$. Part (c) follows from the handshaking lemma and part (b). Part (d) follows directly from part (a) and the fact that $f : V \to V'$ is a bijection. $\square$

A graph property (or a parameter) that is preserved under a graph isomorphism is called an *isomorphism invariant* or simply an *invariant*. For instance, the size, the order and the degree sequence of a graph are examples of invariants. Other examples of invariants include: the number of vertices of a given degree k, the number of circuits of a given length k, the number of connected components, connectivity, being bipartite. Many other invariants will be added as we look at more concepts of graph theory. Remember that when two graphs are isomorphic, they are two different drawings of the same graph so they share the same structural properties and the same building blocks. Properties that are not isomorphism invariants are mostly "cosmetic" and not structural. For example, vertex and edge labeling are not graph invariants. Unfortunately, no complete check list of invariants to determine if two graphs are isomorphic exists. We can have two graphs sharing all the properties mentioned above but still not isomorphic. However, having a list of main isomorphism invariants is handy to show that two graphs are **not** isomorphic. If a graph G has one isomorphism invariant not satisfied by another graph G', then the two graphs are not isomorphic.

Example 9.51. Consider the following graphs:

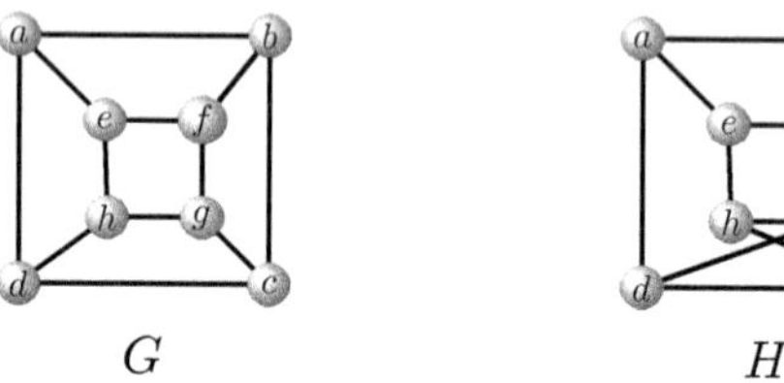

$$G \qquad\qquad\qquad H$$

The two graphs have the same number of vertices (8), the same number of edges (12) and the same degree sequence (namely, $(3, 3, 3, 3, 3, 3, 3, 3)$). Graph G is bipartite but H is not (for instance, (c, h, g, f, b, c) is a circuit of odd length in H). We conclude that G and H are not isomorphic.

Determining if two graphs are isomorphic is not an easy task in general, especially when the two graphs share the same basic structural characteristics like the order, the size and the degree sequence. If no obvious structural difference is detected between two graphs, we try to construct a graph isomorphism between their sets of vertices. This amounts to finding a bijection between the sets of vertices that preserves adjacency. If the graphs have n vertices each, there are $n!$ different bijections between the sets of vertices. The problem is that $n!$ becomes large very rapidly and it is unpractical to check each bijection for the adjacency preserving property.

The following theorem gives a characterization of a graph isomorphism in terms of the adjacency matrices of the two graphs.

Theorem 9.17. *Let $G = (V, E)$ and $G' = (V', E')$ be two graphs of the same order n, $f : V \to V'$ a bijection. Then f is a graph isomorphism (and hence $G \cong G'$) if and only if the adjacency matrix of G with respect to the ordering $v_1, v_2, \ldots, v_n$ of its vertices is the same as the adjacency matrix of G' with respect to the ordering $f(v_1), f(v_2), \ldots, f(v_n)$ of its vertices.*

Proof. Let $M = [a_{ij}]$ be the adjacency matrix of G with respect to the ordering $v_1, v_2, \ldots, v_n$ of its vertices and let $M' = [a'_{ij}]$ be the adjacency matrix of G' with respect to the ordering $f(v_1), f(v_2), \ldots, f(v_n)$ of its vertices. The result follows from the following observation:

$$a_{ij} = 1 \Leftrightarrow \{v_i, v_j\} \in E \Leftrightarrow \{f(v_i), f(v_j)\} \in E' \Leftrightarrow a'_{ij} = 1.$$

$\square$

Remark 9.8. Since adjacency matrices are symmetric, we only need to compare upper (or lower) parts of the matrices M and M' to determine if $M = M'$ in Theorem 9.17.

The above theorem proposes to choose a bijection $f : V \to V'$ carefully so that the adjacency of vertices is preserved, and then prove that f is indeed a graph isomorphism using adjacency matrices.

Example 9.52. Let us go back to the pair of graphs given in the introduction of this section and prove that the graphs are isomorphic by explicitly exhibiting an isomorphism.

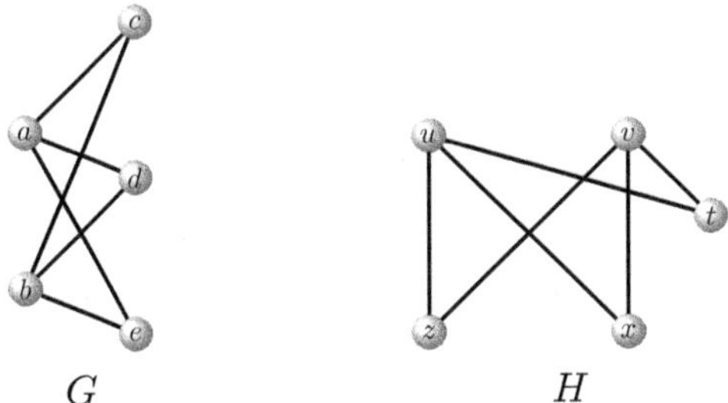

G H

The first step is to carefully choose the bijection between the set of vertices so that degrees and adjacency are preserved. Consider the bijection: $a \to u$, $b \to v$, $c \to z$, $d \to x$, $e \to t$ between the vertices of G and H. We prove that this is indeed a graph isomorphism using Theorem 9.17. Let A be the adjacency matrix of G with respect to the enumeration a, b, c, d, e of its vertices and B the adjacency matrix of H with respect to the enumeration u, v, z, x, t of its vertices. The upper parts of A and B are equal as shown below, and so $A = B$.

$$A = \begin{array}{c} \\ a \\ b \\ c \\ d \\ e \end{array}\begin{array}{c} a\ b\ c\ d\ e \\ \begin{bmatrix} 0 & 0 & 1 & 1 & 1 \\ & 0 & 1 & 1 & 1 \\ & & 0 & 0 & 0 \\ & & & 0 & 0 \\ & & & & 0 \end{bmatrix}\end{array}, \quad B = \begin{array}{c} \\ u \\ v \\ z \\ x \\ t \end{array}\begin{array}{c} u\ v\ z\ x\ t \\ \begin{bmatrix} 0 & 0 & 1 & 1 & 1 \\ & 0 & 1 & 1 & 1 \\ & & 0 & 0 & 0 \\ & & & 0 & 0 \\ & & & & 0 \end{bmatrix}\end{array}.$$

By Theorem 9.17, the bijection is a graph isomorphism and $G \cong H$.

Example 9.53. Consider the following three graphs.

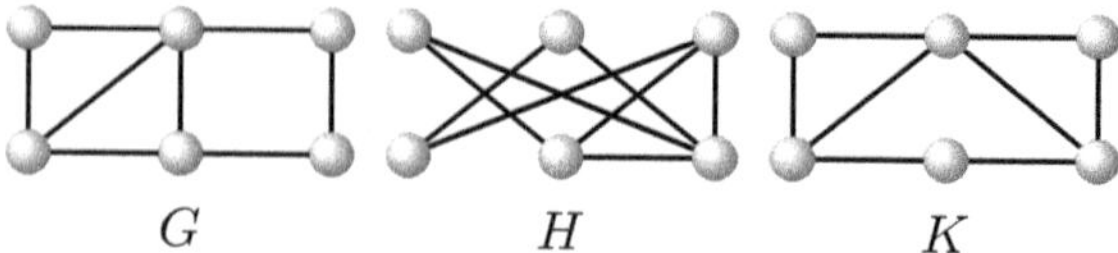

All three graphs have the same order (6), the same size (8) and the same degree sequence $(2, 2, 2, 3, 3, 4)$. So we have to dig a bit deeper to find a structural difference. In graph G, there are two adjacent vertices of degree 2 each. This is not the case in K. This is a enough to conclude that G and K are not isomorphic. Since no obvious structural between G and H can be detected, we try to prove that $G \cong H$. We start by labeling the vertices of the two graphs as indicated in the diagram:

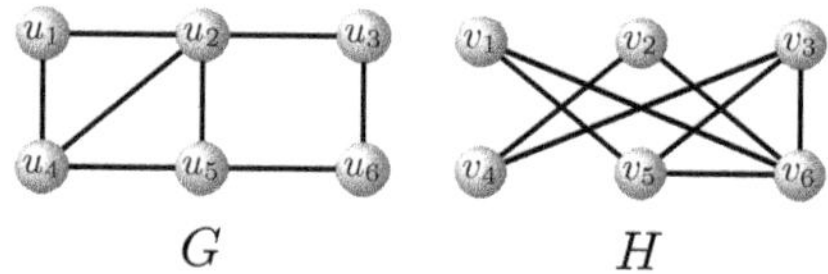

Any graph isomorphism between G and H must associate u_2 with v_6 since these are the only vertices of degree 4 in the graphs. In graph G, vertex u_1 is of degree 2 and is adjacent to one vertex of degree 3 (u_4) and to another of degree 4 (u_2). The same is true for vertex v_1 in H. So we can associate u_1 with v_1. Since u_4 is a neighbor of u_1 in G and v_5 is a neighbor of v_1 (with the same degree) in H, we can associate u_4 with v_5. The other vertex of degree 3 in G, namely u_5, must be associated with v_3. Two choices are left for the image of vertex u_3, namely v_2 and v_4. Note that u_2 and u_3 are neighbors in G but v_4 and v_6 are not in H. So, u_3 is associated with v_2 and consequently u_6 is associated with v_4. We have then constructed the bijection: $u_1 \to v_1$, $u_2 \to v_6$, $u_3 \to v_2$, $u_4 \to v_5$, $u_5 \to v_3$, $u_6 \to v_4$. The upper parts of the adjacency matrix M of G with respect to $u_1, u_2, u_3, u_4, u_5, u_6$ and the adjacency matrix M' of H with respect to $v_1, v_6, v_4, v_5, v_3, v_2$ are as follows:

$$M = \begin{array}{c} \\ u_1 \\ u_2 \\ u_3 \\ u_4 \\ u_5 \\ u_6 \end{array} \begin{array}{c} u_1\ u_2\ u_3\ u_4\ u_5\ u_6 \\ \begin{bmatrix} 0 & 1 & 0 & 1 & 0 & 0 \\ & 0 & 1 & 1 & 1 & 0 \\ & & 0 & 0 & 0 & 1 \\ & & & 0 & 1 & 0 \\ & & & & 0 & 1 \\ & & & & & 0 \end{bmatrix} \end{array}, \quad M' = \begin{array}{c} \\ v_1 \\ v_6 \\ v_2 \\ v_5 \\ v_3 \\ v_4 \end{array} \begin{array}{c} v_1\ v_6\ v_2\ v_5\ v_3\ v_4 \\ \begin{bmatrix} 0 & 1 & 0 & 1 & 0 & 0 \\ & 0 & 1 & 1 & 1 & 0 \\ & & 0 & 0 & 0 & 1 \\ & & & 0 & 1 & 0 \\ & & & & 0 & 1 \\ & & & & & 0 \end{bmatrix} \end{array}.$$

Since $M = M'$, $G \cong H$.

9.9.1 *Exercises*

(1) Which of the following pairs of graphs, if any, are isomorphic:

(a) K_3 and C_3
(b) K_3 and W_3
(c) C_4 and W_3
(d) K_{2n} and $K_{n,n}$, $n \geq 2$

(2)(a) How many non-isomorphic (simple) graphs with three vertices are there? Make a drawing for each one of these graphs.

$\star$ (b) How many non-isomorphic (simple) graphs with four vertices are there? Make a drawing for each one of these graphs.

$\star$(3) How many non-isomorphic 3-regular graphs of order 6 are there?

(4) Draw three graphs of order 6 each with the same number of edges and the same degree sequence but no pair of which are isomorphic.

(5) In each case, determine if the given graphs are isomorphic. If you say they are, label the vertices and exhibit a bijection between the set of vertices and prove it is an isomorphism using Theorem 9.17. If you say that the graphs are not isomorphic, give a graph invariant satisfied by one but not the other.

$\star$ (a)

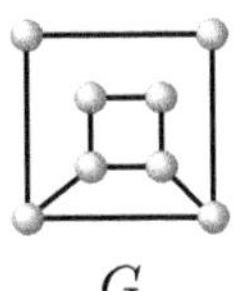

G

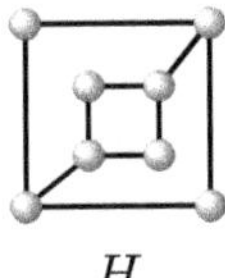

H

(b)

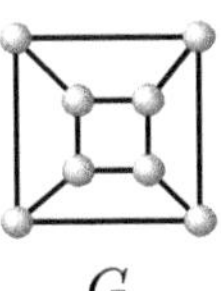

G

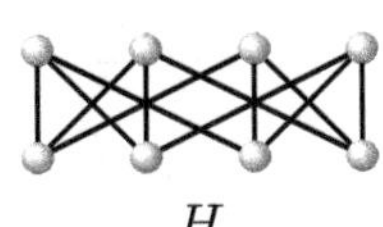

H

$\star$ (c)

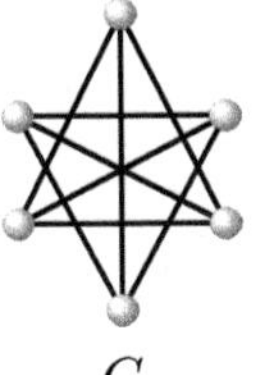

G

H

$\star$ (d)

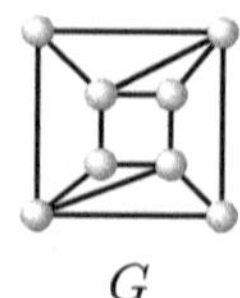 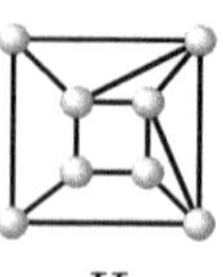

(e)

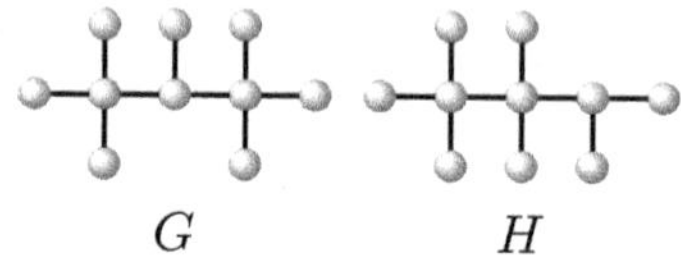

(f)

 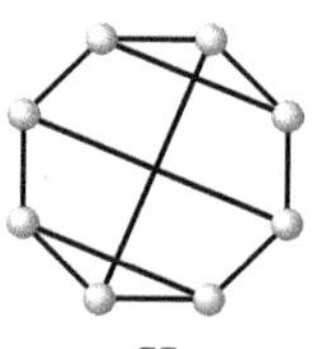

(6) In each case, three graphs are given. Only two of the graphs are isomorphic. For the isomorphic pair, label the vertices and exhibit a graph isomorphism. For the odd graph, specify a structural property not satisfied by the isomorphic pair.

(a)

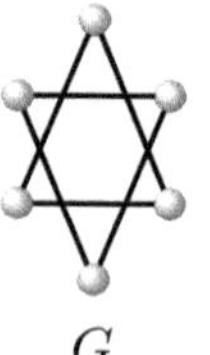 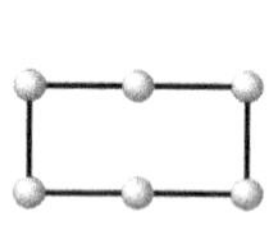 

$\star$ (b)

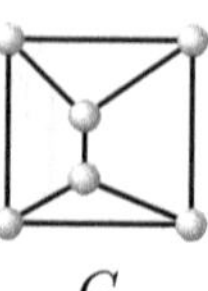 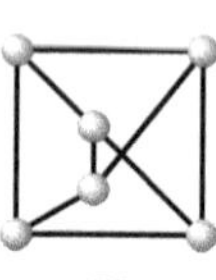 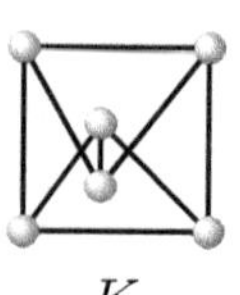

(c)

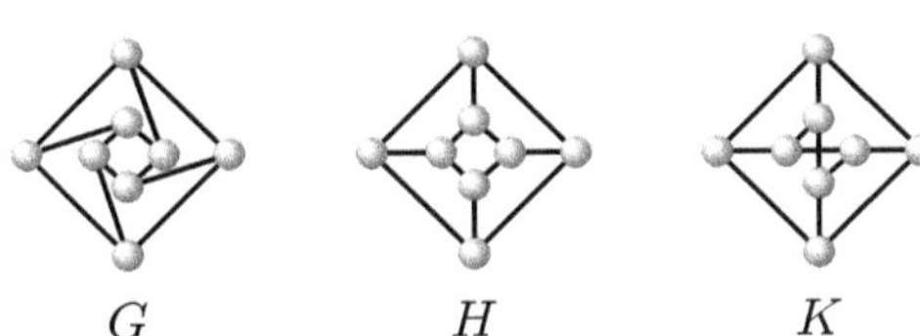

$G \qquad\qquad H \qquad\qquad K$

$\star$(7) Prove that any bijection on the vertex set of the complete graph K_n is a graph isomorphism between K_n and itself.

(8)(a) Let $\mathcal{S}$ be the set of all simple graphs. Show that the binary relation $\cong$ (isomorphism of graphs) is an equivalence relation on $\mathcal{S}$.

(b) Consider the following simple graph G:

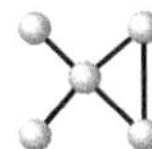

Which of the following graphs are in the equivalence class of G with respect to the equivalence relation defined in part (a)?

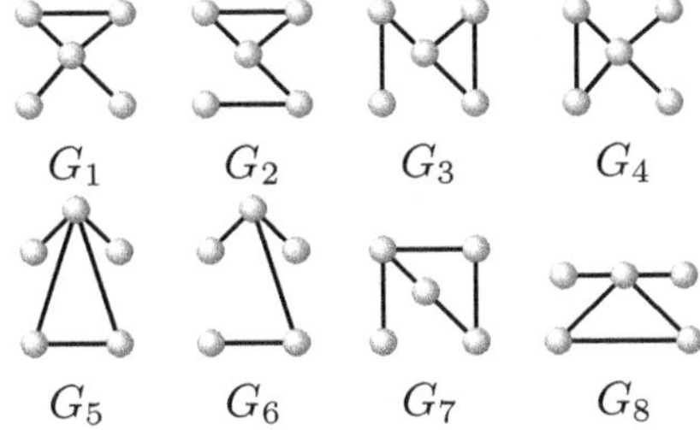

$G_1 \qquad G_2 \qquad G_3 \qquad G_4$

$G_5 \qquad G_6 \qquad G_7 \qquad G_8$

$\star$(9) Let G and G' be two isomorphic graphs. Prove that G is bipartite if and only if G' is bipartite.

(10) Let G and G' be two isomorphic graphs. Prove that G is connected if and only G' is connected.

(11)(a) Prove that two graphs G and H are isomorphic if and only of their complements $\overline{G}$ and $\overline{H}$ are isomorphic.

(b) Use part (a) to determine if the two graphs given below are isomorphic:

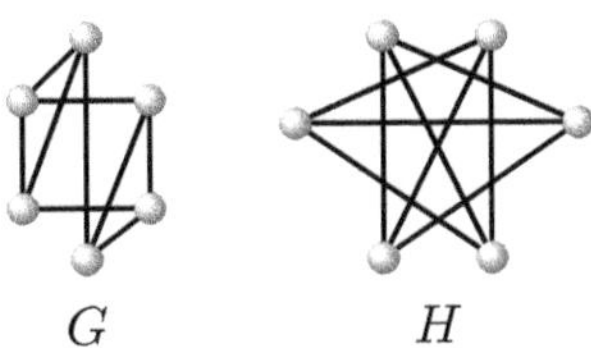

$G \qquad\qquad H$

$\star$(12) A graph G is called *self-complementary* if $G \cong \overline{G}$ where $\overline{G}$ is the complement of G.

    (a) Prove that if G is a simple self-complementary graph of order n, then $n(n-1)$ must be a multiple of 4.

    (b) For each of the following values of the integer n, determine if there Is there a self-complementary graph of order n? If you say such a graph exists, give a possible drawing: (i) $n = 4$; (ii) $n = 5$; (iii) $n = 6$.

9.10   Maximum flow in a network, the Ford–Fulkerson algorithm and maximum bipartite matching

In planning any type of transportation (trucks travelling on highways, electrical current through wires, oil flowing from a well to a refinery through a network of pipes, etc....), an important question to ask is how to maximize the flow while respecting the limitation of channels' capacities (roads, pipes, wires, ...). In this section, we introduce one of the earliest methods to solve the maximum flow problem in a network: the Ford–Fulkerson algorithm. The algorithm will also serve as a useful tool to solve the maximum matching problem in bipartite graphs.

This section is the only part of the book where digraphs are used. Recall that in a digraph, an edge is labeled by an ordered pair (u, v) of vertices to illustrate its direction from u to v.

Definition 9.29. A *flow-network* consists of a *directed* graph $G = (V, E)$, a **flow capacity** function $c : V \times V \to [0, +\infty[$ that assigns to each edge (u, v) of G the maximum flow possible $c(u, v)$ through that edge. If $(u, v) \notin E$, we assume that $c(u, v) = 0$. A flow-network is also characterized by the presence of two special types of vertices: the source and the sink. A **source** is a vertex s such that no edge of G is of the form (v, s) for any $v \in V$ (so no incoming edges to s). A **sink** is a vertex t such that no edge of G is of the form (t, v) for any $v \in V$ (so no outgoing edges from t). Any vertex in G different from the source and the sink is called an *internal* vertex. For simplicity, we assume that the graph G of a flow-network is connected. In particular, for every vertex v there exists a path from the source s to the sink t that goes through v.

The unit of measure of the capacity function in a flow-network depends on the nature of the network. For example, if the network represents oil flowing from a well to a refinery, then the capacity $c(u, v)$ of pipe (u, v) can be measured in litres per minute. If the network represents data flowing through wires, then the capacity $c(u, v)$ of wire (u, v) can be measured in kilobytes per second.

With the notations of Definition 9.29, a flow-network will be denoted by $F = (G, s, t, c)$ where $G = (V, E)$ is the underlying digraph, s is the source, t is the sink and c is the flow capacity function. The following is an illustration of a flow-network with source s and a sink t. The capacity of an edge is represented with an integer along the edge.

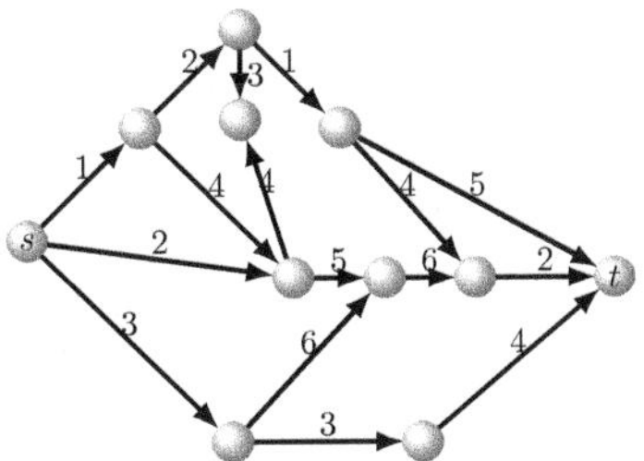

A more formal mathematical definition of the notion of a flow in a network is given next. Later, we impose some assumptions to ease the task of proving some results about the flow.

Definition 9.30. Let $F = (G, s, t, c)$ be a flow-network. A *flow* is a function $f : V \times V \to \mathbb{R}$ satisfying the following three conditions:

- **Capacity constraint:** $f(u, v) \le c(u, v)$ for any $(u, v) \in V \times V$: the flow through the edge (u, v) cannot exceed the capacity of that edge.
- **Flow conservation law:** For any *internal* vertex v, $\sum_{x \in V} f(v, x) = 0$.
- **Skew-symmetry:** For any $(u, v) \in V \times V$, $f(u, v) = -f(v, u)$.

The **value** of a flow f on the network $F = (G, s, t, c)$, denoted by $|f|$, is defined as the total flow out of the source s:

$$|f| = \sum_{v \in V} f(s, v).$$

The notion of a flow in a network introduced in Definition 9.30 is known in the literature as the *net flow*. It can be negative, zero, or positive. Intuitively, the flow on an edge (u, v) of a network is the amount of material carried on that edge from u toward v. So it is natural to assume that a flow takes non-negative values. For an internal vertex $v \in V$, define the *total positive net flow entering v* as being $f^+(v) = \sum_{u \in V;\, f(u,v)>0} f(u,v)$. Similarly, the *total positive net flow leaving* vertex v is defined as $f^-(v) = \sum_{u \in V;\, f(v,u)>0} f(v,u)$. The *total net flow* at vertex v is $f^-(v) - f^+(v)$.

Theorem 9.18. *Let $F = (G, s, t, c)$ be a flow-network with $G = (V, E)$ and let f be a flow on F.*

(a) If (u, v) and (v, u) are not in E, then $f(u, v) = f(v, u) = 0$.
(b) For an internal vertex $v \in V$, $f^-(v) = f^+(v)$.

Proof. If (u, v) and (v, u) are not in E, then $c(u, v) = c(v, u) = 0$. By the capacity constraint property, $f(u, v) \leq c(u, v) = 0$ and $f(v, u) \leq c(v, u) = 0$. By the Skew-symmetry property, $f(u, v) = -f(v, u) \geq 0$. So $0 \leq f(u, v) \leq 0$ which implies that $f(u, v) = 0$. This also shows that $f(v, u) = -f(u, v) = 0$. This proves part (a). For part (b), let v be an internal vertex. The flow conservation law at v can be expressed as $\sum_{f(v,x)>0} f(v, x) + \sum_{f(v,y)<0} f(v, y) = 0$. By the skew-symmetry property, the latter equation can be written as $\sum_{f(v,x)>0} f(v, x) - \sum_{f(y,v)>0} f(y, v) = 0$, or equivalently, $f^-(v) = f^+(v)$. $\qquad\square$

When representing a flow f in a network with a drawing, we use the notation "$m : n$" on edge (u, v) to indicate that the capacity of the edge is $c(u, v) = n$ and the value of the flow from u to v along that edge is $m = f(u, v)$. Only flows on edges of positive capacity are shown. If edge (u, v) is shown with a positive flow $f(u, v)$, then the assumption is that the flow from v to u is $-f(u, v)$ but these negative flows are omitted.

Example 9.54. Consider the following flow-network.

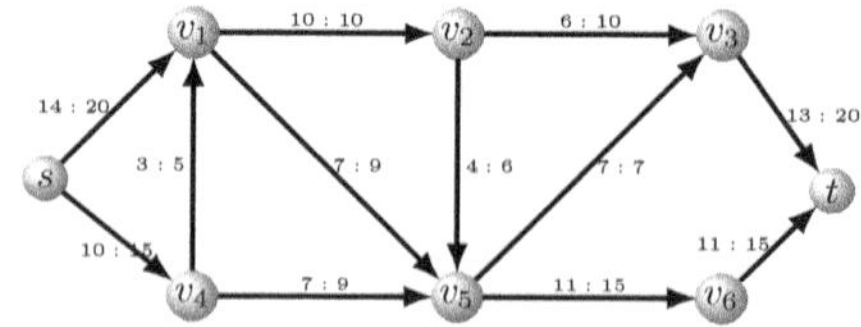

The flow through every edge does not exceed the capacity of the edge. One can easily verify that at every internal vertex, the total flow in is the same as the total flow out. The diagram represents indeed a flow in the network. The value of the flow in this network is $|f| = 10 + 14 = 24$.

Definition 9.31. Let $F = (G, s, t, c)$ be a flow-network and let f be a flow on F. For any subsets X and Y of vertices of G, define $f(X, Y)$ as being the real number $\sum_{u \in X} \sum_{v \in Y} f(u, v)$. If $X = \{u\}$ (respectively, $Y = \{v\}$), we write $f(u, Y)$ (respectively $f(X, v)$) for $f(X, Y)$. The expression $f(X, Y)$ is called an *implicit summation* (for obvious reason).

Theorem 9.19. *Let f be a flow on a network $F = (G, s, t, c)$ with $G = (V, E)$. Then:*

(a) For any subsets X, Y of V, $f(X, Y) = -f(Y, X)$.
(b) For any internal vertex $u \in V$, $f(u, V) = f(V, u) = 0$.
(c) For any subset X of V, $f(X, X) = 0$.
(d) If X, Y are disjoint subsets of V, then $f(X \cup Y, Z) = f(X, Z) + f(Y, Z)$ and $f(Z, X \cup Y) = f(Z, X) + f(Z, Y)$ for any subset Z of V.

Proof.

(a) Follows from the skew-symmetry property.
(b) Follows from the flow conservation law at vertex u and the skew-symmetry property.
(c) By part (a), $f(X, X) = -f(X, X)$ and so $f(X, X) = 0$.
(d) Let X, Y, Z be three subsets of V with $X \cap Y = \emptyset$. Then:

$$f(X \cup Y, Z) = \sum_{u \in X \cup Y} \sum_{v \in Z} f(u, v)$$

$$= \sum_{u \in X} \sum_{v \in Z} f(u, v) + \sum_{u \in Y} \sum_{v \in Z} f(u, v) \text{ (Since } X \cap Y = \emptyset)$$

$$= f(X, Y) + f(Y, Z).$$

The relation $f(Z, X \cup Y) = f(Z, X) + f(Z, Y)$ is proven similarly. $\square$

We are now ready to prove the first interesting result about flow-networks. Let us first recall that the conservation law at an internal vertex says that nothing is stored at that junction of the network. Intuitively, this implies that the total flow pushed out of the source in the network must end up at the sink. This is precisely what is claimed in the next theorem.

Theorem 9.20. *Let f be a flow on a network $F = (G, s, t, c)$ with underlying graph $G = (V, E)$. Then the value $|f|$ of the flow is equal to the total flow into the sink: $|f| = \sum_{u \in V} f(u, t)$.*

Proof. Using the implicit summation notation, we need to prove that $|f| = f(V, t)$. Since the subsets $X = \{s\}$ and $Y = V \backslash \{s\}$ form a partition of V, $0 = f(V, V) = f(Y \cup X, V) = f(Y, V) + f(s, V)$ (using parts (a) and by part (d) of Proposition 9.19). So:

$$f(s, V) = -f(Y, V) = -f(V \backslash \{s\}, V). \tag{9.2}$$

Let $X' = \{t\}$ and $Y' = V \backslash \{s, t\}$. Then X' and Y' are disjoint with $X' \cup Y' = V \backslash \{s\}$. Part (d) of the previous proposition implies that $f(V, X' \cup Y') = f(V, X') + f(V, Y')$. In other words:

$$f(V, V \backslash \{s\}) = f(V, t) + f(V, V \backslash \{s, t\}). \tag{9.3}$$

We conclude that

$$
\begin{aligned}
|f| &= \sum_{v \in V} f(s, v) \quad \text{(Definition of the flow value)} \\
&= f(s, V) \quad \text{(Using the implicit summation notation)} \\
&= -f(V \backslash \{s\}, V) \quad \text{(By 9.2)} \\
&= f(V, V \backslash \{s\}) \quad \text{(By part (a) of Proposition 9.19)} \\
&= f(V, t) + f(V, V \backslash \{s, t\}) \quad \text{(By 9.3)} \\
&= f(V, t) - f(V \backslash \{s, t\}, V) \quad \text{(By part (b) of Proposition 9.19)} \\
&= f(V, t) \quad \text{(Since } f(V \backslash \{s, t\}, V) = 0 \text{ by the conservation law)} \quad \square
\end{aligned}
$$

We now state the main problem we deal with in this section.

The Maximum flow problem. Given a flow-network $F = (G, s, t, c)$, construct a flow f on F of maximum possible value. More notations are needed to solve the maximum flow problem.

Definition 9.32. Consider a flow-network $F = (G, s, t, c)$ with underlying graph $G = (V, E)$ and a flow f.

(a) A *cut* in F is a pair (A, B) where A is a non-empty proper subset of V and $B = \overline{A} = V \backslash A$ is the complement of A in V. In other words, a cut is simply a partition of V consisting of only two subsets.

(b) If (S, T) is a cut such that the source s is in S and the sink t is in T, then (S, T) is called a *source-sink cut* or simply *st*-cut.

(c) The *capacity* of a cut (A, B), denoted by $c(A, B)$, is the sum of capacities of all edges (u, v) with $u \in A$ and $v \in B$: $c(A, B) = \sum_{u \in A, v \in B} c(u, v)$.

(d) An *st*-cut (S, T) is called *minimum* if its capacity $c(S, T)$ is the smallest over all possible *st*-cuts of the network.

(e) The *flow across* a cut (A, B) is $f(A, B) = \sum_{u \in A, v \in B} f(u, v)$.

Example 9.55. Consider the flow-network of Example 9.54 above:

- The value of the flow is $|f| = 10 + 14 = 24$ and the flow into the sink t is $13 + 11 = 24$. This is expected by Theorem 9.20.
- Let $S = \{s, v_1, v_2, v_4, v_5\}$, $T = \{v_3, v_6, t\}$. Then (S, T) is an st-cut with capacity $c(S, T) = c(v_2, v_3) + c(v_5, v_3) + c(v_5, v_6) = 10 + 7 + 15 = 32$. The flow across this cut is $f(S, T) = f(v_2, v_3) + f(v_5, v_3) + f(v_5, v_6) = 6 + 7 + 11 = 24$.
- Let $S' = \{s, v_1, v_2, v_3\}$, $T' = \{v_4, v_5, v_6, t\}$. Then (S', T') is another st-cut with capacity $c(S', T') = c(s, v_4) + c(v_1, v_5) + c(v_2, v_5) + c(v_3, t) = 15 + 9 + 6 + 20 = 50$. Notice that the capacity of edge (v_1, v_2) is not part of $c(S', T')$ nor is the capacity of edge (v_4, v_1). The flow across the cut (S', T') is $f(S', T') = f(s, v_4) + f(v_1, v_4) + f(v_1, v_5) + f(v_2, v_5) + f(v_3, v_5) + f(v_3, t) = 10 - 3 + 7 + 4 - 7 + 13 = 24$ (here we used $f(v_1, v_4) = -3$ and $f(v_3, v_5) = -7$ by the skew-symmetry property).

The fact that the flows across the st-cuts (S, T) and (S', T') in the previous example are both equal to the value of the flow in the network is not a coincidence as shown in the next result.

Theorem 9.21. *For any flow f and any st-cut (S, T) in a flow-network $F = (G, s, t, c)$, we have that $|f| = f(S, T)$.*

Proof. Let the underlying digraph of the network be $G = (V, E)$. By Theorem 9.19 above, we have that $f(S, V) = f(S, S \cup T) = f(S, S) + f(S, T) = f(S, T)$. The same theorem also implies that $f(S, V) = f(s, V) + f(S \setminus \{s\}, V) = f(s, V)$ (since $f(S \setminus \{s\}, V) = 0$ by the conservation rule and the fact that the set $S \setminus \{s\}$ does not contain neither s nor t). We conclude that $|f| = f(s, V) = f(S, V) = f(S, T)$ as required. $\square$

Theorem 9.21 gives us already an upper bound on the maximum flow in a network. More precisely, we have the following.

Corollary 9.4.

(a) *The value of any flow in a flow-network is less than or equal to the capacity of any st-cut in the network.*

(b) *The maximum flow in a flow-network is bounded above by the capacity of a minimum st-cut in the network.*

Proof. Part (b) is a direct consequence of part (a). Let $F = (G, s, t, c)$ be a flow-network with a flow f. If (S, T) is any st-cut in F, then $|f| = f(S, T) = \sum_{u \in S} \sum_{v \in T} f(u, v) \leq \sum_{u \in S} \sum_{v \in T} c(u, v) = c(S, T)$. $\square$

9.10.1 *Residual network*

In order to maximise the value of a flow in a network, it is useful to exploit the unused capacity in some edges to increase the flow but it might also be necessary in some cases to decrease the flow in other edges. One way to represent a decrease of a positive flow $f(u, v)$ from u to v is to introduce an edge (v, u) in a "parallel graph" with a capacity $f(u, v)$ that allows a decrease in the flow of at most $f(u, v)$ units from u to v.

Definition 9.33. Consider a flow-network $F = (G, s, t, c)$ with underlying graph $G = (V, E)$ and a flow f. For vertices u and v of G, the *residual capacity* $c_f(u, v)$ of edge (u, v) is the amount of additional flow that can be pushed from u to v without exceeding the capacity $c(u, v)$. That is: $c_f(u, v) = c(u, v) - f(u, v)$.

Note that if $f(u, v) < 0$, $c_f(u, v) > c(u, v)$. At first glance, this sounds like a violation of the capacity rule. The justification is as follows. Say $f(u, v) = -2$ units, then there is a flow of 2 units from v to u that we can "counter" by pushing 2 units from u to v. These two units are added to the capacity $c(u, v)$ of edge (u, v). For example, if $c(u, v) = 11$ and $f(u, v) = 6$, then $c_f(u, v) = 11 - 6 = 5$. On the other hand, if $c(u, v) = 11$ and $f(u, v) = -3$, then the residual capacity along (u, v) is $c_f(u, v) = 11 - (-3) = 14$. If $f(u, v) = c(u, v)$, then $c_f(u, v) = 0$ and the edge (u, v) is called *saturated*.

Definition 9.34. Given a flow-network $F = (G, s, t, c)$ with underlying graph $G = (V, E)$ and a flow f, the *residual network of F induced by the flow f* is the network $F_f = (G_f, s, t, c_f)$ where the underlying graph is $G_f = (V, E_f)$ with $E_f = \{(u, v) \in V \times V; c_f(u, v) > 0\}$.

So the residual network has the same set of vertices as the original network, the same source and the same sink, but a different set of edges. There are two types of edges in the residual network:

• **The forward edges.** For every non-saturated edge (u, v) in G (i.e, with $c_f(u, v) > 0$), assign a residual edge (u, v) in G_f with capacity $c_f(u, v) = c(u, v) - f(u, v)$. This residual edge serves as an indicator that we can increase the flow along edge (u, v) by an amount of $c_f(u, v)$ without exceeding the capacity $c(u, v)$ of the edge.

• **The backward edges.** For every edge (u, v) in G with $f(u, v) > 0$, assign an edge (v, u) in G_f with capacity $c_f(v, u) = f(u, v)$. This edge serves as an indicator that we can decrease the flow along edge (u, v) by pushing a positive flow of a maximum of $f(u, v)$ units from v to u.

Since each edge in G contributes with at most two edges in G_f, we have $|E_f| \leq 2|E|$.

Example 9.56. The diagram below shows the flow-network F given in Example 9.54 together with its residual network F_f.

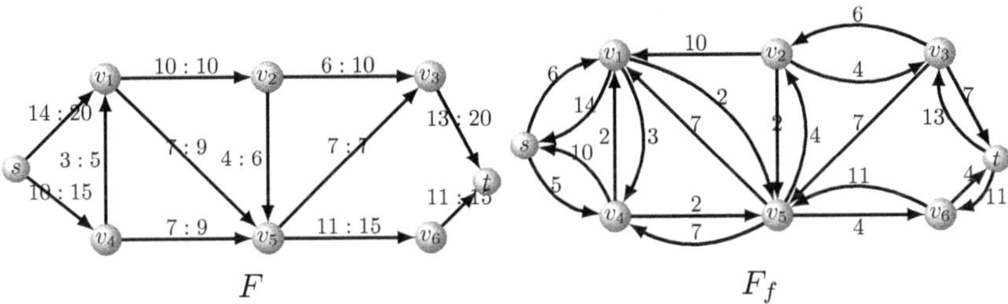

Next, we define the notion of an augmenting path. The existence of such a path in a flow-network means that we can push more flow from the source and consequently the flow is not maximal.

Definition 9.35. Given a flow-network $F = (G, s, t, c)$ and a flow f on F, an *augmenting path* for the flow is a simple *directed* path from s to t in the residual network F_f. By the definition of the residual network, an augmenting path can also be defined as a simple *undirected* path P from s to t in the graph G such that if an edge (u, v) appears in P in the forward direction then it is not saturated ($f(u, v) < c(u, v)$) and if it appears in the backward direction then it has $f(u, v) > 0$. If P is an augmenting path then the *residual capacity* of P (also known as the *bottleneck* of P), denoted by $c_f(P)$, is the maximum additional flow that can be pushed along the edges of P. In other words, $c_f(P)$ is the minimum of residual capacities of all edges on P: $c_f(P) = \min\{c_f(e_i); e_i \text{ is an edge in } P\}$. Notice that $c_f(P) > 0$.

Example 9.57. The path $P = (s, v_1, v_4, v_5, v_6, t)$ in the residual graph of Example 9.56 is an augmenting path of residual capacity $c_f(P) = 2$.

Proposition 9.1. *Let f be a flow on a flow-network $F = (G, s, t, c)$ and let P be an augmenting path in the residual network F_f of F. Define a function*

$f_P : V \times V \to \mathbb{R}$ *as follows:*

$$f_P(u, v) = \begin{cases} c_f(P) & \text{if } (u, v) \in P \\ -c_f(P) & \text{if } (v, u) \in P \\ 0 & \text{otherwise} \end{cases}$$

Then the function $f' = f + f_P : V \times V \to \mathbb{R}$ (defined by $f'(u, v) = f(u, v) + f_P(u, v)$) is a flow on F. Moreover, $|f'| = |f| + c_f(P) > |f|$.

Proof. The fact that f' is a flow is left as an exercise (see Exercise (8) of this section). As for the value of f':

$$|f'| = |f + f_P| = \sum_{v \in V}(f + f_P)(s, v) = \sum_{v \in V}(f(s, v) + f_P(s, v))$$
$$= \sum_{v \in V} f(s, v) + \sum_{v \in V} f_P(s, v) = |f| + c_f(P).$$

The last equality follows from the fact that there is a unique edge (s, w) belonging to the augmenting path P and the value of $f_P(s, w)$ is $c_f(P)$. $\square$

Proposition 9.1 links the existence of an augmenting path to the possibility of increasing the flow in a network.

Proposition 9.2. *Let f be a flow on a network $F = (G, s, t, c)$. If an augmenting path P exists in the residual network F_f, then the flow f is not maximum.*

Proof. If P is an augmenting path in F, then $f' = f + f_P$ is another flow on F with value $|f'| = |f| + c_f(P) > |f|$ since $c_f(P) > 0$ (Proposition 9.1). In particular, f is not maximum. $\square$

9.10.2   The Max-Flow Min-Cut theorem, the Ford–Fulkerson algorithm

By Corollary 9.4 above, we know that the maximum flow value in a network is less than or equal to the capacity of a minimum cut in the network. The Max-Flow Min-Cut theorem shows that the equality actually holds.

Theorem 9.22. *Let $F = (G, s, t, c)$ be a network and let f be a flow on F. The following statements are equivalent.*

(a) $|f| = c(S, T)$ for some st-cut (S, T).
(b) f is a maximum flow.
(c) There is no augmenting path in the residual network F_f.

Proof. Let the underlying graph of the network be $G = (V, E)$. Assume that $|f| = c(S, T)$ for some st-cut (S, T), then f is a maximum flow by Corollary 9.4 above. This proves $(a) \Rightarrow (b)$. The implication $(b) \Rightarrow (c)$ follows from Proposition 9.2. For the implication $(c) \Rightarrow (a)$, assume that no augmenting path exists in F_f and let S be the set of vertices in the network that can reached from s by a directed simple path in F_f and let $T = V \setminus S$. Since there is no directed path from s to t in F_f, (S, T) is an st-cut. Moreover, there is no edge (v, w) in F_f with $v \in S$ and $w \in T$ since otherwise there would be a directed path from s to w in F_f contradicting the fact that $w \in T$. The absence of edge (v, w) in the residual network means that the residual capacity of that edge is zero or equivalently $f(v, w) = c(v, w)$. By Theorem 9.21 above, we know that $|f| = f(S, T) = \sum_{v \in S} \sum_{w \in T} f(v, w) = \sum_{v \in S} \sum_{w \in T} c(v, w) = c(S, T)$. $\square$

Theorem 9.23 (The Max-Flow Min-Cut theorem). *The maximum value of a flow in a network is equal to the minimum of the capacities of all st-cuts in the network.*

Proof. Let f^* be a maximum flow in the network and let δ be the minimum of the capacities of all st-cuts in the network. We know by Corollary 9.4 above that $|f^*| \leq \delta$. On the other hand, Theorem 9.22 tells us that $|f^*| = c(S, T)$ for some st-cut (S, T). Since $\delta \leq c(S, T) = |f^*|$ and the result follows. $\square$

We can now give an algorithm, known as *the Ford–Fulkerson* algorithm (named after Lester Randolph Ford Jr and Delbert Ray Fulkerson), to produce a maximum flow in a network. Set the initial flow to be zero in the network. Using the current network F with flow f, form the residual network F_f and look for an augmenting path in F_f. If such a path P exists, find its bottleneck capacity $c_f(P)$ then adjust the flows on the edges of P in F by increasing the flow on every forward edge by $c_f(P)$ units and decreasing the flow on every backward edge by $c_f(P)$ units. Continue this process until no augmenting path from the source to the sink is possible in the augmented network. At this point, we know that the flow is maximum. The following theorem proves the correctness of the Ford–Fulkerson algorithm in the case where the edge capacities have integer values.

Theorem 9.24. *If all capacities in a flow-network F are integers, then the flow at every iteration of the Ford–Fulkerson algorithm is integer-valued. Moreover, the algorithm produces an integer-valued maximum flow in a finite number of iterations.*

Proof.   Start with the flow $f_0 = 0$. If an augmenting path P exists then its residual capacity $c_f(P)$ is an integer since all the capacities in the network are integers. Adjusting the flow along the edges of P using the bottleneck $c_f(P)$ of P produces a flow f_1 with integer values with $|f_0| < |f_1|$. Continuing this way, we get a sequence $f_0, f_1, f_2, \ldots$ of flows with integer values and such that $|f_0| < |f_1| < |f_2| < \ldots$. This increasing sequence of integers is also bounded by the capacity of any st-cut (Corollary 9.4)in the network, and so it must terminate with a flow f_n for which no augmenting path exits. By Theorem 9.23, f_n must be a maximum flow.   □

Some remarks about the above discussions.

(1) The choice of the term "algorithm" to describe the procedure of Ford-Fulkerun is not very accurate as the procedure does specify the search method used to find an augmenting path.

(2) Theorem 9.24 can be used to prove that a maximum flow exists in a network where edge capacities are positive rational numbers (see Exercise (11.6.4) below).

(3) The Ford–Fulkerson algorithm may not terminate if the capacities are allowed to take irrational values.

Example 9.58. Use Ford–Fulkerson algorithm to find a maximum flow in the network below. Give the value of the maximum flow. As usual, the label "$m : n$" of an edge indicates the capacity n of the edge and the value m of the flow on that edge.

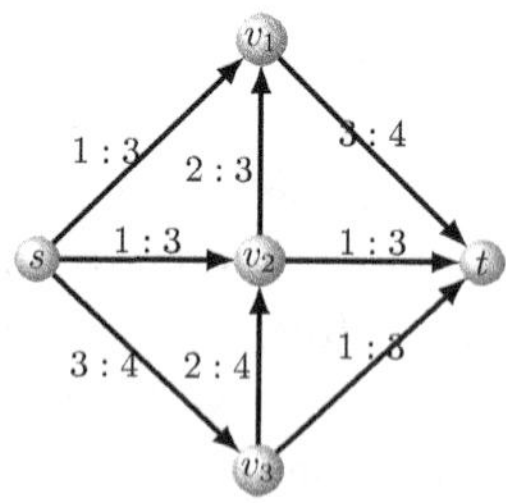

Solution. In the diagram below, the residual graph with an augmenting path in bold is shown on the left and the updated flow-network on the right (obtained by adding or subtracting the residual capacity along the edges of the augmenting path in the previous network).

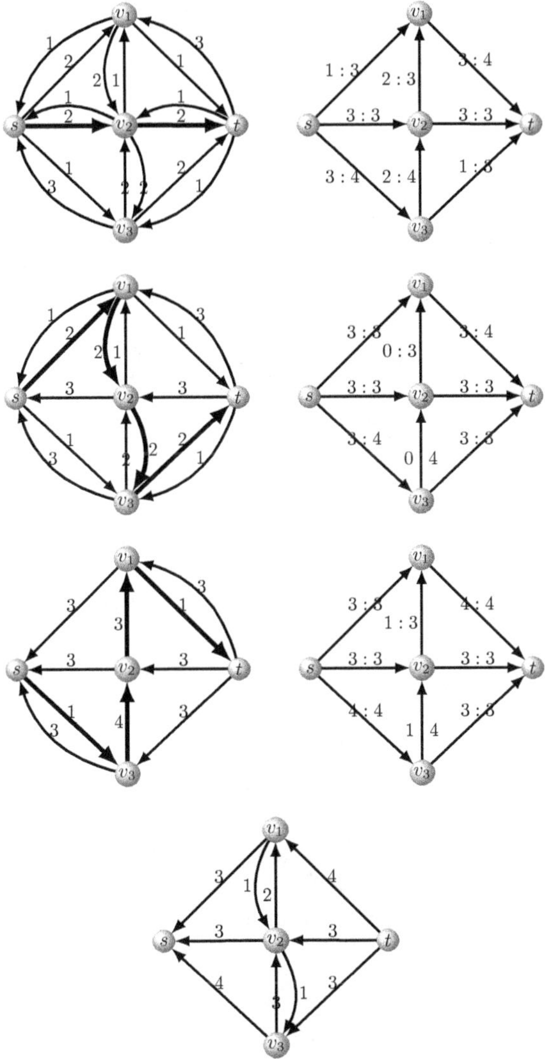

In the last residual network, there is no augmenting path (no arrow leaving the source). This means that the last updated flow is maximum. The value of the maximum flow is $3 + 3 + 4 = 10$.

Example 9.59. Use the Max-Flow Min-Cut theorem (Theorem 9.23 above) to determine the maximum flow in the network of the previous example.

Solution. The Max-Flow Min-Cut theorem tells us that the value of a maximum flow is equal to the minimum capacity of an st-cut. Since there are three internal vertices (beside the source and the sink) in the network, we can count $2^3 = 8$ st-cuts (one st-cut for every subset of $\{v_1, v_2, v_3\}$). The following table shows all 8 cuts of the network together with their capacities.

st-cut (S, T)	Capacity
$S = \{s\}, T = \{v_1, v_2, v_3, t\}$	10
$S = \{s, v_1\}, T = \{v_2, v_3, t\}$	11
$S = \{s, v_2\}, T = \{v_1, v_3, t\}$	13
$S = \{s, v_3\}, T = \{v_1, v_2, t\}$	13
$S = \{s, v_1, v_2\}, T = \{v_3, t\}$	11
$S = \{s, v_1, v_3\}, T = \{v_2, t\}$	14
$S = \{s, v_2, v_3\}, T = \{v_1, t\}$	13
$S = \{s, v_1, v_2, v_3\}, T = \{t\}$	10

The minimum capacity of all possible st-cuts is 10 and it occurs for the st-cuts $(\{s\}, \{v_1, v_2, v_3, t\})$ and $(\{s, v_1, v_2, v_3\}, \{v_3, t\})$. $\Diamond$

The Ford–Fulkerson algorithm does not specify how to choose an augmenting path in the residual network. When we have multiple augmenting paths, a bad choice of the path can increase the number of iterations. This is particularly problematic in a network with large edge capacities. If only few units of flow are added to the edges of the augmenting path in each iteration, it could take a fairly large number of iterations (hence time) for the algorithm to terminate. Exercise (4) below gives an illustration of such a scenario. There are variations of the Ford–Fulkerson algorithm (e.g., the Edmonds–Karp algorithm) where the augmenting paths are chosen in a way to minimize the number of iterations. But we will not discuss these variations in this book.

9.10.3   *Maximum bipartite matching*

In this section, we revisit the notion of maximum matching in bipartite graphs. Recall that a matching in a graph is called *maximum* if it has the largest possible size among all possible matchings of the graph. As it turns out, finding a maximum matching in a bipartite graph is closely related to the Ford–Fulkerson algorithm to find a maximum flow in a network. Given a bipartite graph $G = (V, E)$ with bipartition sets $X = \{x_1, \ldots, x_m\}$ and $Y = \{y_1, \ldots, y_n\}$, start by introducing two new vertices s and t (a source and a sink respectively). Let $V' = V \cup \{s,t\}$ and $E' = \{(x_i, y_j);\ \{x_i, y_j\} \in E\}$ $\cup \{(s, x_i);\ 1 \le i \le m\} \cup \{(y_j, t);\ 1 \le j \le n\}$. Let $G' = (V', E')$ and define the *flow-network associated with* G as being $F_G = (G', s, t, c)$ where the capacity function c is constant with value 1 on every edge in E': $c(u, v) = 1$ for every $(u, v) \in E'$.

Example 9.60. The following shows a bipartite graph G on the left and the associated flow network F_G on the right.

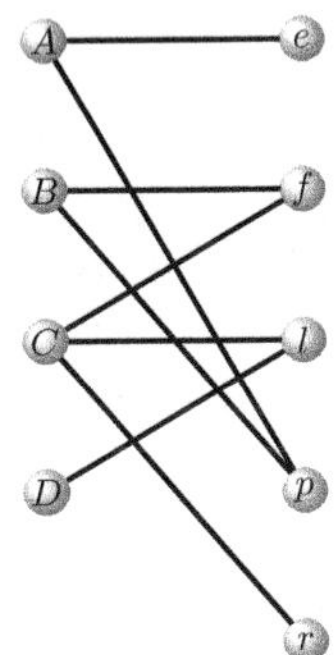
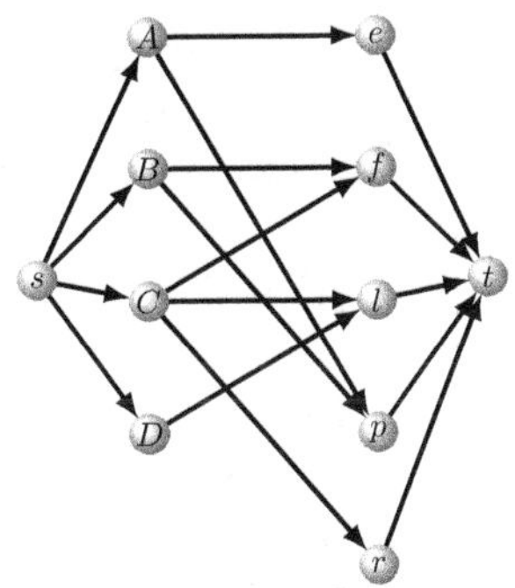

Bipartite Graph G Associated flow network F_G
with every edge capacity equals to 1

We now establish the connection between maximum matching in a bipartite graph and the maximum flow in the associated network.

Theorem 9.25. *Let $G = (V, E)$ be a bipartite graph with bipartition sets X and Y and let F_G be the associated flow-network.*

(1) For any integer-valued flow f of value k on F_G, there exists a matching M_f of size k in G.

(2) Conversely, if M is a matching of size k in G, then there exists an integer-valued flow f_M on F_G with $|f_M| = k$.

(3) The size of a maximum matching in G is equal to the value of a maximum flow on F_G.

Proof.

(1) Assume that f is an integer-valued flow on F_G with value $|f| = k$. Let

$$M_f = \{(x, y);\ x \in X, y \in Y, \text{ and } f(x, y) > 0\}.$$

We prove that M_f is a matching in G. By construction of F_G, each vertex x in X has *at most* one unit of flow entering it through a single edge (s, x). If one unit of flow enters x, one unit of flow must leave x

through a single edge (x, y) with $y \in Y$ (by the conservation rule at x and the fact that the flow is integer-valued). Similarly, one unit of flow must leave y in this case through a single edge (y, t). Therefore, two edges in M_f cannot share an end point and M_f is a matching in G. To prove that $|f|$ is equal to the size of M_f, notice first that $f(X, Y) = \sum_{x \in X} \sum_{y \in Y} f(x, y) = |M_f|$ since $f(x, y) = 1$ for $(x, y) \in M_f$ and $f(x, y) = 0$ for $(x, y) \in E \backslash M_f$. For the st-cut (S, T) where $S = X \cup \{s\}$ and $T = Y \cup \{t\}$, we have that $f(S, T) = f(X, Y) + f(X, t) + f(s, Y) + f(s, t) = f(X, Y) = |M_f|$ since $f(X, t) = f(s, Y) = f(s, t) = 0$. By Theorem 9.21, $|f| = f(S, T) = |M_f|$.

(2) Assume that M is a matching in G of size k. Let $V' = V \cup \{s, t\}$ and define a function $f : V' \times V' \to \mathbb{R}$ as follows. If $(u, v) \in M$, then $f(s, u) = f(u, v) = f(v, t) = 1$ and $f(u, s) = f(v, u) = f(t, v) = -1$. If $(u, v) \notin M$, then $f(u, v) = 0$. It is straightforward to verify that f defines an integer-valued flow on F_G. To see that $|f| = k$, consider the st-cut $S = X \cup \{s\}$, $T = Y \cup \{t\}$ of F_G then $f(S, T)$ has as many non-zero terms as the number of edges in M and each one of these terms is equal to 1. We conclude that $f(S, T) = |M| = k$. By Theorem 9.21 above, $|f| = f(S, T) = k$.

(3) Suppose that M is a maximum matching in G. Let f be the corresponding integer-valued flow on F_G as constructed in the previous part. If $|f|$ is not maximal, then we can apply the Ford-Fulkersen algorithm and produce a maximum flow f' on F_G with $|f| < |f'|$. Moreover, f' is integer-valued by Theorem 9.24. But then the corresponding matching $M_{f'}$ of G has size $|M_{f'}| = |f'| > |f| = |M|$: a contradiction to the fact that M is a maximum matching. We conclude that f is a maximum integer-valued flow in F_G.

$\square$

9.10.4 *Exercises*

(1) The following diagram represents a flow-network with graph $G = (V, E)$, a source s and a sink t. The capacity of each edge is represented by the integer on the edge.

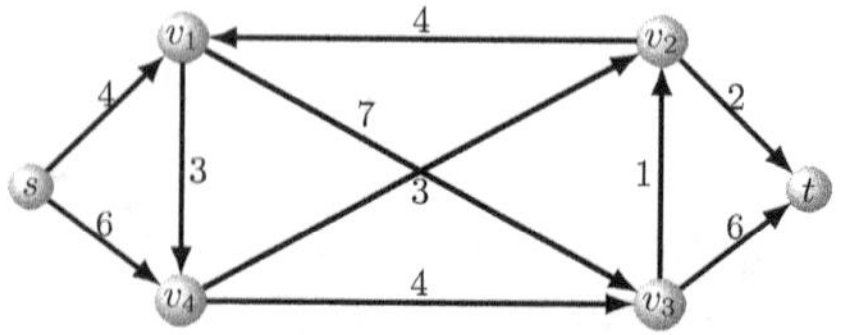

(a) Give two different st-cuts (S, T) and (S', T') on the network and indicate their capacities.

(b) Construct two different flows f and f' in the network using the notation "$m : n$" where m indicates the value of the flow on the edge and n is its capacity.

(c) Give the value of each of the flows f and f' you constructed in the previous part.

(d) With S, S', T, T', f and f' as in the previous parts, compute $f(S, T)$, $f(S', T')$, $f'(S, T)$ and $f'(S', T')$. Verify that $f(S, T) = f(S', T')$ and $f'(S, T) = f'(S', T')$.

(2) The following diagram represents a flow network with graph $G = (V, E)$, a source s, a sink t and a flow f. On each edge, the usual notation "$m : n$" is used represent the capacity $n = c(u, v)$ of the edge and the value $m = f(u, v)$ of the flow on that edge. It is also assumed that $f(v, u) = -f(u, v)$ for any $u, v \in V$.

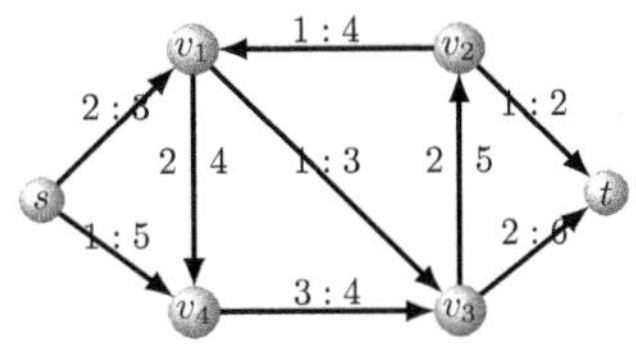

(a) Verify that f is indeed a flow on the network.

(b) Give the value of the flow.

(c) Construct the corresponding residual network.

(d) Conclude that f is not maximum. Justify your answer.

$\star$(3) For the network in Exercise (2):

(a) Apply the Ford–Fulkerson algorithm to find the maximum flow on the network. What is the value of that maximal flow?

(b) Using part (a), give an st-cut in the network of minimum capacity.

$\star$(4) The purpose of this exercise is to show that the choice of an augmenting path can significantly affect the number of iterations in the Ford–Fulkerson algorithm. In the following flow network, each edge has a capacity of 1 or N where $N \geq 2$ is an integer.

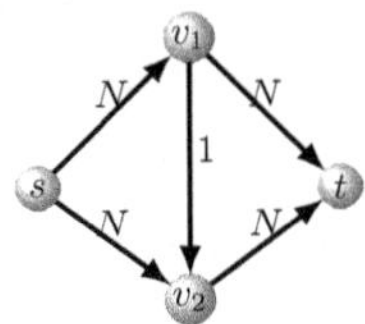

(a) Give all *st*-cuts in the network and their capacities.

(b) Deduce the value of the maximum flow.

(c) Starting with the zero flow, choose two different augmenting paths and use the Ford–Fulkerson algorithm to get to the maximum flow in the network for each of the two paths. How many iterations are required for each one of the two augmented paths?

⋆(5) Consider the following flow network.

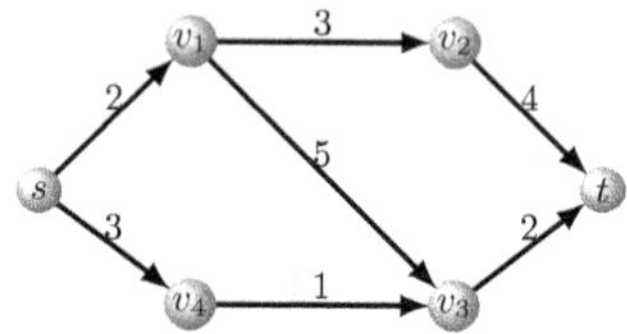

(a) Make a table listing all *st*-cuts of the network and their capacities.

(b) Use the Max-Flow Min-Cut Theorem to find the value of the maximum flow in the network.

(c) Starting with the zero flow, use Ford–Fulkerson algorithm to find the value of the maximum flow in the network.

(6) Repeat the previous exercise using the following flow network:

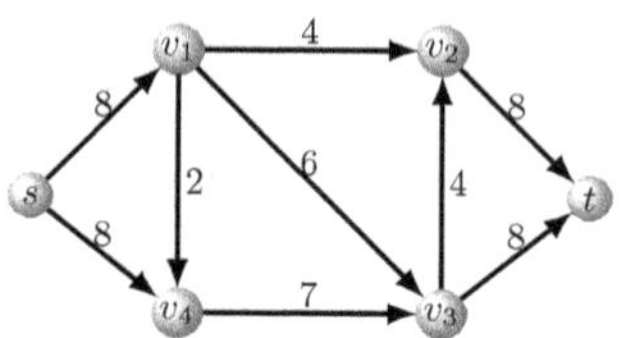

⋆(7) Prove the result of Theorem 9.20 as a corollary of Theorem 9.21.

(8) Prove that the function f' defined in Proposition 9.1 above is a flow on the network F.

⋆(9) In this exercise we use Ford–Fulkerson algorithm to prove that a maximum flow exists in flow network where edge capacities are positive rational numbers (not necessarily integers). Let $F = (G, s, t, c)$ be a flow network with graph $G = (V, E)$ and assume that $c(e_i) = \frac{p_i}{q_i}$ with $p_i \geq 0$, $q_i \geq 1$ are integers for any edge e_i of the network. Let λ be the least common multiple of the q_i's and let F' be the network obtained from F by multiplying each edge capacity with λ.

(a) Prove that if f' is a flow on F', then $f = \frac{1}{\lambda} f'$ is a flow on F with value $|f| = \frac{1}{\lambda}|f'|$.

(b) Prove that if g is a flow on F, then $g' = \lambda g$ is a flow on F' with value $|g'| = \lambda|g|$.

(c) Use the fact that Ford–Fulkerson algorithm terminates and produces a maximum flow on F' to show that a maximum flow on the original flow F exists.

$\star$(10) Let $F = (G, s, t, c)$ be a flow network with integer-valued edge capacities. Let F' be the flow network obtained from F by choosing one edge e in G and increasing its capacity by one unit while leaving the capacities of the other edges unchanged. What is the relation between the values of maximum flows in F and F'?

(11) In each case, use the Ford–Fulkerson algorithm to find a maximum matching in the given bipartite graph G and indicate the size of a maximum matching.

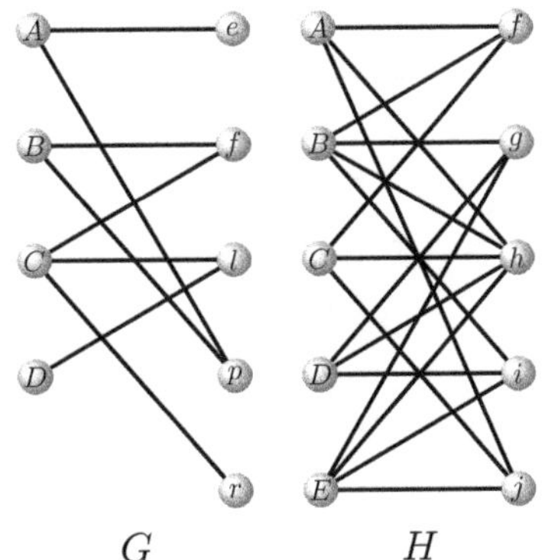

$G \qquad\qquad H$

$\star$(12) Let G be a bipartite graph with bipartition sets X and Y. When the Ford–Fulkerson algorithm is used to find a maximum matching in G, we impose a unit capacity on every edge in the corresponding network $F_G = (G, s, t)$.

(a) Is the assumption of a unit capacity on every edge necessary to find a maximum matching in G?

(b) Assume that we change the capacity of every edge (x, y) with $x \in X$ and $y \in Y$ in the corresponding network of G to ∞ (or just an arbitrary large integer), while keeping the capacities of the other edges at 1, what can be said about a minimal cut (S, T) in the corresponding network?

Chapter 10

More on Graph Theory

Chapter 9 was an introduction to graph theory, its key concepts, its first interesting results, and some of its important applications. The theory is an active domain of research as it extends to many other area of mathematics, computer science and engineering. In this chapter, we continue to explore more interesting concepts and applications of graph theory. We look at the notion of planar graphs, Euler trail and circuits, Hamiltonian paths and circuits and the interesting problem of graph coloring.

10.1 Planar graphs

A well-known problem in graph theory (known as the *three houses and the three utilities problem*), is described as follows. Three houses (labeled A, B and C) are under construction and each house must be connected to the terminal of each of three utilities: water (W), electricity (E) and internet (I). All lines connecting the houses with the services must be at the same depth (in the same plane). The challenge is to connect each house with each of the three utilities in way that the connection lines do not cross anywhere. Using graph theory, the problem asks if we can draw the complete bipartite graph $K_{3,3}$ (representing the houses and the utilities and the connections between them) in the plane in such a way that no two edges cross anywhere except (perhaps) at their endpoints.

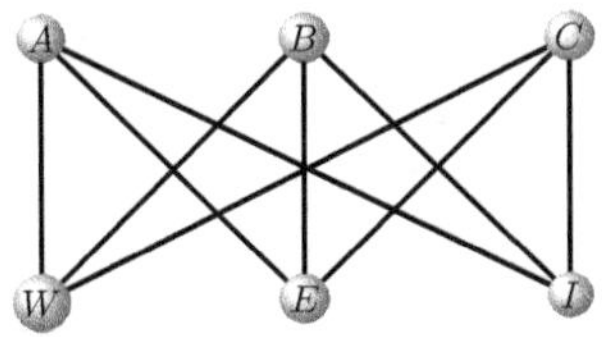

Graphs with no edge crossings are widely used in areas of engineering that require specific designs. For instance, VLSI designs (Very Large Scale Integration) combine thousands of transistors into a single chip and so the lesser the crossings of wires, the better is the design. Another example is in civil engineering where cities try hard to minimize level crossings. The same is true when designing large irrigation canals.

Definition 10.1. A graph is called *planar* if it can be drawn in the plane (like on a paper or a blackboard) in such a way that no two edges cross anywhere except, maybe, at their endpoints. Such a drawing of a planar graph G in the plane is called a *planar embedding* of G. A graph that is drawn in the plane such that no two edges cross at a point different from the end points is called a *plane* graph.

From the definition, it is immediate that every plane graph is planar and every planar graph is isomorphic to a plane graph.

Example 10.1. The complete graph K_4 is planar. The following shows K_4 and three planar embedding G_1, G_2 and G_3 of K_4:

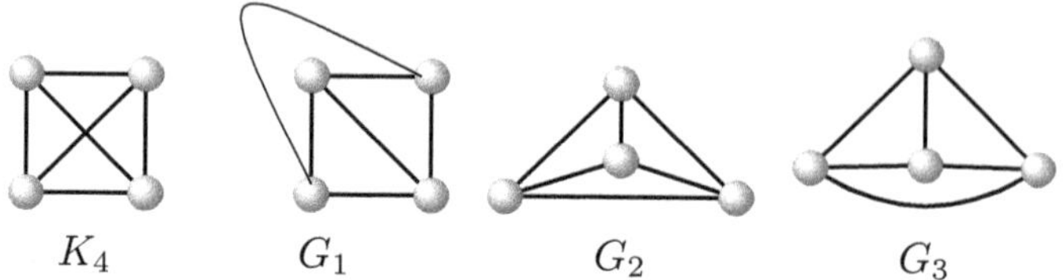

Example 10.2. The graph G in the diagram below is planar. Graph H is one planar embedding of G:

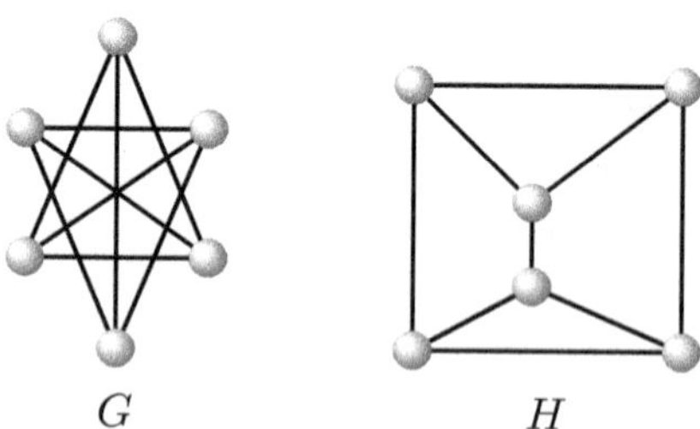

Determining if a graph is planar is not an easy task in general. A characterization of planar graphs was given by Kazimierz Kuratowski. Kuratowski's criteria can be easily stated but not so easy to implement in practice. First, we start with some interesting properties satisfied by planar graphs. In most cases, these properties help proving that a given graph is actually non-planar. First, a definition.

Definition 10.2. Let G be a plane graph. An area of the plane that is bounded by the edges of G and that cannot be subdivided further into smaller subareas is called a *region* or a *face* of G. A face of G is called *finite* (respectively, *infinite*) if it has a finite area (respectively, an infinite area). The edges forming the boundary of a face of G form a closed walk in the graph.

A plane graph divides the plane into a finite number of regions with exactly one infinite region.

Example 10.3. The following planar embedding of the complete graph K_4 divides the plane in four regions (labeled 1, 2, 3 and 4) with region 4 being the infinite one:

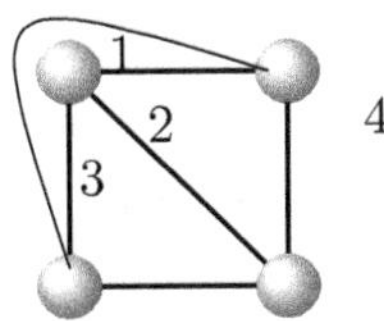

In general, a planar graph has many embeddings in the plane and it is natural to ask if all these embeddings have the same number of faces. Euler settled that question by establishing a relation between the number of faces, the order and the size of the graph. Namely, we have the following.

Theorem 10.1 (Euler's formula). *Let G be a plane connected graph of order n, size e and number of faces f. Then*

$$n - e + f = 2. \tag{10.1}$$

Proof. We use strong induction on e. If $e = 0$, then the graph is K_1 (since it is assumed to be connected). In this case $n = 1$, $e = 0$ and $f = 1$. Relation (10.1) is clearly satisfied in this case. Let $e \geq 1$ and suppose that Euler's formula holds for any connected plane graphs with k edges for any k such that $0 \leq k < e$. Let G be a connected plane graph with n vertices, e edges and f faces. Let $\gamma = \{a, b\}$ be an arbitrary edge of G and let $H = G - \gamma$, the graph obtained from G by removing γ. Then H has n vertices and $e - 1$ edges. If H is connected, then the number of its faces must be $f - 1$. By the induction hypothesis, $n - (e - 1) + (f - 1) = 2$ which implies that $n - e + f = 2$ and Euler's formula is true in this case. If H

is disconnected, then it has two connected components H_1 and H_2 since a single edge of G was removed. Let n_i, e_i and f_i be the number of vertices, edges and faces (respectively) of graph H_i, $i = 1, 2$. Clearly $n_1 + n_2 = n$, $e_1 + e_2 = e - 1$. Since removing edge γ disconnects G, γ must belong to the boundary of the infinite face of G. This means that every face of H_1 and of H_2 is a face of G. Moreover, the infinite face of G is counted twice in the sum $f_1 + f_2$. This implies that $f_1 + f_2 = f + 1$. Since both H_1 and H_2 are connected with size less than e each, the induction hypothesis gives us the two equations: $n_1 - e_1 + f_1 = 2$ and $n_2 - e_2 + f_2 = 2$. Adding these two equations, we get that $(n_1 + n_2) - (e_1 + e_2) + (f_1 + f_2) = 4$. Using the relations $n_1 + n_2 = n$, $e_1 + e_2 = e - 1$ and $f_1 + f_2 = f + 1$, the formula $n - e + f = 2$ follows for G. $\qquad\square$

Corollary 10.1. *All planar embeddings of a connected planar graph have the same number of faces.*

Proof. Let H be a planar embedding of a connected planar graph G. Since H is isomorphic to G, H has the same number of vertices and the same number of edges as G. By Euler's formula, the number of faces of H is $f = e - n + 2$ which is a constant. $\qquad\square$

Example 10.4. Verify Euler's formula for the following plane graph:

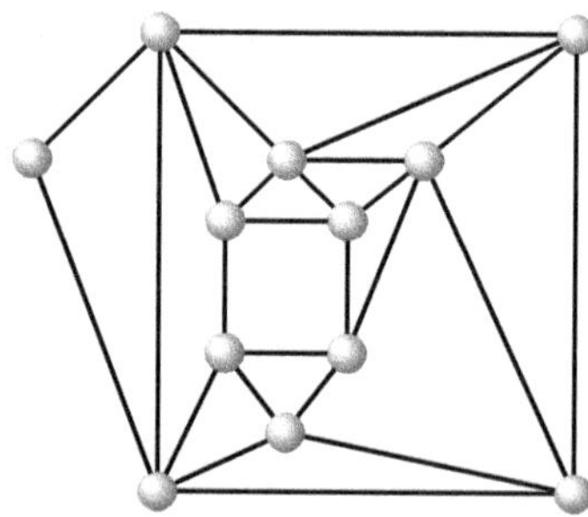

Solution. The graph has $n = 12$ vertices, $e = 25$ edges et $f = 15$ faces, so $n - e + f = 12 - 25 + 15 = 2$ as stated by Theorem 10.1. $\qquad\Diamond$

Definition 10.3. Let G be a plane graph, F a face of G. The **degree** of F, denoted by $\deg(F)$, is defined to be the number of edges of the shortest closed walk of G forming the boundary of F. If an edge is traversed twice when the boundary of F is traced, then it contributes with 2 to the degree of F.

Example 10.5. The following is a plane graph with five faces $F_1, \ldots, F_5$.

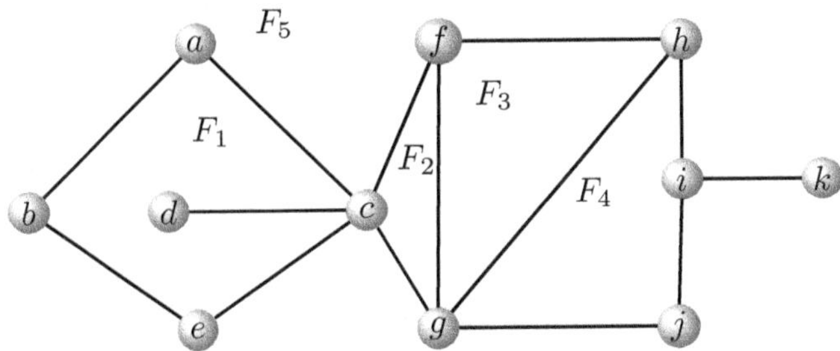

The degrees of the faces are as follows: $\deg(F_1) = 6$ (determined by the closed walk (d, c, a, b, e, c, d)), $\deg(F_2) = 3$ (determined by the closed walk (c, g, f, c)), $\deg(F_3) = 3$ (determined by the closed walk (f, h, g, f)), $\deg(F_4) = 4$ (determined by the closed walk (h, i, j, g, h)) and $\deg(F_5) = 12$ (determined by the closed walk $(a, b, e, c, f, h, i, k, i, j, g, c, a)$).

Adding the degrees of all faces of the previous example gives us 28: twice the number of edges in the graph. This is no coincidence as shown in the following theorem.

Theorem 10.2. *Let G be a connected plane graph. Then the sum of the degrees of all faces of G is equal to two times the number of edges in G.*

Proof. Each edge of G is either a part of the boundary of two different faces (like edge $\{f, g\}$ in the previous example) or occurs twice in the boundary of a single face (like edge $\{c, d\}$ in the previous example). $\qquad\square$

In what follows, we explore a series of corollaries to Euler's formula. Each of these results gives a necessary condition for a graph to be planar.

Corollary 10.2. *Let G be a simple planar graph with order $n \geq 3$ and size e. Then $e \leq 3n - 6$.*

Proof. Assume first that G is connected. Consider a planar embedding E of G, with n vertices, e edges and f faces $F_1, F_2, \ldots, F_f$. Since E is connected with at least three vertices, $\deg(F_i) \geq 3$ for any $i = 1, \ldots, f$ (no faces of degree 1 or 2 since there are no loops and no multiple edges). By Theorem 10.2, $2e = \sum_{i=1}^{f} \deg(F_i) \geq \sum_{i=1}^{f} 3 = 3f$. Euler's formula tells us that $f = 2 + e - n$ and so $2e \geq 3(2 + e - n)$. Consequently $e \leq 3n - 6$. If G is not connected, we can add edges to the planar embedding E of G to obtain a plane simple connected graph E' of order n and size $e' > e$. Since E' is connected, $e' \leq 3n - 6$ and consequently, $e \leq 3n - 6$. $\qquad\square$

Corollary 10.3. *Every simple planar graph has at least one vertex of degree less than or equal to 5.*

Proof. We prove the result by contradiction. Let $\{v_1, \ldots, v_n\}$ be the set of vertices and assume that $\deg(v_i) \geq 6$ for every i. So $\sum_{i=1}^{n} \deg(v_i) \geq \sum_{i=1}^{n} 6 = 6n$. By the handshaking Lemma, we then get that $2e \geq 6n$ or $e \geq 3n > 3n - 6$. This is a contradiction to Corollary 10.2. $\qquad\square$

Corollary 10.4. *Let G be a connected, simple planar graph with order $n \geq 3$, size e and no circuits of length 3. Then $e \leq 2n - 4$.*

Proof. Since there are no circuits of length 3, the degree of every face of G is at least 4. By Theorem 10.2, $2e \geq 4f$ or $f \leq \frac{1}{2}e$. By Euler's formula, we then get that $e - n + 2 \leq \frac{1}{2}e$ which is equivalent to $e \leq 2n - 4$. $\qquad\square$

Important examples of non-planar graphs are $K_{3,3}$ and K_5.

Corollary 10.5. *The complete graph K_5 and the complete bipartite graph $K_{3,3}$ are non-planar.*

Proof. Both K_5 and $K_{3,3}$ are simple and connected. In K_5, the inequality $e \leq 3n - 6$ (of corollary 10.2) is not satisfied since the graph has fives vertices and ten edges. In $K_{3,3}$, there are 6 vertices, 9 edge and no circuit of length 3 since $K_{3,3}$ is bipartite. The inequality $e \leq 2n - 4$ (of corollary 10.4) is not satisfied. $\qquad\square$

A consequence of Corollary 10.5 is that any graph containing K_5 or $K_{3,3}$ as a subgraph is not planar.

10.1.1 *Kuratowski's theorem*

The above corollaries give necessary conditions for a graph to be planar but they are certainly not sufficient. For example, $K_{3,3}$ satisfies the result of Corollary 10.2 but it is not planar. So how can we decide if a given graph is planar? As it turns out, the non-planarity of K_5 and $K_{3,3}$ plays a key role in answering this question. In 1930, the Polish mathematician Kazimierz Kuratowski gave a characterization of planar graphs in the form of a necessary and sufficient condition involving K_5 and $K_{3,3}$. First, some terminology.

Definition 10.4. Let $G = (V, E)$ be a graph, $\{u, v\} \in E$. The operation consisting of removing edge $\{u, v\}$, adding a vertex w to G and connecting w to each of the vertices u and v by one edge (hence w is of degree 2) is called an *elementary subdivision* on edge $\{u, v\}$. The added vertex w is referred to as a *subdivision vertex*. The term "subdivision" comes from the fact that an elementary operation on edge $\{u, v\}$ can be performed by simply "inserting" a vertex w somewhere on the edge between u and v; hence subdividing the edge.

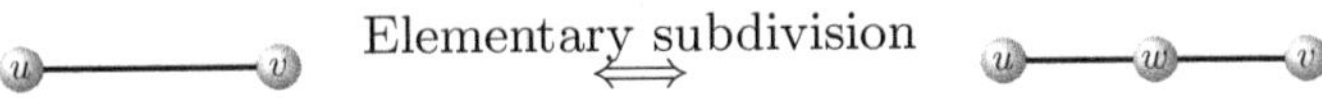

A graph G' is called a *subdivision* of G if $G' \cong G$ or G' can be obtained from G by a finite series of elementary subdivisions on edges of G. We say that the two graphs G_1 and G_2 are *homeomorphic* if they are two subdivisions of the same graph.

Example 10.6. In the following diagram, G_1 and G_2 are homeomorphic as they are subdivisions of graph G:

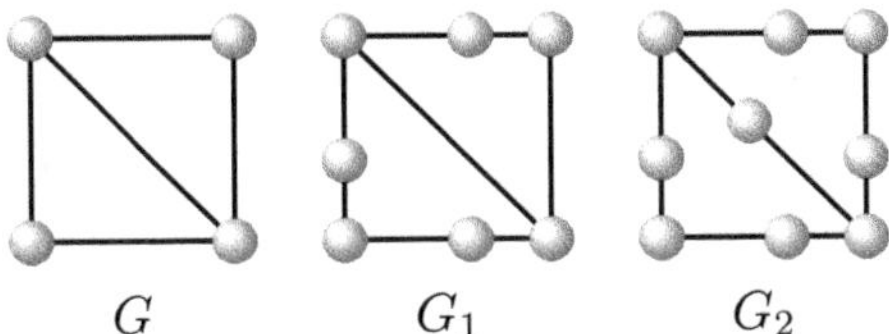

Remark 10.1. It is not hard to see that an elementary subdivision of a graph does not change the planarity of the graph. That is to say, if graphs G and H are homeomorphic then G is planar if and only if H is planar.

We now can state Kuratowski's criteria for planar graphs.

Theorem 10.3 (Kuratowski's theorem). *A graph G is planar if and only if it contains no subgraph homeomorphic to K_5 or to $K_{3,3}$.*

One direction is easy to see: If G has a subgraph H homeomorphic to either K_5 or $K_{3,3}$, then H is not planar and so G is not planar. The proof of the other direction is not that simple and it is very technical. We omit that proof in this book (for a proof of Kuratowski's Theorem, the reader is referred to the book *Applications of Graph Theory* by J. A. Bondy and U. S. R Murty).

Example 10.7. The following diagram shows the Petersen graph P on the left and a subgraph H of P on the right.

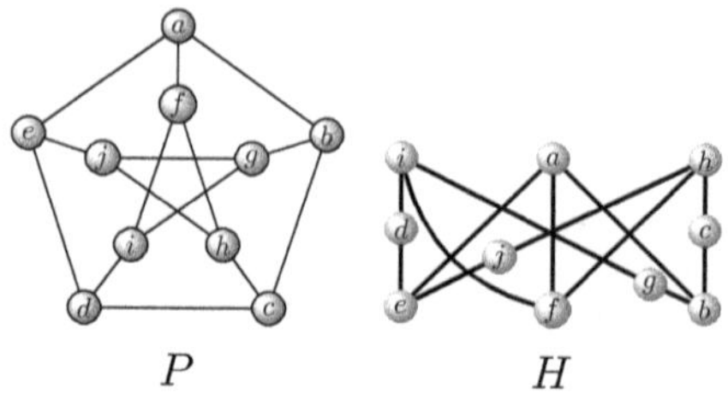

$$P \qquad\qquad H$$

A closer look at H shows that its is a subdivision of $K_{3,3}$. By Kuratowski's Theorem, the Petersen graph is not planar.

10.1.2 *Exercises*

(1) In each case, show that the given graph is planar by drawing a planar embedding of the graph. Verify Euler's formula for each of the graphs.

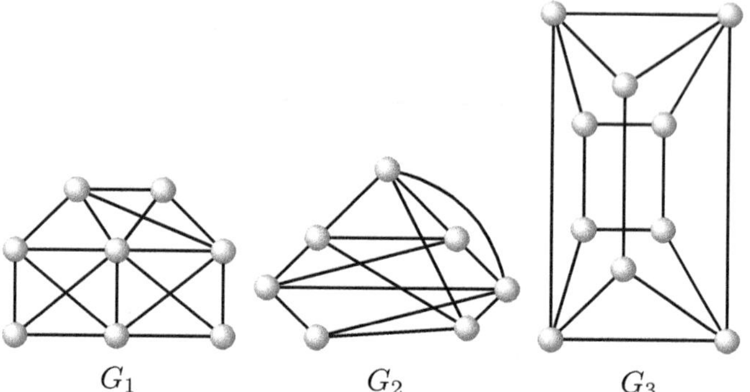

$$G_1 \qquad\qquad G_2 \qquad\qquad G_3$$

$\star$(2) Is it possible for a connected planar graph of order 7 to have 12 edges and 6 faces? If yes, draw such a graph and if no explain why.

(3) How many edges in a connected planar graph if the graph has 5 vertices and 3 faces? Give a possible drawing of such a graph.

$\star$(4) How many vertices in a connected planar graph if the graph has 7 edges and 5 faces? Give a possible drawing of such a graph.

(5) In each case, prove that no simple connected planar graph exists with the given degree sequence.

$\star$ (a) $(4, 4, 5, 6, 6, 7)$ (c) $(4, 4, 4, 5, 5, 5, 6, 6, 6, 7, 7, 7)$

  (b) $(2, 2, 5, 5, 5, 5, 6, 6, 8)$ $\star$ (d) $(6, 6, 6, 6, 6, 7, 7, 7, 9)$

(6) Is it possible for a connected graph G with degree sequence $(2, 2, 3, 4, 4, 5)$ to be planar? If yes, draw such a graph.

⋆(7) Draw two simple connected graphs of degree sequence $(2, 3, 3, 3, 3, 3, 3)$ each but one is planar and the other is not. How many faces must the planar graph have?

(8) Draw two simple connected graphs of degree sequence $(3, 3, 3, 3, 4, 4)$ each but one is planar and the other is not. How many faces must the planar graph have?

⋆(9) Let G be a connected simple planar graph of order $n \geq 3$. Prove that $\sum_{v \in V} (6 - \deg(v)) \geq 12$. (*Hint.* Use Corollary 10.2).

(10) A simple connected planar graph $G = (V, E)$ is such that $\deg(v) \geq 5$ for every vertex $v \in V$. What is the least order G can have? Give a possible drawing of such a graph with the least possible order.

(11) For what values of the integer n is the complete graph K_n planar?

⋆(12) Let $m, n \geq 1$ be integers.

    (a) Prove that $K_{m,n}$ is planar for $m \leq 2$ or $n \leq 2$. How many faces does $K_{m,n}$ have in this case?

    (b) Prove that $K_{m,n}$ is non-planar for $m \geq 3$ and $n \geq 3$.

(13) Let G be a d-regular planar graph of order 10 with $d \geq 1$. If a planar embedding of G has 7 faces, what is the value of d?

(14) Let G be a connected planar graph of order n with e edges and f faces such that the degree of every vertex of G is a multiple of 4. Prove that n and f must have the same parity (either both even or both odd).

(15) Let G be a connected simple planar graph with order at most 11. Prove that G has at least one vertex with degree less than or equal to 4.

⋆(16) Let G be a simple connected planar graph of order $n \geq 4$ and such that $\deg(v) \geq 3$ for every vertex v. Prove that G must have at least four vertices of degree less than or equal to 5 each.

(17) Determine the truth value of each of the following statements. Justify your answer.

    ⋆ (a) Every subgraph of a planar graph is planar.

    (b) Every subgraph of a non-planar graph is non-planar.

    ⋆ (c) K_5 is the only simple connected non-planar graph of order 5.

    (d) If a graph does not contain K_5 or $K_{3,3}$ as a subgraph, then it must be planar.

(18) Let G be a graph and let G' be a subdivision of G. Prove that G is planar if and only if G' is planar.

(19) In each case, determine if the graph is planar. If you say it is, draw a planar embedding of the graph and verify Euler's formula. If you say that the graph is not planar, explain why.

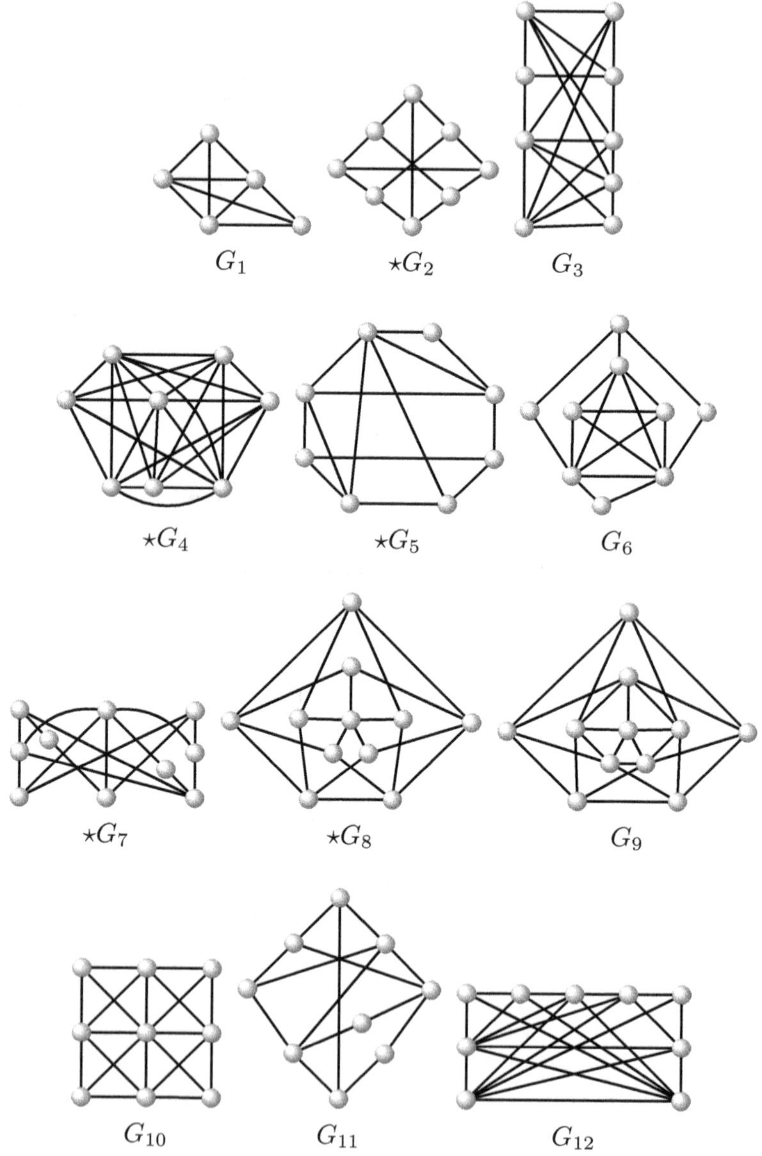

$\star$(20) Let G be a simple, connected, planar graph of order $n \geq 1$. Give two upper bounds on the size e of G in terms of n (*Hint.* One upper bound uses the fact that G is simple and the other uses the fact that G is planar). For what value(s) of n are these two upper bounds equal? Draw all possible non-isomorphic simple connected planar graphs having the two upper bounds equal.

(21) For a simple planar graph of order $n \geq 3$ and size e, Corollary 10.2 above tells us that $e \leq 3n - 6$.

   (a) Give an example of a planar graph with $e = 3n - 6$.

   (b) Give an example of a non-planar graph with $e = 3n - 6$.

$\star$(22) Prove the following generalization of Corollary 10.4. Let G be a simple connected planar graph of order $n \geq 3$ and size $e \geq k$ for some integer $k \geq 3$. If the length of every circuit in G is at least k, then $e \leq \frac{k}{k-2}(n - 2)$.

(23) Euler's formula requires that the graph is connected. There is a generalization of the formula for disconnected graphs. Let n, e and f be the order, the size and the number of faces (respectively) of a graph G with k simple connected and planar components. Prove that $n - e + f = k + 1$.

10.2   Euler trails and Euler circuits

Graph theory was first introduced in the previous chapter using the puzzle of the bridges of Königsberg. In that city, there are two islands A, D and two shores B and C. Seven bridges are built to connect the four regions of the city. Graph G below represents the city and its seven bridges. The

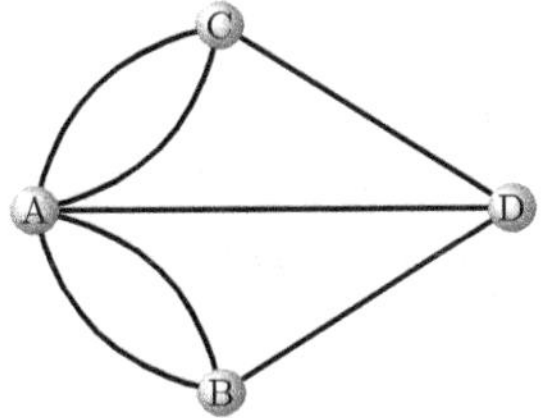

Fig. 10.1   Graph G representing Königsberg and its seven bridges

puzzle is to determine if it possible to start at any region of the city, cross *every* bridge *exactly once* and return to the original region. Clearly, one can attempt to solve this problem by explicitly tracing all possible crossings of the bridges. But you can imagine the complexity of such a task if the scenario is generalized to a city with a larger number of regions and bridges. Euler came up with a surprisingly easy and elegant way to solve the puzzle. The Königsberg bridges puzzle translates to the following problem in graph G: *Is it possible to start at a vertex, cross every edge exactly once and return to the starting vertex?* Such a path is called an Euler circuit.

Definition 10.5. Let G be a connected graph. An *Euler circuit* in G is a closed trail in G traversing each edge of G. An *Euler trail* in G is an *open* trail in G traversing each edge of G. The graph is called Eulerian if it has an Euler circuit.

Recall that a trail in a graph is a walk in the graph with no repeated edges. So an Euler circuit is a closed trail that uses every edge *exactly once*. An Euler trail is an open trail that uses every edge *exactly once*.

Example 10.8. For each the graphs G, H, K and L below, determine if the graph has an Euler circuit, an Euler trail or neither. Note that the edges of H are labeled to help identify trails since H is not simple. The other three graphs are simple and any trail can be identified by the list of vertices it traverses.

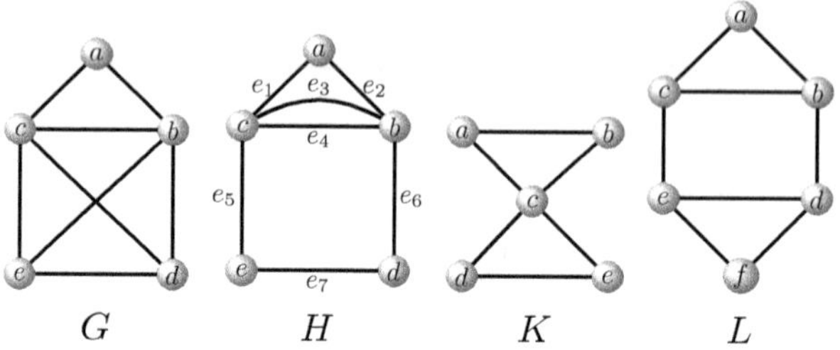

G H K L

Solution. At this point, we don't have any criteria to determine the existence of an Euler trail or an Euler circuit. We proceed by inspection and trial and error. Graph G is not Eulerian but has an Euler trail, namely *dbacbecde*. Graph H is Eulerian: $ae_1ce_5ee_7de_6be_3ce_4be_2a$ is one Euler circuit in H, but the graph has no Euler trail. Similarly, K is Eulerian (*acedcba* is an Euler circuit) but has no Euler trail. Graph L has neither an Euler circuit nor an Euler trail.   ◊

What makes graph L different from the other three in Example 10.8 and what prevents it from having an Euler trail or an Euler circuit? What characteristics allow G to have an Euler trail but not an Euler circuit? As it turns out, the answer to these questions is as simple as writing the degree of each vertex in the graph.

Lemma 10.1. *Let $G = (V, E)$ be a graph such that $\deg(v) \geq 2$ for any $v \in V$. Then G must have a circuit.*

Proof.   The result is clearly true if G is not a simple graph since any loop is a circuit of length one and any pair of multiple edges is a circuit of length two. So we may assume that G is simple. Start at an arbitrary vertex v_0 of G and choose an edge $e_1 = \{v_0, v_1\}$ with $v_1 \neq v_0$ (this is possible since $\deg(v_0) \geq 2$). Similarly, we can choose an edge $e_2 = \{v_1, v_2\}$ with $v_2 \neq v_1$. We repeat this process and get at the kth iteration an edge $e_k = \{v_{k-1}, v_k\}$ with $v_k \neq v_{k-1}$. We obtain this way a trail $T : v_0 v_1 v_2 \ldots v_k \ldots$.

Since there are finitely many vertices in G, the trail T must hit a certain vertex v_i already visited. The part of the trail between the two occurrences of v_i is clearly a circuit in G. $\qquad\square$

Theorem 10.4 (Euler theorem). *Let G be a connected graph. Then:*

(a) G is Eulerian if and only if every vertex in G is of even degree.

(b) G has an Euler trail if and only if it has exactly two vertices u, v of odd degree. Moreover, any Euler trail in G starts at one of the vertices u, v and ends at the other.

Proof.   Assume that G is Eulerian and let C be an Euler circuit in G starting and ending at vertex w. The initial edge of C adds 1 to $\deg(w)$ and each visit of w as an internal vertex of C contributes with 2 to $\deg(w)$. The last edge of C adds another 1 to $\deg(w)$ making $\deg(w)$ even. If v be a vertex of G other than w, then each time we visit v two edges of G are traversed, one to enter and another to leave v. Since all edges incident to v are traversed (by definition of Euler circuit), $\deg(v)$ is even. This proves that every vertex of G has an even degree. Conversely, assume that every vertex of G is of even degree. We proceed by induction on the size m of G to prove that an Euler circuit exists in G. If $m = 0$, then G consists of a single vertex v since it is connected and has no edges. The result is true in this case since the trivial trail $T = v$ is an Euler circuit. Let $m \geq 0$ be an integer and assume and that every connected graph of size k with $0 \leq k \leq m$ and with every vertex of even degree has an Euler circuit. Let $G = (V, E)$ be a connected graph of size $m + 1$ such that $\deg(v)$ is even for every $v \in V$. Note that $\deg(v) \geq 2$ for every $v \in V$ since no vertex of G can have degree 0 (G is connected). By Lemma 10.1, G has a circuit C. If C

uses every edge of G, then we are done. If not, let H be the graph obtained from G by removing all edges of C from G together with any vertex that would have degree 0. It is possible that the graph H is disconnected but every vertex of H remains of even degree (any degree change resulting from removing edges of C is done through removing two distinct adjacent edges from the circuit). By the induction hypothesis, each connected component of H has an Euler circuit since it has fewer edges than G. Moreover, every connected component of H has at least one vertex in common with C since the components are actually formed by removing edges of C. The following is an illustration of a graph G, a circuit C (shown in dashed thick edges), the graph H and its connected components obtained from removing the edges of C from the graph G.

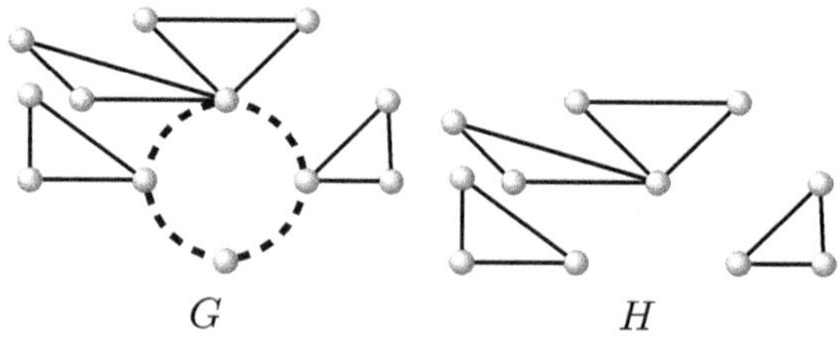

G H

Starting at an arbitrary vertex w of C, travel along the edges of C until a vertex v_1 that lies on a component H_1 of H is met. From vertex w_1, travel along an Euler circuit of H_1 (such a circuit exists by the induction hypothesis) and return to w_1. Once at w_1, continue traveling the circuit C until a second vertex w_2 that lies on another component H_2 is met. From w_2, travel on an Euler circuit of H_2 until you reach back w_2. Continuing this process, we traverse every edge of G exactly once when we return to w since an edge of G is either an edge of C or belongs to one component of H. An Euler circuit has then been traced and the proof is complete by induction. This proves part (a). For part (b), assume G has an Euler trail T with end points u and v. Traveling T starting at u, the first edge out of u contributes with 1 to $\deg(u)$. Every other encounter of vertex u in our path contributes with 2 to $\deg(u)$ since we arrive at u using one edge and exit using another. This shows that $\deg(u)$ is odd. Similarly, $\deg(v)$ is odd. If w is a vertex in G different from u and v, then every time T meets w it goes through two edges incident to w which have not been traced earlier. This shows that $\deg(w)$ is even. Conversely, assume G has exactly two vertices of odd degree u and v. Let H be the graph obtained from G by adding an edge $e = \{u, v\}$ to the set of edges of G (there might be already some edges in G joining u and v). The graph H is connected (since G is) with deg(w) even for any vertex w of H. By part (a), H has an Euler circuit C. Removing edge e from C results in an Euler trail of G starting at one of the vertices u and v and ending at the other. $\square$

Example 10.9. The puzzle of Königsberg's bridges described above can now be solved easily. The graph representing the city and its seven bridges (Figure 10.1 above) has no Euler circuit nor an Euler trail as all four vertices of the graph have odd degrees. This means that it is impossible to start in one region of the city, cross every bridge exactly once and end up in the starting region or in a different region in the city.

Theorem 10.4 gives a surprisingly simple way to determine the existence of an Euler circuit and an Euler trail. Moreover, its proof hints to an algorithm to trace an Euler circuit or an Euler trail when they exist. Given a connected graph G with all vertices of even degree, begin at an arbitrary vertex v_0 and start tracing a trail C_0 along the edges of G until hitting a vertex w with no unused edges to travel out over. Then $w = v_0$ since otherwise deg(w) would be odd) and so C_0 is a circuit at v_0. If C_0 uses all edges of G, then we have traced an Euler circuit. Otherwise, there must be at least one vertex v_1 on C_0 with an even number of unused incident edges. Let G_1 be the graph obtained from G by deleting all edges of C_0 and the resulting isolated vertices. Clearly G_1 has all vertices of even degree and similar to what was done in G, we can trace a circuit C_1 at vertex v_1 in G_1. Rearrange circuit C_0 so it starts and ends at v_1 and join circuits C_0 and C_1 together to get a larger circuit C at v_1. If C uses all edges of G, then we have an Euler circuit. Otherwise, repeat the above process. Since G has finitely many edges, this process must terminate and when it does it produces an Euler circuit. If G has exactly two vertices of odd degree, say v and w, add an edge $e = \{v, w\}$ to the graph and obtain a graph G' with all vertices of even degree. Starting at either v or w, follow the process above to trace an Euler circuit C in G'. Removing edge e from C results in an Euler trail in G starting at one of the vertices u, v and ending at the other.

Example 10.10. A graph G is given below. Determine if G has an Euler circuit or an Euler trail. If it does, trace such a circuit or trail.

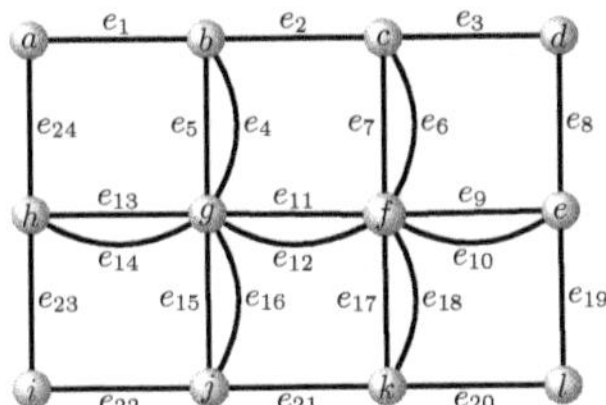

Solution. Note first that each vertex of G has an even degree, so G is Eulerian. We construct an Euler circuit using the algorithm above. Start at vertex a and choose the circuit $C_0 : ae_1be_2ce_3de_8ee_{19}le_{20}ke_{21}je_{22}ie_{23}he_{24}a$ at a. Removing the edges of C_0 and all resulting isolated vertices from G, namely a, d, l and i, gives the following graph G_1:

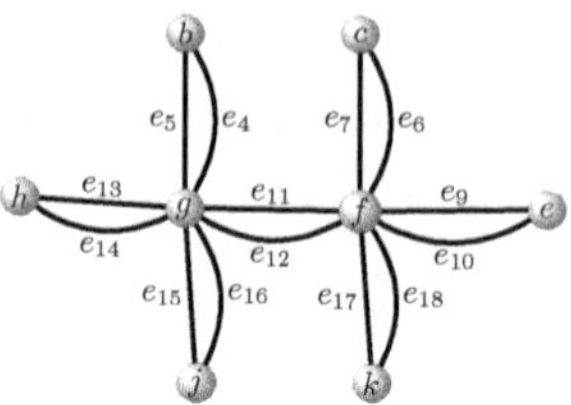

Next, we choose a vertex on C_0 with unused incident edges. Here we have the choice between b, c, e, k, j, h. Choosing vertex b, we start by rearranging circuit C_0 to start and terminate at b:

$$be_2ce_3de_8ee_{19}le_{20}ke_{21}je_{22}ie_{23}he_{24}ae_1b.$$

We then choose the following circuit at b in G_1: $be_5ge_{15}je_{16}ge_4b$. Splicing this circuit with C_0, we get the circuit

$$C_1 : be_2ce_3de_8ee_{19}le_{20}ke_{21}je_{22}ie_{23}he_{24}ae_1be_5ge_{15}je_{16}ge_4b.$$

Removing the edges of circuit $be_5ge_{15}je_{16}ge_4b$ from G_1 together with the resulting isolated vertices, we get the graph G_2 as follows:

The only vertex on circuit $be_5ge_{15}je_{16}ge_4b$ of graph G_1 with unused edges is g. We rearrange C_1 so it starts and ends at g:

$$C_1 : ge_{15}je_{16}ge_4be_2ce_3de_8ee_{19}le_{20}ke_{21}je_{22}ie_{23}he_{24}ae_1be_5g. \qquad (10.2)$$

Now, $ge_{13}he_{14}ge_{11}fe_7ce_6fe_9ee_{10}fe_{18}ke_{17}fe_{12}g$ is an Euler circuit in G_2. Splicing it with circuit (10.2), we get the following Euler circuit of G:

$$ge_{15}je_{16}ge_4be_2ce_3de_8ee_{19}le_{20}ke_{21}je_{22}ie_{23}he_{24}ae_1be_5ge_{13}he_{14}ge_{11}fe_7ce_6fe_9$$
$$ee_{10}fe_{18}ke_{17}fe_{12}g.$$

Example 10.11. Consider the following graph G.

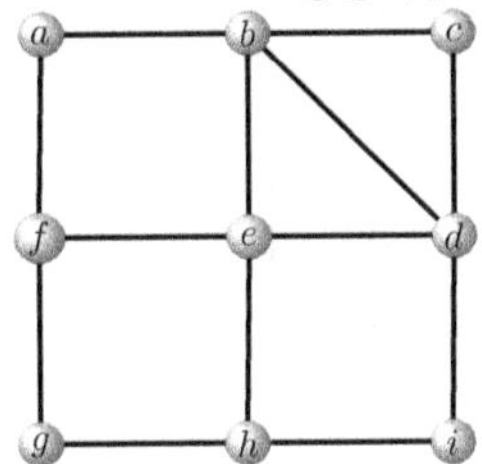

All vertices of G except f and h have even degrees. So G has an Euler trail starting at f (or h) and ending at h (or f). It is easy to verify that $fabedbcdihefgh$ is indeed an Euler trail in the graph.

10.2.1 *Exercises*

(1) In each case, determine if the graph has an Euler trail, an Euler circuit or neither.

(a) K_3	(c) K_5	(e) C_4	(g) W_4	(i) $K_{2,2}$	(k) $K_{3,3}$
(b) K_4	(d) C_3	(f) C_5	(h) W_5	(j) $K_{2,3}$	(l) P_5

(2) For each of the graphs below, determine the value(s) of the integer $n \geq 2$ (if any) for which the graph has: (i) an Euler circuit; (ii) an Euler trail.

$\star$ (a) K_n (b) C_n (c) W_n (d) P_n $\star$ (e) Q_n

$\star$(3) Determine the conditions on the positive integers m and n so that the complete bipartite graph $K_{m,n}$ has: (i) an Euler circuit; (ii) an Euler trail.

(4) In each case, determine if the given graph has an Euler trail, an Euler circuit or neither. Trace such a trail or circuit if the graph has one.

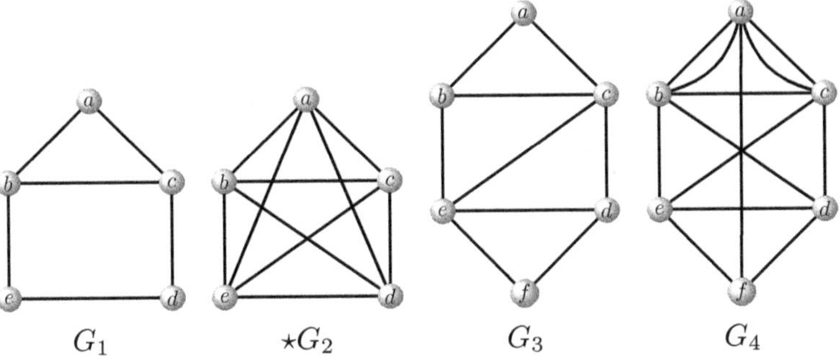

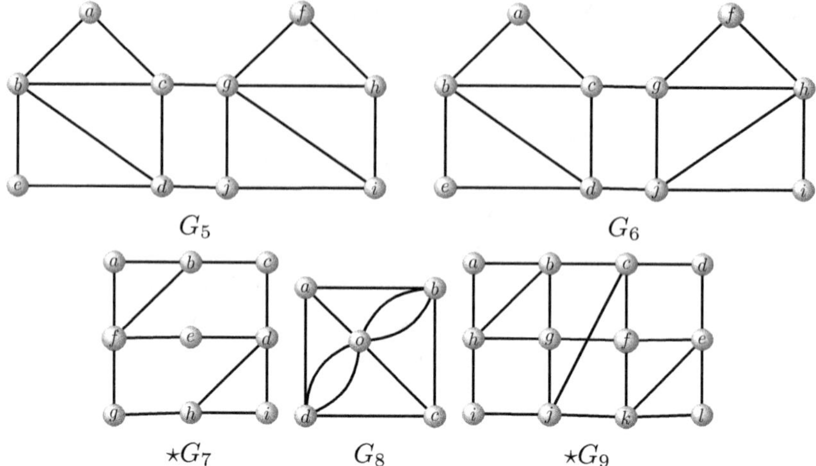

G_5 G_6

$\star G_7$ G_8 $\star G_9$

(5) Prove that the graph G given below is Eulerian. Label the vertices and trace a Euler circuit using the algorithm described in this section. Remove a single edge from G and trace an Euler trail in the resulting graph.

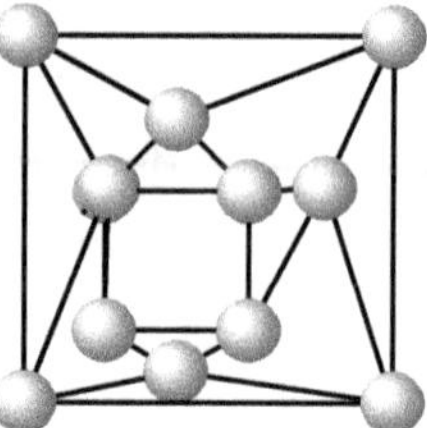

$\star$(6) A city has five regions: three islands A, B, C (shown as circles in the diagram) and two shores (North and South shores). The four regions are connected by 12 bridges (shown as solid straight lines in the diagram).

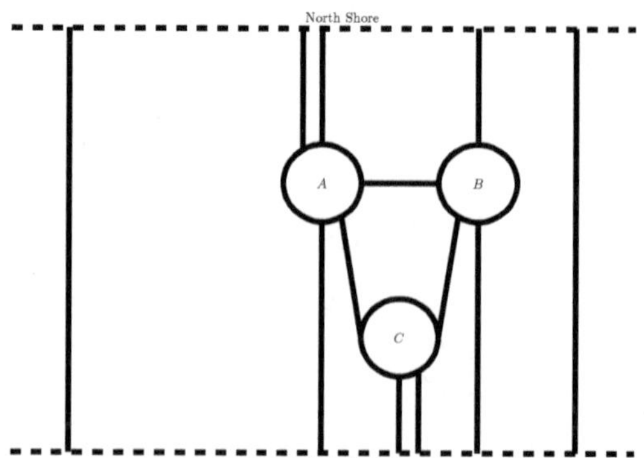

(a) Can you tour the city crossing every bridge exactly once and finishing the tour at the same region of departure? If yes, trace such a tour of the city. If no, can you recommend building one or more bridges between the regions to make such a tour possible?

(b) Can you tour the city crossing every bridge exactly once and finishing your tour at a different region than the region of departure? If yes, trace such a tour of the city. If no, can you recommend building one or more bridges between the regions to make such a tour possible?

(7) In each case, the floor plan of a certain museum is shown. Letters $A, \ldots, I$ represent exhibition rooms in the museum. The small thick rectangles represent doors separating rooms in the museum or leading to the outside. Is it possible to:

  (i) Start in a room, cross every doorway exactly once and return in the same room? If yes, draw such a tour.

 (ii) Start on the outside of the museum, cross every doorway exactly once and end on the outside? If yes, draw such a tour.

(iii) Start in a room, cross every doorway exactly once and end on the outside? If yes, determine if such a tour is possible regardless of the room you started at?

(iv) start in a room, cross every doorway exactly once and finish in a different room? If yes, determine if such a tour is possible between any two rooms of the museum?

Justify your answers using graph theory.

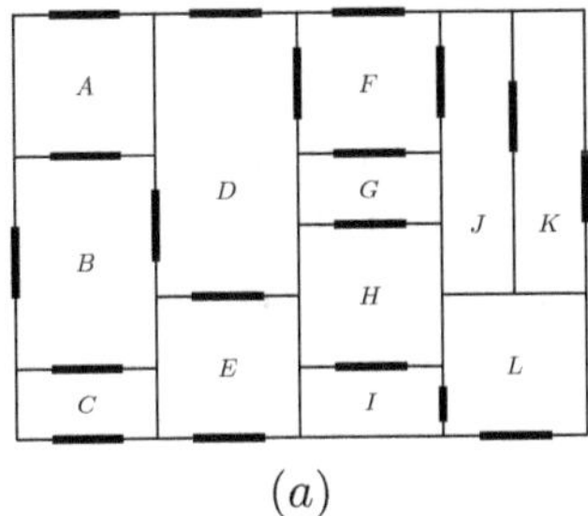

(a)

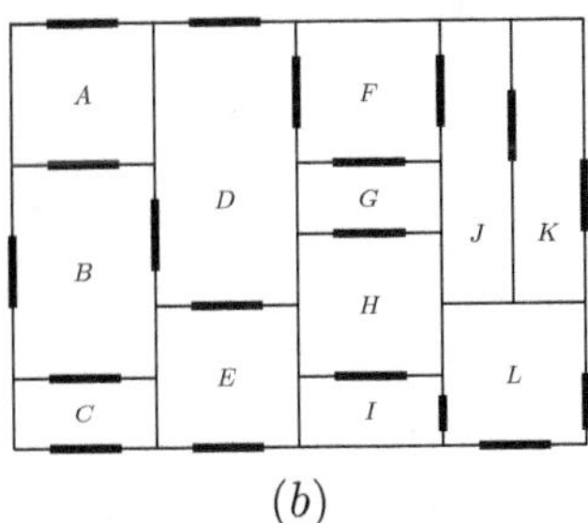

(b)

(8) In each case, determine if the drawing shown can be traced on a paper without lifting the pencil or retracing any line of the drawing.

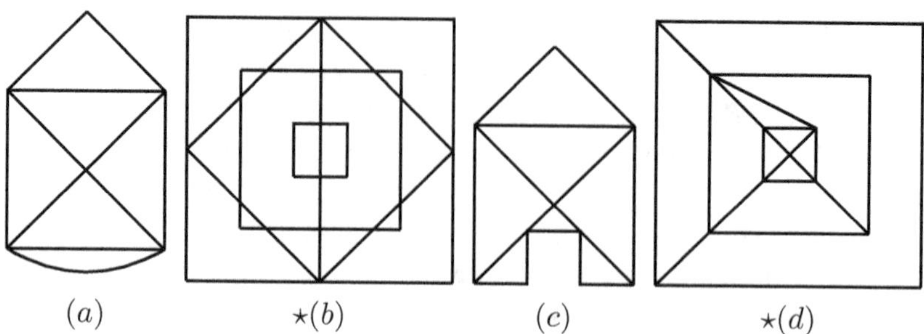

(a) $\star(b)$ (c) $\star(d)$

$\star$(9) A domino tile is a rectangular shape with two sets (possibly empty) of dots on each half of the rectangle. For $0 \leq i \leq j \leq 6$, a tile containing i dots in one half and j dots in the other is denoted by $[i, j]$.

  (a) How many domino tiles $[i, j]$ are there with $0 \leq i < j \leq 6$?

  (b) Is it possible to arrange all domino tiles $[i, j]$ with $0 \leq i < j \leq 6$ end to end in such a way that the numbers of dots on adjacent ends are always the same?

(10) Let G, H be two connected and isomorphic graphs. Prove each of the following statements.

  (a) G is Eulerian if and only if H is Eulerian.

  (b) G has an Euler trail if and only if H has an Euler trail.

$\star$(11) Let G be a connected simple regular d-graph with $d \geq 1$. Prove that the line graph $L(G)$ of G is Eulerian (Recall that the vertices of $L(G)$ are the edges of G and two vertices of $L(G)$ are adjacent if and only if the corresponding edges share a common endpoint in G).

(12) Prove that every connected graph is a subgraph of an Eulerian graph.

$\star$(13) The incidence matrix of a graph G with four vertices v_1, v_2, v_3, v_4 and nine edges $e_1, \ldots, e_9$ is given:

$$I = \begin{array}{c} \\ v_1 \\ v_2 \\ v_3 \\ v_3 \end{array} \begin{array}{c} e_1\ e_2\ e_3\ e_4\ e_5\ e_6\ e_7\ e_8\ e_9 \\ \left[\begin{array}{ccccccccc} 1 & 1 & 1 & 0 & 0 & 1 & 0 & 0 & 0 \\ 1 & 1 & 0 & 1 & 1 & 0 & 1 & 1 & 0 \\ 0 & 0 & 1 & 1 & 1 & 0 & 0 & 0 & 1 \\ 0 & 0 & 0 & 0 & 0 & 1 & 1 & 1 & 1 \end{array} \right] \end{array}$$

  (a) Is G simple? Is G connected? justify.

  (b) Is G Eulerian? Is there an Euler trail in G?

  (c) Give a possible drawing for G.

(14) For each of the following statements, determine if it is true or false. If it is true, prove it and if it is false give a counter-example.

⋆ (a) If a connected graph G of even order $n \geq 4$ has an Euler circuit, then it must have an even size.

  (b) Two connected simple and Eulerian graphs with the same order and the same size must be isomorphic.

⋆ (c) There exists a unique simple connected Eulerian graph of order 4 (up to isomorphism).

  (d) If G and H are two connected Eulerian graphs, then the graph K obtained by joining an arbitrary vertex u in G to an arbitrary vertex v of H with a single edge has an Euler trail.

⋆ (e) If G is a connected simple graph such that $\deg(u) + \deg(v)$ is even for any vertices u, v of G then G must be Eulerian.

(15) Let G be an Eulerian connected simple graph of odd order $n \geq 3$. Prove that G has at least three vertices of the same degree. (*Hint*. Use the pigeonhole principle).

10.3 Hamiltonian circuits and Hamiltonian paths

The puzzle of the bridges of Königsberg described in the previous section asks if one can tour the city by crossing every bridge exactly once. It seems only natural to ask if the tour can be accomplished if one wants to visit every region of the city exactly once. Of course, the Königsberg model is easy to deal with by trial and error but you can imagine that the problem gets complicated quickly as the number of regions increases.

Definition 10.6. Let G be a graph. A *Hamiltonian path* in G is a path that includes every vertex of G. Since by definition, no vertex in a path is repeated, a Hamiltonian path in G uses every vertex of G exactly once. A *Hamiltonian circuit* in G is a circuit that uses every vertex of G exactly once (expect for the initial and the terminal vertices). We say that the graph is *Hamiltonian* if it has a Hamiltonian circuit.

Example 10.12. The cycle C_n $(n \geq 3)$ and the complete graph K_n $(n \geq 1)$ are Hamiltonian graphs (see Exercise (6) below).

If a graph G is Hamiltonian then removing any edge from a Hamiltonian circuit creates a Hamiltonian path in the graph.

Example 10.13. Consider the three graphs:

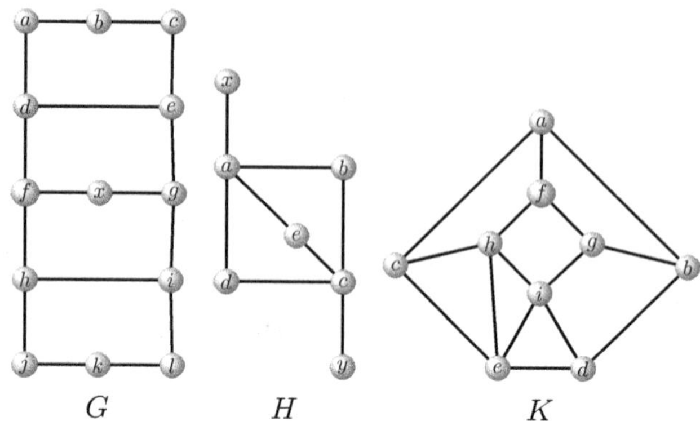

G H K

Graph G has a Hamiltonian path, namely $abcedfxgihjkl$, but no Hamiltonian circuit (we will prove this a bit later). Graph H has neither Hamiltonian circuit nor Hamiltonian path. Graph K has a Hamiltonian circuit, namely $afgbdiehca$. Removing edge $\{a, f\}$ from the circuit gives the Hamiltonian path $fgbdiehca$.

For the purpose of determining if a graph is Hamiltonian, we can restrict ourselves to simple graphs since removing loops or parallel edges between vertices does not change the fact that the graph is Hamiltonian or not (see Exercises (4) below). For the rest of this section, we only consider simple graphs.

The apparent similarity between Hamiltonian and Eulerian graphs is misleading. Unlike an Eulerian path or circuit, there is no necessary and sufficient condition for the existence of a Hamiltonian path or circuit. There are, however, some interesting sufficient conditions that we explore next.

Theorem 10.5 (Ore theorem). *Let G be a simple graph of order $n \geq 3$. If $\deg(u) + \deg(v) \geq n$ for every pair of non-adjacent vertices u and v of G, then G is Hamiltonian.*

Proof. Assume that $\deg(u) + \deg(v) \geq n$ for every pair of non-adjacent vertices u and v of G. We start by proving that G is connected. If it is not, let C_1 and C_2 be two different connected components of G of order n_1 and n_2 respectively. If u_1 is a vertex in C_1 and u_2 is a vertex in C_2, then $\deg(u_1) + \deg(u_2) \geq n$ since u_1 and u_2 are not adjacent. On the other hand, $\deg(u_1) \leq n_1 - 1$ and $\deg(u_2) \leq n_2 - 1$ since G is simple. So $\deg(u_1) + \deg(u_2) \leq n_1 + n_2 - 2 < n$: a contradiction. This proves that G is connected. If G is not Hamiltonian, then we can successively add to G all edges $\{u, v\}$ for every non-adjacent vertices u, v of G and eventually obtain the complete graph K_n. Since K_n is Hamiltonian, we must change from a non-Hamiltonian to a Hamiltonian graph at some stage of this procedure. Let H be the last non-Hamiltonian graph obtained by the process of adding edges between non-adjacent vertices of G. So H is a *maximal* non-Hamiltonian supergraph of G in the sense that if one more edge is added to H, the new graph would be Hamiltonian. If v is any vertex of G, then clearly $\deg_H(v) \geq \deg_G(v)$ since G is a subgraph of H. This means that the condition $\deg(u) + \deg(v) \geq n$ for every pair (u, v) of non-adjacent vertices holds also for H. Since H is not the complete graph, it must have at least one pair (a, b) of non-adjacent vertices. The graph $H + ab$ obtained from H by adding the edge $\{a, b\}$ is now Hamiltonian. Moreover, any Hamiltonian circuit in $H + ab$ must include edge $\{a, b\}$. This means in particular that graph H contains a Hamiltonian path σ between a and b. Relabel the vertices of G if necessary and write $\sigma : u_1 u_2 u_3 \ldots u_{n-1} u_n$ where $u_1 = a$ and $u_n = b$. Consider the two sets:

- $A = \{i \in \mathbb{N};\ 2 \leq i \leq n,\ \text{and } u_i \text{ is adjacent to } a \text{ in } H\}$
- $B = \{i \in \mathbb{N};\ 2 \leq i \leq n,\ \text{and } u_{i-1} \text{ is adjacent to } b \text{ in } H\}$.

If $\{a, u_i\}$ and $\{u_{i-1}, b\}$ are both edges of H for some $i \in \{2, \ldots, n\}$, then $a u_i u_{i+1} \ldots b u_{i-1} u_{i-2} \ldots a$ would be a Hamiltonian circuit in H, which is a contradiction. This shows that $A \cap B = \emptyset$. The cardinality $|A|$ of A is equal to $\deg_H(a)$ since the graph has no loops. Similarly, $|B| = \deg_H(b)$. Since a and b are not adjacent in H, our assumption implies that $\deg_H(a) + \deg_H(b) \geq n$. Now, A and B are two subsets of $\{2, 3, \ldots, n\}$, so $|A \cup B| \leq n - 1$. By the principle of inclusion–exclusion, $|A \cap B| = |A| + |B| - |A \cup B| \geq n - (n-1) = 1$, a contradiction to the fact that $A \cap B = \emptyset$. We conclude that G is Hamiltonian. $\qquad\square$

Example 10.14. Consider the graph G:

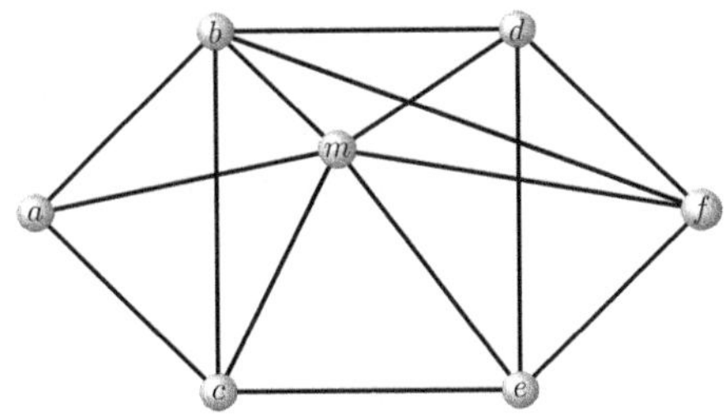

The list of non-adjacent pairs of vertices of G and the sum of their degrees is given in the following table:

Non-adjacent vertices u, v	$\deg(u) + \deg(v)$
a, d	7
a, e	7
a, f	7
b, e	8
c, d	8
c, f	8

The table shows that $\deg(u) + \deg(v) \geq 7$ for every pair of non-adjacent vertices of G. Since G is of order 7, Ore's Theorem implies that G is Hamiltonian. For instance, $abmdfeca$ is a Hamiltonian circuit in G.

Theorem 10.6 (Bondy and Chvátal). *Let G be a graph of order n and let u, v be two non-adjacent vertices of G such that $\deg(u) + \deg(v) \geq n$. Then G is Hamiltonian if and only if the graph $G' = G + \{u, v\}$ (obtained by connecting u and v with one edge) is Hamiltonian.*

Proof. One direction is easy: If G is Hamiltonian then G' is Hamiltonian since any Hamiltonian circuit in G is also one in G'. For the other direction, assume that G' is Hamiltonian but G is not. Then every Hamiltonian circuit in G' must include edge $\{u, v\}$ and therefore there exists a Hamiltonian uv-path in G (obtained by removing edge $\{u, v\}$ from a Hamiltonian circuit in G'). Following the same argument as in the proof of Theorem 10.5, we get a Hamiltonian circuit in G which is a contradiction. We conclude that G is Hamiltonian. $\qquad\square$

Gabriel Dirac gave a sufficient condition for the existence of a Hamiltonian circuit in terms of individual degrees of vertices. Dirac's result is an easy corollary of Ore's Theorem.

Theorem 10.7 (Dirac). *Let G be a graph of order $n \geq 3$. If $\deg(v) \geq \frac{n}{2}$ for every vertex v of G, then G is Hamiltonian.*

Proof. We may assume that G is not isomorphic to the complete graph K_n since the latter is Hamiltonian. Let u, v be two non-adjacent vertices in G, then $\deg(u) + \deg(v) \geq \frac{n}{2} + \frac{n}{2} = n$. The result follows by Theorem 10.5. $\qquad\square$

Example 10.15. For the graph G of order 6 given below, the degree of every vertex is greater than or equal to $\frac{6}{2} = 3$. By Dirac's Theorem, G is

Hamiltonian. For instance, *camdebc* is a Hamiltonian circuit in G.

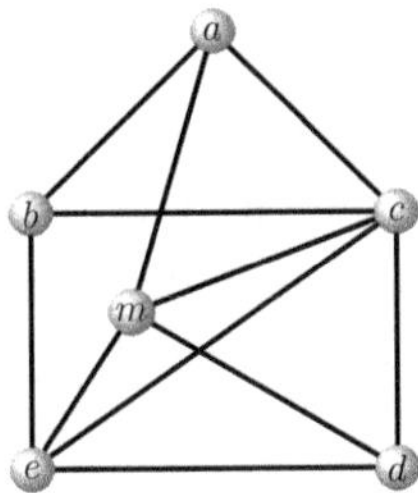

Recall (see Definition 9.15 of Chapter 9) that the Hamiltonian closure $cl(G)$ of a graph G of order n is the graph obtained from G by successively adding edges between non-adjacent vertices u, v of G satisfying $\deg(n)+\deg(v) \geq n$ until this can no longer be done.

Theorem 10.8. *A graph G is Hamiltonian if and only if $cl(G)$ is Hamiltonian.*

Proof. If G is Hamiltonian then so is $cl(G)$ since G is a subgraph of $cl(G)$ with the same set of vertices. If $cl(G)$ is Hamiltonian, then repeated applications of Theorem 10.6 implies that G is Hamiltonian. $\square$

Corollary 10.6. *Let G be a graph of order $n \geq 3$. If $cl(G) = K_n$, then G is Hamiltonian.*

Proof. This is a direct consequence of Theorem 10.8 and the fact that K_n is Hamiltonian. $\square$

Example 10.16. For the following graph G:

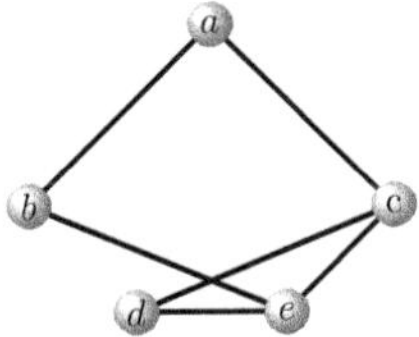

the Hamiltonian closure $cl(G)$ is obtained by adding edges $\{b,c\}$, $\{b,d\}$, $\{a,d\}$ and $\{a,e\}$ respectively, as shown in the following diagram:

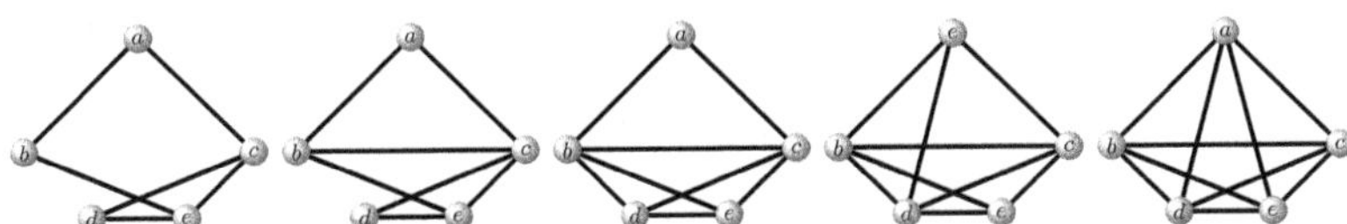

Note that $cl(G) = K_5$, the complete graph of order 5. This means that G is Hamiltonian. For instance, *acdeba* is a Hamiltonian circuit in G.

Determining if a graph is Hamiltonian is often a matter of trial and error. One starts by attempting to trace a Hamiltonian circuit in the graph. If that fails, Theorems 10.5 and 10.7 or some other algorithms can be applied. Given a graph G of order $n \geq 1$, we give in what follows few basic properties that can help to prove that G is not Hamiltonian or to trace a Hamiltonian circuit in G if such a circuit exists.

(1) A Hamiltonian circuit in G, if it exists, must have exactly n edges.

(2) For any vertex v of G, a Hamiltonian circuit in G (if it exists) must include *exactly two* edges incident to v (see Exercise (5) below). In particular, if $\deg(v) = 2$ then any Hamiltonian circuit of G must include the two edges incident to v.

(3) If G has a vertex of degree 1, then it is not Hamiltonian.

(4) For any vertex v of G, once a Hamiltonian circuit we are trying to construct has passed through v, we can delete all unused edges incident to v before we continue to trace the circuit.

(5) A Hamilton circuit in G cannot contain a simple subcircuit which does not include all vertices of G (see Exercise (5) below).

Example 10.17. Prove that graphs G and H of Example 10.13 above are not Hamiltonian.

Solution. In graph G, vertices a, b, c, x, j, k, l are of degree 2 each. The two edges incident to each one of these vertices must be included in any Hamiltonian circuit in the graph, if one exists. Namely, edges $\{a, b\}$, $\{b, c\}$, $\{a, d\}$, $\{c, e\}$, $\{x, f\}$, $\{x, g\}$, $\{h, j\}$, $\{j, k\}$, $\{k, l\}$ and $\{l, i\}$ are all included in any Hamiltonian circuit as illustrated in bold in the following diagram:

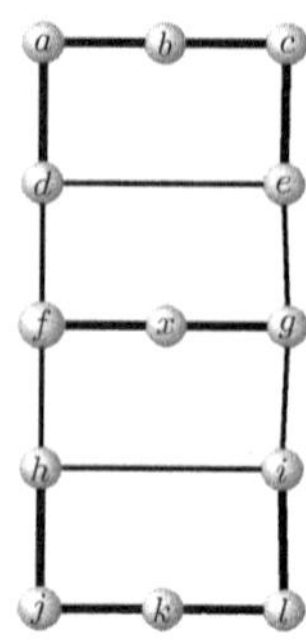

Edge $\{d, e\}$ cannot be part of a Hamiltonian circuit since otherwise *abceda* would be a subcircuit not containing all vertices, a contradiction to property (5) above. So edges $\{d, f\}$ and $\{e, g\}$ must be included in a Hamiltonian circuit by property (2). But now we have the subcircuit *abcegxfda* which does not include all vertices of G, a contradiction to property (5). We conclude that G is not Hamiltonian. Graph H of Example 10.13 is easily seen as non-Hamiltonian as it has vertices of degree 1 (namely x and y). ◊

Example 10.18. Prove that the graph G given below is not Hamiltonian.

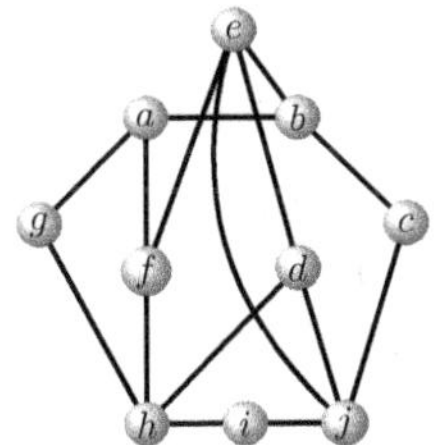

Solution. The graph has three vertices of degree 2: c, g and i. Edges incident to these three vertices, namely $\{a, g\}$, $\{g, h\}$, $\{h, i\}$, $\{i, j\}$, $\{j, c\}$ and $\{c, b\}$, must be included in any Hamiltonian circuit in G if one exists. Edge $\{a, b\}$ cannot be part of any Hamiltonian circuit since otherwise *abcjihga* would be a subcircuit not containing all vertices of G, a contradiction to property (5) above. This implies that edge $\{b, e\}$ must be part of any Hamiltonian circuit (as such a circuit must contain two edges incident to b). The following diagram shows in bold the edges that must be included (so far) in any Hamiltonian circuit with the edge $\{a, b\}$ removed:

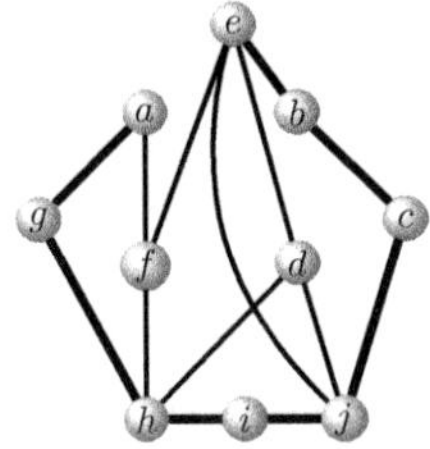

Since every Hamiltonian circuit uses exactly two of the incident edges at any vertex, we can disregard edges $\{j, e\}$, $\{j, d\}$, $\{h, d\}$ and $\{h, f\}$ when trying to trace a Hamiltonian circuit:

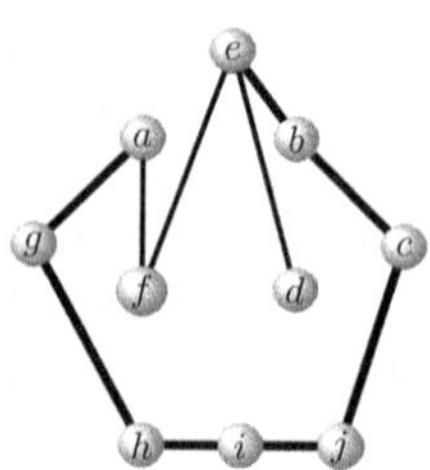

Notice that $\deg(d) = 1$ which makes it impossible for a Hamiltonian circuit to pass through d. We conclude that the graph is not Hamiltonian.   ◊

10.3.1   *Exercises*

⋆(1) Is it possible for an Euler circuit to be a Hamilton circuit at the same time in a graph? Justify your answer.

(2) In each case, draw a simple connected graph G with the given conditions.

   (a) G is Eulerian but not Hamiltonian.
   (b) G is Hamiltonian but not Eulerian.
  ⋆ (c) G is Hamiltonian and has an Euler trail but not Eulerian.

(3) Let $n \geq 3$ be an integer. In each case, prove that the graph is Hamiltonian.

   (a) C_n (b) W_n (c) K_n

(4) Recall that the underlying simple graph $S(G)$ of a graph G is the simple subgraph obtained from G by removing all loops and all but one edge from each set of parallel edges in G. Let G be a graph of order $n \geq 2$. Prove that G is Hamiltonian if and only if $S(G)$ is Hamiltonian.

(5) Let $G = (V, E)$ be a graph of order $n \geq 2$, $v \in V$.

   (a) Prove that any Hamiltonian circuit in G must include *exactly two* edges incident to v.
   (b) Prove that a Hamiltonian circuit in G cannot have a subcircuit which doesn't include all vertices of G.

⋆(6) Let G be a bipartite graph with bipartition sets X and Y. Prove that if G is Hamiltonian, then $|X| = |Y| \geq 2$. For what values of the positive integers m and n the complete bipartite graph $K_{m,n}$ is Hamiltonian?

⋆(7) Let G be a connected graph with degree sequence $(7, 7, 6, 6, 6, 6, 5, 5)$. Prove that G is Hamiltonian.

(8) Let G be 20-regular graph with 380 edges. Prove that G is Hamiltonian.

(9) Let G be k-regular graph of order $2k - 1$, $k \geq 2$. Prove that G is Hamiltonian.

(10) Let $G = (V, E)$ be a graph with $V = \{v_1, \ldots, v_6\}$ and $E = \{e_1, \ldots, e_9\}$. The incidence matrix of G with respect to the ordering $v_1, \ldots, v_6$ and $e_1, \ldots, e_9$ of its vertices and edges is as follows:

$$\begin{bmatrix} 1 & 0 & 0 & 0 & 0 & 1 & 1 & 0 & 1 \\ 0 & 1 & 0 & 0 & 0 & 0 & 0 & 1 & 1 \\ 0 & 1 & 1 & 0 & 0 & 0 & 0 & 0 & 0 \\ 1 & 0 & 1 & 1 & 0 & 0 & 0 & 1 & 0 \\ 0 & 0 & 0 & 1 & 1 & 0 & 1 & 0 & 0 \\ 0 & 0 & 0 & 0 & 1 & 1 & 0 & 0 & 0 \end{bmatrix}.$$

Give a possible drawing of G. Is G is simple? Hamiltonian? If you say it is Hamiltonian, use the given labeling of the vertices to give a Hamiltonian circuit in the graph.

(11) Consider the following graph G:

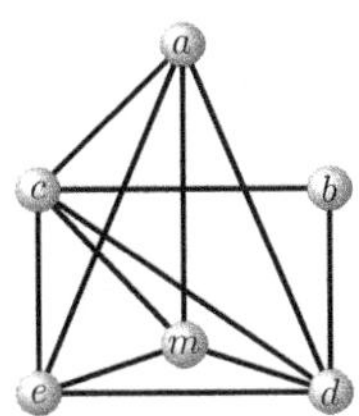

(a) Prove that G is Hamiltonian by exhibiting a Hamiltonian circuit.

(b) Which of Theorems 10.5 and 10.7 can be applied to conclude that G is Hamiltonian?

(12) Give an example of a Hamiltonian graph that does not satisfy either one of the hypotheses of Ore's Theorem or Dirac's Theorem.

⋆(13) Nine scientists (labeled $1, 2, \ldots, 9$) are attending a conference. In the following table, a ✓ in the cell (i, j) indicates that scientists i and j share some common research interest and the symbol × indicates that their research areas do not intersect. Of course, cells (i, i) are left empty.

	1	2	3	4	5	6	7	8	9
1		✓	×	✓	×	✓	✓	×	×
2	✓		×	✓	×	×	✓	×	×
3	×	×		×	×	✓	×	✓	×
4	✓	✓	×		×	✓	✓	✓	✓
5	×	×	×	×		✓	×	✓	✓
6	✓	×	✓	✓	✓		×	×	×
7	✓	✓	×	✓	×	×		✓	×
8	×	×	✓	✓	✓	×	✓		×
9	×	×	×	✓	✓	×	×	×	

The conference organizers would like to sit the nine scientists around a table in such a way that each scientist sits between two other scientists with whom he or she shares some common research interest. Is that possible? If yes, show such arrangement and if no explain why.

(14) In each case use either Ore's or Dirac's Theorem to prove that the graph is Hamiltonian.

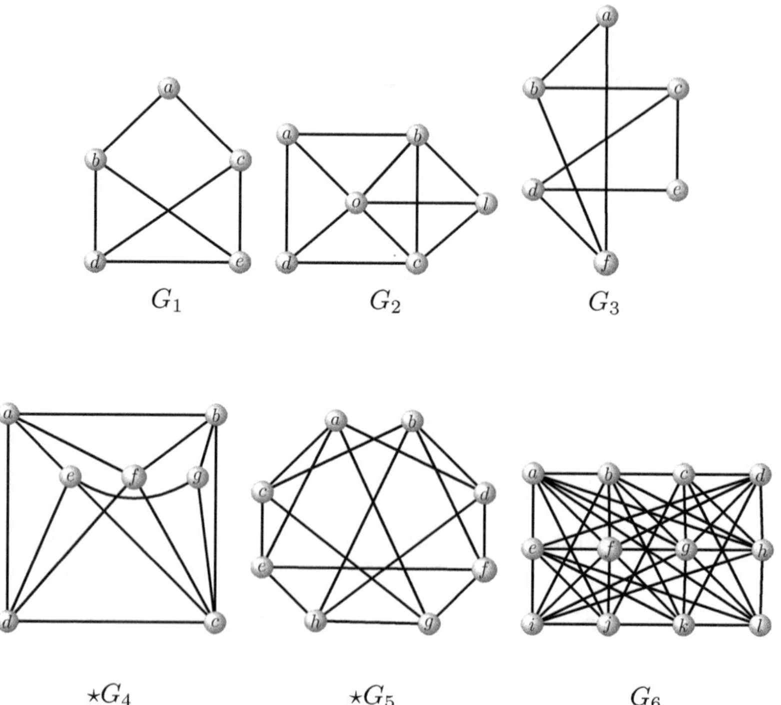

G_1 G_2 G_3

$\star G_4$ $\star G_5$ G_6

(15) For each of the following graphs, determine if it is Hamiltonian. If you say it is, trace a Hamilton circuit. If you say it is not, explain why.

G_1

G_2

G_3

$\star G_4$

G_5

$\star G_6$

G_7

$\star G_8$

G_9

(16) For the graph G given below, prove it is Hamiltonian by finding its Hamiltonian closure.

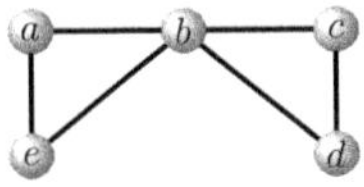

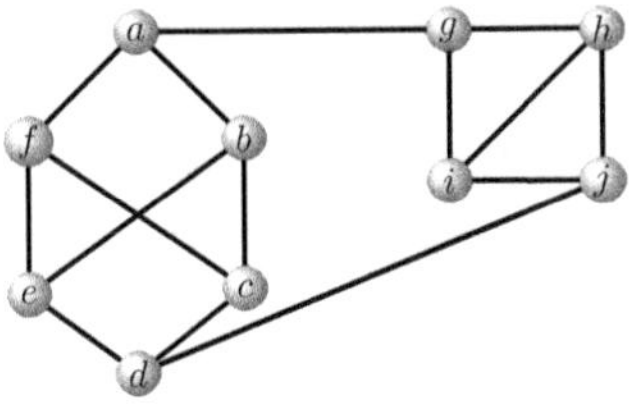

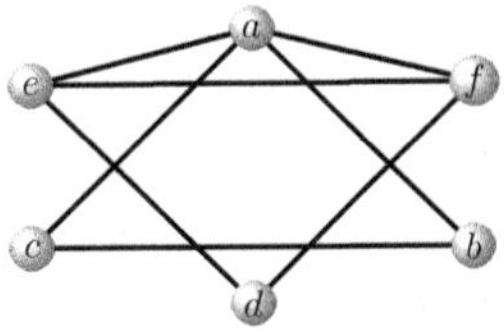

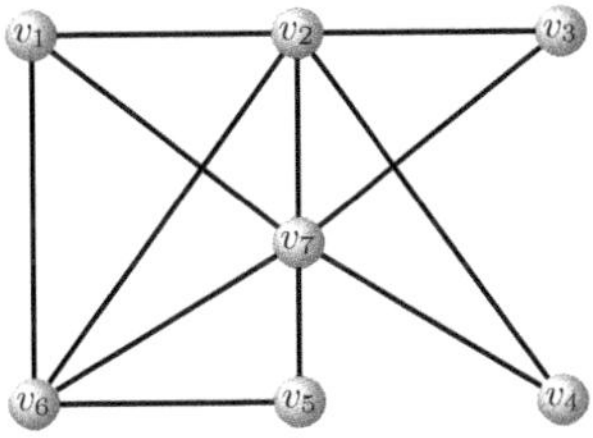

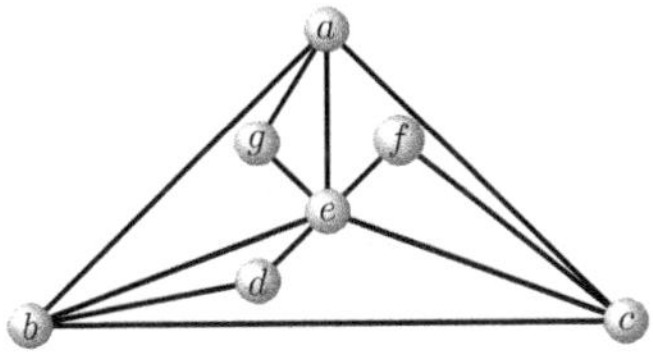

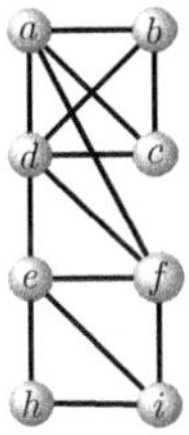

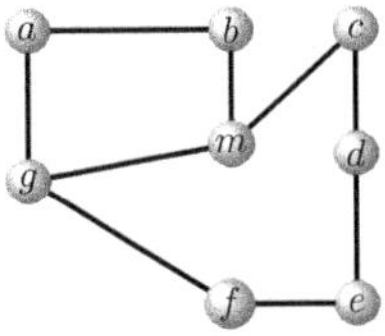

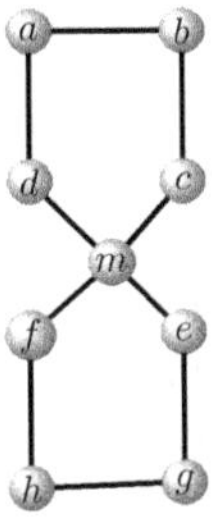

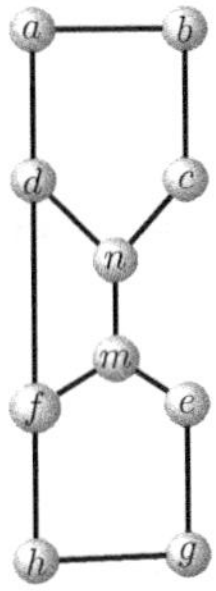

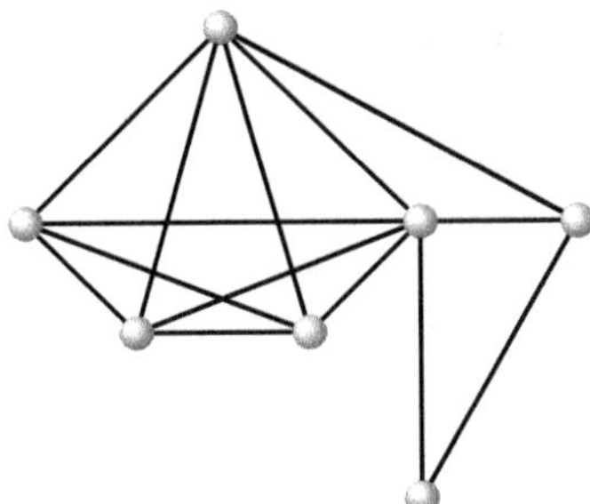

$\star$(17) Prove that the number of different Hamiltonian circuits at a vertex v of the complete graph K_n with $n \geq 3$ is equal to $\frac{(n-1)!}{2}$.

(18) Let G be a graph of order $n \geq 3$ such that every pair u, v of non-adjacent vertices satisfies the relation $\deg(u) + \deg(v) \geq n - 1$. Prove that G contains a Hamiltonian path.

(19) Give an example of a non-Hamiltonian graph G such that the line graph $L(G)$ is both Eulerian and Hamiltonian.

$\star$(20) Prove that if G is an Eulerian graph, then its line graph $L(G)$ is Hamiltonian.

10.4 Graph coloring

In a certain department, there are several working committees and every faculty member has to serve on at least one committee. There are faculty members who serve on multiple committees. Every committee meets once a week for 90 minutes. Obviously, two committees with members in common cannot meet at the same time. The challenge is to come up with the least number of weekly meeting time slots to ensure there are no conflicts. This scheduling problem can be converted into a graph theory problem as follows. Let G be the graph with committees as vertices and two vertices are adjacent if and only if the corresponding committees share at least one common member. Assign a color (in this case, a meeting time) to every vertex in such a way that two adjacent vertices are assigned different colors. The above scheduling problem can then be stated as follows: What is the least number of colors needed to make sure all vertices are colored and no two adjacent vertices are assigned the same colors? Beside the scheduling problem, graph coloring has several interesting applications. It is considered one of the oldest problems in graph theory.

Definition 10.7. Let $G = (V, E)$ be a graph. A *proper coloring* of G is a function $c : V \to \mathbb{N}$ such that $c(u) \neq c(v)$ whenever $\{u, v\} \in E$.

Since we are dealing with graphs of finite orders, we can trivially assign a different color for every vertex to obtain a proper coloring. Of course, we are looking for an optimal solution where the number of colors used is as small as possible. For simplicity, we use positive integers $1, 2, 3, \ldots$ to refer to the colors.

Definition 10.8. Let $G = (V, E)$ be a graph and let $c : V \to \mathbb{N}$ be a proper coloring of G.

(a) For $i \in \mathbb{N}$, the subset $c^{-1}(i)$ of V (of all vertices assigned the same color i) is called a *color class* of G.

(b) A proper coloring of G that uses k given colors is called a *k-coloring* of G. In other words, a k-coloring of G is an onto function $c : V \to \{1, 2, \ldots, k\}$ such that $c(u) \neq c(v)$ whenever $\{u, v\} \in E$. A k-coloring of G partitions V into k color classes.

(c) The *chromatic number* of G, denoted by $\chi(G)$, is the *smallest* integer k for which a k-coloring of G exists. If G is of order n, then clearly $\chi(G) \leq n$.

(d) We say that G is *k-colorable* if there exists a k-coloring of it. It is called *k-chromatic* if $\chi(G) = k$.

Example 10.19. The following diagram shows a three-colorable graph. In fact, we can easily see that the graph is three-chromatic. The diagram shows a three-coloring of the graph using colors 1, 2 and 3.

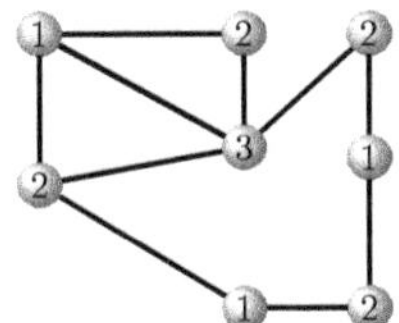

Remark 10.2. If G is not connected, a proper coloring of G consists of a proper coloring of each of the connected components of G. In other words, proper coloring a connected component of G does not affect the proper coloring of the other components. Also, if a vertex of G has a loop, the graph could never be properly colored following Definition 10.8 above. In the context of graph coloring, graphs are assumed loopless. Moreover, replacing any set of multiple edges between a given pair of vertices in G by a single edge does not affect the adjacency of the vertices. That allows us to assume in what follows in this section that all graphs are connected and simple.

Let $G = (V, E)$ be a graph and let $c : V \to \{1, 2, \ldots, k\}$ be a k-coloring of G. The following observations follow immediately from the above definitions.

(a) The fact that a graph coloring is a *function* from V to $\mathbb{N}$ implies in particular that every vertex is assigned one and only one color.

(b) Every color class is an independent set of vertices. That is, no two vertices in the same class are neighbors.

(c) $\chi(G) \le k$.

(d) G is 1-colorable if and only if $E = \emptyset$ (the graph has no edges).

Theorem 10.9. *Let $G = (V, G)$ be a simple connected graph of order $n \ge 2$. Then $\chi(G) = 2$ if and only if G is bipartite.*

Proof. Assume $\chi(G) = 2$. Let $f : V \to \{1, 2\}$ be a proper 2-coloring of G and let X and Y be the two color classes of f. Clearly $\{X, Y\}$ is a partition of V and every edge of G has one endpoint in X and another in Y since no vertices of the same color are adjacent. We conclude that G is bipartite with bipartition (X, Y). Conversely, assume G is bipartite and let (X, Y) be a bipartition of G. The function $f : V \to \{1, 2\}$ defined by

$$f(v) = \begin{cases} 1 & \text{if } v \in X \\ 2 & \text{if } v \in Y \end{cases}$$

is clearly a proper 2-coloring of G. This shows that $\chi(G) \le 2$. On the other hand, $\chi(G) \ge 2$ since G is connected and of order at least 2. We conclude that $\chi(X) = 2$. $\qquad\square$

Theorem 10.10. *As usual, let K_n and C_n denote the complete graph and the cycle graph of order n, respectively. Then:*

(a) For $n \ge 1$, $\chi(K_n) = n$

(b) For $n \ge 3$, $\chi(C_n) = \begin{cases} 2 & \text{if } n \text{ is even} \\ 3 & \text{if } n \text{ is odd} \end{cases}$

Proof.

(a) In K_n, two distinct vertices are adjacent. A proper coloring of K_n must assign different colors to different vertices. The result follows.

(b) If n is even, the cycle C_n is bipartite and the result follows from Theorem 10.9. If n is odd, relabel the vertices of C_n if necessary so that $C_n = v_1 v_2 \ldots v_n v_1$. Assign color 1 to any vertex v_k with k odd and color 2 to any vertex v_k with k even. Vertices v_1 and v_n are adjacent but share the same color (since n is odd). A third color is then needed for v_n to have a proper coloring of the cycle. $\qquad\square$

The following is a basic but very useful result.

Theorem 10.11. *If H is a subgraph of a graph G then $\chi(H) \le \chi(G)$.*

Proof. The proof is left as an exercise (see Exercise (1) below). $\qquad\square$

Example 10.20. Consider the following graph G:

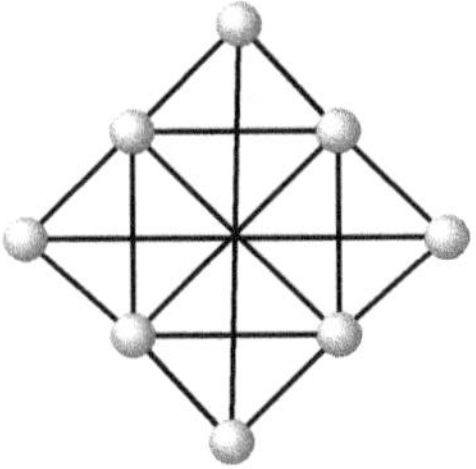

Since G contains K_4 as a subgraph (can you see it?), $\chi(G) \geq 4$. In addition, the following diagram shows a four-coloring of G, which implies that $\chi(G) = 4$.

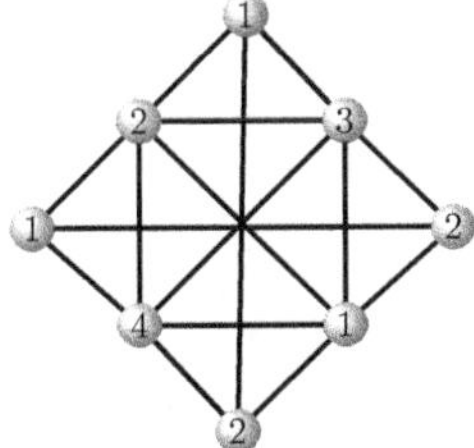

Example 10.21. Find the chromatic number of each of the following graphs.

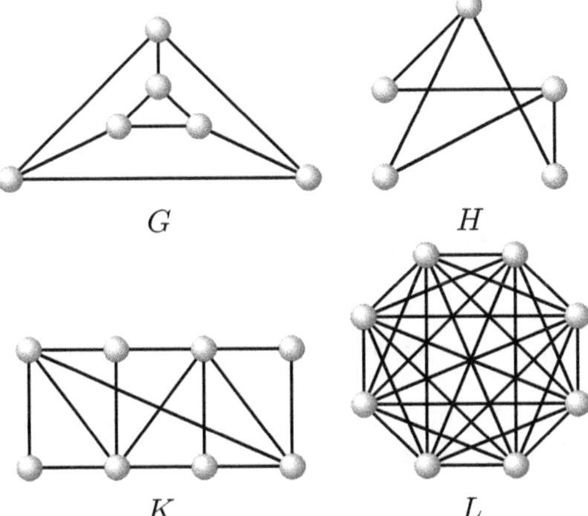

Solution. Graph G contains a triangle as a subgraph, its chromatic number is then at least 3. We prove next that $\chi(G) = 3$. Notice that G consists of an inner and an outer triangle. Start by coloring the vertices of the outer triangle with three colors 1, 2 and 3. Then alternate the same colors on the vertices of the inner triangle so that no two adjacent vertices share the same color:

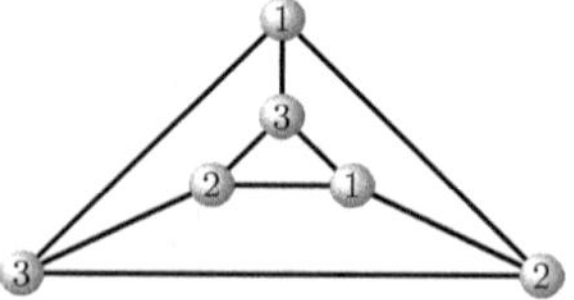

For graph H, start by labeling the vertices as follows:

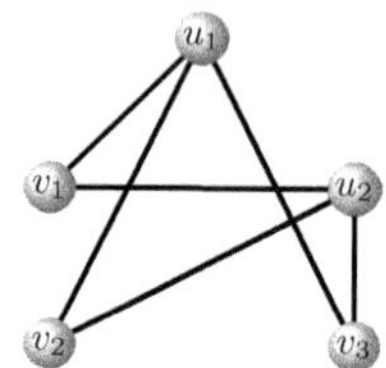

Clearly, H is a bipartite graph with bipartition sets $V_1 = \{v_1, v_2, v_3\}$ and $V_2 = \{u_1, u_2\}$. We conclude that $\chi(H) = 2$. Graph K contains triangles as subgraphs, so $\chi(K) \geq 3$. The following shows a three-coloring of K which proves that the chromatic number of K is indeed 3:

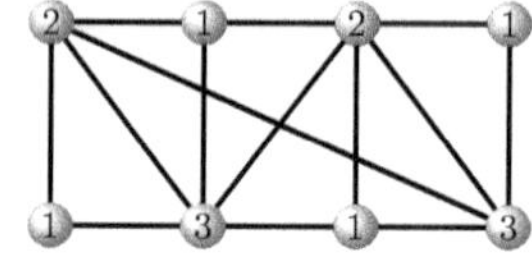

Graph L is the complete graph on 8 vertices (K_8), so $\chi(L) = 8$. $\Diamond$

The following result gives an upper bound on the chromatic number of a graph.

Theorem 10.12. *For any graph G:*

$$\chi(G) \leq \Delta_G + 1. \tag{10.3}$$

where Δ_G is the maximum degree of G.

Proof. We proceed by induction on the order n of G. If $n = 1$, then $G = K_1$ (since G is assumed simple and connected) and so $\chi(G) = 1$ and $\Delta_G = 0$. The inequality holds in this case. Let $n \geq 1$ be an integer and assume that inequality (10.3) holds for any graph of order less than or equal to n. Let G be a graph of order $n + 1$ and let v be an arbitrary vertex in G. Let $H = G - v$ be the graph obtained from G by removing v and its incident edges. By the induction hypothesis, $\chi(H) \leq \Delta_H + 1$.

In particular, there exists a proper coloring c of H using at most $\Delta_H + 1$ colors. Since $\deg_G(v) \leq \Delta_G$, the $\deg_G(v)$ neighbors of v in G use up to Δ_G different colors of c. If $\Delta_H = \Delta_G$, then there exists at least one color α of the $\Delta_H + 1$ colors not used by the neighbors of v. Assigning color α to vertex v gives us a proper coloring of G using at most $\Delta_H + 1 = \Delta_G + 1$ colors. This shows that $\chi(G) \leq \Delta_G + 1$ in this case. If $\Delta_H \leq \Delta_G - 1$, then assigning a new color to vertex v (other than the $\Delta_H + 1$ colors) gives a proper coloring of G with at most $(\Delta_H + 1) + 1 = \Delta_H + 2$ colors. Since $\Delta_H + 2 \leq (\Delta_G - 1) + 2 = \Delta(G) + 1$, we also get that $\chi(G) \leq \Delta_G + 1$ in this case. The result follows by the principle of induction. $\qquad\square$

Note that the equality in (10.3) is achieved in the following graphs:

- The cycle C_n with $n \geq 3$ odd: $\chi(C_n) = 3$ and $\Delta_{C_n} = 2$.
- The complete graph K_n, $n \geq 1$: $\chi(K_n) = n$ and $\Delta_{K_n} = n - 1$.

As it turns out, these are the only graphs where the equality in (10.3) is achieved. For any other graph G, $\chi(G) \leq \Delta_G$ as stated by the following theorem.

Theorem 10.13 (Brooks). *If G is a (connected) graph other than the complete graph or the cycle graph of odd order, then $\chi(G) \leq \Delta_G$.*

We omit the proof as it involves many cases some of which are beyond the scope of this book.

10.4.1 *The chromatic polynomial*

Given a simple connected graph G and a positive integer λ, we turn our attention in this section to the following problem: How many different λ-coloring of G are there? Here two λ-colorings c_1 and c_2 of G are considered different if $c_1(v) \neq c_2(v)$ for at least one vertex v of G. The number of λ-colorings of G is a function, called the *chromatic function* of G, denoted by $f_G(\lambda)$. As we will prove later, $f_G(\lambda)$ is a polynomial function in λ known as the *chromatic polynomial* of G. The problem of determining the number of λ-colorings was introduced by George Birkhoff in 1912 who was interested in the Four Color Problem (proving that every map can be colored using only four colors in a way that no two regions sharing a border have the same color). While they did not contribute directly to the solution of the Four Color Problem, chromatic polynomials found their way in other applications since they hold interesting properties that can be used to extract information about the graph.

Remark 10.3. Let $f_G(\lambda)$ be the chromatic function of a graph G.

(1) If $\lambda < \chi(G)$, then clearly $f_G(\lambda) = 0$.
(2) $\chi(G) = \lambda$ if and only if λ is the smallest integer such that $f_G(\lambda) \neq 0$.
(3) If $G_1, G_2, \ldots, G_k$ are the connected components of G, then $f_G(\lambda) = f_{G_1}(\lambda) \cdot f_{G_2}(\lambda) \cdots f_{G_k}(\lambda)$. This is a direct consequence of the product principle of combinatorics and the fact that coloring one component can be done independently from other components.

The last part of Remark 10.3 allows us to focus solely on connected graphs.

Example 10.22. Consider the empty graph G of order $n \geq 1$ (this is the graph with n vertices but no edges). If we have λ colors available, then there are λ ways to properly color each vertex since no two vertices are adjacent. So $f_G(\lambda) = \lambda^n$. Since the smallest positive integer for which $\lambda^n \neq 0$ is $\lambda = 1$, $\chi(G) = 1$.

Example 10.23. Consider the complete graph K_n $(n \geq 1)$ and label its vertices as $v_1, \ldots, v_n$. Let λ be a non-negative integer and assume we have λ colors available to properly color K_n. We can color v_1 with any of the λ colors, v_2 with any of the remaining $\lambda - 1$ colors and so on. By the product rule, the chromatic function of K_n is $f_{K_n}(\lambda) = \lambda(\lambda-1)(\lambda-2) \cdots (\lambda-n+1)$. Since $f_{K_n}(i) = 0$ for any $i \in \{0, 1, 2, \ldots, n-1\}$ and $f_{K_n}(n) = n! \neq 0$, $\chi(K_n) = n$.

Example 10.24. For the path graph P_n of order $n \geq 1$ with vertices labeled as $v_1, \ldots, v_n$:

$$v_1 \text{ ——— } v_2 \text{ ——— } v_3 \text{ — — — } v_n$$

If there are λ colors available, we can start by assigning any of the λ colors to vertex v_1. Vertex v_2 can be assigned any of the $\lambda - 1$ remaining colors. Since we can assign the same color to vertex v_3 as the one assigned to v_1, there are $\lambda - 1$ coloring possibilities for vertex v_3. Continuing with this reasoning, we conclude that the chromatic function of P_n is

$$f_{P_n}(\lambda) = \lambda \underbrace{(\lambda - 1)(\lambda - 1) \cdots (\lambda - 1)}_{n-1 \text{ times}} = \lambda (\lambda - 1)^{n-1}.$$

In particular, $f_{P_n}(1) = 0$ and $f_{P_n}(2) = 2 \neq 0$ which implies that $\chi(P_n) = 2$.

Example 10.25. Find the chromatic function of the graph G given in the following diagram.

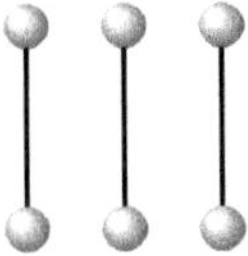

Solution. The graph has three connected components, each of which is the complete graph K_2 with chromatic function $\lambda(\lambda - 1)$. By the last part of Remark 10.3 above, $f_G(\lambda) = (\lambda(\lambda - 1))^3 = \lambda^3(\lambda - 1)^3$. In particular, $\chi(G) = 2$. $\diamond$

The fact that $f_G(\lambda)$ is a polynomial in each of the above examples is not a coincidence. In what follows, we look at some techniques that will allow us to prove that this is the case for any graph. These techniques will also help to compute the chromatic polynomial (and hence the chromatic number) of a graph by breaking it into two "simpler" parts. First a definition.

Definition 10.9. Let G be a graph and let $e = \{u, v\}$ be an edge in G (which is not a loop). Let $G \backslash e$ be the subgraph of G obtained by deleting e while keeping vertices u and v. Let G/e be the graph obtained from G by deleting e, fusing vertices u and v into one vertex w that we join to all vertices which were adjacent in G to either u or v (or both) and replacing all resulting parallel edges with a single edge. We say that graph $G \backslash e$ is obtained by *deletion* and graph G/e by *contraction*.

Example 10.26. The following diagram shows a graph G, the deletion $G - e$ and the contraction G/e with the edge $e = \{b, d\}$.

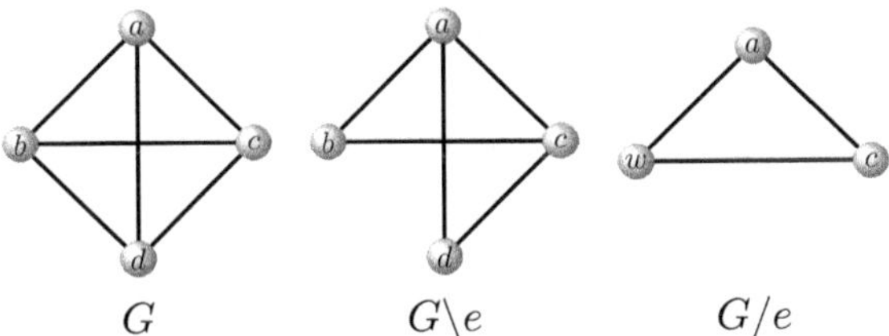

The following result (known as the Deletion-Contraction property) is an important tool in graph coloring both from both a theocratical and practical point of view. Using this property, together with an induction reasoning, we prove that the chromatic function of a graph is indeed a polynomial. The property also provides a mechanical procedure for the computation of the chromatic polynomial of certain graphs.

Theorem 10.14 (Deletion-Contraction property). *Let $G = (V, E)$ be a simple connected graph. Then for any edge e of G, we have:*

$$f_G(\lambda) = f_{G \backslash e}(\lambda) - f_{G/e}(\lambda). \tag{10.4}$$

Proof. Write $e = \{u, v\}$. Assume we have λ colors available. Let $c : V \to \{1, 2, \ldots, \lambda\}$ be a λ-coloring of $G \backslash e$. If $c(u) \neq c(v)$ (u, v have different colors), then c is clearly a λ-coloring of G. If $c(u) = c(v)$, then assigning the common color of u and v to the newly created vertex combining u and v in G/e gives a λ-coloring of G/e. So any λ-coloring of $G \backslash e$ is either a λ-coloring of G or a λ-coloring of G/e (and obviously cannot be both at the same time). We conclude that $f_{G \backslash e}(\lambda) = f_G(\lambda) + f_{G/e}(\lambda)$ and relation (10.4) follows. $\square$

We can now prove that the chromatic function of any graph is a polynomial.

Theorem 10.15. *The chromatic function of a graph G of order n is a polynomial in λ of degree n that we call the chromatic polynomial of G and we denote by $P_G(\lambda)$.*

Proof. We use induction on the size m of G. For $m = 0$, G is the empty graph and the result follows from Example 10.22 above. Let $m \geq 0$ and assume that the result is true for any graph of size k with $0 \leq k \leq m$. Let G be a graph of order n and of size $m + 1$. If e is an edge in G, then $G \backslash e$ is of order n and size m and G/e is of order $n - 1$ and of size at most m. By the induction hypothesis, $f_{G \backslash e}(\lambda)$ is a polynomial of degree n and $f_{G/e}(\lambda)$ is a polynomial of degree $n - 1$. By the Deletion-Contraction Property of Theorem 10.14, $f_G(\lambda) = f_{G \backslash e}(\lambda) - f_{G/e}(\lambda)$ is a polynomial of degree n. $\square$

The following is a consequence of Theorem 10.14.

Theorem 10.16. *For $n \geq 3$, the chromatic polynomial of the cycle graph C_n is $(\lambda - 1)^n + (-1)^n(\lambda - 1)$.*

Proof. We use induction on n. If $n = 3$, C_n is isomorphic to K_3 and the chromatic polynomial in this case is $\lambda(\lambda - 1)(\lambda - 2)$ by Example 10.23 above. On the other hand, $(\lambda - 1)^3 + (-1)^3(\lambda - 1) = (\lambda - 1)^3 - (\lambda - 1) = (\lambda - 1)((\lambda - 1)^2 - 1) = (\lambda - 1)(\lambda^2 - 2\lambda) = \lambda(\lambda - 1)(\lambda - 2)$. The result is then true for $n = 3$. Let $n \geq 3$ and assume that $P_{C_n}(\lambda) = (\lambda - 1)^n + (-1)^n(\lambda - 1)$. Now, consider the cycle C_{n+1} and pick any edge e in it. The graph $C_{n+1} \backslash e$ is the path P_{n+1} and so $P_{C_{n+1} \backslash e}(\lambda) = \lambda(\lambda - 1)^n$ by Example 10.24. Contracting edge e from C_{n+1} leaves us with the cycle C_n, so $P_{C_{n+1}/e}(\lambda) =$

$(\lambda - 1)^n + (-1)^n(\lambda - 1)$ by the induction hypothesis. By Theorem 10.14,
$P_{C_{n+1}}(\lambda) = P_{C_{n+1}\backslash e}(\lambda) - P_{C_{n+1}/e}(\lambda) = \lambda(\lambda-1)^n - (\lambda-1)^n - (-1)^n(\lambda-1) = \lambda(\lambda-1)^n - (\lambda-1)^n + (-1)^{n+1}(\lambda-1) = (\lambda-1)^n(\lambda-1) + (-1)^{n+1}(\lambda-1) = (\lambda-1)^{n+1} + (-1)^{n+1}(\lambda-1)$. $\qquad\square$

Successive applications of Theorem 10.14 reduces the problem of computing the chromatic polynomial of a graph G to that of a simpler graph since both $G\backslash e$ and G/e have fewer edges than G.

Example 10.27. Use the Deletion-Contraction property of Theorem 10.14 to find the chromatic polynomial of the graph G given below. What is $\chi(G)$?

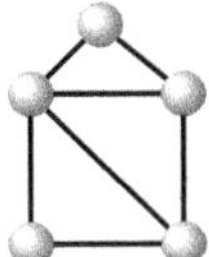

Solution. In the following diagram, a succession of graphs is shown. In each step, the graphs represent the Deletion-Contraction property of Theorem 10.14. The operations $-$ and $+$ stand for the difference and the sum of the chromatic polynomials of the corresponding graphs in each case.

$$= ((\lambda - 1)^5 + (-1)^5(\lambda - 1)) - \lambda(\lambda - 1)^3$$
$$+ \lambda(\lambda - 1)^2 - \lambda^2(\lambda - 1)(\lambda - 2)$$
$$+ \lambda(\lambda - 1)(\lambda - 2)$$
$$= \lambda(\lambda - 1)(\lambda - 2)^3$$

Since $\lambda = 3$ is the smallest integer for which $P_G(\lambda) \neq 0$, $\chi(G) = 3$. $\qquad\Diamond$

The coefficients of the chromatic polynomial of a graph give many interesting properties about the graph. The following theorem explores some of these properties.

Theorem 10.17. *Let $G = (V, E)$ be a graph of order n, size m and with chromatic polynomial $P_G(\lambda)$.*

(1) *$P_G(\lambda)$ is a monic polynomial (i.e., its leading coefficient is 1) of degree n with integer coefficients and constant term 0.*

(2) *If we write $P_G(\lambda) = \lambda^n + a_{n-1}\lambda^{n-1} + \cdots + a_1\lambda$, then:*

    (i) *$a_{n-1} = -m$.*

    (ii) *a_i has the opposite sign of a_{i+1} for $i = 1, \ldots, n-2$.*

    (iii) *If j is the smallest integer such that $a_j \neq 0$, then j is the number of connected components of G.*

(3) *If $m \geq 1$ (so G is not the empty graph), then the sum of the coefficients of $P_G(\lambda)$ is zero.*

Proof.

(1) Left as an Exercise (see Exercise (18) below).

(2) Write $P_G(\lambda) = \lambda^n + a_{n-1}\lambda^{n-1} + \cdots + a_1\lambda$.

    (i) We proceed by induction on the size m of G. If $m = 0$, the result is true since the chromatic polynomial of the empty graph is λ^n in which the coefficient of λ^{n-1} is zero. Let $m \geq 0$ and assume that the result is true for every graph of size k with $0 \leq k \leq m$. Let G be a graph of order n and size $m + 1$ and let e be an edge in G. Graph $G \backslash e$ is of order n and size m and graph G/e is of order $n - 1$ and size $m' \leq m$. By part (1), we know that $P_{G \backslash e}(\lambda)$ and $P_{G/e}(\lambda)$ are monic of degree n and $n - 1$ respectively, each with 0 as constant term. Moreover, the induction hypothesis tells us that the coefficient of λ^{n-1} in $P_{G \backslash e}(\lambda)$ is $-m$ and the coefficient of λ^{n-2} in $P_{G/e}(\lambda)$ is $-m'$. Write $P_{G \backslash e}(\lambda) = \lambda^n - m\lambda^{n-1} + \cdots + a_1\lambda$ and $P_{G/e}(\lambda) = \lambda^{n-1} - m'\lambda^{n-2} + \cdots + b_1\lambda$. By the Deletion-Contraction property, $P_G(\lambda) = P_{G \backslash e}(\lambda) - P_{G/e}(\lambda) = \lambda^n - m\lambda^{n-1} + \cdots + a_1\lambda - (\lambda^{n-1} - m'\lambda^{n-2} + \cdots + b_1\lambda) = \lambda^n - (m + 1)\lambda^{n-1} + (a_{n-2} + m')\lambda^{n-2} + \cdots + (a_1 - b_1)\lambda$. The coefficient of λ^{n-1} in $P_G(\lambda)$ is then $-(m + 1)$, as required.

    (ii) Left as an Exercise (see Exercise (18) below).

    (iii) By induction on the size m. Since the empty graph of order n has n connected components (each consists of a single vertex) and its chromatic polynomial is λ^n, the result is true for $m = 0$. Let $m \geq 0$ and assume the result is true for any graph of size m. Let G be a graph of order n and size $m+1$ and let j be the number of connected components of G. For any edge e of G, the graph G/e has j connected components and the graph $G\backslash e$ has either j or $j + 1$ connected components. If $G\backslash e$ has j connected components, then the induction hypothesis tells us that $P_{G\backslash e}(\lambda) = \lambda^n + \alpha_{n-1}\lambda^{n-1} + \cdots + \alpha_j\lambda^j$ and $P_{G/e}(\lambda) = \lambda^{n-1} + \beta_{n-2}\lambda^{n-2} + \cdots + \beta_j\lambda^j$ for some non-zero integers α_j and β_j. By the Deletion-Contraction property, $P_G(\lambda) = P_{G\backslash e}(\lambda) - P_{G/e}(\lambda) = \lambda^n + (\alpha_{n-1}-1)\lambda^{n-1} + \cdots + (\alpha_j - \beta_j)\lambda^j$. By part (ii), α_j and β_j have opposite signs, which means that $(\alpha_j - \beta_j) \neq 0$ and j (the number of connected components in G) is the smallest integer such that the coefficient of λ^j in $P_G(\lambda)$ is not zero. The result is true is this case. If the number of connected components of $G\backslash e$ is $j + 1$, then using the same argument as before we can write $P_G(\lambda) = P_{G\backslash e}(\lambda) - P_{G/e}(\lambda) = \lambda^n + \alpha_{n-1}\lambda^{n-1} + \cdots + \alpha_{j+1}\lambda^{j+1} - \left(\lambda^{n-1} + \beta_{n-2}\lambda^{n-2} + \cdots + \beta_j\lambda^j\right) = \lambda^n + (\alpha_{n-1} - 1)\lambda^{n-1} + \cdots + \alpha_{j+1}\lambda^{j+1} - \beta_j\lambda^j$. The result is also true in this case since $\beta_j \neq 0$. The result follows by the principle of induction.

(3) Since $m \geq 1$, there exists at least one edge in G and so a minimum of two colors are needed to properly color the graph. In other words, $P_G(1) = 0$. Write $P_G(\lambda) = \lambda^n + \alpha_{n-1}\lambda^{n-1} + \cdots + \alpha_1\lambda$, then the relation $P_G(1) = 0$ implies that $1 + \alpha_{n-1} + \alpha_{n-2} + \cdots + \alpha_1 = 0$, as required. $\qquad\square$

Example 10.28. None of the polynomials $p_1(x) = \lambda^4 - 4\lambda^3 + 2\lambda^2 - 2\lambda + 3$, $p_2(x) = \lambda^4 - 4\lambda^3 + 2\lambda^2 - 3\lambda$ and $p_3(x) = \lambda^4 - 4\lambda^3 + 2\lambda^2 + 3\lambda$ can be the chromatic polynomial of a graph. Polynomial $p_1(x)$ has a non-zero constant term. The sum of the coefficients in $p_2(x)$ is not equal to zero. The coefficients of $p_3(x)$ do not alternate in sign.

Example 10.29. The chromatic polynomial of a certain graph G is given:

$$P_G(\lambda) = \lambda^8 - 12\lambda^7 + 66\lambda^6 - 214\lambda^5 + 441\lambda^4 - 572\lambda^3 + 423\lambda^2 - 133\lambda.$$

(a) Is G connected?
(b) What is the order of G?
(c) What is the size of G?
(d) Is G bipartite?

Solution. (a) Since $a_1 = -133 \neq 0$, the number of connected components of G is 1. So G is connected.

(b) The order of G is the degree of $P_G(\lambda)$: 8.

(c) The size of G is $-a_7 = 12$.

(d) The chromatic number of G is the smallest positive integer i such that $P_G(i) \neq 0$. Since $P_G(1) = 0$ and $P_G(2) = 2 \neq 0$, $\chi(G) = 2$. In particular, G is bipartite. $\hfill \Diamond$

Example 10.30. To avoid signal interference, telecommunication companies must assign different frequencies to towers that are in range of each others. In a certain region, there are 8 towers labeled $T_1, \ldots, T_8$. The following pairs (i, j) represent towers T_i and T_j which are in range of one another: $(1, 2)$, $(1, 5)$, $(1, 6)$, $(2, 3)$, $(2, 5)$, $(2, 6)$, $(3, 4)$, $(3, 6)$, $(3, 7)$, $(4, 7)$, $(4, 8)$, $(5, 6)$ and $(7, 8)$. What the minimum number of frequencies the company needs in order to avoid any signal interference?

Solution. Construct the graph G with tours $T_1, \ldots, T_8$ as vertices, and edges joining towers which are in the range of one another.

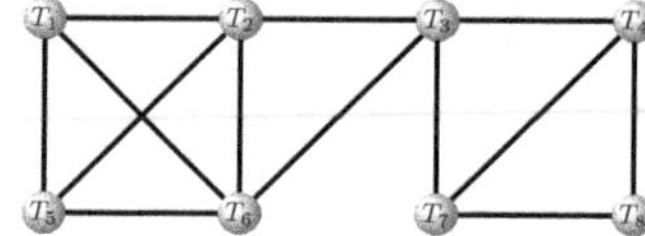

Then the minimum number of frequencies needed to avoid interferences between towers is the chromatic number of the graph G constructed above (the color of a vertex represents the frequency assigned to the corresponding tower). Note that G contains a copy of K_4 shown in bold in the diagram:

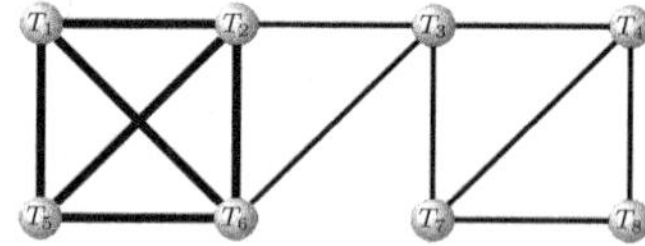

This shows in particular that $\chi(K_4) = 4 \leq \chi(G)$. Since G is not a complete graph nor a cycle of odd length, Brooks theorem tells us that $\chi(G) \leq \Delta_G = 4$. We conclude that $\chi(G) = 4$. $\hfill \Diamond$

10.4.2   *Coloring planar graphs: Five and four-color theorems*

A famous problem in graph theory is the four-color problem: Can we color every map using at most four colors in a way that no two regions sharing a border have the same color? This can be turned into a graph coloring problem as follows: represent each region with a vertex and declare two vertices adjacent if the corresponding regions share a border. Then the four-color problem asks if we have a proper four-coloring of the graph. We assume that the graph corresponding to any map coloring is simple and planar.

The four-color problem was first proposed by Francis Guthrie in 1852 in an attempt to color the map of all counties of England. Many false proofs appeared in the literature since then and it took more than a century to finally have an affirmative solution to the problem. But even after the publication in 1976 of what is considered widely as a proof (by Kenneth Ira Appel and Wolfgang Haken), some mathematicians still dispute its validity today because of its heavy computational nature (using computers to analyze nearly two thousands cases).

One attempt at a proof of the four-color problem was given by Alfred Bray Kempe in 1879. That proof was widely accepted at the time and it wasn't until a decade later that Percy John Heawood proved that Kempe's proof had a flaw. While Heawood was not able to fix Kempe's proof, he gave a proof of the result but for five colors rather than four.

Theorem 10.18 (Five-color theorem). *If G is a simple planar graph, then $\chi(G) \leq 5$.*

Proof. We proceed by induction on the order n of the graph. The result is clearly true if $n \leq 5$. Let $n \geq 6$ and assume that every simple planar graph of order $k < n$ has a five-coloring. Let G be a simple planar graph of order n. By Corollary 10.3 above, we can choose a vertex v of G with $\deg(v) \leq 5$. If $\deg(v) < 5$, the graph $G - v$ obtained from G by removing v and all its incident edges is simple and planar of order $n - 1$. By the induction hypothesis, $G - v$ has a five-coloring. Assigning to vertex v any color not used by its neighbors in G (at most 4) results in a five-coloring of G. The result is true in this case. Assume next that $\deg(v) = 5$ and let $v_1, \ldots, v_5$ be the five neighbors of v in G. If $\{v_i, v_j\}$ is an edge of G for every pair (i, j) with $i \neq j$, then G would have a subgraph isomorphic to K_5 and G would not be planar by Kuratowski's theorem. We may then assume that v_1, v_2 are non-adjacent neighbors of v. Let H be the graph obtained contracting v_1, v_2 in the graph $G - v$ (so we start by removing v and all its incident edges from G and then we merge v_1, v_2 into one vertex w adjacent to any vertex of $G - v$ which is a neighbor of either v_1 or v_2). Clearly H is simple and planar and of order $n - 2$. By the induction hypothesis, H has a five-coloring. This produces a five-coloring in $G - v$ in which vertices v_1, $v2$ have the same color (the color assigned to the merged vertex w). One color is now free and can be assigned to v to produce a five-coloring of G. The result follows by mathematical induction. $\qquad\square$

We finish this section with the four-color theorem. As mentioned above, the proof of the theorem relies heavily on computational analysis of possible cases. We omit the proof as it is beyond the scope of this book.

Theorem 10.19 (Four-color theorem). *If G is a planar and simple graph, then $\chi(G) \le 4$.*

10.4.3 Exercises

All graphs in the following set of exercises are assumed to be simple and connected, unless otherwise indicated.

$\star$(1) If H is a subgraph of a graph G, prove that $\chi(H) \le \chi(G)$.

(2) For a graph G, prove that $\chi(G) \ge 3$ if and only if G has a simple cycle of odd length.

(3) If G is a graph with n connected components $C_1, \ldots, C_n$, prove that $\chi(G) = \max\{\chi(C_i),\ i = 1, \ldots, n\}$.

$\star$(4) Let G be a graph of order n. Prove that $\chi(G) = n$ if and only if $G = K_n$.

$\star$(5) Find the chromatic number of the Petersen graph:

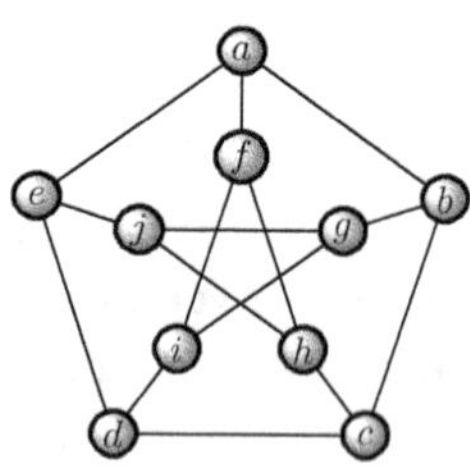

$\star$(6) Prove that in a graph G of order $n \ge 1$ there exists at least one vertex v with $\deg(v) \ge \chi(G) - 1$.

(7) If G is a graph with n connected components, prove that λ^n is a factor of $P_G(\lambda)$.

(8) Let G be a graph and let $G' = G + v$ be the graph obtained from G by adjoining a vertex v to G and join it to each vertex of G by an single edge. Prove that $P_{G'}(\lambda) = \lambda \cdot P_G(\lambda - 1)$.

$\star$(9) Find the chromatic polynomial of the wheel graph W_n. What is $\chi(W_n)$?

(10) Let $k \geq 2$ be an integer. A graph $G = (V, E)$ is called k-partite if there exists a partition $V_1, V_2, \ldots, V_k$ of V such that for every $i \in \{1, \ldots, k\}$, no two vertices in V_i are adjacent and every edge in E is of the form $\{v_i, v_j\}$ where $v_i \in V_i$, $v_j \in V_j$ with $i \neq j$. Bipartite graphs are simply 2-partite.

   (a) Draw possible k-partite graphs for $k = 3$ and $k = 4$.

   (b) Prove that if G is k-partite, then $\chi(G) = k$ (*Hint.* Use induction on k).

(11) In each case, write the chromatic polynomial of the given graph.

   (a) K_5 (b) P_5 (c) C_5 (d) W_5 $\star$ (e) $K_{2,5}$

(12) In each case, find the number of different proper λ-coloring of the graph G for the given value of λ (*Hint.* Use your answers to Exercise (11)).

   (a) $G = K_5$, $\lambda = 8$ (d) $G = W_5$, $\lambda = 9$

   (b) $G = P_5$, $\lambda = 7$ (e) $G = K_{2,5}$, $\lambda = 5$

   (c) $G = C_5$, $\lambda = 8$

(13) In each case, use the Deletion-Contraction property (Theorem 10.14) to find the chromatic polynomial of the graph. Determine the chromatic number of the graph.

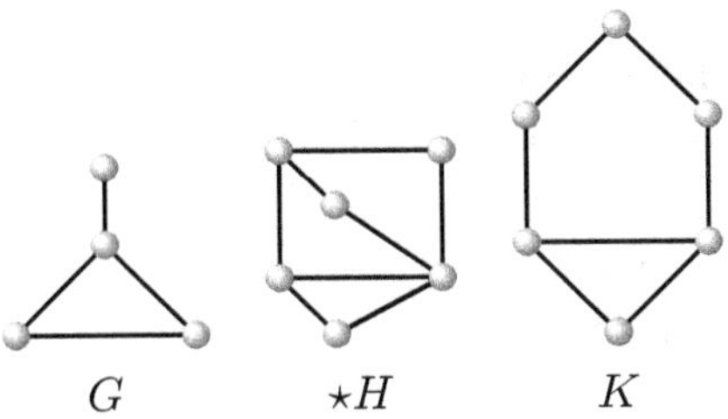

(14) In each case, determine the chromatic number χ and provide a possible

χ-coloring of the graph.

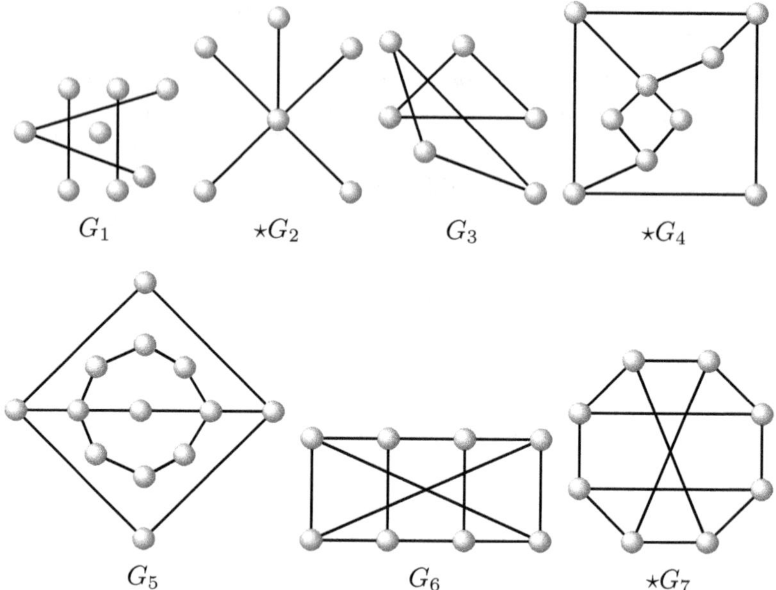

(15) In each case, the given polynomial cannot be the chromatic polynomial of a graph. Explain why.

$\star$ (a) $2\lambda^4 - 3\lambda^3 + 2\lambda^2 - \lambda$

   (b) $\lambda^4 + 4\lambda^3 - 2\lambda^2 - 3\lambda$

   (c) $\lambda^4 - 3\lambda^3 + 4\lambda^2 - 3\lambda + 1$

$\star$ (d) $\lambda^4 - 3\lambda^3 + 2\lambda^2 - 5\lambda$

(16) Can $\lambda^2(\lambda - 1)(\lambda - 2)(\lambda - 3)(\lambda - 4)$ be the chromatic polynomial of a *planar* graph? if you say yes, draw such a graph and if you say no explain why.

(17) In each case, the chromatic polynomial $P_G(\lambda)$ of a graph G is given. Find the order n, the size m, the chromatic number χ and the number ω of connected components of G.

   (a) $\lambda^8 - 5\lambda^7 + 9\lambda^6 - 7\lambda^5 + 2\lambda^4$.

$\star$ (b) $\lambda^9 - 10\lambda^8 + 44\lambda^7 - 122\lambda^6 + 183\lambda^5 - 186\lambda^4 + 135\lambda^3 - 55\lambda^2 + 10\lambda$.

   (c) $P_G(\lambda) = \lambda^{10} - 15\lambda^9 + 105\lambda^8 - 455\lambda^7 + 1353\lambda^6 - 2861\lambda^5 + 4275\lambda^4 - 4305\lambda^3 + 2606\lambda^2 - 704\lambda$.

(18) Let G be a graph of order n, size m and with chromatic polynomial $P_G(\lambda)$. Use induction on m to prove that $P_G(\lambda)$ is a monic polynomial (i.e., its leading coefficient is 1) with integer coefficients alternating in sign and that the constant term of $P_G(\lambda)$ is 0.

(19) Two graphs are called *chromatically equivalent* if they have the same chromatic polynomial.

(a) Give an example of two chromatically equivalent graphs.
(b) True or False: Two chromatically equivalent graphs must have the same order, the same size, the same chromatic number and the same number of connected components.
(c) True or False: Two chromatically equivalent graphs must be isomorphic.

(20) A department of mathematics offers seven graduate courses labeled as $C_1, \ldots, C_7$. When planning the schedule for final exams in these courses, the department has to take into account if there are students taking more than one of these courses at the same time. In the table below, a $\times$ symbol in row i and column j indicates that courses C_i and C_j have at least one student in common.

	C_1	C_2	C_3	C_4	C_5	C_6	C_7
C_1		$\times$	$\times$		$\times$	$\times$	$\times$
C_2	$\times$			$\times$	$\times$		$\times$
C_3	$\times$			$\times$		$\times$	$\times$
C_4		$\times$	$\times$			$\times$	
C_5	$\times$	$\times$				$\times$	$\times$
C_6	$\times$		$\times$	$\times$	$\times$		$\times$
C_7	$\times$	$\times$	$\times$		$\times$	$\times$	

Using graph coloring, find the minimum number of time slots needed to schedule all 7 finals such that no conflict occurs.

$\star$(21) Eight chemical products labeled $C_1, \ldots, C_8$ are to be transported in trucks from a warehouse to the lab of a certain pharmaceutical company. Due to their structures, some of these products cannot be transported in the same truck at the same time as shown in the following table:

Product	cannot be transported in the same truck at the same time with
C_1	C_2, C_3, C_7
C_2	C_1, C_3, C_5, C_7, C_8
C_3	C_1, C_2, C_5, C_7, C_8
C_4	C_6, C_8
C_5	C_2, C_3, C_6, C_7, C_8
C_6	C_4, C_5
C_7	C_1, C_2, C_3, C_5, C_8
C_8	C_2, C_3, C_4, C_5, C_7

Using graph coloring, find the minimal number of trucks needed to transport all the products at the same time. Label the smallest number of trucks needed as $T_1, T_2, \ldots$ and give a possible distribution of the 8 products into the labeled trucks.

(22) Let k be a positive integer. A graph G is called *k-critical* if $\chi(G) = k$ and $\chi(G - v) < k$ for any vertex v of G. We say that a graph is *critical* if it is k-critical for some positive integer k. So k-critical graphs are "barely" k-chromatic in the sense that removing any vertex decreases the chromatic number.

(a) Give a description of 1-critical and of 2-critical graphs.

(b) Give two non-isomorphic 3-critical graphs.

(c) Prove that every critical graph is connected.

(d) For each of the following graphs, determine if it is critical.

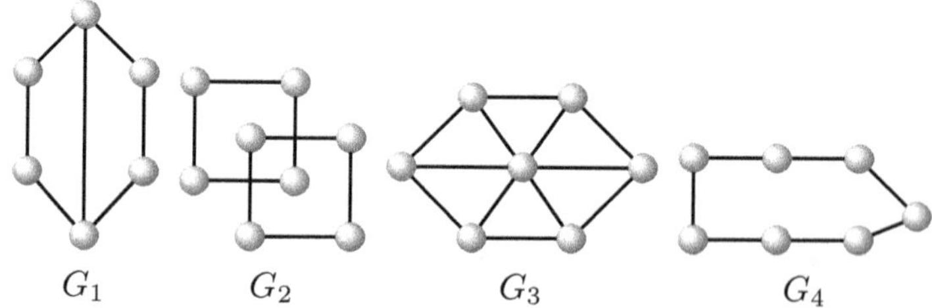

G_1 G_2 G_3 G_4

Chapter 11

Trees

11.1 Introduction

From a computer science perspective, not all graphs are created equal. A family of graphs that plays a central role in digital inquires and information theory is known as *Trees*. The notion of a tree graph was first introduced in mid 19th-century by Gustave Kirchoff as part of his work on electrical networks. Later, they were extensively studied by Arthur Cayley (1824–1887) who was interested in their structures for the purpose of counting the number of structural isomers of the saturated hydrocarbons $C_k H_{2k+2}$ where k is a positive integer. One of the most interesting results in graph theory is actually attributed to Cayley's work on trees and became widely known as *Cayley's formula*. The formula states that the number of labeled trees (a graph in which every vertex is assigned a unique name or label) with n vertices is equal to n^{n-2}. With the digital revolution in the second half of the twentieth century, trees became increasingly essential in numerous applications and their theory found its way in coding, data structure, electronic queries, storing digital information and others.

11.2 Basic definitions and terminology

Definition 11.1. A graph without cycles is called a *forest*. A connected forest is called a *tree*. In other words, a tree is a connected graph with no cycles. Vertices of a tree graph are referred to as *nodes*.

Two quick facts can be drawn from the above definition. The first is that every forest (hence every tree) must be a simple graph. This is true since the presence of loops or multiple edges creates cycles in the graph. The second is that any connected component of a forest is a tree.

Example 11.1. Consider the following five graphs:

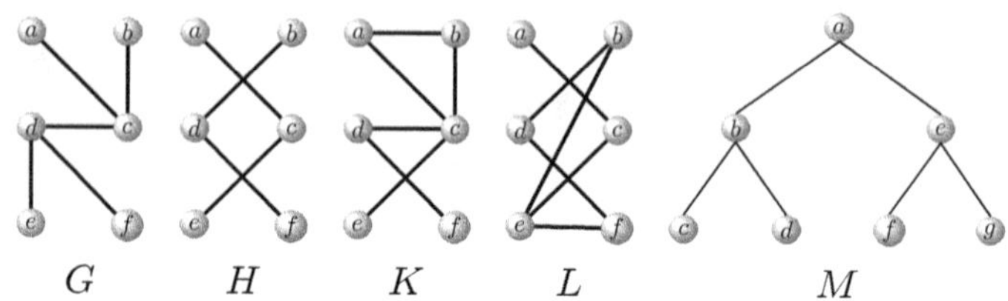

$$G \qquad H \qquad K \qquad L \qquad M$$

Graphs G and M are trees (hence forests) since they are connected with no cycles. Graph H is a forest but not a tree since it has no cycles but it is not connected. Graph K is not a forest since it has a cycle $(abca)$, hence not a tree. Similarly, graph L is not a forest since it has a cycle $(befdb)$.

The following result is used in many textbooks as an alternative definition of trees.

Theorem 11.1. *A graph G is a tree if and only if there exists a unique path between any pair of distinct nodes of G.*

Proof.   Assume first that G is a tree and let x, y be two distinct nodes of G. Since G is connected, there exists an xy-path in G. For the uniqueness of such a path, assume that P_1 and P_2 are two different xy-paths in G. The path P obtained by following P_1 from x to y and then P_2 from y to x is a cycle at node x. This is a contradiction since G has no cycles. Conversely, assume that there exists a unique path between any pair of distinct nodes of G. By definition, G is connected. Assume that G has a cycle C and let u, v be two distinct nodes on C. Clearly, C can be broken into two distinct paths: one from u to v and another from v to u. This contradicts the uniqueness of a path between u and v. We conclude that G is connected with no cycles, hence a tree. $\qquad\qquad\square$

Theorem 11.1 allows us to choose a direction for the edges in a tree T by specifying a *root* in the tree, called a *designated root*, as follows. Choosing any node r in T, we know that there exists a unique path from r to any other node v of T. Edges of T are then directed away from the designated root r.

Definition 11.2. A tree, together with a designated root and edges directed away from the root, is a directed graph that we call a *rooted tree*.

With the understanding that edges are directed away from the designated root in a rooted tree, there is no need to include arrows on the edges. It should be also noted that changing the designated root in a tree would produce a different rooted tree as illustrated in the following example.

Example 11.2. In the diagram below, H and K are two different rooted trees obtained from graph G by designating nodes a and c as roots, respectively.

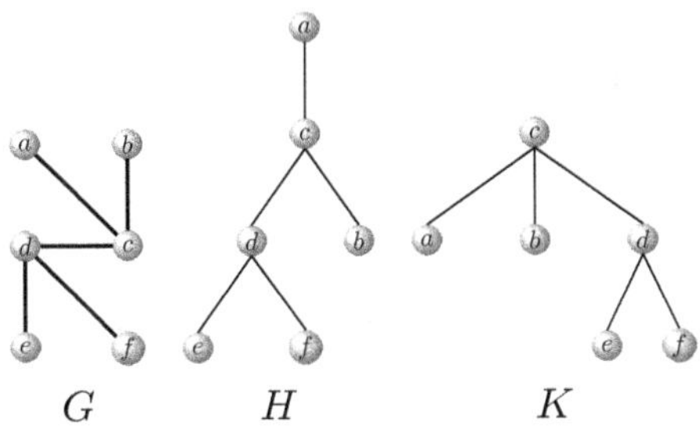

Rooted trees have natural links to traditional family trees that start at the first ancestor known to the family.

Definition 11.3. Let T be a rooted tree with a designated root labeled a.

(1) The *level* (also known as the *depth*) of a node v in T, denoted by $l(v)$, is defined to be the length of the unique path from a to v. The level of the root a is 0.

(2) A node v is called the *parent* of the node u if (u, v) is a directed edge in T. In other words, v is the vertex immediately before u on the path from the root to v. If v is the parent of u, we say that u is a *child* of v. Note that a node can have many children but at most one parent (see Exercise (5) below for uniqueness of a parent).

(3) If v is a node in T different from the root a, then any node on the path from a to v (excluding v itself) is called an *ancestor* of v. If u is an ancestor of v, we say that v is a *descendant* of u. So the descendants of node u are all nodes having u as an ancestor.

(4) Nodes v_1 and v_2 are called *siblings* if they have the same parent.

(5) A node with no children is called a *leaf*. A node with at least one child is called an *internal node* (or interior node). The root of a tree is a leaf if and only if the tree consists of one node only.

(6) The *height* (also known as the *depth*) of T is the length of the longest path from the root to a leaf in T. If T has only one node, its height is defined to be 0.

(7) If v is an internal node in T, the subtree rooted at v is the subgraph $T' = (V', E')$ of T with V' consisting of v and all of its descendants and E' is the set of all edges incident to the descendants of v.

In theory, a tree may have *infinitely* many levels, and any level (other than level 0) may have an infinite number of nodes. However, in all what follows we only consider finite trees (so every internal node has a finitely many descendants).

Example 11.3. Consider the following rooted tree T.

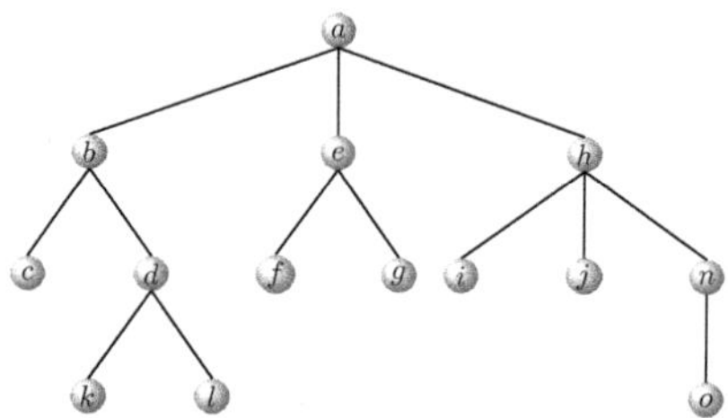

Node a is the root. Nodes b, e and h are siblings with parent a. Same is true for nodes i, j and n with parent h. Node h is an ancestor of nodes i, j, n and o. Nodes k and l are siblings and are descendants of b and d. Nodes b, e and h are at level 1. Nodes c, d, f, g, i, j and n are at level 2. Nodes o, k and l are at level 3. Nodes b, h and d are internal (among others) while c, k, l and o are leaf nodes. The height of the rooted tree is 3 with (a, b, d, k) being one of the longest paths from the root to a leaf node in T. The following is the subtree of T rooted at b:

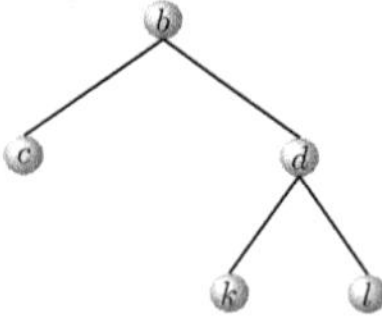

Definition 11.4. Let $k \geq 0$ be an integer. We say that a rooted tree T is k-ary if every internal node of T has *at most* k children. In particular, a 2-ary tree is called a *binary* tree and a 3-ary tree is called a *ternary* tree. A k-ary rooted tree is called *complete* if every node is either a leaf or has exactly k children. A tree of height $h \geq 1$ is called *balanced* if the level of every leaf is either h or $h - 1$. A k-ary complete tree of height h in which all leaves appear at level h is called a *full k-ary* tree.

It is worth mentioning here that the above definitions can differ from one textbook to another. For example, in some textbooks a k-ary tree is called complete if every node is either a leaf or has exactly k children and all leaves are at the same level. One should also note that if T is a complete tree of order $n \geq 3$, then T is not complete as a graph (so T is not isomorphic to K_n).

Example 11.4. Consider the following rooted trees:

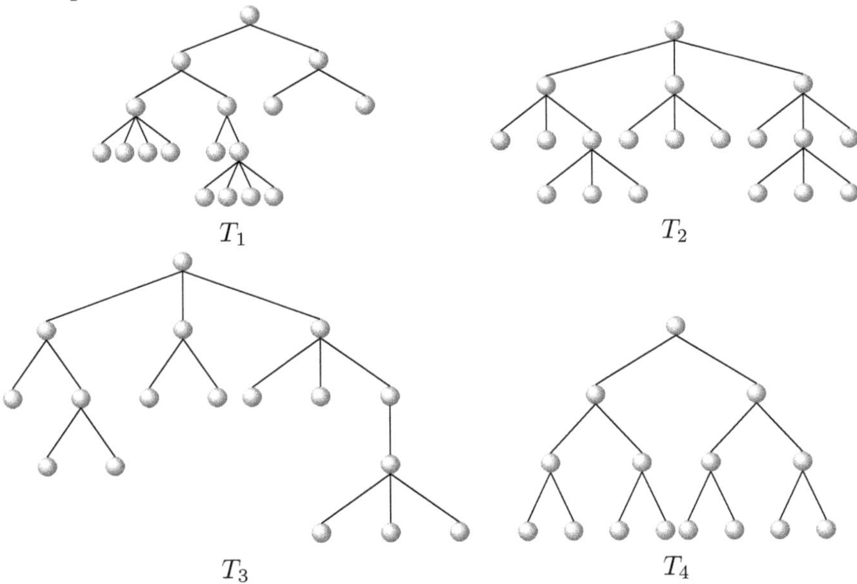

T_1 is 4-ary of height 4, not balanced since some leaves appear at depth 2 and not complete (hence not full). Tree T_2 is ternary of height 3, it is balanced since every leaf appears at level 2 or 3 and complete since every node is either a leaf or has three children. It is not full since there are leaves at level 2. Tree T_3 is ternary of height 4 but neither balanced (leaves at level 2) nor complete (there are internal nodes with only two children). Tree T_4 is a full binary tree of height 3 (every internal node has exactly two children and every leaf is at level 3).

Definition 11.5. An *ordered* rooted tree is a rooted tree in which the children of each internal node are ordered, typically as child 1, child 2 and so on from left to right. In an ordered binary tree, if a node has two children then we refer to them as the *left child* and the *right child*. If v is a node with two children in a binary rooted tree, the subtree rooted at the left (respectively right) child is called the left (respectively right) subtree at v.

Ordered rooted trees are usually drawn in a way that the order of siblings of every internal node is implicitly embedded in the drawing.

Example 11.5. Consider the following binary tree:

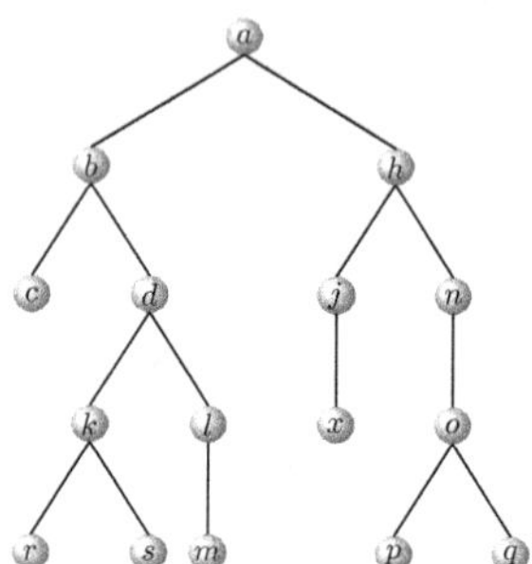

The left child of node h is j and the right child is n. The left and right subtrees of h are shown in the following diagram:

The left subtree of h The right subtree of h

11.2.1   *Exercises*

(1) In each case, determine if the graph is a forest, a tree or neither.

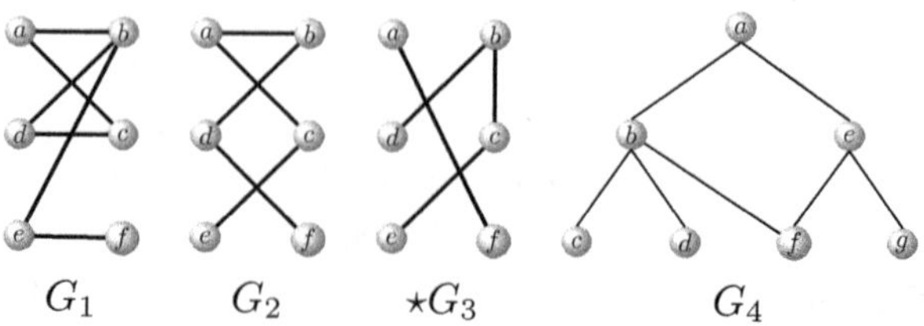

G_1 G_2 $\star G_3$ G_4

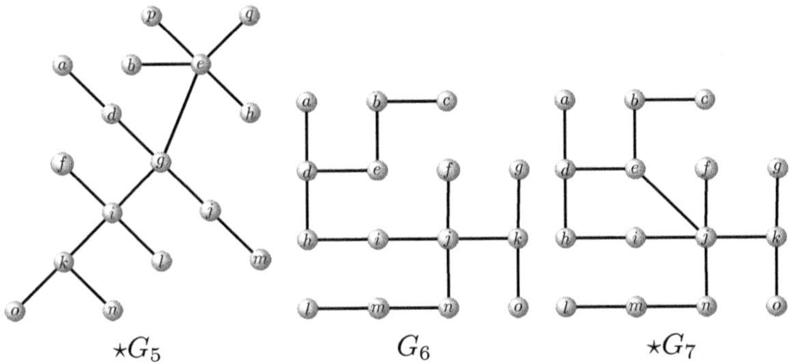

$\star G_5$ G_6 $\star G_7$

(2) for what value(s) of the integer n is the complete graph K_n a tree?

(3) for what value(s) of the integers m and n is the complete bipartite graph $K_{m,n}$ a tree?

(4) For each of the following trees, give the smallest positive integer k such that the tree is k-ary and the height h of the tree. Determine if the tree is balanced, complete or full.

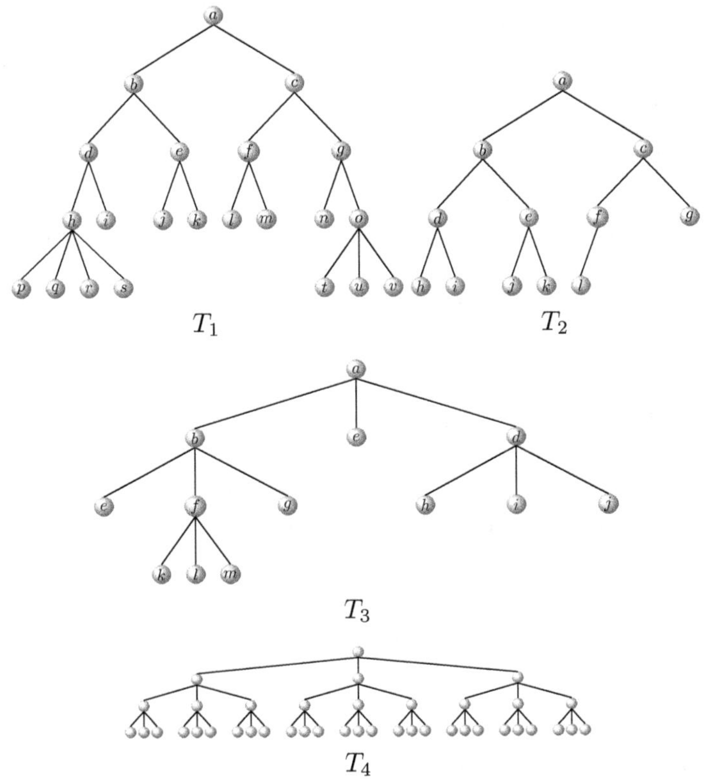

T_1 T_2

T_3

T_4

⋆(5) Show that in a rooted tree T, every node other than the root has a unique parent.

(6) Consider the following tree T:

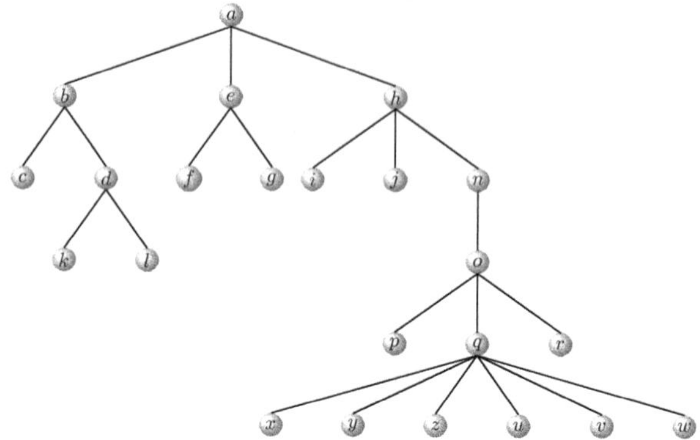

(a) List all the leaf nodes of T.

(b) List all the internal nodes of T at height 2.

⋆ (c) Which node is the parent of o?

(d) What is the height of node q?

⋆ (e) What is the height of T?

(f) List all siblings of node v.

⋆ (g) List all descendants of node n.

(h) List all ancestors of node q other than the root of the tree.

⋆ (i) Give two nodes at level three which are not siblings.

(j) Give two descendants of o which are not siblings.

(k) What is the smallest positive integer m such that T is m-ary?

(7) For the tree T given in the Exercise (6), draw the subtree rooted at d, h and o.

(8) Prove that two siblings in a rooted tree have at the same height.

(9) Prove that every tree is a bipartite graph.

(10) Draw all non-isomorphic trees of order 3, 4 and 5.

(11) Draw a full ternary rooted ordered tree of height 3.

(12) In each case, draw a 4-ary ordered rooted tree satisfying the given property.

(a) Unbalanced of height 3.

⋆(b) Balanced but not full.

(c) Complete but unbalanced.

(d) Balanced but not complete.

⋆(e) Complete but not full of height 3.

(f) Full of height 2.

(13) Let T be a rooted tree and let V be its set of nodes. Define a binary relation R on V as follows: $u\,R\,v$ if and only if u and v are at the same level.

   (a) Prove that R is an equivalence relation on V.

   (b) For the tree T given in Exercise (6), give the partition of V corresponding to the equivalence relation of part (a).

$\star$(14) Let G be a connected graph. Prove that G is a tree if and only if every edge of G is a bridge. (Recall that an edge e is called a bridge in G if $G - e$ is a disconnected graph).

11.3   First results about trees

Now that we have covered the basic terminology of trees, we start exploring some interesting facts about them.

Theorem 11.2. *Let $m \geq 1$ be an integer. If T is an m-ary tree of height h, then the number l of leaf nodes of T satisfies $l \leq m^h$.*

Proof. We proceed by induction on h. If $h = 0$, then T consists of the root only which is a leaf and the inequality $l \leq m^h$ is true in this case. Let $h \geq 0$ and assume that the result is true for any m-ary tree of height at most h. Let T be a tree of height $h + 1$. Deleting all edges in T connecting the root with nodes at level 1 creates at most m subtrees (since the root has at most m children) each of which is m-ary of height at most h. Moreover, the number l of leaf nodes in T is the sum of all leaves in the subtrees. By the induction hypothesis, the number of leaf nodes in each subtree is at most m^h. We conclude that $l \leq m(m^h) = m^{h+1}$. The result follows by the principal of mathematical induction. $\qquad\qquad\square$

Theorem 11.3. *For any integer $n \geq 1$, a tree with n nodes has $n-1$ edges.*

Proof. There are several ways to prove this result. A common way is to use mathematical induction on the number of nodes or the number of edges of the tree (see Exercise (16) below). There is however a much simpler proof of the result based on a natural one-to-one correspondence between the nodes other than the root and the edges of the tree. With the exception of the root, every node is the terminal point of a unique edge (due to the fact that there are no circuits in a tree). This creates a one-to-one correspondence between nodes other than the root and the set of edges in the tree. The result follows. $\qquad\square$

Corollary 11.1. *Any tree with at least two nodes must have at least two nodes of degree 1.*

Proof. Let T be a tree with n nodes, $n \geq 2$. If $\deg(v) \geq 2$ for every node v of T, then the sum of the degrees of all the nodes in T is greater than or equal to $2n$. By Theorem 11.3 and the handshaking lemma, we get that $2(n-1) \geq 2n$ which is a contradiction. If there is exactly one node of degree 1, then the sum of the degrees of all the nodes in T in this case would be at least $2(n-1)+1$. Again, Theorem 11.3 and the handshaking lemma imply that $2(n-1) \geq 2(n-1)+1$ which is also a contradiction. The result follows. $\qquad\square$

If T is a tree, we denote the number of internal nodes and leaf nodes in T by i and l respectively. If the tree is of order n, then $n = i + l$ since every node is either internal or a leaf and cannot be both.

Theorem 11.4. *Let $m \geq 2$ be an integer, T an m-ary tree of order n and height h with l leaf nodes and i internal nodes.*

(a) If T is complete, then $n = mi + 1$.
(b) If T is full, then:

    (i) The number of nodes at level j is m^j.
    (ii) $l = m^h$.
    (iii) $n = \dfrac{m^{h+1}-1}{m-1}$.
    (iv) $i = \dfrac{m^h-1}{m-1}$.

Proof.

(a) Since the tree is m-ary and complete, every internal node has m children. The number of children in the tree is then mi. Now, every node is a child except for the root node. The result follows.
(b) Assume now that T is a full tree. For part (i), we proceed by induction on j. If $j = 0$, then there is only one node at level j, namely the root. The result is true in this case since $m^0 = 1$. Let $j \geq 0$ and assume that level j contains m^j nodes. Since each of these nodes has m children (as the tree is full), level $j + 1$ contains $m(m^j) = m^{j+1}$ nodes. The result follows by the induction principle. Part (ii) follows from part (i) using $j = h$ and the fact that all leaf nodes in a full tree are at level h. For part (iii), we use the fact that $n = mi + 1$ (since T is in particular a

complete m-ary tree) and $i = n - l = n - m^h$ (part (ii)) to get that $n = m(n - m^h) + 1$. This can be rearranged as $(m-1)n = m^{h+1} - 1$ or $n = \frac{m^{h+1}-1}{m-1}$. Using the relation $n = l + i$ and parts (ii) and (iii), we get that $i = n - l = \frac{m^{h+1}-1}{m-1} - m^h = \frac{m^{h+1}-1-m^h(m-1)}{m-1} = \frac{m^h-1}{m-1}$ and part (iv) follows. $\qquad\square$

Having bounds on the height and the order of a tree is often helpful in applications. The next result addresses bounds for some trees. Recall that if x is a real number then the ceiling of x, denoted by $\lceil x \rceil$, is the smallest integer greater than or equal to x.

Theorem 11.5. *Let T be an m-ary tree of order n with height h and l leaf nodes. Then:*

(a) $h \geq \lceil \log_m(l) \rceil$
(b) *If T is complete and balanced, then:*

   (i) $h = \lceil \log_m(l) \rceil$
   (ii) $\frac{m^h-1}{m-1} < n \leq \frac{m^{h+1}-1}{m-1}$

Proof.

(a) By Theorem 11.2, we know that $l \leq m^h$. The result follows by taking $\log_m$ on both sides of the inequality and using the definition of the ceiling function.
(b) Assume that T is complete and balanced. Let T' be the tree obtained from T by removing all leaf nodes at level h together with their incident edges. Then T' is an m-ary full tree of height $h - 1$. Theorem 11.4 tells us that T' has exactly m^{h-1} leaf nodes. Since T is of height h, it must have at least one leaf node at level h and consequently T has more leaf nodes than T'. We then get that $m^{h-1} < l \leq m^h$. Taking $\log_m$ gives $h-1 < \log_m(l) \leq h$. By definition of the ceiling function, $h = \lceil \log_m(l) \rceil$. This proves part (i). Part (ii) is left as an exercise (see Exercise (17) below). $\qquad\square$

11.3.1   *Exercises*

(1) How many nodes in a tree with 2987 edges?
$\star$(2) How many internal nodes in a complete 5-ary tree T with 256 nodes? How many leaf nodes in T?

(3) How many nodes in a complete ternary tree T with 12 internal nodes? How many leaf nodes in T?

(4) Give a lower bound on the height of a 5-ary tree with 87 leaf nodes.

$\star$(5) Give bounds on the order of a complete and balanced 4-ary tree of height 5.

(6) A forest F consists of k connected components and a total of n vertices. Express the total number of edges in F in terms of n and k.

(7) Let T be a complete m-ary tree with n nodes. Express the number of internal nodes and the leaf nodes in terms of m and n.

(8) For each of the sequences below, determine if it can be the degree sequence of a tree. If you say it is, give a possible drawing of such a tree. If you say it is not, explain why.

$\star$ (a) $(5, 5, 5, 4, 4, 3)$

$\star$ (b) $(5, 1, 1, 1, 1, 1, 1)$

(c) $(3, 3, 2, 1, 1)$

$\star$ (d) $(4, 4, 3, 3, 3, 2, 2, 1, 1, 1, 1, 1, 1, 1)$

(e) $(6, 4, 4, 3, 1, 1, 1, 1, 1, 1, 1, 1, 1, 1, 1)$

$\star$(9) Let T be a full ternary tree of height 8 with n nodes, e edges, i internal nodes, and l leaf nodes. Let n_k be the number of nodes at level k in T. Find the value of each of the following.

(a) n (b) e (c) l (d) i (e) n_2 (f) n_5

(10) Let T be a full 4-ary tree of height h with n nodes, e edges, i internal nodes, and 1024 leaf nodes. Let n_k be the number of nodes at level k in T. Find the value of each of the following.

(a) h (b) n (c) e (d) i (e) n_3 (f) n_5

(11) Let m be a positive integer. In each case determine if there exists a full m-ary tree T with the given height h and the number l of leaf nodes. If you say such a tree exists, give a possible drawing of the tree.

(a) $h = 3, l = 27$ (b) $h = 4, l = 32$

$\star$(12) A tree T has 9 nodes of degree 1, two nodes of degree 2, one node of degree 3 and the rest of the nodes are of degree 4 each. How many nodes of degree 4 does T have?

$\star$(13) A complete k-ary tree T of height 3 has 46 leaf nodes. Determine all possible values of k.

(14) Joe shared a photo on a social media platform with 6 of his friends. Each one of the 6 friends either shares it with six other friends (who have never received it before) or does not share it with anyone, and so on. We assume that no person receives the photo from more than one friend. If the photo was shared 5460 times in total: (a) How many people (including Joe) shared the photo? (b) How many people did not share the photo? Answer these questions by modeling the situation with an m-ary tree for a certain positive integer m.

(15) Corollary 11.3 above states that any tree of order $n \geq 2$ must have at least two pendent nodes (nodes of degree 1). Describe trees of order $n \geq 2$ having *exactly* two pendent nodes. Justify your answer.

(16) Give a proof of Theorem 11.3 using induction on the number of nodes of the tree.

$\star$(17) Let T be an m-ary tree of order n and height h Prove that if T is complete and balanced, then $\frac{m^h - 1}{m - 1} < n \leq \frac{m^{h+1} - 1}{m - 1}$.

$\star$(18) Prove that the chromatic polynomial of a tree with n nodes is $\lambda(\lambda - 1)^{n-1}$ (*Hint.* Use induction on n, the cancellation-contraction property and the fact that a tree has at least one node of degree 1). What is the chromatic number of a tree?

(19) Theorem 11.3 states that a tree with n nodes has $n - 1$ edges. Prove the converse: a simple connected graph G of order n with $n - 1$ edges must be a tree.

(20) Prove that a tree is a planar graph (*Hint.* Use induction on the number of the nodes of the tree).

11.4 Traversal of trees

Trees are often used in computer science to store data and later retrieve it and process it. It is therefore important to have procedures for visiting nodes in the tree and to search for particular parts of the stored data. Depending on the purpose of visiting the nodes, the order in which the nodes are traversed is important and needs to be taken into account. In this section, we assume that all trees are rooted and ordered.

Definition 11.6. Given a rooted and ordered tree T, a *traversal* of T is a systematic procedure to visit (and process) every node of T exactly once.

There are several ways of traversing a tree and each traversal is a recursive procedure that depends on three main components of the tree: the left subtree, the root and right subtree. The following definition explores three common tree traversals. In each case, the traversal creates a certain ordering on the nodes, following the order in which they are visited. A tree with only one node (the root) has only one traversal. Preorder, postorder and inorder traversals are defined for arbitrary trees.

Definition 11.7. Let T be an ordered rooted tree with root v and order $n \geq 2$. Let $T_1, T_2, \ldots, T_k$ be the subtrees of T from left to right.

(a) The *preorder traversal* of T consists of visiting the root v first, then the nodes of T_1 in preorder, then the nodes of T_2 in preorder and continuing this way until the nodes of T_k are visited in preorder. In a preorder traversal of T, the parent node is visited before the children and siblings are traversed from left to right order.

(b) The *postorder traversal* of T consists of visiting, in postorder, the nodes of T_1 then the nodes of T_2 until the nodes of T_k are visited. We finish by visiting the root. In a postorder traversal of T, children nodes are visited before the parent node and siblings are traversed from left to right order.

(c) The *inorder traversal* of T consists of visiting the nodes of T_1 in inorder, then the root v, then the nodes of T_2 in inorder and so on until the nodes of T_k are visited in inorder

Example 11.6. Consider the following ternary tree T:

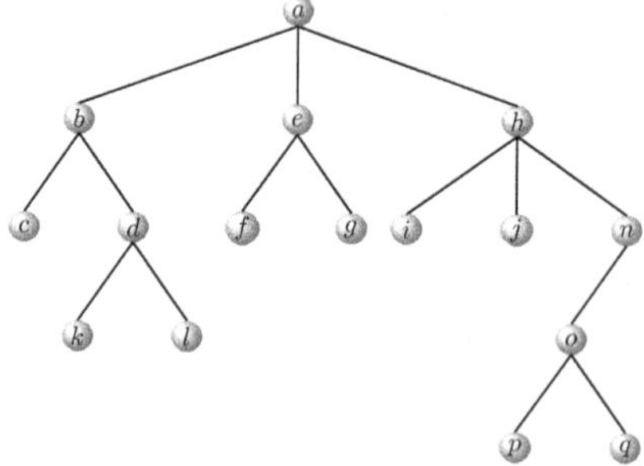

Give the order of the nodes of T following a preorder traversal, a postorder and an inorder traversal of the tree.

Solution. In preorder traversal we start by visiting the root a, then the left subtree followed by the middle and the right subtrees each in preorder.

The following diagram explains the succession of traversals and finishes by
the order of nodes of T:

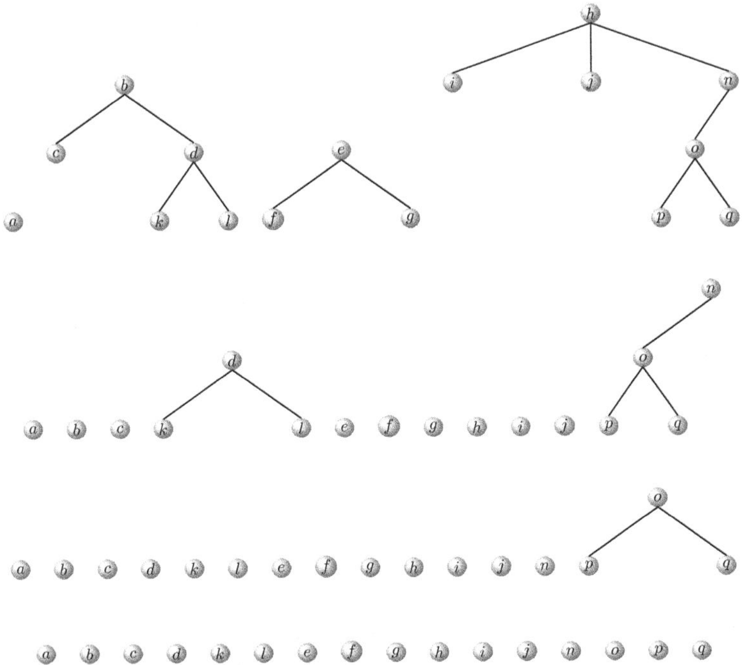

The following is the order on nodes of T following the preorder traversal of
the tree:

$$a\ b\ c\ d\ k\ l\ e\ f\ g\ h\ i\ j\ n\ o\ p\ q.$$

In postorder traversal we start by visiting the left subtree followed by
the middle and the right subtrees each in postorder and then we finish
by visiting the root a. The following diagram explains the succession of
traversals and finishes by the final order in which the postorder visits the
nodes of T:

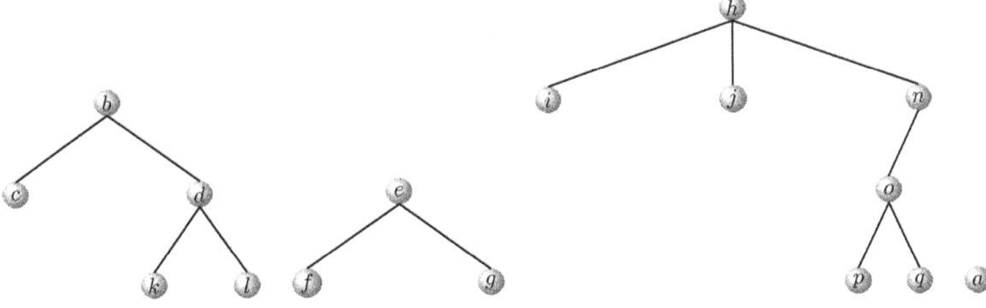

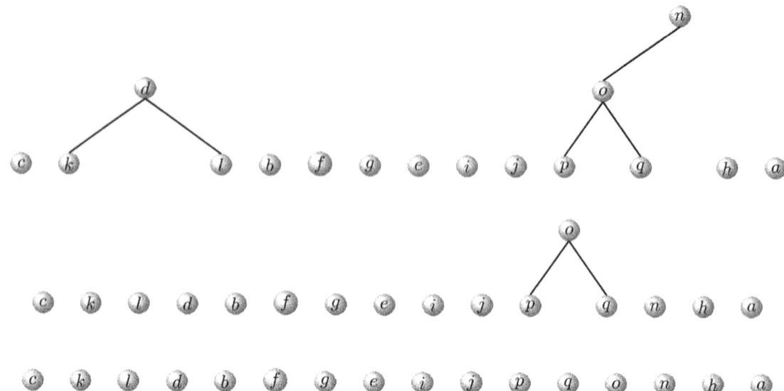

The following is the order on nodes of T following the postorder traversal of the tree:

$$c\ k\ l\ d\ b\ f\ g\ e\ i\ j\ p\ q\ o\ n\ h\ a.$$

For the inorder traversal of the tree, we give the order of the nodes but we leave out the details:

$$c\ b\ k\ d\ l\ a\ f\ e\ g\ i\ h\ j\ p\ o\ q\ n.$$

$\Diamond$

Example 11.7. For the binary tree T below, traverse the tree in inorder fashion and give the corresponding order on the nodes of T.

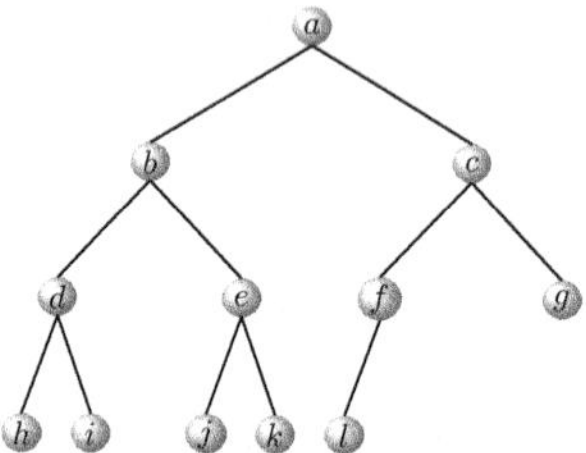

Solution. Traversing the left subtree first, followed by the root and then the right subtree (all in inorder) yields the following order on the nodes:

$$h\ d\ i\ b\ j\ e\ k\ a\ l\ f\ c\ g.$$

$\Diamond$

Remark 11.1. When traversing a tree, we could encounter an empty left or right subtree. In this case, we have to be careful how we approach such a situation in each of the three modes of traversals explored above. See Exercise (1) below for an example.

One way to have a total ordering on the nodes of an ordered rooted tree is to label the nodes according to their levels and their placements in that level from left to right. This way of ordering the nodes resembles the ordering of words in the dictionary (known as the lexicographic order).

Definition 11.8 (Universal address system). Let T be an ordered rooted tree with root r. The *universal address system* is a procedure to label the nodes of T and then use that labeling to define a total order on the nodes. The universal address system is defined recursively as follows.

1. Assign label 0 to r.
2. Assign labels $1, 2, 3, \ldots$ to all nodes at level 1 (children of r) from left to right respectively.
3. If v is an internal node at level n labeled α and if $v_1, \ldots, v_k$ are all the children of v from left to right respectively, then child v_i is assigned label $\alpha \cdot i$ for each $i \in \{1, 2, \ldots, k\}$.

The universal address system is used to define a total order on the nodes of T as follows: If u and v are nodes of T labeled $\alpha_1 \cdot \alpha_2 \cdot \ldots \cdot \alpha_s$ and $\beta_1 \cdot \beta_2 \cdot \ldots \cdot \beta_t$ respectively, then define $u < v$ if either one of the following two scenarios occurs:

(a) $\alpha_1 < \beta_1$;
(b) There exists i such that $\alpha_j = \beta_j$ for $j = 1, 2, \ldots, i-1$ and $\alpha_i < \beta_i$.

This total order on the nodes can be obtained by traversing the tree in preorder fashion.

Example 11.8. The following shows an ordered rooted tree T with nodes labeled using the universal address system:

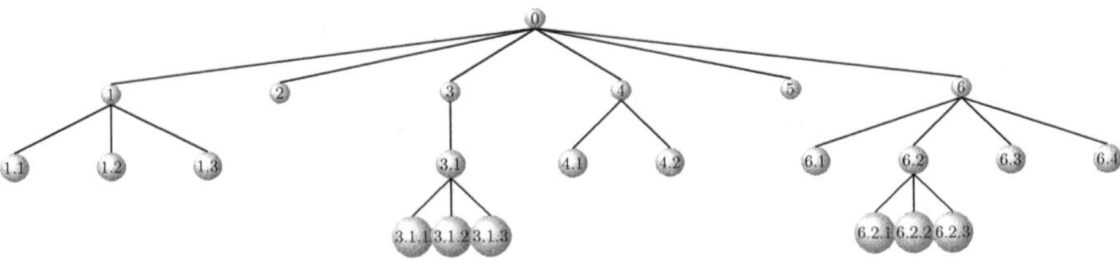

The corresponding total ordering of the nodes of T is the following: $0 < 1 < 1.1 < 1.2 < 1.3 < 2 < 3 < 3.1 < 3.1.1 < 3.1.2 < 3.1.3 < 4 < 4.1 < 4.2 < 5 < 6 < 6.1 < 6.2 < 6.2.1 < 6.2.2 < 6.2.3 < 6.3 < 6.4$

11.4.1 *Binary search tree*

Given a totally ordered set S of data elements, an important question in computer science is to find ways to store and later retrieve elements in S. Binary trees can be very useful to achieve this. The idea is to store the elements of S in the nodes of a rooted ordered binary tree in such a way that the total order defined on S is "compatible" with the order of the tree.

Definition 11.9. A *binary search tree* is a rooted binary tree T with a total order $\leq$ on the set V of its nodes such that for any node $v \in V$, $v_L \leq v$ for any node v_L on the left subtree of v and $v_R \geq v$ for any node v_R on the right subtree of v. We say in this case that T is a binary search tree on the set V.

Example 11.9. Let $V = \{a, b, c\}$ ordered with the usual lexicographic alphabetic order: $a < b < c$. The following are all the possible different search trees on V:

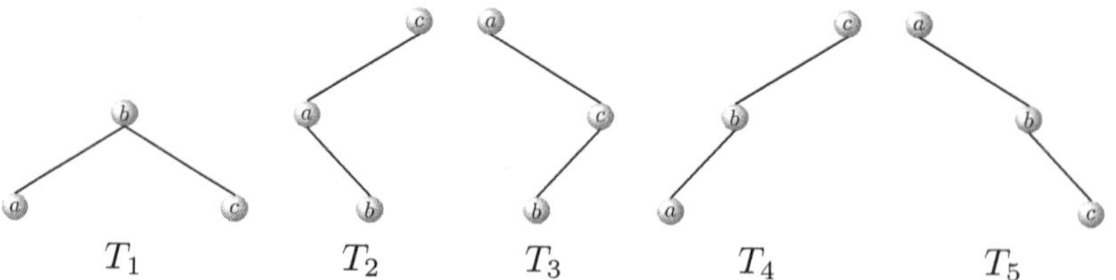

Given a non-empty totally ordered set V, a binary search tree T on V can be formed using the following steps.

(1) Line up the elements of V in a list of arbitrary order: $v_1, v_2, \ldots, v_n$.
(2) For $k = 1$, set the binary search tree to be just the node v_1.
(3) For $2 \leq k \leq n$, compare v_k with v_1:

  (3.1) If $v_k < v_1$, start at v_1 and move to the left subtree of v_1.

    (3.1.1) If the left subtree of v_1 is empty, add v_k as a left child of v_1.
    (3.1.1) Otherwise, repeat step 3 using the left subtree.

  (3.2) If $v_k > v_1$, start at v_1 and move to the right subtree of v_1.

    (3.2.1) If the right subtree is empty, add v_k as a right child of v_1.
    (3.2.1) Otherwise, repeat step 3 using the right subtree.

Example 11.10. Let $V = \{4, 6, 8, 7, 1, 3, 9, 2, 5\}$ ordered with the natural ordering on integers. Construct a binary search tree on V.

Solution. We start by using the following order on the elements of V:

$$4\ 6\ 8\ 7\ 1\ 3\ 9\ 2\ 5.$$

The steps of constructing the search tree are illustrated below:

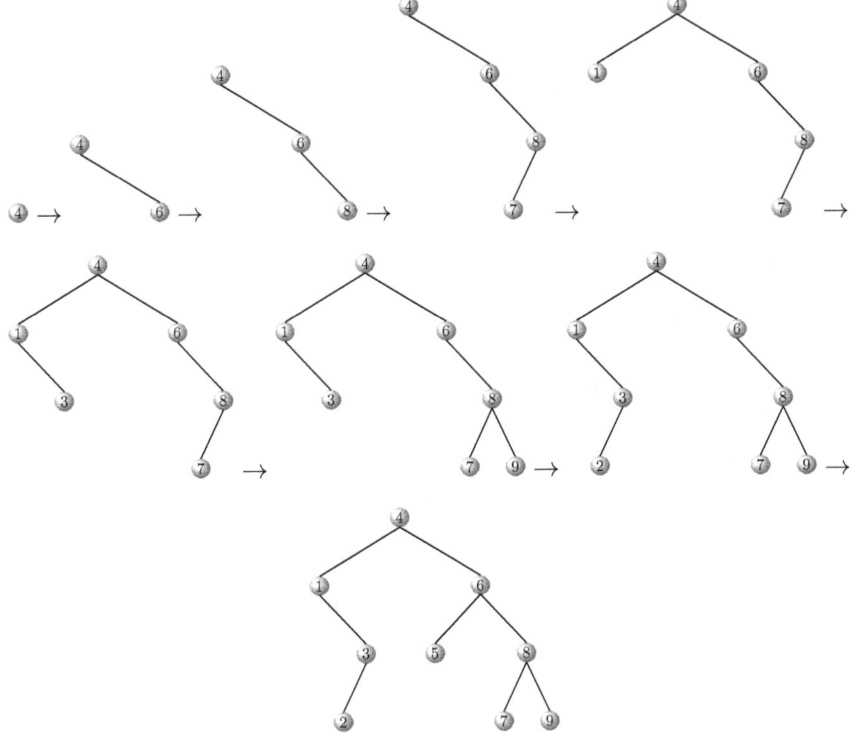

If the binary search tree on a totally ordered set V is given, then an inorder traversal of the tree allows us to sort the elements of V in lexicographic order. Searching for an element x in V is also easy using the binary search tree. This is similar to inserting a new node in the tree. Start at the root node r of the tree, and when you are at a node v in the tree compare the sought value x with v:

- If $v = x$, the element x is found.
- If $x > v$, proceed to the right subtree of v. If no such subtree exists, x is not on the tree.
- If $x < v$, proceed to the left subtree of v. If no such subtree exists, x is not on the tree.

Example 11.11. The following shows a binary search tree for the data in the set $S = \{\chi, \varpi, \zeta, \rho, \omega, \eta, \nu, \xi\}$

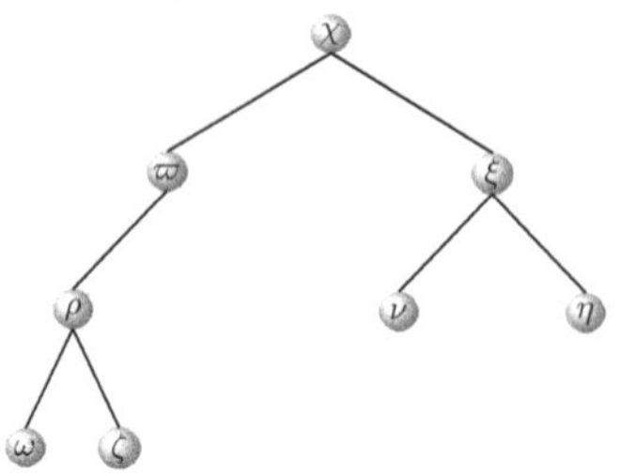

(a) Sort the data in an increasing order.

(b) Give the path followed in the tree to access data item ζ.

Solution.   (a) This is accomplished by performing an inorder traversal of the tree: $\omega \ \rho \ \zeta \ \varpi \ \chi \ \nu \ \xi \ \eta$.

(b) Starting at the root χ, we move to the left since $\zeta < \chi$. We compare ζ with the new node we arrive at: ϖ. Since $\zeta < \varpi$, we move to the left at ϖ. The new node we encounter is ρ. Since $\zeta > \rho$, we move to the right at ρ and we reach ζ. The search path that leads to ζ is shown in the diagram below.

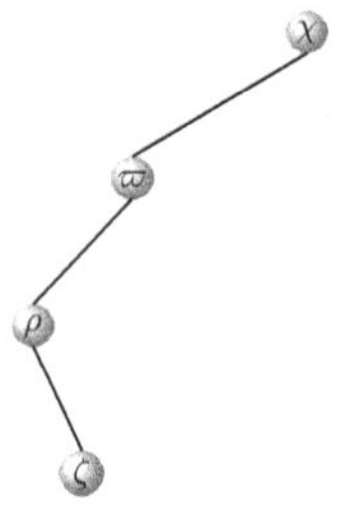

$$\Diamond$$

An important ordering in many areas of Mathematics is that of polynomials in several variables. Recall that a monomial in n variables $x_1, x_2, \ldots, x_n$ is an expression of the form $x_1^{a_1} x_2^{a_2} \cdots x_n^{a_n}$ where each exponent a_i is a non-negative integer. The lexicographic ordering on the set of monomials with respect to the order $x_1 > x_2 > \ldots > x_n$ on the variable is defined as follows: if $\mathbf{x} = x_1^{a_1} x_2^{a_2} \cdots x_n^{a_n}$ and $\mathbf{y} = x_1^{b_1} x_2^{b_2} \cdots x_n^{b_n}$ are two monomials, then $\mathbf{x} > \mathbf{y}$ if and only if the leftmost non-zero component of the vector $(a_1 - b_1, a_2 - b_2, \ldots, a_n - b_n)$ is positive. For example, using lexicographic ordering on two variables x and y with $x > y$, we have: $1 < y < y^2 < y^3 < x < xy < x^2 y$.

Example 11.12. Consider the set

$$V = \{xy^2z, x^2y^2z^2, z^3, x, y^2z, x^3y^2z^8, x^2y^9z^2\}$$

of monomials in three variables x, y and z. Form the binary search tree T on the set V ordered with the monomial lexicographic order with $x > y > z$. Traverse T in inorder fashion and verify that the resulting list of elements of V is the same as the ordering of the monomials using the lexicographical order of the previous part. Finally, give the path in T to access the element y^2z of V.

Solution. The binary search tree is constructed step by step as shown in the diagram below:

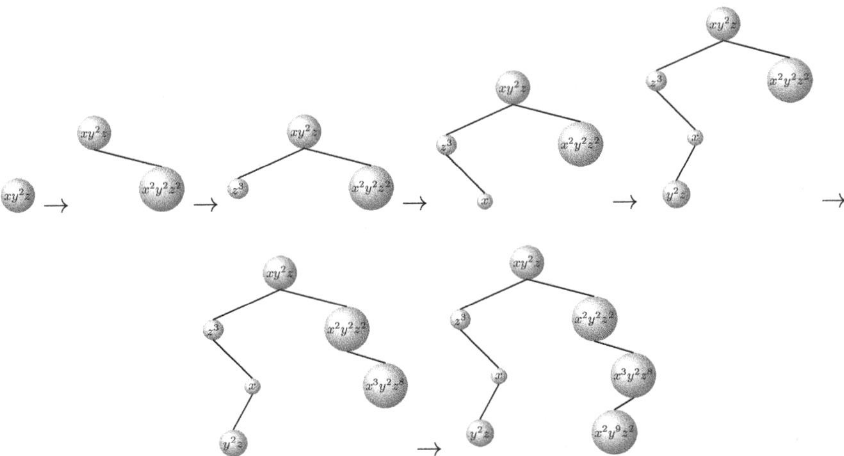

Traversing the above tree in inorder fashion leads to the following ordering on the monomials in V: $z^3 < y^2x < x < xy^2z < x^2y^2z^2 < x^2y^9z^2 < x^3y^2z^8$ which is exactly the lexicographic ordering on these monomials with $x > y > z$. For the path to access the element y^2z in T, start at the root xy^2z, move to the left since $y^2z < xy^2z$. The next node we encounter is z^3. Since $z^3 < y^2z$, we move to the right and we encounter node x. Since $y^2z < x$, we move to the left. Since the new node we encounter is y^2z, we are done:

$\Diamond$

11.4.2   *Exercises*

(1) The following diagram shows three binary rooted trees T_1, T_2 and T_3.

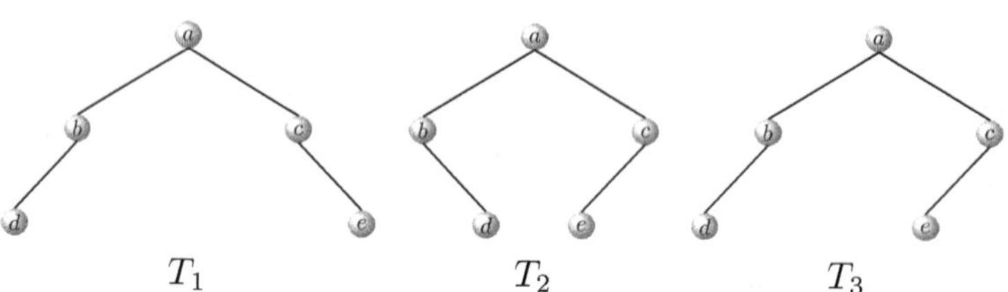

The trees are isomorphic (as graphs) but their traversals could yield different orders on their nodes. For each of the trees, give the order on the nodes corresponding to: (i) a preorder traversal; (ii) a postorder traversal; (iii) an inorder traversal.

(2) For each of the following binary trees, give the order in which the nodes appear when the tree is traversed using: (i) a preorder traversal (ii) a postorder traversal (iii) an inorder traversal.

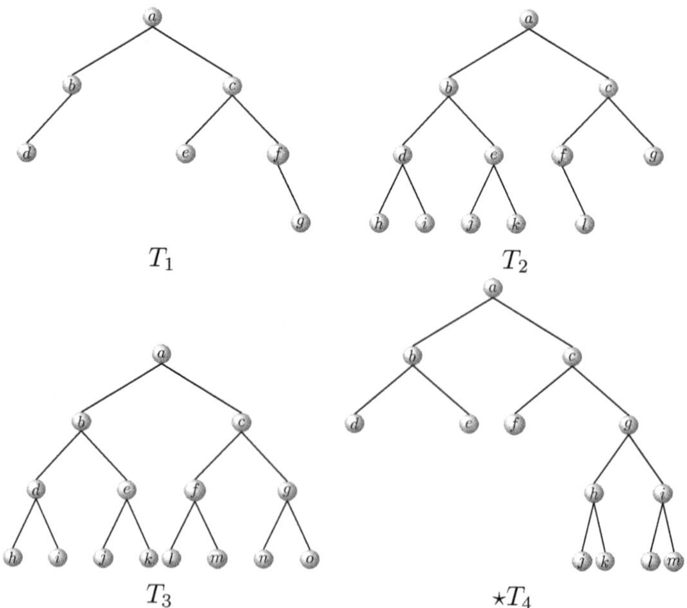

(3) Consider the following ternary tree T:

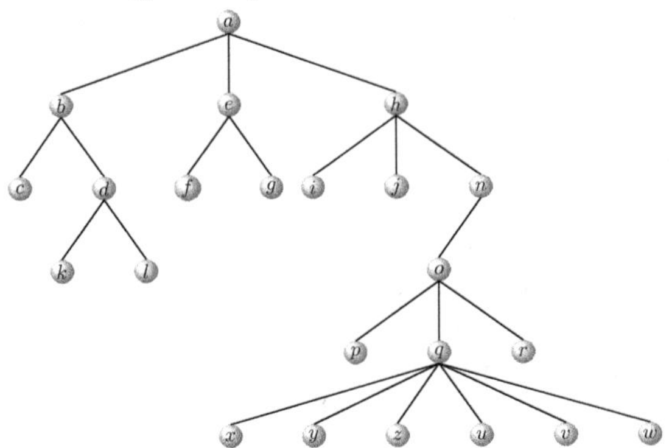

Give the order in which the nodes of T appear when T is traversed using: (i) a preorder traversal (ii); a postorder traversal and (iii) an inorder traversal.

(4) For each of the following ordered rooted binary trees, label the nodes according to the universal address system. Give the total lexicographic ordering on the nodes that corresponds to that labeling.

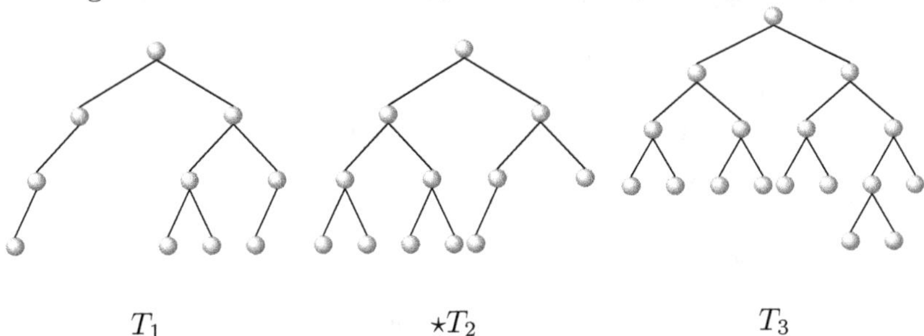

T_1 $\star T_2$ T_3

(5) For each of the following ordered rooted ternary trees, label the nodes according to the universal address system. Give the total lexicographic ordering on the nodes that corresponds to that labeling.

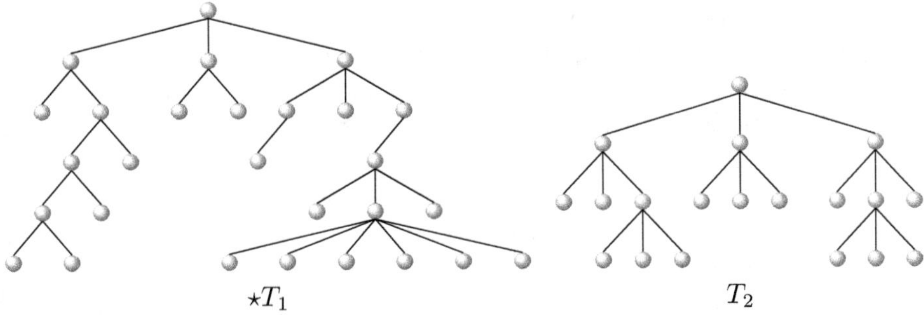

$\star T_1$ T_2

(6) The nodes of an ordered rooted tree T are labeled using the universal address system. The node in T with the *largest* address is v with universal address 4.4.4.5.7.

  $\star$ (a) Find the level of v.

  (b) Find the height of T.

  $\star$ (c) Find the number of siblings of v.

  $\star$ (d) Find the addresses of all of the siblings of v.

  (e) Find the address of the parent node of v.

  (f) Find the address of five of the ancestors of v.

  $\star$ (g) Find the smallest possible number of nodes in T.

  (h) Find all the other addresses that *must* appear in T.

(7) Consider the set $S = \{12, 3, 5, 13, 4, 6, 1, 19, 10, 14, 2\}$ with the standard order $<$ on the integers. Form the binary search tree T for the set S. Traverse T in inorder fashion and verify that the output is the list the elements of S ordered from the smallest to the largest.

$\star$(8) The set S consists of the following (unsorted) English word: *capital, binary, mathematics, mathematician, when, also, river, rather, or, capsule, ordered, ordering, myth, rhythm.*

  (a) Form the binary search tree T for the set S ordered with the standard alphabetical (lexicographic) dictionary order.

  (b) Traverse T in inorder fashion and verify that the output is the list the elements of S is the natural order.

  (c) Give the path followed in T to access the word *rhythm*.

(9) Consider the set

$$S = \{y^2zt, xy^2z^2t, t^3, x, z, y^2, xt^2, x^3yz^4t^5, z^5t^2, x^2y^9z^2t^4\}$$

of monomials in four variables x, y, z and t. Form the binary search tree T for the set S ordered with the lexicographic with $x > y > z > t$. Traverse T in inorder fashion and verify that the output is the list the elements of S is the natural order.

$\star$(10) Beside the lexicographic ordering, there are other ways to order monomials. The graded lexicographic ordering, denoted by $>_{grlex}$, is another ordering of interest for monomials. For the monomial $\mathbf{x} = x_1^{a_1} x_2^{a_2} \cdots x_n^{a_n}$, define the total degree of $\mathbf{x}$, denoted by $|\mathbf{x}|$, as being $|\mathbf{x}| = \sum_{i=1}^{n} a_i$. Given two monomials $\mathbf{x} = x_1^{a_1} x_2^{a_2} \cdots x_n^{a_n}$ and $\mathbf{y} = x_1^{b_1} x_2^{b_2} \cdots x_n^{b_n}$, then $\mathbf{x} >_{grlex} \mathbf{y}$ if $|\mathbf{x}| > |\mathbf{y}|$ or $|\mathbf{x}| = |\mathbf{y}|$ and $\mathbf{x} >_{lex} \mathbf{y}$ where $>_{lex}$ is the regular lexicographic ordering. Redo Exercise (9) with the set S ordered with the graded lexicographic with $x > y > z > t$.

11.5 Modeling with trees

The particular structure of trees makes them very useful in modeling real life problems. In this section, we explore few applications of trees.

11.5.1 *Chemistry*

Trees are ideal to represent the structure of some types of chemical compounds. Saturated hydrocarbons are chemical compounds formed by Carbon (C) and Hydrogen (H) atoms such that all of the carbon atoms are connected by a single bond and each carbon atom is bonded directly to four other atoms (hydrogen or carbon). In these compounds, we have a maximum number of hydrogen atoms for a given number of carbon atoms. If there are n atoms of carbon in a saturated hydrocarbon, then it can be shown (see Exercise (2) below) that there must be $2n + 2$ atoms of hydrogen in the structure, hence the chemical representation C_nH_{2n+2}. A saturated hydrocarbon is represented with a tree where each atom (C or H) is a node and each bond between atoms is an edge in the tree. Since each carbon atom has four electrons available for bonding and each hydrogen atom has a single electron, $\deg(C) = 4$ and $\deg(H) = 1$ in every tree representation of C_nH_{2n+2}. The first graph in the following diagram represents CH_4 (methane), the second and third graphs are two representations of C_4H_{10}. The second graph is known as Butane and third graph is known as 2-methyl propane. Clearly, all three graphs are trees as they are connected with no cycles. Notice that the trees representing the Butane and the 2-methyl propane are not isomorphic. For instance, in the tree representing the 2-methyl propane, there is a node C adjacent to three nodes of degree 4 each. No such a node in the tree representing the Butane. We say that these two representations are chemical *isomers* for the saturated hydrocarbon C_4H_{10}. They have the same number of atoms of Carbon and Hydrogen, but they differ in the way the atoms are bond together.

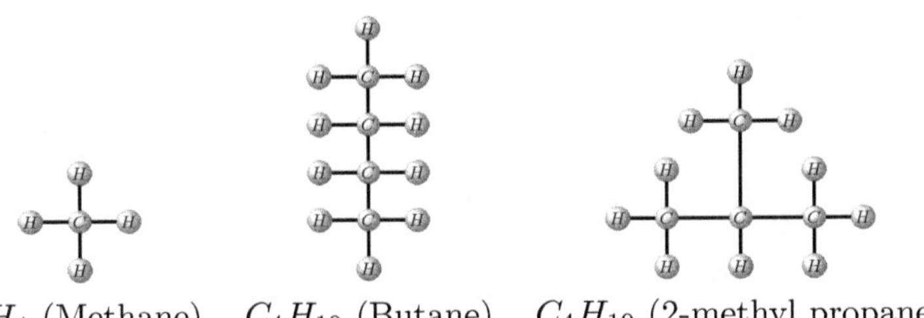

CH_4 (Methane) C_4H_{10} (Butane) C_4H_{10} (2-methyl propane)

In general, to determine the number of different isomers of the saturated hydrocarbon C_nH_{2n+2} one has to determine how many non-isomorphic trees with n nodes of degree 4 and $2n+2$ nodes of degree 1 are there. The fact that every graph representation of C_nH_{2n+2} is indeed a tree is left as an exercise (see Exercise (1) below).

11.5.2 *Digital arithmetic expressions*

Trees, in particular binary trees, are widely used in computer science. Arithmetic Expressions like addition $(+)$, subtraction $(-)$, multiplication $(*)$, division $(/)$ and exponentiation $(\uparrow)$ are called binary operations since they take two operands (input) and produce a third element (output). Notice that some of these operations are not commutative. For example $a - b$ is not the same as $b - a$. The order of operands must then be taken into account. This is where ordered rooted trees come in handy. If $\circ$ is an arithmetic operation and a, b are two numbers, then the number $a \circ b$ can be represented by a binary tree with $\circ$ as the root and operands a and b as left and right child respectively:

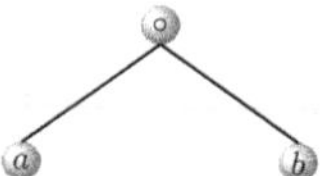

Example 11.13. The following diagram shows binary tree representations of basic arithmetic operations. Addition and multiplication are commutative while subtraction, division and exponentiation are not.

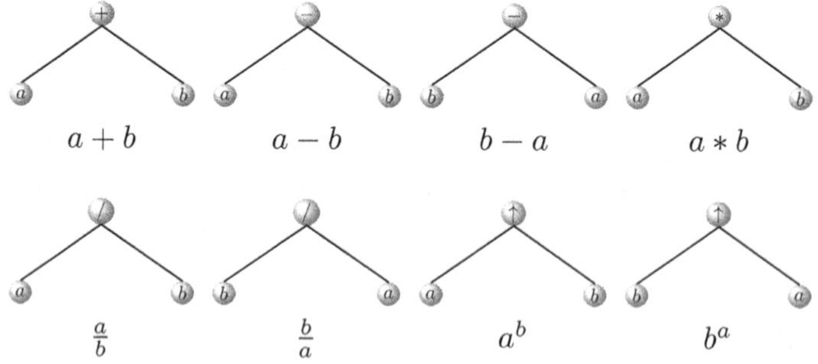

Complex arithmetic expressions involve operands which contain one or more operations each. For example, the expression $\alpha = \left(a - \frac{b}{2}\right) * \left(7 + a^b\right)$ consists of multiplying two operands, each of which is an expression itself with some arithmetic operations. In this case, the tree representing α is built, from the bottom up, out of subtrees of individual operands.

Example 11.14. To construct the binary tree of the expression $\alpha = \left(a - \frac{b}{2}\right) * \left(7 + a^b\right)$, we start by constructing the subtree of the left operand $\left(a - \frac{b}{2}\right)$. First, we construct the tree of the division $\frac{b}{2}$ and then the tree of the subtraction $\left(a - \frac{b}{2}\right)$:

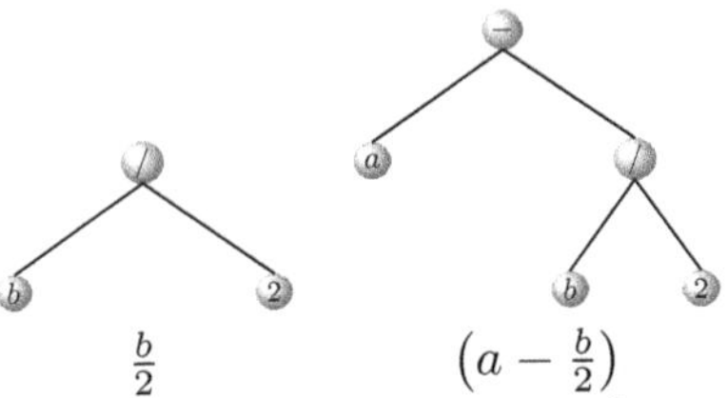

$$\frac{b}{2} \qquad\qquad \left(a - \frac{b}{2}\right)$$

Next, we form the subtree of the right operand $\left(7 + a^b\right)$:

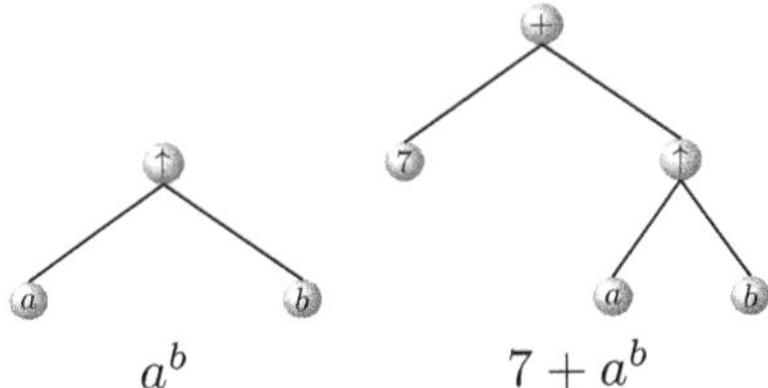

$$a^b \qquad\qquad 7 + a^b$$

Combining the two subtrees with the multiplication operation, we get the binary tree representing the original expression:

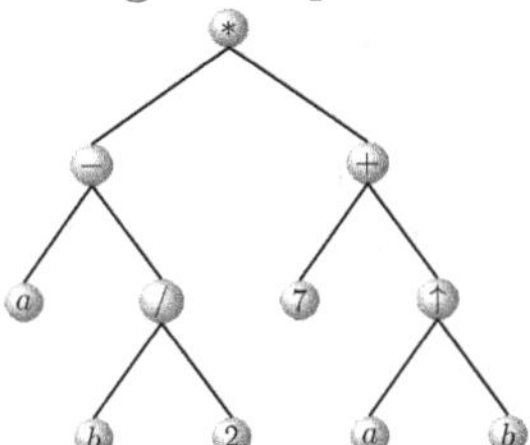

Example 11.15. Find the arithmetic expression represented by the following binary tree.

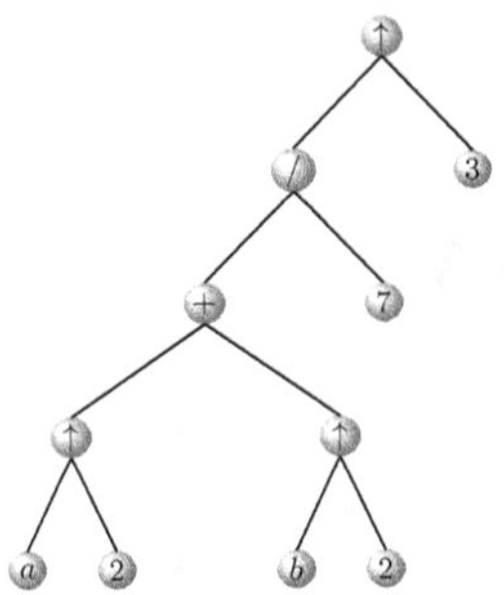

Solution. Starting at the left-most tree from the bottom, recursively evaluate every subtree. The required expression is the final value obtained at the root of the tree, namely $\left(\frac{a^2+b^2}{7}\right)^3$:

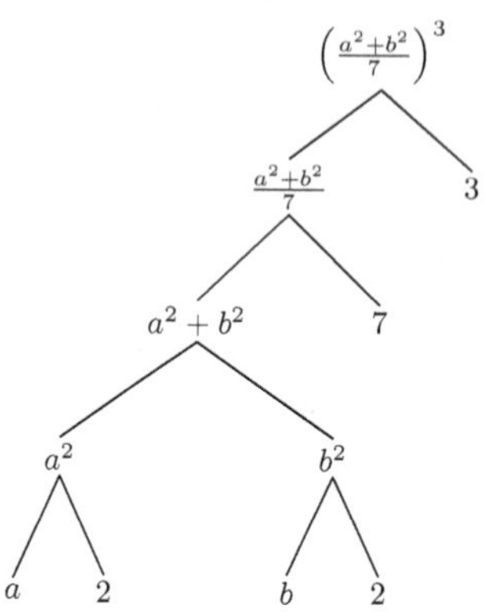

$\Diamond$

11.5.3   *Prefix, postfix and infix notations*

When a computer program evaluates an arithmetic expression, it does so by performing the binary operations present with respect to the provided hierarchy of operations. For example, parentheses precede exponentiation which precedes multiplication and division (multiplication and division have the same hierarchy), which in turn precedes addition and subtraction (which have the same hierarchy). A computer program evaluates an arithmetic expression by working the operations in each parenthesized component from left to right in the hierarchal order stored in the program.

Example 11.16. An arithmetic expression is shown in the first row of the following diagram. The integers in the following row represent the order in which the operations are performed by a computer program.

$$z + (\, x - y \uparrow 2\,) \, * \, (\, z + u - v \uparrow 2\,) / (\, x \uparrow 2 + y \uparrow 3\,)$$

$$11 \quad\quad 8 \quad 7 \quad\quad 10 \quad\quad 6 \quad 5 \quad 4 \quad 9 \quad\quad 2 \quad 3 \quad 1$$

The expression value is $z + (x - y^2)\dfrac{z+u-v^2}{x^2+y^3}$.

    The standard notation we use for binary operations, such as $a + b$, where the operation symbol lies between the two operands is known as the *infix* notation. For a more complex expression, an infix form requires the use of many parentheses to respect the hierarchy of operations. A computer program must then do a lot of back and forth scanning and storing of intermediate values in order to compute the expression. That usually takes time and storage space. To avoid the use of parentheses, the polish mathematician Jan Lukasiewicz introduced in 1920 two alternative notations that became later known as the *Polish notations* in his honor. The first Polish notation is known as the *prefix notation* (also referred to simply as the Polish notation) in which the binary operation symbol precedes the two operands, so we write $\circ ab$ instead of $a \circ b$. The second Polish notation is known as the *postfix notation* (also referred to as the reverse Polish notation) in which the binary operation symbol follows the two operands, so we write $ab\circ$ instead of $a \circ b$.

Example 11.17. In infix notation, the arithmetic expression $\frac{a}{b} + cd - e^2$ is written as: $(a/b) + (c * d) - (e \uparrow 2)$. In prefix notations, (a/b), $(c * d)$ and $(e \uparrow 2)$ are represented as $/ab$, $*cd$ and $\uparrow e2$ respectively. So the prefix notation of $(a/b) + (c*d) - (e \uparrow 2)$ is $- + /ab * cd \uparrow e2$. In postfix notations, (a/b), $(c * d)$ and $(e \uparrow 2)$ are represented as $ab/$, $cd*$ and $e2 \uparrow$ respectively. So the postfix notation of $(a/b) + (c * d) - (e \uparrow 2)$ is $ab/cd * +e2 \uparrow -$.

In prefix notation, the operator symbol precedes the two operands. This means that in order to evaluate an expression written in prefix form, we read the expression from right to left. When we encounter an operator, we perform the corresponding operation using the two operands immediately to the right of the operator. In postfix notation, the operator symbol follows the two operands. To evaluate an expression written in postfix form, we read the expression from left to right. When we encounter an operator, we perform the corresponding operation using the two operands immediately to the left of the operator. In both cases, the new element we obtain is a new operand for the next operator (if one exists).

Remark 11.2. The above procedure to evaluate an expression written in either prefix or postfix notation describes how to proceed when encountering a *binary* operator. If we encounter a *unary* operator, we follow the same procedure except that we perform the corresponding operation on the single operand immediately before or after depending on what notation we are using.

Example 11.18. Find the value of the following expression written in prefix notation: $- 9 * * 2 1 / \uparrow 2 2 2$.

Solution. Reading the expression from right to left, the steps of evaluating the expression are shown in the following diagram:

$$- 9 * * 2 1 / \underbrace{\uparrow 2 2}_{2^2=4} 2; \quad - 9 * * 2 1 \underbrace{/ 4 2}_{\frac{4}{2}=2}; \quad - 9 * \underbrace{* 2 1}_{2\times 1=2} 2;$$

$$- 9 \underbrace{* 2 2}_{2\times 2=4}; \quad \underbrace{- 9 4}_{9-4=\boxed{5}}.$$

$\Diamond$

Example 11.19. Find the value of the following expression written in postfix notation: $12\ 3\ 2\ * \ -\ 3\ 1 \uparrow\ 3\ /\ -$.

Solution. Reading the expression from left to right, the steps of evaluating the expression are shown in the following diagram:

$$12 \underbrace{3\ 2\ *}_{3\times 2=6} -\ 3\ 1\uparrow\ 3\ /\ -; \quad \underbrace{12\ 6\ -}_{12-6=6}\ 3\ 1\ \uparrow\ 3\ /\ -; \quad 6 \underbrace{3\ 1\ \uparrow}_{3^1=3}\ 3\ /\ -;$$

$$6 \underbrace{3\ 3\ /}_{\frac{3}{3}=1}\ -; \quad \underbrace{6\ 1\ -}_{6-1=\boxed{5}}.$$

$\Diamond$

Converting an expression from the infix notation (standard) to a prefix or a postfix notation could be overwhelming and time consuming. This is where binary trees come in handy. If T is the binary tree representing the expression, then a preorder (respectively an inorder, respectively a postorder) traversal of T leads to the prefix (respectively the infix, respectively the postfix) notation of the expression.

Example 11.20. Give the infix, prefix and the postfix forms of the arithmetic expression $\left(a - \frac{b}{2}\right) * \left(7 + a^b\right)$.

Solution. The binary tree T representing the expression was constructed in Example 11.14 as follows:

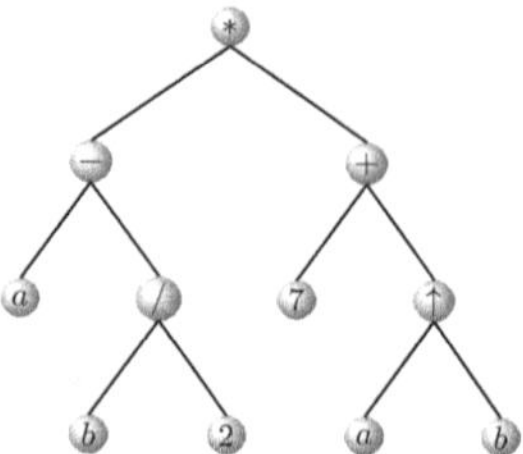

A preorder traversal of the tree yields the order $* - a / b\, 2 + 7 \uparrow a\, b$. This is the prefix notation of the expression. An inorder traversal of the tree is $a - b / 2 * 7 + a \uparrow b$. This is the infix notation of the expression. Finally, a postorder traversal of the tree gives the postfix notation of the expression: $a\, b\, 2 / - 7\, a\, b \uparrow + *$. $\qquad\qquad\Diamond$

Like arithmetic expressions, well-formed formulas of propositional logic can be represented using binary trees. In this case, the operations are the logic connectives and the operands are the well-formed logic subformulas. Similar to arithmetic expressions, well-formed compound propositions can be expressed in prefix, infix and postfix forms in order to avoid (or minimize) the use of parentheses. Expressions containing sets and set operations (like intersection, union,...) can also be represented using prefix, infix and postfix notations. Notice that the negation connective ($\neg$) and the complement of a set A ($\overline{A}$ or A') are unary operations (apply to one operand only) that can be represented inside a binary tree using a node (for the operation) with one child (the operand).

Example 11.21. Consider the following well-formed formula:

$$\phi : ((p \vee q) \to \neg(\neg p \wedge (q \leftrightarrow r))).$$

Draw an ordered rooted binary tree representing ϕ. Express ϕ using prefix, infix and postfix notations.

Solution. Like before, the binary tree is constructed from the bottom up using subtrees of intermediate operations.

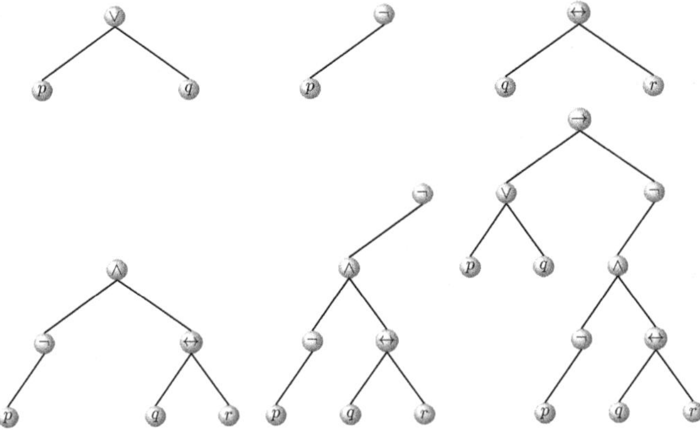

Traversing the tree in preorder yields the prefix notation: $\to \ \vee\, p\, q\, \neg\ \wedge \ \neg\, p\ \leftrightarrow\ q\, r$. Traversing the tree in inorder yields the infix notation: $p \vee$

$q \to p \neg \wedge q \leftrightarrow r \neg$. Finally, traversing the tree in postorder yields the postfix notation: $p\, q \vee p \neg q\, r \leftrightarrow \wedge \neg \to$. $\Diamond$

Example 11.22. A, B and C are three subsets of a universal set $\mathcal{U}$. For the subset $X = (A \cup C) \setminus (A \cap (B \cup C))$ of $\mathcal{U}$, draw an ordered rooted binary tree representing X and express X using prefix, infix and postfix notations.

Solution. The ordered rooted binary tree representing X is:

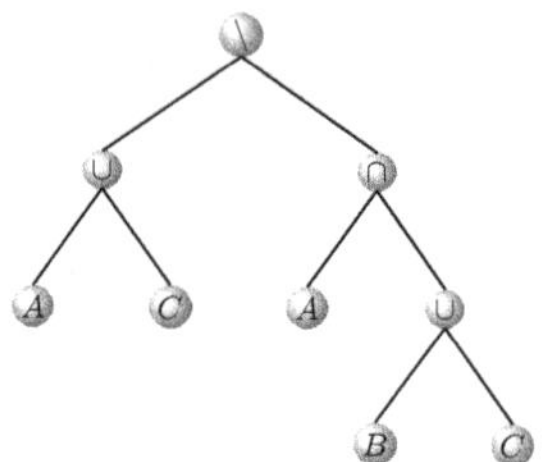

Traversing the tree in preorder yields the prefix notation: $\setminus \cup A\, C \cap A \cup B\, C$. Traversing the tree in inorder yields the infix notation: $A \cup C \setminus A \cap B \cup C$. Finally, traversing the tree in postorder yields the postfix notation: $A\, C \cup A\, B\, C \cup \cap \setminus$. $\Diamond$

11.5.4 *Coding*

When you click the "save" button after working on a document (text, image, audio, ...), a convenient interpretation of what happens next is to imagine that your computer stores the file in the form of a (long) finite binary string.

$$\boxed{0\,1\,1\,1\,1\,1\,1\,1\,1\,1\,1\,0\,0\,0\,0\,0\,1\,0\,0\,0\,0\,0\,1\,0\,0\,0\,0\,0\,0\,0\,0\,0\,1\,0\,0\,0\,1\,0\,0}$$

Before it is converted into a digital form, information usually comes in a *raw form* that we call *source data*. Source data can be a text, an image, an audio or a video. A source data is usually a series of characters from a finite set $\mathcal{A} = \{\alpha_1, \ldots, \alpha_r\}$ that we call the *alphabet*. The process of creating a digital form (binary strings) to represent the data source for storage or transmission is called *encoding*. Encoding an input file with alphabet $\mathcal{A} = \{\alpha_1, \ldots, \alpha_r\}$ is achieved by assigning a binary string c_i to each character α_i in the alphabet. The set $\mathcal{C} = \{c_1, c_2, \ldots, c_r\}$ thus obtained is called a *binary code* and each c_i is called a *codeword*. For example, if $\mathcal{A} = \{X, Y, Z, W\}$, then $\mathcal{C} = \{00110, 0, 1010, 111\}$ is a binary code for $\mathcal{A}$ with codewords 00110, 0, 1010 and 111. By a *data compression algorithm*, we usually mean a process through which we can represent data in a compact and digital form that uses less space to store or less time to transmit over a network than the original form. This is usually done by reducing unwanted noise (or redundancy) in the data to a certain degree where it is still possible to recover it in an acceptable form. Of course, a compression algorithm is only efficient if we are able to reverse the process and retrieve the original sequence of characters from the encoded digital form. This process is called *decoding* or *decompressing*. In the literature, the word *compression* is often used to indicate the compression and the decompression processes of data.

Example 11.23. The extended version of the ASCII code assigns binary codes of length 8 each for each character (letter from the English alphabet, punctuation, numbers and other symbols for a total of 256 characters). For example, the (capital) letter "J" is assigned the codeword 01001010, the letter "o" is assigned the codeword 01101111 and 01100101 is the codeword assigned for the letter "e". The ASCII code is an example of a *fixed length code* as it assigns the same length of 1 byte (=8 bits) for each codeword. However, fixed length codes are not usually the most efficient ones for compression purposes. To save storage space, it would make more sense to assign shorter codewords for characters that appear more frequently in a source data and longer ones for less frequent characters. For example, some studies suggest that in a typical English text, the letter "e" appears more frequently than other letters (this is not exactly true if we include other non-letter characters, like the space character which is a bit more frequent than "e") followed by the letter "t", whereas the letters "q" and "z" appear at the bottom of the list of most frequent letters. So, if we are compressing an English text, a good compression strategy would be to assign shorter codewords for "e" and "t" and longer ones for "q" and "z". In this case, our code is called a *variable length code*.

11.5.4.1 *Using binary trees to represent binary codes*

Binary trees are often used to represent binary codes. To represent the binary code $C = \{c_1, \ldots, c_n\}$ with a tree, start with codeword c_1 and read its bits one at a time from left to right. If 0 is read, create a left branch from the root of the tree with a new node at its end. Draw a right branch if 1 is read. Move to the next bit in c_1. If it is 0, move down one edge along the left branch and record a new node at the end of the new edge. Do the same but along the right branch if the second bit is 1. Repeat the process until all bits of the codeword are read and mark the corresponding character. Starting from the root node, repeat the same process for the other codewords.

Example 11.24. Consider the alphabet $\Sigma = \{A,\ B,\ C,\ D,\ E\}$. The binary tree for the code $\mathcal{C} = \{00,\ 01,\ 10,\ 110,\ 111\}$ of Σ is the following.

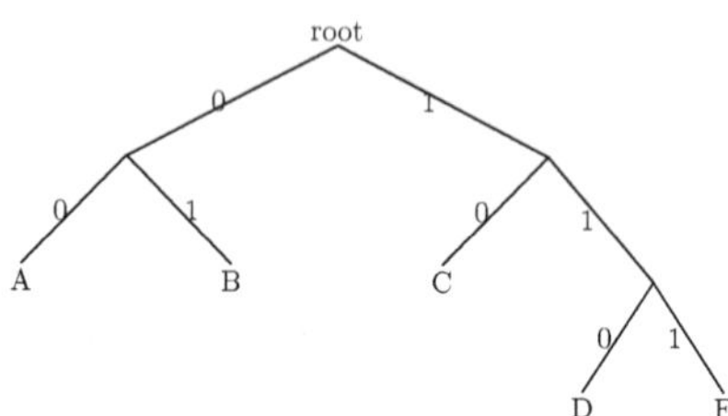

Variable length codes are very practical from compression perspective, but they don't always achieve optimal results. For a code to be efficient, one must include in its design a unique way of decoding. For example, consider the variable code $\mathcal{C} = \{0,\ 01,\ 10,\ 1001,\ 1\}$ for the alphabet $\{a, b, c, d, e\}$. The word encoded as 101001 can be decoded in more than one way: "cd", "ead" or "ccb". This is certainly not a well-designed code because of this decoding ambiguity. A closer look at $\mathcal{C}$ shows that the problem lies in the fact that some codewords are prefix (or start) of other codewords. For example, the codeword 10 is a prefix of the codeword 1001. One way to avoid the confusion is to include a symbol to indicate the end of a codeword, but this risks to be costly considering the number of times the end symbol must be included in the encoded file.

Definition 11.10. A code $\mathcal{C}$ with the property that no codeword is a prefix of some other codeword in $\mathcal{C}$ is called a *prefix-free* code (also known as *instantaneous* codes).

Prefix-free codes are *uniquely decodable*: given an encoded digital message M, there is unique way to decode M using a prefix-free code.

Example 11.25. The code $\mathcal{C} = \{10,\ 01,\ 000,\ 1111\}$ is prefix-free since no codeword in $\mathcal{C}$ is a prefix of another codeword.

The task of determining if a given code is prefix-free could be challenging for large codes. Using the associated binary tree is one way to simplify the task. A code is prefix-free if all the alphabet characters are associated with leaf nodes on the associated binary tree. If one character is located at an internal node, the code is not prefix-free.

Example 11.26. Determine if the code: $A \to 00$, $B \to 010$, $C \to 011$, $D \to 10$, $E \to 110$, $F \to 111$ and $G \to 11$ on the alphabet $\mathcal{A} = \{A, B, C, D, E, F, G\}$ is prefix-free.

Solution. We construct the binary tree for the code:

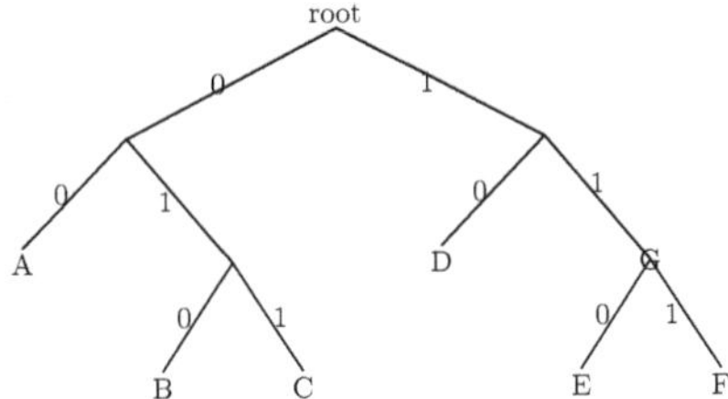

Since character G corresponds to an internal node in the tree, the above code is not prefix-free. $\Diamond$

Given an encoded binary message M on a prefix-free binary code $\mathcal{C}$, the algorithm to decode M using the binary tree T associated with $\mathcal{C}$ is described in the following steps.

(1) Starting at the root of T, move to the left branch if the first bit in M from the left is 0. Move to the right branch if that bit is 1.
(2) Repeat the previous step using adjacent bits in M (read from left to right) until you reach a leaf node.
(3) Record the letter corresponding to the leaf node.
(4) Return to the root of T and repeat the previous steps using the remaining bits in M.
(5) Finish when all bits in M are read.

Example 11.27. Consider the code $\mathcal{C} = \{00, 01, 10, 110, 111\}$ for the alphabet $\{A, B, C, D, E\}$ of Example 11.24. The tree associated with $\mathcal{C}$ (given in Example 11.24 above) shows that the code is prefix-free. Suppose we want to decode the binary message $M = 01000011010111$. Starting at the root of the tree, we move left using the first 0 in M and then we move right using the second 1. A leaf with label "B" is reached. We record the letter "B" and return to the root. At this point, we continue with the string 000011010111. From the root, we move left (for the first 0 in the new string) and then left again along the same branch (for the second 0 in the new string). The leaf labeled "A" is reached. We record "A" next to the letter "B" previously found. Continuing in this manner, we find that the message 01000011010111 decodes as "BAADCE".

11.5.5   *Exercises*

(1) Prove that a graph representing the saturated hydrocarbon structure $C_n H_{2n+2}$ must be a tree (*Hint.* Use Exercise (19) of Section 11.3 above).

$\star$(2) Use graph theory to prove that if a saturated hydrocarbon has n carbon atoms, then it must have $2n + 2$ hydrogen atoms.

(3) How many isomers are there for the Pentane $C_5 H_{12}$? Give a tree representation for each isomer.

(4) Let a and b be two positive integers. In each case, draw an ordered rooted binary tree representing the given arithmetic expression.

(a) $\left(a - \frac{2}{b}\right) + 3$

$\star$ (b) $a - \left(\frac{2}{b} + 3\right)$

$\star$ (c) $a^2 + b^2 - \left(\frac{2}{b} + \frac{a}{2}\right) + 3$

(d) $a^b * \left(2 + \frac{a}{b}\right)$

(e) $a^{\left(b*2+\frac{a}{b}\right)}$

$\star$(5) Determine the value of the arithmetic expression represented by the following binary tree:

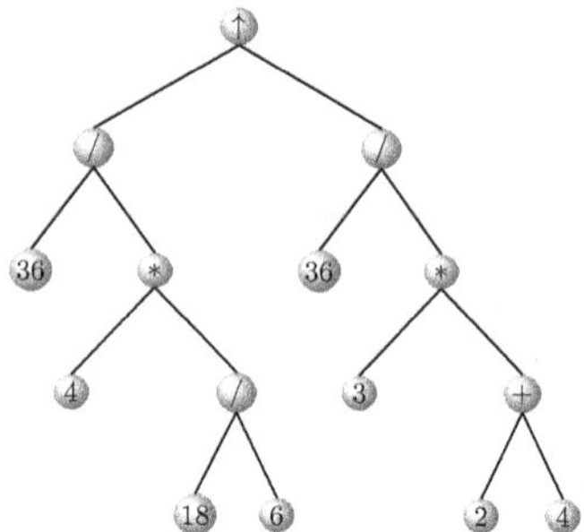

(6) Let a and b be positive real numbers. Write the arithmetic expression represented by the following binary tree using standard parenthesized notation. Then write your answer in infix, prefix and postfix notations.

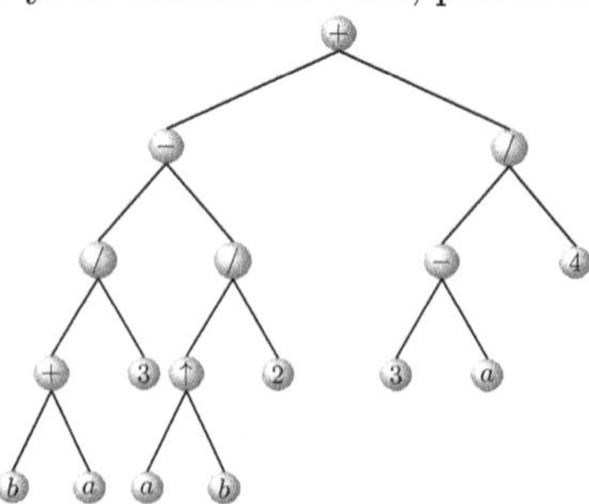

(7) In each case, give the numerical value of the arithmetic expression written in prefix form.

$\star$ (a) $* - 5 / 6\,2\,2$ (c) $- * \uparrow 2\,2 * 5\,2\,6$
  (b) $- + * 4\,5 / 3\,2\,9$ $\star$ (d) $\uparrow\uparrow\uparrow 2\,1\,1 + - 6\,1\,0$

(8) In each case, the expression is given in postfix form. Give the value of the expression.

  (a) $6\,2 / 5 + \uparrow 5\,2 -$ $\star$ (c) $1\,3 * 2 \uparrow 9 - 5 + 9\,3 / *$
$\star$ (b) $3\,2 * 2 \uparrow 6\,4 - 8\,4 / * -$ (d) $5\,4 - 2 \uparrow 1 \uparrow 1\,1 \uparrow\uparrow$

(9) Consider the algebraic expression $\alpha = (a + (b + c)d)x^{y^{z^{t}}}$ where a, b, c, d, x, y, z, t are positive integers. Draw the binary tree representing α and write it in prefix, infix and postfix forms.

(10) For each of the following well-formed formulas of propositional logic, draw an ordered rooted binary tree representing the formula and then express it using prefix, infix and postfix notations.

  (a) $((\neg A \leftrightarrow B) \wedge C)$ $\star$ (c) $((A \leftrightarrow B) \to (A \wedge (B \vee C)))$
  (b) $\neg((A \vee B) \to (B \wedge \neg C))$ (d) $(A \leftrightarrow ((B \vee C) \to (C \to D)))$

(11) A, B, C and D are four subsets of a universal set $\mathcal{U}$. The notation E' refers to the *complement* of E in $\mathcal{U}$. In each case, draw an ordered rooted binary tree representing the given subset of $\mathcal{U}$. Express the subset using prefix and postfix notations.

  (a) $(A \backslash B) \cup (B \backslash C)$ (c) $A \backslash (B' \backslash (C \backslash D))$
  (b) $((A \cup B) \cap D) \cup (B \cap (C \cup D))$ (d) $((A \backslash B) \cup C) \backslash (D \cap (C \cup D'))$

(12) Given a set S of symbols and a set O of *binary* operators, we can talk about *well-formed expressions* in prefix, infix and postfix notations. The definition of well-formed expression in postfix notation is given recursively as follows:

  (1) Any element of S is a well-formed expression in postfix notation.
  (2) If ϕ and ψ are any two well-formed expressions in postfix notation and $\circ$ is a binary operator (element of O), then $\phi\psi\circ$ is a well-formed expression in postfix notation.
  (3) Nothing else is a well-formed expression in postfix notation.

  (a) Give a recursive definition of a well-formed expression over S using operations from O in infix notation.
  (b) Give a recursive definition of a well-formed expression over S using operations from O in prefix notation.

(c) Let $S = \{a, b, c, d\}$ and $O = \{\natural, \top, \Diamond, \circ\}$. The first column in the table below contains some expressions using elements from S and O. Complete the table by inserting a check mark ($\checkmark$) in the column labeled "prefix" (respectively infix, postfix) if the expression is a well-defined expression in prefix notation (respectively infix, postfix notations) and an $\times$ if it is not.

	prefix	infix	postfix
$ab\natural\top cd$	$\times$		
$\Diamond c\Diamond c\natural\top \circ cdab$			
$\circ\natural \circ \top\Diamond aabbcd$	$\checkmark$		
$aab\natural\top cd \circ \Diamond$			
$ab\natural\top cd$			
$\Diamond \circ aab\natural\top c$			
$a\top d\natural d\Diamond d \circ b$			
$bc\top d\natural da\Diamond d \circ b\top c$			$\times$

(13) On the set P of all positive integers, define the following binary operations: $a \circ b = min\{a, b\}$, $a\Delta b = max\{a, b\}$ and $a\Diamond b = gcd(a, b)$ (here gcd stands for the *greatest common divisor*). Addition, multiplication and exponentiation of integers are denoted by $+$, $*$ and $\uparrow$, respectively.

(a) Give the value of each of the following expressions written in prefix notation:

    (1) $\Diamond \circ \Delta 2345$ $\star$ (3) $\Delta + \circ * \Delta 549182$

    (2) $\uparrow + * \circ 4511\Diamond 12$ (4) $\Diamond\Delta \circ \circ \circ 23454\,(26)$

(in part (4), (26) represents the integer twenty six).

(b) Give the value of each of the following expressions written in postfix notation:

    (1) $436 \circ \circ 2 * 8\Delta$ (3) $21 \uparrow 4 * 8 + 4 \circ 3\Diamond$

    $\star$ (2) $13154 \circ \Diamond\Delta +$ (4) $11 \uparrow 111111\Diamond\Delta \circ \circ \circ \Diamond$

$\star$(14) The binary tree representing a prefix-free binary code $\mathcal{C}$ is full of height 5. How many characters are there in the alphabet of $\mathcal{C}$?

$\star$(15) Let α, β, γ and κ be four bits (elements of the set $\{0, 1\}$). Determine the values of α, β, γ and κ so that the binary code: $A \to 01$, $B \to \alpha 10\beta 1$, $C \to 10111$, $D \to \gamma 1\kappa$, $E \to 110110$ and $F \to 1010$ with alphabet $\mathcal{A} = \{A, B, C, D, E\}$ is prefix-free.

(16) Consider the binary code $A \to 00$, $B \to 010$, $C \to 011$, $D \to 10$, $E \to 110$ and $F \to 111$ of the alphabet $\{A, B, C, D, E, F\}$. Construct the binary tree of the code and show that the code is prefix-free. Decode the message $M = 11100100100001100110$.

$\star$(17) The following is the binary tree of a binary code $\mathcal{C}$ on the alphabet $\mathcal{A} = \{A, B, C, D, E, F, G, H, I, J\}$:

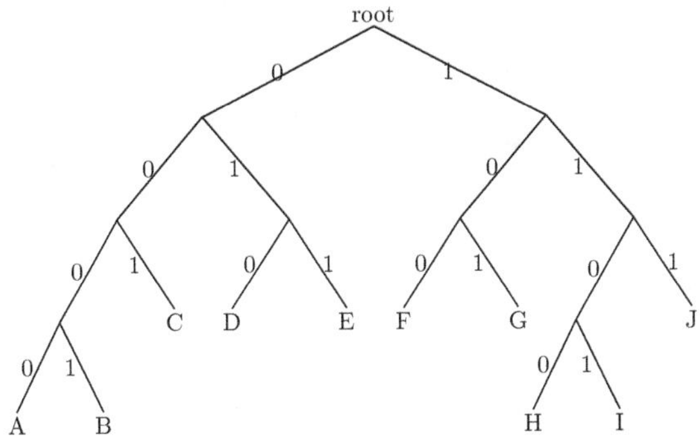

(a) Give the codeword for each of the alphabet characters.
(b) Encode the word IJACDFHDDCIIBBEFE using the tree.
(c) Decode each of the following encoded messages:

   (i) 0001101000
   (ii) 000110000001110111100
   (iii) 1110001101000011001000111111111

11.6 Spanning subtrees and graph search

Given a graph $G = (V, E)$, recall that a subgraph H of G is called a *spanning subgraph* if it has the same set of vertices as G.

Definition 11.11. Given a graph G, a *spanning tree* of G is a spanning subgraph of G which is also a tree.

If a graph G contains a spanning tree T, then it is possible to find a path between any two distinct vertices of G using the edges of T only. This suggests that T "generates" or "spans" G; hence the name *spanning tree*.

Example 11.28. The following diagram shows a graph G and two spanning trees T_1 and T_2 with edges drawn in thick lines.

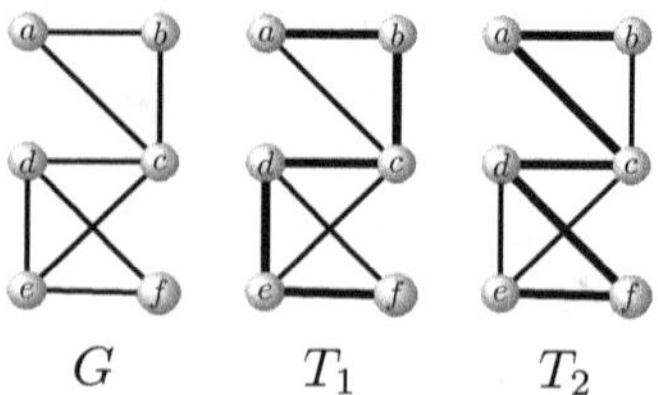

$$G \qquad\qquad T_1 \qquad\qquad T_2$$

It is not hard to see that a graph has a spanning tree if and only if
its underlying simple graph (obtained from G by removing all loops and
replacing any set of parallel edges with a single edge) has a spanning tree.
This suggests that we can assume that the graph is simple when we look
for a spanning tree. For the rest of this section, all graphs considered are
assumed simple.

Theorem 11.6. *A graph is connected if and only if it has a spanning tree.*

Proof. Since every tree is connected (by definition), a graph having a
spanning tree must be connected. Conversely, Let G be a connected graph.
If G has no cycles, then it is a tree and we are done. If G has cycles, then the
subgraph G_1 obtained by removing an edge from a cycle of G is connected
(removing an edge from a cycle in G will not disconnect G) and contains
all the vertices of G. If G_1 has no cycles, then it is a spanning tree of G
and we are done. If G_1 has cycles, then remove an edge from one cycle of
G_1 and obtain a spanning connected subgraph G_2 of G_1. Continuing this
way, we obtain a sequence $G_1, G_2, \ldots$ of spanning connected subgraphs of
G. Since G has a finite number of edges, this sequence of subgraphs must be
finite. The last subgraph we obtain in the sequence is a connected spanning
subgraph of G with no cycles; hence a spanning tree of G. $\qquad\square$

Corollary 11.2. *Let G be a connected graph of order n and size e. Then*
$e \geq n - 1.$

Proof. Left as an Exercise (see Exercise (2) below). $\qquad\square$

The proof of Theorem 11.6 gives an algorithm to construct a spanning
tree in a connected graph G: identify a cycle in G (if one exists) and remove
an edge from that cycle to obtain a spanning subgraph of G. Repeat the
process until no cycles remain. The last subgraph obtained is a spanning
tree.

Example 11.29. Construct a spanning tree of the wheel W_6:

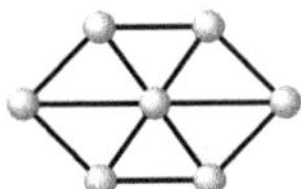

Solution. The following diagram shows a sequence $H \to K$ where H, K are subgraphs of W_6. In subgraph H, a cycle (in bold edges) and an edge e of that cycle (indicated by a dashed line) are identified. Subgraph K is $H - e$.

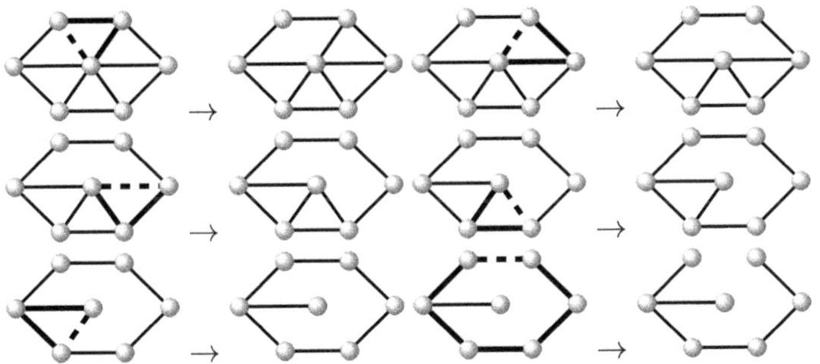

The last graph in the above diagram is a spanning tree. $\Diamond$

Searching for a spanning tree in a graph could be an overwhelming task for large graphs. The technique of successively removing edges from cycles in the graph is not very practical in general. In the next section, we describe two algorithms that can generate a spanning tree in a connected graph in a more efficient way.

11.6.1 *Graph search*

In many applications of graph theory, we need to "explore" the graph at hand. Exploring a graph could take many forms depending on what we are trying to achieve. In some cases, we could be interested in visiting every vertex and exploring edges and paths. In other cases, we could be interested in starting at a certain vertex and looking for the shortest possible path (in terms of the number of edges) to reach another vertex in the graph. A good practical example of graph search is the suggestion of friends by some social media platforms. In this case, the problem is represented by a graph with individual accounts as vertices and edges connecting vertices which are "friends". Suggesting a friend for you consists of starting at the vertex s corresponding to your account and searching through the graph for vertices reachable from s through simple paths of length 2 (two edges).

In this section, we look at two very well-known search algorithms. The first one (Depth-First Search) goes as deep as possible in a path starting at a given vertex while the second (Breadth-First Search) explores vertices reachable from a given vertex "layer by layer".

11.6.1.1 *Depth-First search (DFS)*

Let $G = (V, E)$ be a connected graph. As the name suggests, the Depth-First Search algorithm (DFS for short) is a search algorithm in G that goes "as deep as possible" at every stage. The algorithm produces a spanning tree in G when it terminates. The DFS algorithm starts by choosing an arbitrary vertex v of G. The *current* tree consists of the single vertex v. The algorithm then adds new edges (x, y) (from x to y) to the tree in such a way that: (i) x is the last vertex added to the current tree and (ii) y is not a vertex in the current tree. These two conditions insure that the new graph we obtain at every stage is connected and has no cycle in it; hence a tree. After adding a new edge (x, y), it is possible that no new edge can be added to the current tree. In this case, the algorithm backtracks to vertex x and if possible, chooses a new edge (x, z) satisfying (i) and (ii). If that is not possible, the algorithm backtracks to the parent of vertex x and tries to add a new edge and so on. The backtracking continues until the original vertex v is reached and no new edges can be added. When that happens, the resulting graph has visited every vertex of the graph (otherwise, adding a new edge would be possible) and since it is a tree, it is a spanning tree of G.

Example 11.30. For the graph G given below, apply the DFS algorithm to find a spanning tree.

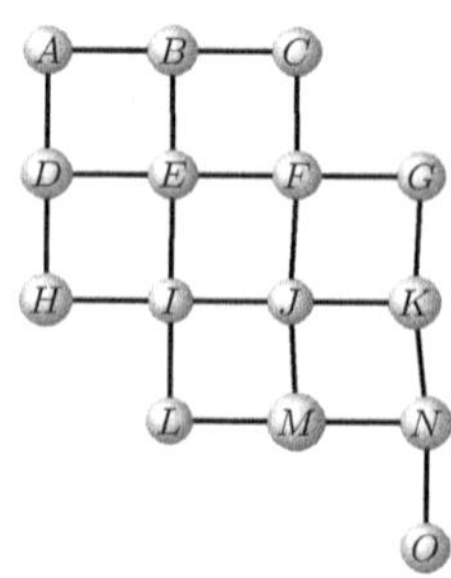

Solution. Start the DFS at vertex A. The thick edges in the diagram are the new edges added to the tree using the DFS algorithm. The label on an edge indicates the order in which the edge was added to the tree.

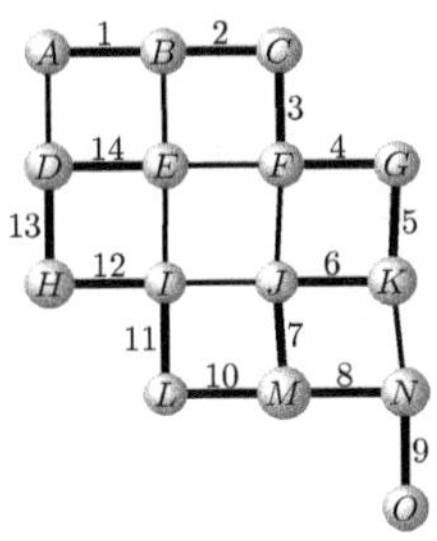

Removing all unlabeled edges from te above diagram gives us a spanning tree of the graph:

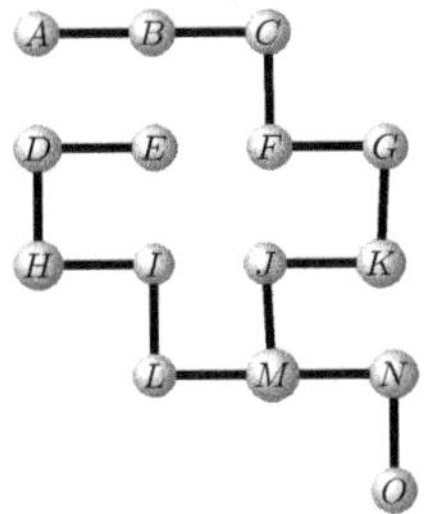

$\Diamond$

11.6.1.2 *Breadth-First search*

In a nutshell, the BFS starts at a given vertex, explores all its neighbors first before moving on to vertices on other levels. The algorithm uses a queue data structure to keep track of the order of vertices to visit and to make sure that every vertex is visited exactly once. The vertices of the graph are visited in the same order they appear in the queue. The steps of the BFS algorithm are as follows:

(1) Initiate all vertices to the statue *unvisited.*
(2) Choose an initial vertex v of the graph, place it in the front of the queue and change its status to *waiting.* Initialize the tree T to be the one consisting of vertex v only.
(3) Keep performing the following steps until the queue is empty:

  (3.1) Remove the first vertex from the front of the queue and mark its statue as *visited.*

(3.2) Add all neighbors of the removed vertex having status *unvisited* to the end of queue and change their status to *waiting*.

(3.3) Add all edges (x, y) to the tree T where x is the visited vertex and y is a waiting vertex.

Like the DFS, the BFS produces a spanning tree in the graph when the queue is empty. We will not prove the correctness of this algorithm.

Example 11.31. Use the BFS to produce a spanning tree in the following graph.

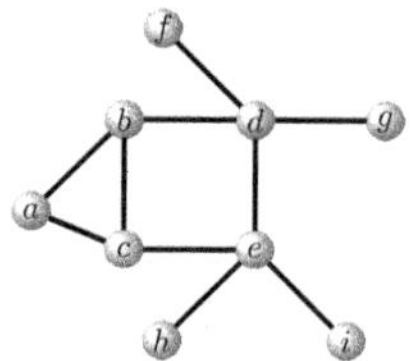

Solution. We start by marking the statue of every vertex in G as *unvisited*. Choose vertex a as a starting vertex for the BFS. Place a at the front of the queue and change its status to *waiting*. The tree generated at this stage consists of the node a. The following shows the queue on the left and the current tree on the right.

Next, we remove a from the queue and we change its statute to *visited*. Locate the neighbors of a with statue *unvisited* (namely b and c), place them at the back of the queue and change their status to *waiting*. Add edges (a, b) and (a, c) to the tree.

The next vertex waiting at the front of the queue is b. Remove it from the queue and change its statue to *visited*. The only neighbor of b with statue *unvisited* is d (neighbor a has statue *visited* and c has statue *waiting*). Insert d at the back of the queue and change its statue to *waiting*. The new tree is obtained by adding edge (b, d) to the previous tree.

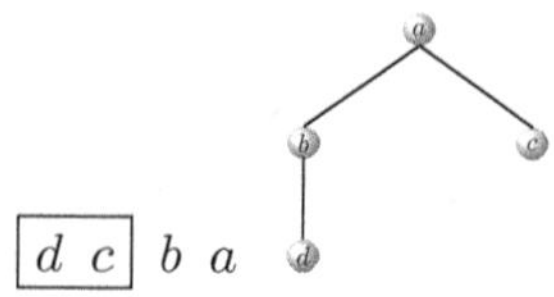

The next vertex waiting at the front of the queue is c. Remove it from the queue and change its statute to *visited*. The only neighbor of c with statue *unvisited* is e (neighbor a has statue *visited* and d has statue *waiting*). Insert e at the back of the queue and change its statue to *waiting*. The new tree is obtained by adding edge (c, e) to the previous tree:

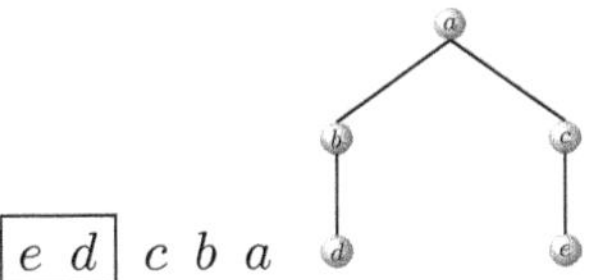

Next, remove vertex d from the queue, change its statue to *visited*. Place the neighbors of d with statue *unvisited*, namely f and g, at the back of the queue and change their statue to *waiting*. Add edges (d, f) and (d, g) to the previous tree:

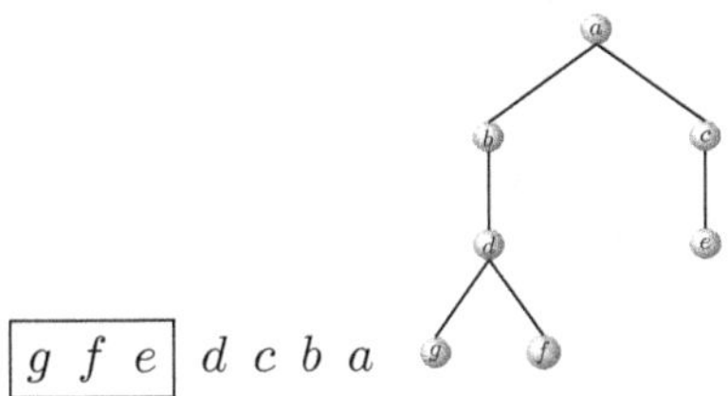

Remove e from the front of the queue, change its statue to *visited* and place its neighbors with statue *unvisited*, namely h and i, at the back of the queue and change their statue to *waiting*. Add edges (e, h), (e, i) to the previous tree:

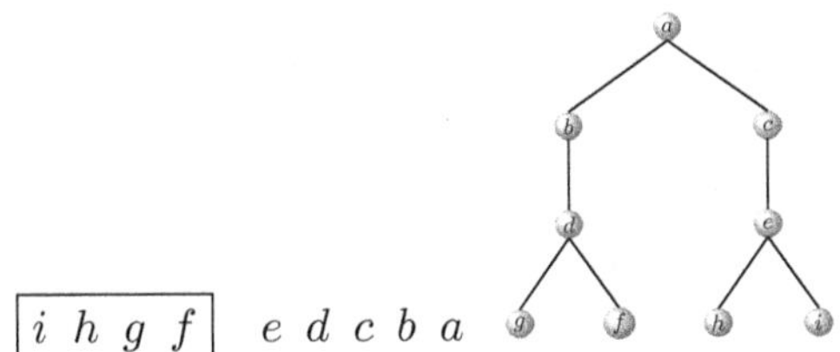

Next, remove f from the front of the queue, change its statue to *visited*. No neighbors of f exist with statue *unvisited*:

$$\boxed{i\ h\ g}\ f\ e\ d\ c\ b\ a$$

No new edges added to the last tree in this case. The same scenario repeats for vertices g, h and i:

$$\boxed{}\ i\ h\ g\ f\ e\ d\ c\ b\ a.$$

The queue is now empty and the spanning tree of G generated by the BFS starting at vertex a is the last tree in the above procedure:

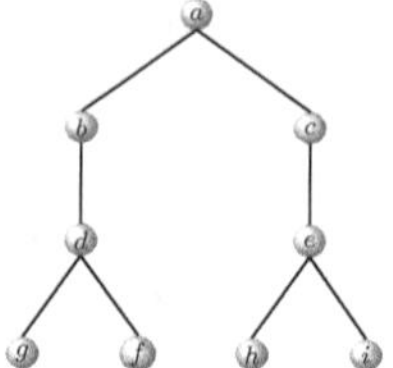

$\Diamond$

11.6.2   *Search algorithms and graph connectivity*

Given the adjacency matrix of a graph G with respect to a certain ordering $v_1, v_2, \ldots, v_n$ of its vertices, how can we determine if G is connected? By Theorem 11.6, this question is equivalent to finding a spanning tree in G. We can then use either the DFS or the BFS to try to find a spanning tree. The following theorem tells us that if either one of the DFS or the BFS terminates by visiting all vertices, then the graph is connected.

Theorem 11.7. *Let G be a graph and let v be any vertex in G.*

(a) The DFS and the BFS algorithms visit all vertices that are reachable from v.

(b) Starting at v, if the DFS or the BFS terminates by visiting all vertices of G, then G is connected. Otherwise, it is disconnected.

Proof. For part (a), we proceed by induction on the level $k \geq 0$ from the source v. If $k = 0$, the result is clearly true since at level 0, there is only vertex v. Let $k \geq 0$ and assume that the DFS and the BFS algorithms visit all vertices at level k from v. Since every vertex at level $k + 1$ from v is connected to one vertex at level k from v, the algorithms must then visit all vertices at level $k + 1$. The result follows by the principle of induction. Part (b) follows easily from part (a). $\square$

Example 11.32. Let $G = (V, E)$ be a graph of order 6. The adjacency matrix of G with respect to the ordering $v_1, v_2, v_3, v_4, v_5, v_6$ of its vertices is given as follows:

$$
A = \begin{array}{c} \\ v_1 \\ v_2 \\ v_3 \\ v_4 \\ v_5 \\ v_6 \end{array}
\begin{array}{c} v_1\ v_2\ v_3\ v_4\ v_5\ v_6 \\
\left[\begin{array}{cccccc}
0 & 1 & 1 & 1 & 0 & 1 \\
1 & 0 & 0 & 1 & 0 & 1 \\
1 & 0 & 0 & 1 & 1 & 0 \\
1 & 1 & 1 & 0 & 0 & 1 \\
0 & 0 & 1 & 0 & 0 & 1 \\
1 & 1 & 0 & 1 & 1 & 0
\end{array}\right]
\end{array}
$$

Determine if G is connected.

Solution. We use a breadth-first search starting at vertex v_1. It is the same procedure as in the previous example, except that we get the neighbors of the current vertex from the adjacency matrix rather than the drawing of the graph. The BSF finishes by producing the following spanning tree:

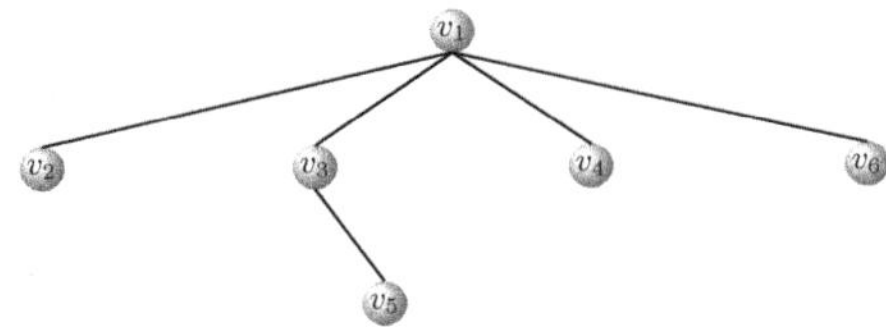

The fact that a spanning tree is generated shows that the graph is connected.

$\Diamond$

11.6.3 *Minimal spanning tree*

Definition 11.12. A *weighted* graph is a graph $G = (V, E)$ with a map $w : E \to \mathbb{R}$ called the *weight function*. If $e \in E$, we call $w(e)$ the *weight* of edge e. If H is a subgraph of G, the *weight* of H, denoted by $w(H)$, is defined as the sum of the weights of all edges of H.

Definition 11.13. Let $G = (V, E)$ be a connected weighted graph with a weight function $w : E \to \mathbb{R}$. A *minimal spanning tree* is a spanning tree of G such that $w(T)$ is minimal.

A graph can have more than one minimal tree, but finding a minimal tree is not an easy task in general. In this section, we describe two algorithms to find a minimal tree. We will not, however, prove the correctness of these algorithms.

11.6.3.1 *Kruskal's algorithm*

Let $G = (V, E)$ be a weighted connected graph of order n, size m and a weight function $w : E \to \mathbb{R}$. Kruskal's Algorithm starts by arranging the edges of G in a non-decreasing fashion according to their weights. Edges are examined for inclusion in the minimal spanning tree one at a time. The criteria of including a new edge in the tree is simple: edge e is added to the tree if it does not form a cycle with the other edges already in the tree. More precisely, the steps of Kruskal's algorithm are as follows.

(1) Choose an edge e_1 of G such that $w(e_1)$ is the smallest of the set $\{w(e); e \in E\}$, mark it as *treated* and include it in the tree.
(2) Assume edges $e_1, e_2 \ldots, e_{k-1}$ have been marked *treated*, choose a non-treated edge e_k such that:

    (i) e_k does not form a cycle with any subset of $\{e_1, e_2 \ldots, e_{k-1}\}$ and
    (ii) $w(e_k)$ is the smallest among all non-treated edges satisfying (i).

    Mark e_k as *treated* and add it to the tree.
(3) Repeat Step (2) until the tree has exactly $n - 1$ treated edges. This would be a minimal spanning tree of G.

In the above algorithm, we used the word "tree" loosely. A more appropriate term is "forest" as the algorithm could add edges not incident to any of the previously added edges.

Example 11.33. Use Kruskal's algorithm to find a minimal spanning tree of the weighted graph G below. Give the weight of the minimal tree.

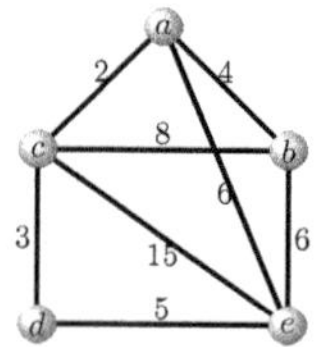

Solution. We start by sorting the edges of in a non-decreasing fashion as shown in the following table.

Edge	$\{a,c\}$	$\{c,d\}$	$\{a,b\}$	$\{d,e\}$	$\{b,e\}$	$\{a,e\}$	$\{c,b\}$	$\{c,e\}$
Weight	2	3	4	5	6	6	8	15

The minimal spanning tree starts with edge $\{a, c\}$. The next edge in the table is $\{c, d\}$ and since edges $\{a, c\}$, $\{c, d\}$ do not form a cycle, we include $\{c, d\}$ in the tree. We also include the next edges $\{a, b\}$ and $\{d, e\}$ since the four edges $\{a, c\}$, $\{c, d\}$, $\{a, b\}$ and $\{d, e\}$ do not form cycles. Since the graph is of order 5 and we already have 4 edges, we stop the process and edges $\{a, c\}$, $\{c, d\}$, $\{a, b\}$ and $\{d, e\}$ form a minimal spanning tree:

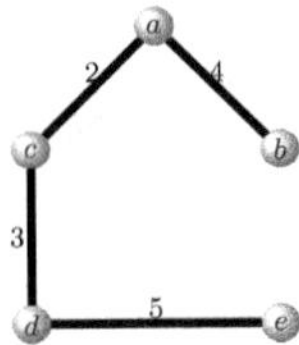

The weight of the minimal spanning tree is $2 + 3 + 4 + 5 = 14$.

$\Diamond$

Example 11.34. Use Kruskal's algorithm to find a minimal spanning tree of the graph G given below. Give the weight of the minimal tree.

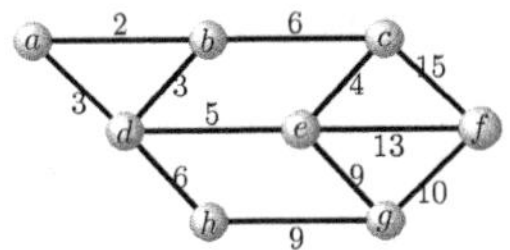

Solution. We sort the edges of G in a non-decreasing fashion according to their weights: $\{a, b\}$, $\{a, d\}$, $\{b, d\}$, $\{c, e\}$, $\{d, e\}$, $\{b, c\}$, $\{d, h\}$, $\{e, g\}$, $\{h, g\}$, $\{g, f\}$, $\{e, f\}$, $\{c, f\}$. Start by including edge $\{a, b\}$ in the tree. Then edge $\{a, d\}$ is treated and included in the tree. The next edge to be considered is $\{b, d\}$, but since it forms a cycle with the edges previously treated we skip it and move to treat $\{c, e\}$. We continue treating edges according to their appearance in the above weight order. The tables below show the edges of G with their weights at the second row. A $\checkmark$ symbol in the third row indicates that the corresponding edge is included in the minimal spanning tree and an $\times$ symbol indicates that the edge is skipped using Kruskal's algorithm.

$\{a, b\}$	$\{a, d\}$	$\{b, d\}$	$\{c, e\}$	$\{d, e\}$	$\{b, c\}$
2	3	3	4	5	6
$\checkmark$	$\checkmark$	$\times$	$\checkmark$	$\checkmark$	$\times$

$\{d, h\}$	$\{e, g\}$	$\{h, g\}$	$\{g, f\}$	$\{e, f\}$	$\{c, f\}$
6	9	9	10	13	15
$\checkmark$	$\checkmark$	$\times$	$\checkmark$	$\times$	$\times$

The minimal spanning tree in G is shown in the following diagram.

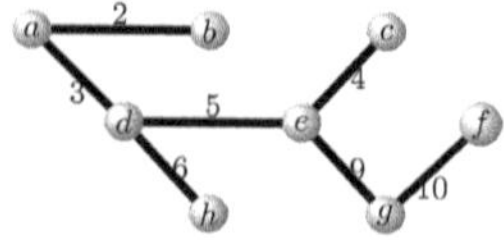

The weight of this minimal spanning tree is $2 + 3 + 4 + 5 + 6 + 9 + 10 = 39$.
$\Diamond$

11.6.3.2   *Prim's algorithm*

Let $G = (V, E)$ be a connected weighted graph. For Kruskal's algorithm, edges can be added to the tree without necessarily being incident to any of the previously added edges as long as no cycles are formed. Starting at an arbitrary vertex v of G, Prim's algorithm builds a tree out of v by adding one edge of minimal weight at a time that links the tree to a new vertex in the graph. At each stage, Prim's algorithm adds one edge to the current tree $T = (A, F)$ using the following steps.

(1) Start at an arbitrary vertex v of G, set $A = \{v\}$ and $F = \emptyset$.
(2) Choose an edge $e_1 = \{v, u\}$ of minimal weight incident to v. Set $A = \{v, u\}$, $F = \{e_1\}$.
(3) At every iteration of the algorithm add a new edge $\{x, y\}$ of minimal possible weight such that $x \in A$, $y \notin A$, and $\{x, y\}$ does not create a cycle with the edges already added to the tree. Add node y to A and edge $\{x, y\}$ to F.
(4) Continue this process until $A = V$.

Example 11.35. Use Prim's algorithm to construct a minimal spanning tree T of the graph given in Example 11.34 above.

Solution. We start at vertex a, add one edge at a time following the steps in Prim's algorithm. The table below shows the edges added to T, the set A of nodes of the current tree and the remaining vertices $V \backslash A$ at every iteration.

Step	New edge added	A	$V \backslash A$
Initial		$\{a\}$	$\{b, c, d, e, f, g, h\}$
1	$\{a, b\}$	$\{a, b\}$	$\{c, d, e, f, g, h\}$
2	$\{b, d\}$	$\{a, b, d\}$	$\{c, e, f, g, h\}$
3	$\{d, e\}$	$\{a, b, d, e\}$	$\{c, f, g, h\}$
4	$\{e, c\}$	$\{a, b, d, c, e\}$	$\{f, g, h\}$
5	$\{d, h\}$	$\{a, b, c, d, e, h\}$	$\{f, g\}$
6	$\{h, g\}$	$\{a, b, c, d, e, h, g\}$	$\{f\}$
7	$\{g, f\}$	$\{a, b, c, d, e, f, g, h\}$	$\emptyset$

The minimal spanning tree obtained by Prim's algorithm is shown in the
following diagram.

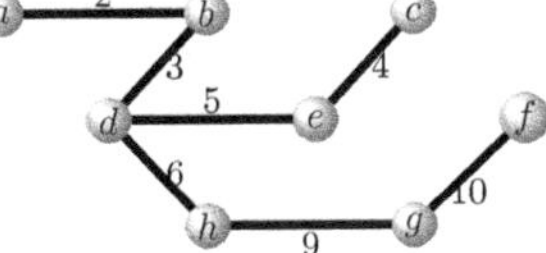

The weight of he minimal spanning tree is $2 + 3 + 5 + 4 + 6 + 9 + 10 = 39$. $\Diamond$

11.6.4 *Exercises*

(1) In each case, construct a spanning tree for the given graph starting at
vertex a using: (i) Depth-First Search; (ii) Breadth-First Search.

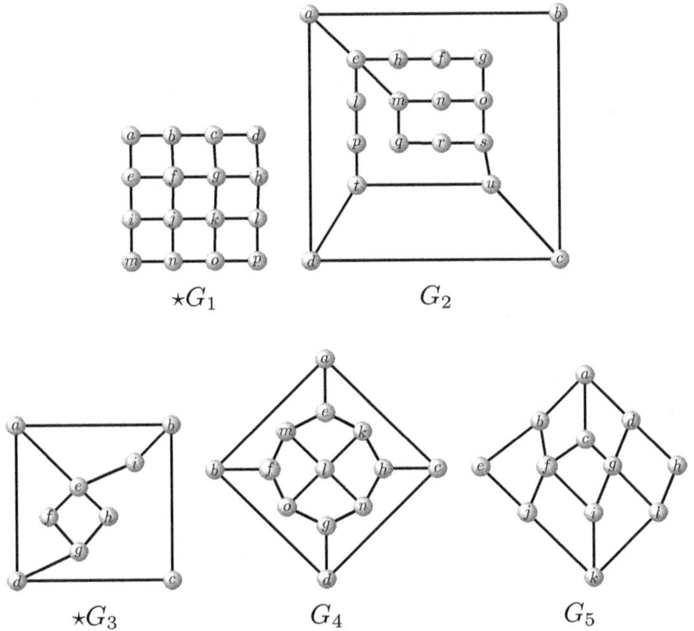

$\star G_1$ G_2

$\star G_3$ G_4 G_5

(2) Prove Corollary 11.2.

$\star$(3) Let T be a spanning tree of a graph G. If e is an edge of G which is not in T, prove that the subgraph $T + e$ of G obtained by adding e to T is not a tree.

(4) Find a minimal spanning tree of the weighted graph G given below using: (i) Kruskal's algorithm (ii) Prim's algorithm. Give the weight of the minimal tree.

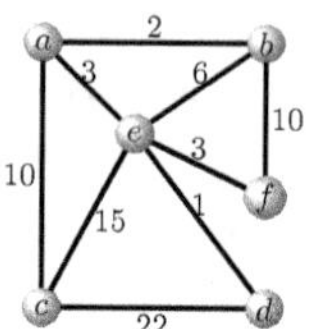

(5) Let $G = (V, E)$ be a weighted connected graph with a constant weight function (so $w(e_i) = w(e_j)$ for any $e_i, e_j \in E$). Prove that every spanning tree of G is a minimal spanning tree.

$\star$(6) A bus company in Ontario, Canada is launching a new service to connect the following five cities in the province: Ottawa, Bellville, Toronto, Windsor and Kingston. The following table gives the distances between these cities (in Km).

City	Ottawa	Kingston	Bellville	Toronto	Windsor
Ottawa	-	220	235	500	700
Kingston	220	-	85	280	480
Bellville	235	85	-	188	540
Toronto	500	280	188	-	330
Windsor	700	480	540	330	-

(a) Draw a weighted graph representing these five cities and their distances apart.

(b) The company charges 0.42 dollar/km. Trace the cheapest bus route that visits every city exactly once. What is the cheapest cost to do that tour?

(7) Somewhere in the Atlantic Ocean, a country consists of a federation of 9 islands that we label using the letters a, b, c, d, e, f, g, h and i. People can move from one island to another using a boat service offered by a local company. In the following weighted graph, each of the nine islands is represented by a vertex and each boat line between islands is represented by an edge. The weight of each edge represents the cost of using the corresponding boat service (one way).

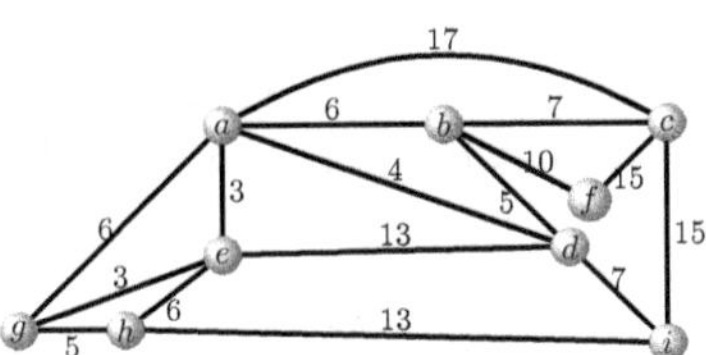

In order to deal with some financial difficulties, the boat company has to cut some lines. But the cut has to be done in a way to minimize the cost for citizens and to ensure that any island in the country can still be reached using boat lines from any other island. What lines should the company keep in order to make sure the requirements are met. What would be the minimal cost of visiting all nine islands?

(8) The following is the *weight matrix* of a weighted graph $G = (V, E)$ with respect to the ordering $v_1, v_2, v_3, v_4, v_5, v_6$ of its vertices. The entry m_{ij} in the matrix represents the weight of the edges $\{v_i, v_j\}$ of G. The symbol $-$ in the ith row and the jth column indicates that there exists no edge between vertices v_i and v_j:

$$W = \begin{bmatrix} - & 15 & 9 & 4 & 3 & - \\ 15 & - & 14 & - & - & 8 \\ 9 & 14 & - & - & - & 9 \\ 4 & - & - & - & 3 & 5 \\ 3 & - & - & 3 & - & 6 \\ - & 8 & 9 & 5 & 6 & - \end{bmatrix}$$

(a) Use Kruskal's algorithm to prove that G is connected. Give a minimal spanning tree in G. What is the weight of a minimal tree?

(b) Repeat the previous part using Prim's algorithm.

$\star$(9) The following diagram shows seven points in the Cartesian plane with integer coordinates.

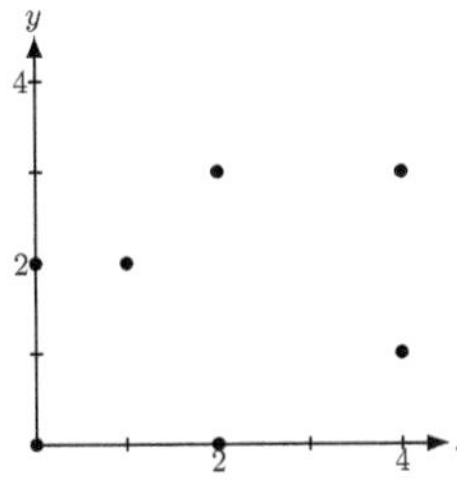

Join each pair of points with a straight line and find the path in the Cartesian plane that connect all seven points and which has a minimal total length. What the minimal total length?

(10) Let $G = (V, E)$ be a weighted graph with weight function w.

    (a) Modify Kruskal's algorithm to get a maximal spanning tree in G.

    (b) Use the modified Kruskal algorithm to find a maximal spanning tree for the graph given in Exercise (4) above and give the weight of that maximal spanning tree.

$\star$(11) Let G be a connected weighted graph with a cycle C and a weight function w. Assume that C has an edge e such that $w(e) > w(f)$ for any other edge f in C. Let $G' = G - e$. Prove that G' is connected and that any minimal spanning tree of G' is also a minimal spanning tree of G.

(12) Let $G = (V, E)$ be a connected weighted graph with a weight function w satisfying $w(e_i) \neq w(e_j)$ for all edges e_i, e_j with $i \neq j$. Prove that G has a unique minimal spanning tree (*Hint.* Use a proof by contradiction).

Appendix: Solutions to Suggested Exercises

1.2

(1)(b) This is a proposition.

   (c) Not a proposition. Just an integer.

   (d) This is a proposition.

   (k) This is not a proposition.

(2)(a) (i) The sum of the squares of two integers is not always a perfect square (ii) The sum of the squares of two integers is always a perfect square.

   (e) (i) There exists a real number that satisfies the equation $x^2 - 1 = 0$ (ii) There exists a real number that satisfies the equation $x^2 + 1 = 0$.

(3)(d) $(p \leftrightarrow (\neg q \wedge \neg r))$ where p: Joe fails the course, q: Joe does not do the effort, r: The professor warns Joe. The main logic connective is the biconditional $\leftrightarrow$.

(4)(b) Exclusive or. You cannot do both at the same time.

(5)(f) True. It is an implication of the form $\mathbf{F} \to \mathbf{T}$.

   (g) False. It is a biconditional of the form $\mathbf{F} \leftrightarrow \mathbf{T}$.

(6) The only possibility that $(B \to \neg A)$ is false is that the hypothesis B is true and the conclusion $\neg A$ is false, which means A and B are both true. From this we get the truth value of all the propositions below: (d) True; (i) False; (j) True.

(7) Consider the implication: If $1 + 1 = 0$, then $1 + 2 = 3$. This is a true implication since it is of the form $\mathbf{F} \to \mathbf{T}$. Its converse is If $1 + 2 = 3$, then $1 + 1 = 0$ which is false since it is of the form $\mathbf{T} \to \mathbf{F}$.

(8)(b) The contrapositive is: If a function is not continuous on the interval $[a, b]$, then it is not differentiable on that interval. The converse is: If a function is continuous on the interval $[a, b]$, then it is differentiable on that interval.

(9)(d) The truth tables of the formulas are as follows.

p	q	$(p \to q)$	$(q \to p)$	$((p \to q) \wedge (q \to p))$	$(p \leftrightarrow q)$
T	T	F	T	T	T
T	F	F	T	F	F
F	T	T	F	T	F
F	F	T	T	T	T

(12)(c) $s \wedge t = 110000$, $s \vee t = 110111$, $s \oplus t = 000111$, $st = 110100110000$.

(13)(c) True: for any bits b_1, b_2 and b_3, $(b_1 \wedge b_2) \wedge b_3 = b_1 \wedge (b_2 \wedge b_3)$.

   (g) False. A counterexample: $(1 \oplus 1)1 = 11$ but $(1 \oplus 1)(1 \oplus 1) = 00$.

1.3

(1)(b) This is a wff. The proof is illustrated in the table below where each of the formulas shown in the column on the left is a wff and the justification is in the second column.

		Justification
1.	A, B, C	Rule (1)
2.	$\neg A, \neg C$	Line 1 and Rule (2)
3.	$(\neg A \wedge B)$	Lines 1, 2 and Rule (3.1)
4.	$\neg(\neg A \wedge B)$	Line 3 and Rule (2)
5.	$(\neg C \vee B)$	Lines 1, 2 and Rule (3.2)
6.	$(A \leftrightarrow (\neg C \vee B))$	Lines 1, 5 and Rule (3.4)
7.	$(\neg(\neg A \wedge B) \to (A \leftrightarrow (\neg C \vee B)))$	Lines 4, 6 and Rule (3.3)

The main connective of the formula is the implication $\to$.

   (e) Not a wff: missing parentheses in $(A \to B \wedge C)$.

(3)(c) The truth table of the formula

$$\phi : ((P \to (Q \to R)) \to ((P \to Q) \to (P \to R)))$$

is the following.

P	Q	R	$(Q \to R)$	$(P \to Q)$	$(P \to (Q \to R))$	$((P \to Q) \to R)$	ϕ
T	T	T	T	T	T	T	T
T	T	F	F	T	F	F	T
T	F	T	T	F	T	T	T
T	F	F	T	F	T	T	T
F	T	T	T	T	T	T	T
F	T	F	F	T	T	F	F
F	F	T	T	T	T	T	T
F	F	F	T	T	T	F	F

Since the truth table contains both **T**'s and **F**'s, it is a contingency. The valuations that satisfy it are: $(\mathbf{T}, \mathbf{T}, \mathbf{T})$, $(\mathbf{T}, \mathbf{T}, \mathbf{F})$, $(\mathbf{T}, \mathbf{F}, \mathbf{T})$, $(\mathbf{T}, \mathbf{F}, \mathbf{F})$, $(\mathbf{F}, \mathbf{T}, \mathbf{T})$ and $(\mathbf{F}, \mathbf{F}, \mathbf{T})$.

(g) The truth table of the formula is the following.

A	B	$\neg B$	$(A \vee B)$	$(\neg B \to A)$	$\neg(\neg B \to A)$	$((A \vee B) \leftrightarrow \neg(\neg B \to A))$
T	**T**	**F**	**T**	**T**	**F**	**F**
T	**F**	**T**	**T**	**T**	**F**	**F**
F	**T**	**F**	**T**	**T**	**F**	**F**
F	**F**	**T**	**F**	**F**	**T**	**F**

Since the table contains **F**'s only, the formula is a contradiction.

(4)(c) Let $\phi_1 : \neg(Q \wedge R)$, $\phi_2 : ((\neg P \to Q) \wedge (\neg R \to P))$, $\phi_3 : (P \vee \neg R)$ and $\phi_4 : (P \to (Q \to \neg R))$. The truth table of the set $\{\phi_1, \phi_2, \phi_3, \phi_4\}$ is as follows.

P	Q	R	$\neg P$	$\neg R$	$(Q \wedge R)$	$(\neg P \to Q)$	$(\neg R \to P)$	$(Q \to \neg R)$	ϕ_1	ϕ_2	ϕ_3	ϕ_4
T	**T**	**T**	**F**	**F**	**T**	**T**	**T**	**F**	**F**	**T**	**T**	**F**
T	**T**	**F**	**F**	**T**	**F**	**T**	**T**	**T**	**F**	**T**	**T**	**T**
T	**F**	**T**	**F**	**F**	**F**	**T**	**T**	**T**	**F**	**T**	**T**	**T**
T	**F**	**F**	**F**	**T**	**F**	**T**	**T**	**T**	**F**	**T**	**T**	**T**
F	**T**	**T**	**F**	**F**	**T**	**T**	**T**	**F**	**F**	**T**	**F**	**T**
F	**T**	**F**	**T**	**T**	**F**	**T**	**F**	**T**	**T**	**F**	**T**	**T**
F	**F**	**T**	**T**	**F**	**F**	**F**	**T**	**T**	**T**	**F**	**F**	**T**
F	**F**	**F**	**T**	**T**	**FF**	**F**	**T**	**T**	**F**	**T**	**T**	**T**

The set is coherent. The following valuations satisfy all formulas in the set: $(\mathbf{T}, \mathbf{T}, \mathbf{F})$, $(\mathbf{T}, \mathbf{F}, \mathbf{T})$ and $(\mathbf{T}, \mathbf{F}, \mathbf{F})$.

(5)(b) α does not satisfy the formula since the latter has the form of an implication with a true hypothesis and a false conclusion.

(d) α satisfies the formula since the latter has the form of an implication with a false hypothesis.

1.4

(2)(b) $((R \vee (\neg P \wedge Q))$

(3)(c) A snow storm is coming is sufficient for the temperature not to drop below the freezing degree if and only if the wind is not blowing.

(4) Here is one possible answer.

(c) $(H \to (C \vee R)) \wedge \neg G$ where H: *Joe wears his hat*, C: *Joe enters the construction site*, R: *Joe changes the roof*, G: *Joe wears his safety glasses on site.*

(6) Consider the following atoms:

- O: There is a power outage
- L: The emergency lights turn on
- E: The emergency exist is clear
- S: All students leave the auditorium safely

The following is a translation of the above argument into symbols of propositional logic.

$$(O \to L)$$
$$(E \to (S \vee L))$$
$$((S \to E) \wedge \neg(E \to S))$$
$$\underline{(S \to \neg L)}$$
$$\therefore (E \vee \neg O)$$

(7) Consider the following atoms:

- U: The user enter a valid password
- L: The user is granted access
- A: The user account is activated
- E: The user pays a fee
- S: The user account is locked

The following is a translation of the above argument into symbols of propositional logic.

$$(\neg U \to (\neg L \vee E))$$
$$((A \wedge L) \to (E \vee S))$$
$$\underline{(E \leftrightarrow (\neg U \wedge \neg S))}$$
$$\therefore (U \to E)$$

1.5

(1) A possible proposition that any inhabitant can say is *"I am a knight"*.

(4) If A is a knight, then *"B is a knight but I am not"* is true and so both *"B is a knight"* and *"I am not"* are true. This is a contradiction since *"I am not"* is false in this case. If A is a knave, then *"B is a knight but I am not"* is false and so one of *"B is a knight"* and *"I am not"* must be false. Since *"I am not"* is true in this case, *"B is a knight"* is false and B is a knave. We conclude that both A and B are knaves.

(7) If A is a knave, then *"B is a knave"* is false and B is a knight. So *"A and C are of the same nature"* is true and C is a knave. If A is a knight, then B is a knave and the sentence *"A and C are of the same*

nature" is false. So A and C are of opposite natures and C is a knave. We conclude that C is a knave.

(10) If A is a knight, then *"B is a knight and there is no hotel on this island"* is true and so B is a knight and there is no hotel on this island. In particular B speaks true and *"There is no hotel on this island only if A is a knave"* is true. This is a contradiction, since the latter is an implication of the form $(\mathbf{T} \to \mathbf{F})$. If A is a knave, then *"B is a knight and there is no hotel on this island"* is false and so either B is a knave or there is a hotel on the island. If B is a knave, then *'There is no hotel on this island only if A is a knave"* is false and so A is a knight in particular. This is a contradiction. If there is a hotel on the island, then *"There is no hotel on this island only if A is a knave"* is true (an implication with false hypothesis) and so B is a knight, and no contradiction in this case. We conclude that A is a knave, B is a knight and there is a hotel on the island.

(12) If B is a knave, then *"A is a knave unless C is knight"* is false and so both *"A is a knave"* and *"C is knight"* are false. This is a contradiction since A speaks false in this case so he is a knave. If B is a knight, then one of *"A is a knave"* and *"C is knight"* must be true. Since A is a knight (speaking the truth in this case), C is a knight. There is no contradiction in this case. We conclude that all three inhabitants are knights.

1.6

(3) We use truth tables:

(a)

p	q	r	$(q \wedge r)$	$(p \vee q)$	$(p \vee r)$	$(p \vee (q \wedge r))$	$((p \vee q) \wedge (p \vee r))$
T	T	T	T	T	T	T	T
T	T	F	F	T	T	T	T
T	F	T	F	T	T	T	T
T	F	F	T	T	T	T	T
F	T	T	T	T	T	T	T
F	T	F	F	T	F	F	F
F	F	T	F	F	T	F	F
F	F	F	F	F	F	F	F

(b)

p	q	r	$(q \vee r)$	$(p \wedge q)$	$(p \wedge r)$	$(p \wedge (q \vee r))$	$((p \wedge q) \vee (p \wedge r))$
T	T	T	T	T	T	T	T
T	T	F	T	T	F	T	T
T	F	T	T	F	T	T	T
T	F	F	F	F	F	F	F
F	T	T	T	F	F	F	F
F	T	F	T	F	F	F	F
F	F	T	F	F	F	F	F
F	F	F	F	F	F	F	F

(4)

P	Q	R	$(P \vee Q)$	$(P \rightarrow R)$	$(Q \rightarrow R)$	$((P \vee Q) \rightarrow R)$	$((P \rightarrow R) \wedge (Q \rightarrow R))$
T	T	T	T	T	T	T	T
T	T	F	T	F	F	F	F
T	F	T	T	T	T	T	T
T	F	F	T	F	T	F	F
F	T	T	T	T	T	T	T
F	T	F	T	T	F	F	F
F	F	T	F	T	T	T	T
F	F	F	F	T	T	T	T

(5)(c)

A	B	C	$(B \wedge C)$	$(A \rightarrow B)$	$(A \rightarrow C)$	ϕ	ψ
T	T	T	T	T	T	T	T
T	T	F	F	T	F	F	F
T	F	T	F	F	T	F	F
T	F	F	F	F	F	F	F
F	T	T	T	T	T	T	T
F	T	F	F	T	T	T	T
F	F	T	F	T	T	T	T
F	F	F	F	T	T	T	T

The formulas are equivalent since they have the same truth tables.

(6)(c)

$$((A \leftrightarrow B) \wedge \neg A) \equiv ((A \wedge B) \vee (\neg A \wedge \neg B)) \wedge \neg A) \text{ (Material equivalence (2))}$$
$$\equiv ((A \wedge B \wedge \neg A) \vee (\neg A \wedge \neg B \wedge \neg A)) \text{ (Distributivity)}$$
$$\equiv (\mathbf{F} \vee (\neg A \wedge \neg B \wedge \neg A)) \text{ (Contradiction)}$$
$$\equiv (\neg A \wedge \neg B \wedge \neg A)) \text{ (Or-simplification (false))}$$
$$\equiv (\neg A \wedge \neg B) \text{ (And-simplification)}$$
$$\equiv \neg(A \vee B) \text{ (DeMorgan)}$$

(f) Let $\phi : ((S \to (\neg P \to S)) \to P)$, then:

$$\phi \equiv (\neg(S \to (\neg P \to S)) \vee P) \text{ (Implication elimination)}$$
$$\equiv ((S \wedge \neg(\neg P \to S)) \vee P) \text{ (Negation of the implication)}$$
$$\equiv ((S \wedge (\neg P \wedge \neg S)) \vee P) \text{ (Negation of the implication)}$$
$$\equiv ((S \wedge \neg P \neg S) \vee P) \text{ (Associativity of } \wedge)$$
$$\equiv (\mathbf{F} \vee P) \text{ (Contradiction)}$$
$$\equiv P \text{ (Or-simplification (false))}$$

(8) The "nor" connective of p and q stands for $(\neg p \wedge \neg q)$. Its truth table is:

p	q	$\neg p$	$\neg q$	$(p \downarrow q)$
T	T	F	F	F
T	F	F	T	F
F	T	T	F	F
F	F	T	T	T

$(p \downarrow q) \equiv (\neg p \wedge \neg q) \equiv \neg(p \vee q)$ (DeMorgan). So $(p \vee q) \equiv \neg(p \downarrow q)$.

(11) $(p \to (p \leftrightarrow q)) \equiv (\neg p \vee (p \leftrightarrow q)) \equiv (\neg p \vee (p \wedge q) \vee (\neg p \wedge \neg q)) \equiv (\neg(p \wedge \neg(p \wedge q)) \vee (\neg p \wedge \neg q)) \equiv \neg((p \wedge \neg(p \wedge q)) \wedge \neg(\neg p \wedge \neg q))$.

(13)(a) We identify the rows in the truth table where the formula is true. Each one of these rows corresponds to a minterm. We then connect all these minterms with $\vee$: $((\neg A \wedge B \wedge C) \vee (\neg A \wedge \neg B \wedge C))$ is a logic formula equivalent to ψ.

(b) Using DeMorgan law: $((\neg A \wedge B \wedge C) \vee (\neg A \wedge \neg B \wedge C)) \equiv (\neg(A \vee \neg B \vee \neg C) \vee \neg(A \vee B \neg C))$.

(c) Using DeMorgan law: $((\neg A \wedge B \wedge C) \vee (\neg A \wedge \neg B \wedge C)) \equiv \neg(\neg(\neg A \wedge B \wedge C) \wedge \neg((\neg A \wedge \neg B \wedge C))$.

(14)(b) $(((A \to B) \wedge (B \to C)) \to (A \to C)) \equiv (\neg((A \to B) \wedge (B \to C)) \vee (A \to C)) \equiv (\neg(A \to B) \vee \neg(B \to C) \vee (A \to C)) \equiv ((A \wedge \neg B) \vee (B \wedge \neg C) \vee (\neg A \vee C)) \equiv (((A \vee B) \wedge (A \vee \neg C) \wedge (\neg B \vee B) \wedge (\neg B \vee \neg C)) \vee (\neg A \vee C)) \equiv (((A \vee B) \wedge (A \vee \neg C) \wedge \mathbf{T} \wedge (\neg B \vee \neg C)) \vee (\neg A \vee C)) \equiv (((A \vee B) \wedge (A \vee \neg C) \wedge (\neg B \vee \neg C)) \vee (\neg A \vee C)) \equiv ((A \vee B \vee \neg A \vee C) \wedge (A \vee \neg C \vee \neg A \vee C) \wedge (\neg B \vee \neg C \vee \neg A \vee C)) \equiv (\mathbf{T} \wedge \mathbf{T} \wedge \mathbf{T}) \equiv \mathbf{T}$.

(15)(d) The function f is not continuous and $f(0) = 0$.

1.7

(1)(a) False. The completely branched truth tree of the formula $(A \vee B)$ has two open branches but the formula is not a contradiction.

(d) True. In this case, no valuation satisfies $\neg\phi$, so ϕ is true for any valuation.

(g) True. If the set $\{\phi\}$ is incoherent, then there exists no valuation that satisfies ϕ. So ϕ is a contradiction.

(2)(a)

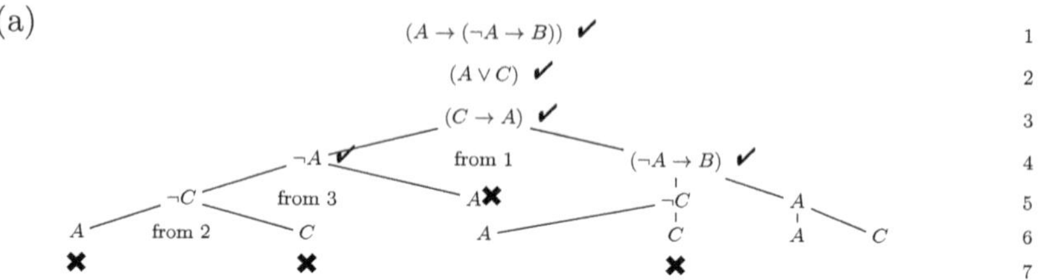

There are open branches. The set is satisfiable. Following the open branches, we get the following valuations of $\{A, B, C\}$ that satisfy $\mathcal{E}$: $(\mathbf{T}, \mathbf{T}, \mathbf{T})$, $(\mathbf{T}, \mathbf{T}, \mathbf{F})$, $(\mathbf{T}, \mathbf{F}, \mathbf{T})$ and $(\mathbf{T}, \mathbf{F}, \mathbf{F})$

(3)(c)

$$\neg((X \to (Y \to Z)) \to ((\neg X \to Y) \to Z)) \checkmark \qquad 1$$

$(X \to (Y \to Z))$ $\checkmark$	from 1	2
$\neg((\neg X \to Y) \to Z)$ $\checkmark$	from 1	3
$(\neg X \to Y)$ $\checkmark$	from 3	4
$\neg Z$ $\checkmark$	from 3	5
$\neg X$ / Y from 2 $(Y \to Z)$		6
$\neg\neg X$ / Y $\neg\neg X$ from 4 Y		7

There are open branches. The formula is not a tautology. Following the open branches, we get the following valuations of $\{X, Y, Z\}$ for which the formula is flase: $(\mathbf{T}, \mathbf{T}, \mathbf{F})$, $(\mathbf{T}, \mathbf{F}, \mathbf{F})$ and $(\mathbf{F}, \mathbf{T}, \mathbf{F})$.

(4)(b) Assume that $(((A \to B) \wedge (C \to D)) \to ((A \to D) \vee (C \to B)))$ is not a tautology. Then there exists a valuation such that $((A \to B) \wedge (C \to D))$ is true and $((A \to D) \vee (C \to B))$ is false. This implies that $(A \to B)$ and $(C \to D)$ are both false, and so A and C are true and B and D are false. In particular $(A \to B)$ is false and so is $((A \to B) \wedge (C \to D))$. This is a contradiction. We conclude that $(((A \to B) \wedge (C \to D)) \to ((A \to D) \vee (C \to B)))$ is a tautology

1.8

(1)(a) False. The existence of a valuation that satisfies every premise is not sufficient for the argument to be valid. That valuation must also satisfies the conclusion.

(c) False. The fact that the conclusion is a contradiction does not necessarily mean that the argument is invalid. We still need a valuation that satisfy all premises. For example, the argument with premises $((A \to (B \vee C)))$, $((\neg A \wedge B))$, A and conclusion $(B \wedge \neg B)$ is valid with a contradiction as a conclusion.

(d) False. We could still have a counter-example in this case. The argument could still be invalid.

(g) True. If the set $\{p_1, \ldots, p_n\}$ is incoherent, then there could not be a counter-example.

(2)(a) We use the method of truth tree. We start a tree with all the formulas in $\mathcal{E}$ at the root.

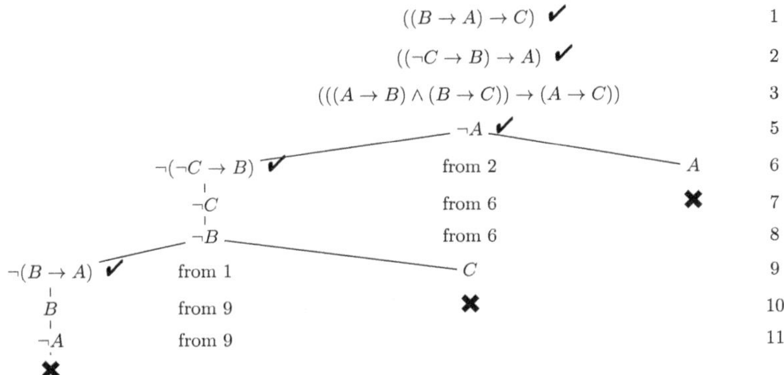

Since all the branches are closed (even before breaking all compound formulas), there exists no valuation that satisfies all formulas in the set. We conclude that $\mathcal{E}$ is incoherent.

(b) Notice that the set $\mathcal{E}$ of logic formulas given in the previous part consists of all the premises and the negation of the conclusion of the argument. Since $\mathcal{E}$ is incoherent, there exists no valuation that satisfies all the premises but not the conclusion. In other words, there is no counter-example of the argument and the latter is valid.

(3)(b) Label the premises as $p_1 : (B \to \neg A)$, $p_2 : \neg(\neg A \vee \neg C)$, $p_3 : (\neg C \to (B \to \neg C)$ and $p_4 : (A \to C)$. Label the conclusion as $\phi : (A \wedge \neg B)$. The truth tables of these formulas are as follows.

A	B	C	$\neg A$	$\neg B$	$\neg C$	$(\neg A \vee \neg C)$	$(B \to \neg C)$	p_1	p_2	p_3	p_4	ϕ
T	T	T	F	F	F	F	F	F	T	T	T	F
T	T	F	F	F	T	T	T	F	F	T	F	F
T	F	T	F	T	F	F	T	T	T	T	T	T
T	F	F	F	T	T	T	T	T	F	T	F	T
F	T	T	T	F	F	T	F	T	F	T	T	F
F	T	F	T	F	T	T	T	T	F	T	T	F
F	F	T	T	T	F	T	T	T	F	T	T	F
F	F	F	T	T	T	T	T	T	F	T	T	F

The argument is valid: there is only one valuation that satisfies all the premises at the same time: $(v(A) = \mathbf{T}, v(B) = \mathbf{F}, v(C) = \mathbf{T})$ and that valuation satisfies also the conclusion.

(4)(b)

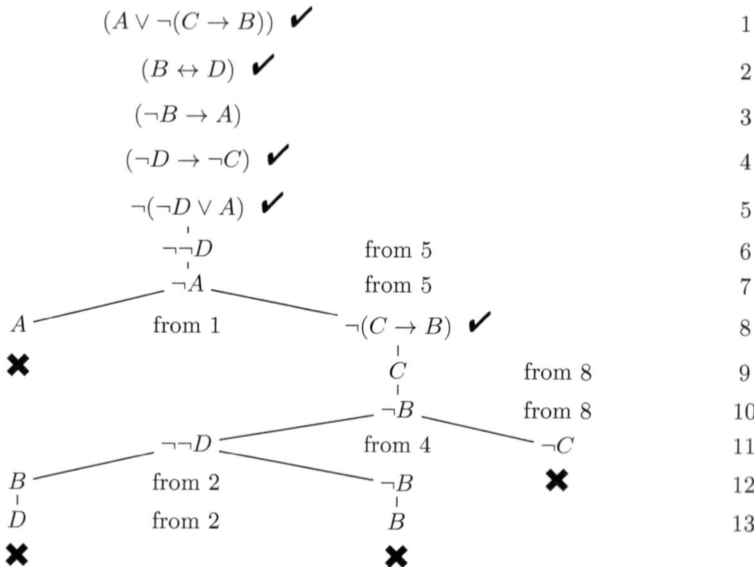

There are no open branches. The argument is valid.

(5) Assume that there exists a counter-example α. Then $v(D \to \neg A) = \mathbf{F}$ and so $v(D) = v(A) = \mathbf{T}$. This implies that $v(B \vee A) = \mathbf{T}$ and so $v(\neg(B \vee A)) = \mathbf{F}$. This is a contradiction since $\neg(B \vee A)$ is one of premises).

(6)(b) The argument translates as follows using symbols of propositional logic:

$$\frac{\begin{array}{c} (J \to (P \vee A)) \\ ((P \wedge A) \to R) \end{array}}{(J \to R)}$$

where J: *Joe calls for a meeting*, P: *Phil attends*, A: *Amanda attends*, and R: *There is a huge argument*. Using a truth tree, we branch the tree with all premises and the negation of the conclusion at the root.

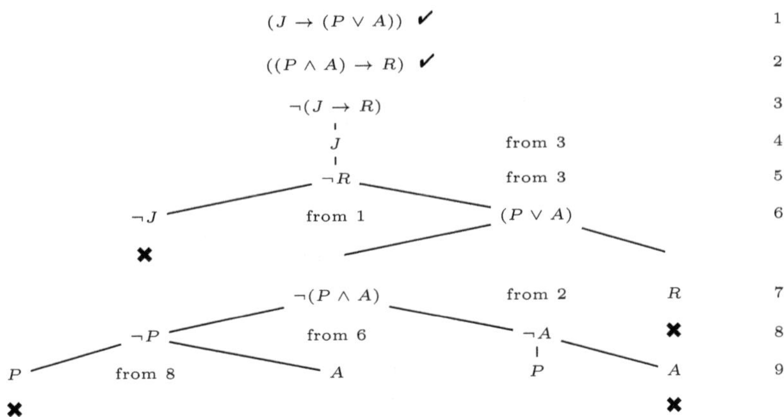

There are open branches. The argument is not valid.

(d) Consider the following logic atoms: *L*: *Joe wants to get his lifeguard certificate*, *C*: *Joe finishes his course*, *G*: *Joe reduces his gym time*, *E*: *Joe has a low self-esteem*. The argument translates as follows using symbols of propositional logic:

$$(L \to C)$$
$$(G \to C)$$
$$(G \to E)$$
$$\frac{(E \to \neg C)}{\neg L}$$

Using a truth tree: we branch the tree with all premises and the negation of the conclusion at the root.

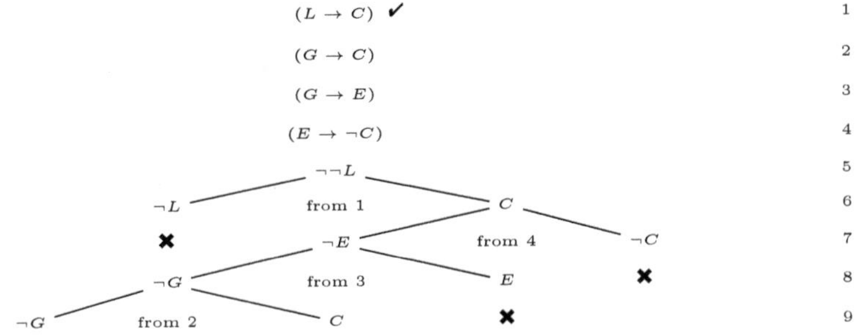

There are open branches. The argument is not valid.

(7)(a) The argument translates as follows into symbols of propositional logic:

$$(B \to S)$$
$$(\neg S \vee \neg L)$$
$$(\neg L \to (\neg W \vee P))$$
$$(P \to W)$$
$$(P \to B)$$
$$\neg B$$
$$\overline{(S \to P)}$$

(b) We use a truth tree to determine the validity of the argument.

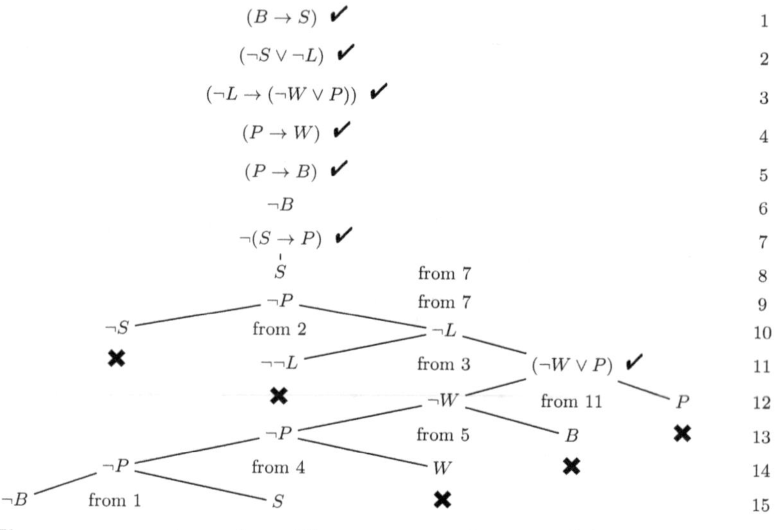

There are open branches. The argument is not valid.

1.9

(2) (a) Disjunction Syllogism; (d) $\vee$-introduction; (g) Proof by cases.

(3)(a)

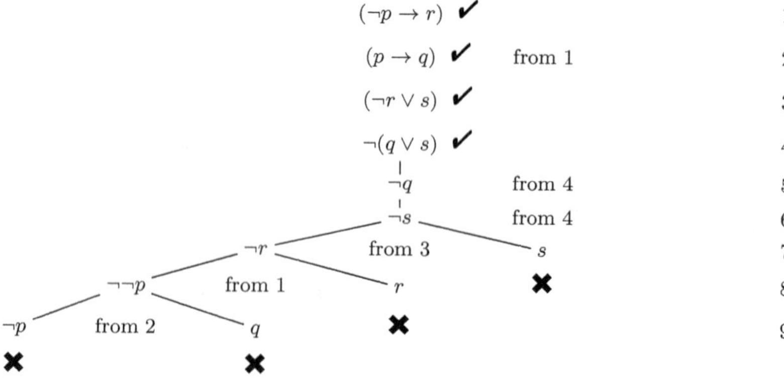

All branches are closed, the argument is valid.

(b)

	Step	Justification
1.	$(\neg p \to r)$	Premise
2.	$(p \to q)$	Premise
3.	$(\neg r \lor s)$	Premise
4.	$(r \to s)$	Logic equivalence, 3
5.	$(\neg p \to s)$	Hypothetical Syllogism, 1,4
6.	$(\neg q \to \neg p)$	Logic equivalence, 2
7.	$(\neg q \to s)$	Hypothetical Syllogism, 5,6
8.	$(q \lor s)$	Logic equivalence, 7

(4)(b) Using a truth tree:

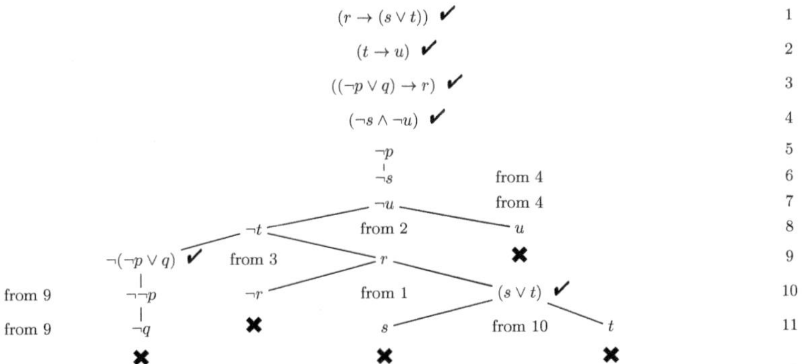

All branches are closed. The argument is valid. The formal proof of
the argument is as follows.

	Step	Justification
1.	$(r \to (s \lor t))$	Premise
2.	$(t \to u)$	Premise
3.	$((\neg p \lor q) \to r)$	Premise
4.	$((\neg s \land \neg u) \to r)$	Premise
5.	$\neg p$	To prove p by contradiction
6.	$(\neg p \lor q)$	$\lor$-introduction, 5
7.	r	Modus ponens, 3,6
8.	$(s \lor t)$	Modus ponens, 1,7
9.	$\neg s$	$\land$-elimination, 2
10.	t	Disjunctive Syllogism, 8, 9
11.	u	Modus ponens, 2,10
12.	$\neg u$	$\land$-elimination, 2
13.	$(u \land \neg u)$	$\land$-introduction, 11, 12
14.	$\mathbf{F}$	Logic equivalence, 13
15.	$(\neg p \to \mathbf{F})$	$5 - 14$
16.	$\neg\neg p$	Contradiction, 15
17.	p	Logic equivalence, 16

(5)(a) Let

- W: The war is over
- K: The Klingon Kingdom surrenders
- N: King Namnar dies
- S: Prince Shawra takes the thrown

The translation of the argument into logic symbols is:

$$(W \to K)$$
$$((K \vee W) \leftrightarrow N)$$
$$\underline{(\neg K \leftrightarrow (N \vee \neg S))}$$
$$\therefore (K \vee \neg S)$$

(b) Using a truth tree:

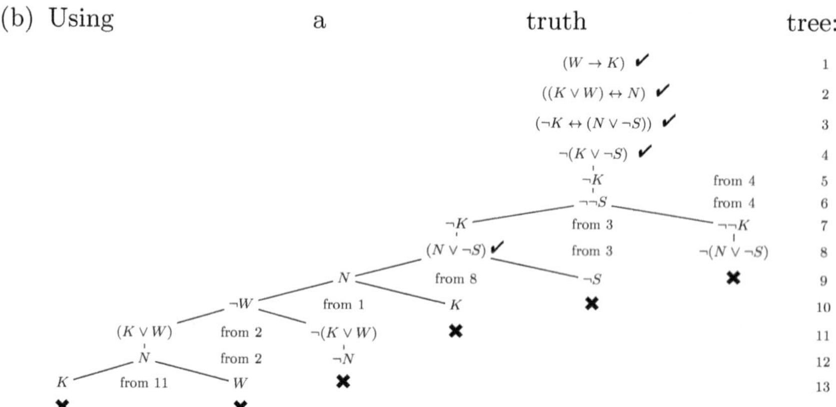

All branches all closed. The argument is valid.

(c)

1. $(W \to K)$		Premise
2. $((K \vee W) \leftrightarrow N)$		Premise
3. $(\neg K \leftrightarrow (N \vee \neg S))$		Premise
4. $\neg(K \vee \neg S)$		Hypothesis, to use the contradiction rule
5. $(\neg K \wedge \neg\neg S)$		Logical equivalence (De Morgan), 4
6. $(\neg K \wedge S)$		Logical equivalence (Double negation), 5
7. $\neg K$		$\wedge$-elimination, 6
8. $((\neg K \to (N \vee \neg S)) \wedge ((N \vee \neg S) \to \neg K))$	Logical equivalence, 3
9. $(\neg K \to (N \vee \neg S))$		$\wedge$-elimination, 7
10. $(N \vee \neg S)$		Modus Ponens, 3, 9

	11. $(S \to N)$	Logical equivalence (Material implication), 10
	12. S	$\wedge$-elimination, 6
	13. N	Modus ponens, 11, 12
	14. $(K \vee W)$	Modus ponens, 2, 13
	15. W	Disjunctive Syllogism, 7, 14
	16. K	Modus ponens, 1, 15
	17. $(\neg K \wedge K)$	$\wedge$-Introduction, 7, 16
	18. $\mathbf{F}$	Logical equivalence (Contradiction), 17
	19. $(\neg(K \vee \neg S) \to \mathbf{F})$	Implication, $4 - 18$
	20. $\neg\neg(K \vee \neg S)$	Contradiction, 19
	21. $(K \vee \neg S)$	Logical equivalence (Double negation), 20

1.10

(1) A DNF equivalent to ϕ is $((A \wedge \neg B \wedge C) \vee (A \wedge \neg B \wedge \neg C) \vee (\neg A \wedge B \wedge C) \vee (\neg A \wedge B \wedge C) \vee (\neg A \wedge \neg B \wedge \neg C))$. A CNF equivalent to ϕ is $((\neg A \vee \neg B \vee \neg C) \wedge (\neg A \vee \neg B \vee C) \wedge (A \vee B \vee \neg C)$.

(2)(b) (i) The truth table of the formula $\phi : ((A \to B) \leftrightarrow (\neg A \to (B \wedge C)))$ is the following.

A	B	C	$\neg A$	$(A \to B)$	$(B \wedge C)$	$(\neg A \to (B \wedge C))$	ϕ
T	T	T	F	T	T	T	T
T	T	F	F	T	F	T	T
T	F	T	F	F	F	T	F
T	F	F	F	F	F	T	F
F	T	T	T	T	T	T	T
F	T	F	T	T	F	F	F
F	F	T	T	T	F	F	F
F	F	F	T	T	F	F	F

From the table, we get the following (complete) DNF equivalent to the formula: $((A \wedge B \wedge C) \vee (A \wedge B \wedge \neg C) \vee (\neg A \wedge B \wedge C))$.

(ii) Using a truth tree:

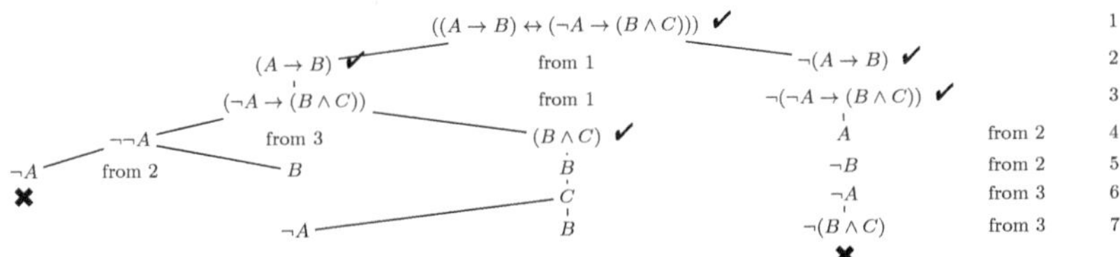

Following the open branches in the tree, we get the following DNF equivalent to the original formula: $((A \wedge B) \vee (\neg A \wedge B \wedge C) \vee (B \wedge C))$.

(iii) Let $\phi : ((A \rightarrow B) \leftrightarrow (\neg A \rightarrow (B \wedge C)))$. Using algebraic manipulations, we have:

$$\begin{aligned}
\phi &\equiv (((A \rightarrow B) \wedge (\neg A \rightarrow (B \wedge C))) \\
&\quad \vee (((\neg(A \rightarrow B) \wedge \neg(\neg A \rightarrow (B \wedge C))))) \\
&\equiv (((\neg A \vee B) \wedge (\neg\neg A \vee (B \wedge C))) \\
&\quad \vee ((A \wedge \neg B) \wedge (\neg A \wedge \neg(B \wedge C)))) \\
&\equiv (((\neg A \vee B) \wedge (A \vee (B \wedge C))) \\
&\quad \vee ((A \wedge \neg B \wedge \neg A \wedge (\neg B \vee \neg C))) \\
&\equiv ((\neg A \wedge A) \vee (\neg A \wedge B \wedge C) \vee (B \wedge A) \vee (B \wedge B \wedge C) \vee \mathbf{F}) \\
&\equiv ((\neg A \wedge B \wedge C) \vee (A \wedge B) \vee (B \wedge C))
\end{aligned}$$

(4) Since $(p \wedge \mathbf{T}) \equiv p$, $(A \wedge \neg B) \equiv ((A \wedge \neg B) \wedge (C \vee \neg C) \equiv ((A \wedge \neg B \wedge C) \vee (A \wedge \neg B \wedge \neg C))$. Similarly, $(B \wedge \neg C) \equiv ((A \wedge B \wedge \neg C) \vee (\neg A \wedge B \wedge \neg C))$. We conclude that $((A \wedge \neg B \wedge C) \vee (A \wedge \neg B \wedge \neg C) \vee (A \wedge B \wedge \neg C) \vee (\neg A \wedge B \wedge \neg C))$ is the complete DNF equivalent to ϕ.

(6) Here is one possibility $\phi : ((A \wedge B) \vee C)$.

2.2

(1)(b) Not well-defined. We need to specify what an expert is.

   (e) Well-defined. This is the set of the integers $-30, -29, -1, \ldots, 0, 1, \ldots, 29, 30$.

   (g) Not well-defined. We need to specify what a large integer means.

(2)(b) $\{-10, -9, -8, -7, -6, -5, -4, -3, -2, -1, 0, 1, 2\}$.

   (f) Note that any multiple of 3 and 5 at the same time is just a multiple of 15: $\{0, 15, 30, 45, 60, 75, 90\}$.

(3)(b) $\mathcal{U} = \mathbb{R}$, $A = \{x \in \mathcal{U}; x(x^2 - 1)(x^2 - 2) = 0\}$.

   (d) $\mathcal{U} = \mathbb{N}$, A is the set of all non-negative integer powers of 2.

(4)(c) "$x \in A$" is true since $x^3 = (\sqrt[3]{2})^3 = 2 \in \mathbb{N}$ and 3 is an integer.

   (f) "$x \in A$" is false since $|(-2)^2 - 10| = 6 > 5$.

(5)(a) False   (d) True   (h) False

(6)(b) False: $\{1\} \in \{1, \{1\}\}$ but $\{1\} \notin \{1\}$.

   (f) True. Both sets are empty.

(7)(b) Empty since the equation $2n^2 - n + 3 = 0$ has no real roots.

   (f) Not empty, $\{\emptyset\} \in A$.

(8)(b) Finite of cardinality 10.

   (c) Infinite.

   (g) Infinite.

(9)(a) 4, 46, 6304, 119214946.

   (c) $(1,0)$, $(0,1)$, $(2,1)$, $(2,0)$.

(10)(b) (1) **Basis clause.** $1, 3 \in A$.

   (2) **Recursive clause.** If $x, y \in A$, then $x + y \in A$.

   (3) **Terminal clause.** Nothing else is in A unless it is obtained from the Basis and Recursive Clauses.

(11)(b) This is the set of all natural numbers which are not divisible by 5.

(12)(b) This is the set of all non-positive integer multiples of 3.

(13)(a) A is the set of all binary strings of the form $\underbrace{11\ldots1}_{n \text{ times}}\underbrace{00\ldots0}_{n \text{ times}}$ where $n \geq 0$ is a integer.

   (d) A is the set of all binary strings having an even number of 0's.

(15) Here are 10 distinct elements of the set A: $(0,0)$, $(1,0)$, $(0,1)$, $(0,2)$, $(2,0)$, $(3,0)$, $(0,3)$, $(1,1)$, $(2,1)$, $(2,2)$. It looks like A is the set of all point with integer coordinates in the first quadrant of the Cartesian plane.

2.3

(1) A power set contains at least the empty set, so it can never be empty.

(2)(a) This is a subset of A: $\emptyset$ is a subset of any set.

   (d) Not a subset of A: $a \notin A$.

   (h) This is a subset of A: it is actually A itself.

(3)(b) If $x \in A$, then $x = 3^n$ for a certain even integer n. Write $n = 2k$ where $k \in \mathbb{Z}$, then $x = 3^{2k} = (3^2)^k = 9^k \in B$. This proves that $A \subseteq B$. Conversely, if $x \in B$, then there exists an integer k such that $x = 9^k = (3^2)^k = 3^{2k} \in A$. Thus proves that $B \subseteq A$.

   (d) If $x \in A$, then $x = 12s + 3t$ for some $s, t \in \mathbb{Z}$. So $x = 3(4s + t) \in B$ since $k = 4s + t \in \mathbb{Z}$. This shows that $A \subseteq B$. Conversely, if $x \in B$ then $x = 3k$ for some $k \in \mathbb{Z}$. In particular, $x = 12(0) + 3k \in A$.

(4)(b) False, $1 \notin A$, so $\{1\} \not\subseteq A$.

   (e) True, $\{1, 2\}$ is an element of A.

   (i) False, $1 \notin A$.

(5) The elements of $\wp(A)$ are $\emptyset$, $\{a\}$, $\{b\}$, $\{c\}$, $\{d\}$, $\{a,b\}$, $\{a,c\}$, $\{a,d\}$, $\{b,c\}$, $\{b,d\}$, $\{c,d\}$, $\{a,b,c\}$, $\{a,b,d\}$, $\{a,c,d\}$, $\{b,c,d\}$ and $\{a,b,c,d\}$.

(8) Such a set does not exist since otherwise we would have: $|A| = \frac{\ln(32766)}{\ln(2)} \approx 14.999912$ (Not an integer)

(10)(a) False since $\wp(\emptyset) = \{\emptyset\}$ is not empty.

(c) True since the emptyset is a subset of any set.

(f) True since $|\wp(\{\emptyset, \{\emptyset\}\})| = 2^2 = 4$, so $|\wp(\wp(\{\emptyset, \{\emptyset\}\}))| = 2^4 = 16$.

(11)(d) True since $\{-22, 2, 26\} \subseteq A : -22 = 3(-2)^3 + 2$, $2 = 3(0)^3 + 2$ and $26 = 3(2)^3 + 2$.

(f) False: $\{-22\} \in \wp(A)$ (since $-22 \in A$) but $\{-22\} \notin \wp(\mathbb{N})$.

(12)(c) False: the interval $[-3, 3]$ is not a subset of $\mathbb{Q}$.

(13) In each case determine if the given set can be realized as the power set of a set A. If you say it is, determine the set A.

(b) This is not a powerset since it does not contain the empty set.

(14)(b) $\{x \in \mathbb{R}; 2 \le x < 2\}$.

(e) $\{x \in \mathbb{R}; x \le -1\}$.

(g) $\{x \in \mathbb{R}; 2 \le x\}$.

(18)(a) Assume that $A \subseteq B$ and $B \subseteq C$. Let $x \in \mathcal{U}$. If $x \in A$, then $x \in B$ (since $A \subseteq B$) and so $x \in C$ (since $B \subseteq C$). This shows that $A \subseteq C$.

(19)(c) 00001001

(e) 11011001

(20)(a) $\{a, e, f, g\}$

(c) $\{a, b, d, e, f, g\}$

2.4

(1)(d) False since $4 \in A \cap C$, so $4 \notin A \oplus C$.

(h) False since $0 \notin A$.

(k) False since $1 \in A \cap \overline{B}$ (as $1 \in A$ and $1 \notin B$) so $1 \notin A \oplus \overline{B}$.

(2)(a) $\{e, g\}$.

(e) $\overline{A \cup C} = \{d\}$, so $\overline{A \cup C} \cap B = \{d\}$

(g) $(B \backslash C) = \{a, d, j\}$, so $A \backslash (B \backslash C) = \{b, e, f, g\}$.

(4)(b) $(A \cap B) \backslash D$.   (e) $(A \cup B) \backslash C$.

(5)(b) $B \cap (A \cup C)$

(6)(b) $3 \in C \backslash A$ (h) $E \in C \backslash D \ne \emptyset$

(e) $\mathcal{U} = A \cup B$

(7) Let $A = \{1,2\}$, $B = \{2,3\}$ and $C = \{1,3,4\}$. Then $A \cap B = \{2\} \neq \emptyset$, $A \cap C = \{1\} \neq \emptyset$ and $A \cap B = \{3\} \neq \emptyset$ but $A \cap B \cap C = \emptyset$.

(9)(c) $A \oplus \mathcal{U} = (A \cup \mathcal{U}) \setminus (A \cap \mathcal{U}) = \mathcal{U} \setminus A = \overline{A}$.

(10) $(A \cup B) \setminus (A \cap B) = (A \cup B) \cap \overline{(A \cap B)}$ (elimination of set difference) $= (A \cup B) \cap (\overline{A} \cup \overline{B})$ (De Morgan) $= (A \cap \overline{A}) \cup (A \cap \overline{B}) \cup (B \cap \overline{A}) \cup (B \cap \overline{B})$ (distributivity) $= \emptyset \cup (A \setminus B) \cup (B \setminus A) \cup \emptyset$ (elimination of $\setminus$ and the exclusion law) $= (A \setminus B) \cup (B \setminus A)$ (Identity law).

(12)(a)

$$
\begin{aligned}
(A \cap B) \cup \overline{(\overline{A} \cup B)} &= (A \cap B) \cup (\overline{\overline{A}} \cap \overline{B}) \text{ (De Morgan)}\\
&= (A \cap B) \cup (A \cap \overline{B}) \text{ (Double complement)}\\
&= (A \cup A) \cap (A \cup \overline{B}) \cap (B \cup A) \cap (B \cup \overline{B}) \text{ (Distributivity)}\\
&= A \cap (A \cup \overline{B}) \cap (B \cup A) \cap \mathcal{U}\\
&\quad \text{(Idempotent and Complementation laws)}\\
&= A \cap (A \cup (\overline{B} \cap B)) \text{ (Distributivity)}\\
&= A \cap (A \cup \emptyset) \text{ (Exclusion law)}\\
&= A \cap A \text{ (Identity law)}\\
&= A \text{ (Idempotent law)}
\end{aligned}
$$

(b) Let $\phi : ((A(x) \wedge B(x)) \vee \neg(\neg A(x) \vee B(x)))$. We need to prove that $\phi \equiv A(x)$ or equivalently that the corresponding logic expression $(\phi \leftrightarrow A(x))$ is a tautology. We use a truth table:

$A(x)$	$B(x)$	$\neg A(x)$	$(A(x) \wedge B(x))$	$(\neg A(x) \vee B(x))$	$\neg(\neg A(x) \vee B(x))$	ϕ
T	T	F	T	T	F	T
T	F	F	F	F	T	T
F	T	T	F	T	F	F
F	F	T	F	T	F	F

(13)(b) Let $A = \{1\}$, $B = C = \{1,2\}$. Then $A \setminus B = \emptyset$, $A \setminus C = \emptyset$, $B \setminus C = \emptyset$, $A \setminus (B \setminus C) = (\{1\}, A \setminus B) \setminus C = \emptyset$ and $(A \setminus B) \setminus (A \setminus C) = \emptyset$.

(14)(a) False. Let $A = \{1\}$, $B = \{1,2\}$. Then $A \setminus B = \emptyset$, but $A \neq B$

(e) True: If there exists $x \in A \cup B$ such that $x \ /B$, then $x \in A$ and so $x \in A \cap B$ since $A \subseteq A \cap B$. But that means in particular that $x \in B$: a contradiction.

(15) (a) Assume first that $A \subseteq B$. If $x \in A$, then $x \in A \cap B$ and so $A \subseteq A \cap B$. The inclusion $A \cap B \subseteq A$ is also true. So $A \cap B = A$. Conversely, if $A \cap B = A$, then for any $x \in A$, $x \in A \cap B$ and therefore $x \in B$. This shows that $A \subseteq B$.

(e) We use properties of set operations.

$$
\begin{aligned}
A\setminus(A\cap B) &= A\cap\overline{A\cap B}\;\text{(Elimination of set difference)}\\
&= A\cap(\overline{A}\cup\overline{B})\;\text{(De Morgan)}\\
&= (A\cap\overline{A})\cup(A\cap\overline{B})\;\text{(Distributivity)}\\
&= \emptyset\cup(A\cap\overline{B})\;\text{(Exclusion law)}\\
&= A\cap\overline{B}\;\text{(Identity law)}\\
&= A\setminus B\;\text{(Elimination of set difference)}
\end{aligned}
$$

(18) (a) Assume first that $A\cap(B\cup C) = (A\cap B)\cup C$. If $x\in C$, then $x\in(A\cap B)\cup C$ and so $x\in A\cap(B\cup C)$. In particular, $x\in A$ and so $C\subseteq A$. Conversely, assume that $C\subseteq A$. Then $A\cup C = A$ and so $A\cap(B\cup C) = (A\cup C)\cap(B\cup C) = (A\cap B)\cup C$ (by the distributivity property).

(19) (b)

$$
\begin{aligned}
A\setminus(B\cap C) &= A\cap\overline{B\cap C}\;\text{(Elimination of set difference)}\\
&= A\cap(\overline{B}\cup\overline{C})\;\text{(DeMorgan)}\\
&= (A\cap\overline{B})\cup(A\cap\overline{C})\;\text{(Distributivity)}\\
&= (A\setminus B)\cup(A\setminus C)\;\text{(Elimination of set difference)}
\end{aligned}
$$

(d)

$$
\begin{aligned}
\overline{A\cap(\overline{A}\cup B)}\cup B &= \overline{A}\cup\overline{(\overline{A}\cup B)}\cup B\;\text{(De Morgan)}\\
&= \overline{A}\cup(\overline{\overline{A}}\cap\overline{B})\cup B\;\text{(De Morgan)}\\
&= \overline{A}\cup(A\cap\overline{B})\cup B\;\text{(Double complementation)}\\
&= ((\overline{A}\cup A)\cap(\overline{A}\cup\overline{B}))\cup B\;\text{(Distributivity)}\\
&= (\mathcal{U}\cap(\overline{A}\cup\overline{B}))\cup B\;\text{(Complementation law)}\\
&= (\overline{A}\cup\overline{B})\cup B\;\text{(Identity law)}\\
&= (\overline{A}\cup\mathcal{U})\;\text{(Complementation law)}\\
&= \mathcal{U}\;\text{(Domination law)}
\end{aligned}
$$

(20) $\bigcap_{i=0}^{\infty} A_i = \{0\}$ and $\bigcup_{i=0}^{\infty} A_i = \mathbb{N}$.

(23)(a) Assume that $A\subseteq B$. If $x\in\wp(A)$, then $X\subseteq A$ and so $X\subseteq B$ or equivalently, $X\in\wp(B)$. This proves that $\wp(A)\subseteq\wp(B)$.

(25) One of the two sets must be empty since otherwise $A\times B\neq\emptyset$.

(27)(c) The elements of the set $B \times A$ are: $(\alpha, (a, 1))$, $(\alpha, (b, 2))$, $(\alpha, (c, 3))$, $(\beta, (a, 1))$, $(\beta, (b, 2))$, $(\beta, (c, 3))$.

(d) The elements of the set $A \times B$ are $(a, 1)$, $(a, 2)$, $(b, 1)$, $(b, 2)$. The elements of the set $A \times B \times C$ are $((a, 1), \alpha)$, $((a, 1), \beta)$, $((a, 2), \alpha)$, $((a, 2), \beta)$, $((b, 1), \alpha)$, $((b, 1), \beta)$, $((b, 2), \alpha)$, $((b, 2), \beta)$.

(29) Let $A = \{1\}$, $B = \{2\}$ and $C = \{3\}$. Then $(1, (2, 3)) \in A \times (B \times C)$ but $(1, (2, 3)) \notin (A \times B) \times C$ since $1 \notin A \times B$. We conclude that $A \times (B \times C) \neq (A \times B) \times C$.

(30)(b) $(0, 5) \in I \times J$ since $0 \in I$ and $5 \in J$.

(e) $(1, 3) \in I \times J$ since $1 \in I$ and $3 \in J$.

(g) $(\frac{9}{2}, 5) \notin I \times J$ since $5 \notin J$.

(32)(c) $A = \{a, b\}$, $B = \{0, 1, 2\}$.

(33)(b) False since $(a, b) \in \{(a, b), (c, d)\}$ but $(a, b) \notin \{a, b, c, d\}$.

(d) False since $(a, b) \in \{(a, b), (c, d)\}$ but $(a, b) \notin \{(a, c), (b, d)\}$.

(f) False since $\{(a, c), (b, d)\} \not\subseteq \{a, c\} \times \{b, d\}$. For example, $(a, c) \in \{(a, c), (b, d)\}$ but $(a, c) \notin \{a, c\} \times \{b, d\}$.

(36)(b) $A = B = \{1\}$.

(h) $A = B = \{1, 2, 3, 4, 5, 6\}$.

(38) The elements of $A \times B \times C$ are (a, b, c), (a, b, d), (a, c, c), (a, c, d), (b, b, c), (b, b, d), (b, c, c), (b, c, d).

(40) For $x \in \mathcal{U}$: $(x, y) \in A \times (B \cup C) \Leftrightarrow x \in A$ and $y \in B \cup C \Leftrightarrow x \in A$ and $(y \in B$ or $y \in C) \Leftrightarrow (x \in A$ and $y \in B)$ or $(x \in A$ and $y \in C) \Leftrightarrow (x, y) \in A \times B$ or $(x, y) \in A \times C \Leftrightarrow (x, y) \in (A \times B) \cup (A \times C)$.

2.5

(1)(b) $\{\{1\}, \{2\}, \{3\}\}$, $\{\{1, 2\}, \{3\}\}$, $\{\{1, 3\}, \{2\}\}$, $\{\{1\}, \{2, 3\}\}$, $\{\{1, 2, 3\}\}$

(3)(a) Not a partition. It contains the empty set as a block.

(d) Not a partition: $\{\rho, \phi, \xi\} \cup \{\alpha, \beta\} \cup \{\gamma\} \neq A$.

(4)(b) $\{\{1\}, \{2\}, \{3\}, \{4\}, \{5, 6, 7\}\}$.

(7) $\{1, 2, 3\} \cap \{4, 5\} = \emptyset$, $\{1, 2, 3\} \cap \{6, 7\} = \emptyset$, $\{4, 5\} \cap \{6, 7\} = \emptyset$ and $\{1, 2, 3\} \cup \{4, 5\} \cup \{6, 7\} = \{1, 2, 3, 4, 5, 6, 7\}$. So, $\{\{1, 2, 3\}, \{4, 5\}, \{6, 7\}\}$ is a partition of A. A possible refinement of $\mathcal{F}$ consisting of five blocks is:

$$\{\{1, 2\}, \{4, 5\}, \{3\}, \{6\}, \{7\}\}.$$

(8) Clearly $I \cap J = I \cap \{0\} = J \cap \{0\} = \emptyset$. Moreover $I \cup J \cup \{0\} = \mathbb{R}$. This show that $\{I, J, \{0\}\}$ is a partition of $\mathbb{R}$. A possible refinement of $\{I, J, \{0\}\}$ consisting of five blocks is $\{]-\infty, -1[, [-1, 0[, \{0\},]0, 1[, [1, \infty[\}$.

(11) By definition, members of the family $\mathcal{G}$ are non-empty. Since $B = A \cap B$ (as $B \subseteq A$), $B = (\cup_{i \in I} A_i) \cap B$ (as $A = \cup_{i \in I} A_i$) and so $= \cup_{i \in I}(B \cap A_i)$ by distributivity property. Moreover, if $i, j \in I$, then $(B \cap A_i) \cap (B \cap A_j) = B \cap (A_i \cap A_j) = B \cap \emptyset$ (since $\mathcal{F}$ est une partition de A) $= \emptyset$. We conclude that $\{B \cap A_i;\ i \in I$ and $B \cap A_i \neq \emptyset\}$ is a partition of B.

(14) Since $A_i \neq \emptyset$ and $B_j \neq \emptyset$ for any $i \in I$ and any $j \in J$, $A_i \times B_j \neq \emptyset$ For any $(i, j) \in I \times J$. If $i, j \in I$ and $r, s \in J$ with $(i, j) \neq (r, s)$ then $(A_i \times B_r) \cap (A_j \times B_s) = (A_i \cap A_j) \times (B_r \cap B_s) = \emptyset$. Moreover, $\cup_{(i,j) \in I \times J}(A_i \times B_j) = (\cup_{i \in I} A_i) \times (\cup_{j \in J} B_j) = A \times B$. This proves that $\{A_i \times B_j;\ (i, j) \in I \times J\}$ of subsets of $A \times B$ is a partition of $A \times B$.

2.6

(1) No, $(\mathbb{R}, +, \times)$ is not a Boolean algebra. For example, the distributive law $x + yz = (x + y)(x + z)$ does not hold in general for real numbers x, y and z.

(3) Let $X, Y, Z \in \wp(A)$.

  (B1) $X \cap Y$ and $X \cup Y$ are elements of $\wp(A)$.

  (B2) $X \cap Y = Y \cap X$ and $X \cup Y = Y \cup X$.

  (B3) $X \cap (Y \cup Z) = (X \cap Y) \cup (X \cap Z)$ and $X \cup (Y \cap Z) = (X \cup Y) \cap (X \cup Z)$

  (B4) $X \cap A = A \cap X = X$ and $X \cup \emptyset = \emptyset \cup X = X$. This shows that the identity element of $\cup$ is $\emptyset$ and that of $\cap$ is A.

  (B5) Let $\overline{X}$ be the complement of X in A. So $X \cap \overline{X} = \emptyset$ and $X \cup \overline{X} = A$.

(4) Assume that u and v are both units elements of a Boolean algebra B. By definition of the unit element, $u \cdot v = v$ (since u is a unit element) and $u \cdot v = u$ (since v is a unit element). We conclude that $u = v$.

(5)(b) We have that $b' + b = 1$ $b'b = 0$. By the uniqueness of the complement (part (a)), $(b')' = b$.

(7) For any $b \in B$:

$$
\begin{aligned}
b + 1 &= (b + 1)1 && \text{(Axiom (B4))} \\
&= (b + 1)(b + b') && \text{(Axiom (B5))} \\
&= b + 1 \cdot b' && \text{(Axiom (B3))} \\
&= b + b' && \text{(Axiom (B4))} \\
&= 1 && \text{(Axiom (B5))}
\end{aligned}
$$

(9) Assume first that $a + b = b$. Then $ab = a(a + b) = aa + ab = a + ab$ (since $aa = a$) $= a(1 + b)$ (Distributivity) $= a(1)$ (Absorption laws) $= a$.

Conversely, assume that $ab = a$. Then $a + b = ab + a = (a+1)b = (1)b$ (Absorption laws) $= b$.

(11) We know that any Boolean algebra must contain at least the two distinct identity elements 0 and 1. Assume that B is a Boolean algebra containing exactly three pairwise distinct elements, namely 0, 1 and a. Then $a' \neq 0$ since otherwise $a = 1$. Similarly $a' \neq 1$. Then $a' = a$. But that would mean that $aa' = aa = a = 0$: a contradiction. So no Boolean algebra contains exactly three pairwise distinct elements.

(14) By the uniqueness of the complement element, we need to show two things: $(ab) + (a' + b') = 1$ and $(ab)(a' + b') = 0$.

$$
\begin{aligned}
(ab) + (a' + b') &= (ab + a') + b' &&\text{(Associative law)} \\
&= (a + a')(b + a') + b' &&\text{(Axiom (B3))} \\
&= (1)(b + a') + b' &&\text{(Axiom (B5))} \\
&= (b + a') + b' &&\text{(Axiom (B4))} \\
&= (a' + b) + b' &&\text{(Axiom (B2))} \\
&= a' + (b + b') &&\text{(Associativity laws)} \\
&= a' + 1 &&\text{(Axiom (B5))} \\
&= 1 &&\text{(Boundless laws).}
\end{aligned}
$$

$$
\begin{aligned}
(ab)(a' + b') &= (ab)a' + (ab)b' &&\text{(Axiom (B3))} \\
&= (ba)a' + (ab)b' &&\text{(Axiom (B2))} \\
&= b(aa') + a(bb') &&\text{(Associativity laws} \\
&= b(0) + a(0) &&\text{(Axiom (B5))} \\
&= 0 + 0 &&\text{(Boundless laws)} \\
&= 0 &&\text{(Axiom (B4)).}
\end{aligned}
$$

(15)(a) $(a + b + c')(a' + b + c')$ (c) $(a + b)'(a + b + c')$

(18)(a) $(a + (ab)')'ab' = a'((ab)')'ab'$ (De Morgan) $= a'(ab)ab'$ (Double complement) $= (a'a)(bab')$ (Associativity) $= (0)(bab')$ (Axiom (B5)) $= 0$ (Boundless law).

(e) $abc' + abc + a'bc + a + c = ab(c + c') + a'bc + a + c$ (Distributivity) $= ab + a'bc + a + c$ (Axiom (B4)) $= b(a + a'c) + a + c$ (Distributivity) $= b(a + a')(a + c) + a + c$ (Distributivity) $= b(a + c) + a + c$ (Axiom (B5)) $= (b + 1)(a + c)$ (Distributivity) $= (1)(a + c) = a + c$ (Axiom ((B4)).

2.7

(1)(d) Both SOP and POS

   (e) Neither

(2)(c) $xyz' + yzt + xyzt' + xt = xyz'(t + t') + (x + x')yzt + xyzt' + x(y + y')(z + z')t = xyz't + xyz't' + xyzt + x'yzt + xyzt' + xyzt + xyz't + xy'zt + xy'z't$.

(3)(c) $x(yz + xy'z')' + (yz)(x + y') = xyz + xxy'z' + yzx + yzy'$ (Distributivity) $= xyz + xy'z' + yzx$ (since $xx = x$, $yy' = 0$) $= xyz + xy'z'$ (since $xyz + yzx = xyz + xyz = xyz$). This is already in CSOP.

(4)(b) $xy(x'yz + y'z')'$

(5)(a) The CSOP form of f is $xyzt + xyzt' + xyz't + xy'zt' + xy'z't + x'y'z't' + x'yzt' + x'yz't' + x'yz't$. Six groupings of 1's are needed in the map as shown below. and find a minimal form of the expression.

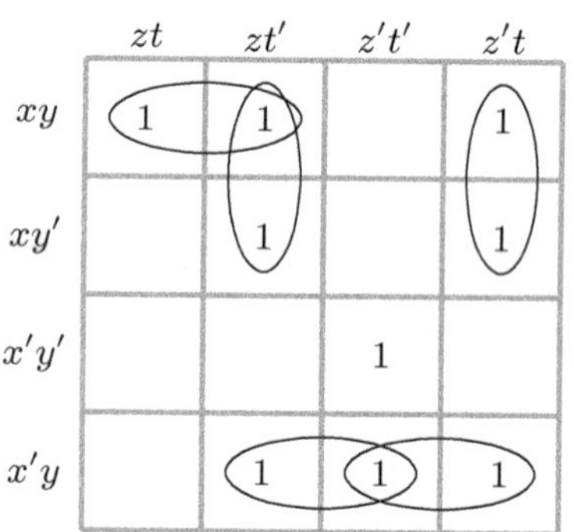

Summing up the contribution of every group in the map, we get the following minimal form of the expression: $xyz + xzt' + xz't + x'z't' + x'yt' + x'yz'$.

(7) $x'yzt$, $x'yz't'$, $x'y'zt'$ and $xyzt'$.

(8)(b) The Karnaugh map of the form is.

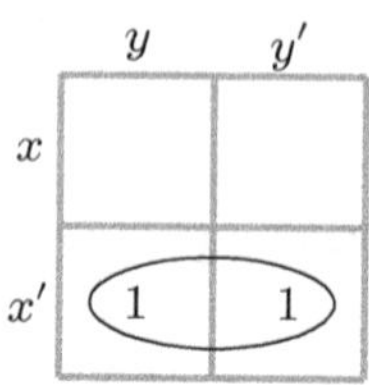

(9)(b) The Karnaugh map for the expression together with the appropriate groupings in each map are given in the following diagram:

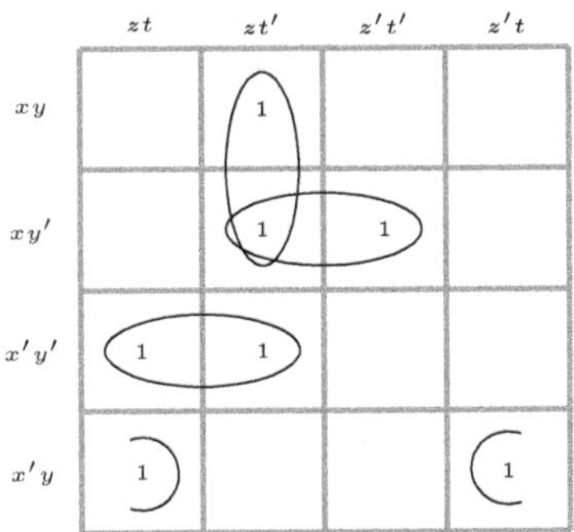

Summing up the contribution of every group in the map, we get the following minimal form of the expression: $xz' + x'y + x'yz$.

(10)(a) The Karnaugh map corresponding to the CSOP is:

Summing up the contribution of every group in the map, we get the following minimal form of the expression: $xzt' + xy't' + x'y'z + x'yt$.

2.8

(1)(a)

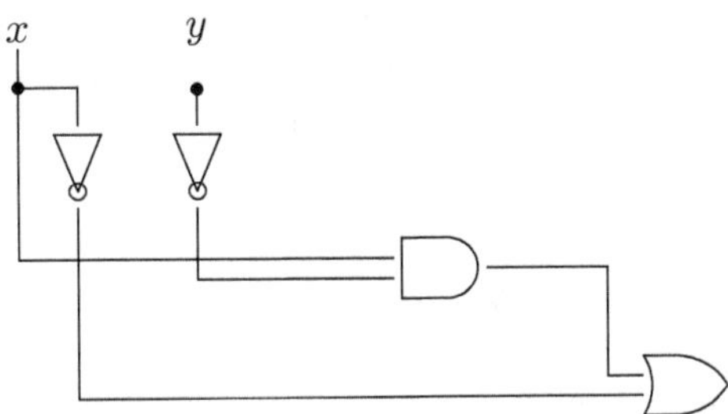

(3) The output of the circuit is the function:

$$f(x, y, z, t) = [((x' + y' + z) + x'yz' + xy'z + xyz + (x' + y' + z))'.$$

Using Boolean algebra laws, the function simplifies to xyz'. The following logic circuit uses fewer gates and produces the same output as the original one.

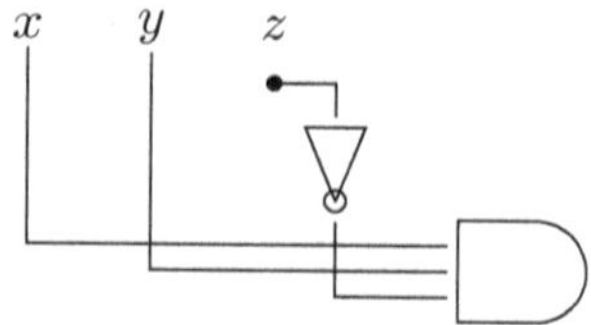

3.2

(3) Assume that n is a multiple of 3, then $n = 3k$ for some $k \in \mathbb{Z}$ and so $n^2 + 2n = (3k)^2 + 2(3k) = 9k^2 + 6k = 3(3k^2 + 2k)$ is a multiple of 3.

(5) Let m, n be two integers of different parities. Without loss of generality, we can assume that m is even and n is odd. So $m = 2k$ and $n = 2s + 1$ for some integers k, s. Therefore, $mn = 2k(2s + 1) = 2(2ks + k)$ is even.

(7) We need to prove the implication: "if $n^5 + 7$ is even, then n is odd". The best approach is to proceed by indirect proof. The contrapositive of the implication is: "if n is even, then $n^5 + 7$ is odd". So assume that n is even and write $n = 2k$ for some $k \in \mathbb{Z}$. Then $n^5 + 7 = (2k)^5 + 7 = 32k^5 + 7 = 2(16k^53) + 1$ is odd.

(12) We proceed by a direct proof. Assume that x and y are two rational numbers, and write $x = \frac{a}{b}$, $y = \frac{c}{d}$ for some integers a, b, c and d with $b \neq 0$ and $d \neq 0$. Then $mx + ny = m\frac{a}{b} + n\frac{c}{d} = \frac{mad + abc}{bd}$ is a rational number for any integers m and n.

(15) Let $x \geq 0$ and $y \geq 0$ be real numbers. We have that $\left(\frac{x+y}{2}\right)^2 - xy = \frac{x^2 + y^2 + 2xy}{4} - xy = \frac{x^2 + y^2 - 2xy}{4} = \left(\frac{x-y}{2}\right)^2 \geq 0$. We conclude then that $\left(\frac{x+y}{2}\right)^2 \geq xy$ and so $\frac{x+y}{2} \geq \sqrt{xy}$ since x, y are non-negative.

(18) The best approach is to proceed by indirect proof. The contrapositive of the implication is: *If $\sqrt{x}$ is a rational number, then x is a rational number.* Assume that $\sqrt{x}$ is a rational number and write $\sqrt{x} = \frac{a}{b}$ for some integers a, b with $b \neq 0$. Squaring both sides, we get $x = \frac{a^2}{b^2}$ which is a rational number.

(21) The best approach is to proceed by indirect proof. The contrapositive of the implication is: *If a and b are not relatively prime then so are a^2 and b^2.* Assume that a and b are not relatively prime then there exists a common divisor $d \geq 2$ of a and b. But then d is also a common divisor of a^2 and b^2. Therefore a^2 and b^2 are not relatively prime.

(23) The best approach is to proceed by indirect proof. The contrapositive of the implication is: *If each of m, n is a perfect square, then mn is a perfect square.* Assume that m, n are perfect squares and write $m = a^2$

and $n = b^2$ for some integers a and b. Then $mn = a^2b^2 = (ab)^2$ is a perfect square.

3.3

(2) Assume first that n is a multiple of 6 and write $n = 6k$ for some integer k. Then $n = 2(3k) = 3(2k)$, which means that n is both even and a multiple of 3. Conversely, assume that n is even and a multiple of 3 and write $n = 2s = 3t$. In particular, $3t$ is even and hence t is even. Write $t = 2k$ for some integer k, then $n = 3t = 6k$ is a multiple of 6.

(6) If $x < y$, then $x - \frac{x+y}{2} = \frac{x-y}{2} < 0$ and consequently, $x < \frac{x+y}{2}$. Conversely, if $x < \frac{x+y}{2}$ then $x - \frac{x+y}{2} = \frac{x-y}{2} < 0$ which implies that $x - y < 0$ and so $x < y$.

(8) Assume that $A \backslash (A \backslash B) = B$. Then $B \subseteq A$ since $A \backslash (A \backslash B) \subseteq A$. Conversely, if $B \subseteq A$, then $A \backslash (A \backslash B) = A \cap \overline{(A \backslash B)} = A \cap \overline{A \cap \overline{B}} = A \cap (\overline{A} \cup \overline{\overline{B}}) = A \cap (\overline{A} \cup B) = (A \cap \overline{A}) \cup (A \cap B) = \emptyset \cup B$ (since $A \cap \overline{A} = \emptyset$ and $B \subseteq A$) $= B$.

3.4

(3) Assume by contradiction that r is the smallest positive rational number. Write $r = \frac{a}{b}$ for some positive integers a and b. Then $\frac{r}{2} = \frac{a}{2b}$ is a positive rational number smaller than r: a contradiction.

(5) Assume by contradiction that $\sqrt{3}$ is rational and write $\sqrt{3} = \frac{a}{b}$ for some positive integers a and b. Furthermore, we can assume that the only positive common divisor of a and b is 1. Squaring both sides of the equation $\sqrt{3} = \frac{a}{b}$, we get that $a^2 = 3b^2$. Since 3 is a prime number dividing a^2, 3 must divide a by the hint. Write $a = 3k$ for some integer k, then the equation $a^2 = 3b^2$ simplifies to $3k^2 = b^2$. Again, 3 is a prime number dividing b^2, so 3 must divide b. This is a contradiction since 3 is a common divisor of a and b.

(7) Assume by contradiction that the result is false. Then the number of objects in every box must be at most 1. Therefore the total number of objects is at most n. This is a contradiction since we have $n+1$ objects.

(11) Assume that there exists a perfect square of the form $4n + 2$ where n is an integer. Then there exists an integer k such that $4n + 2 = k^2$. In particular k^2 is even and so k is even. Write $k = 2t$ for some integer t. The equation $4n + 2 = k^2$ becomes $4n + 2 = 4t^2$ or $2n + 1 = 2t^2$ after

dividing by 2. This is a contradiction since $2n + 1$ is odd and $2t^2$ is even. We conclude that no perfect square of the form $4n + 2$ exists.

(15) Assume by contradiction that $B \cap (A \backslash C) \neq \emptyset$ and let $x \in B \cap (A \backslash C)$. Then $x \in B$, $x \in A$ and $x \notin C$. This implies that $x \in (A \cap B) \backslash C$ which is a contradiction since $A \cap B \subseteq C$.

(17) Note first that $\frac{2+3\sqrt{2}}{3-2\sqrt{2}} = \frac{(2+3\sqrt{2})(3+2\sqrt{2})}{(3-2\sqrt{2})(3+2\sqrt{2})} = \frac{18+13\sqrt{2}}{1} = 18 + 13\sqrt{2}$. Assume by contradiction that $\frac{2+3\sqrt{2}}{3-2\sqrt{2}}$ is rational. Then there exist two integers a and b such that $b \neq 0$ and $18 + 13\sqrt{2} = \frac{a}{b}$ and therefore $\sqrt{2} = \frac{\frac{a}{b}-18}{13}$. But $\frac{\frac{a}{b}-18}{13}$ is rational number which means that $\sqrt{2}$ is rational. This is a contradiction.

(20) Assume by contradiction that $2 \leq x \leq 3$ and $x^2 - 5x + 6 > 0$. Notice that $x^2 - 5x + 6 = (x-2)(x-3)$ so the fact that $x^2 - 5x + 6 > 0$ implies that either $x - 2 > 0$ and $x - 3 > 0$ or $x - 2 < 0$ and $x - 3 < 0$. From the first case we get that $x > 3$ and from the second case $x < 2$. We get a contradiction in both cases by the assumption that $2 \leq x \leq 3$. We conclude that $x^2 - 5x + 6 \leq 0$.

3.5

(2) If n is even then $n = 2k$ for some integer k and so $n^3+3n+5 = 8k^3+6n+5 = 2(4k^3+3n+2)+1$ is odd. If n is odd then $n = 2k+1$ for some integer k and so $n^3 + 3n + 5 = 8k^3 + 12k^2 + 12k + 9 = 2(4k^3 + 6k^2 + 6k + 4) + 1$ is odd. Since the result is true in all possible cases, it is true for any integer n.

(4) There are three possible values for the remainder r of the division of an integer n by 3: $r = 0, 1$ or 2. If $r = 0$, then $n = 3k$ for some integer k and so $n^2 + n + 8 = 9k^2 + 3k + 8 = 3(3k^3 + k + 2) + 2$ is not divisible by 3. If $r = 1$, then $n = 3k + 1$ for some integer k and so $n^2 + n + 8 = 9k^2 + 9k + 10 = 3(3k^3 + 3k + 3) + 1$ is not divisible by 3. If $r = 2$, then $n = 3k + 2$ for some integer k and so $n^2 + n + 8 = 9k^2 + 15k + 14 = 3(3k^3 + 5k + 4) + 2$ is not divisible by 3.

(9) Consider the following cases. If A is a knight, then he speaks truth and so B is a knight. Therefore the statement "Either A is a knave and C is a knight, or A is a knight and C is a knave" is true. Since "A is a knave and C is a knight" is false (as A is a knight), "A is a knight and C is a knave" is true and C is a knave. If A is a knave, then he speaks false and so B is a knave. Therefore the statement "Either A is a knave and C is a knight, or A is a knight and C is a knave" is false. Therefore

both "A is a knave and C is a knight" and "A is a knight and C is a knave" are false. Since A is a knave, "C is a knight" must be false so C is a knave. We conclude that C is a knave.

(12) Let x, y and z be three arbitrary real numbers. There are six possibilities: $x \leq y \leq z$, $x \leq z \leq y$, $x \leq y \leq z$, $x \leq y \leq z$, $x \leq y \leq z$ and $x \leq y \leq z$. If $x \leq y \leq z$, then $x \oplus (y \oplus z) = x \oplus z = z$ and $(x \oplus y) \oplus z = z$. The result is true in this case. Similar conclusion is drawn in the other cases.

3.6

(2) We use weak induction.

Base step. For $n = 1$, $\sum_{k=1}^{n}(2k-1)^2 = 1^1 = 1$ and $\frac{n(2n-1)(2n+1)}{3} = 1$ and the result is true in this case.

$\sum_{k=1}^{n}(2k-1)^2 = 1^2 + 3^3 + 5^2 + \cdots + (2n-1)^2 = \frac{n(2n-1)(2n+1)}{3}$

Inductive step. Let $n \geq 1$ be an integer.

Induction hypothesis. Assume that $\sum_{k=1}^{n}(2k-1)^2 = 1^2 + 3^3 + 5^2 + \cdots + (2n-1)^2 = \frac{n(2n-1)(2n+1)}{3}$.

Conclusion. We need to prove that $\sum_{k=1}^{n+1}(2k-1)^2 = 1^2 + 3^3 + 5^2 + \cdots + (2n-1)^2 + (2n+1)^2 = \frac{(n+1)(2n+1)(2n+3)}{3}$.

$$\sum_{k=1}^{n+1}(2k-1)^2 = \sum_{k=1}^{n}(2k-1)^2 + (2n+1)^2$$

$$= \frac{n(2n-1)(2n+1)}{3} + (2n+1)^2 \text{ (By the induction hypothesis)}$$

$$= \frac{n(2n-1)(2n+1) + 3(2n+1)^2}{3}$$

$$= \frac{(2n+1)(n(2n-1) + 3(2n+1))}{3}$$

$$= \frac{(2n+1)(2n^2 + 5n + 3)}{3}$$

$$= \frac{(2n+1)(n+1)(2n+3)}{3}$$

(5) We use weak induction.

Base step. For $n = 1$, $2 \cdot 2 + 3 \cdot 2^2 + 4 \cdot 2^3 + \cdots + (n+1) \cdot 2^n = 2 \cdot 2 = 4$ and $n \cdot 2^{n+1} = 4$. The result is true in this case.

Inductive step. Let $n \geq 1$ be an integer.

Induction hypothesis. Assume that $2 \cdot 2 + 3 \cdot 2^2 + 4 \cdot 2^3 + \cdots + (n+1) \cdot 2^n = n \cdot 2^{n+1}$

Conclusion. We need to prove that $2 \cdot 2 + 3 \cdot 2^2 + 4 \cdot 2^3 + \cdots + (n+1) \cdot 2^n + (n+2) \cdot 2^{n+1} = (n+1) \cdot 2^{n+2}$.

$$2 \cdot 2 + 3 \cdot 2^2 + 4 \cdot 2^3 + \cdots + (n+1) \cdot 2^n + (n+2) \cdot 2^{n+1}$$
$$= n \cdot 2^{n+1} + (n+2) \cdot 2^{n+1} \text{ (By the induction hypothesis)}$$
$$= 2^{n+1}(2n+2)$$
$$= 2^{n+1}2(n+1)$$
$$= (n+1)2^{n+2}$$

(8) We use weak induction.

Base step. For $n = 0$, $S_n = a$ and $a\frac{1-r^{n+1}}{1-r} = a\frac{1-r}{1-r} = a$ (since $r \neq 1$. The result is true in this case.

Inductive step. Let $n \geq 0$ be an integer.

Induction hypothesis. Assume that $S_n = a\frac{1-r^{n+1}}{1-r}$.

Conclusion. We need to prove that $S_{n+1} = a\frac{1-r^{n+2}}{1-r}$:

$$S_{n+1} = a + ar + \cdots + ar^n + ar^{n+1}$$
$$= a\frac{1-r^{n+1}}{1-r} + ar^{n+1} \text{ (By the induction hypothesis)}$$
$$= a\frac{1-r^{n+1} + r^{n+1}(1-r)}{1-r}$$
$$= a\frac{1-r^{n+2}}{1-r}.$$

The expression $1 - \frac{1}{3} + \frac{1}{9} - \frac{1}{27} + \cdots + \frac{(-1)^n}{3^n}$ is a geometric sum of $n+1$ terms with first term 1 and ration $-\frac{1}{3}$. We conclude that $1 - \frac{1}{3} + \frac{1}{9} - \frac{1}{27} + \cdots + \frac{(-1)^n}{3^n} = \frac{1-\left(\frac{1}{3}\right)^{n+1}}{1-\frac{1}{3}} = \frac{3}{2}\left(\frac{3^{n+1}-1}{3^{n+1}}\right).$

(12) We use weak induction.

Base step. For $n = 0$, $2^{3n+1} + 5 = 7$ is divisible by 7. The result is true in this case.

Inductive step. Let $n \geq 0$ be an integer.

Induction hypothesis. Assume that $2^{3n+1} + 5$ is divisible by 7. That is $2^{3n+1} + 5 = 7k$ for some integer k

Conclusion. We need to prove that $2^{3(n+1)+1} + 5$ is divisible by 7.

$$2^{3(n+1)+1} + 5 = 2^{3n+4} + 5 = 27 \cdot 2^{3n+1} + 5$$
$$= 8(7k-5) + 5 \text{ (By the induction hypothesis)}$$
$$= 56k - 35 = 7(8k-5) : \text{ divisible by 7}$$

(1) We use weak induction.

Base step. For $n = 0$, $10^{n+2} + 11^{2n+1} = 111$ is divisible by 111 . The result is true in this case.

Inductive step. Let $n \geq 0$ be an integer.

Induction hypothesis. Assume that $10^{n+2} + 11^{2n+1}$ is divisible by 111. That is $10^{n+2} + 11^{2n+1} = 111k$ for some integer k

Conclusion. We need to prove that $10^{n+3} + 11^{2n+3}$ is divisible by 111.

$$10^{n+3} + 11^{2n+3} = 10 \cdot 10^{n+2} + 11^2 \cdot 11^{2n+1}$$

$$= 10(111k - 11^{2n+1}) + 121 \cdot 11^{2n+1} \ \text{(By the induction hypothesis)}$$

$$= 111(10k + 11^{2n+1}) : \ \text{divisible by 111}$$

(17) We start by checking the inequality for some initial values of $n \geq 0$:

n	$2n+1$	2^n	$2n+1 < 2^n$
0	1	1	False
1	3	2	False
2	5	4	False
3	7	8	True
4	9	16	True

We prove that $2n + 1 < 2^n$ for $n \geq 3$ using weak induction.

Base step. For $n = 3$, the result is true according to the above table.

Inductive step. Let $n \geq 3$ be an integer.

Induction hypothesis. Assume that $2n + 1 < 2^n$.

Conclusion. We need to prove that $2(n + 1) + 1 < 2^{n+1}$.

$$2(n + 1) + 1 = (2n + 1) + 2$$
$$< 2^n + 2 \ \text{(By the induction hypothesis)}$$
$$< 2^n + 2^n \ \text{since } n \geq 3$$
$$= 2 \cdot 2^n = 2^{n+1}.$$

(20) We use weak induction.

Base step. For $n = 1$, the result is true since both sides are equal to $1 + x$.

Inductive step. Let $n \geq 1$ be an integer.

Induction hypothesis. Assume that $(1 + x)^n \geq 1 + nx$.

Conclusion. We need to prove that $(1 + x)^{n+1} \geq 1 + (n + 1)x$.

Multiplying the inequality $(1 + x)^n \geq 1 + nx$ by $1 + x$, we get that $(1 + x)^{n+1} \geq (1 + nx)(1 + x)$ since $1 + x > 0$. Now, $(1 + nx)(1 + x) = 1 + x + nx + nx^2 = 1 + (n + 1)x + nx^2 \geq 1 + (n+1)x$. The result follows by the principle of mathematical induction.

(27) $a_3 = 2a_2 - a_1 = 3$, $a_4 = 2a_3 - a_2 = 4$ and $a_5 = 2a_4 - a_3 = 5$. Looks like $a_n = n$ for any $n \geq 1$. We prove this observation using strong induction.

Base step. For $n = 1$ and $n = 2$, the result is true by the given information.

Inductive step. Let $n \geq 2$ be an integer.

Induction hypothesis. Assume that $a_k = k$ pour tout k tel que $1 \leq k \leq n$.

Conclusion. We need to prove that $a_{n+1} = n + 1$. Since $n + 1 \geq 3$, $a_{n+1} = 2a_n - a_{n-1}$. By the induction hypothesis, $a_n = n$ and $a_{n-1} = n - 1$ and therefore, $a_{n+1} = 2n - (n - 1) = n + 1$. The result follows by the principle of mathematical induction.

(31) We use strong induction.

Base step. For $n = 0$, $4^0 - 3^0 = 0 = a_0$. For $n = 1$, $4^n - 3^n = 1 = a_1$. The result is true for $n = 0$ and $n = 1$.

Inductive step. Let $n \geq 1$ be an integer.

Induction hypothesis. Assume that $a_k = 4^k - 3^k$ for any integer k with $0 \leq k \leq n$.

Conclusion. We need to prove that $a_{n+1} = 4^{n+1} - 3^{n+1}$.

$$
\begin{aligned}
a_{n+1} &= 7a_n - 12a_{n-1} \\
&= 7(4^n - 3^n) - 12(4^{n-1} - 3^{n-1}) \text{ (By the induction hypothesis)} \\
&= 7 \cdot 4^n - 7 \cdot 3^n - 3 \cdot 4 \cdot 4^{n-1} - 3 \cdot 4 \cdot 3^{n-1} \\
&= 7 \cdot 4^n - 7 \cdot 3^n - 3 \cdot 4^n - 4 \cdot 3^n \\
&= 4 \cdot 4^n - 3 \cdot 3^n \\
&= 4^{n+1} - 3^{n+1}.
\end{aligned}
$$

(34) We use strong induction.

Base step. $f_3 = f_2 + f_1 = f_0 + 2f_1 = 2$. Since $\left(\frac{1+\sqrt{5}}{2}\right)^{3-2} < 2$, the result is true for $n = 3$. For $n = 4$, note that $f_4 = f_3 + f_2 = 3$ and $\left(\frac{1+\sqrt{5}}{2}\right)^{4-2} = \frac{3+\sqrt{5}}{2} < 3$. The result is also true in this case.

Inductive step. Let $n \geq 4$ be an integer.

Induction hypothesis. Assume that $f_k \geq \left(\frac{1+\sqrt{5}}{2}\right)^{k-2}$ for any integer k with $3 \leq k \leq n$.

Conclusion. We need to prove that $f_{n+1} \geq \left(\frac{1+\sqrt{5}}{2}\right)^{n-1}$.

$$
\begin{aligned}
f_{n+1} &= f_n + f_{n-1} \\
&\geq \left(\frac{1+\sqrt{5}}{2}\right)^{n-2} + \left(\frac{1+\sqrt{5}}{2}\right)^{n-3} \quad \text{(by the induction hypothesis)} \\
&= \left(\frac{1+\sqrt{5}}{2}\right)^{n-3}\left(\frac{1+\sqrt{5}}{2}+1\right) \\
&= \left(\frac{1+\sqrt{5}}{2}\right)^{n-3}\left(\frac{3+\sqrt{5}}{2}\right) \\
&= \left(\frac{1+\sqrt{5}}{2}\right)^{n-3}\left(\frac{1+\sqrt{5}}{2}\right)^{2} \\
&= \left(\frac{1+\sqrt{5}}{2}\right)^{n-1}.
\end{aligned}
$$

(37) The proof follows directly from the well-ordering principle. Assume that the truth of $P(n_0)$ and $P(k)$ for any integer k with $n_0 \leq k \leq n$ imply the truth of $P(n+1)$. We need to prove that $P(n)$ is true for any $n \geq n_0$. Let A be the set of all integers $n > n_0$ such that $P(n)$ is false. If $A \neq \emptyset$, then A must have a smallest element, say m_0, by the well-ordering principal. By the minimality of m_0, we have that $m_0 - 1 \notin A$ and so $P(m_0 - 1)$ is true. So we have that $P(n_0), P(n_0 + 1), \ldots, P(m_0 - 1)$ are all true. By our assumption, $P(m_0)$ must be true. This is a contradiction to the fact that $m_0 \in A$. We conclude that $A = \emptyset$ and $P(n)$ is true for any $n \geq n_0$.

(41) Note that 22 cannot be written under the form $5s + 7t$ for some non-negative integers s and t. On the other hand, $23 = 3 \cdot 5 + 1 \cdot 7$, $24 = 2 \cdot 5 + 2 \cdot 7$, $25 = 5 \cdot 5 + 0 \cdot 7$. Looks like any amount greater than or equal to 23 can be formed using just 5-cent and 7-cent stamps. We prove this using strong induction. Some rough work is needed before we proceed with the formal proof: note that $n + 1 = (n - 4) + 5$. In order to use the induction hypothesis, we must have that $n - 4 \geq 23$ or $n \geq 27$. So, in the base case, we include the values $n = 23, 24, 24, 25, 26$ and 27.

Base step. $23 = 3 \cdot 5 + 1 \cdot 7$, $24 = 2 \cdot 5 + 2 \cdot 7$, $25 = 5 \cdot 5 + 0 \cdot 7$, $26 = 1 \cdot 5 + 3 \cdot 7$ and $27 = 4 \cdot 5 + 1 \cdot 7$.

Inductive step. Let $n \geq 27$ be an integer.

> *Induction hypothesis.* Assume that $n = 5s + 7t$ for some non-negative integers s and t.
>
> *Conclusion.* We need to prove that $n + 1 = 5m + 7n$ for some non-negative integers m and n.

$$n + 1 = (n - 4) + 5$$
$$= (5a + 7b) + 5 \; a, b \in \mathbb{N}$$

(By the induction hypothesis and the fact that $n - 4 \geq 23$)

$$= 5(a + 1) + 7b \; a + 1, b \in \mathbb{N}.$$

3.7

(1)(d) Here are 5 possible elements of S: $0 \in S$, $2 \in S$ (base elements), $0 + 2^2 = 4 \in S$, $0 + 2^0 = 1 \in S$ and $2^2 + 4 = 8 \in S$.

(2)(b) Here are 5 possible elements of S: $0 \in S$ (base element), $00 \in S$, $10 \in S$, $000 \in S$ and $110 \in S$.

(4)   *Base step.* $1, 2 \in S$.

    Recursion step. If $x \in S$, then $x + 5 \in S$.

    Restriction step. Nothing else is in S unless it is obtained from the Basic and Recursive steps.

(5)(b) 1. $1 \in S, 01 \in S$ (Base step)

    2. $1011 \in S$ (Line 1, Recursive step ($x = 1$, $y = 01$, so $xyx = 1011 \in S$)

    3. $1011011011 \in S$ (Line 2, Recursive step ($x = 1011 \in S$, $y = 01 \in S$, so $xyx = 1011011011 \in S$)

(7)(a)   *Base step.* $1 \in S$.

    Recursion step. If $x \in S$, then $0x0 \in S$.

    Restriction step. Nothing else is in S unless it is obtained from the Basic and Recursive steps.

  (b)   *Base step.* $\lambda \in S$.

    Recursion step. If $x \in S$, then $x00 \in S$.

    Restriction step. Nothing else is in S unless it is obtained from the Basic and Recursive steps.

  (c)   *Base step.* $00 \in S$.

    Recursion step. If $x \in S$, then $x0 \in S$.

    Restriction step. Nothing else is in S unless it is obtained from the Basic and Recursive steps.

  (d)   *Base step.* $\lambda \in S$.

    Recursion step. If $x \in S$, then $0x1 \in S$.

Restriction step. Nothing else is in S unless it is obtained from the Basic and Recursive steps.

(8) *Base step.* The empty word λ is in S.
Recursion step. If $x \in S$, and $\alpha \in \Sigma$ then $x\alpha \in S$.
Restriction step. Nothing else is in S except elements in the base step or can be produced by the recursion step.

(11) *Base step.* The empty string λ is in S.
Recursion step. If $x \in S$, then $x00$, $x01$, $x10$ and $x11 \in S$.
Restriction step. Nothing else is in S except elements in the base step or can be produced by the recursion step.

(15) *Base step.* $1 \in S$, $10 \in S$.
Recursion step. If $x \in S$, then $1x00$, $x01$, $x10$ and $x11 \in S$.
Restriction step. Nothing else is in S except elements in the base step or can be produced by the recursion step.

(17) We use structural induction.

Base step. The base element 0 is divisible by 3.
Recursion step. Assume that $x \in \mathbb{Z}$ is divisible by 3. Then $x = 3k$ for some $k \in \mathbb{Z}$ and so $x + 3 = 3(k + 1)$ and $x - 3 = 3(k - 1)$ are both divisible by 3.

(19) We use structural induction.

Base step. Since $0 \leq 2(0)$, the result is true for the base element $(0, 0)$.
Recursion step. Let $(a, b) \in S$ be such that $a \leq b$. Then $a \leq 2(b + 1)$ and $a + 2 \leq 2b + 2 = 2(b + 1)$. This implies that the result is true for the elements $(a, b + 1)$, and $(a + 2, b + 1)$.

We conclude $a \leq 2b$ for any $(a, b) \in S$.

(25) Let S be the subset of $\mathbb{N}$ defined recursively as follows:

Base step. $2 \in S$.
Recursion step. If $x \in S$, then $x + 3 \in S$.
Restriction step. Nothing else is in S except elements in the base step or can be produced by the recursion step.

Let E be the subset of $\mathbb{N}$ of integers having a remainder of 2 upon division by 3. We prove that $E = S$. Note first that a typical element of E is of the form $x = 3k + 2$ for some $k \in \mathbb{N}$. We use weak induction on k to prove the inclusion $E \subseteq S$.

Base step. For $k = 0$, $3k + 2 = 23 \in S$ by the Base step of S.
Inductive step. Let $k \geq 0$ be an integer.

Induction hypothesis. Assume that $3k + 2 \in S$.

Conclusion. We need to prove that $3(k+1) + 2 \in S$. Note that $3(k+1) + 2 = 3k + 2 + 3 \in S$ by the Recursion step of S and the fact that $3k + 2 \in S$ (induction hypothesis).

For the inclusion $S \subseteq E$, we use structural induction.

Base step. $2 \in E$ since the remainder of 2 upon division by 3 is 2.
Recursion step. Let $x \in S$ be also an element of E. So $x = 3k + 2$ for some $k \in \mathbb{N}$. Then $x + 3 = 3(k+1) + 2$ has a remainder of 2 upon division by 3 and so $x + 3 \in E$.

We conclude that $E = S$.

(26) Let S be the subset of $\mathbb{N}$ defined recursively as follows:

Base step. $a \in S$.
Recursion step. If $x \in S$, then $x^a \in S$.
Restriction step. Nothing else is in S except elements in the base step or can be produced by the recursion step.

We prove that $E = S$. Note first that a typical element of E is of the form a^{a^n} for some $n \in \mathbb{N}$. we use weak induction on k to prove the inclusion $E \subseteq S$.

Base step. For $n = 0$, $a^{a^n} = a \in S$ by the Base step of S.
Inductive step. Let $n \geq 0$ be an integer.

Induction hypothesis. Assume that $a^{a^n} \in S$.
Conclusion. We need to prove that $a^{a^{n+1}} \in S$. Note that $a^{a^{n+1}} = \left(a^{a^n}\right)^a \in S$ by the Recursion step of S and the fact that $a^{a^n} \in S$ (induction hypothesis).

For the inclusion $S \subseteq E$, we use structural induction.

Base step. $a \in E$ since $a = a^{a^0}$.
Recursion step. Let $x \in S$ be also an element of E. Then $x = a^{a^n}$ for some $n \in \mathbb{N}$ and so $x^a = \left(a^{a^n}\right)^a = a^{a^{n+1}} \in E$.

We conclude that $E = S$.

4.2

(1)(b) *Joe* and *Patrick* are constants, *"finished his first year with a higher average than"* is a binary predicate.

(3)(b) $P(91)$ is false since there exists no integer k such that $3^k = 91$.

  (f) $P\left(\frac{1}{2187}\right)$ is true since $3^{-7} = \frac{1}{2187}$.

(5)(b) The equation can be rewritten as $(x+1)^2 + (y-1)^2 = 9$. This is the equation of a circle centered at the point $(-1, 1)$ and of radius 3.

(6) The inequality $x^2 + 3x \leq 2x + 6$ is equivalent to $(x+3)(x-2) \leq 0$. The solution set is $-3 \leq x \leq 2$.

   (a) If $\mathcal{U} = \mathbb{N}$, the solution set is $\{0, 1, 2\}$.

   (b) If $\mathcal{U} = \mathbb{N}$, the solution set is $\{-3, -2, -1, 0, 1, 2\}$.

   (c) If $\mathcal{U} = \mathbb{R}$, the solution set is the interval $[-3, 2]$.

(8)(c) True: $D(11, 1)$ is false and so $\neg D(11, 1)$ is true. On the other hand, $D(-2, 23)$ is false. This means that $(\neg D(11, 1) \leftrightarrow D(-2, 23))$ is false and so $\neg\, (\neg D(11, 1) \leftrightarrow D(-2, 23))$ is true.

(9)(c) False since $C(0, \pi)$ reads "$\cos 0 \leq \sin \pi$ which is not true as $\cos 0 = 1 > \sin \pi = 0$.

   (h) True: $A\left(\frac{\pi}{4}\right) = \sin\left(\frac{\pi}{4}\right) \cos\left(\frac{\pi}{4}\right) = \frac{1}{2} > 0$. So $A\left(\frac{\pi}{4}\right)$ is true and thus $\neg A\left(\frac{\pi}{4}\right)$ is false. On the other hand, $C\left(\frac{\pi}{6}, \frac{\pi}{3}\right)$ is false since $\cos\left(\frac{\pi}{6}\right) = \sin\left(\frac{\pi}{3}\right) = \frac{\sqrt{3}}{2}$. Also, $S(0, 0)$ is true since $\sin^0 + \cos^2 0 = 1 \leq \frac{3}{2}$ and so $\neg S(0, 0)$ is false. We conclude that $\left(\neg S(0, 0) \vee C\left(\frac{\pi}{6}, \frac{\pi}{3}\right)\right)$ is false and so $\left(\neg A\left(\frac{\pi}{4}\right) \leftrightarrow \left(\neg S(0, 0) \vee C\left(\frac{\pi}{6}, \frac{\pi}{3}\right)\right)\right)$ is true.

(10)(d) Note that $(\neg P(x) \rightarrow Q(x))$ is logically equivalent to $(\neg\neg P(x) \vee Q(x) \equiv (P(x) \vee Q(x))$. So the required truth set is $T_P \cup T_Q = \mathbb{R}$.

4.3

(1)(a) False: the statement says that there exists a real number x larger than any other real number. Clearly, no such real number exist.

   (d) True: the statement says that for any real number x, there exists a real number y less than or equal to x. For instance, $y = x - 1$ is one possibility.

(2)(b) This is false. For $x = -1 \in \mathcal{U}$, $Q(x)$ is true but $P(x)$ is false, so $(Q(x) \rightarrow P(x))$ is false.

   (f) This is false. For $x = -1$, $\neg P(x)$ is true and $R(x)$ is false which means that the implication $((\neg P(x) \vee Q(y)) \rightarrow R(x))$ is false for any choice of y in $\mathcal{U}$.

(3)(a) This is true, For example $D(1, 1)$ is true. In fact $D(1, x)$ is true for any $x \geq 1$.

   (d) This is false. $D(1, 2)$ is true but $P(1)$ is false.

   (f) This is true. If $x \neq 0$, then x divides other integers. For instance, x divides x.

(5)(c) This is true. Take $x = 2$, then $(\neg P(x) \wedge R(x))$ is true and so $(\exists x)(\neg P(x) \wedge R(x))$ is true. Since $R(3)$ is true, $(R(3) \rightarrow (\exists x)(\neg P(x) \wedge R(x)))$ is true.

   (e) This is false. Note first that $l\left(2^5 + 1, 3^3 + 6\right)$ is false. On the other hand, $(\exists m)(\neg q(m) \wedge r(m))$ is true ($m = 2$ is such an integer). so the biconditional $\left(l\left(2^5 + 1, 3^3 + 6\right) \leftrightarrow (\exists m)(\neg q(m) \wedge r(m))\right)$ is false.

(6)(c) False. Let $m = 0$ then there exists no integer n such that $m^2 + n^2 = 3$ since otherwise $n = \sqrt{3}$ is not an integer.

(7)(b) False. If a real number x exists such that $xy + z = x + yz$ for any real numbers y and z, then setting $y = z = 2$ gives $2x + 2 = x + 4$ and so $x = 2$. Setting $y = z = 0$, we get that $x = 0$. So, on one hand, $x = 0$ and on the other $x = 2$ which is absurd.

(8)(b) False. If x is a real number satisfying $x < \frac{1}{n}$ for any $n \in \mathbb{N}^+$, then taking the limit as n approaches infinity we get that $x < 0$: a contradiction.

(e) True. For any $n \in \mathbb{N}^+$, we can always find a positive real number smaller that $\frac{1}{n}$.

(9)(d)

$$\neg(\exists x)(\exists y)(\forall z)\left((\forall u)(P(u) \vee Q(x)) \rightarrow (P(y) \wedge Q(z))\right)$$
$$\equiv (\forall x)\neg(\exists y)(\forall z)\left((\forall u)(P(u) \vee Q(x)) \wedge (P(y) \rightarrow Q(z))\right)$$
$$\equiv (\forall x)(\forall y)\neg(\forall z)\left((\forall u)(P(u) \vee Q(x)) \wedge (P(y) \rightarrow Q(z))\right)$$
$$\equiv (\forall x)(\forall y)(\exists z)\neg\left((\forall u)(P(u) \vee Q(x)) \wedge (P(y) \rightarrow Q(z))\right)$$
$$\equiv (\forall x)(\forall y)(\exists z)\left(\neg(\forall u)(P(u) \vee Q(x)) \vee \neg(P(y) \rightarrow Q(z))\right)$$
$$\equiv (\forall x)(\forall y)(\exists z)\left((\exists u)\neg(P(u) \vee Q(x)) \vee (P(y) \wedge \neg Q(z))\right)$$
$$\equiv (\forall x)(\forall y)(\exists z)\left((\exists u)(\neg P(u) \wedge \neg Q(x)) \vee (P(y) \wedge \neg Q(z))\right)$$

(10)(b) Let $\phi : (\exists x)(\forall y)\left((\forall t)P(x,t) \vee Q(y,t)\right)$. Then:

$$\neg\phi \equiv (\forall x)\neg(\forall y)\left((\forall t)P(x,t) \vee Q(y,t)\right)$$
$$\equiv (\forall x)(\exists y)\neg\left((\forall t)P(x,t) \vee Q(y,t)\right)$$
$$\equiv (\forall x)(\exists y)\left(\neg(\forall t)P(x,t) \wedge \neg Q(y,t)\right)$$
$$\equiv (\forall x)(\exists y)\left((\exists t)\neg P(x,t) \wedge \neg Q(y,t)\right)$$

(11)(b)

$$\neg(\exists n)(\forall m)\left((m < n) \rightarrow (\exists k)(n = m + k)\right)$$
$$\equiv (\forall n)\neg(\forall m)\left((m < n) \rightarrow (\exists k)(n = m + k)\right)$$
$$\equiv (\forall n)(\exists m)\neg\left((m < n) \rightarrow (\exists k)(n = m + k)\right)$$
$$\equiv (\forall n)(\exists m)\left((m \geq n) \wedge \neg(\exists k)(n = m + k)\right)$$
$$\equiv (\forall n)(\exists m)\left((m \geq n) \wedge (\forall k)\neg(n = m + k)\right)$$
$$\equiv (\forall n)(\exists m)\left((m \geq n) \wedge (\forall k)(n \neq m + k)\right)$$

(14)(a) The following gives a step-by-step proof that ϕ is a wff.

    (a) $A(z)$, $B(y)$, $C(x,z)$ and $D(x,z)$ are all wffs by rule (1).

    (b) $(\forall z)C(x,z)$ is a wff by line 1 and rule (3.1).

    (c) $((\forall z)C(x,z) \to B(y))$ is a wff by lines 1 and 2 and rule (2.4).

    (d) $(\forall y)((\forall z)C(x,z) \to B(y))$ is a wff by line 3 and rule (3.1).

    (e) $(\exists x)(\forall y)((\forall z)C(x,z) \to B(y))$ is a wff by line 4 and rule (3.2).

    (f) $(\forall x)D(x,z)$ is a wff by line 1 and rule (3.1).

    (g) $(A(z) \leftrightarrow (\forall x)D(x,z))$ is a wff by lines 1 and 6 and rule (2.5).

    (h) $(\exists z)(A(z) \leftrightarrow (\forall x)D(x,z))$ is a wff by line 7 and rule (3.2).

    (i) $((\exists x)(\forall y)((\forall z)C(x,z) \to B(y)) \wedge (\exists z)(A(z) \leftrightarrow (\forall x)D(x,z)))$ is a wff by lines 5 and 8 and rule (2.2).

  (b) $(\forall z)C(x,z)$, $(\forall x)D(x,z)$, $((\forall z)C(x,z) \to B(y))$ and $(A(z) \leftrightarrow (\forall x)D(x,z))$ are four non-atomic subformulas of ϕ.

4.4

(1)(a) Joe has a son.

  (d) If Mich is a son of Joe, then there exists someone who is a son of Joe.

(2)(b) $(R(j,c) \to (\forall x)R(j,x))$.

  (d) $(\forall x)(\exists y)R(x,y)$.

(3)(a) $(\exists x)(R(x) \wedge I(x))$.

  (e) $(\exists x)(R(x) \wedge ((\forall y)((C(y,x) \to G(y,-2))))$.

  (h) $\neg(\exists x)(\exists y)((\exists n)(I(n) \wedge C(x,n)) \wedge (\exists q)(R(q) \wedge C(y,q)) \wedge G(x,y))$.
Another equivalent way is the following:
$(\forall x)(\forall y)(((\exists n)(I(n) \wedge C(x,n)) \wedge (\exists t)(R(t) \wedge C(y,t))) \to \neg G(x,y))$

(4)(a) No dear can hunt after Simba.

  (d) Yogi is the only bear who can hunt any lion.

  (f) There exists an animal that can be hunted by any lion that can outrun it.

  (h) There is a bear other than Yogi that can hunt any animal as long as it can outrun it.

(5) We use the following predicates: $E(x)$: "x is even", $O(x)$: "x is odd", $P(x)$: "x is prime", $D(x,y)$: "x divides y."

  (a) $(\forall n)(P(n) \to O(n))$.

  (d) $(\forall n)((P(n) \wedge (n \neq 2)) \to O(n))$.

  (g) $(\forall n)(P(n) \to (\forall m)(D(m,n) \to ((m = 1) \vee (m = n))))$

(5) (b)

(9) (d)

(11) The statement says that on every row, we can find an entry which is either less than or equal to -5 or is the square of a natural number. Matrices C, E and F satisfy this condition.

(13)(a) There exist two stars adjacent in the same row. This is false.

(d) No triangle is adjacent to a square in a column. This is false.

4.5

(1)(e)

Quantifier	Scope
$(\exists x)$	$(\exists x)(\forall y)(A(x) \vee (\exists z)(A(y) \vee B(z)))$
$(\forall y)$	$(\forall y)(A(x) \vee (\exists z)(A(y) \vee B(z)))$
$(\exists z)$	$(\exists z)(A(y) \vee B(z))$

There are two occurrences of the variable x and they are both bound by $(\exists x)$. There are two occurrences of the variable y and they are both bound by $(\forall y)$. There are two occurrences of the variable z and they are both bound by $(\exists z)$. The wff is closed since it contains no free variables.

(g)

Quantifier	Scope
$(\forall x)$	$(\forall x)(\forall y)(\neg B(x) \wedge A(y))$
$(\forall y)$	$(\forall y)(\neg B(x) \wedge A(y))$
$(\exists x)$	$(\exists x)A(x)$

There are four occurrences of the variable x. The first two are bound by $(\forall x)$ and the last two bound by $(\exists x)$. There are three occurrences of the variable y. The first two are bound by $(\forall y)$ and the third is free. The wff is open since it contains a free variable.

(2)(b)

Quantifier	Scope
$(\forall x)$	$(\forall x)(A(x, y) \rightarrow (\forall z)((\exists x)B(z, x) \wedge A(x, z)))$
$(\forall z)$	$(\forall z)((\exists x)B(z, x) \wedge A(x, z))$
$(\exists x)$	$(\exists x)B(z, x)$

There are five occurrences of the variable x. The first two and the last are bound by $(\forall x)$. The third and fourth occurrences are bound by $(\exists x)$. So all occurrences of x are bound and x is a bound variable. There is only one occurrence of the variable y and it is a free occurrence. The variable y is free. There are three occurrences of the variable z and they are all bound by $(\forall z)$. The variable z is bound. The wff is open since it contains a free variable.

(4) (b) $\phi(x/t) = \phi$, $\phi(y/t) = (P(a) \rightarrow (\forall x)(Q(x,z) \rightarrow Q(t,z)))$ and
$\phi(x/t) = (P(a) \rightarrow (\forall x)(Q(x,t) \rightarrow Q(y,t)))$.

(d) $\phi(x/t) = ((\forall x)(A(x) \rightarrow B(x,y)) \rightarrow ((\exists x)A(x) \rightarrow B(t,z)))$,
$\phi(y/t) = ((\forall x)(A(x) \rightarrow B(x,t)) \rightarrow ((\exists x)A(x) \rightarrow B(x,z)))$ and
$\phi(z/t) = ((\forall x)(A(x) \rightarrow B(x,y)) \rightarrow ((\exists x)A(x) \rightarrow B(x,t)))$.

4.6

(1)(a) ϕ is false in this interpretation. For instance, take $x = 0 \in \mathbb{R}$. Then $(\exists y)P(x,y)$ means: "there exists $y \in \mathbb{R}$ such that $x + y = 1$" which is true ($y = 1$). On the other hand, the statement $Q(f(x,2),-1)$ means $(2x)^2 - 1 = 3$ or $x = \pm 1$. This is false since $x = 0$.

(2)(b) ϕ is true in this interpretation. It states the following: *There exists a country x such that for any country y in the same continent as x and that shares a land border with x, y has a smaller area than Canada.* This is true for any country x since Russia is the only country with a larger area than Canada, but in this case the implication $((P(x,y) \wedge E(x,y)) \rightarrow Q(y,c))$ remains true since its hypothesis is false.

(3)(a) False. If ϕ is satisfiable, then it is true for some interpretation but it could be false in some other interpretation. For example, the wff $(\forall x)P(x)$ is true in the interpretation with universe $\mathcal{U} = \{2^n;\ n$ is a positive integer$\}$ and $P(x)$: "x is even". But it is false in the interpretation with universe $\mathcal{U} = \mathbb{N}$ and $P(x)$: "x is even". So $(\forall x)P(x)$ is satisfiable but invalid.

(e) True since if there exists an interpretation of ϕ which is a model of ϕ, then ϕ would be satisfiable. "x is even" is a countermodel of ϕ.

(4)(b) Let I be an arbitrary model of the wff $(\exists x)(P(x) \wedge Q(x))$ with universe $\mathcal{U}$. Then there exists an element $a \in \mathcal{U}$ such that $(P(a) \wedge Q(a))$ is true. That implies that there exists an element $a \in \mathcal{U}$ such that $P(a)$ is true and $Q(a)$ is true. In other words, I is a model for $((\exists x)P(x) \wedge (\exists x)Q(x))$. We conclude that $((\exists x)(P(x) \wedge Q(x)) \rightarrow ((\exists x)P(x) \wedge (\exists x)Q(x)))$ is valid.

(5)(d) Let I be the interpretation with universe $\mathcal{U} = \mathbb{R}$, $P(x,y)$: "$xy^2 \geq 0$", $f(x,y) = xy$ and $Q(x,y)$: "$x^2 + y^2 = 1$" and $y = 1$ y is a free variable) Then $\neg P(f(x,x),y)$ means: "$x^2 < 0$" which is false for any $x \in \mathbb{R}$. This means that the implication $(\neg P(f(x,x),y) \rightarrow Q(x,y))$ is true for any $x \in \mathbb{R}$. The interpretation I is then a model for $(\forall x)(\neg P(f(x,x),y) \rightarrow Q(x,y))$.

(6)(a) Model: $\mathcal{U} = \{1\}$ (singleton), $M(x)$: "x is even", $P(x, y)$: "$x + y \leq 2$. Countermodel: $\mathcal{U} = \{1, 2\}$ (singleton), $M(x)$: "x is even", $P(x, y)$: "$x + y = 0$.

4.7

(1)(c) Let $x = y = 1$, then $L(x, y) = 1$ and the statement is true.

(e) Let $x = 1$, then for any integer y, x is a divisor of y and so $D(x, y)$ is true.

(g) Let x, y be two integers. Assume that x is prime and that x is not a divisor of y. If d is any common divisor of x and y, then d is equal to either 1 or x (as being a divisor of x). But since x is not a divisor of y, $d = 1$. So $\gcd(x, y) = 1$ and $G(x, y)$ is true.

(3) The statement $(\forall a)\left((\exists k)\left(a = 3^{2^k}\right) \to (\exists b)\left((a < b) \wedge P(b)\right)\right)$ says that for any integer of the form $a = 3^{2^k}$ for some integer k, we can find a prime number b greater than a. So, it is enough to prove that the integer 1853020188851841 is of the form $a = 3^{2^k}$ for some integer k. Trial and error shows that $1853020188851841 = 3^{2^5}$ and the result follows. Another way to proceed is to write $3^{2^k} = 1853020188851841$ and then take ln on both sides: $2^k = \frac{\ln(1853020188851841)}{\ln(3)}$. Taking ln a second time gives $k = 5$.

(4)(c) Choose $x = -1$ (or any negative real number). Then for any real number y, if $y > 0$ then automatically $y > x$ and so the conjunction $((y > 0) \wedge (x \geq y))$ is false. In particular $(((y > 0) \wedge (x \geq y)) \to (y > y^2))$ is true.

(5)(a) The statement says that there exists a *unique* integer x such that $100 \leq x^2$ which is false since, for example, both $x = 10$ and $x = 11$ satisfy that condition.

(8) The formal proof of the argument is the following

1. $(\forall x)(\neg P(x) \to Q(x))$	Premise
2. $(\exists x)\neg P(x)$	Premise
3. $(\forall x)(Q(x) \to R(x))$	Premise
4. $(\forall x)(S(x) \to \neg R(x))$	Premise
5. $\neg P(c),\ c \in \mathcal{U}$	(EI), Line 2
6. $(\neg P(c) \to Q(c))$	(UI), Line 1
7. $Q(c)$	Modus ponens, Lines 5, 6
8. $(Q(c) \to R(c))$	(UI), Line 3
9. $R(c)$	Modus ponens, Lines 7, 8
10. $(S(c) \to \neg R(c))$	(UI), Line 4
11. $\neg S(c)$	Modus tollens, Lines 9, 10
10. $(\exists x)\neg S(x)$	(EG), Line 11

5.2

(1)(a) This is a function since every University student is assigned a unique student number. The domain is the set of all students at Oxford University (past and present), the codomain is $\mathbb{N}$.

(e) The rule is not a function since the elements -1 and 1 have no image. If we restrict the rule to the domain $]-\infty, -1[\cup]-1, 1[\cup]1, \infty[$ then we obtain a function.

(g) The rule is not a function since every element has two different images. If we assign 2^x to every real number x, then we get a function.

(2)(a) False. The real numbers $-\sqrt{3}, \sqrt{3}$ don't have images.

(d) False. For example, $f(-1) = \frac{1}{2} \notin \mathbb{Z}$.

(f) False. The element $\sqrt{3} \in]-\sqrt{3}, \sqrt{3}]$ and $f\left(\sqrt{3}\right)$ is not a real number.

(5)(b) This is a well-defined function. To see this, let $\frac{a}{b}$, $\frac{c}{d}$ be two rational numbers such that $\frac{a}{b} = \frac{c}{d}$. Then $ad = bc$ and consequently $(ad)^3 = (bc)^3$ which is equivalent to $a^3 d^3 = b^3 c^3$. This implies that $\frac{a^3}{b^3} = \frac{c^3}{d^3}$ or $f\left(\frac{a}{b}\right) = f\left(\frac{c}{d}\right)$.

(d) This is a well-defined function. Clearly, a family member can have more than one hobby.

(7)(c) This is a partial function. For instance, $f(0) = -4 \notin \mathbb{N}$. The domain of f is the set of all integers greater than or equal to 4.

(e) This is a partial function. For instance, $f(1) = \left(\frac{1}{3}, \frac{1}{4}\right) \notin \mathbb{Z} \times \mathbb{Z}$. The domain of f is the set of all integers n such that both $\frac{n}{3}$ and $\frac{n}{4}$ are integers. That is equivalent to say that n is a multiple of 12.

(9) Let $y = f(x) = \frac{-x+2}{3x-2}$. Then $-x + 2 = y(3x - 2)$. Isolating x in this equation, we get that $x = \frac{2(1+y)}{1+3y}$. In order for y to be a real number (and so the image is defined), $y \neq -\frac{1}{3}$. In other words, every real number is an image except $-\frac{1}{3}$. We conclude that $\text{Im}(f) = \mathbb{R} \setminus \{-\frac{1}{3}\}$.

(13) Consider the function $f : \mathbb{R} \to \mathbb{Z}$ defined piecewise as follows:

$$f(x) = \begin{cases} 1 & \text{for } x \leq 0 \\ 2 & \text{for } 0 < x \leq 1 \\ 3 & \text{for } x \geq 1 \end{cases}$$

Clearly, the image of f is the set $\{1, 2, 3\}$.

(14)(b) True: $h(-2) = 10 = f(-2)$, $h(-1) = 7 = f(-1)$, $h(0) = 4 = f(0)$, $f(3) = -5 = f(3)$ and $h(4) = -8 = f(4)$.

(e) False: $k(-2) = 0 \neq f(-2)$.

(15) Any extension of g to a function $f : A \to B$ must fix the same values as g for the elements a, b and c of A. Therefore, we must only assign

values for $f(d)$ and $f(e)$ in B. There are six possible choices for $f(d)$ and for each of these choices, there are six possible choices for $f(e)$. Therefore, there are $6 \times 6 = 36$ possible extensions of g.

(19) Let $a \in A$. Then:

$$
\begin{aligned}
a \in f^{-1}(T_1 \cup T_2) &\Leftrightarrow f(a) \in T_1 \cup T_2 \\
&\Leftrightarrow f(a) \in T_1 \text{ or } f(a) \in T_2 \\
&\Leftrightarrow a \in f^{-1}(T_1) \text{ or } a \in f^{-1}(T_2) \\
&\Leftrightarrow a \in f^{-1}(T_1) \cup f^{-1}(T_2).
\end{aligned}
$$

This shows that $f^{-1}(T_1 \cup T_2) = f^{-1}(T_1) \cup f^{-1}(T_2)$.

(24)(d) False:

$$
k(x) = (x-1)\left(U(x-1) - U(x-3)\right) + xU(x-3) = (x-1)U(x-1) + U(x-3).
$$

(25)(c) This is an odd function: $f_3(-x) = -2(-x)^3 + 4(-x) = 2x^3 - 4x = -f_3(x)$.

(g) This is an even function: $f_7(-x) = \frac{|-x|}{|-x|^3+1} = \frac{|x|}{|x|^3+1} = f_7(x)$.

(26)(a) True: For any $x \in \mathbb{R}$, $(f+g)(-x) = f(-x) + g(-x) = f(x) + g(x) = (f+g)(x)$.

(g) False: $f(x) = x^2$ is even and $g(x) = x$ is odd. Then $(f+g)(x) = x^2+x$ is not even since $f(-1) = 0$ but $f(1) = 2$.

5.3

(3)(e) 5

(i) 2

(4) (a) $n+1$

(6) Let A, B be two subsets of a universal set $\mathcal{U}$. Let $x \in \mathcal{U}$. If $\chi_{A \cup B}(x) = 0$, then $x \notin A \cup B$ and therefore $x \notin A$ and $x \notin B$. Consequently, $\chi_A(x) = \chi_B(x) = 0$ and $\chi_A(x) + \chi_B(x) - \chi_A(x) \cdot \chi_B(x) = 0$. Conversely, assume that $\chi_A(x) + \chi_B(x) - \chi_A(x) \cdot \chi_B(x) = 0$. Then $\chi_A(x) + \chi_B(x) = \chi_A(x) \cdot \chi_B(x)$. If $\chi_A(x) \cdot \chi_B(x) = 1$, then $\chi_A(x) = \chi_B(x) = 1$ and therefore $\chi_A(x) + \chi_B(x) = 2$: a contradiction. Therefore, $\chi_A(x) \cdot \chi_B(x) = 0$ and consequently $\chi_A(x) + \chi_B(x) = 0$. But the latter equation is only possible if $\chi_A(x) = \chi_B(x) = 0$, or equivalently, $x \notin A \cup B$ and therefore $\chi_{A \cup B}(x) = 0$. We conclude that $\chi_{A \cup B} = 0 \Leftrightarrow \chi_A + \chi_B - \chi_A \cdot \chi_B = 0$ and the result follows.

(10) If $x \in \mathbb{Z}$, then $\lfloor x \rfloor = x$ and $\lfloor -x \rfloor = -x$. Therefore, $\lfloor x \rfloor + \lfloor -x \rfloor = 0$ in this case. If $x \notin \mathbb{Z}$, then $\lfloor x \rfloor \le x < \lfloor x \rfloor + 1$. This implies that $-\lfloor x \rfloor - 1 < -x \le -\lfloor x \rfloor$ and therefore, $\lfloor -x \rfloor = -\lfloor x \rfloor - 1$. We conclude

that $\lfloor x \rfloor + \lfloor -x \rfloor = -1$ in this case. In conclusion:

$$\lfloor x \rfloor + \lfloor -x \rfloor = \begin{cases} 0 & \text{if } x \in \mathbb{Z} \\ -1 & \text{if } x \notin \mathbb{Z} \end{cases}$$

(14) Let $m = \lfloor x \rfloor$. Then $m \leq x < m + 1$. Adding n to all sides, we get: $m+n \leq x+n < m+n+1$. This implies that $\lfloor x+n \rfloor = m+n = \lfloor x \rfloor + n$.

(15) Find each of the following integers.

  (b) 6 since $-85 = -7(13) + 6$

  (e) 53 since $2211 = (53)(41) + 38$

(17)(c) First notice that $\frac{3}{7} = 0.42857142857...$ with a periodical part consisting of six digits. Since $1543 \pmod 6 = 1$, the answer is 2.

5.4

(1)(a) This is not a binary operation on $\mathbb{N}$ since, for instance, $2 \star 3 = -1 \notin \mathbb{N}$.

  (e) This is not a binary operation on $\mathbb{R} \backslash \mathbb{Q}$ since, for instance, $\sqrt{2}\sqrt{2} = 2 \notin \mathbb{R} \backslash \mathbb{Q}$.

(2)(c) S is not closed for the binary operation: $12 \star 39 = 156 \notin S$.

  (e) S is closed for the binary operation: If $x, y, a, b \in \mathbb{Z}$, then $(x + a, y + b) \in \mathbb{Z} \times \mathbb{Z}$.

(3)(a) The operation is commutative: $a \star b = \frac{ab}{3} = \frac{ba}{3} = b \star a$. It is associative: $(a \star b) \star c = \frac{ab}{3} \star c = \frac{\frac{ab}{3}c}{3} = \frac{abc}{9}$. On the other hand, $a \star (b \star c) = a \star \frac{bc}{3} = \frac{a\frac{bc}{3}}{3} = \frac{abc}{9}$. So $(a \star b) \star c = a \star (b \star c)$. The operation has $e = 3$ as an identity element: For any $a \in \mathbb{Q}$, $a \star 3 = \frac{3a}{3} = a$.

  (f) The operation is commutative: $a \star b = a+b+2 = b+a+2 = b \star a$. It is associative: $(a \star b) \star c = (a+b+2) \star c = a+b+2+c+2 = a+b+c+4$. On the other hand, $a \star (b \star c) = a \star (b+c+2) = a+b+c+2+2 = a+b+c+4$. So $(a \star b) \star c = a \star (b \star c)$. The operation has $e = -2$ as an identity element: For any $a \in \mathbb{Q}$, $a \star -2 = a + (-2) + 2 = a$.

(9)(b) Let $S \in \wp(A)$. If S is invertible with respect to $\cup$, then there exists $X \in \wp(A)$ such that $S \cup X = X \cup S = \emptyset$ (since $\emptyset$ is the identity element of $\cap$). But that is only possible if $X = S = \emptyset$. Therefore, $\emptyset$ is the only invertible element of $\wp(A)$ with respect to the $\cup$.

(11) Assume $a \in A$ is invertible. Then a^{-1} is the unique element of A satisfying: $a \star a^{-1} = a^{-1} \star a = e$. By the definition of the inverse element, a^{-1} is invertible and $(a^{-1})^{-1} = a$.

(12) Let $A = \{1, 2, 3, 4, 5\}$. Consider the binary operation $\star$ on A defined by $x \star y = \max\{x, y\}$, where $\max\{x, y\}$ stands for the maximum of x and y.

(a) The Cayley Table is as follows:

$\star$	1	2	3	4	5
1	1	2	3	4	5
2	2	2	3	4	5
3	3	3	3	4	5
4	4	4	4	4	4
5	5	5	5	5	5

(c) Let $x, y, z \in A$. We need to prove that $\max(\max(x, y), z) = \max(x, \max(y, z))$. Notice that there are six possible cases to consider depending on the relative positions of the three integers: $x \le y \le z$, $x \le z \le y$, $y \le x \le z$, $y \le z \le x$, $z \le x \le y$, and $z \le y \le x$. We prove the result in the case $x \le y \le z$. The others cases are proved similarly. Since $x \le y \le z$, $\max(\max(x, y), z) = \max(y, z) = z$ and $\max(x, \max(y, z)) = \max(x, z) = z$.

(d) For each element of A, determine if it is invertible, and if it is give its inverse.

(17)(a) From the table, we have that $(2 \star 2) \star 3 = 1 \star 3 = 3$. On the other hand, $2 \star (2 \star 3) = 2 \star 1 = 2$. We conclude that $(2 \star 2) \star 3 \ne 2 \star (2 \star 3)$ and the operation $\star$ is not associative.

(b) From the table, we can see that 1 is an identity element.

(e) Both 2 and 3 have two inverses each.

(18)(a) Let $x, y \in A$. There are two possibilities: $x + y < n$ and $x + y \ge n$. If $x + y < n$, then $x \star y = x + y \in A$. If $x + y \ge n$, then $x \star y = x + y - n$. Note that since $0 \le x \le n - 1$ and $0 \le y \le n - 1$, $0 \le x + y \le 2n - 2$. So if $x + y - n > n - 1$, then $x + y > 2n - 1$: a contradiction. So $x \star y \in A$ for any $x, y \in A$ and $\star$ is a binary operation on A.

(b) For any $x \in A$, since $x + 0 = x < n$, we have that $x \star 0 = x + 0 = x$. So 0 is an identity element of $\star$.

(e) Let $a \in A$. Since $a + (n - a) = n \ge n$, $a \star (n - a) = a + (n - a) - n = 0$. This proves that a is invertible and $n - a$ is an inverse of a.

5.5

(1)(a) This is not a valid recursively defined function since $f(1) = 2f(-2) + 1$ and $f(-2)$ is not defined. In other words, both steps are not enough to compute $f(1)$ and therefore f is not a well-defined function on $\mathbb{N}$.

(e) This is a valid recursively defined function since we can determine the value of $f(n)$ for any $n \ge 0$.

(3) $f(2) = \frac{f(1)}{f(0)} + 3(2) = \frac{3}{1} + 6 = 9$. To find $f(4)$, we first need to find $f(3)$:

$f(3) = \frac{f(2)}{f(1)} + 3(3) = \frac{9}{3} + 9 = 12$. So $f(4) = \frac{f(3)}{f(2)} + 3(4) = \frac{12}{9} + 12 = \frac{40}{3}$.

To find $f(6)$, we need $f(5)$ first: $f(5) = \frac{f(4)}{f(3)} + 3(5) = \frac{\frac{40}{3}}{12} + 15 = \frac{145}{9}$.

So $f(6) = \frac{f(5)}{f(4)} + 3(6) = \frac{\frac{145}{9}}{\frac{40}{3}} + 18 = \frac{461}{24}$. Finally: $f(7) = \frac{f(6)}{f(5)} + 3(7) =$

$\frac{\frac{145}{9}}{\frac{40}{3}} + 18 = \frac{25743}{1160}$.

(4)(b) Note that $f(0) = 5^0 = 1$ and $f(n+1) = 5^{n+1} = 5(5^n) = 5f(n)$.
Consider the function $g : \mathbb{N} \to \mathbb{N}$ defined recursively as follows.

> *Base step.* $g(0) = 1$.
> *Recursion step.* If $n \geq 1$, then $g(n) = 5g(n-1)$.

We prove that $f(n) = g(n)$ for any $n \geq 0$ using induction on n. For $n = 0$, $g(0) = 1$ by definition and $f(0) = 1$ so the base step is true. For the inductive step, let $n \geq 0$ and assume that $f(n) = g(n)$. We need to prove that $f(n+1) = g(n+1)$. By the recursive definition of g, $g(n+1) = 5g(n) = 5f(n)$ (by the induction hypothesis) $= 5(5^n) = 5^{n+1} = f(n+1)$. By the principle of induction, $f(n) = g(n)$ for any $n \geq 0$ and the above recursive definition of f is indeed correct.

(f) Note that $f(0) = 0$ and $f(n+1) = 3^{n+1} - 2^{n+1} = 3(3^n) - 2(2^n) = 3(3^n - 2^n) + 3(2^n) - 2(2^n) = 3f(n) + 2^n$. Consider the function $g : \mathbb{N} \to \mathbb{N}$ defined recursively as follows.

> *Base step.* $g(0) = 0$.
> *Recursion step.* If $n \geq 0$, then $g(n+1) = 3g(n) + 2^n$.

We prove that $f(n) = g(n)$ for any $n \geq 0$ using induction on n. For $n = 0$, $g(0) = 0$ by definition and $f(0) = 0$ so the base step is true. For the inductive step, let $n \geq 0$ and assume that $f(n) = g(n)$. We need to prove that $f(n+1) = g(n+1)$. By the recursive definition of g, $g(n+1) = 3g(n) + 2^n = 3(3^n - 2^n) + 2^n$ (by the induction hypothesis) $= 3^{n+1} - 3(2^n) + 2^n = 3^{n+1} - 2(2^n) + 2^n = 3^{n+1} - 2^{n+1} = f(n+1)$. By the principle of induction, $f(n) = g(n)$ for any $n \geq 0$ and the above recursive definition of f is indeed correct.

(6) Trying some values for $n \geq 0$, it looks like $f(n) = 2n + 5$ for any $n \geq 0$. Let $g : \mathbb{N} \to \mathbb{N}$ defined by $g(n) = 2n + 5$. We use mathematical induction to prove that $g(n) = f(n)$ for any $n \geq 0$. The base case is clear since $g(0) = f(0) = 5$. Let $n \geq 0$ and assume that $f(n) = g(n)$. Then $f(n+1) = 2 + f(n)$ (by the recursive definition of f) $= 2 + g(n)$ (by the induction hypothesis) $= 2 + (2n + 5) = 2(n+1) + 5 = g(n+1)$. By the principle of induction, $f(n) = g(n) = 2n + 5$ for any $n \geq 0$.

(8) Let $B(n)$ be the balance in Joe's bank account at the end of year n. Then at the end of the following year (year $n + 1$), the balance is $B(n + 1) = B(n) + 0.11B(n)$ where $0.11B(n)$ represents the interest accumulated at the end of year $n+1$. So $B(n+1) = B(n)+0.11B(n) = 1.11B(n)$. In addition, $B(1) = 5000.00$.

(10)(c) $A(0, 1) = 1 + 1 = 2$

(g) $A(1, 3) = A(0, A(1, 2)) = A(1, 2) + 1 = A(0, A(1, 1)) + 1 + 1 = A(1, 1) + 2 = A(0, A(1, 0)) + 2 = (A(1, 0) + 1) + 2 = A(1, 0) + 3 = A(0, 1) + 3 = (1 + 1) + 3 = 5.$

(11) For $n = 0$, $A(1, n) = A(0, 1) = 1 + 1 = 2$ and $n + 2 = 2$. The base case is true. Let $n \leq 0$ and assume that $A(1, n) = n + 2$. Then $A(1, n + 1) = A(0, A(1, n)) = A(1, n) + 1 = (n + 2) + 1$ (by the induction hypothesis) $= (n + 1) + 2$. By the induction principle, $A(1, n) = n + 2$ for $n \geq 0$.

(14)(c)

$$f(n) = \begin{cases} 1 & \text{if } n = 0 \\ f(n - 1) + 2n + 1 & \text{if } n \geq 1 \end{cases}$$

(i) Here are the values of $f(n)$ for $n = 0, 1 \ldots, 8$: 1, 4, 9, 16, 25, 36, 49, 64, 81.

(ii) Here are the values of $g(n)$ for $n = 0, 1 \ldots, 7$: 3, 5, 7, 9, 11, 13, 15, 17.

(iii) Here are the values of $g(n)$ for $n = 0, 1 \ldots, 6$: 2, 2, 2, 2, 2, 2 2. We notice that $g(n) = 2$ for any $n \geq 0$.

(iv) From parts (a) and (b), a suitable suggestion for a general closed-form expression for $f(n)$ is a second degree polynomial: $f(n) = An^2 + Bn + C$ for some $A, B, C \in \mathbb{N}$.

(v) Since we have three unknowns (A, B and C), we use the three equations: $f(0) = 1$, $f(1) = 4$ and $f(2) = 9$ to determine the values of these three coefficients. From $f(0) = 1$, we get that $C = 1$. From $f(1) = 4$, we get: $A + B + C = 4$ or $A + B = 3$. From $f(2) = 9$, we get: $4A + 2B + C = 9$ or $2A + B = 4$. Combining the two last equations, we obtain that $A = 1$ and $B = 2$. So a good candidate for a closed-form expression of the function is the expression $n^2 + 2n + 1 = (n + 1)^2$.

(vi) Consider the function $g(n) = (n + 1)^2$ on $\mathbb{N}$. We prove (using induction) that $f(n) = g(n)$ for any $n \geq 0$. The property is clearly true for $n = 0$. Let $n \geq 0$ and suppose that $f(n) = g(n)$. Then $f(n + 1) = f(n) + 2(n + 1) + 1 = g(n) + 2n + 3$ (by the induction hypothesis) $= (n + 1)^2 + 2n + 3 = n^2 + 4n + 4 =$

$(n + 2)^2 = g(n + 1)$. We conclude that $f(n) = g(n)$ for all $n \in \mathbb{N}$ and the expression $(n + 1)^2$ is a closed-form formula for the function f.

5.6

(2) Here are all the bijective functions $f : A \to B = A$:

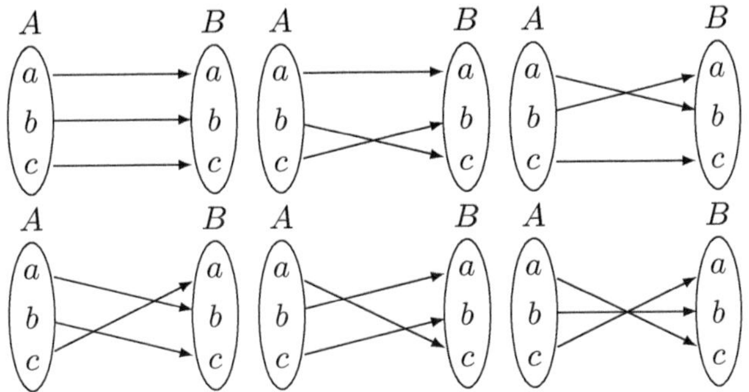

(5)(a) This is a bijective function. If $m, n \in \mathbb{Z}$ are such that $f(m) = f(n)$, then $m - 5 = n - 5$ and consequently, $m = n$. The function is injective. If $m \in \mathbb{Z}$, let $n = m + 5 \in \mathbb{Z}$. Then $f(n) = (m + 5) - 5 = m$ and so f is surjective.

(f) This is a bijective function. Let $m, n \in \mathbb{Z}$ be such that $f(m) = f(n)$, then $(-1)^m + m = (-1)^n + n$. If m is odd and n is even, the latter equation becomes $-1 + m = 1 + n$ or $m = n + 2$ which means that m is even: a contradiction. We also obtain a contradiction if m is even and n is odd. So the equation $(-1)^m + m = (-1)^n + n$ is only possible if m and n have the same parity (both even or both odd. If both m and n are even, the equation $(-1)^m + m = (-1)^n + n$ becomes $1 + m = 1 + n$ which implies that $m = n$. If both m and n are odd, the equation $(-1)^m + m = (-1)^n + n$ becomes $-1 + m = -1 + n$ which also implies that $m = n$. This shows that f is injective. For the surjectivity, fix an integer $m \in \mathbb{Z}$. If m is even, let $n = m + 1$ then $m + 1$ is odd and $f(n) = (-1)^{m+1} + m + 1 = -1 + m + 1 = m$. If m is odd, let $n = m - 1$ then $m - 1$ is even and $f(n) = (-1)^{m-1} + m - 1 = 1 + m - 1 = m$. This shows that any $m \in \mathbb{Z}$ is an image and f is surjective. We conclude that f is bijective.

(7)(c) This is not injective: $f(-1) = f(1) = 0$. Fix $y \in \mathbb{R}$. Assume there exists $x \in \mathbb{R}$ such that $f(x) = 1$. If $x \geq 0$, then $x - x^2 = 1$ or $x^2 - x + 1 = 0$. This is a contradiction since the latter equation has no real roots. If $x < 0$, then $-x - x^2 = 1$ or $x^2 + x + 1 = 0$. This is also a contradiction since the latter equation has no real roots. We conclude that $1 \in \mathbb{R}$ is not an image and the function is not surjective.

(d) The function is injective: If $a, b \in \mathbb{R}$ are such that $f(a) = f(b)$ then $\sqrt[3]{a}+1 = \sqrt[3]{b}+1 \Rightarrow \sqrt[3]{a} = \sqrt[3]{b} \Rightarrow a = b$. If $y \in \mathbb{R}$, let $x = (y-1)^3 \in \mathbb{R}$. Then $f(x) = \sqrt[3]{x} + 1 = \sqrt[3]{(y-1)^3} + 1 = y - 1 + 1 = y$. So f is surjective. We conclude that f is bijective.

(8)(c) The function is not injective: $f(2) = 2$ and $f(5) = \frac{5-1}{2} = 2$. The function is surjective. To see this, let $m \in \mathbb{Z}$ fixed. If m is even, then $m = f(m)$. If m is odd, let $n = 2m + 1 \in \mathbb{Z}$, then $f(n) = \frac{2m+1-1}{2}$ (since $2m + 1$ is odd) $= m$. So every element in $\mathbb{Z}$ is an image and the function is surjective.

(9)(a) The function is injective: If $w_1, w_2 \in \Sigma^*$ are such that $f(w_1) = f(w_2)$, then $w_1^R = w_2^R$. Taking the reverse again, we get that $\left(w_1^R\right)^R = \left(w_2^R\right)^R$ and so $w_1 = w_2$. The function is also surjective: for $w \in \Sigma^*$, let $t = w^R \in \Sigma^*$. Then $f(t) = t^R = \left(w^R\right)^R = w$. The function is a bijection.

(d) The function is not injective: $f(01) = f(10) = 1$ but 01 and 10 are different strings. The function is surjective: given any $n \in \mathbb{N}$, let $w = \underbrace{00\ldots0}_{n \text{ times}} \in \Sigma^*$, then $f(w) =$ number of 0's in w=n.

(12) Consider the function $f : \mathbb{N} \to \mathbb{N}$ defined piecewise as follows:

$$f(n) = \begin{cases} n + 1 & \text{if } n \text{ is even} \\ n - 1 & \text{if } n \text{ is odd} \end{cases}$$

Let $m, n \in \mathbb{Z}$ such that $f(m) = f(n)$. If m is even and n is odd, then the equations $f(m) = f(n)$ translates to $m+1 = n-1$ and so $n = m+2$ is even: a contradiction. We also get a contradiction if m is odd and n is even. We conclude that m and n must have the same parity. If m and n are both even, then $f(m) = f(n)$ translates to $m+1 = n+1$ and so $m = n$. We get the same conclusion if m and n are both odd. We conclude that if $f(m) = f(n)$ then $m = n$ and f is injective. To prove that f is surjective, let $m \in \mathbb{Z}$ be arbitrary. If m is even, let $n = m+1$. Then n is odd and $f(n) = n - 1 = m$. If m is odd, let $n = m - 1$. Then n is even and $f(n) = n + 1 = m$. We conclude that any integer in $\mathbb{Z}$ is an image and the function is surjective.

(16) The function is not injective: $f(1, 2) = f(2, 4) = \frac{1}{2}$ but $(1, 2) \neq (2, 4)$ as elements of $\mathbb{Z} \times \mathbb{Z}^*$. The function is surjective: If $m \in \mathbb{Q}$, then $m = \frac{m}{1} = f(m, 1)$ with $(m, 1) \in \mathbb{Z} \times \mathbb{Z}^*$.

(18) Let $t \in [-1, 1]$, so $-1 \le t \le 1$. Multiplying all sides by $\frac{b-a}{2}$ (a positive real number), we get that $-\frac{b-a}{2} \le \frac{b-a}{2}t \le \frac{b-a}{2}$. Adding $\frac{b+a}{2}$ to all sides, we get $-\frac{b-a}{2} + \frac{b+a}{2} \le \frac{b-a}{2}t + \frac{b+a}{2} \le \frac{b-a}{2} + \frac{b+a}{2}$. Simplifying, we get: $a \le \frac{b-a}{2}t + \frac{b+a}{2} \le b$. This proves that $f(t) \in [a, b]$ for any $t \in [-1, 1]$. Let $s, t \in [-1, 1]$ such that $f(s) = f(t)$. Then $\frac{b-a}{2}s + \frac{b+a}{2} = \frac{b-a}{2}t + \frac{b+a}{2}$ and therefore $s = t$. The function is injective. For $m \in [a, b]$, let $t = \frac{2m}{b-a} - \frac{b+a}{b-a}$. Since $a \le m \le b$, $\frac{2a}{b-a} \le \frac{2m}{b-a} \le \frac{2b}{b-a}$ and so $\frac{2a}{b-a} - \frac{b+a}{b-a} \le \frac{2m}{b-a} - \frac{b+a}{b-a} \le \frac{2b}{b-a} - \frac{b+a}{b-a}$. This shows that $\frac{a-b}{b-a} \le t \le \frac{b-a}{b-a}$ or $-1 \le t \le 1$. So $t \in [-1, 1]$. Moreover, $f(t) = \frac{b-a}{2}t + \frac{b+a}{2} = \frac{b-a}{2}\left(\frac{2m}{b-a} - \frac{b+a}{b-a}\right) + \frac{b+a}{2} = m$. So f is surjective. Hence, the expression $f(t) = \frac{b-a}{2}t + \frac{b+a}{2}$ defines a bijection from $[-1, 1]$ to $[a, b]$.

(22) Consider the function $f : A \to \wp(A)$ defined by $f(x) = \{x\}$. If $x, y \in A$ are such that $f(x) = f(y)$ then $\{x\} = \{y\}$ and therefore $x = y$. The function is injective. It is not surjective as, for example, $\emptyset \in \wp(A)$ is not an image.

(24) Let $m \in \mathbb{Z}$ be an arbitrary integer. If $m < 0$, then $m = f(m)$ by definition. If $m \ge 0$, let $n = m + 1 \in \mathbb{Z}$. Then $n \ge 0$ and so $f(n) = n - 1 = m$. We conclude that for each $m \in \mathbb{Z}$, there exists $n \in \mathbb{Z}$ such that $f(n) = m$. The function is surjective.

(25)(b) True since $|B| > |A|$.

   (f) True since $|A \times B| = 20 > |\wp(A)| = 2^4 = 16$.

(28) Let $B, C \in \wp(A)$ be such that $\phi(B) = \phi(B)$. Then $\chi_B = \chi_C$. For $x \in A$, $x \in B \Leftrightarrow \chi_B(x) = 1 \Leftrightarrow \chi_C(x) = 1 \Leftrightarrow x \in C$. This proves that $B = C$ and the function ϕ is injective. Let $f : A \to \{0, 1\}$ be a function and consider the subset $B = \{x \in A;\ f(x) = 1\}$ of A. For any $x \in A$, $\chi_B(x) = 1 \Leftrightarrow x \in B \Leftrightarrow f(x) = 1$. This proves that $f = \chi_B = \phi(B)$. The function ϕ is then surjective, hence bijective.

(33)(b) Assume that f is injective. By part (a), we only need to show that the inclusion $f^{-1}(f(X)) \subset X$ is true. If $a \in f^{-1}(f(X))$, then $f(a) \in f(X)$. Then there exists $x \in X$ such that $f(a) = f(x)$. Since f is injective, this implies that $a = x$. In particular $a \in X$.

5.7

(1)(a) 2

   (f) 4

(2)(b) 22

   (f) $(g \circ f)(a - 1) = g(f(a - 1)) = g(-2a + 5) = (-2a + 5)^2 - 3 = 4a^2 - 20a + 22$.

(h) $(f \circ f \circ g)(a) = (f \circ f)(g(a)) = (f \circ f)(a^2 - 3) = f(f(a^2 - 3)) = f(-2(a^2 - 3) + 3) = f(-2a^2 + 9) = -2(-2a^2 + 6) + 3 = 4a^2 - 9.$

(3) Let $x \in \mathcal{H}$

   (a) $(f \circ f)(x) = f(f(x)) =$ the father of the father of $x=$ paternal grand father of $x.$

   (f) $(f \circ m \circ m)(x) = f(m(m(x))) =$ the father of the maternal grand mother of $x=$maternal great grand father of the mother of $x.$

(5) Let $x \in \mathbb{R}.$

   (c) $(g \circ h)(x) = g(h(x)) = g\left(\frac{x}{x^2+1}\right) = -2\left(\frac{x}{x^2+1}\right) + 3 = \frac{3x^2 - 2x + 3}{x^2+1}.$

   (e) $f^2(x) = f(f(x)) = f(x^2 + 1) = (x^2 + 1)^2 + 1 = x^4 + 2x^2 + 2.$

   (g)

$$(f \circ g \circ h)(x) = (f \circ g)(h(x)) = f(g(h(x)))$$

$$= f\left(g\left(\frac{x}{x^2 + 1}\right)\right)$$

$$= f\left(-2\frac{x}{x^2 + 1} + 3\right)$$

$$= f\left(\frac{3x^2 - 2x + 3}{x^2 + 1}\right)$$

$$= \left(\frac{3x^2 - 2x + 3}{x^2 + 1}\right)^2 + 1$$

$$= \frac{10x^4 - 13x^3 - 24x^2 - 12x + 10}{x^4 + 2x^2 + 1}.$$

(6)(b) Let $h(x) = |x|$, $g(x) = \frac{x}{1+x^2}$. Then for any $x \in \mathbb{R}$, $(g \circ h)(x) = g(h(x)) = g(|x|) = \frac{|x|}{1+|x|^2} = f(x).$

(9) Assume that $g \circ f$ is surjective and let $x \in C$. Then there exists $a \in A$ such that $c = g \circ f(a)$ (since $g \circ f$ is surjective). Let $b = f(a) \in B$. Then $c = g \circ f(a) = g(f(a)) = g(b)$. This proves that g is surjective.

(12) Assume that f and g are two non-decreasing functions from $\mathbb{R}$ to $\mathbb{R}$. We prove that $f \circ g$ is non-decreasing. Let $x, y \in \mathbb{R}$ be such that $x \leq y$. Then $g(x) \leq g(y)$ (since g is non-decreasing) and so $f(g(x)) \leq f(g(y))$ (since f is non-decreasing). This shows that $(f \circ g)(x) \leq (f \circ g)(y)$ and $f \circ g$ is non-decreasing.

(15) It suffices that the function is bijective. If $x, y \in \mathbb{R}\backslash\{\frac{3}{2}\}$ are such that $f(x) = f(y)$, then $\frac{x+1}{2x-3} = \frac{y+1}{2y-3}$. Cross multiplying gives: $2xy - 3x + 2y - 3 = 2xy + 2x - 3y - 3$ which simplifies to $5x = 5y$ or $x = y$. For $y \in \mathbb{R}\backslash\{\frac{1}{2}\}$, let $x = \frac{3y+1}{2y-1}$. Then $x \in \mathbb{R}$ since $y \neq \frac{1}{2}$ and it is easy to

verify that $f(x) = y$. This shows that f is surjective. We conclude that f is bijective and hence invertible. Moreover, $f^{-1}(x) = \frac{3x+1}{2x-1}$.

(16)(a) Let $a \neq b \in A$. Then $f(\{a\}) = f(\{b\}) = 1$ but $\{a\} \neq \{b\}$. So f is not injective, hence not invertible.

(d) The function is injective: If $f\left(\left[\begin{smallmatrix} a & b \\ c & d \end{smallmatrix}\right]\right) = f\left(\left[\begin{smallmatrix} x & y \\ z & t \end{smallmatrix}\right]\right)$, then $\left[\begin{smallmatrix} a & c \\ b & d \end{smallmatrix}\right] = \left[\begin{smallmatrix} x & z \\ y & t \end{smallmatrix}\right]$ and therefore $x = a$, $y = b$, $c = z$ and $d = t$. This proves that $\begin{bmatrix} a & b \\ c & d \end{bmatrix} = \begin{bmatrix} x & y \\ z & t \end{bmatrix}$. Let $B = \begin{bmatrix} x & y \\ z & t \end{bmatrix} \in \mathbb{M}_{22}$ at let $A = \begin{bmatrix} x & z \\ y & t \end{bmatrix} \in \mathbb{M}_{22}$. Clearly $f(A) = B$ and F is surjective. This show that f is bijective and hence invertible. The inverse function of f is the function f itself.

(19)(b) The property is true for $n = 1$ since f is surjective. Assume that $n \geq 1$ and that f^n is surjective. Then $f^{n+1} = f^n \circ f$ is surjective by the induction hypothesis and Theorem 5.8.

(20)(a) Assume that $g \circ f$ is injective and f is surjective. Let $a, b \in B$ such that $g(a) = g(b)$. Since f is surjective, there exist $x, y \in A$ such that $a = f(x)$ and $b = f(y)$. The equation becomes $g(f(x)) = g(f(y))$ or $(g \circ f)(x) = (g \circ f)(y)$. Since $g \circ f$ is injective, we get that $x = y$ and so $a = f(x) = f(y) = b$. The function g is injective.

(27) Let $C = \mathrm{Im}(f) \subseteq B$, let $g : A \to C$ defined by $g(x) = f(x)$ for any $x \in A$ and $h : C \to B$ be the inclusion map of C. For any $x \in A$, $(h \circ g)(x) = h(g(x)) = g(x) = f(x)$. We conclude that $h \circ g = f$.

6.2

(2) Assume that 1 and $1'$ are two identity elements for the multiplication of integers. Then $1 \cdot 1' = 1$ (since $1'$ is the identity element for $\cdot$) and $1 \cdot 1' = 1'$ (since 1 is the identity element for $\cdot$). We conclude that $1 = 1' = 1 \cdot 1'$.

(4)(a) Since $(-m) + m = 0$, the additive inverse of $(-m)$ is m. In other words, $-(-m) = m$

(d) $(m + n) + (-n - m) = ((m + n) + (-n)) + (-m)$ (associativity of $+$) $= (m + (n + (-n))) + (-m)$ (associativity of $+$) $= (m + 0) + (-m)$ (since $n + (-n) = 0$) $= m + (-m)$ (Since $m + 0 = m$) $= 0$ (Since $m + (-m) = 0$). By the uniqueness of the additive inverse, we conclude that $-(m + n) = -n - m$.

(5)(a) We need to show that $x(-y)$ is the additive inverse of xy:
$$xy + x(-y) = x(y + (-y)) \text{ (Axiom A7)} = x0 \text{ (Axiom A9)} = 0$$
(Theorem 1.2).

(9) Assume that $x < y$, then there exist $t \in \mathbb{Z}^+$ such that $y = x + t$. If $z > 0$, then $z \in \mathbb{Z}^+$ and we have: $yz = (x + t)z = xz + tz$ (Axiom A7), so $xz < yz$ since $tz \in \mathbb{Z}^+$.

(11)(a)

$$
\begin{aligned}
(a + b)(c + d) &= (a + b)c + (a + b)d && \text{(distributivity)} \\
&= (ac + bc) + (ad + bd) && \text{(distributivity)} \\
&= ac + (bc + (ad + bd)) && \text{(associativity of $+$)} \\
&= ac + ((bc + ad) + bd) && \text{(associativity of $+$)} \\
&= ac + ((ad + bc) + bd) && \text{(commutativity of $+$)} \\
&= (ac + (ad + bc)) + bd && \text{(associativity of $+$)} \\
&= ((ac + ad) + bc) + bd && \text{(associativity of $+$)} \\
&= (ac + ad) + (bc + bd) && \text{(associativity of $+$)} \\
&= ac + ad + bc + bd && \text{(associativity of $+$)}
\end{aligned}
$$

(13) Fix an integer k and let $\Sigma_k = \{n \in \mathbb{Z}; n \geq k\}$. The purpose of this exercise is to prove that the well-ordering principle (axiom WOP above) can be extended to the set Σ_k. Let T be a nonempty subset of Σ_k.

  (a) Since T is a nonempty subset of Σ_k, we can choose an element α in T. The integer $\alpha - k$ belongs to T'. In particular, $T' \neq \emptyset$. Moreover, every element of T' has the form $n - k$ for some $n \in T$ and since $T \subseteq \Sigma_k$, $n \geq k$ and consequently $n - k \geq 0$. We conclude that T' is a nonempty subset of $\mathbb{N}$.

  (b) By the well-ordering principle, T' has a smallest element e.

  (c) First note that $e = n - k$ for some $n \in T$ and so $e + k = n \in T$. For $m \in T$, $m - k \in T'$ and so $m - k \geq e$ or equivalently, $m \geq e + k$. This proves that $e + k$ is the smallest element of T.

6.3

(1)(a) False: $a = 2$, $b = 3$ are two non-zero integers with both $a \mid b$ and $b \mid a$ false.

  (d) False: $4 \mid 12$ and $6 \mid 12$ but $4 \cdot 6 = 24 \nmid 12$.

  (g) True: If $a \mid b$ and $c \mid d$, then there exist $m, n \in \mathbb{Z}$ such that $b = ma$ and $d = nc$. Multiplying the two equations together, we get that $bd = (mn)ac$ and so $ac \mid bd$.

  (i) False: $6 \mid 2 \cdot 9 = 18$ but $6 \nmid 2$ and $6 \nmid 9$.

(4) Assume that $m \mid n$ and $n \mid m$. In particular, $m \neq 0$ and $n \neq 0$. There exist integers a, b such that $n = am$ and $m = bn$. This implies that $n = abn$ and therefore $1 = ab$ (simplifying by n). Since a, b are integers, this implies that either $a = b = 1$ or $a = b = -1$ and so either $m = n$ or $m = -n$.

(6)(c) False: $p = 2$ is a prime number, but $p + 1 = 3$ is not composite.

  (e) False: $p = 3$ and $q = 5$ are prime numbers, but $pq + 1 = 16$ is not a prime number.

(8)(d) $3^{2022} - 5^{4044} = (3^{1022})^2 - 5^{511})^2 = ((3^{1022}) - 5^{511}))((3^{1022}) + 5^{511}))$: not prime.

(9)(f) 111000111 is not prime: divisible by 3 as the sum of the digits is 6 divisible by 3.

(12) $(3, 5)$, $(5, 7)$, $(11, 13)$, $(17, 19)$, $(29, 31)$, $(41, 43)$, $(59, 61)$ and $(71, 73)$.

(15)(f) $544 = 2^5 \times 17$

  (h) $4851 = 3^2 \times 7^2 x 11$.

(19) Assume that $p = 2^n - 1$ is a prime number with $n \geq 2$. If n is not prime, then there exists $a, b \in \mathbb{N}$ such that $n = ab$ with $1 < a < n$ and $1 < b < n$. So

$$2^n - 1 = 2^{ab} - 1$$
$$= (2^a)^b - 1 = (2^a - 1)((2^a)^{b-1} + (2^a)^{b-2} + \cdots + 2^a + 1)$$
$$= (2^a - 1)(2^{a(b-1)} + 2^{a(b-2)} + \cdots + 2^a + 1).$$

Given that $a \geq 2$, we have that $2^a - 1 > 1$ and $2^{a(b-1)} + 2^{a(b-2)} + \cdots + 2^a + 1 > 1$. This is a contradiction to the assumption that p is prime. We conclude that n must be prime.

(21)(c) $169 \leq 181 \leq 196$, so $13 \leq \sqrt{181} \leq 14$. Primes less than or equal to 13 are $2, 3, 5, 7, 13$ and none of them divides 181. So 181 is prime.

6.4

(3)(a) $q = 7$, $r = 4$: $123 = 7 \times 17 + 4$.

  (e) $q = -39$, $r = 6$: $-462 = (-39) \times 12 + 6$

(5) Write $n = 2k + 1$ for some integer k. Then $n^4 = (2k + 1)^4 = 16k^4 + 32k^3 + 24k^2 + 8k + 1 = 8(2k^4 + 4k^3 + 3k^2 + k) + 1$ and the result follows.

(7)(e) $1010101101 = 1 \cdot 2^9 + 0 \cdot 2^8 + 1 \cdot 2^7 + 0 \cdot 2^6 + 1 \cdot 2^5 + 0 \cdot 2^4 + 1 \cdot 2^3 + 1 \cdot 2^2 + 0 \cdot 2^1 + 1 \cdot 2^0 = 685$

(8)(c) 11100001

  (f) 11000111110

  (i) 11000000111001

(9)(c) $2 \cdot 8^3 + 3 \cdot 8^2 + 4 \cdot 8^2 + 5 \cdot 8^0 = 1253$

(10)(c) 4525

(11)(c) $q(x) = x^2 - 3x + 1$, $r(x) = 3x - 2$: $q(x)g(x) + r(x) = (x^2 - 3x + 1)(x^2 + 1) + 3x - 2 = x^4 - 3x^3 + 2x^2 - 1 = f(x)$.

(13) Let $n \in \mathbb{Z}$. If the remainder of the division of n by 8 is 7, then $n = 8k + 7$ for some integer k. So $n = 4(2k) + 4 + 3 = 4s + 3$ with $s = 2k \in \mathbb{Z}$. This shows that the remainder of the division of n by 4 is 3.

6.5

(1)(e) $\gcd(a, a^2) = |a|$

(h) $\gcd(-2a, 3a) = 2|a|$

(2)(a) False: 1, 2, 4 are relatively prime, but they are not pairwise relatively prime since $\gcd(2, 4) = 2 \neq 1$.

(d) True. By the definition of pairwise relatively prime integers.

(4) The divisors of 35 are: 1, 5, 7 and 35. From the set of all positive integers less than 35, we remove all integers multiples of 5 and multiples 7 and we get 1, 2, 4, 6, 8, 9, 11, 12, 13, 16, 17, 18, 19, 22, 23, 24, 26, 27, 29, 31, 32, 33 and 34.

(5)(f) $285 = 3 \cdot 5 \cdot 19$, $465 = 3 \cdot 5 \cdot 31$. So, $\gcd(285, 465) = 3 \cdot 5 = 15$ and $\operatorname{lcm}(285, 465) = 3 \cdot 5 \cdot 19 \cdot 31 = 8835$.

(h) $2010 = 2 \cdot 3 \cdot 5 \cdot 67$, $4095 = 3^2 \cdot 5 \cdot 7 \cdot 13$. So, $\gcd(2010, 4095) = 3 \cdot 5 = 15$ and $\operatorname{lcm}(296, 666)(2010, 4095) = 2 \cdot 3^2 \cdot 5 \cdot 7 \cdot 13 \cdot 67 = 548730$.

(i) $616 = 2^3 \cdot 7 \cdot 11$, $255255 = 3 \cdot 5 \cdot 7 \cdot 11 \cdot 13 \cdot 17$. So, $\gcd(616, 255255) = 7 \cdot 11 = 77$ and $\operatorname{lcm}(616, 255255) = 2^3 \cdot 3 \cdot 5 \cdot 7 \cdot 11 \cdot 13 \cdot 17 = 2042040$.

(6)(c)

$$\begin{aligned} 1320 &= 15 \cdot 84 + \underline{60} \\ 84 &= 1 \cdot 60 + \underline{24} \\ 60 &= 2 \cdot 24 + \underline{12} \\ 24 &= 2 \cdot 12 + \underline{0} \end{aligned}$$

$\gcd(1320, 84) = 12$. Working backwards, we get that $12 = 60 - 2 \cdot 24 = 60 - 2(84 - 60) = 3 \cdot 60 - 2 \cdot 84 = 3(1320 - 15 \cdot 84) - 2 \cdot 84 = 3 \cdot 1320 + (-47) \cdot 84$.

(e)

$$\begin{aligned} 255255 &= 414 \cdot 616 + \underline{231} \\ 616 &= 2 \cdot 231 + \underline{154} \\ 231 &= 1 \cdot 154 + \underline{77} \\ 154 &= 2 \cdot 77 + \underline{0} \end{aligned}$$

$\gcd(255255, 616) = 77$. Working backwards, we get that $77 = 231 - 154 = 231 - (616 - 2 \cdot 231) = 3 \cdot 231 - 616 = 3 \cdot (255255 - 414 \cdot 616) - 616 = 3 \cdot 255255 + (-1243) \cdot 616$.

(7)(c) $\gcd(56, 168, 256) = 8$, $\mathrm{lcm}(56, 168, 256) = 301056$

(8) We use the relation $ab = gl$.

  (b) $a = \frac{gl}{b} = \frac{22 \cdot 6468}{66} = 2156$.

(9)(e) $128 = 2^7$, so the divisors of the 128 are of integers of the form 2^k for $0 \le k \le 7$: $2^0 = 1$, $2^1 = 2$, $2^2 = 4$, $2^3 = 8$, $2^4 = 16$, $2^5 = 32$, $2^6 = 64$, $2^7 = 128$.

  (j) $2197 = 13^3$, so the divisors of the 2197 are of integers of the form 13^k for $0 \le k \le 3$: $13^0 = 1$, $13^1 = 13$, $13^2 = 169$, $13^3 = 2197$.

(11) If such integers a, b exist, then 3 would be a common divisor of a and b. So $3 \mid a + b = 23456701$. This is a contradiction since 23456701 is not divisible by 3. So no such integers exist.

(15) Let $g = \gcd(m + n, m - n)$. Then g is a common divisor of $m + n$ and $m - n$ and so it is a divisor of both $(m + n) + (m - n) = 2m$ and $(m + n) - (m - n) = 2n$. Therefore, $g \mid \gcd(2m, 2n) = 2\gcd(m, n) = 2$ since $\gcd(m, n) = 1$. This shows that $g = \gcd(m + n, m - n)$ is either 1 or 2.

(21) Assume first that $\gcd(n, p) = 1$. If $\gcd(n, p^k) > 1$, then $\gcd(n, p^k)$ has a prime divisor q. Therefore q is a common prime divisor of n and p^k. By Euclid' Lemma, $q \mid p$ and therefore, $q = p$. But this is a contradiction since $\gcd(n, p) = 1$. For the converse, assume $\gcd(n, p^k) = 1$. Then there exist $a, b \in \mathbb{Z}$ such that $an + bp^k = 1$. In particular, $an + (bp^{k-1})p = 1$ and $\gcd(n, p) = 1$.

(22) The statement is false. Consider the equation $2x + y = 2$. Then $x = -1$ and $y = 4$ satisfy the equation but $\gcd(2, 1) = 1 \ne 2$.

(26) By induction on n. The result is trivially true for $n = 1$. Let $n \ge 1$ be an integer and assume the result is true for n. If $p \mid a_1 a_2 \cdots a_{n+1}$, then $p \mid a_1 a_2 \cdots a_n$ or $p \mid a_{n+1}$ by Corollary 6.3. If $p \mid a_{n+1}$, then we are done. If $p \mid a_1 a_2 \cdots a_n$, then $p \mid a_i$ for some $i \in \{1, \ldots, n\}$ by the induction hypothesis, The result follows by the principle of induction.

6.6

(5) Assume that $a \equiv b \pmod{n}$ and d is a positive divisor of n. Write $a - b = ns$ and $n = td$ for some $s, t \in \mathbb{Z}$. Then $a - b = (st)d$ and so $a \equiv b \pmod{d}$.

(9) For any integer n, $n \equiv 0, 1, 2, 3$ or $4 \pmod 5$. So, $n^4 \equiv 0^4 = 0, 1^4 =$

$1, 2^4 = 16, 3^4 = 81$ or $4^4 = 256 \pmod 5$. But each of the integers 16, 81 and 256 is congruent to $1 \pmod 5$. The result follows.

(11)(b) Note that $3^4 = 81 \equiv 1 \pmod{80}$, so $3^{257} = (3^4)^{64} \cdot 3 \equiv 1^{64} \cdot 3 = 3 \pmod{80}$.

(12)(d) Since $\gcd(8, 12) = 4 \mid 16$, the congruence has a total of four non-congruent solutions. By inspection, $x = 8$ is a solution of the congruence $8x \equiv 16 \pmod{12}$ since $8 \times 8 - 16 = 48$ is divisible par 12. The non-congruent solutions are 8, $8 + \frac{12}{4} = 11$, $8 + 2\frac{12}{4} = 14$ and $8 + 3\frac{12}{4} = 17$.

  (f) The congruence has a unique solution modulo 7 since $\gcd(17, 7) = 1$. By inspection, $x = 1$ is a solution and any other solution is congruent to 1 modulo 7.

  (i) The congruence has no solution since $\gcd(37, 148) = 37$ which does not divide 38.

  (j) Since $\gcd(91, 143) = 13 \mid 78$, the congruence has a total of 13 non-congruent solutions. By inspection, $x = 4$ is a solution. The non-congruent solutions are 4, $4 + \frac{143}{13} = 15$, $4 + 2\frac{143}{13} = 26$, $4 + 3\frac{143}{13} = 37$, $4 + 4\frac{143}{13} = 48$, $4 + 5\frac{143}{13} = 59$, $4 + 6\frac{143}{13} = 70$, $4 + 7\frac{143}{13} = 81$, $4 + 8\frac{143}{13} = 92$, $4 + 9\frac{143}{13} = 103$, $4 + 10\frac{143}{13} = 114$, $4 + 11\frac{143}{13} = 125$ and $4 + 12\frac{143}{13} = 136$.

(14) Since $ac \equiv bc \pmod n$, there exists an integer k such that $ac - bc = kn$. This means that $n \mid a - b$ since $\gcd(c, n) = 1$ and $n \mid (a - b)c$. This shows that $a \equiv b \pmod n$.

(18) We start by rewriting each of the congruences under the form $x \equiv a \pmod n$. Multiplying the first congruence by 2, we get that $x \equiv 2 \pmod 3$ since $4 \equiv 1 \pmod 3$. Multiplying the second congruence by 2, we get that $x \equiv 4 \pmod 3$ since $6 \equiv 1 \pmod 5$. Multiplying the third congruence by 3, we get that $x \equiv 15 \equiv 4 \pmod{11}$ since $12 \equiv 1 \pmod{12}$. Since 3, 5 and 11 are pairwise relatively prime, the Chinese remainder theorem tells us that the system has a unique solution modulo $n = 3 \cdot 5 \cdot 11 = 165$. Let $m_1 = \frac{165}{3} = 55$, $m_2 = \frac{165}{5} = 33$ and $m_3 = \frac{165}{11} = 15$. For the congruence $55x \equiv 1 \pmod 3$, $x_1 = 1$ is the unique solution modulo 3. For the congruence $33x \equiv 1 \pmod 5$, $x_2 = 2$ is the unique solution modulo 5. For the congruence $15x \equiv 1 \pmod{11}$, $x_3 = 3$ is the unique solution modulo 11. So $x = m_1 a_1 x_1 + m_2 a_2 x_2 + m_3 a_3 x_3 = 554$ is a solution of the system of three congruence relations. Any other solution is congruent to 554 modulo 165. We conclude that the smallest positive solution to the system of congruences is $x = 554 - 3 \cdot 165 = 59$.

(20) Let n be the number of the marbles in the box before it was dropped. The information given by the kid gives the following system of congruences: $n \equiv 3 \,(\text{mod } 5)$, $n \equiv 6 \,(\text{mod } 11)$, $n \equiv 8 \,(\text{mod } 13)$ and $n \equiv 10 \,(\text{mod } 17)$. Since $5, 11, 13$ and 17 are pairwise relatively prime, the Chinese remainder theorem tells us that the system has a unique solution modulo $n = 5 \cdot 11 \cdot 13 \cdot 17 = 12155$. Let $m_1 = \frac{12155}{5} = 2431$, $m_2 = \frac{12155}{11} = 1105$, $m_3 = \frac{12155}{13} = 935$ and $m_4 = \frac{12155}{17} = 715$. For the congruence $2431x \equiv 1 \,(\text{mod } 5)$, $x_1 = 1$ is the unique solution modulo 5. For the congruence $1105x \equiv 1 \,(\text{mod } 11)$, $x_2 = 9$ is the unique solution modulo 11. For the congruence $935x \equiv 1 \,(\text{mod } 13)$, $x_3 = 12$ is the unique solution modulo 13. For the congruence $715x \equiv 1 \,(\text{mod } 17)$, $x_4 = 1$ is the unique solution modulo 17. So $x = m_1 a_1 x_1 + m_2 a_2 x_2 + m_3 a_3 x_3 + m_4 a_4 x_4 = 163873$ is a solution of the system of three congruence relations. Any other solution is congruent to 26873 modulo 12155. We conclude that the smallest positive solution to the system of congruences is $x = 163873 - 13 \cdot 12155 = 5858$.

6.7

(3)(e) $\phi(65) = \phi(5 \cdot 13) = \phi(5)\phi(13) = (5 - 1) \cdot (13 - 1) = 48$ since $\gcd(5, 13) = 1$.

(i) $\phi(143) = \phi(11 \cdot 13) = \phi(11)\phi(13) = (11 - 1) \cdot (13 - 1) = 120$ since $\gcd(11, 13) = 1$.

(o) $\phi(20200) = \phi(2^3 \cdot 5^2 \cdot 101) = (2^3 - 2^2) \cdot (5^2 - 5^1) \cdot (101 - 1) = 8000$.

(r) $\phi(58697100) = \phi(2^2 \cdot 3^2 \cdot 5^2 \cdot 7^2 \cdot 11^3) = (2^2 - 2^1) \cdot (3^2 - 3^1) \cdot (5^2 - 5^1) \cdot (7^2 - 7^1) \cdot (11^3 - 11^2) = 12196800$.

(5) Let $n \geq 2$ be an even integer, and let $n = 2^r p_1^{t_1} p_2^{t_2} \cdots p_k^{p_k}$ be its prime decomposition with $r \geq 1$ and $p_i \geq 3$ is odd for all i. Then $2n = 2^{r+1} p_1^{t_1} p_2^{t_2} \cdots p_k^{p_k}$ and $\phi(2n) = (2^{r+1} - 2^r)(p_1^{t_1} - p_1^{t_1 - 1}) \cdots (p_k^{t_k} - p_k^{t_k - 1}) = 2(2^r - 2^{r-1})(p_1^{t_1} - p_1^{t_1 - 1}) \cdots (p_k^{t_k} - p_k^{t_k - 1}) = 2\phi(n)$.

(10)(e) We have that $9^{18} \equiv 1 \,(\text{mod } 19)$ by Fermat's theorem. By the division algorithm, $1989 = 119(18) + 9$, so $9^{1989} \equiv (9^{18})^{119} \cdot 9^9 \equiv 9^9 = 387420489 \equiv 1 \,(\text{mod } 19)$.

(h) We know that $3^{16} \equiv 1 \,(\text{mod } 17)$. Since $1000 = 62(16) + 8$, $3^{1000} \equiv (3^{16})^{62} \cdot 3^8 \equiv 3^8 = 6561 \equiv 16 \,(\text{mod } 17)$.

(j) $128^{128} = (2^7)^{128} = 2^{896} = 2^{648} \cdot 2^{248} \equiv 2^{248} \,(\text{mod } 703)$ since $2^{648} = 2^{18 \cdot 36} \equiv 1 \,(\text{mod } 703 = 19 \cdot 37)$ by Euler's Corollary. Now, $2^{10} = 1024 \equiv 321 \,(\text{mod } 703)$. So $2^{20} \equiv 321^2 = 103041 \equiv 403 \,(\text{mod } 703)$, $2^{40} \equiv 403^2 = 162409 \equiv 16 \,(\text{mod } 703)$, $2^{160} \equiv 16^4 = 65536 \equiv 157 \,(\text{mod } 703)$. We conclude that $2^{248} = (2^{40})^2 \cdot 2^{160} \cdot 2^8 \equiv 16^2 \cdot 157 \cdot 2^8 = 10289152 \equiv 44 \,(\text{mod } 703)$.

(12) Since 101 is a prime that does not divide 7, Fermat's little theorem tells us that $7^{100} \equiv 1 \pmod{101}$. Now, $7^{34567892123465412002} = \left(7^{100}\right)^{34567892123465412} \cdot 7^2 \equiv 7^2 = 49 \pmod{101}$.

(13)(b) Since 7 is a prime that does not divide 10, Fermat's little theorem tells us that $10^6 \equiv 1 \pmod 7$. By part (a), we know that $10^{100} \equiv 4 \pmod 6$ and so $10^{100} = 6k + 4$ for some integer $k \geq 1$. So, $10^{10^{100}} = 10^{6k+4} = (10^6)^k \cdot 10^4 \equiv 10^4 \pmod 7$. Now, $10 \equiv 3 \pmod 7$, so $10^4 \equiv 3^4 = 81 \equiv 4 \pmod 7$. So, $10^{10^{100}} \equiv 4 \pmod 7$.

(17) Using Euler's theorem, we have that $m^{\phi(n)} \equiv 1 \pmod n$ and $n^{\phi(m)} \equiv 1 \pmod m$. But $m^{\phi(n)} \equiv 0 \pmod m$ and $n^{\phi(m)} \equiv 0 \pmod n$. So $m^{\phi(n)} + n^{\phi(m)} \equiv 0 + 1 = 1 \pmod m$ and $m^{\phi(n)} + n^{\phi(m)} \equiv 1 + 0 = 1 \pmod n$. Since $\gcd(m, n) = 1$, this show that $m^{\phi(n)} + n^{\phi(m)} \equiv 1 \pmod{mn}$. If p and q are distinct prime numbers, then $\gcd(p, q) = 1$, $\phi(p) = p - 1$, $\phi(q) = q - 1$ and so $p^{q-1} + q^{p-1} \equiv 1 \pmod{pq}$.

6.8

(1)(a) PRRQ   (d) RABJHQ

(2) Using the decryption function $D_k(x) = (x - 3) \pmod{26}$ to decrypt the numerical message, we get:

$$18\,04\,11\,11 \quad 24\,14\,20\,17 \quad 07\,14\,20\,18\,04$$

which corresponds to the plaintext message: SELL YOUR HOUSE.

(4) $f_{a,b}(x) = ax + b \pmod{26}$ where $a, b \in \{0, 1, \ldots, 25\}$.

(a) Note that the congruence relation $y \equiv ax + b \pmod{26}$ can be written in the form $ax \equiv b - y \pmod{26}$. We know that this congruence has a unique solution $x \in \{0, 1, \ldots, 25\}$ if and only if $\gcd(a, 26) = 1$. In this case, $f(x) = ax + b \pmod{2)6}$ is a valid encryption function since we could solve for x in $ax \equiv b - y \pmod{26}$.

(b) There are $\phi(26) = 12$ possible values for the coefficient a in the affine cipher $f_{a,b}(x) = ax + b \pmod{26}$. For each of these values, there are 26 possible values for the integer $b \in \{0, 1, \ldots, 25\}$. Therefore, there are $12 \cdot 26 = 312$ affine ciphers.

(c) (i) H (ii) G (iii) A (iv) P (v) C

(d) Since $26 + (-5)(5) = 1$, $(-5)(5) \equiv 1 \pmod{26}$ or equivalently, $(21)(5) \equiv 1 \pmod{26}$. If $y = 5x + 7 \pmod{26}$, then $y - 7 = 5x$

(mod 26) and so $x = 21(y-7) = 21y - 147 \equiv 21y - 17$ (mod 26). So, the decryption function of the affine cipher is $g_{5,7}(x) = 21x - 17$ (mod 26).

(e) The numerical version of the message HMHUWZU is

$$07\,12\,07\,20\,22\,25\,20.$$

Using the decryption function $g_{5,7}(x) = 21x - 17$ (mod 26), the above numerical message becomes:

$$00\,01\,00\,13\,03\,14\,13$$

which translates to the plaintext ABANDON.

(6) We know that $\phi(n) = (p-1)(q-1)$ and that $\gcd(\phi(n), e) = 1$. We use the Euclidean algorithm to find integers c and d such that $c\phi(n) + de = 1$. Then (n, d) would be the private key of the cryptosystem.

(b) $\phi(n) = (23-1)(37-1) = 792$. Using the Euclidean algorithm: $(20)(792) + (-337)(47) = 1$. So $d = -337 \equiv 455$ (mod 792) and the private key is $(n, d) = (851, 455)$.

(d) $\phi(n) = (179-1)(211-1) = 37380$. Using the Euclidean algorithm, $(-38)(37380) + (10219)(139) = 1$ and the private key is $(n, d) = (37769, 10219)$.

(7) We start by writing n as the product of two primes p and q.

(b) $n = 55 = 5 \cdot 11$, so $p = 5$, $q = 11$, $\phi(n) = 40$. Using the Euclidean algorithm, $(3)(40) + (-17)(7) = 1$ and so $d = -17 \equiv 23$ (mod 40). The decryption scheme is then $m = c^{23}$ (mod 55).

(d) $n = 221 = 13 \cdot 17$, so $p = 13$, $q = 17$, $\phi(n) = 192$. Using the Euclidean algorithm, $(7)(192) + (-79)(17) = 1$ and so $d = -79 \equiv 113$ (mod 192). The decryption scheme is then $m = c^{113}$ (mod 221).

(10) Assume that n and $\phi(n)$ in a RSA cryptosystem are known. We need to find primes p and q such that $n = pq$. Since $\phi(n) = (p-1)(q-1)$, $\phi(n) = pq - (p+q) + 1 = n - p - q + 1$. Solving for q in this last equation, we get that $q = (n+1) - \phi(n) - p$. Therefore, $n = p((n+1) - \phi(n) - p)$ or $p^2 - (n+1-\phi(n))p + n = 0$. This is a quadratic equation in p that we can solve.

7.2

(1)(d) $R_1 = \emptyset$, $R_2 = \{(1,a)\}$, $R_3 = \{(1,b)\}$, $R_4 = \{(2,a)\}$, $R_5 = \{(2,b)\}$, $R_6 = \{(1,a),(1,b)\}$, $R_7 = \{(1,a),(2,a)\}$, $R_8 = \{(1,a),(2,b)\}$, $R_9 = \{(1,b),(2,a)\}$, $R_{10} = \{(1,b),(2,b)\}$, $R_{11} = \{(2,a),(2,b)\}$,

$R_{12} = \{(1,a),(1,b),(2,a)\}$, $R_{13} = \{(1,a),(2,a),(2,b)\}$, $R_{14} = \{(1,b),(2,a),(2,b)\}$, $R_{15} = \{(1,a),(1,b),(2,b)\}$, $R_{16} = \{(1,a),(1,b),(2,a),(2,b)\}$.

(2)(a) False: $3(0) - 2(1) = -2 \neq -1$

   (e) True: $3(1) - 2(2) = -1$

   (h) True: $3(3) - 2(5) = -1$

(4)(b) R_2 is a relation on the set A, $\mathrm{Dom}(R_2) = \mathrm{Ran}(R_2) = A$.

   (d) R_5 is a relation on the set B, $\mathrm{Dom}(R_5) = \mathrm{Ran}(R_5) = B$.

(5)(a) $R = \{(C_1, C_2), (C_1, C_3), (C_3, C_2), (C_4, C_2)\}$.

(7)(b) $(-1,1)$, $(-1,2)$, $(-1,3)$, $(-1,6)$, $(0,1)$, $(0,2)$, $(0,3)$, $(0,6)$, $(1,1)$, $(1,2)$, $(1,3)$, $(2,1)$, $(2,2)$, $(2,3)$, $(3,1)$, $(3,2)$, $(3,3)$, $(4,1)$, $(4,2)$.

   (d) $(0,12)$, $(1,12)$, $(2,12)$, $(3,12)$, $(4,12)$.

(8)(b) In R: $(-4)(2) = (-8)(1)$.

   (e) Not in R: $(\sqrt{2})\left(\frac{1}{\sqrt{2}}\right) \neq (-1)(1)$.

(9)(a) $R(1) = \{1, 3, 4\}$

   (c) $R(\{1,2\}) = A$

(10)(a) $R(-1) = \{y \in \mathbb{R};\ 4(-1)^2 + 9y^2 = 1\} = \{y \in \mathbb{R};\ 9y^2 = -3\} = \emptyset$

   (e) $R\left(\frac{1}{8}\right) = \{y \in \mathbb{R};\ 4\left(\frac{1}{8}\right)^2 + 9y^2 = 1\} = \{y \in \mathbb{R};\ y^2 = \frac{15}{144}\} = \{\pm\frac{\sqrt{15}}{12}\}$.

   (g) $R(\{1,2\}) = \{y \in \mathbb{R};\ (1,y) \in R \text{ or } (2,y) \in R\} = R(1) \cup R(2) = \emptyset \cup \emptyset = \emptyset$.

(11)(b) If $y \in R(A_1 \cap A_2)$, then there exists $x \in A_1 \cap A_2$ such that $(x,y) \in R$. Since $x \in A_1$ and $x \in A_2$, we have that $(x,y) \in R(A_1) \cap R(A_2)$.

7.3

(1) R consists of the following elements: $(1,1)$, $(1,2)$, $(1,3)$, $(1,4)$, $(1,9)$, $(1,12)$, $(2,2)$, $(2,4)$, $(2,12)$, $(3,3)$, $(3,9)$, $(3,12)$, $(4,4)$, $(4,12)$, $(9,9)$, and $(12,12)$. The adjacency matrix of R is the following:

$$M_R = \begin{array}{c c} & \begin{array}{c c c c c c} 1 & 2 & 3 & 4 & 9 & 12 \end{array} \\ \begin{array}{c} 1 \\ 2 \\ 3 \\ 4 \\ 9 \\ 12 \end{array} & \left[\begin{array}{c c c c c c} 1 & 1 & 1 & 1 & 1 & 1 \\ 0 & 1 & 0 & 1 & 0 & 1 \\ 0 & 0 & 1 & 0 & 1 & 1 \\ 0 & 0 & 0 & 1 & 0 & 1 \\ 0 & 0 & 0 & 0 & 1 & 0 \\ 0 & 0 & 0 & 0 & 0 & 1 \end{array}\right] \end{array}$$

and the digraph of R is:

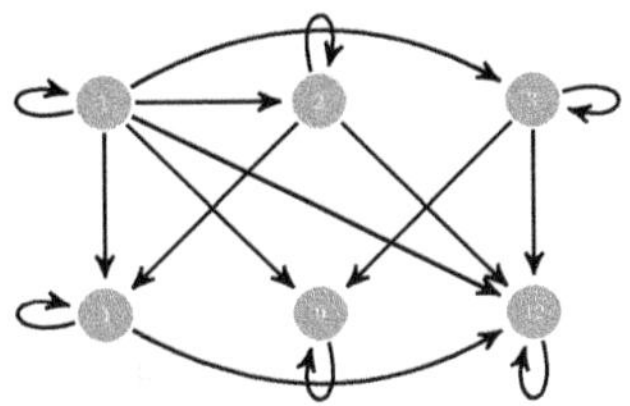

(3)(b) The elements of S are: (a, x), (a, y), (a, z), (a, w), (b, y), (b, z), (b, w), (c, y), (d, x), (d, z) and (d, w).

(4)(c) The Boolean matrix representation of R is:

$$
M_R = \begin{array}{c} \\ -3 \\ -2 \\ -1 \\ 0 \\ 1 \\ 4 \\ 6 \\ 9 \\ 12 \end{array}
\begin{array}{c} \begin{array}{ccccccccc} -3 & -2 & -1 & 0 & 1 & 4 & 6 & 9 & 12 \end{array} \\
\left[\begin{array}{ccccccccc}
1 & 1 & 1 & 1 & 0 & 0 & 0 & 0 & 0 \\
1 & 1 & 1 & 1 & 1 & 0 & 0 & 0 & 0 \\
1 & 1 & 1 & 1 & 1 & 0 & 0 & 0 & 0 \\
1 & 1 & 1 & 1 & 1 & 0 & 0 & 0 & 0 \\
0 & 1 & 1 & 1 & 1 & 1 & 0 & 0 & 0 \\
0 & 0 & 0 & 0 & 1 & 1 & 1 & 0 & 0 \\
0 & 0 & 0 & 0 & 0 & 1 & 1 & 1 & 0 \\
0 & 0 & 0 & 0 & 0 & 0 & 1 & 1 & 1 \\
0 & 0 & 0 & 0 & 0 & 0 & 0 & 1 & 1
\end{array}\right]
\end{array}
$$

and the digraph of R is:

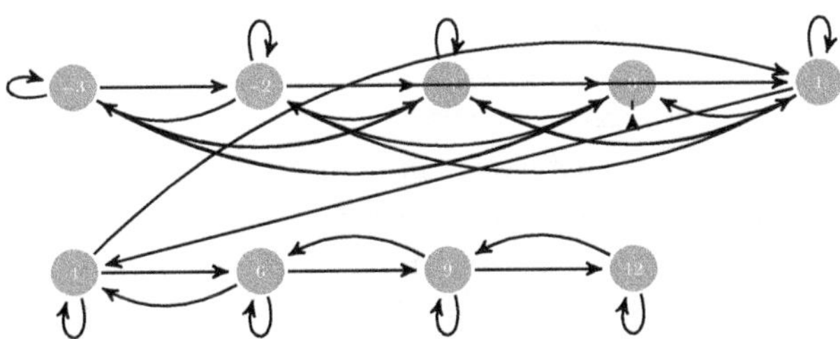

(6) Write $A = [a_{ij}]$ where a_{ij} is the entry on the ith row and the jth column of A. For $i \in \{1, 2, \ldots, m\}$ and $j \in \{1, 2, \ldots, n\}$, the entry c_{ij} on the ith row and the jth column of $A \odot I_n$ is given by $c_{ij} = (a_{i1} \wedge b_{1j}) \vee \ldots \vee (a_{ij} \wedge b_{jj}) \vee \ldots (a_{in} \wedge b_{nj})$ where b_{kl} is the entry at row k and column l of the identity matrix I_n. Since $b_{kl} = 0$ for $k \neq l$ and $b_{kl} = 1$ for $k = l$, we conclude that $c_{ij} = (a_{ij} \wedge b_{jj}) = (a_{ij} \wedge 1) = a_{ij}$. This shows that $A \odot I_n = A$. The proof that $I_m \odot A = A$ is done similarly.

(9)(b) $A \vee B = \begin{bmatrix} 1 & 1 \\ 1 & 0 \end{bmatrix}$

(e) Undefined

(h) $(A \odot B) \odot (A \odot C) = \begin{bmatrix} 0 & 1 & 0 \\ 0 & 1 & 0 \end{bmatrix}$

(10)(b) $A \odot B = \begin{bmatrix} 1 & 1 & 0 & 0 \\ 1 & 1 & 1 & 1 \\ 1 & 1 & 1 & 1 \\ 1 & 1 & 1 & 1 \end{bmatrix}$, $B \odot A = \begin{bmatrix} 1 & 1 & 0 & 1 \\ 0 & 1 & 1 & 1 \\ 1 & 1 & 1 & 1 \\ 1 & 1 & 0 & 1 \end{bmatrix}$

7.4

(2)(b) R_2 is function. It is injective but not surjective.

(3)(b) R_2 is a function from $\mathbb{R}^*$ to $\mathbb{R}^*$. For every $x \in \mathbb{R}^*$, $(x, \frac{1}{x}) \in R_2$. So the existence condition is satisfied. If $x, y, z \in \mathbb{R}^*$ are such that $(x, y) \in R_2$ and $(x, z) \in R_2$ then $xy = xz = 1$ and therefore $y = z$ since $x \neq 0$. The uniqueness condition is also satisfied. If $x, y, z \in \mathbb{R}^*$ are such that $(x, z) \in R_2$ and $(y, z) \in R_2$ then $xz = yz = 1$ and therefore $x = y$ since $y \neq 0$. The function is then injective. If $y \in \mathbb{R}^*$, let $x = \frac{1}{y} \in \mathbb{R}^*$. Then $xy = 1$ and therefore, $(x, y) \in R_2$. The function is then surjective.

(d) R_4 is a function from $\mathbb{Q}$ to $\mathbb{Q}$. For every $x \in \mathbb{Q}$, $(x, |x|) \in R_4$. So the existence condition is satisfied. If $x, y, z \in \mathbb{Q}$ are such that $(x, y) \in R_4$ and $(x, z) \in R_4$ then $y = z = |x|$. The uniqueness condition is also satisfied. The function is not injective since If $x, y, z \in \mathbb{Q}$ are such that $(-1, 1) \in R_4$ and $(1, 1) \in R_4$ but $1 \neq -1$. The function is not surjective. For example, there exists no $x \in \mathbb{Q}$ such that $(x, -1) \in R_4$ since otherwise this would mean that $|x| = -1$.

(4)(c) Not a function: both $(1, 2)$ and $(1, 3)$ are elements of the relation.

(5)(b) The relation is a function since every row of the matrix contains a unique 1 which means that both the existence and the uniqueness conditions are satisfied. It is not injective since (c, x) and (d, x) are in the relation. It is not surjective since there exists no element $u \in A$ such that (u, y) is in the relation.

7.5

(1)
- $R \cap S = \{(1, a), (1, c), (3, d)\}$.
- $R \cup S = \{(1, a), (1, b), (1, c), (1, d), (2, a), (2, b), (2, d), (3, a), (3, c), (3, d)\}$.
- $R \backslash S = \{(1, b), (2, b), (3, c)\}$.
- $S \backslash R = \{(1, d), (2, a), (2, d), (3, a)\}$.

- $\neg R = \{(1,d),(2,a),(2,c),(2,d),(3,a),(3,b)\}$.
- $\neg S = \{(1,b),(2,b),(2,c),(3,b),(3,c)\}$.
- $R^{-1} = \{(a,1),(b,1),(c,1),(b,2),(c,3),(d,3)\}$.
- $S^{-1} = \{(a,1),(c,1),(d,1),(a,2),(d,2),(a,3),(d,3)\}$.

(3) For $a,b \in A$:

- $a\,R^2\,b$ if and only if there exists $c \in A$ such that $a\,R\,c$ and $c\,R\,b$. This means that a is a parent of c and c is a parent b. Therefore, a is a grandparent of b. So, $a\,R^2\,b$ if and only if a is a grandparent of b.
- $a\,R^{-1}\,b \Leftrightarrow b\,R\,a$. So $a\,R^{-1}\,b$ if and only if a is a child of b.

(4)
- $a\,(R \cap S)\,b$ if and only if a has both direct and indirect flights from Ottawa to b.
- $a\,(R \cup S)\,b$ if and only if a has either direct or indirect flights from Ottawa to b.
- $a\,(R \backslash S)\,b$ if and only if a has only direct flights from Ottawa to b.
- $a\,(S \backslash R)\,b$ if and only if a has only indirect flights from Ottawa to b.

(7) Let $a,b \in \mathbb{Z}$. Then:

- $a\,R^{-1}\,b \Leftrightarrow b\,R\,a \Leftrightarrow b-a$ is even $\Leftrightarrow a-b$ is even $\Leftrightarrow a\,R\,b$. We conclude that $R^{-1} = R$.
- $a\,\neg R\,b \Leftrightarrow a\,\not{R}\,b \Leftrightarrow a-b$ is odd.

(8)(a) False: $1R^{-1}3 \Leftrightarrow 3\,R\,1 \Leftrightarrow 1 = 2^k(3)$ for some $k \in \mathbb{Z}$ which is not true.

(c) False: $1\,\not{R}\,3$ since $3 \neq 2^k \cdot 1$ for any $k \in \mathbb{Z}$. So $1\,(R \not{\cap} S)\,3$.

(h) True: enough that one of $\frac{1}{2}R\frac{1}{16}$ and $\frac{1}{2}S\frac{1}{16}$ is true and they are both true.

(9)(b) True: $c\,S^{-1}\,a \Leftrightarrow a\,S\,c$ which is true.

(c) False: $a\,\not{R}\,b$ since $3 \neq 2^k \cdot 1$ for any $k \in \mathbb{Z}$. So $1\,(R \not{\cap} S)\,3$.

(g) False: $d\,(R \cap S)^{-1}\,b \Leftrightarrow b\,(R \cap S)\,d$ which is false since $b\,R\,d$ is false.

(12)(a) Let $a \in A$. If $a \in \mathrm{Dom}\,(R \cap S)$, then $(a,b) \in R \cap S$ for some $b \in B$ and so $(a,b) \in R$ and $(a,b) \in S$ for some $b \in B$. This shows that $a \in \mathrm{Dom}(R)$ and $a \in \mathrm{Dom}(S)$. The result follows.

(14) Let $A = B = \{1,2,3,4\}$, $R = \{(1,1),(1,2),(2,3),(2,4),(3,4)\}$, $S = \{(1,2),(2,2),(4,3)\}$. Then $\mathrm{Ran}(R) = \{1,2,3,4\}$, $\mathrm{Ran}(S) = \{2,3\}$. So, $\mathrm{Ran}(R) \cap \mathrm{Ran}(S) = \{2,3\}$. On the other hand, $R \cap S = \{(1,2)\}$ and so $\mathrm{Ran}(R \cap S) = \{2\} \neq \mathrm{Ran}(R) \cap \mathrm{Ran}(S)$.

(15) Let $a \in A$, $b \in B$ be two arbitrary elements.

(c) $(b,a) \in (R \cap S)^{-1} \Leftrightarrow (a,b) \in R \cap S \Leftrightarrow (a,b) \in R$ and $(a,b) \in S \Leftrightarrow (b,a) \in R^{-1}$ and $(b,a) \in S^{-1} \Leftrightarrow (b,a) \in R^{-1} \cap S^{-1}$.

(16)(a)

$$M_{R^{-1}} = (M_R)^T = \begin{array}{c} \\ x \\ y \\ z \\ w \end{array}\begin{array}{cccc} a & b & c & d \\ \end{array}\begin{bmatrix} 1 & 0 & 0 & 1 \\ 1 & 0 & 0 & 0 \\ 1 & 1 & 0 & 0 \\ 0 & 0 & 1 & 1 \end{bmatrix}, \quad M_{S^{-1}} = (M_S)^T = \begin{array}{c} \\ x \\ y \\ z \\ w \end{array}\begin{array}{cccc} a & b & c & d \\ \end{array}\begin{bmatrix} 0 & 0 & 1 & 1 \\ 0 & 0 & 0 & 0 \\ 0 & 1 & 0 & 0 \\ 1 & 0 & 0 & 0 \end{bmatrix}$$

$$M_{R \cap S} = M_R \wedge M_S = \begin{array}{c} \\ a \\ b \\ c \\ d \end{array}\begin{array}{cccc} x & y & z & w \\ \end{array}\begin{bmatrix} 0 & 0 & 0 & 0 \\ 0 & 0 & 1 & 0 \\ 0 & 0 & 0 & 0 \\ 1 & 0 & 0 & 0 \end{bmatrix}, \quad M_{R \cup S} = M_R \vee M_S = \begin{array}{c} \\ a \\ b \\ c \\ d \end{array}\begin{array}{cccc} x & y & z & w \\ \end{array}\begin{bmatrix} 1 & 1 & 1 & 1 \\ 0 & 0 & 1 & 0 \\ 1 & 0 & 0 & 1 \\ 1 & 0 & 0 & 1 \end{bmatrix}$$

$$M_{R \setminus S} = \begin{array}{c} \\ a \\ b \\ c \\ d \end{array}\begin{array}{cccc} x & y & z & w \\ \end{array}\begin{bmatrix} 1 & 1 & 1 & 0 \\ 0 & 0 & 0 & 0 \\ 0 & 0 & 0 & 1 \\ 0 & 0 & 0 & 1 \end{bmatrix}, \quad M_{S \setminus R} = \begin{array}{c} \\ a \\ b \\ c \\ d \end{array}\begin{array}{cccc} x & y & z & w \\ \end{array}\begin{bmatrix} 0 & 0 & 0 & 1 \\ 0 & 0 & 0 & 0 \\ 1 & 0 & 0 & 0 \\ 0 & 0 & 0 & 0 \end{bmatrix}$$

(19) Let $a, b \in P$. Then $a\,(S \circ R)\,b$ if and only if there exists $c \in P$ such that $a\,R\,c$ and $c\,S\,b$. So a is a daughter of c and c is a sibling of b, which makes b an aunt or an uncle of a. Therefore, $a\,(R \circ R)\,b$ if and only if b is an aunt or an uncle of a. Now, $a\,(R \circ S)\,b$ if and only if there exists $c \in P$ such that $a\,S\,c$ and $c\,R\,b$. So a is a sibling of c and c is a daughter of b, which makes b a parent of a. Therefore, $a\,(S \circ S)\,b$ if and only if b is a parent of a.

(22)(a) False. $1\,(R \circ S)\,-1$ if and only if there exists $c \in \mathbb{Z}$ such that $1\,S\,c$ and $c\,R\,-1$. But $c\,\not{R}\,-1$ since otherwise $-1 = c^2$.

  (h) True. $0\,R\,0$ and $0\,S\,0$ and so $0\,(S \circ R)\,0$.

  (i) True. $-1\,R\,1$ (since $1 = (-1)^2$ and $1\,S\,1$ (since $1^2 = 1$), so $-1\,(R \circ R)$.

  (l) True. $-\sqrt[16]{2}\,S\,\sqrt{2}$ (since $\sqrt{2} = (-\sqrt[16]{2})^8$ and $\sqrt{2}\,S\,16$ (since $(\sqrt{2})^8 = 16$), so $-\sqrt[16]{2}\,(S \circ S)\,16$.

(24)(a)

$(a, c) \in (S \cup T) \circ R$

$\Leftrightarrow (\exists b \in B)((a, b) \in R \wedge (b, c) \in S \cup T)$

$\Leftrightarrow (\exists b \in B)((a, b) \in R \wedge ((b, c) \in S \vee (b, c) \in T))$

$\Leftrightarrow (\exists b \in B)((a, b) \in R \wedge (b, c) \in S)$

$\vee ((a, b) \in R \wedge (b, c) \in T))$

$\Leftrightarrow (\exists b \in B)((a, b) \in R \wedge (b, c) \in S)$

$\vee (\exists b \in B)((a, b) \in R \wedge (b, c) \in T))$

$\Leftrightarrow (a, c) \in S \circ R \vee (a, c) \in T \circ R$

$\Leftrightarrow (a, c) \in (S \circ R) \cup (T \circ R).$

(26) Note first that $R^{-1} = \{(b, a), (d, a), (a, b), (b, d)\}$. The adjacency matrix of R^{-1} is

$$
M_{R^{-1}} = \begin{array}{c} \\ a \\ b \\ c \\ d \end{array}
\begin{array}{c} a\ b\ c\ d \\ \begin{bmatrix} 0\ 1\ 0\ 0 \\ 1\ 0\ 0\ 1 \\ 0\ 0\ 0\ 0 \\ 1\ 0\ 0\ 0 \end{bmatrix} \end{array}
$$

So the adjacency matrix of $R^{[-2]}$ is

$$
M_{R^{[-2]}} = \begin{array}{c} a \\ b \\ c \\ d \end{array}
\begin{array}{c} a\ b\ c\ d \\ \begin{bmatrix} 0\ 1\ 0\ 0 \\ 1\ 0\ 0\ 1 \\ 0\ 0\ 0\ 0 \\ 1\ 0\ 0\ 0 \end{bmatrix} \end{array}
\odot
\begin{array}{c} a \\ b \\ c \\ d \end{array}
\begin{array}{c} a\ b\ c\ d \\ \begin{bmatrix} 0\ 1\ 0\ 0 \\ 1\ 0\ 0\ 1 \\ 0\ 0\ 0\ 0 \\ 1\ 0\ 0\ 0 \end{bmatrix} \end{array}
=
\begin{array}{c} a \\ b \\ c \\ d \end{array}
\begin{array}{c} a\ b\ c\ d \\ \begin{bmatrix} 1\ 0\ 0\ 1 \\ 1\ 1\ 0\ 0 \\ 0\ 0\ 0\ 0 \\ 0\ 1\ 0\ 0 \end{bmatrix} \end{array}
$$

So $R^{[-2]} = \{(a, a), (a, d), (b, a), (b, b), (d, b)\}$. For $R^{[-3]}$:

$$
M_{R^{[-3]}} = \begin{array}{c} a \\ b \\ c \\ d \end{array}
\begin{array}{c} a\ b\ c\ d \\ \begin{bmatrix} 1\ 0\ 0\ 1 \\ 1\ 1\ 0\ 0 \\ 0\ 0\ 0\ 0 \\ 0\ 1\ 0\ 0 \end{bmatrix} \end{array}
\odot
\begin{array}{c} a \\ b \\ c \\ d \end{array}
\begin{array}{c} a\ b\ c\ d \\ \begin{bmatrix} 0\ 1\ 0\ 0 \\ 1\ 0\ 0\ 1 \\ 0\ 0\ 0\ 0 \\ 1\ 0\ 0\ 0 \end{bmatrix} \end{array}
=
\begin{array}{c} a \\ b \\ c \\ d \end{array}
\begin{array}{c} a\ b\ c\ d \\ \begin{bmatrix} 1\ 1\ 0\ 0 \\ 1\ 1\ 0\ 1 \\ 0\ 0\ 0\ 0 \\ 1\ 0\ 0\ 1 \end{bmatrix} \end{array}
$$

So $R^{[-3]} = \{(a, a), (a, b), (b, a), (b, d), (d, a), (d, d)\}$.

(28) The adjacency matrices M_R and M_S of R and S are as follows:

$$
M_R = \begin{array}{c} a_1 \\ a_2 \\ a_3 \\ a_4 \end{array}
\begin{array}{c} b_1\ b_2\ b_3 \\ \begin{bmatrix} 0\ 1\ 1 \\ 1\ 0\ 1 \\ 1\ 1\ 0 \\ 0\ 0\ 1 \end{bmatrix} \end{array},
\quad
M_S = \begin{array}{c} b_1 \\ b_2 \\ b_3 \end{array}
\begin{array}{c} c_1\ c_2\ c_3 \\ \begin{bmatrix} 1\ 0\ 1 \\ 1\ 1\ 0 \\ 0\ 1\ 0 \end{bmatrix} \end{array}
$$

The adjacency matrix of the relation $S \circ R$ is $M_R \odot M_S$:

$$
\begin{array}{c}
\begin{array}{ccc} b_1 & b_2 & b_3 \end{array} \\
\begin{array}{c} a_1 \\ a_2 \\ a_3 \\ a_4 \end{array}
\begin{bmatrix} 0 & 1 & 1 \\ 1 & 0 & 1 \\ 1 & 1 & 0 \\ 0 & 0 & 1 \end{bmatrix}
\end{array}
\odot
\begin{array}{c}
\begin{array}{ccc} c_1 & c_2 & c_3 \end{array} \\
\begin{array}{c} b_1 \\ b_2 \\ b_3 \end{array}
\begin{bmatrix} 1 & 0 & 1 \\ 1 & 1 & 0 \\ 0 & 1 & 0 \end{bmatrix}
\end{array}
=
\begin{array}{c}
\begin{array}{ccc} c_1 & c_2 & c_3 \end{array} \\
\begin{array}{c} a_1 \\ a_2 \\ a_3 \\ a_4 \end{array}
\begin{bmatrix} 1 & 1 & 0 \\ 1 & 1 & 1 \\ 1 & 1 & 1 \\ 0 & 1 & 0 \end{bmatrix}
\end{array}
$$

The elements of $S \circ R$ are: (a_1, c_1), (a_1, c_2), (a_2, c_1), (a_2, c_2), (a_2, c_3), (a_3, c_1), (a_3, c_2), (a_3, c_3) and (a_4, c_2).

(30)(a) True. $a\,S\,c$ and $c\,R\,d$, so $a\,(R \circ S)\,d$.

(d) False. the statement is equivalent to $a\,(R \circ S)\,b$ but there exists no $z \in A$ such that $a\,S\,z$ (so $z = c$ or $z = d$) and $z\,R\,b$.

(e) True. the statement is equivalent to $d\,R\,z$ and $z\,S\,b$ for some $z \in A$. The element $z = d$ satisfies these two conditions.

(32)(a)

$$
M_{S \circ R} = M_R \odot M_S =
\begin{array}{c}
\begin{array}{cccc} x & y & z & w \end{array} \\
\begin{array}{c} a \\ b \\ c \\ d \end{array}
\begin{bmatrix} 1 & 1 & 1 & 0 \\ 0 & 0 & 1 & 0 \\ 0 & 0 & 0 & 1 \\ 1 & 0 & 0 & 1 \end{bmatrix}
\end{array}
\odot
\begin{array}{c}
\begin{array}{ccc} 1 & 2 & 3 \end{array} \\
\begin{array}{c} x \\ y \\ z \\ w \end{array}
\begin{bmatrix} 0 & 0 & 0 \\ 0 & 0 & 1 \\ 1 & 0 & 0 \\ 1 & 0 & 0 \end{bmatrix}
\end{array}
=
\begin{array}{c}
\begin{array}{ccc} 1 & 2 & 3 \end{array} \\
\begin{array}{c} a \\ b \\ c \\ d \end{array}
\begin{bmatrix} 1 & 0 & 1 \\ 1 & 0 & 1 \\ 1 & 0 & 1 \\ 1 & 0 & 1 \end{bmatrix}
\end{array}
$$

The elements of $S \circ R$ are $(a, 1)$, $(a, 3)$, $(b, 1)$, $(b, 3)$, $(c, 1)$, $(c, 3)$, $(d, 1)$ and $(d, 3)$.

(35) Let $m, n \in \mathbb{Z}$. Then $(m, n) \in R^{[2]}$ if and only if there exists $z \in \mathbb{Z}$ such that $(m, z) \in R$ and $(z, n) \in R$, which implies that $m + r$ and $r + n$ are both odd and so $m + r + r + n = m + n + 2r$ is even. Since $2r$ is even, $m + n$ must be even. This means that m and n have the same parity (both odd or both even). The converse is also true: assume that m and n have the same parity. If m and n are both even, let $z = 1 \in \mathbb{Z}$, then both $m + z$ and $z + n$ are odd and therefore $(m, n) \in R^{[2]}$. If m and n are both odd, let $z = 0 \in \mathbb{Z}$, then both $m + z$ and $z + n$ are odd and therefore $(m, n) \in R^{[2]}$. We conclude that $(m, n) \in R^{[2]}$ if and only if m and n have the same parity. If $(m, n) \in R^{[3]}$, then there exists $z \in \mathbb{Z}$ such that $(m, z) \in R$ and $(z, n) \in R^{[2]}$. So $m + z$ is odd, z and n have the same parity. If z is odd, then n is odd and m is even (since $m + z$ is odd and z is odd) and so $m + n$ is odd. If z is even, then n is even and m is odd (since $m + z$ is odd and z is even) and so $m + n$ is odd. We conclude that $(m, n) \in R$. Conversely, if $(m, n) \in R$, then $m + n$ is odd. If we choose $z = n \in \mathbb{Z}$, then $m + z$ is odd and n, z have the same parity. This

means that $m\,R\,z$ and $z\,R^{[2]}\,n$. In other words, $(m,n) \in R^{[2]} \circ R = R^{[3]}$. We conclude that $R^{[3]} = R$. Now, $R^{[4]} = R^{[3]} \circ R = R \circ R = R^{[2]}$, $R^{[5]} = R^{[4]} \circ R = R^{[2]} \circ R = R^{[3]}$ and in general, $R^{[n]} = R^{[2]}$ if n is even and $R^{[n]} = R$ if n is odd.

7.6

(2) The adjacency matrix of R is as follows:

$$
M_R = \begin{array}{c} \\ a \\ b \\ c \\ d \\ e \end{array}
\begin{array}{c} a\ b\ c\ d\ e \\ \left[\begin{array}{ccccc} 0 & 0 & 0 & 1 & 0 \\ 1 & 0 & 1 & 0 & 0 \\ 0 & 0 & 0 & 0 & 0 \\ 0 & 1 & 0 & 0 & 1 \\ 0 & 1 & 1 & 0 & 0 \end{array}\right] \end{array}
$$

The adjacency matrix of R^2 and R^3 are as follows:

$$
M_{R^2} = \begin{array}{c} a \\ b \\ c \\ d \\ e \end{array}
\begin{array}{c} a\ b\ c\ d\ e \\ \left[\begin{array}{ccccc} 0 & 1 & 0 & 0 & 1 \\ 0 & 0 & 0 & 1 & 0 \\ 0 & 0 & 0 & 0 & 0 \\ 1 & 1 & 1 & 0 & 0 \\ 1 & 0 & 1 & 0 & 0 \end{array}\right] \end{array},
\qquad
M_{R^3} = \begin{array}{c} a \\ b \\ c \\ d \\ e \end{array}
\begin{array}{c} a\ b\ c\ d\ e \\ \left[\begin{array}{ccccc} 1 & 1 & 1 & 0 & 0 \\ 0 & 1 & 0 & 0 & 1 \\ 0 & 0 & 0 & 0 & 0 \\ 1 & 0 & 1 & 1 & 0 \\ 0 & 0 & 0 & 1 & 0 \end{array}\right] \end{array}
$$

The digraphs of R^2 and R^3 are as follows:

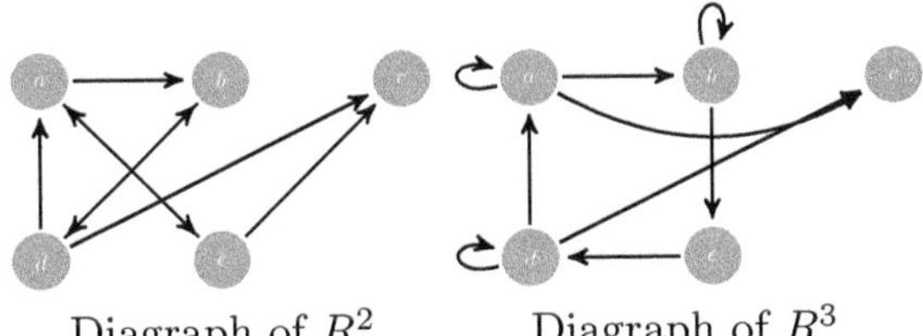

Diagraph of R^2 Diagraph of R^3

(3) The adjacency matrix of R is:

$$
M_R = \begin{array}{c} a \\ b \\ c \\ d \\ e \end{array}
\begin{array}{c} a\ b\ c\ d\ e \\ \left[\begin{array}{ccccc} 0 & 0 & 0 & 1 & 0 \\ 1 & 0 & 1 & 1 & 1 \\ 0 & 0 & 0 & 0 & 0 \\ 0 & 1 & 0 & 0 & 1 \\ 0 & 1 & 1 & 0 & 0 \end{array}\right] \end{array}
$$

The adjacency matrices of R^n, $n = 2, 3, 4, 5$ are as follows:

$$M_{R^2} = \begin{array}{c} \\ a \\ b \\ c \\ d \\ e \end{array}\begin{array}{c} a\ b\ c\ d\ e \\ \left[\begin{array}{ccccc} 0 & 1 & 0 & 0 & 1 \\ 0 & 1 & 1 & 1 & 1 \\ 0 & 0 & 0 & 0 & 0 \\ 1 & 1 & 1 & 1 & 1 \\ 1 & 0 & 1 & 1 & 1 \end{array}\right] \end{array}, \quad M_{R^3} = \begin{array}{c} \\ a \\ b \\ c \\ d \\ e \end{array}\begin{array}{c} a\ b\ c\ d\ e \\ \left[\begin{array}{ccccc} 1 & 1 & 1 & 1 & 1 \\ 1 & 1 & 1 & 1 & 1 \\ 0 & 0 & 0 & 0 & 0 \\ 1 & 1 & 1 & 1 & 1 \\ 0 & 1 & 1 & 1 & 1 \end{array}\right] \end{array}$$

$$M_{R^4} = \begin{array}{c} \\ a \\ b \\ c \\ d \\ e \end{array}\begin{array}{c} a\ b\ c\ d\ e \\ \left[\begin{array}{ccccc} 1 & 1 & 1 & 1 & 1 \\ 1 & 1 & 1 & 1 & 1 \\ 0 & 0 & 0 & 0 & 0 \\ 1 & 1 & 1 & 1 & 1 \\ 1 & 1 & 1 & 1 & 1 \end{array}\right] \end{array}, \quad M_{R^5} = \begin{array}{c} \\ a \\ b \\ c \\ d \\ e \end{array}\begin{array}{c} a\ b\ c\ d\ e \\ \left[\begin{array}{ccccc} 1 & 1 & 1 & 1 & 1 \\ 1 & 1 & 1 & 1 & 1 \\ 0 & 0 & 0 & 0 & 0 \\ 1 & 1 & 1 & 1 & 1 \\ 1 & 1 & 1 & 1 & 1 \end{array}\right] \end{array}$$

Therefore,

$$M_{R^\infty} = R \vee R^2 \vee R^3 \vee R^4 \vee R^5 = \begin{array}{c} \\ a \\ b \\ c \\ d \\ e \end{array}\begin{array}{c} a\ b\ c\ d\ e \\ \left[\begin{array}{ccccc} 1 & 1 & 1 & 1 & 1 \\ 1 & 1 & 1 & 1 & 1 \\ 0 & 0 & 0 & 0 & 0 \\ 1 & 1 & 1 & 1 & 1 \\ 1 & 1 & 1 & 1 & 1 \end{array}\right] \end{array}$$

(a) $R^2(a) = \{b, d\}$

(b) $R^3(a) = A$

(f) $R^2(e) = \{a, c, d, e\}$

(g) $R^4(e) = A$

(k) $R^\infty(c) = \emptyset$

(6) Since A has four elements, $M_{R^\infty} = M_R \vee M_{R^2} \vee M_{R^3} \vee M_{R^4}$.

(a)

$$M_{R^2} = \begin{array}{c} \\ a \\ b \\ c \\ d \end{array}\begin{array}{c} a\ b\ c\ d \\ \left[\begin{array}{cccc} 0 & 0 & 0 & 0 \\ 0 & 1 & 0 & 1 \\ 1 & 0 & 1 & 0 \\ 0 & 0 & 0 & 0 \end{array}\right] \end{array}, \quad M_{R^3} = \begin{array}{c} \\ a \\ b \\ c \\ d \end{array}\begin{array}{c} a\ b\ c\ d \\ \left[\begin{array}{cccc} 0 & 0 & 0 & 0 \\ 1 & 0 & 1 & 0 \\ 0 & 1 & 0 & 1 \\ 0 & 0 & 0 & 0 \end{array}\right] \end{array}, \quad M_{R^4} = \begin{array}{c} \\ a \\ b \\ c \\ d \end{array}\begin{array}{c} a\ b\ c\ d \\ \left[\begin{array}{cccc} 0 & 0 & 0 & 0 \\ 0 & 1 & 0 & 1 \\ 1 & 0 & 1 & 0 \\ 0 & 0 & 0 & 0 \end{array}\right] \end{array}$$

Therefore, $M_{R^\infty} = \begin{array}{c} \\ a \\ b \\ c \\ d \end{array}\begin{array}{c} a\ b\ c\ d \\ \left[\begin{array}{cccc} 0 & 0 & 0 & 1 \\ 1 & 1 & 1 & 1 \\ 1 & 1 & 1 & 1 \\ 0 & 0 & 0 & 0 \end{array}\right] \end{array}$

(10)(a) For $n = 1$, $a\,R\,b \Leftrightarrow b = a + 1 \Leftrightarrow a = b - 1$. The property is true in this case. Let $n \geq 1$ and assume that $a\,R^n\,b$ if and only if $a = b - n$. Then $a\,R^{n+1}\,b$ if and only if there exists $c \in \mathbb{Z}$ such that $a\,R\,c$ and $c\,R^n\,b$. By the induction hypothesis, this is equivalent to say that $c = b - n$ and $a = c - 1$, which in turns is equivalent to $a = (b - n) - 1 = b - (n + 1)$. The result is then true for any $n \geq 1$.

   (b) For $a, b \in \mathbb{Z}$, $a\,R^{\infty}\,b \Leftrightarrow a\,R^n\,b$ for some integer $n \geq 1 \Leftrightarrow a = b - n$ for some integer $n \geq 1$.

7.7

(2)(a) R is reflexive (every vertex has a loop), not symmetric since $b\,R\,c$ but $c\,\not{R}\,b$, not transitive since $a\,R\,b$ and $b\,R\,c$ but $a\,\not{R}\,c$, not irreflexive since $a\,R\,a$, not asymmetric since $a\,R\,b$ and $b\,R\,a$ are true at the same time, not antisymmetric since $a\,R\,b$ and $b\,R\,a$ but $a \neq b$.

(4)

		Reflexive	Symmetric	Transitive	Irreflexive	Antisymmetric	Asymmetric
R_1		Yes	No	No	No	No	No
R_2		No	No	No	Yes	Yes	Yes
R_3		No	No	No	Yes	No	No
R_4		Yes	Yes	Yes	No	Yes	No

(5)(a) R_1 is not reflexive as no one shakes hands with himself. The relation is symmetric since if m_i shook hand with m_j, then m_j also shook hand with m_i. The relation is not transitive as it is very possible that m_i shook hand with m_j and m_j shook hand with m_k but m_i did not shake hand with m_k.

   (e) Clearly R_5 is not reflexive. It is symmetric since if m_i met m_j for the first time at the conference, then m_j met m_i for the first time at the conference. The relation is not transitive: If m_i met m_j for the first time at the conference and m_j met m_k for the first time at the conference, it is not automatic that m_i met m_k for the first time at the conference.

(6)(c) R_4 is not reflexive since x is not a cousin of x. It is symmetric since if x is a cousin of y, then y is a cousin of x. It is not transitive since if x and y are cousins and y and z are cousins, it does not imply that x and z are cousins.

   (d) R_5 is clearly reflexive as any human x has the same height and the same eye color as x. It is symmetric since if x and y have either the same height or the same eye color, then y and x have either the same height or the same eye color. It is not transitive since if x and y have either the same height or the same eye color and y and z have either the same height or the same eye color, it does not mean that x and z have either the same height or the same eye color.

(7)(a) The relation is reflexive (entries on the main diagonal are all 1's), it is symmetric (the transpose of its adjacency matrix is the same as the adjacency matrix) and it is transitive (every time $x\,R\,y$ and $y\,R\,z$, we have that $x\,R\,z$). It is not irreflexive (since it is reflexive), not antisymmetric ($a\,R\,b$ and $b\,R\,a$ but $a \neq b$) and not asymmetric (both (a,b) and (b,a) are in R).

(d) The relation is not reflexive (entries on the main diagonal are not all 1's), it is symmetric (the transpose of its adjacency matrix is the same as the adjacency matrix) and it is not transitive ($a\,R\,e$ and $e\,R\,a$ but $a\,\not{R}\,a$). It is not irreflexive (since $d\,R\,d$), not antisymmetric ($(a,e) \in R$ and $(e,a) \in R$ but $a \neq e$) and not asymmetric ($(a,e) \in R$ and $(e,a) \in R$ at the same time).

(8)(b) The relation is reflexive: $A \cup X = A \cup X$ for any $A \in \wp(\mathcal{U})$. It is symmetric: if $A \cup X = B \cup X$ then $B \cup X = A \cup X$ for any $A, B \in \wp(\mathcal{U})$. It is transitive: for any $A, B, C \in \wp(\mathcal{U})$ if $A\,R\,B$ and $B\,R\,C$, then $A \cup X = B \cup X$ and $B \cup X = C \cup X$ and therefore $A \cup X = C \cup X$ which means that $A\,R\,C$.

(d) The relation is reflexive: $A \backslash A = \emptyset$ for any $A \in \wp(\mathcal{U})$. It is not symmetric: let $A = \{1\}$, $B = \{1,2\}$ then $A \backslash B = \emptyset$ but $B \backslash A = \{2\} \neq emptyset$. It is transitive: for any $A, B, C \in \wp(\mathcal{U})$ if $A\,R\,B$ and $B\,R\,C$, then $A \backslash B = B \backslash C = \emptyset$, which means that $A \subseteq B$ and $B \subseteq C$ and therefore $A \subseteq C$ which means that $A \backslash C = \emptyset$ and $A\,R\,C$.

(9)(a) The relation is not reflexive: $(1)(1) = 1 \neq 0$. It is symmetric: if $xy = 0$ then $yx = 0$. It is not transitive: $1\,R\,0$, $0\,R\,2$ but $1\,\not{R}\,2$.

(c) The relation is reflexive: $|m - m| = 0 \leq 3$ for any $m \in \mathbb{Z}$. It is symmetric: if $|m - n| \leq 3$, then $|n - m| = |m - n| \leq 3$ for any $m, n \in \mathbb{Z}$. It is not transitive: $3\,R\,0$, $0\,R\,-1$ but $3\,\not{R}\,-1$.

(f) The relation is not reflexive: $2\,\not{R}\,2$ since $\gcd(2,2) = 2 \neq 1$. It is symmetric: if $\gcd(m,n) = 1$, then $\gcd(n,m) = 1$ for any $m, n \in \mathbb{N} \backslash \{0\}$. It is not transitive: $\gcd(3,4) = 1$ and $\gcd(4,9) = 1$ but $\gcd(3,9) = 3 \neq 1$.

(13)(b) Consider the following relation on $\mathbb{R}$: $x\,R\,y \Leftrightarrow x + y = 1$. R is not reflexive since, for example $1\,/\,1$ ($1 + 1 = 1 \neq 1$), R is symmetric (for any $x, y \in \mathbb{R}$: if $x + y = 1$, then $y + x = 1$), R is not transitive since $\frac{1}{3}\,R\,\frac{2}{3}$ and $\frac{2}{3}\,R\,\frac{1}{3}$ but $\frac{1}{3}\,\not{R}\,\frac{1}{3}$ ($\frac{1}{3} + \frac{1}{3} \not\leq 1$).

(14) Let $x, y, z \in A$.

(d) If $(x, y) \in R \cup S$, then $(x, y) \in R$ or $(x, y) \in S$. Since R and S are symmetric, $(y, x) \in R$ or $(y, x) \in S$ and so $(y, x) \in R \cup S$. This shows that the relation $R \cup S$ is symmetric.

(e) If $(x, y) \in R \cap S$ and $(y, z) \in R \cap S$, then $(x, y) \in R$, $(x, y) \in S$, $(y, z) \in R$ and $(y, z) \in S$. Since R and S are transitive, $(x, z) \in R$ and $(x, z) \in S$ and so $(x, z) \in R \cap S$. This shows that the relation $R \cap S$ is transitive.

(18) Let $A = \{a, b, c, d, e\}$ and consider the relations $R = \{(a, b), (c, d)\}$ and $S = \{(b, c), (d, e)\}$ on A. Clearly, R and S are both transitive but $S \circ R = \{(a, c), (c, e)\}$ is not.

(20) Let R be a symmetric relation on A. We use induction to prove that R^n is symmetric for any $n \geq 1$. The property is clearly true for $n = 1$ since $R^1 = R$ is symmetric. Let $n \geq 1$ and assume that R^n is symmetric. Let $x, y \in A$ such that $x \, R^{n+1} \, y$. Then there exists $z \in A$ such that $x \, R \, z$ and $z \, R^n \, y$. Since R is symmetric, $z \, R \, x$ and by the induction hypothesis, R^n is symmetric and so $y \, R^n \, z$. The conditions $y \, R^n \, z$ and $z \, R \, x$ imply that $y \, (R \circ R^n) \, x$ or equivalently, $y \, R^{n+1} \, x$ and R^{n+1} is symmetric.

(22) Let $x, y, z \in A$ such that $x \, (R \circ S) \, y$ and $y \, (R \circ S) \, z$. Then there exists $a, b \in A$ such that $x \, S \, a$, $a \, R \, y$, $y \, S \, b$ and $b \, R \, z$. The conditions $a \, R \, y$, $y \, S \, b$ imply that $a \, (S \circ R) \, b$. Since $S \circ R \subseteq R \circ S$, we get that $a \, (R \circ S) \, b$. Then there exists $c \in A$ such that $a \, S \, c$ and $c \, R \, b$. Since R and S are transitive, $x \, S \, c$ and $c \, R \, z$. This shows $x \, (R \circ S) \, z$ and so $R \circ S$ is transitive.

(23)(c) Assume R is antisymmetric. Let $(x, y) \in R \cap R^{-1}$, then $(x, y) \in R$ and $(x, y) \in R^{-1}$ and so $(x, y) \in R$ and $(y, x) \in R$. Since R is antisymmetric, $x = y$ and therefore, $(x, y) \in I_A$. This shows that $R \cap R^{-1} \subseteq I_A$. For the converse, assume that $R \cap R^{-1} \subseteq I_A$ and let $x, y \in A$ be such that $(x, y) \in R$ and $(y, x) \in R$. Then $(x, y) \in R$ and $(x, y) \in R^{-1}$ and so $(x, y) \in I_A$. This shows that $x = y$ and R is antisymmetric.

7.8

(3)(a) If R is reflexive, then $I_A \subseteq R$ and so $r(R) = R \cup I_A = R$. Conversely, if $r(R) = R$, then $R \cup I_A = R$ and so $I_A \subseteq R$ which means that R is reflexive.

(b) If R is symmetric, then $R^{-1} = R$ and so $s(R) = R \cup R^{-1} = R$. Conversely, if $s(R) = R$, then $R \cup R^{-1} = R$ and so $R^{-1} \subseteq R$ which means that R is symmetric.

(c) If R is transitive, then it is the smallest transitive relation on A containing R and so $t(R) = R$ by definition of $t(R)$. Conversely, if $R = t(R)$, then R is in particular transitive since $t(R)$ is transitive.

(4)(a)   − $r(R) = \{(a,a), (a,e), (b,b), (c,a), (c,c), (d,d), (e,e)\}$.
     − $s(R) = \{(a,a), (a,e), (e,a), (b,b), (c,a), (a,c)\}$.
     − $t(R) = \{(a,a), (a,c), (a,e), (b,b), (c,a), (c,e)\}$.

(6)(b) $r(R) = R$ since R is reflexive (a divides a for any $a \in \mathbb{Z}\backslash\{0\}$, $s(R)$ is the relation "a divides b or b divides a", $t(R) = R$ since R is transitive (a divides b and b divides c imply that a divides c).

(e) $r(R) = R$ since R is reflexive ($m + m = 2m$ is even for any integer m), $s(R) = R$ since R is symmetric (if $m + n$ is even, then $n + m$ is even), and $t(R) = R$ since R is transitive (if $a\,R\,b$ and $b\,R\,c$, then $a + b$ and $b + c$ are even and so $(a+b) + (b+c) = a + c + 2b$ is even. But that means that $a + c$ is even).

(7)(b)

$$M_{r(R)} = \begin{array}{c} \\ a \\ b \\ c \end{array}\begin{array}{c} a\ b\ c \\ \begin{bmatrix} 1\ 1\ 0 \\ 1\ 1\ 0 \\ 1\ 0\ 1 \end{bmatrix}\end{array} \quad M_{s(R)} = \begin{array}{c} \\ a \\ b \\ c \end{array}\begin{array}{c} a\ b\ c \\ \begin{bmatrix} 0\ 1\ 1 \\ 1\ 0\ 0 \\ 1\ 0\ 1 \end{bmatrix}\end{array} \quad M_{t(R)} = \begin{array}{c} \\ a \\ b \\ c \end{array}\begin{array}{c} a\ b\ c \\ \begin{bmatrix} 1\ 1\ 0 \\ 1\ 1\ 0 \\ 1\ 1\ 1 \end{bmatrix}\end{array}$$

(8)(c)

$$r(R) = \begin{array}{c} \\ a \\ b \\ c \\ d \end{array}\begin{array}{c} a\ b\ c\ d \\ \begin{bmatrix} 1\ 0\ 1\ 0 \\ 1\ 1\ 1\ 0 \\ 1\ 0\ 1\ 0 \\ 1\ 0\ 0\ 1 \end{bmatrix}\end{array} \quad s(R) = \begin{array}{c} \\ a \\ b \\ c \\ d \end{array}\begin{array}{c} a\ b\ c\ d \\ \begin{bmatrix} 1\ 1\ 1\ 1 \\ 1\ 1\ 1\ 0 \\ 1\ 1\ 1\ 0 \\ 1\ 0\ 0\ 1 \end{bmatrix}\end{array} \quad t(R) = \begin{array}{c} \\ a \\ b \\ c \\ d \end{array}\begin{array}{c} a\ b\ c\ d \\ \begin{bmatrix} 1\ 1\ 1\ 0 \\ 1\ 1\ 1\ 0 \\ 1\ 1\ 1\ 0 \\ 1\ 1\ 1\ 1 \end{bmatrix}\end{array}$$

(9)(b)

$$r(R) = \begin{array}{c} \\ a \\ b \\ c \\ d \\ e \end{array}\begin{array}{c} a\ b\ c\ d\ e \\ \begin{bmatrix} 1\ 0\ 1\ 0\ 1 \\ 1\ 1\ 1\ 0\ 0 \\ 0\ 0\ 1\ 1\ 0 \\ 0\ 0\ 0\ 1\ 0 \\ 0\ 0\ 0\ 1\ 1 \end{bmatrix}\end{array} \quad s(R) = \begin{array}{c} \\ a \\ b \\ c \\ d \\ e \end{array}\begin{array}{c} a\ b\ c\ d\ e \\ \begin{bmatrix} 1\ 1\ 1\ 0\ 1 \\ 1\ 0\ 1\ 0\ 0 \\ 1\ 1\ 0\ 1\ 0 \\ 0\ 0\ 1\ 1\ 1 \\ 1\ 0\ 0\ 1\ 0 \end{bmatrix}\end{array} \quad t(R) = \begin{array}{c} \\ a \\ b \\ c \\ d \\ e \end{array}\begin{array}{c} a\ b\ c\ d\ e \\ \begin{bmatrix} 1\ 0\ 1\ 1\ 1 \\ 1\ 0\ 1\ 1\ 1 \\ 0\ 0\ 0\ 1\ 0 \\ 0\ 0\ 0\ 1\ 0 \\ 0\ 0\ 0\ 1\ 0 \end{bmatrix}\end{array}$$

(11)(b) Suppose that R is symmetric. If $(x, y) \in r(R) = R \cup I_A$, then either $x\,R\,y$ or $x = y$ and so either $y\,R\,x$ or $y = x$. This means that $(y, x) \in R \cup I_A = r(R)$ and thus $r(R)$ is symmetric. Next, let $x, y \in A$ such that $x\,t(R)\,y$. By definition of $t(R)$, there exists an integer $n \geq 1$ such that $(x, y) \in R^n$. By definition of R^n, there exist elements $y_0, y_1, \ldots, y_n \in A$ such that $y_0 = x$, $y_n = y$ and $y_k\,R\,y_{k+1}$ for every

k. Let $z_k = y_{n-k}$. Then $z_0 = y_n = y$, $z_n = y_0 = x$ and since $y_k\,R\,y_{k+1}$ for every k and R is symmetric, $y_{n-k}\,R\,y_{n-k-1}$ for every $k = 0, \ldots, n-1$. This means that $z_k\,R\,z_{k+1}$. By definition, we get that $(y, x) \in R^n$ and thus $(y, x) \in t(R)$.

(12) Let $A = \{a, b, c\}$ and consider the relation $R = \{(a, b), (b, c), (a, c)\}$ on A. Then R is transitive but $s(R)$ is not since both (a, b) and (b, a) are in $s(R)$ but $(a, a) \notin A$.

(13)(b) Note that $R \subseteq r(R) \subseteq tr(R)$. So $tr(R)$ is a transitive relation containing R. Therefore, $tr(R)$ must contain $t(R)$ since the latter is the smallest transitive relation containing R. By Exercise (11), $tr(R)$ is also reflexive (since $r(R)$ is). Therefore, $rt(R) \subseteq tr(R)$ since $rt(R)$ is the smallest reflexive relation on A containing $t(R)$. On the other hand, $rt(R)$ is transitive by Exercise (11) (since $t(R)$ is). We also have that $R \subseteq t(R) \subseteq rt(R)$, so $r(R) \subseteq rt(R)$ since $rt(R)$ is reflexive and contains R. Since $rt(R)$ is a transitive relation containing $r(R)$, $tr(R) \subseteq rt(R)$. We conclude that $tr(R) = rt(R)$.

7.9

(2) R is not reflexive as $(5, 5) \notin R$. The relation is not symmetric since (for instance) $(2, 5) \in R$ but $(5, 2) \notin R$. Finally, R is not transitive since (for instance) $(2, 4) \in R$, $(4, 3) \in R$ but $(2, 3) \notin R$.

(3)(b) The relation is not an equivalence relation on A. It is not symmetric: $(1, 2) \in R$ but $(2, 1) \notin R$. It is not transitive since $(4, 1) \in R$ and $(1, 2) \in R$ but $(4, 2) \notin R$.

(5)(b) The relation is not an equivalence relation on A. It is not reflexive (no one is a cousin of himself). It is not transitive since if x is a cousin of y and y is a cousin of z, it does mean that x is a cousin of z.

(e) The relation is not an equivalence relation on A. It is clearly reflexive and symmetric, but not transitive: let $f(x) = x - 1$, $g(x) = x^2 - 1$ and $h(x) = x + 1$ then $f\,R\,g$ (since $f(1) = g(1) = 0$, $g\,R\,h$ (since $g(-1) = h(-1) = 0$ but $f\,\not R\,h$ since $f(1) \neq h(1)$ et $f(-1) \neq h(-1)$.

(7)(a) The relation is an equivalence relation on A. It is reflexive (all entries on the main diagonal of M_R are 1's), symmetric ($M_R^T = M_R$) and transitive: when applying Warshall's algorithm, we find that $W_4 = M_R$.

(10) The answer is 4. First note that $a\,\not R\,b$ since otherwise $[a] = [b]$ but we know this is not case (the classes are of different cardinalities).

Similarly, $a \not{R} c$ and $b \not{R} c$. So $[a]$, $[b]$ and $[c]$ are three distinct elements of the quotient set A/R of A. Since these three classes take $6+1+4 = 11$ elements of A, there exists one more equivalence class consisting of a single element of A.

(13)(a) The relation is reflexive since for every $x \in \Sigma_4$, x and x start and end with same digit. If $x, y \in \Sigma_n$ are such that $x\,R\,y$, then x and y have the same first digit and the same last digit and so y and x have the same first digit and the same last digit and so $y\,R\,x$. The relation is symmetric. If $x, y, z \in \Sigma_4$ are such that $x\,R\,y$ and $y\,R\,z$, then x and y have the same first digit and the same last digit and y and z have the same first digit and the same last digit. This implies that x and z have the same first digit and the same last digit and so $x\,R\,z$. The relation is transitive and hence it is an equivalence relation.

(b) Note first that there are $2^4 = 16$ elements in Σ_4. the partition of Σ_4 is determined by the equivalence classes of R: $[1111] = \{1111, 1101, 1011, 1001\}$, $[1110] = \{1111, 1100, 1010, 1000\}$, $[0111] = \{0111, 0101, 0011, 0001\}$ and $[0110] = \{0110, 0100, 0010, 0000\}$.

(c) The relation is not an equivalence relation since it is not transitive: $1111\,R\,1100$ and $1100\,R\,0000$ but $1111\,\not{R}\,0000$.

(17)(a) For any $a \in \mathbb{Z}$, we have that $3a + 5a = 8a$ which is even. This shows that $a\,R\,a$ and the relation is reflexive. For $a, b \in \mathbb{Z}$ if $a\,R\,b$ then $3a + 5b$ is even and so $3b + 5a = 3a + 5b + 2a - 2b = 3a + 5b + 2(a - b)$ is even since both $3a + 5b$ and $2(a - b)$ are even. This shows that R is symmetric. Finally, let $a, b, c \in \mathbb{Z}$ be such that $a\,R\,b$ and $b\,R\,c$, then both $3a + 5b$ and $3b + 5c$ are even. Therefore, $(3a + 5b) + (3b + 5c) = 3a + 5c + 8b$ is even and consequently, $3a + 5c$ is even since $8a$ is even. This shows that R is transitive and hence an equivalence relation on $\mathbb{Z}$.

(b) Let $a \in \mathbb{Z}$. If a is even, then $3a + 5(0)$ is even and so $a \in [0]$. If a is odd, then $3a + 5(1)$ is even (as the sum of two odd integers) and so $a \in [1]$. Since $0\,\not{R}\,1$ (as $3(0) + 5(1) = 5$ is odd), $[0] \neq [1]$ and these are the only two classes in the quotient set $\mathbb{Z}/R$.

(20)(a) For any $a \in \mathbb{R}^*$, $|a|a = a|a|$ and R is reflexive. If $a, b \in \mathbb{R}^*$ are such that $a\,R\,b$, then $|a|b = a|b|$ and so $|b|a = b|a|$ which means that $b\,R\,a$. The relation is symmetric. Let $a, b, c \in \mathbb{R}^*$ such that $a\,R\,b$ and $b\,R\,c$. Then $|a|b = a|b|$ and $|b|c = b|c|$. Multiplying the second equation by a (which is non-zero) gives: $a|b|c = ab|c|$. Using the first equation we get that $|a|bc = ab|c|$. Finally, simplifying by b (which is non-zero) gives: $|a|c = a|c|$. Therefore, $a\,R\,c$ and the relation is transitive. We conclude that R is an equivalence relation on $\mathbb{R}^*$.

(2) Let $a \in \mathbb{R}^*$. If $a > 0$ then $a \in [1]$ since $|a| \cdot 1 = a \cdot |1| = a$ in this case. If $a < 0$, then $a \in [-1]$ since $|a| \cdot (-1) = a \cdot |-1| = a$ in this case as well. We conclude that the partition of $\mathbb{R}^*$ induced by R consists of two equivalence classes: $[1] = \{a \in \mathbb{R}^*; a > 0\}$ and $[-1] = \{a \in \mathbb{R}^*; a < 0\}$.

(23)(b) R consists of the elements $(1,1)$, $(2,2)$, $(3,3)$, $(4,4)$, $(5,5)$, $(6,6)$, $(1,2)$, $(2,1)$, $(1,3)$, $(3,1)$, $(2,3)$, $(3,2)$, $(4,5)$, $(5,4)$, $(4,6)$, $(6,4)$, $(5,6)$ and $(6,5)$.

(24)(a) This is not a well-defined function: $[1]_3 = [4]_3$ but $[1]_4 \neq [4]_4$.

(c) This is a well-defined function: If $[x]_{18}, [y]_{18} \in \mathbb{Z}_{18}$ are such that $[x]_{18} = [y]_{18}$ then $x - y = 18k$ for some $k \in \mathbb{Z}$. But then $[x]_6 = [y]_6$ since $x - y = 18k = 6(3k)$.

(f) This is a well-defined function: If $[x]_4 = [y]_4$, then $x - y = 4k$ for some $k \in \mathbb{Z}$ and so $2^x = 2^{y+4k} = 2^y \cdot 2^{4k} = 2^y \cdot 16^k \equiv 2^y \pmod 3$ since $16 \equiv 1 \pmod 3$.

(25)(b) The relation is reflexive since $x^2 + y = x^2 + y$ for any $(x,y) \in \mathbb{R}^2$. If $(x,y),(a,b) \in \mathbb{R}^2$ are such that $(x,y)\,S\,(a,b)$, then $x^2 + y = a^2 + b$ and so $a^2 + b = x^2 + y$. This shows that $(a,b)\,S\,(x,y)$ and S is symmetric. For the transitivity, let $(x,y),(a,b),(c,d) \in \mathbb{R}^2$ such that $(x,y)\,S\,(a,b)$ and $(a,b)\,S\,(c,d)$, then $x^2 + y = a^2 + b$ and $a^2 + b = c^2 + d$. So $x^2 + y = c^2 + d$. This shows that $(a,b)\,S\,(c,d)$ and S is transitive. We conclude that S is an equivalence relation on $\mathbb{R}^2$. For the equivalence classes, fix $(a,b) \in \mathbb{R}^2$, then $[(a,b)] = \{(x,y) \in \mathbb{R}^2; x^2 + y = a^2 + b\}$. The equation $x^2 + y = k$ (where $k \in \mathbb{R}$) represents the parabola $y = -x^2 + k$ in $\mathbb{R}^2$ obtained by shifting the parabola $y = -x^2$ vertically by a scale of k.

(30)(a) For $x \in A$, $\frac{x}{x} = 1 \in \mathbb{Q}$. The relation is reflexive. If $x, y \in A$ are such that $x\,R\,y$ then $\frac{x}{y} \in \mathbb{Q}$ and so $\frac{y}{x} = \frac{1}{\frac{x}{y}} \in \mathbb{Q}$. This proves that $y\,R\,x$ and R is symmetric. If $x, y, z \in A$ are such that $x\,R\,y$ and $y\,R\,z$ then $\frac{x}{y} \in \mathbb{Q}$ and $\frac{y}{z} \in \mathbb{Q}$. We then get that $\frac{x}{z} = \frac{x}{y}\frac{y}{z} \in \mathbb{Q}$ and so $x\,R\,z$. The relation is an equivalence relation on A.

(b) $[1] = \{x \in A; x\,R\,1\} = \{x \in A; \frac{x}{1} = x \in \mathbb{Q}\} = \mathbb{Q}^*$.

(c) Let $x = a + b\sqrt{2} \in A$ with $b \neq 0$. Then $x = b\left(\frac{a}{b} + \sqrt{2}\right)$ and so $\frac{x}{\frac{a}{b}+\sqrt{2}} = b \in \mathbb{Q}$. This shows that $x\,R\,\frac{a}{b} + \sqrt{2}$ and so $[x] = [\frac{a}{b} + \sqrt{2}]$.

(32) Assume the condition: $(\forall x \in A)(\exists y \in A)((x,y) \in R)$ holds, we need to show that R is reflexive. Let $x \in A$ then there exists $y \in A$ such that $(x,y) \in R$. Since R is symmetric, $(y,x) \in R$. Since R is transitive,

$(x, x) \in R$ and so R is reflexive and hence an equivalence relation on A. Conversely, if R is an equivalence relation on A then for any $x \in A$, $(x, x) \in R$ since R is reflexive. The condition $(\forall x \in A)(\exists y \in A)\,((x, y) \in R)$ holds.

(34) Let $A = \{a, b, c\}$ and consider the following relations on A:

- $R = \{(a, a), (b, b), (c, c), (a, b), (b, a)\}$
- $S = \{(a, a), (b, b), (c, c), (b, c), (c, b)\}$

Clearly, R and S are equivalence relations on A but $R \cup S$ is not since $(a, b), (b, c) \in R \cup S$ but $(a, c) \notin R \cup S$ (so $R \cup S$ is not transitive).

(36) Let $a, b, x, y \in \mathbb{Z}$ be such that $[a] = [b]$ and $[x] = [y]$. Then there exist $s, t \in \mathbb{Z}$ such that $a - b = sn$ and $x - y = tn$. So $ax - by = (b + sn)x - b(x - tn) = bx + snx - bx + btn = (sx + bt)n$. This shows that $ax - by$ is a multiple of n and so $ax\,R\,by$ or equivalently, $[ax] = [by]$. We conclude that $[a] \otimes [x] = [b] \otimes [y]$ and the operation $\otimes$ well-defined on $\mathbb{Z}_n$.

(40) $r(R)$ consists of the pairs $(1, 1)$, $(2, 2)$, $(3, 3)$, $(2, 1)$ and $(2, 3)$, $tr(R)$ consists of $(1, 1)$, $(2, 2)$, $(3, 3)$, $(2, 1)$, $(2, 3)$ and $str(R)$ consists of $(1, 1)$, $(2, 2)$, $(3, 3)$, $(2, 1)$, $(1, 2)$, $(2, 3)$, $(3, 2)$. Note that $str(R)$ is not transitive as $(1, 2), (2, 3) \in str(R)$ but $(1, 3) \notin str(R)$.

(41) We compute $tsr(R)$.

(c) $r(R)$ consists of the pairs (a, a), (b, b), (c, c), (d, d), (a, b), (b, c), (c, d), $sr(R)$ consists of (a, a), (b, b), (c, c), (d, d), (a, b), (b, a), (b, c), (c, b), (c, d), (d, c). Finally, $tsr(A) = A \times A$.

7.10

(3)(a) This is not a partial order on $\mathbb{Z}$. It is not antisymmetric since $1\,R\,0$ and $0\,R\,1$ at the same time. It is also not transitive: $0\,R-1$ and $-1\,R-2$ but $0\,\not{R}\,-2$.

(b) This is a partial order on $\mathbb{Z}$. It is reflexive by definition. It is antisymmetric since we cannot have $|m| < |n|$ and $|m| < |n|$ at the same time. It is transitive since if $m\,R\,n$ and $n\,R\,p$ then $(m = n$ or $|m| < |n|)$ and $(n = p$ or $|n| < |p|)$. This is equivalent to $(m = n$ and $n = p)$ or $(m = n$ and $-n-_{\text{i}}-p-)$ or $(|m| < |n|$ and $n = p)$ or $(|m| < |n|$ and $|n| < |p|)$. This leads to $m = p$ or $|m| < |p|$, or equivalently, $m\,R\,p$.

(e) This is a partial order on $\mathbb{Z}$. Note first that $m^3 - n^3 = (m - n)(m^2 - nm + n^2)$ and that $m^2 - nm + n^2 > 0$ for any $m, n \in \mathbb{Z}$ (the discriminant is $\Delta = (-n)^2 - 4(1)(n^2) = -3n^2 < 0$). This means that $m\,R\,n \Leftrightarrow m^3 \le n^3 \Leftrightarrow m^3 - n^3 \le 0 \Leftrightarrow (m-n)(m^2 - nm + n^2) \le 0 \Leftrightarrow m - n \le 0 \Leftrightarrow m \le n$. We know already that the natural order of numbers is a partial order

(5)(a) This is a partial order on A. The relation is reflexive as every entry on the main diagonal is 1. It is antisymmetric as for every $x, y \in A$, if $(x, y) \in R$ then $(y, x) \notin R$. The relation is transitive: when we apply Warshall's algorithm, we find that $W_4 = R$.

(7)(a) 000101, 0110101, 100101, 10011, 10101, 111, 1110000, 1110101010.

(8)(b) The arrangement of the monomials in increasing order is as follows: z^7, yz^5, $y^3 z^3$, xy^5, $x^2 yz^3$, $x^3 z^2$, $x^3 yz$, $x^3 y^2$, $x^4 y$, x^5. .

(10) Arranging the terms of f in increasing order, we get:

$$f = -xy + xyz^2 + xy^3 z^2 + z^2 w^2 + 2xyz^2 w^2 + 3w^3 xyz^2 - 5w^3 xy^2 z^2.$$

(12)(a) $(0, 2, 6)$, $(1, 0, 6)$, $(1, 4, 0)$, $(1, 8, 0)$, $(2, 3, 9)$, $(2, 4, 1)$, $(3, 4, 7)$. .

(13) Let $(a, b) \in A \times B$. Then $a\,R\,a$ and $b\,S\,b$ since R and S are reflexive. So $(a, b)\,T\,(a, b)$. The relation T is reflexive. If $(a, b), (c, d) \in A\times$ are such that $(a, b)\,T\,(c, d)$ and $(c, d)\,T\,(a, b)$ then $a\,R\,c$, $b\,S\,d$, $c\,R\,a$ and $d\,S\,b$. This shows that $a = c$ and $b = d$ since R and S are antisymmetric. We conclude that $(a, b) = (c, d)$ and the relation T is antisymmetric. For the transitivity of T, let $(a, b), (c, d), (z, t) \in A \times B$ such that $(a, b)\,T\,(c, d)$ and $(c, d)\,T\,(z, t)$. Then $a\,R\,c$, $b\,S\,d$, $c\,R\,z$ and $d\,S\,t$. By the transitivity of R and S we get that $a\,R\,z$ and $b\,S\,t$. This means that $(a, b)\,T\,(z, t)$ and T is transitive. We conclude that T is a partial order on $A \times B$.

(15)(b)

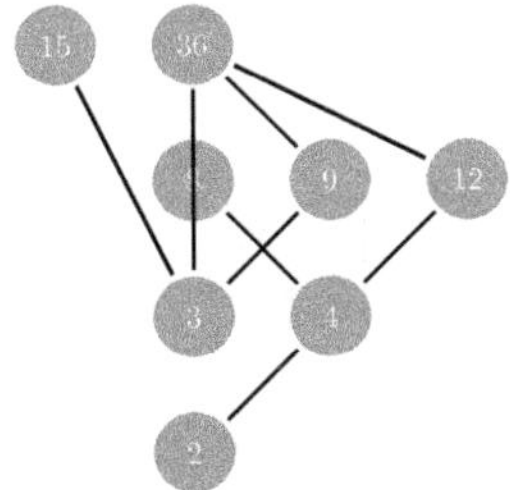

(16)(c) $a \nsubseteq b$ and $b \nsubseteq a$. So a, b are not comparable elements of $(\wp\,(\wp(A)))$.

(d) $a \subseteq b$, so a, b are comparable elements of $(\wp\,(\wp(A)))$.

(18) First we prove that $\preccurlyeq$ is a partial order on $P \times P$. Let $(a, b) \in P \times P$, then $(a, b) \preccurlyeq (a, b)$ since $ab \le ba$ and so $\preccurlyeq$ is reflexive. Let (a, b), $(c, d) \in P \times P$ such that $(a, b) \preccurlyeq (c, d)$ and $(c, d) \preccurlyeq (a, b)$. Therefore, $ad \le bc$ and $ad \le bc$ which implies that $ad = bc$. Since $\gcd(a, b) = 1$ and $\mid bc$, then $a \mid c$. Similarly, $c \mid a$, $d \mid b$ and $b \mid d$. Since the integers a, b, c, d

are positive, $a = c$ and $b = d$ and so $(a,b) = (c,d)$. The relation is antisymmetric. Let (a,b), (c,d), $(e,f) \in P \times P$ such that $(a,b) \preccurlyeq (c,d)$ and $(c,d) \preccurlyeq (e,f)$. Then, $ad \leq bc$ and $cf \leq de$. For the first inequality we get that $a \leq \frac{bc}{d}$ and from the second we get that $f \leq \frac{de}{c}$. Therefore, $af \leq \left(\frac{bc}{d}\right)\left(\frac{de}{c}\right) = be$. This proves that $(a,b) \preccurlyeq (e,f)$ and the relation is transitive. Thus, $\preccurlyeq$ is a partial order on $P \times P$. Given (a,b) and $(c,d) \in P \times P$, either $ad \leq bc$ or $bc \leq ad$. This means that either $(a,b) \preccurlyeq (c,d)$ or $(c,d) \preccurlyeq (a,b)$ and the relation is a total partial order on $P \times P$.

(19) One possible answer is $A = \{1, 2, 4, 8, 16, 32, 64\}$. It is easy to verify that every two elements of A are comparable with the divisibility relation.

(21) Let k^m and k^n be two elements of $\mathcal{G}_k$ where $m, n \in \mathbb{N}$. Since the standard order $\leq$ is a total order on $\mathbb{N}$, either $m \leq n$ or $n \leq n$. Therefore, either $k^m | k^n$ (if $m \leq n$) or $k^n | k^m$ (if $n \leq m$). This shows that the divisibility relation is a total order on the subset $\mathcal{G}_k$ of $\mathbb{N}$.

7.11

(1)(a) g, h, i.   (c) No least element.
   (b) a, e, f.   (d) a.

(2)(b) 11, 48, 60, 72.
   (d) No greatest element.
(4)(a) $\{2\}$, $\{3\}$, $\{4\}$, $\{5\}$.
   (c) No least element.

(5)(b) a (f) j (i) c (j) j

(6)(a) $\inf(X) = \emptyset$, $\sup(X) = \{a, b\}$.
   (d) $\inf(X) = \emptyset$, $\sup(X) = \{a, b, d\}$.
   (f) $\inf(X) = \{b, d\}$, $\sup(X) = \{a, b, c, d\}$.
(8)(b) Assume that g, g' are two least upper bounds of B. Since g is in particular an upper bound of B, we must have $g' \preccurlyeq g$. Similarly, $g \preccurlyeq g$ and therefore $g = g'$ since $\preccurlyeq$ is antisymmetry. The least upper element of B (if it exists) is unique.
(9)(a) Interval $[0, 1[$ with the standard order $\leq$ on real numbers.
   (c) The poset $\{2, 3, 4\}$ with the divisibility relation of integers.
(10) Note that $1 \in S$ and for every integer $n \geq 1$, we have that $\frac{a}{n} \leq 1$. So 1 is a greatest element of S. If $a \in S$ is a least element of S, then $a \in S$ and $a \leq \frac{1}{n}$ for any $n \geq 1$. But that implies that $a \leq 0$ since otherwise

there exists $n \geq 1$ such that $a > \frac{1}{n}$. This is a contradiction and S has no least element.

(14) The set $\mathbb{R}$ with the standard order on the reals is a totally ordered poset which is not well-ordered because, for example, the non-empty interval $]0, 1]$ has no least element with respect to $\leq$.

(15) Let A be a non-empty finite poset. We only prove that A has a maximal element, the proof that it has a minimal element is identical. Pick an arbitrary element $a_1 \in A$. If a_1 is maximal then we are done. If a_1 is not maximal, then there exists an element $a_2 \in A$ such that $a_1 \prec a_2$. If a_2 is maximal then we are done. If a_2 is not maximal, then there exists an element $a_3 \in A$ such that $a_2 \prec a_3$. Since $\prec$ is transitive, we get that $a_1 \prec a_3$ and so $a_1 \neq a_3$. Since A is finite, this process must terminate and when it does, we get a sequence $a_1 \prec a_2 \prec a_3 \prec \ldots \prec a_n$ of distinct elements of A. The element a_n is a maximal element of A.

(17)(c) This is not a lattice. For example, each of the elements d, e, f is a lower bound of $\{b, c\}$ but none of them is $\inf\{b, c\}$.

   (f) This is a lattice since $\inf\{x, y\}$ and $\sup\{x, y\}$ exist for any $x, y \in A$.

(18)

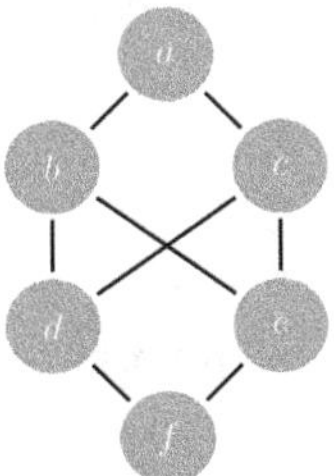

In this poset, each of the elements b, c and e is an upper bound for the subset $\{d, e\}$ but the subset $\{d, e\}$ has noleast upper bound. The poset represented by the Hasse diagram is not a lattice. Clearly, a is the greatest element and f is the least element of the poset.

(20)(b) $\sup\{6, 16\} = \mathrm{lcm}(6, 16) = 48 \notin A$.

(21)(c) We have that $a \preccurlyeq (a \vee b)$ (since $a \vee b$ is an upper bound of $\{a, b\}$) and $a \preccurlyeq a$ (since $\preccurlyeq$ is reflexive), so $a \preccurlyeq a \wedge (a \vee b)$. On the other hand, $a \wedge (a \vee b) \preccurlyeq a$ since $a \wedge (a \vee b)$ is a lower bound for $\{a, a \vee b\}$. We conclude that $a \wedge (a \vee b) = a$ by the fact that $\preccurlyeq$ is antisymmetric.

(23) Let B be a non-empty finite subset B of A and assume that $|B| = n$. If $n = 1$, then $B = \{a\}$ and clearly $\inf(B) = \sup(B) = a$ in this case. The result is true for $n = 1$. Assume that $n \geq 1$ and that $\inf(B)$ and $\sup(B)$ exist for any finite subset of A of cardinality n. Let B be a

finite subset of A of cardinality $n + 1$ and write $B = C \cup \{a\}$ for some subset C of A of cardinality n and some element $a \in A \backslash C$. By the induction hypothesis, $p = \inf(C)$ and $q = \sup(C)$ exist. Since A is a lattice, $p \wedge a$ and $q \vee a$ exist. For $x \in B$, either $x = a$ or $x \in C$. If $x = a$, then $p \wedge a \preccurlyeq a = x$. If $x \in C$, then $p \preccurlyeq x$ and so $p \wedge a \preccurlyeq p \preccurlyeq x$. This shows that $p \wedge a$ is a lower bound for B. On the other hand, if l is any lower bound of B then l is also a lower bound for C and so $l \leq p$. Since $l \preccurlyeq a$, $l \preccurlyeq p \wedge a$. This shows that $p \wedge a = \inf(B)$. Next we show that $\sup(B) = q \vee a$. For $x \in B$, if $x = a$ then $x \preccurlyeq q \vee a$ and if $x \in C$ then $x \preccurlyeq q \preccurlyeq q \vee a$. This shows that $q \vee a$ is an upper bound of B. If λ is any upper bound of B then λ is also an upper bound for C and so $q \leq \lambda$. Since $a \preccurlyeq \lambda$, $a \vee q \preccurlyeq \lambda$. This shows that $q \vee a = \sup(B)$.

(24)(a) Not bounded: there is a least element (1) but no greatest element.

 (c) Bounded: 1 is the least element and 12 is the greatest element.

 (d) Bounded: $\emptyset$ is the least element and $\mathbb{N}$ is the greatest element.

 (g) Not bounded: there is a least element (0) but no greatest element.

(27)(a) (i) Let $a, b, c \in \mathbb{N} \backslash \{0\}$. Without loss of generality, we may assume that $a \leq b \leq c$. Then $a \vee (b \wedge c) = a \vee b = b$ and $(a \vee b) \wedge (a \vee c) = b \wedge c = b$. Also, $a \wedge (b \vee c) = a \wedge c = a$ and $(a \wedge b) \vee (a \wedge c) = a \vee a = a$.

 (b) Consider the poset $(A = \{1, 2, 3, 4, 12\}, |)$ (with the divisibility inherited from $\mathbb{N}$). Notice that $2 \vee (3 \wedge 4) = 2 \vee 1 = 2$, but $(2 \vee 3) \wedge (2 \vee 4) = 12 \wedge 4 = 4$. The poset is not distributive.

(28)(b) $a \wedge (b \vee c) = a \wedge x = a$. On the other hand, $(a \wedge b) \vee (a \wedge c) = y \vee y = y$.

(30)(a) Assume that y, z are two complements of the element $x \in A$. Then $y = y \vee 0 = y \vee (z \wedge x) = (y \vee z) \wedge (y \vee x) = (y \vee z) \wedge 1 = y \vee z$. Similarly, we can establish that $z = y \vee z$. Therefore, $y = z$ and the complement in unique when it exists.

 (b) We have that $0 \vee 0' = 1$ by the definition of the complement element. On the other hand, $0 \vee 0' = 0'$ since 0 is the least element of A. We conclude that $0' = 1$. The proof of the fact that $1' = 0$ is similar.

(30) Consider the poset $D_{12} = \{1, 2, 3, 4, 6, 12\}$ with the divisibility relation. Assume that the element 6 of D_{12} has a complement x'. Then, in particular, $\gcd(x', 6) = 1$. Since 1 is the only element of D_{12} satisfying $\gcd(x', 6) = 1$, we must have that $x' = 1$. But this is absurd since $\text{lcm}(x', 6) = 12$ but $\text{lcm}(1, 6) = 6$.

(34)(a) $d \, R \, e \, R \, c \, R \, b \, R \, a$.

 (b) $h \, R \, f \, g \, R \, e \, R \, c \, R \, d \, R \, b \, R \, a$.

8.2

(1) We proceed by induction on $n \geq 2$. The result is true for $n = 2$ as proved in Theorem 8.1. Let $n \geq 2$ and assume that the result is true for n sets. Let $A = A_1 \cup \cdots \cup A_n$, then $A \cap A_{n+1} = (A_1 \cap A_n) \cup \cdots \cup (A_n \cap A_{n+1}) = \emptyset$ and so A and A_{n+1} are disjoint. By the proof of case $n = 2$ and the associativity of the union of sets, we have that $|A_1 \cup A_2 \cup \ldots A_{n+1}| = |A_1 \cup A_2 \cup \ldots A_n| + |A_{n+1}|$. By the induction hypothesis, $|A_1 \cup A_2 \cup \ldots A_n| = |A_1| + \cdots + |A_n|$. We conclude that $|A_1 \cup A_2 \cup \ldots A_{n+1}| = |A_1| + \cdots + |A_n| + |A_{n+1}|$ and the result is true.

(2)(c) There are 40 ways to draw a value card and 4 ways to draw a Jack and these events are mutually exclusive. By the sum principle, there are $40 + 4 = 44$ ways to draw a value card or a Jack.

(4) A sum of 6 is obtained by the following outcomes $(1,5)$, $(5,1)$, $(2,4)$, $(4,2)$ and $(3,3)$. A sum of 9 is obtained by the following outcomes $(3,6)$, $(6,3)$, $(4,5)$ and $(5,4)$. By the sum principle, there are $5 + 4 = 9$ ways we can get a sum of 6 or 9 on the faces of the dice.

(5)(a) There are 18 ways to pick the flavor. For each of these ways, there are 6 ways to choose the toping, 3 ways to pick the cone size and 4 ways to pick the cone type. By the product principle, there are $18 \cdot 6 \cdot 3 \cdot 4 = 1296$ ways to get an ice cream cone.

(7)(b) By the product principle, there are $26 \cdot 25 \cdot 24 \cdot 23 \cdot 22 = 7893600$ words with five letters such that no letter is repeated.

(d) Let Σ be the set of all words with five letters and no occurrence of the letter A. Let Ω be the set of all words with five letters and exactly one occurrence of the letter A. Clearly, $|\Sigma| = 25^5$. To compute $|\Omega|$, notice that there are five possible positions of the letter A in the word. For each one of these positions, there are 25^4 ways to choose the other letters. By the product principle, there are $5 \cdot 25^4 = 1953125$ words with five letters containing one occurrence of A. By the sum rule, there are $25^5 + 5 \cdot 25^4$ words with 5 letters and *at most* one occurrence of the letter A.

(8)(b) If none of the letters are repeated, then there are $26 \cdot 25 \cdot 24 \cdot 10 \cdot 10 \cdot 10 = 15600000$ licence plates.

(c) If none of the letters and none of the digits are repeated, then there are $26 \cdot 25 \cdot 24 \cdot 10 \cdot 9 \cdot 8 = 11232000$ different licence plates.

(9)(b) There are $10 \cdot 26 \cdot 10 \cdot 26 \cdot 10 = 676000$ postal codes with K as the first letter and the same number with H as the first letter. By the sum principle, there are $676000 + 676000 = 2704000$ postal codes with first letter either K or H.

(13) We can think of a photo as an ordered list $x_1 x_2 \ldots x_7$ where each x_i represents one person. There are 7 possible choices for x_1. For each of these choices, there are 6 possible choices for x_2, and so on. By the product principle, there are $7 \cdot 6 \cdots 2 \cdot 1 = 5040$ ways to take a photo of the seven persons.

(16) Note first that there is a total of $26 + 10 + 5 = 41$ characters. Let Σ, Γ and Δ be the set of possible passwords having 5, 6 and 7 characters, respectively. Clearly these sets are pairwise disjoint.

   (a) By the sum principle, there are $|\Sigma \cup \Gamma \cup \Delta| = |\Sigma| + |\Gamma| + |\Delta| = 41^5 + 41^6 + 41^7 = 199620234323$ possible passwords.

(17) Let O be the set of surveyed people who said they shop online at JoeMart and S the set of surveyed people who said they shop in store. We have that $|O| = 550$, $|S| = 580$ and $|G \cap H| = 200$. The set of people who shop at JoeMart (online or in store) is $|O \cup S| = |O| + |H| - |O \cap S| = 550 + 580 - 200 = 930$. Therefore, there are $1000 - 930 = 70$ among the surveyed people who do not shop at JoeMart at all.

(22) Let L, C and D be the set of surveyed students who took linear algebra, calculus II and discrete math, respectively. By the inclusion–exclusion principle, $|L \cup C \cup D| = |L| + |C| + |D| - |L \cap C| - |L \cap D| - |C \cap D| + |L \cap C \cap D| = 80 + 32 + 63 - 17 - 25 - 16 + 9 = 126$. Since there are 21 students who took none of the three courses, $n = 126 + 21 = 147$.

(25) By the principle of product, there are 4^9 passwords of length 9 using only the four letters A, B, C and D. Let X, Y, Z and W be the sets of passwords not containing A, B, C, and D respectively the number of passwords containing at least one appearance of each of the four letters is $4^9 - |X \cup Y \cup Z \cup W|$. By the product principle, $|X| = |Y| = |Z| = |W| = 3^9$, $|X \cap Y| = |X \cap Z| = |X \cap W| = |Y \cap Z| = |Y \cap W| = |Z \cap W| = 2^9$, $|X \cap Y \cap Z| = |X \cap Y \cap W| = |X \cap Z \cap W| = |Y \cap Z \cap W| = 1^9 = 1$ and finally, $|X \cap Y \cap Z \cap W| = 0$. By the inclusion–exclusion principle, $|X \cup Y \cup Z \cup W| = |X| + |Y| + |Z| + |W| - |X \cap Y| - |X \cap Z| - |X \cap W| - |Y \cap Z| - |Y \cap W| - |Z \cap W| + |X \cap Y \cap Z| + |X \cap Y \cap W| + |X \cap Z \cap W| + |Y \cap Z \cap W| - |X \cap Y \cap Z \cap W| = 4 \cdot 3^9 - 6 \cdot 2^9 + 4$. We conclude that the number of passwords with at least one appearance of each of the four letters is $4^9 - (4 \cdot 3^9 - 6 \cdot 2^9 + 4) = 186480$.

8.3

(2) There are 7 remainders in the division by 7, namely $0, 1, 2, 3, 4, 5$ and 6. We use the pigeonhole principle with the objects being the 132

integers, the boxes are the 7 remainders in the division by 7. Each of the 132 integers is placed in the box corresponding to its remainder upon division by 7. By the pigeonhole principle, there exists at least one box (remainder) containing at least $\lceil \frac{132}{7} \rceil = 19$ integers. These 19 integers have the same remainder in the division by 7.

(5) Write $S = \{a_1, a_2, \ldots, a_n\}$. For each $k \in \{1, 2, \ldots, n\}$, let $s_k = a_1 + a_2 + \cdots + a_k$. Use the pigeonhole principle with objects being the integers $s_1, \ldots, s_n$ and boxes the $n - 1$ possible remainders upon division by n. Since there are n objects to be placed in n boxes, there is at least one box (remainder) containing at least two objects, say s_i and s_j. Since s_i and s_j have the same remainder upon division by n, their difference is a multiple of n. We can clearly assume that $i < j$, so $s_j - s_i = a_{i+1} + \cdots + a_j$. The subset $\{a_{i+1}, \ldots, a_j\}$ of S satisfies the required condition.

(9) We use the pigeonhole principle with the objects being the number n of students registered in the course, the boxes are the 10 possible grades and each student is placed in the box corresponding to his/her grade obtained in the class. To ensure that at least 9 students receive the same grade, we must have $8 < \frac{n}{10} \leq 9$. So $n \geq 10 \cdot 8 + 1 = 81$.

(15) Note first that there are 15 prime numbers in A: 2, 3, 5, 7, 11, 13, 17, 19, 23, 29, 31, 37, 41, 43 and 47. Let P be the set of these 15 prime numbers. If a is one of the n selected numbers in A, place a in the box corresponding to the smallest prime number divisor of a in P. For example, 8 is placed in box "2", 17 is placed in box "17" and so on. By the pigeonhole principle, there exists at least one box containing at least $\lceil \frac{n}{15} \rceil$ integers. Since $n \geq 16$, $\lceil \frac{n}{15} \rceil \geq 2$. That means that at least two of the n integers, say a and b, have the same smallest prime factor. In other words, a and b are not relatively prime.

8.4

(2) There are $P(4, 3) = \frac{4!}{(4-3)!} = 24$ 3-permeations of the objects a, b, c, d. The full list is: abc, abd, acd, acb, adb, adc, bac, bad, bcd, bca, bda, bdc, cab, cad, cba, cbd, cda, cdb, dab, dac, dba, dbc, dca, dcb.

(3)(b) This corresponds to the number of 5-permutations without repetition of 15 objects: $P(15, 5) = \frac{15!}{(15-5)!)} = 360360$ possible outcomes.

(5) There are $12! = 479001600$ ways of making a necklace that can be worn on one side only, and $\frac{12!}{2} = 239500800$ ways of making a necklace that can be worn on both sides.

(6)(c) $P(n,5) = 32P(n-1,3) \Rightarrow \frac{n!}{(n-5)!} = 32\frac{(n-1)!}{(n-4)!} \Rightarrow n = \frac{32}{n-4} \Rightarrow n^2 - 4n - 32 = 0 \Rightarrow (n-8)(n+4) = 0 \Rightarrow n = 8.$

(8)(a) If there are no restrictions on seating arrangements, then we are simply looking at the number of permutations without repetition of 12 persons: there are $12! = 479001600$ such permutations.

  (b) If the women are sitting together, we treat all 5 women as one "person". Together with the 7 men, there are 8! ways of arranging all 12 persons. For each one of these arrangements, there are 7! ways to permute the women and 5! to permute the women. We conclude that there are $8!7!5! = 24385536000$ arrangements with all women sitting together.

  (c) There are $5! = 120$ arrangements for all the women on the right of the row. For each of these arrangements, there are $7! = 5040$ ways to arrange all men on the left of the row. By the product rule, there are $120 \cdot 5040 = 604800$ ways to arrange all women on the right and all men on the left of the row. A similar number of ways for arranging all women on the left and all men on the right of the row. We conclude that there are 1209600 arrangements with all the men are sitting together and the women are all sitting together.

(10) Notice that the word *"COVERING"* is formed by 8 *different* letters. Permuting all the letters of the word corresponds to a permutation, without repetition, of 8 objects. There are $8! = 40320$ different words. Now, if we require that the letter "I" immediately precedes the letter "N", then we can treat the string "IN" as one symbol and we then have $7! = 5040$ words with "I" immediately preceding "N".

(12) Notice first that the letters of the word *"SAVOURY"* are pairwise distinct.

  (b) This is the number of 4-permutations of 7 objects: $P(7,4) = \frac{7!}{(7-4)!} = 840$ such words.

  (e) We treat the three vowels present in the word (A, O and U) as one object. There $5! = 120$ ways to permute the five objects AOU, S, V, R and Y. For each one of these permutations, there are $3! = 6$ ways to permute the three vowels. By the product principle, there are $120 \cdot 6 = 720$ such words.

(16) In addition to the five different colors $C_1, \ldots, C_5$, introduce a sixth one, say Black B, to represent unlit projectors. A signal corresponds to a permutation with repetitions allowed of five objects from the set $\{C_1, \ldots, C_5, B\}$. Therefore, there are $6^5 = 7776$ possible signals.

(19) This amounts to the number of different words we obtain by permuting all the letters of the word:

$$\underbrace{B\cdots B}_{5}\underbrace{G\cdots G}_{4}\underbrace{R\cdots R}_{6}\underbrace{Y\cdots Y}_{7}$$

There are $\frac{22!}{(5!4!6!7!)} = 10450944000$ different arrangements.

(22)(a) We use induction. For $n = 1$, $\frac{(2n)!}{2^n} = 1$ is an integer. Let $n \geq 1$ and assume that $\frac{(2n)!}{2^n}$ is an integer. Then $\frac{(2(n+1))!}{2^{n+1}} = \left(\frac{(2n+2)(2n+1)}{2}\right)\left(\frac{(2n)!}{2^n}\right) = (n+1)(2n+1)\left(\frac{(2n)!}{2^n}\right)$: this is an integer by the induction hypothesis.

(b) For this part, we use a combinatorial argument. Consider the n distinct symbols $x_1, x_2, \ldots, x_n$. The number of different words obtained by permuting all symbols of the word $x_1 x_1 x_2 x_2 \ldots x_n x_n$ is $\frac{(2n)!}{\underbrace{2 \cdot 2 \cdots 2}_{n}} = \frac{(2n)!}{2^n}$ and we know that this has to be an integer.

(24) If $r > n$, the answer is clearly 0. If $r \leq n$, then every distribution of the r objects in the n boxes with every box containing at most one object corresponds to an injective function $f : \{1, 2, \ldots, r\} \to \{1, 2, \ldots, n\}$. There are $\frac{n!}{(n-r)!}$ such functions.

8.5

(1)(c) $\binom{50}{42} = \frac{50!}{42!8!} = 536878650$

(2)(a) $\binom{n+2}{2} = 15$ implies that $\frac{(n+2)!}{2!n!} = 15$ which simplifies to $\frac{(n+2)(n+1)}{2} = 15$ or $n^2 + 3n - 28 = 0$. The roots of this quadratic equation are $n = 4$ and $n = -7$. Of course the latter is to be rejected as it makes $n + 2 < 0$ and the solution is $n = 4$.

(c) $\binom{16}{n+2} = \binom{16}{n}$ is equivalent to $\frac{16!}{(n+2)!(14-n)!} = \frac{16!}{n!(16-n)!}$ which simplifies to $(n+1)(n+2) = (16-n)(15-n)$ or $34n - 238 = 0$. The solution is $n = 7$.

(6) There are $\binom{10}{5} = \frac{10!}{5!5!} = 252$ ways to choose the 5 students. For each one of these choices, there are $4! = 24$ ways to place the five students around a table. By the product principle, there are $252 \cdot 24 = 6048$ ways to place five students around the table.

(7)(b) Such a subset can be written under the form $Y \cup Z$ where $Y \subseteq B$ is of cardinality 4 and Z is a subset of $A \backslash B$. There are $\binom{9}{4} = \frac{9!}{4!5!} = 126$ ways to choose Y and 2^{11} ways of choosing Z (since $|A \backslash B| = 11$). By the product principle, there are $126 \cdot 2^{11}$ such subsets.

(9)(c) This amounts to choosing exactly 18 or exactly 19 or exactly 20 from the first 20 questions. In the first case, there are $\binom{20}{18}\binom{20}{17} = 216600$

ways to choose the questions on the exam. In the second case, there are $\binom{20}{19}\binom{20}{16} = 96900$ ways to choose the questions on the exam. In the last case, there are $\binom{20}{20}\binom{20}{15} = 15504$ ways to choose the questions on the exam. By the sum rule, there are $216600 + 96900 + 15504 = 329004$ ways to choose the questions on the exam.

(11) If Joe is on the committee, then Jessica is not. So there are $\binom{11}{4} = 330$ possible committees with Joe on it but not Jessica. Similarly, there are 330 committees with Jessica on but not Joe. By the sum principle, there are 660 possible committees.

(12)(a) First we choose 7 persons out of 15: there are $15|choose7 = 6435$ ways to do that. For each one of these ways, there are $\binom{7}{3}$ ways the fill the three positions in Mathematics, $\binom{4}{2}$ ways to fill the two positions in Computer Science, $\binom{2}{1}$ ways the fill the position in English and $\binom{1}{1}$ way to fill the Geography position. By the product rule, there are $15|choose7 \cdot \binom{7}{3} \cdot \binom{4}{2} \cdot \binom{2}{1} \cdot \binom{1}{1} = 2702700$ ways to fill seven positions if there are no restrictions.

(b) If exactly three women must be hired, then there are $\binom{7}{3} \cdot \binom{8}{4} = 2450$ ways to choose the seven candidates among the 15 who applied. Next, and like in the previous part, there are $\binom{7}{3} \cdot \binom{4}{2} \cdot \binom{2}{1} \cdot \binom{1}{1} = 420$ ways to fill in the four subjects with these chosen 7 candidates. By the product principle, there are $2450 \cdot 420 = 1029000$ ways to fill the seven positions in this case.

(13)(e) There are three possible cases: (i) Ron is on the committee but Rebecca is not; (ii) Ron is not on the committee but Rebecca is on the committee; (iii) Both Ron and Rebecca are not on the committee. In case (i), there are $\binom{4}{1} = 4$ ways to choose the politician (other than Ron) and for each of these ways, there are $\binom{14}{2} = 91$ to choose the two biologists (other than Rebecca) and $\binom{18}{2} = 153$ to choose the other two members of the committee. Thus, there are $4 \cdot 91 \cdot 153 = 55692$ possible committees. Arguing similarly, we find that the total number of possible committees in case (ii) is $\binom{4}{2} \cdot \binom{14}{1} \cdot \binom{18}{2} = 12852$. Finally, the number of possible committees in case (iii) is $\binom{4}{2} \cdot \binom{14}{2} \cdot \binom{18}{2} = 83538$. By the sum principle, there are $55692 + 55692 + 83538 = 194922$ possible committees.

(16)(a)

$$\binom{n}{k}\binom{k}{r} = \frac{n!}{k!(n-k)!} \cdot \frac{k!}{r!(k-r)!}$$
$$= \frac{n!}{k!(n-k)!(k-r)!}.$$

On the other hand,

$$\binom{n}{r}\binom{n-r}{k-r} = \frac{n!}{r!(n-r)!} \cdot \frac{(n-r)!}{(k-r)!(n-r-(k-r))!}$$
$$= \frac{n!}{k!(n-k)!(k-r)!}.$$

Hence $\binom{n}{k}\binom{k}{r} = \binom{n}{r}\binom{n-r}{k-r}$.

(b) Imagine forming a committee of k members out of a group of n people. The committee must have an executive subcommittee of r persons. There are two ways to form the committee and its executive subcommittee. First choose k persons out of n: there are $\binom{n}{k}$ ways to do that. For each of these ways, there are $\binom{k}{r}$ ways of choosing the executive subcommittee. By the product rule, there are $\binom{n}{k} \cdot \binom{k}{r}$ ways of choosing the committee and its executive subcommittee. A second way consists of first choosing the r members of the executive committee out of n persons: there are $\binom{n}{r}$ ways to do that. For each of these ways there are $\binom{n-r}{k-r}$ of choosing the other $k-r$ members of the committee. By the product rule, there are $\binom{n}{r} \cdot \binom{n-r}{k-r}$ ways to form the committee. The result follows.

(21)(a) This is the number of derangements of 8 objects: $D_8 = 8!\left(\frac{1}{2!} - \frac{1}{3!} + \frac{1}{4!} - \frac{1}{5!} + \frac{1}{6!} - \frac{1}{7!} + \frac{1}{8!}\right) = 14833$.

(b) At least three guests get the wrong coat is equivalent to exactly three guests get the wrong coat, or exactly four guests get the wrong coat, ..., or exactly eight get the wrong coat. All these events are mutually exclusive. To compute the number of ways exactly k guests get the wrong coat ($3 \leq k \leq 8$), we start by choosing these k guests: there are $\binom{8}{k}$ ways to do that. For each one of these ways, there are D_k ways of giving each of the k guests the wrong coat since the other $8-k$ coats are given to the right guests. By the sum rule, the number of ways at least three guests get the wrong coat at the end of the party is $\sum_{k=3}^{8} \binom{8}{k} D_k = 40291$.

8.6

(1)

$$(x + 2y)^{11} = \sum_{k=0}^{11} \binom{11}{k} x^{10-k}(2y)^k = \sum_{k=0}^{11} 2^k \binom{11}{k} x^{11-k} y^k$$

To find the coefficient of $x^2 y^9$, we set $k = 9$ in $2^k \binom{11}{k} x^{11-k}$ and we get the coefficient $2^2 \cdot \binom{11}{9} = 220$.

(4) Set $x = 1$, $y = -1$ in the binomial theorem, we get that

$$0 = (1-1)^n = \sum_{k=0}^{n} \binom{n}{k}(1)^{n-k}(-1)^k = \binom{n}{0} - \binom{n}{1} + \binom{n}{2} - \cdots + (-1)^n \binom{n}{n}.$$

(6)(b) $\binom{12}{2,3,3,4} = \frac{12!}{2!3!3!4!} = 277200$

(7)(b) Using the coefficients on the fifth row of the Pascal triangle, we get that $(x + y + z)^4 = x^4 + 4x^3 y + 6x^2 y^2 + 4xy^3 + y^6$

(9) Using the multinomial theorem: $\left(3x^2 + 2y - \frac{1}{y^2}\right)^{14} = \sum_{a+b+c=14} \frac{14!}{a!b!c!}(3x^2)^a (2y)^b (-y^{-2})^c = \sum_{a+b+c=14} \frac{14!}{a!b!c!}(-1)^c 3^a 2^b x^{2a} y^{b-2c}$. To obtain the coefficient of $x^6 y^2$, we need that $2a = 6$, $b - 2c = 2$ and $a + b + c = 14$. This gives $a = 3$, $b = 8$ and $c = 3$. The required coefficient is $\frac{14!}{3!8!3!}(-1)^3 3^3 2^8 - 415134720$.

(11) We use the multinomial theorem.

(b) The coefficient of $x^2 y z^2 w^3$ in the expansion of $(-2x + 3y + 2z - w)^8$ is $(-2)^2 3 2^2 (-1)^3 \frac{8!}{2!1!2!3!} = -80640$.

(12) Using the binomial theorem:

$$\sum_{k=0}^{40} \binom{40}{k} 15^k = \sum_{k=0}^{40} \binom{40}{k} 15^k (1)^{40-k}$$
$$= (15 + 1)^{40}$$
$$= \left(2^4\right)^{40} = 2^{160}.$$

We conclude that $n = 2$.

(13)(c) By the multinomial theorem: $(3x - 2y - 4z + 6w)^9 = \sum_{a+b+c+d=9} \frac{9!}{a!b!c!d!} 3^a (-2)^b (-4)^c 6^d x^a y^b z^c w^d$. The sum of all the coefficients in this expansion is $\sum_{a+b+c+d=9} \frac{9!}{a!b!c!d!} 3^a (-2)^b (-4)^c 6^d$ and this can be obtained by setting $x = y = z = w = 1$. This gives that:

$$\sum_{a+b+c+d=9} \frac{9!}{a!b!c!d!} 3^a (-2)^b (-4)^c 6^d = (3 - 2 - 4 + 6)^9 = 19683.$$

8.7

(1)(a) $\binom{10+24-1}{24} = \binom{33}{24} = 38567100$ ways.

  (b) If we want at least two squares of each sort, then it remains to choose $36-20 = 16$ chocolate squares: there are $\binom{10+16-1}{16} = \binom{33}{24} = 2042975$ ways to do that.

(2)(b) This amounts to finding all integer solutions to the equation $y_1 + y_2 + y_3 + y_4 + y_5 + y_6 = 12$. There are $\binom{12+6-1}{12} = \binom{17}{12} = 6188$ ways to do that.

(5)(c) If each kid is given one of the three fruits, then the problem amounts to distributing the remaining 1 apple, 3 bananas and 2 oranges to 6 kids. There are $\binom{1+6-1}{1} \cdot \binom{3+6-1}{3} \cdot \binom{2+6-1}{2} = 6 \cdot 56 \cdot 21 = 7056$ ways of distributing the fruits to the six kids in this case (by the product rule).

(6) Let $y_i = x_i - 2$ for $i = 1, 2, 3, 4$. Then the equation is equivalent to $y_1 + y_1 + y_3 + y_4 = 29$ with $0 \le y_i \le 8$ for each i. If we ignore the upper bound on the variables, then there are $\binom{29+4-1}{29} = 4960$ integer solutions to the equation. Let A_i be the set of all integer solutions of $y_1 + y_2 + y_3 + y_4 = 29$ with $y_i \ge 9$. Then the required number is $4960 - |A_1 \cup A_2 \cup A_3 \cup A_4| = 4060 - |A_1| - |A_2| - |A_3| - |A_4| + |A_1 \cap A_2| + |A_1 \cap A_3| + |A_1 \cap A_4| + |A_2 \cap A_3| + |A_2 \cap A_4| + |A_3 \cap A_4| - |A_1 \cap A_2 \cap A_3| - |A_1 \cap A_2 \cap A_4| - |A_1 \cap A_3 \cap A_4| - |A_2 \cap A_3 \cap A_4| + |A_1 \cap A_2 \cap A_3 \cap A_4|$. Now $|A_1| = |A_2| = |A_3| = |A_4| = \binom{20+4-1}{20} = 1771$, $|A_1 \cap A_2| = |A_1 \cap A_3| = |A_1 \cap A_4| = |A_2 \cap A_3| = |A_2 \cap A_4| = |A_3 \cap A_4| = \binom{11+4-1}{11} = 364$, $|A_1 \cap A_2 \cap A_3| = |A_1 \cap A_2 \cap A_4| = |A_1 \cap A_3 \cap A_4| = |A_2 \cap A_3 \cap A_4| = \binom{2+4-1}{2} = 10$, and $|A_1 \cap A_2 \cap A_3 \cap A_4| = 0$ since no integer solution of $y_1 + y_1 + y_3 + y_4 = 29$ satisfies $y_i \ge 9$ for each i (note that $4 \times 9 = 36 > 29$). We conclude that the number of integer solutions of the equation $x_1 + x_2 + x_3 + x_4 = 37$ such that $2 \le x_i \le 10$ for each $i = 1, 2, 3, 4$ is $4960 - 4 \cdot 1771 + 6 \cdot 364 - 4 \cdot 10 + 0 = 20$.

(7) This is the number of non-negative integer solutions to the equation $a_1 + a_2 + \cdots + a_7 = 21$: there are $\binom{21+7-1}{21} = 296010$ such solutions.

(9)(b) There are $\binom{13+6-1}{13}$ ways to distribute the 13 blue balloons in the six boxes. For each of these ways, there is a similar number of ways to distribute the 13 red balloons in the six boxes. By the product rule, there are $\binom{13+6-1}{13}^2 = 73410624$ ways of distributing the 26 balloons in the boxes in this case.

  (d) By the product rule, there are $\binom{26}{2} \cdot \binom{24}{4} \cdot \binom{20}{5} \cdot \binom{15}{5} \cdot \binom{10}{5} \cdot \binom{5}{5} = 325 \cdot 10626 \cdot 15504 \cdot 3003 \cdot 252 = 40518448303132800$ ways to do that.

(11)(a) Let $\alpha = (\alpha_1, \alpha_2, \ldots, \alpha_k)$ be a composition of n into k summands.
Then $\alpha_i \geq 1$ is an integer and $\alpha_1 + \alpha_2 + \cdots + \alpha_r = n$. The number of
non-negative integer solutions to the equation $\alpha_1 + \alpha_2 + \cdots + \alpha_k = n$
is $\binom{k+n-1}{n}$. If we require that $\alpha_i \geq 1$ for each i, then the number of
positive integer solutions of the equation $\alpha_1 + \alpha_2 + \cdots + \alpha_k = n$ is
$\binom{k+(n-k)-1}{n-k} = \binom{n-1}{n-k}$.

(b) For positive integer n, a composition has exactly k summands with
$1 \leq k \leq n$. By the sum rule, the number of all compositions of n is
$\sum_{k=1}^{n} \binom{n-1}{n-k} = \sum_{k=1}^{n} \binom{n-1}{k-1} = 2^{n-1}$.

(13) Start by creating five spaces between the four women and the ends
of the row: $\square\,W\,\square\,W\,\square\,W\,\square\,W\square$. There are $\binom{12}{5} = 792$ ways of
choosing five men at the ends of the row and and in every space between
the women. For each of these ways, there are $\binom{7+5-1}{5} = 462$ ways of
distributing the 7 remaining men among the 5 possible spaces. There
are 9! ways to permute the men in there positions and 4! to permute
the women in there positions. By the product principle, there are
$792 \cdot 462 \cdot 9! \cdot 4! = 3186701844480$ ways to have the required arrangement.

9.2

(1)(a) Directed. there is no indication that roads are two ways.

(c) Directed. The relation "preys on" is not symmetric.

(2) Of course, many scenarios are possible. Below we give one possible
scenario. As usual, V is the set of vertices.

(a) V is a group of six persons in a gathering. Two vertices are adjacent
if and only if the corresponding persons shook hands.

(4) The following is a possible graph representation of the office space.

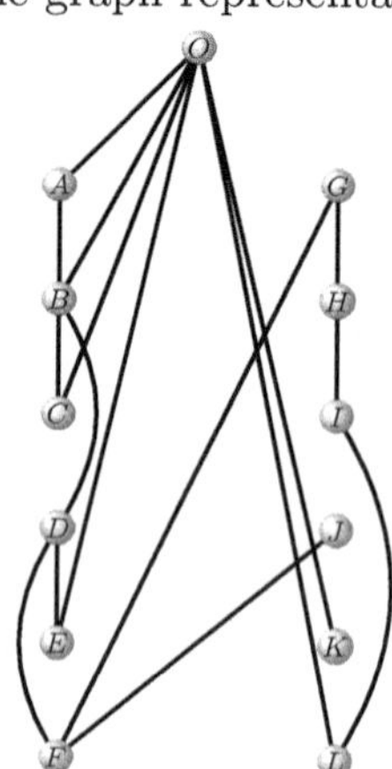

(6) Recall that the binary relation R on the set A is symmetric if and only if $a\,R\,b$ implies $b\,R\,a$ for any $a, b \in A$.

    (d) The relation is not symmetric. For example $3\,R\,2$ since $2(3)-3(2) = 0$ but $2\,\not{R}\,3$ since $2(2) - 3(3) \neq 0$. The relation can be represented by the following digraph. Note also that $a\,R\,a$ if and only if $2a - 3a = 0$ which implies that $a = 0$. Since $0 \notin A$, none of the vertices in the diagraph has a loop.

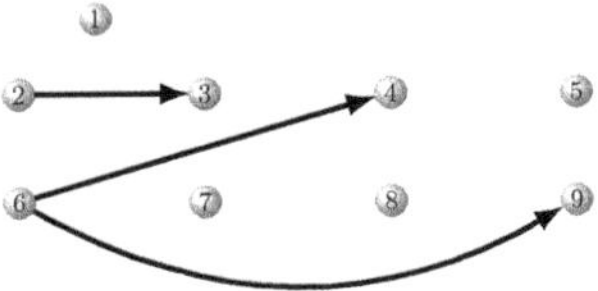

(8)(b)

$$
\begin{array}{c c}
 & \begin{array}{c c c c c} E & C & D & A & B \end{array} \\
\begin{array}{c} E \\ C \\ D \\ A \\ B \end{array} &
\left[\begin{array}{c c c c c}
1 & 1 & 2 & 0 & 0 \\
1 & 1 & 0 & 1 & 1 \\
2 & 0 & 0 & 3 & 0 \\
0 & 1 & 3 & 0 & 3 \\
0 & 1 & 0 & 3 & 1
\end{array}\right]
\end{array}
$$

(10) We know that the matrix is of format 6×8 since the graph has six vertices and eight edges. The incidence matrix relative to the ordering A, B, C, D, E, F of the vertices and the ordering $a_1, a_2, a_3, a_4, a_5, a_6, a_7, a_8$ of its edges is as follows.

$$
\begin{array}{c c}
 & \begin{array}{c c c c c c c c} a_1 & a_2 & a_3 & a_4 & a_5 & a_6 & a_7 & a_8 \end{array} \\
\begin{array}{c} A \\ B \\ C \\ D \\ E \\ F \end{array} &
\left[\begin{array}{c c c c c c c c}
1 & 1 & 0 & 0 & 0 & 0 & 0 & 0 \\
1 & 0 & 1 & 0 & 0 & 0 & 1 & 0 \\
0 & 1 & 1 & 1 & 1 & 0 & 0 & 0 \\
0 & 0 & 0 & 1 & 0 & 1 & 0 & 1 \\
0 & 0 & 0 & 0 & 1 & 1 & 1 & 0 \\
0 & 0 & 0 & 0 & 0 & 0 & 0 & 1
\end{array}\right]
\end{array}
$$

(11)(c) Since the adjacency matrix is 4×4, G is of order 4. The graph is not simple since the adjacency matrix is not boolean (entries not consisting of 0's and 1's only). The following is a possible drawing of G. The given adjacency matrix is with respect to the ordering

A, B, C, D of the vertices.

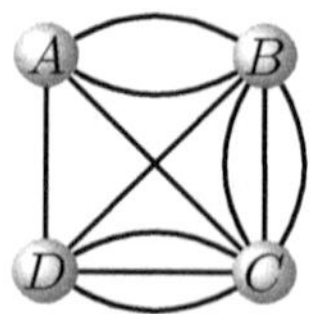

(14) Write $V = \{v_1, v_2, \ldots, v_n\}$. A simple graph with V as the set of vertices is determined uniquely by its set of edges E. Since every edge of G is of the form $\{v_i, v_j\}$ with $i \neq j$, graphs with V as set of vertices are in one-to-one correspondences with the set $\mathcal{E} = \{\{v_i, v_j\}, i \neq j\}$. Since the cardinality of $\mathcal{E}$ is $\binom{n}{2}$, there are $2^{\binom{n}{2}} = \frac{n!}{2!(n-2)!} = \frac{n(n-1)}{2}$ subsets of $\mathcal{E}$. We conclude that there are $2^{\frac{n(n-1)}{2}}$ different simple graphs with vertex set V.

9.3

(3) The Petersen graph has 10 vertices, each of degree 3. We can easily count 15 edges in the graph. So the sum of the degrees of all vertices is $10 \times 3 = 30$, which is twice the number of edges.

(5) The graph $K_{m,n}$ has m vertices of degree n each and n vertices of degree m each. If e is the number of edges in $K_{m,n}$, the handshaking lemma tells us: $m(n) + n(m) = 2e$ giving that $e = mn$.

(6)(d) The degree vertex of G is $(5, 4, 3, 3, 3, 3, 1)$.

(8) The sum of degrees of all vertices is $7 + 5 + 4 + 4 + 3 + 2 + 1 = 26$. By the handshaking lemma, this is twice the number of edges. Therefore, the of size 13 (there are 13 edges in the graph). The graph cannot be simple since it has seven vertices and one vertex of degree 7 (in a simple graph of order n, the degree of any vertex is at most $n - 1$).

(10) Let n by the number of vertices in G. By the handshaking lemma we have $4n = 2 \times 16$ giving that $n = 8$. Here is a possible drawing for such a graph:

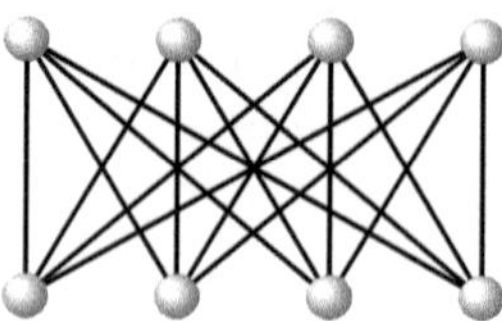

(12) Let n be the number of vertices of degree 4. By the handshaking lemma, $(7 \times 1) + (3 \times 2) + (7 \times 3) + (4n) = 2 \times 21 = 42$. Solving for n gives $n = 2$. So G has $7 + 3 + 7 + 2 = 19$ vertices.

(15) Represent each computer with a vertex in a graph G. Two vertices are adjacent if and only if the corresponding computers are connected with a cable. If n is the number of computers in the office, then the degree of every vertex in G is $n-1$. The handshaking lemma gives that $n(n-1) = 2 \times 66.$ simplifying gives: $n^2 - n - 132 = 0$ which has two roots $n = -11$, $n = 12$. We conclude that there are 12 computers in the office.

(18) Let x, y and z be the number of vertices of degree 2, 3 and 4, respectively. We have:

$$\begin{cases} x + y + z = 12 & \text{(since there are 12 vertices)} \\ y = 2z & \text{(G has twice as many vertices} \\ & \text{of degree 3 as vertices of degree 5)} \\ 2x + 3y + 4z = 32 & \text{(by the handshaking lemma)} \end{cases}$$

Solving this linear system of equations gives $x = 6$, $y = 4$ and $z = 2$.

(20) Let $G = (V, E)$ be a graph of order n. Write $V = V_E \cup V_O$ where V_E and V_O are the sets of vertices of G of even and odd degrees, respectively. By the handshaking lemma, $\sum_{v \in V_E} + \sum_{v \in V_O} = 2e$ where e is the number of edges in G. This means in particular that $\sum_{v \in V_O} = 2e - \sum_{v \in V_E}$ is even as $2e$ is even and $\deg(v)$ is even for any $v \in V_E$. But since $\deg(v)$ is odd for any $v \in V_O$, $|V_O|$ must be even.

(21)(c) Using the same terminology as in part (b), the equation $nd = 2e$ implies that d divides $2e$ but since it is an odd integer, d must divide e.

(d) As usual, let n be the number of vertices and e the number of edges. By the handshaking lemma, $6n = 2e$ which implies that $e = 3n$.

(23)(a) Since the complete graph has a single edge between any two vertices, we can get an edge by choosing any two distinct vertices. So the number of edges in K_n is equal to $\binom{n}{2} = \frac{n!}{(n-2)!2!} = \frac{n(n-1)}{2}$.

(b) Let e be the number of edges in K_n. Recall that $\deg(v) = n - 1$ for every vertex in K_n. By the handshaking lemma: $n(n-1) = 2e$ which implies that $e = \frac{n(n-1)}{2}$.

(27) Let G be a graph of order $n \geq 2$.

(a) Assume by contradiction that G has a vertex u of degree 0 and a vertex v of degree $n - 1$. In particular, vertex v is adjacent to all other vertices in G, But this is a contradiction since u is of degree 0 and cannot be adjacent to any vertex.

(b) From part (a), we have $n - 1$ possibilities for the degrees of the n

vertices of the graph. By the pigeonhole principal, there must be at least one pair of distinct vertices of G with the same degree.

(c) This is a direct application of part (b) using a graph in which each person is represented by a vertex and each handshake with an edge.

(29)

(1)(e) The sequence is graphical. The following is a realization of it:

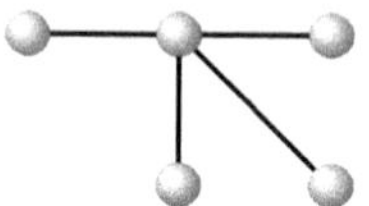

(f) The sequence is not graphical. There are five terms in the sequence with the first term greater than 5.

(g) The sequence is not graphical. Assume by contradiction that it is, and let G be a simple graph having $(3,3,3,1)$ as the degree sequence. Label the vertices v_1, v_2, v_3 and v_4 with $\deg(v_1) = \deg(v_2) = \deg(v3) = 3$ and $\deg(v_4) = 1$. As $\deg(v_1) = 3$, v_1 is adjacent to all other vertices. Connect v_1 with the other three vertices with one edge. This makes vertices v_1 and v_4 "saturated". Vertex v_2 can be joined by a single edge to v_3. Any other edge in the graph would be either incident to v_1 or v_4 or would be a parallel edge between v_2 and v_3. This is a contradiction.

(30)(a)

$$l_0 = (4,4,3,2,2,1) \text{ is graphical} \Leftrightarrow$$
$$l_1 = (3,2,1,1,1) \text{ is graphical} \Leftrightarrow$$
$$l_2 = (1,0,0,1) \leftrightarrow (1,1,0,0) \text{ is graphical (rearranged)} \Leftrightarrow$$
$$l_3 = (0,0,0) \text{ is graphical.}$$

We conclude that the sequence $(4,4,3,2,2,1)$ is graphical. Reversing the sequence above leads to the realization G_0 of the sequence:

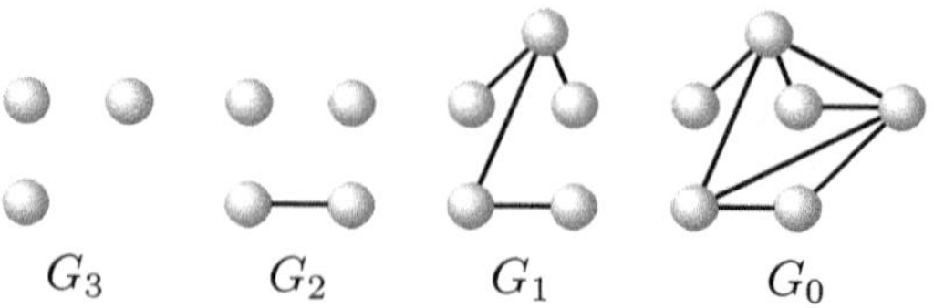

$G_3 \qquad G_2 \qquad G_1 \qquad G_0$

(d)

$$l_0 = (8, 7, 6, 6, 5, 3, 2, 2, 2, 1) \text{ is graphical} \Leftrightarrow$$
$$l_1 = (6, 5, 5, 4, 2, 1, 1, 1, 1) \text{ is graphical} \Leftrightarrow$$
$$l_2 = (4, 4, 3, 1, 0, 0, 1, 1) \Leftrightarrow$$
$$(4, 4, 3, 1, 1, 1, 0, 0) \text{ is graphical (rearranged)} \Leftrightarrow$$
$$l_3 = (3, 2, 0, 0, 1, 0, 0) \Leftrightarrow$$
$$(3, 2, 1, 0, 0, 0, 0) \text{ is graphical (rearranged)} \Leftrightarrow$$
$$l_4 = (1, 0, -1, 0, 0, 0) \text{is graphical.}$$

The last sequence contains a negative integer, $(8, 7, 6, 6, 5, 3, 2, 2, 2, 1)$ is therefore not graphical.

9.4

(1) Write $G = (V, E)$ and $G_i = (V_i, E_i)$ for $i = 1, 2, 3, 4$. G_1 is a subgraph of G: $V_1 \subseteq V$ and $E_1 \subseteq E$. Graph G_2 is not a subgraph of G as it has one edge $\{a, b\}$ which are not in G. Graph G_3 is a subgraph of G: $V_3 \subseteq V$ and $E_3 \subseteq E$. Graph G_4 is not a subgraph of G as edges $\{a, d\}$ and $\{b, c\}$ are not edges in G.

(4)

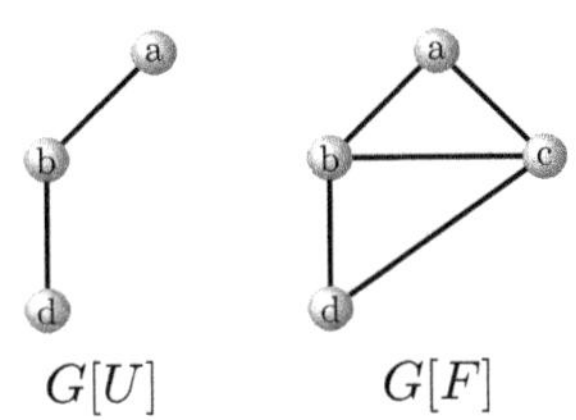

$$G[U] \qquad G[F]$$

(5)

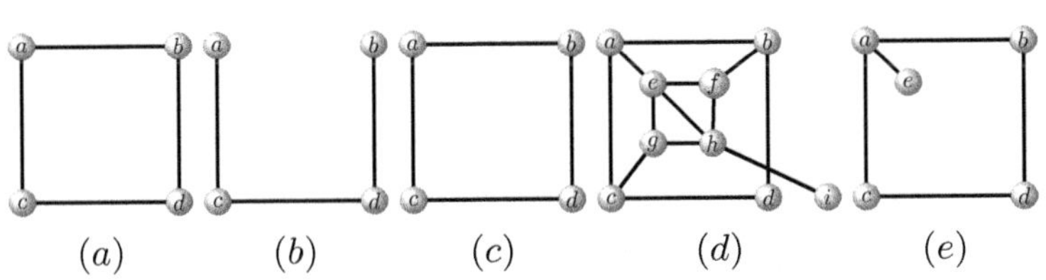

$$(a) \qquad (b) \qquad (c) \qquad (d) \qquad (e)$$

(7) R is reflexive since every graph is a subgraph of itself. R is not symmetric since if H_1 is a subgraph of H_2, H_2 is not necessarily a subgraph of H_1. For example, $H_1 = (\{a, b\}, \{\{a, b\}\})$ is subgraph of $H_2 = (\{a, b, c\}, \{\{a, b\}, \{a, c\}\})$ but H_2 is not a subgraph of H_1. The relation R is antisymmetric: If $H_1 = (V_1, E_1)$ and $H_2 = (V_2, E_2)$ are

two subgraphs of G such that $H_1 \, R \, H_2$ and $H_2 \, R \, H_1$ then $V_1 \subseteq V_2$, $E_1 \subseteq H_2$ (since H_1 is a subgraph of H_2) and $V_2 \subseteq V_1$, $E_2 \subseteq H_1$ (since H_2 is a subgraph of H_1). We conclude that $V_1 = V_2$ and $V_2 = V_1$ and so $H_1 = H_2$. The relation is transitive: let $H_i = (V_i, E_i)$, $i = 1, 2, 3$, be three subgraphs of G such that $G_1 \, R \, G_2$ and $G_2 \, R \, G_3$. Then $V_1 \subseteq V_2 \subseteq V_3$ and $E_1 \subseteq E_2 \subseteq E_3$. Moreover, if $\{a, b\}$ is an edge in E_1, then $a, b \in V_1$ (since H_1 is a subgraph of H_2). We conclude that H_1 is a subgraph of H_3. In other words, $H_1 \, R \, H_3$.

(8)(a) A spanning subgraph of G is completely determined by the subset of edges chosen from the collection of all edges of G (since it has the same set of vertices as G, by definition). There are 2^m possible subsets of the set of edges of G, so there are 2^m possible spanning subgraphs of G.

(e) Such a spanning subgraph is completely determined by the choice of the k edges: there are $\binom{n}{2} = \frac{n(-1)}{2}$ edges in K_n which means a total of $\binom{\frac{n(-1)}{2}}{k}$ spanning subgraphs of K_n with exactly k edges.

(9)(b) Let V_1 and V_2 by the bipartition of the set of vertices of $K_{m,n}$. So every edge in $K_{m,n}$ joins one vertex in V_1 to another in V_2 and no two vertices in either one of the sets V_1 and V_2 are adjacent. Assume that $\{a, b, c\}$ is a clique of size 3 in $K_{m,n}$. Without loss of generality, we can assume that $a, b \in V_1$ and $c \in V_2$. Since a, b are adjacent vertices (as part of the clique), we obtain a contradiction (no two vertices in V_1 are adjacent). So no cliques of size 3. Clearly there are cliques of size 2 in $K_{m,n}$. We conclude that the clique number of $K_{m,n}$ for $m, n \geq$ is 2.

(10)(b) Here are three cliques of size 3: $\{a, b, c\}$, $\{c, d, h\}$ and $\{b, e, f\}$ together with their corresponding induced subgraphs (shown in bold in the graph).

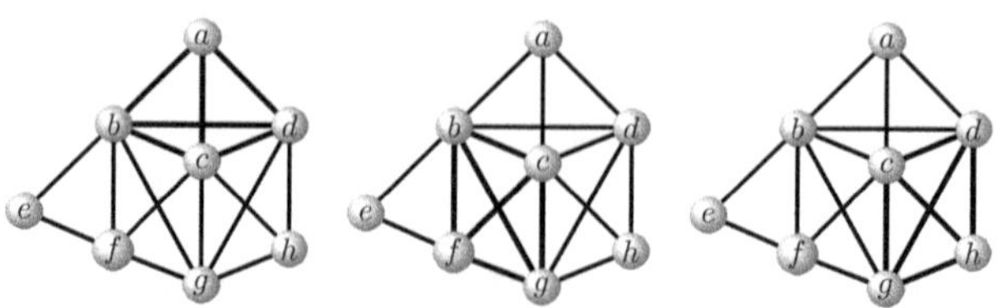

(12) Since every pair of vertices in K_n are adjacent, the independence number of K_n is 1.

(14)

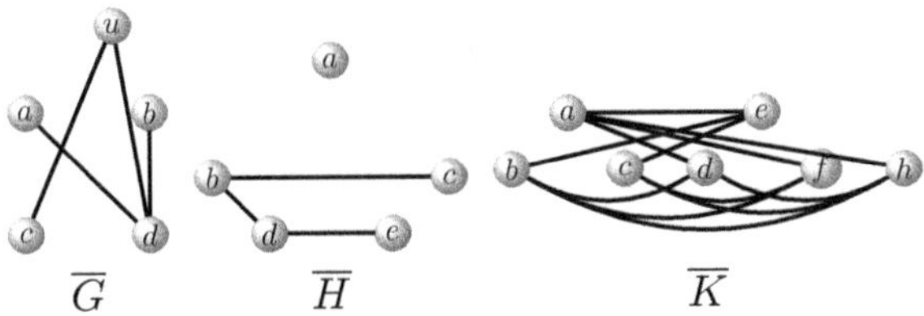

(17)(a) Here is an illustration of $C_3 + C_4$:

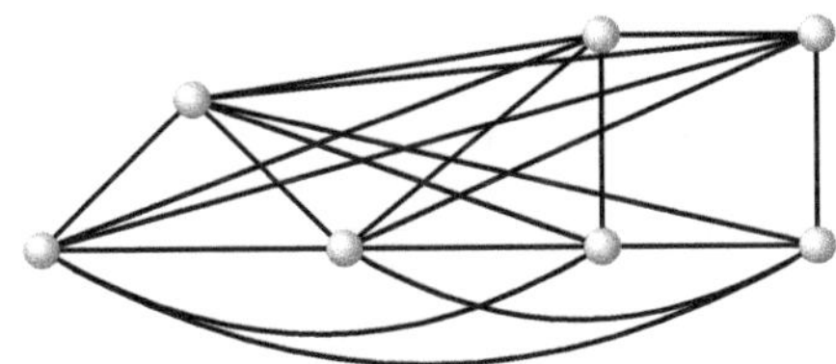

(b) We claim that $K_m + K_n = K_{m+n}$. First note that, like K_{m+n}, $K_m + K_n$ has $m + n$ vertices by definition of the join. Let u, v be any two distinct vertices of $K_m + K_n$. If both u and v belong to K_m, then they are adjacent in $K_m + K_n$. The same conclusion is drawn if both vertices belong to K_n. If u belongs to K_m and v belongs to K_n, then they are joined by a single edge in $K_m + K_n$ by definition of the join. We conclude that every pair of distinct vertices in $K_m + K_n$ are joined by a single edge in $K_m + K_n$ and so the latter is just K_{m+n}.

(c) Recall that W_n is obtained from C_n by adding a vertex in the center and connecting it with one edge to every other vertex of C_n. Since a single vertex is represented by K_1, the wheel can be realized as $K_1 + C_n$.

(20) Since $\deg(v) = d$, there are d edges $e_1, \ldots, e_d$ of G incident to v. Each one of these edges is a vertex in $L(G)$ and these vertices are pairwise adjacent in $L(G)$ since they share vertex v. Then each pair of distinct elements of the set $\{e_1, \ldots, e_d\}$ is an edge in $L(G)$. There are $\binom{d}{2} = \frac{d(d-1)}{2}$ such pairs.

(23) By successively adding an edge between pairs of non-adjacent vertices with the sum of degree greater than or equal to 6 (the order of the graph) until all such pairs are joined by an edge, we obtain the following

Hamiltonian closure of the graph:

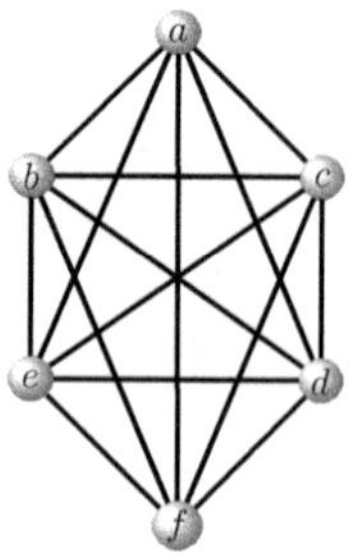

9.5

(1)(c) False: a path is open.

    (e) True: In the following simple graph, The walk (a, c, d, f, c, b) is a trail as all the edges are distinct, but it is not a path as the vertex c is visited twice.

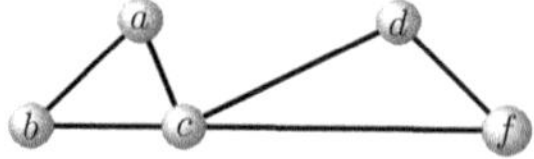

    (g) False: the graph of part (e) has no bridges but it has an articulation point (vertex c).

(2) Note first that since the Petersen graph is simple, we can identify every walk in it with the list of its vertices.

    (a) A path of length 5 is a, b, c, d, i, g. A path of length 6 is a, b, c, d, i, g, j. A path of length 8 is $a, b, c, d, e, j, h, fi, g$. A path of length 9 is $a, f, i, g, j, h, c, b, a, e$.

    (c) $\operatorname{dis}(a, f) = 1$, $\operatorname{dis}(a, g) = 2$, $\operatorname{dis}(b, i) = 2$, $\operatorname{dis}(c, e) = 3$, $\operatorname{dis}(h, i) = 2$, $\operatorname{dis}(d, j) = 2$ and $\operatorname{dis}(a, g) = 2$.

(4)(b) In the graph G below, $\delta(G) = 3$, so $\kappa(G) \leq \kappa'(G) \leq 3$. Since G has no bridges (no removal of a single edge would disconnect the graph), $\kappa'(G) \geq 2$. Note that $\{e_1 = \{b, e\}, e_2 = \{c, h\}\}$ is an edge-cut of the graph, so $\kappa'(G) \geq 2$. On the other hand, the removal of vertices b and c would disconnect the graph, so $\kappa(G) \leq 2$ and so $\kappa(G) = 2$ since $\kappa(G) \leq \kappa'(G) = 2$. We conclude that $\kappa(G) = \kappa'(G) = 2 < \delta(G) = 3$.

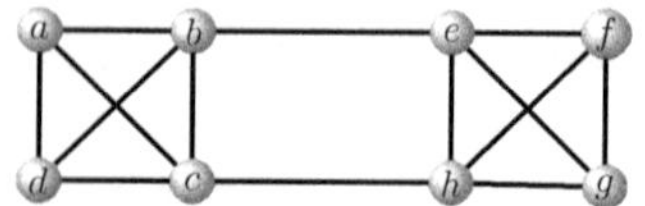

(d) In the graph G below, $\delta(G) = 3$, so $\kappa(G) \leq \kappa'(G) \leq 3$. Since G has no bridges (no removal of a single edge would disconnect the graph), $\kappa'(G) \geq 2$. Removing the two edges $\{b, c\}$ and $\{d, c\}$ would disconnect the graph, so $\kappa'(G) = 2$. On the other hand, $\kappa(G) = 1$ since removing vertex c disconnects the graph. So $1 = \kappa(G) < \kappa'(G) = 2 < \delta(G) = 3$.

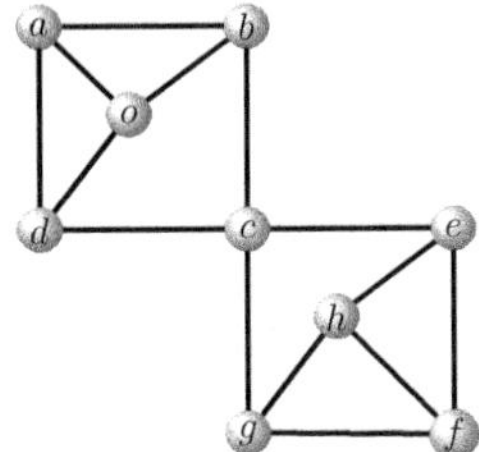

(7)(a) Deleting one vertex from P_n $(n \geq 2)$ will disconnect the graph. Similarly, deleting an edge of P_n disconnects the path. So $\kappa(P_n) = \kappa'(P_n) = 1$.

(b) Deleting one vertex (or one edge) from C_n produces a path. Deleting a second vertex (or a second edge) will disconnect the graph. So $\kappa(C_n) = \kappa'(C_n) = 2$.

(c) Removing the central vertex from the wheel W_n gives us the cycle graph C_n $(n \geq 3)$. By part (b), $\kappa(W_n) = \kappa(C_n) + 1 = 3$. For the edge connectivity, notice that removing one or two edges from W_n will not disconnects the graph. Let x be the central vertex of W_n and let $\{u, v\}$ and $\{v, w\}$ be two edges W_n where the vertices u, v and w are different from x. Removing the edges $\{u, v\}$, $\{v, w\}$ and $\{v, x\}$ from W_n would isolate vertex v and disconnects W_n. Therefore, $\kappa'(W_n) = 2$.

(10)(a) If $v \in V$ then by convention, v is connected to itself by the empty path (path of length zero), so $v\,R\,v$ and R is reflexive. Let $u, v \in V$ such that $u\,R\,v$. Then there exists a uv-path $P = uv_1v_2 \ldots v_k v$ in G. Traversing P is the reverse order gives the vu-path $P' = vv_k \ldots v_2 v_1 u$. So $v\,R\,u$ and R is symmetric. Now, let $u, v, w \in V$ such that $u\,R\,v$ and $v\,R\,w$. Then there exist a uv-path and a vw-path in G. Thus there exists a uw-path in G and $u\,R\,w$. So R is transitive. We conclude that R is an equivalence relation on the set V of vertices of G.

(b) Let X be an be an equivalence class of the relation R and let $v \in X$. By definition of the relation R, $X = C(v)$ and therefore $G[X] = G[C(v)]$ is the connected component containing v.

(11) Assume by contradiction that the edge $e = \{u, v\}$ is a bridge of G. Then $G - e$ is disconnected with exactly two connected components C_1 and C_2. Moreover, we can assume that u belongs to C_1 and v to C_2. So C_1 is a connected graph with all vertices of even degree except for vertex u of odd degree. By the Handshaking Theorem $\deg(u) = 2|E(C_1)| - \sum_{x \neq u} \deg(x)$: this is a contradiction since the righthand is even.

(15) Since $\deg(v) \geq 2$, vertex v must have at least two neighbors, say x and y. Since there is a unique path p from x to y, p must then be the path xvy. This means in particular that there no xy-path in $G - v$ and so $G - v$ is disconnected.

(17) Assume that G is not connected. We prove that $\overline{G}$ must be connected. Let u, v be any two vertices in $\overline{G}$. If u and v belong to the same connected component of G, choose a vertex w in a different connected component of G. Then both u and v are adjacent to w in $\overline{G}$ (by definition of the complement of a graph) and therefore there exists a path from u to v in $\overline{G}$. If u and v belong to different connected components of G, then they are not adjacent in G and so they must be adjacent in $\overline{G}$. In particular there exists a uv-path in $\overline{G}$. We conclude that a uv-path always exists in $\overline{G}$ and $\overline{G}$ is connected.

9.6

(3)(a) Graphs S_2, S_3 and S_4 are shown in the diagram below.

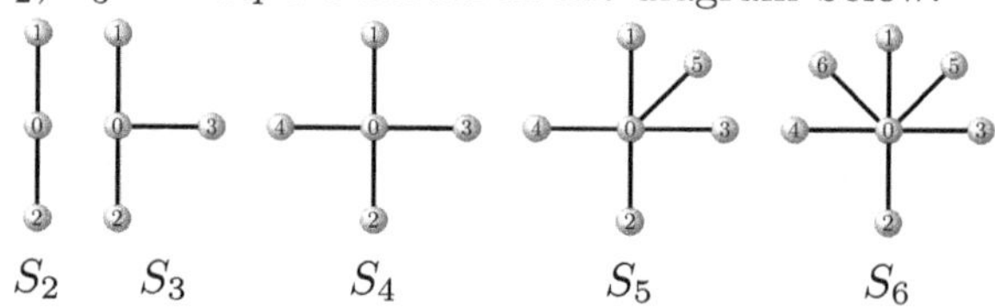

(b) Note that in the star graph S_n, there is one vertex of degree n and every other vertex is of degree 1. This shows in particular that there are no cycles in S_n and in particular no odd cycles. This means that S_n is bipartite. Let $V_1 = \{0\}$ and $V_2 = \{1, 2 \ldots, n\}$. Then (V_1, V_2) is a bipartition of S_n.

(4)(a) K_1 is not bipartite since there is no bipartition of a singleton. K_2 is bipartite with the obvious bipartition. For $n \geq 3$, the complete graph K_n is not bipartite since every pair of vertices are adjacent (so no bipartition of vertices is possible).

(d) The wheel graph W_n (for $n \geq 3$) is not bipartite as it has triangles (cycles of length three) with one vertex being the center of the wheel.

(5) Let V_e and V_o be the sets of vertices of Q_n with an even number and an odd number of 0's, respectively. Since every binary string (vertex in Q_n) has either an odd or an even number of 0's, $V = V_e \cup V_o$. Moreover, it is clear that $V_e \cap V_o = \emptyset$ since no binary string can have both an even and odd number of 0's at the same time. Next, assume by contradiction that $x, y \in V_e$ are adjacent. Then x and y differ in exactly one position. This means that a 0 in one of them is replaced by 1 in the other. This is a contradiction to the fact that x and y both have an even number of 0's. We reach the same conclusion if we assume that $x, y \in V_o$ are adjacent. So every edge of Q_n joins one vertex of V_e to another in V_o. We conclude that (V_e, V_o) is a bipartition of Q_n and Q_n is bipartite.

(6)(a) By induction on $|E|$. If $|E| = 1$, then $\sum_{v \in V_1} \deg(v) = \sum_{v \in V_2} \deg(v) = 1$ and the result is true. Let $k \geq 1$ and assume that the result is true for every bipartite graph with k edges. Let G be a bipartite graph with bipartition (V_1, V_2) and $k+1$ edges. Delete an edge $e = \{x, y\}$ of G and let $G' = G - e$. Graph G' is bipartite as a subgraph of a bipartite graph (Corollary 9.2) of order k. Moreover, (V_1, V_2) is also a bipartition of G' as G and G' has the same set of vertices. By the induction hypothesis,

$$\sum_{v \in V_1} \deg_{G'}(v) = \sum_{v \in V_2} \deg_{G'}(v) = k. \tag{A.1}$$

If $v \neq x$ is a vertex of V_1, then $\deg_{G'}(v) = \deg_G(v)$. Similarly, If $v \neq y$ is a vertex of V_2, then $\deg_{G'}(v) = \deg_G(v)$. Moreover, $\deg_G(x) = \deg_{G'}(x) + 1$ and $\deg_G(y) = \deg_{G'}(y) + 1$. So

$$\sum_{v \in V_1} \deg_G(v) = \deg_G(x) + \sum_{v \in V_1 \setminus \{x\}} \deg(v)$$

$$= \deg_{G'}(x) + 1 + \sum_{v \in V_1 \setminus \{x\}} \deg_{G'}(v)$$

$$= \sum_{v \in V_1} \deg_{G'}(v) + 1$$

$$= \sum_{v \in V_2} \deg_{G'}(v) + 1 \text{ (by A.1)}$$

$$= \deg_{G'}(y) + \sum_{v \in V_2 \setminus \{y\}} \deg(v) + 1$$

$$= \deg_G(y) + \sum_{v \in V_2 \setminus \{y\}} \deg_G(v) = \sum_{v \in V_2} \deg_G(v)$$

(b) By part (a), $\sum_{v \in V_1} \deg(v) = \sum_{v \in V_2} \deg(v)$. But $\deg(v) = d$ for every vertex v of G. So the equality of part (a) becomes $d|V_1| = d|V_2|$ and so $|V_1| = |V_2|$.

(9)(a) The following diagram shows a bipartition of the graph:

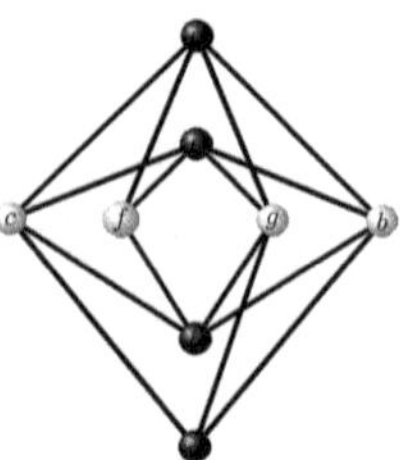

(b) The adjacency matrix of G relative to the ordering a, b, c, d, e, f, g, h of the vertices is as follows:

$$
\begin{array}{c}
\begin{array}{cccccccc} a & b & c & d & e & f & g & h \end{array} \\
\begin{array}{c} a \\ b \\ c \\ d \\ e \\ f \\ g \\ h \end{array}
\left[
\begin{array}{cccccccc}
0 & 1 & 1 & 0 & 0 & 1 & 1 & 0 \\
1 & 0 & 0 & 1 & 1 & 0 & 0 & 1 \\
1 & 0 & 0 & 1 & 1 & 0 & 0 & 1 \\
0 & 1 & 1 & 0 & 0 & 0 & 1 & 0 \\
0 & 1 & 1 & 0 & 0 & 1 & 1 & 0 \\
1 & 0 & 0 & 0 & 1 & 0 & 0 & 1 \\
1 & 0 & 0 & 1 & 1 & 0 & 0 & 1 \\
0 & 1 & 1 & 0 & 0 & 1 & 1 & 0
\end{array}
\right]
\end{array}
$$

(c) Since $V_1 = \{a, e, d, h\}$ and $V_2 = \{b, c, f, g\}$ form a bipartition of the graph, let us write the adjacency matrix of G relative to the ordering a, e, d, h, b, c, f, g of the vertices:

$$
\begin{array}{c}
\begin{array}{cccccccc} a & e & d & h & b & c & f & g \end{array} \\
\begin{array}{c} a \\ e \\ d \\ h \\ b \\ c \\ f \\ g \end{array}
\left[
\begin{array}{cccccccc}
0 & 0 & 0 & 0 & 1 & 1 & 1 & 1 \\
0 & 0 & 0 & 0 & 1 & 1 & 1 & 1 \\
0 & 0 & 0 & 0 & 1 & 1 & 0 & 1 \\
0 & 0 & 0 & 0 & 1 & 1 & 1 & 1 \\
1 & 1 & 1 & 1 & 0 & 0 & 0 & 0 \\
1 & 1 & 1 & 1 & 0 & 0 & 0 & 0 \\
1 & 1 & 0 & 1 & 0 & 0 & 0 & 0 \\
1 & 1 & 1 & 1 & 0 & 0 & 0 & 0
\end{array}
\right]
\end{array}
$$

Let $\mathbf{0} = \begin{bmatrix} 0 & 0 & 0 & 0 \\ 0 & 0 & 0 & 0 \\ 0 & 0 & 0 & 0 \\ 0 & 0 & 0 & 0 \end{bmatrix}$ and $B = \begin{bmatrix} 1 & 1 & 1 & 1 \\ 1 & 1 & 1 & 1 \\ 1 & 1 & 0 & 1 \\ 1 & 1 & 1 & 1 \end{bmatrix}$, then the adjacency matrix above (with respect to the ordering a, e, d, h, b, c, f, g) has the form

$$
\begin{bmatrix} \mathbf{0} & B \\ B^T & \mathbf{0} \end{bmatrix}
$$

(10)(a) If (V_1, V_2) is a bipartition of the graph with $|V_1| = k$ and $|V_2| = n-k$, then $\deg(v) \leq n-k$ for every $v \in V_1$ and $\deg(v) \leq k$ for every $v \in V_2$. We conclude that

$$\sum_{v \in V} \deg(v) = \sum_{v \in V_1} \deg(v) + \sum_{v \in V_2} \deg(v)$$
$$\leq \underbrace{(n-k) + \cdots + (n-k)}_{k \text{ times}} + \underbrace{k + \cdots + k}_{n-k \text{ times}}$$
$$= k(n-k) + (n-k)k$$
$$= 2k(n-k).$$

By the Handshaking Lemma, $\sum_{v \in V} \deg(v) = 2e$. We conclude that $2e \leq 2k(n-k)$ or $e \leq k(n-k)$.

(b) Consider the function of one variable k, $f(k) = k(n-k)$. This is a parabola (in the variable k) which intersects the horizontal axis k in two points $k = 0$ and $k = n$. The maximum value of the function is then reached at the midpoint $\frac{n}{2}$. Replacing in $f(k)$ we get

$$f\left(\frac{n}{2}\right) = \frac{n}{2}\left(n - \frac{n}{2}\right) = \frac{n^2}{4}.$$

So $e \leq k(n-k) \leq \frac{n^2}{4}$.

(c) Consider the simple graph G:

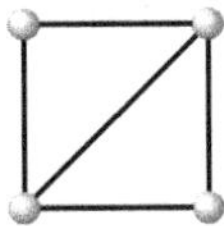

The order of G is $n = 4$ and the size is $e = 5$. Clearly, $e > \frac{n^2}{4}$. The graph is not bipartite as it has cycles of length 3.

(11)(c) The following is 4-partite simple graph of order 12. The graph is not bipartite as it has circuits of length three (for example a, d, g, a).

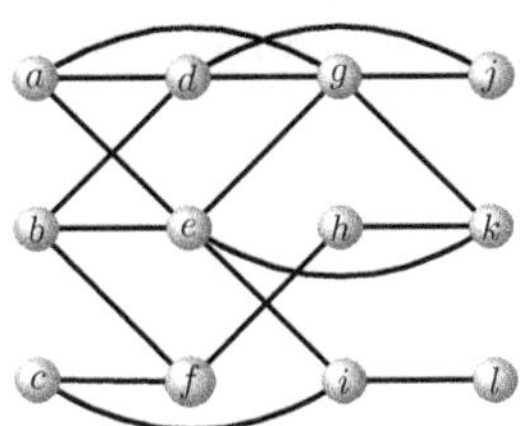

The corresponding 4-partition of the vertices is $V_1 = \{a, b, c\}$, $V_2 = \{d, e, f\}$, $V_3 = \{g, h, i\}$ and $V_4 = \{j, k, l\}$.

9.7

(1) The number of nodes covered by a matching M of size r is $2r$.

(2) In graph H, the subset of edges is not a matching since edges $\{a,b\}$ and $\{a,c\}$ of the subset are incident at vertex a. In graph L, the subset of edges $M = \{\{a,b\},\{f,c\}\}$ is a matching. It is maximal since if any other edge of L is added to M, the subset is no longer a matching. It is not maximum since there is a larger matching (for instance, $\{\{a,e\},\{c,f\},\{b,d\}\}$ is a matching of order 3). Clearly, the matching is not perfect.

(4) Graph G has perfect matching whose edges are shown in bold in the following diagram.

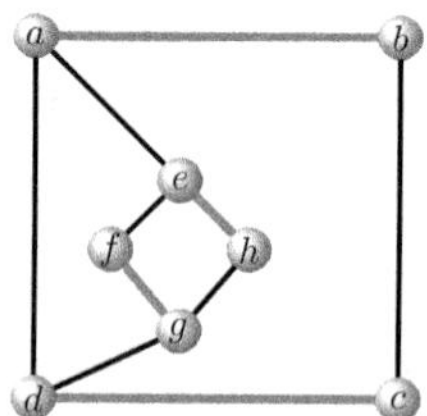

Any perfect matching in K must cover vertices e and d and so must contain edges $\{c,e\},\{c,d\}$, a contradiction. So the graph does not contain a perfect matching. The following diagram shows a maximum matching of K:

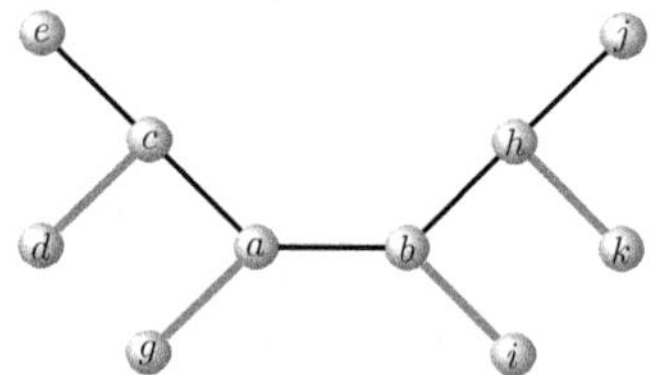

Graph N has a perfect matching as shown in the diagram below.

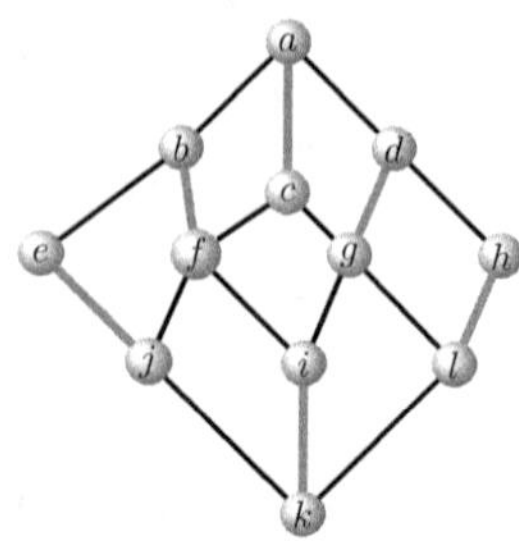

(5)(a) Let M be a maximum matching of the graph G. If M is not maximal, then M is strictly contained in another matching M'. In particular $|M| < |M'|$ (M' has more edges than M), a contradiction to the fact that $|M|$ is the largest cardinality of a matching.

(b) The diagram below shows a matching M in a graph using bold edges:

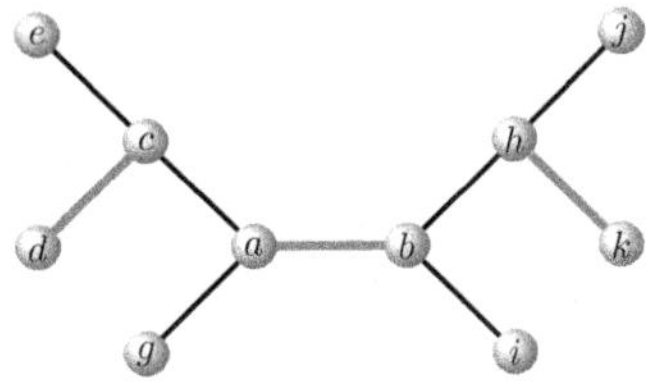

Adding any of the remaining edges of the graph to M does not result in a matching. So M is maximal. However, it is not maximum since the graph has a matching of size 4 formed by the edges $\{c, d\}$, $\{a, g\}$, $\{b, i\}$, $\{h, k\}$.

(9) If n is odd, there is no perfect matching. Assume n is even. Label the vertices of C_n as $v_1, v_2, \ldots, v_n$. We clearly have the two perfect matchings

$$M_1 = \{\{v_1, v_2\}, \{v_3, v_4\}, \ldots, \{v_{n-1}, v_n\}\}$$
$$M_2 = \{\{v_2, v_3\}, \{v_4, v_5\}, \ldots, \{v_n, v_1\}\}$$

of C_n. If M is any perfect matching of C_n, then M either contains $\{v_1, v_2\}$ or not. If $\{v_1, v_2\} \in M$, then $\{v_i, v_{i+1}\} \in M$ for any odd i. This shows that $M = M_1$. Similarly, if $\{v_1, v_2\} \notin M$ then $M = M_2$. We conclude that C_n has only two perfect matchings, namely M_1 and M_2.

9.8

(1) The condition is not sufficient. To see this, consider the following bipartite graph G with bipartition sets $X = \{A, B, C, D\}$ and $Y = \{e, f, l, p\}$:

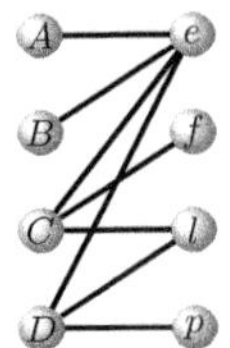

For the subset $U = \{A, B\}$ of X, $N(U) = \{e\}$ and so $|N(U)| < |U|$. The graph does not have a perfect matching of X onto Y by Hall's Theorem. Hence the graph does not have a perfect matching, but the condition $|X| = |Y|$ is satisfied.

(4) Assume the upper left corner is white. Let W and B be the sets of white and black squares on the truncated board, respectively. Notice first that $|W| = |B|$. Consider the bipartite graph G with bipartition sets W and B. Two vertices are joined by an edge if and only the corresponding squares share a side. The truncated chessboard can be covered by rectangular 2×1 tiles as required if and only if there exists a perfect matching M in G. Then by definition of a perfect matching in a bipartite graph, M is a perfect matching of X onto Y and a perfect matching of Y onto X. It is not hard to construct such a matching in G and so the required tiling exists.

(6)(c) Let $U = \{a, b, c\}$, then $N(U) = \{f, l\}$. Since $|U| > |N(U)|$, the graph has no perfect matching of X onto Y.

(7) Consider the bipartite graph G with bipartition sets $R = \{R_1, R_2, \ldots, R_{13}\}$ corresponding to the 13 rows and $V = \{\text{Ace}, 1, 2, \ldots, J, Q, K\}$ corresponding to the 13 possible values in the deck. A vertex $R_i \in R$ is joined by an edge with a vertex $v \in V$ if and only if row R_i contains a card with value v. A perfect matching in G corresponds to a choice of 13 cards, one from each row, such that the cards have 13 different values. It suffices then to prove that a perfect matching exists in G. Let U be a non empty subset of R (so a collection of row numbers). The total number of cards in all the rows in U is $4|U|$ cards since every row has four cards. Since there are at most four cards of each value in the deck, there must be at least $|U|$ different values in the cards of the rows of U. This means that $|U| \leq |N(U)|$. Hall's Theorem tells us that a perfect matching exists in G since $|R| = |V| = 13$.

9.9

(2) For a simple graph of order n, the degree of every vertex is at most $n - 1$. By the Handshaking Lemma, there are at most $\frac{n(n-1)}{2}$ edges in the graph.

  (b) For $n = 4$, there are at most $\frac{4(4-1)}{2} = 6$ edges. There is one such graph with 0 edges, one graph with one edge, two graphs with two edges, three graphs with three edges, two graphs with four edges, one graph with five edges and only one with six edges. A total of 11 non-isomorphic graphs shown in the diagram below.

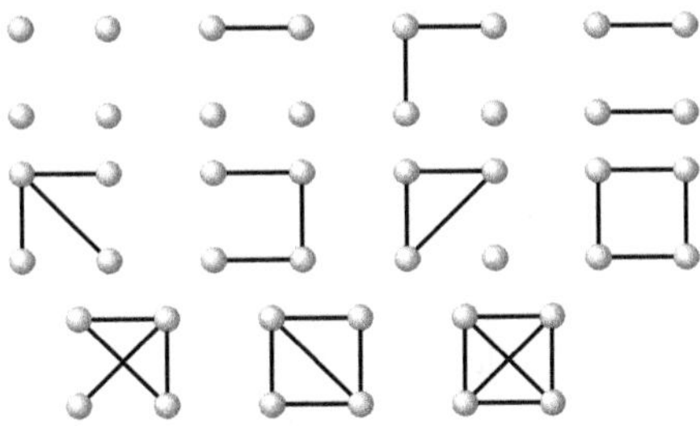

(3) A 3-regular graph on 6 vertices must have 9 edges. There are only two such graphs, one is bipartite (on the right) and another is not.

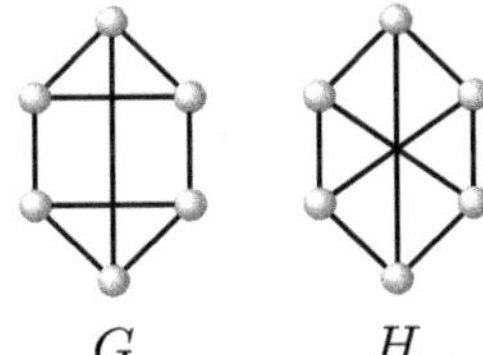

$$G \qquad\qquad H$$

(5)(a) $G \not\cong H$: In G the four vertices of degree 3 form a simple circuit of length 4. This is not the case in H.

(c) $G \cong H$. To see this, we start by labeling the vertices:

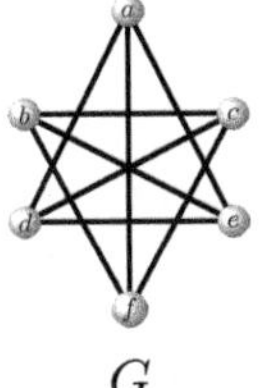

$$G \qquad\qquad H$$

The bijection $a \mapsto u_6$, $b \mapsto u_1$, $c \mapsto u_5$, $d \mapsto u_4$, $e \mapsto u_3$, $f \mapsto u_2$ is an isomorphism. We prove this using the adjacency matrix of G with respect to the enumeration a, b, c, d, e, f of its vertices and the adjacency matrix of H with respect to the enumeration $u_6, u_1, u_5, u_4, u_3, u_2$ of its vertices:

$$
\begin{array}{c}
\begin{array}{cccccc} a & b & c & d & e & f \end{array} \\
\begin{array}{c} a \\ b \\ c \\ d \\ e \\ f \end{array}
\begin{bmatrix}
0 & 0 & 0 & 1 & 1 & 1 \\
0 & 0 & 1 & 0 & 1 & 1 \\
0 & 1 & 0 & 1 & 0 & 1 \\
1 & 0 & 1 & 0 & 1 & 0 \\
1 & 1 & 0 & 1 & 0 & 0 \\
1 & 1 & 1 & 0 & 0 & 0
\end{bmatrix}
\end{array}
\qquad
\begin{array}{c}
\begin{array}{cccccc} u_6 & u_1 & u_5 & u_4 & u_3 & u_2 \end{array} \\
\begin{array}{c} u_6 \\ u_1 \\ u_5 \\ u_4 \\ u_3 \\ u_2 \end{array}
\begin{bmatrix}
0 & 0 & 0 & 1 & 1 & 1 \\
0 & 0 & 1 & 0 & 1 & 1 \\
0 & 1 & 0 & 1 & 0 & 1 \\
1 & 0 & 1 & 0 & 1 & 0 \\
1 & 1 & 0 & 1 & 0 & 0 \\
1 & 1 & 1 & 0 & 0 & 0
\end{bmatrix}
\end{array}
$$

(d) $G \not\cong H$ since in H, there is a simple circuit of length three with each vertex of degree 4 (shown with bold edges). No such a circuit in G.

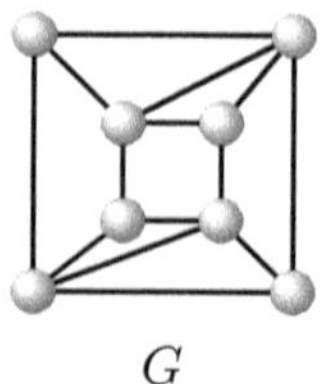
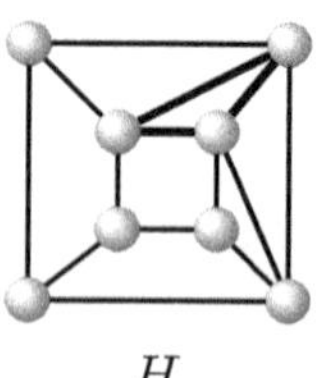

$$G \qquad\qquad H$$

(6)(b) G and K have simple circuits of length three each. No such circuit in H. So $G \not\cong H$ and $H \not\cong K$. We prove next that $G \cong K$. First we label the vertices.

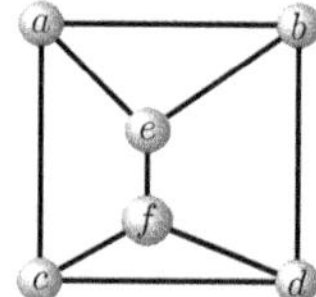
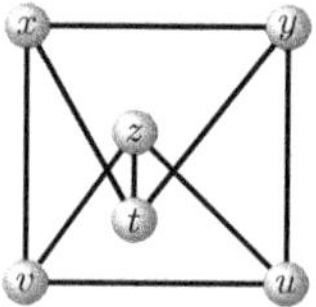

Consider the bijection $a \mapsto x, b \mapsto y, c \mapsto v, d \mapsto u, e \mapsto t, f \mapsto z$. We prove that this an isomorphism of graphs by showing the adjacency matrix of H with respect to the ordering a, b, c, d, e, f of its vertices is equal to the adjacency matrix of K with respect to the ordering x, u, z, t, y, v of its vertices:

$$
\begin{array}{c}
\begin{array}{cccccc} a & b & c & d & e & f \end{array} \\
\begin{array}{c} a \\ b \\ c \\ d \\ e \\ f \end{array}
\left[
\begin{array}{cccccc}
0 & 1 & 1 & 0 & 1 & 0 \\
1 & 0 & 0 & 1 & 1 & 0 \\
1 & 0 & 0 & 1 & 0 & 1 \\
0 & 1 & 1 & 0 & 0 & 1 \\
1 & 1 & 0 & 0 & 0 & 1 \\
0 & 0 & 1 & 1 & 1 & 0
\end{array}
\right]
\end{array}
\qquad
\begin{array}{c}
\begin{array}{cccccc} x & y & v & u & t & z \end{array} \\
\begin{array}{c} x \\ y \\ v \\ u \\ t \\ z \end{array}
\left[
\begin{array}{cccccc}
0 & 1 & 1 & 0 & 1 & 0 \\
1 & 0 & 0 & 1 & 1 & 0 \\
1 & 0 & 0 & 1 & 0 & 1 \\
0 & 1 & 1 & 0 & 0 & 1 \\
1 & 1 & 0 & 0 & 0 & 1 \\
0 & 0 & 1 & 1 & 1 & 0
\end{array}
\right]
\end{array}
$$

(7) The adjacency matrix of the complete graph K_n with respect to any ordering of the vertices consists of 0's on the main diagonal and 1's everywhere else. This means that any bijection $f : V(K_n) \to V(K_n)$ is an isomorphism of graphs.

(9) Let $f : V(G) \to V(G')$ be a graph isomorphism. Assume that G is bipartite and let $V(G) = V_1 \cup V_2$ be a bipartition of $V(G)$. In particular, if $\{x, y\} \in E(G)$ then $u \in V_1$ and $v \in V_2$ (or vice versa). Let $U_1 = f(V_1)$ and $U_2 = f(V_2)$. If $v \in V(G')$, let $u = f^{-1}(v) \in V(G)$. Since (V_1, V_2) is a bipartition of $V(G)$, $u \in V_1$ or $u \in V_2$ and so $v \in U_1$ or $v \in U_2$. This shows that $V(G') = U_1 \cup U_2$. If $v \in U_1 \cap U_2$ then $v = f^{-1}(u) \in$

$V_1 \cap V_2 = \emptyset$ which is a contradiction. This shows that (U_1, U_2) is a partition of $V(G')$. Moreover, for $v = f(x)$, $w = f(y) \in V(G')$: if $\{v, w\} \in E(G')$ then $\{x, y\} \in E(G)$ (since f is an isomorphism) and so $x \in V_1$ and $y \in V_2$ (or vice versa). This shows that $u \in U_1$ and $v \in U_2$ (or vice versa). So G' is bipartite with U_1, U_2 as bipartition sets. By the symmetry property of graph isomorphism we can conclude that if G' is bipartite, then G is bipartite and the result is true.

(12)(a) Let G be a self-complementary graph of order n. Let e be the number of edges of G. Then the number of edges in the complement $\overline{G}$ of G is $\binom{n}{2} - e$ since the number of edges of K_n is $\binom{n}{2}$. Since G is self-complementary, G and $\overline{G}$ must have the same number of edges: $e = \binom{n}{2} - e$ which implies that $e = \frac{1}{2}\binom{n}{2} = \frac{n(n-1)}{4}$. In particular, $n(n-1)$ must be a multiple of 4.

(i) Notice that $4(4-1) = 12$ is a multiple of 4. The following diagram shows a graph G of order 4 together with its complement $\overline{G}$. Notice that $G \cong \overline{G}$ since they are both isomorphic to the path P_4, so G is self-complementary.

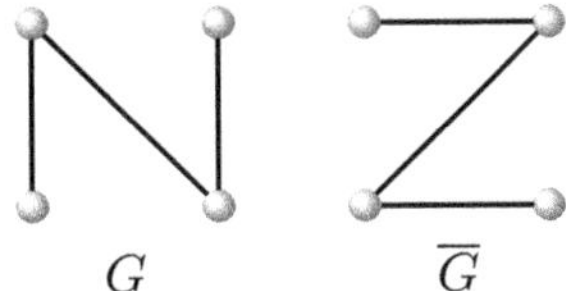

(ii) Notice first that $5(5-1) = 20$ is a multiple of 4. The following diagram shows a graph G of order 5 together with its complement $\overline{G}$. Notice that $G \cong \overline{G}$ since they are both isomorphic to the cycle C_5, so G is self-complementary.

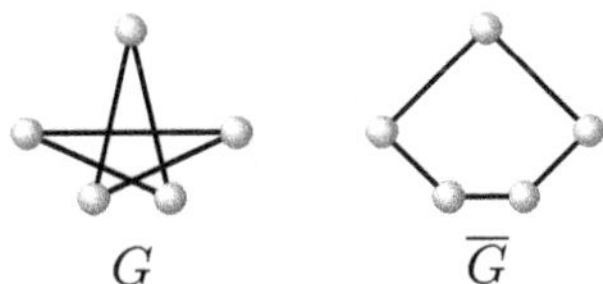

(iii) Since $6(6-1) = 30$ is not a multiple of 4, no self-complementary graph of order 6 exists.

9.10

(3)(a) In the following diagram, we show iterations of the Ford–Fulkerson algorithm. In each step we show the residual network (on the left), an augmenting path using bold edges and the updated flow-network (on the right).

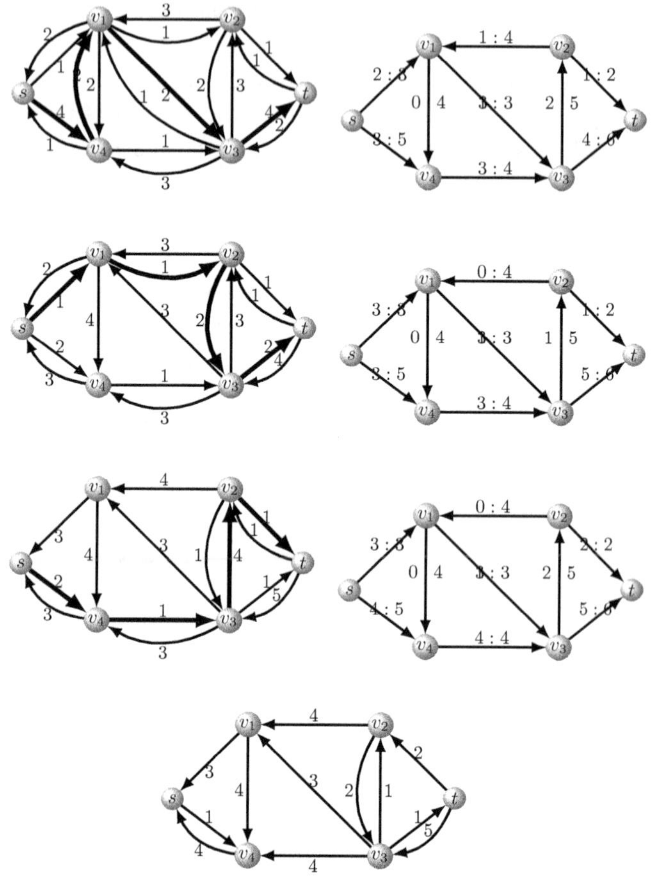

Since the last residual network does not have an augmenting path, the flow in the last updated flow-network is maximum and its value is $|f| = 3 + 4 = 7$.

(b) By the Max-Flow Min-Cut theorem, the value of a maximum flow in the network is equal to the minimum capacity of an st-cut. So, we look for an st-cut of capacity 7. For the st-cut $S = \{s, v_1, v_4\}$, $T = \{v_2, v_3, t\}$, the capacity is $c(v_1, v_3) + c(v_4, v_3) = 7$. So (S, T) is an st-cut with minimum capacity in the network.

(4)(a) The following table shows all st-cuts of the network together with their capacities.

st-cut (S, T)	Capacity
$S = \{s\}, T = \{v_1, v_2, t\}$	$2N$
$S = \{s, v_1\}, T = \{v_2, t\}$	$2N + 1$
$S = \{s, v_2\}, T = \{v_1, t\}$	$2N$
$S = \{s, v_1, v_2\}, T = \{t\}$	$2N$

(b) The minimum capacity of all possible st-cuts is $2N$. By the Max-Flow Min-Cut theorem, the maximum flow in the network is $2N$.

(c) Using the augmented path $p_1 : s - v_2 - t$, the Ford–Fulkerson algorithm reaches the maximum flow (of $2N$ in just one step). Starting with the augmenting path $s - v_1 - v_2 - t$ and using augmented paths of three edges at every iteration of the Ford–Fulkerson, the flow increases by exactly 1 unit each iteration. It takes $2N$ iterations before the algorithm reaches the maximum flow.

(5)(a) The following table shows all st-cuts of the network together with their capacities.

st-cut (S, T)	Capacity
$S = \{s\}, T = \{v_1, v_2, v_3, v_4, t\}$	5
$S = \{s, v_1\}, T = \{v_2, v_3, v_4, t\}$	11
$S = \{s, v_2\}, T = \{v_1, v_3, v_4, t\}$	9
$S = \{s, v_3\}, T = \{v_1.v_2, v_4, t\}$	7
$S = \{s, v_4\}, T = \{v_1.v_2, v_3, t\}$	3
$S = \{s, v_1, v_2\}, T = \{v_3, v_4, t\}$	9
$S = \{s, v_1, v_3\}, T = \{v_2, v_4, t\}$	8
$S = \{s, v_1, v_4\}, T = \{v_2, v_3, t\}$	9
$S = \{s, v_2, v_3\}, T = \{v_1, v_4, t\}$	11
$S = \{s, v_2, v_4\}, T = \{v_1, v_3, t\}$	7
$S = \{s, v_3, v_4\}, T = \{v_1, v_2, t\}$	4
$S = \{s, v_1, v_2, v_3\}, T = \{v_4, t\}$	9
$S = \{s, v_1, v_2, v_4\}, T = \{v_3, t\}$	10
$S = \{s, v_1, v_3, v_4\}, T = \{v_2, t\}$	5
$S = \{s, v_2, v_3, v_4\}, T = \{v_1, t\}$	8
$S = \{s, v_1, v_2, v_3, v_4\}, T = \{t\}$	6

(b) We see that the minimum capacity of an st-cut is 3. By the Max-Flow Min-Cut theorem, the value of a maximum flow is 3.

(c) In the following diagram, the residual network is shown on the left with an augmenting path shown in bold edges and the corresponding updated flow on the right.

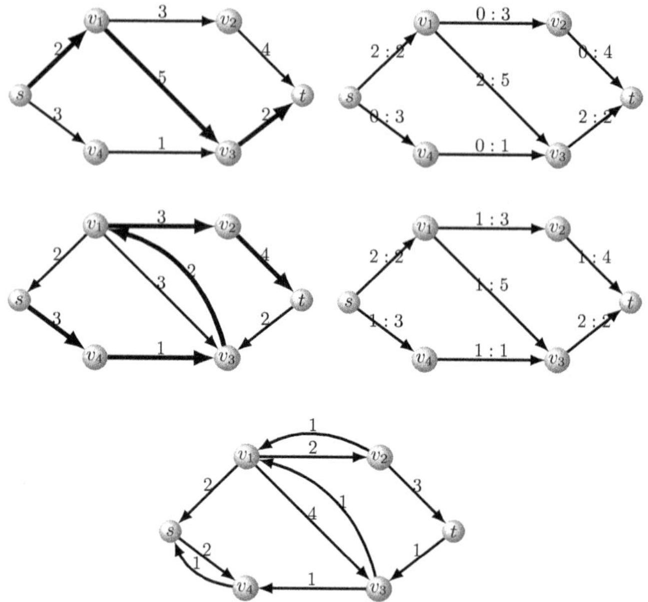

Notice that the last residual network contains no augmenting path. The previous network has then a maximum flow with value $2+1=3$ as expected from part (a).

(7) Assume that $|f| = f(S,T)$ for any st-cut in the network, we need to show that the value of the flow is $|f| = \sum_{v \in V} f(v,t)$. Let $S = V \backslash \{t\}$ and $T = \{t\}$, then (S,T) is an st-cut, and so $|f| = f(S,T) = \sum_{v \in S} f(v,t)$.

(9)(a) Assume that f' is a flow on F'. We prove that $f = \frac{1}{\lambda} f'$ is a flow on F. If $(u,v) \in V \times V$, then $f(u,v) = \frac{1}{\lambda} f'(u,v) \le \left(\frac{1}{\lambda}\right)(\lambda)\, c(u,v) = c(u,v)$ which proves the capacity constraints for f. For $u \in V$, $\sum_{v \in V} f(u,v) = \sum_{v \in V} \frac{1}{\lambda} f'(u,v) = \frac{1}{\lambda}\left(\sum_{v \in V} f'(u,v)\right) = \frac{1}{\lambda}(0) = 0$. This proves the flow conservation rule. Finally, if $(u,v) \in V \times V$, then $f(v,u) = \frac{1}{\lambda} f'(v,u) = \frac{1}{\lambda}(-f'(u,v)) = -\frac{1}{\lambda} f'(u,v) = -f(u,v)$. The skew-symmetry is verified for f. We conclude that f is a flow on F. Moreover, $|f| = \sum_{v \in V} f(s,v) = \sum_{v \in V} \left(\frac{1}{\lambda}\right) f'(s,v) = \frac{1}{\lambda} \sum_{v \in V} f'(s,v) = \frac{1}{\lambda}|f'|$.

(b) Let g be a flow on F and $g' = \lambda g$. We prove that g' is a flow on F'. If $(u,v) \in V \times V$, then $g'(u,v) = \lambda g(u,v) \le \lambda c(u,v) = c'(u,v)$ which proves the capacity constraints for g'. For $u \in V$, $\sum_{v \in V} g'(u,v) = \sum_{v \in V} \lambda g(u,v) = \lambda\left(\sum_{v \in V} g(u,v)\right) = \lambda(0) = 0$. This proves the flow conservation rule for g'. Finally, if $(u,v) \in V \times V$, then $g'(v,u) = \lambda g(v,u) = \lambda(-g(u,v)) = -\lambda g(u,v) = -g'(u,v)$. The skew-symmetry is verified for g'. We conclude that g' is a flow on F'. Moreover, $|g'| = \sum_{v \in V} g'(s,v) = \sum_{v \in V} (\lambda) g(s,v) = \lambda \sum_{v \in V} g(s,v) = \lambda |g|$.

(c) If we multiply all the capacities of edges in F by λ, we get a new flow-network F' with integer capacities. We know that the Ford–Fulkerson algorithm applied to F' will terminate and produce a maximum flow f' on F'. Let $f = \frac{1}{\lambda} f'$. Then f is a flow on F of value $|f| = \frac{1}{\lambda}|f'|$ by part (a). Assume that f is not a maximum flow on F, then there exists a flow g on F such $|f| < |g|$. By part (b), λg is a flow on F' with value $|g'| = \lambda |g|$. Moreover:

$$|f| < |g| \Rightarrow \sum_{v \in V} f(s,v) < \sum_{v \in V} g(s,v)$$

$$\Rightarrow \lambda \sum_{v \in V} f(s,v) < \lambda \sum_{v \in V} g(s,v)$$

$$\Rightarrow \sum_{v \in V} (\lambda f)(s,v) < \sum_{v \in V} (\lambda g)(s,v)$$

$$\Rightarrow \sum_{v \in V} f'(s,v) < \sum_{v \in V} (\lambda g)(s,v)$$

$$\Rightarrow |f'| < |\lambda g|.$$

This is a contradiction since f' is a maximum flow in F'. We conclude that f is a maximum flow in F. So the Ford–Fulkerson algorithm terminates and produces a maximum flow on F.

(10) Notice first that since all capacities of the new network are integers, we know that a maximum flow f' exists in F'. Assume that f is a maximum flow in the original network F. Clearly, f is still an integer-valued flow in F' and so $|f| \leq |f'|$. But we can say more. By the Max Flow Min-Cut Theorem, there exists an st-cut (S,T) in G such that $|f| = c(S,T)$. If $e = (u,v)$ with $u \in S$ and $v \in T$, then the capacity of (S,T) as an st-cut of F' is $c(S,T) + 1$. In this case, $|f'| = c(S,T) + 1 = |f| + 1$. If e is not of the form (u,v) with $u \in S$ and $v \in T$, then the capacity of (S,T) as an st-cut of F' is the same an st-cut in F. In this case, $|f| = c(S,T) = |f'|$. We conclude that $|f'|$ is either equal to $|f|$ or to $|f| + 1$.

(12)(a) No, edges (x,y) with $x \in X$ and $y \in Y$ can have any capacity $N \geq 1$. As long as the capacity of every edge (s,x) where s is the sink and $x \in X$ is equal to 1, any flow on an edge (s,x) must have a value of 1. Since the Ford–Fulkerson algorithm produces integral flows, a flow of 1 unit on an edge (s, x) leaves at exactly one edge (x, y). The Ford–Fulkerson algorithm will terminate with a maximum flow, which corresponds to a maximum bipartite matching on G.

(b) Let (S, t) be a minimal cut in the corresponding network. If S contains a vertex of $x \in X$ and T contains a vertex $y \in Y$, then $c(S, T)$ could not be minimal since $c(x, y) = \infty$. Therefore either $x \notin S$ for any $x \in X$ or $y \notin T$ for any $y \in T$. In other words, either $S \cap X = \emptyset$ or $T \cap Y = \emptyset$.

10.1

(2) No since this would violate Euler's formula: $n - e + f = 7 - 12 + 6 = 1 \neq 2$.

(4) Let n, e and f be the number of vertices, edges and faces of the graph, respectively. Using Euler's formula: $n - e + f = 2 \Rightarrow n = e - f + 2 = 7 - 5 + 2 = 4$. Here a possible drawing of such a graph:

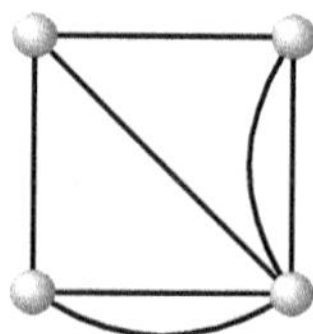

(5)(a) The graph has $n = 6$ vertices $e = \frac{4+4+5+6+6+7}{2} = 16$ edges. The inequality $e \leq 3n - 6$ is not satisfied.

(d) The graph has no vertex of degree less than or equal to 5.

(7) The following diagram shows a simple connected planar graph of degree sequence $(2, 3, 3, 3, 3, 3, 3)$.

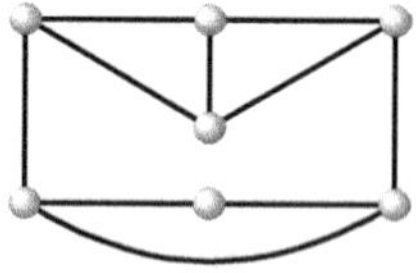

The graph in this case has $n = 7$ vertices, $e = \frac{2+3+3+3+3+3+3}{2} = 10$ edges. The number of faces is $f = 2 - n + e = 5$. To construct a non-planar simple connected graph, Kuratowski's Theorem tells us that the graph must have a subdivision of either K_5 or $K_{3,3}$ as a subgraph. The following shows a simple graph of degree sequence $(2, 3, 3, 3, 3, 3, 3)$

which is a subdivision of $K_{3,3}$, hence non-planar:

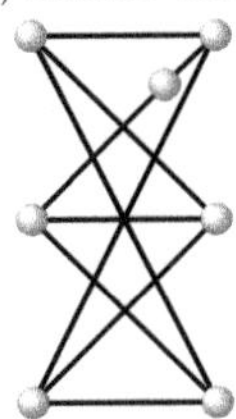

(9) Let n be the order of G and e be its size. Since G is simple, connected and planar we know that $e \leq 3n - 6$. By the handshaking lemma,

$$\sum_{v \in V} \deg(v) = 2e \leq 6n - 12$$

$$\Rightarrow 6n - \sum_{v \in V} \deg(v) \geq 12$$

$$\Rightarrow \sum_{v \in V} 6 - \sum_{v \in V} \deg(v) \geq 12$$

$$\Rightarrow \sum_{v \in V} (6 - \deg(v)) \geq 12.$$

(12)(a) Without loss of generality, we may assume that $m \leq 2$. If $m = 1$, then clearly $K_{1,n}$ is planar for any $n \geq 1$ with only one face, the infinite one. If $m = 2$, then $K_{2,n}$ is planar for any n. To see this, let $X = \{x_1, x_2\}$ and $Y = \{y_1, y_2, \ldots, y_n\}$ be the partition sets. Line up the vertices from Y vertically in a line and place x_1 on the left of that line and x_2 on the right. Connect x_1 and x_2 with one edge with each of the y_i's. This creates a planar embedding of $K_{2,n}$ with n faces.

(b) For $m \geq 3$ and $n \geq 3$, the graph $K_{m,n}$ contains $K_{3,3}$ as a subgraph and therefore it is non-planar.

(16) Let e be the number of edges of G, H the set of vertices of G with degree at most 5. Let $m = |H|$. We need to prove that $m \geq 4$. By the handshaking Lemma,

$$2e = \sum_{v \in H} \deg(v) + \sum_{v \notin H} \deg(v) \geq 3m + 6(n - m) = 6n - 3m.$$

Since G is planar, we know that $e \leq 3n-6$ or equivalently, $2e \leq 6n-12$. We conclude that $6n - 3m \leq 6n - 12$ or equivalently $m \geq 4$.

(17)(a) The statement is true. Assume G is a planar graph and let H be a subgraph of G. Consider a planar embedding G' of G. Removing all edges and vertices from G' that are not in H yields a planar embedding of H.

(c) False. We can draw a simple connected graph with degree sequence $(3, 3, 4, 5, 5)$ which does not satisfy the inequality $e \leq 3v - 6$, hence non-planar but different from K_5 (whose degree sequence is $(4, 4, 4, 4, 4)$).

(19)(a) G_2 is not planar. To see this, let us start by labeling the vertices:

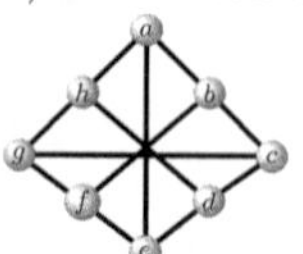

Consider the following Subgraph of G_2:

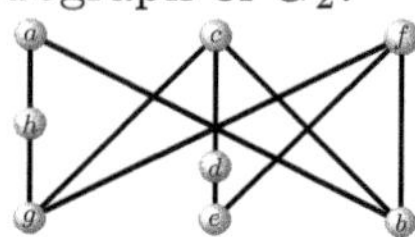

Since this subgraph is a subdivision of $K_{3,3}$, G_2 is a non-planar by Kuratowski's Theorem.

(b) In a planar simple graph, there exists at least one vertex of degree at most 5. We conclude that graph G_4 is non-planar since the minimal degree of a vertex in the graph is 6.

(c) Graph G_5 is planar. The following diagram shows a planar embedding of G_5:

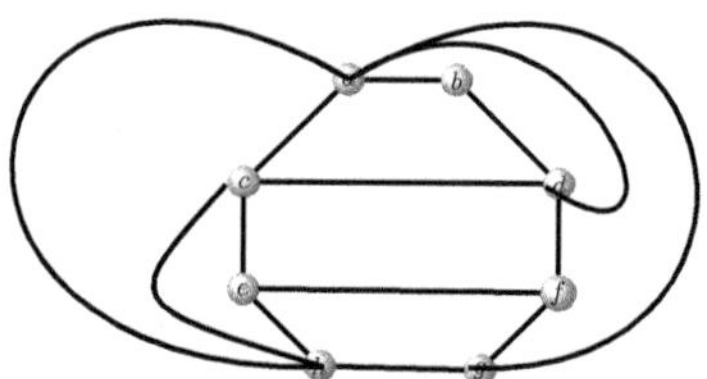

The graph has $n = 8$ vertices, $e = 14$ vertices and $f = 8$ faces. So $n - e + f = 8 - 14 + 8 = 2$ and Euler's formula is satisfied by G_5.

(d) The graph G_7 is non-planar as it contains the following subdivision of $K_{3,3}$ as a subgraph:

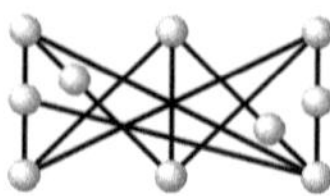

(e) The graph G_8 is simple connected with $n = 11$ vertices and $e = 20$ edges. Moreover, there are no triangles in G_8 (simple circuits of length 3). The inequality $e \leq 2v - 4$ is not satisfied and so G_8 is non-planar.

(20) Since G is simple, $\deg(v) \leq n - 1$ for any vertex v of G. By the handshaking Lemma, $2e = \sum_{v \in V(G)} \deg(v) \leq n(n-1)$. This implies that $e \leq \frac{n(n-1)}{2}$. Since G is also planar, $e \leq 3n - 6$. Let us now check algebraically when these two upper bounds are equal:

$$\frac{n(n-1)}{2} = 3n - 6$$
$$\Rightarrow \frac{n^2 - n - 2(3n-6)}{2} = 0$$
$$\Rightarrow n^2 - 7n + 12 = 0$$
$$\Rightarrow n = 3,\ n = 4.$$

The graph must then be of order 3 or 4. The following diagram shows all possible non-isomorphic simple connected planar graphs with the two upper bounds equal:

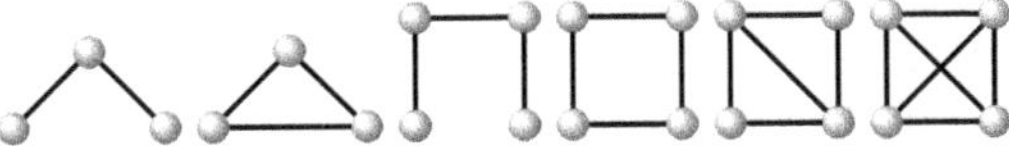

(22) Let f be the number of faces of G. Since $e \geq k$ and no circuit of length less than k exists in G, the degree of every face of G is at least k. Since the sum of the degrees of all faces of G is equal to two times the number of edges in G, we get that $2e \geq kf$ or $f \leq \frac{2}{k}e$. Combining this with the value of f from Euler's formula $f = e - n + 2$, we get that $e - n + 2 \leq \frac{2}{k}e$ which is equivalent to $e \leq \frac{k}{k-2}(n-2)$.

10.2

(2)(a) The complete graph K_2 has exactly two vertices, each of degree 1. So K_2 has an Euler trail but no Euler circuit. For $n \geq 3$, the degree of every vertex in K_n is $n - 1$. So K_n is Eulerian if and only if $n \geq 3$ is odd. The only complete graph with an Euler trail is K_2.

(e) We have seen that in the cube graph Q_n, every vertex represents a binary string of length n. Two vertices are adjacent if and only if the corresponding binary strings differ by exactly one bit. Now, given a binary string s of length n there are n binary strings of length n that differ from s by exactly one bit. So the degree of every vertex in Q_n is exactly n. We conclude that Q_n will have an Euler circuit if and only if n is even. No cube graph will have an Euler trail.

(3) Let m, n be two positive integers. The complete bipartite graph $K_{m,n}$ has two bipartition sets $X = \{x_1, \ldots, x_n\}$ and $\{y_1, \ldots, y_m\}$ with

$\deg(x_i) = m$ for any $i = 1, \ldots, n$ and $\deg(y_j) = n$ for any $j = 1, \ldots, m$. So $K_{m,n}$ has an Euler circuit if and only if both m and n are even integers. The graph $K_{m,n}$ has an Euler trail and if only if there exist exactly two vertices of odd degrees. It is not hard to verify that this happens in either one of these cases: (i) $m = n = 1$; (ii) $m = 2$ and n is odd; (iii) $n = 2$ and m is odd.

(4)   • Graph G_2 has all vertices of even degree. So G_2 has an Euler circuit but no Euler trail. Here is one Euler circuit: *abcdecadbea*.

      • Graph G_7 has exactly two vertices of odd degree, namely b and h. So G_7 has an Euler trail and no Euler circuit. Here is one Euler trail: *bcdefbafghidh*.

      • Graph G_9 has all vertices of even degree. So G_9 has an Euler circuit but no Euler trail. Here is one Euler circuit: *abcdefcjihgfkelkjgbha*.

(6) We construct the graph G with each region of the city (an island or a shore) is a vertex and two vertices are adjacent if and only if there a bridge between the corresponding regions. The North and South shores are indicated by the vertices NS and SS respectively in the graph.

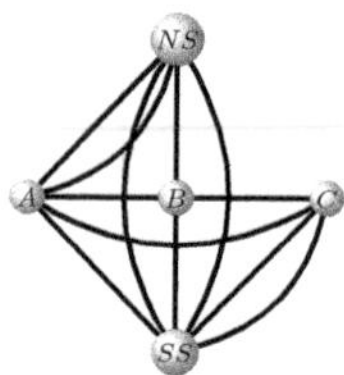

(a) The problem is equivalent to finding an Euler circuit in G. Since the graph has vertices of odd degree (namely, A and NS), such a circuit is not possible and therefore such a tour in the city is not possible. Building a bridge between island A and the North shore will make such a tour possible in the city.

(b) The problem is equivalent to finding an Euler trail in G. Since the graph has exactly two vertices of odd degree (namely, A and NS), such a tour possible is possible in the city (only between island A and the North shore).

(8) For each drawing, we consider a graph having each intersection of lines as a vertex and each line between them as edges. The question asks for the existence of an Euler circuit or an Euler trail in the graph.

      • Each vertex in the graph corresponding to the drawing (a) has an even degree. An Euler circuit exists and we can retrace the drawing with the conditions specified.

- There are more than two vertices of odd degree in the graph corresponding to the drawing (d). Therefore, no Euler circuits and no Euler trail exist in that graph. The drawing cannot be retraced on a paper with the specified conditions.

(9)(a) The number of tiles is the number of ways of choosing two elements out of 7: $\binom{7}{2} = \frac{7!}{2!5!} = 21$.

(b) Form the graph G with vertices $0, 1, \ldots, 7$. Two vertices i, j are connected with an edge if and only if $[i, j]$ is a domino tile with $0 \leq i < j \leq 6$. Clearly G is the complete graph K_7 as it is connected simple with degree of each vertex equals to 6. Arranging all domino tiles $[i, j]$ with $0 \leq i < j \leq 6$ end to end in such a way that the numbers of dots on adjacent ends are always the same corresponds to an Euler trail (or an Euler circuit) in G. Since G has all vertices of even degree, such an arrangement is then possible.

(11) Let G be a connected simple regular d-graph with $d \geq 1$. If $e = \{u, v\}$ is an edge of G, then e in incident to $d - 1$ edges at vertex u and $d - 1$ edges at v. So the degree of e as a vertex in $L(G)$ is $2(d - 1)$. Since every vertex has an even degree in the (connected) graph $L(G)$, the line graph is Eulerian.

(13)(a) G is not simple since edges e_1 and e_2 are parallel (have the same endpoints). The graph G is connected. The matrix tells us there is a path between any two distinct vertices (in fact, every pair of distinct vertices are adjacent).

(b) From the incidence matrix, we get that $\deg(v_1) = 4$, $\deg(v_2) = 6$, $\deg(v_3) = 4$ and $\deg(v_4) = 4$. Since G is connected with every vertex of even degree, the graph G Eulerian. The graph does not have an Euler trail since every vertex is of even degree.

(c) The following is a possible drawing for G:

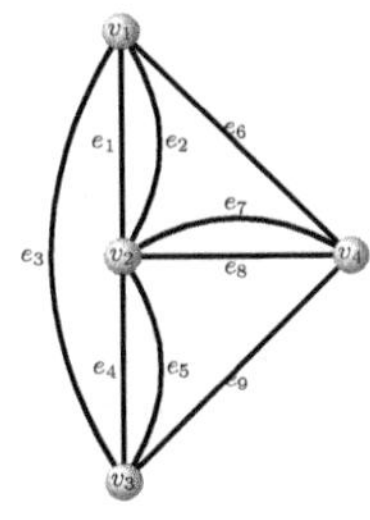

(14)(a) False. Here is a counter-example:

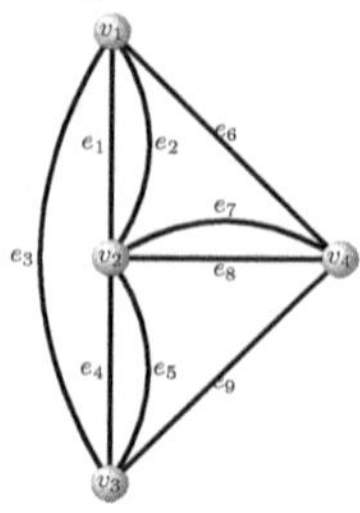

(c) True. Since the graph is connected and simple, $1 \leq \deg(v) \leq 3$ for any vertex v of G. Since the graph is Eulerian, the degree of every vertex is even. That implies that each vertex is of degree 2. The only possibility is then the cycle graph C_4.

(e) False. A counter-example is K_2, the complete graph of order 2.

10.3

(1)

(1) This is possible. The cycle graph C_3 is at the same time an Euler and a Hamiltonian circuit.

(2)(c) The following graph has a Hamiltonian circuit, namely $abcda$ but no Euler circuit since it has vertices of odd degree. Moreover, the graph has an Euler trail since it has exactly two vertices of odd degree.

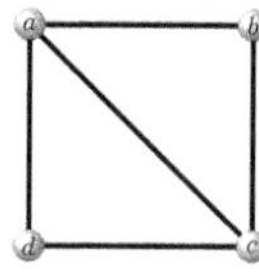

(6) Let $X = \{x_1, \ldots, x_m\}$ and $Y = \{y_1, \ldots, y_n\}$ and assume that the bipartite graph G is Hamiltonian. First note that $|X| \geq 2$ and $|Y| \geq 2$ since otherwise $m = 1$ or $n = 1$ and the graph has a vertex of degree 1 and therefore it cannot be Hamiltonian. Let C be a Hamiltonian circuit in G. Rearranging the vertices if necessary, we can assume that $C = x_1 y_1 x_2 y_2 \ldots x_m y_n x_1$. In order for C to cross every vertex exactly one, m must be the same as n. As a consequence, the complete bipartite graph $K_{m,n}$ is Hamiltonian if and only if $m = n$ for $m, n \geq 2$.

(7) The graph is of order $n = 8$ and from the degree list we see that $\deg(v) \geq 4 = \frac{n}{2}$. By Dirac Theorem, G is Hamiltonian.

(9) First note that since $k \geq 2$, $2k-1 \geq 3$ so the graph is of order at least 3. Moreover, $k - \frac{2k-1}{2} = \frac{1}{2} > 0$ which means that $\deg(v) = k > \frac{2k-1}{2}$ for every vertex v of G. Dirac Theorem tells us that G is Hamiltonian.

(12) Consider the cycle graph C_5 of order 5. The degree of every vertex of C_5 is 2 which means that Dirac's Theorem hypothesize is not satisfied. Also, the sum of degrees of any pair of non-adjacent vertices is 4 which also means that Ore's Theorem hypothesize is not satisfied. However, the cycle C_5 is Hamiltonian.

(14) The problem can be represented by the following graph with each label (scientist) is a vertex and two vertices are joined by an edge if and only if the corresponding scientists share some common research interest:

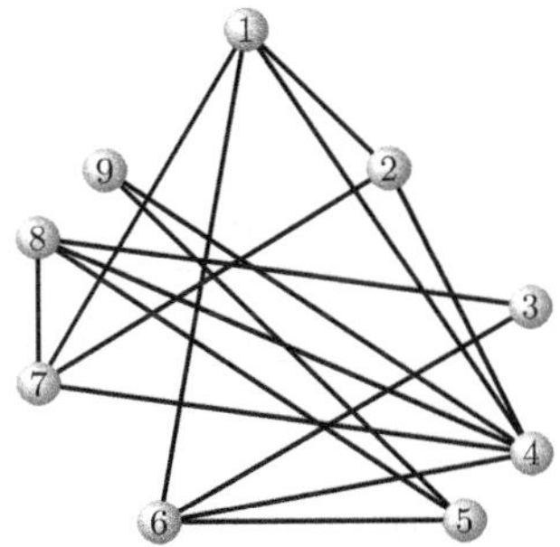

Arranging the nine scientists around a table in such a way that each scientist sits between two other scientists with some common research interest corresponds to a Hamiltonian circuit in the graph. Such an arrangement is possible since the graph has the following Hamiltonian circuit: 1783659421.

(14)(d) Graph G_4 is of order 7. Since $\deg(g) = 3 < \frac{7}{2}$, Dirac's Theorem does not apply. The list of non-adjacent pairs of vertices of G and the sum of their degrees is given in the following table:

Pair $\{u, v\}$ of non-adjacent pairs of vertices	$\deg(u) + \deg(v)$
$\{a, g\}$	7
$\{a, c\}$	9
$\{b, d\}$	8
$\{b, e\}$	8
$\{d, g\}$	7
$\{e, f\}$	8
$\{f, g\}$	8

The table shows that $\deg(u) + \deg(v) \geq 6$ for every pair of non-adjacent vertices of G_4. Ore's Theorem implies that G_4 is Hamiltonian.

(e) Graph G_5 is of order 8. Since $\deg(v) \geq 4 = \frac{8}{2}$ for every vertex v of the graph, Dirac's Theorem implies that G_5 is Hamiltonian.

(15)   • Graph G_4 is not Hamiltonian. Since $\deg(v_3) = \deg(v_4) = \deg(v_5) = 2$, any Hamiltonian circuit in G_5 must include all edges incident to these vertices, namely: $\{v_2, v_3\}$, $\{v_3, v_7\}$, $\{v_2, v_4\}$, $\{v_4, v_7\}$, $\{v_5, v_6\}$ and $\{v_5, v_7\}$. But then, the circuit would include more two edges incident to vertex v_7, a contradiction.

       • Graph G_6 is Hamiltonian. Here is a Hamiltonian circuit: $acbdehifa$.

       • Graph G_8 is not Hamiltonian. Since $\deg(c) = \deg(d) = \deg(f) = \deg(e) = 2$, any Hamiltonian circuit in G_8 must include all edges incident to these vertices. In particular, $\{c, m\}$, $\{d, m\}$, $\{f, m\}$ and $\{e, m\}$ must be included in any Hamiltonian circuit. But then, the circuit would include more two edges incident to vertex m, a contradiction.

(17) Fix a vertex v of the complete graph K_n, $n \geq 3$. By the property of K_n, we can move from v to any other vertex of K_n along a single edge. A Hamiltonian circuit at v has the form $C = vw_1w_2 \ldots w_{n-1}v$. Note that the reverse circuit $C' = vw_{n-1} \ldots w_2w_1v$ is the same as C since it uses the same edges. Since there are $(n-1)!$ permutations of $w_1, \ldots, w_{n-1}$, there are $\frac{(n-1)!}{2}$ different Hamiltonian circuits at a vertex v.

(20) Suppose G is Eulerian and let C be an Eulerian circuit in G. We can certainly relabel the edges and vertices of G and write $C = v_0e_1v_1e_2v_2 \ldots e_{m-1}v_{m-1}e_mv_0$ where $e_1, \ldots, e_m$ are all the distinct edges of G (the vertices $v_0, v_1, \ldots$ appearing in C are not necessarily distinct). Recall that the vertices of $L(G)$ are the edges $e_1, \ldots, e_m$ of G and two vertices of $L(G)$ are adjacent if and only if the corresponding edges of G share a common endpoint. This implies that in $L(G)$: e_1, e_2 are adjacent, e_2, e_3 are adjacent, ..., e_{m-1}, e_m are adjacent and e_m, e_1 are adjacent. This shows that $e_1e_2 \ldots e_me_1$ is a Hamiltonian circuit in $L(G)$ and $L(G)$ is Hamiltonian.

10.4

(1) Let $G = (V, E)$ and $H = (V'E')$. Then $V' \subseteq V$ and $E' \subseteq E$. Let $c : V \to \mathbb{N}$ be a proper coloring of G, then the restriction $c' : V' \to \mathbb{N}$ of c to V' (defined by $c'(v) = c(v)$ for any $v \in V'$) provides a proper coloring of H since any two adjacent vertices in H are also adjacent in G. This means in particular that H can be properly colored with $\chi(G)$ colors. By definition of the chromatic number, $\chi(H) \leq \chi(G)$.

(4) We know already that if $G = K_n$, then $\chi(G) = n$. For the converse, assume that G is a graph of order $n \geq 2$ such that $\chi(G) = n$. If $G \neq K_n$, then there exists a pair of non-adjacent vertices u, v of G. We can then

color u, v with the same color and have an $(n-1)$-coloring of G. That would mean $\chi(G) \le n-1$: a contradiction. We conclude that every pair of vertices in G are adjacent and so $G = K_n$.

(5) Note that the Petersen graph is not bipartite as it has circuits of odd length. So $\chi(P) \ge 3$. On the other hand, P is neither complete nor a cycle graph of odd length, so $\chi(G) \le \Delta_G = 3$ by Brooks Theorem. We conclude that $\chi(G) = 3$.

(6) Assume by contradiction that $\deg(v) < \chi(G) - 1$ for every vertex v of G. Let v be a vertex of G with maximal degree Δ_G, then $\Delta_G = \deg(v) < \chi(G) - 1$. But we know that $\chi(G) \le \Delta_G + 1$. This implies that $\Delta_G < \chi(G) - 1 \le (\Delta_G + 1) - 1 = \Delta_G$: This is clearly absurd. We conclude that there exists at least one vertex v with $\deg(v) \ge \chi(G) - 1$.

(9) Recall that the way we defined W_n: Add one vertex v at the center of the cycle graph C_n and join it with a single edge to each vertex of C_n, which makes W_n of order $n+1$. If there are λ colors available to color W_n, we have λ ways to color the central vertex v. So there are $\lambda - 1$ colors available to color the vertices of C_n and this can be done in $(\lambda - 2)^n + (-1)^n(\lambda - 2)$ ways since the chromatic polynomial of the cycle C_n with k colors available is $(k-1)^n + (-1)^n(k-1)$. We conclude that the chromatic polynomial of W_n is $P_{W_n}(\lambda) = \lambda((\lambda-2)^n + (-1)^n(\lambda-2))$. The chromatic number of W_n is the smallest integer λ such that $P_G(\lambda) \ne 0$. That depends on the parity of n. If n is even, then $P_{W_n}(\lambda) = \lambda((\lambda-2)^n + (\lambda-2)) = \lambda(\lambda-2)((\lambda-2)^{n-1}+1)$. In this case, $P_{W_n}(0) = P_{W_n}(1) = P_{W_n}(2) = 0$ but $P_{W_n}(3) = 6 \ne 0$. So $\chi(W_n) = 3$. If n is odd, then $P_{W_n}(\lambda) = \lambda((\lambda-2)^n - (\lambda-2)) = \lambda(\lambda-2)((\lambda-2)^{n-1} - 1)$. In this case, $P_{W_n}(0) = P_{W_n}(1) = P_{W_n}(2) = P_{W_n}(3) = 0$ but $P_{W_n}(\lambda)(4) = 8(2^{n-1} - 1) \ne 0$. So $\chi(W_n) = 4$ in this case. In summary:

$$\chi(W_n) = \begin{cases} 3 & \text{if } n \text{ is even} \\ 4 & \text{if } n \text{ is odd} \end{cases}$$

(11)(e) Let $X = \{x_1, x_2\}$ and $Y = \{y_1, \ldots, y_5\}$ be the bipartition sets of $K_{2,5}$. Assume we have λ colors available. Since vertices x_1 and x_2 are not adjacent, they could be colored using the same color or two different colors. Assume first that x_1 and x_2 have the same color. In this case, there is λ ways of choosing that common color for x_1 and x_2. Once the common color for x_1 and x_2 is chosen, there are $(\lambda-1)^5$ ways of coloring vertices $y_1, \ldots, y_5$ (since they are non-adjacent). So there are $\lambda(\lambda-1)^5$ ways to color the vertices of $K_{2,5}$ in this case.

Assume next that x_1 and x_2 have different colors. In this case there are $\lambda(\lambda - 1)$ ways to choose the colors for x_1 and x_2 and for each one of the ways there are $(\lambda - 2)^5$ ways to color the vertices of Y. In total, there are $\lambda(\lambda - 1)(\lambda - 2)^5$ ways to color the vertices of $K_{2,5}$ in this case. Since these two possibilities are mutually exclusive, the total number of λ-coloring of $K_{2,5}$ is $\lambda(\lambda - 1)^5 + \lambda(\lambda - 1)(\lambda - 2)^5$. Since $K_{2,5}$ is still bipartite, $\chi(K_{2,5}) = 2$.

(13)(b) A succession of graphs is shown. In each step, the graphs represent the Deletion-Contraction property. The operations $-$ and $+$ stand for the difference and the sum of the chromatic polynomials corresponding to the graphs in each case.

$$= (\lambda - 2)P_{P_5}(\lambda) - (\lambda - 3)P_{P_4}(\lambda) + (\lambda - 3)P_{P_3}(\lambda)$$
$$= (\lambda - 2)\lambda(\lambda - 1)^4 - (\lambda - 3)\lambda(\lambda - 1)^3 + (\lambda - 3)\lambda(\lambda - 1)^2$$
$$= \lambda(\lambda - 1)^2 \left((\lambda - 1)^2(\lambda - 2) - (\lambda - 1)(\lambda - 3) + (\lambda - 3) \right)$$
$$= \lambda(\lambda - 1)^2(\lambda - 2)(\lambda^2 - 3\lambda + 4)$$

Since $P_G(1) = P_G(2) = 0$ and $P_G(3) = 48 \neq 0$, $\chi(G) = 3$.

(14) • Graph G_2 is bipartite with the central vertex in one bipartition set and every other vertex of the graph is in the other. So $\chi(G_2) = 2$. The following is a 2-coloring of the graph:

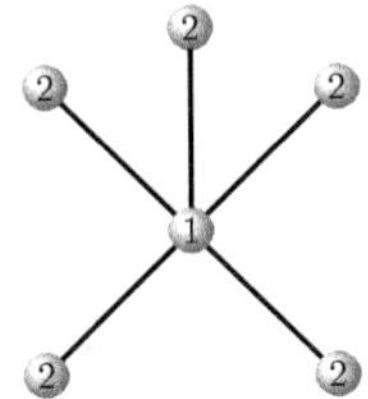

• Notice first that G_4 contains the cycle C_7 as a subgraph, so $\chi(C_7) = 3 \leq \chi(G_4)$. On the other hand, Brooks theorem tells us that $\chi(G_4) \leq \Delta = 4$. We conclude that $\chi(G_4)$ is either 3 or 4. The following diagram shows that we can get away with three colors and so $\chi(G_4) = 3$:

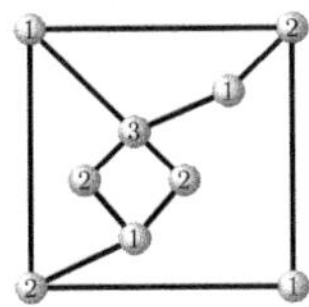

• Graph G_7 is not bipartite (there are simple circuit of odd length), so $\chi(G_7) \geq 3$. On the other hand, G_7 is clearly not the complete graph, nor it is the cycle graph of odd length. Brooks theorem tells us that $\chi(G_6) \leq 3$. We conclude that $\chi(G_7) = 3$. The following shows a proper three-coloring of G_6:

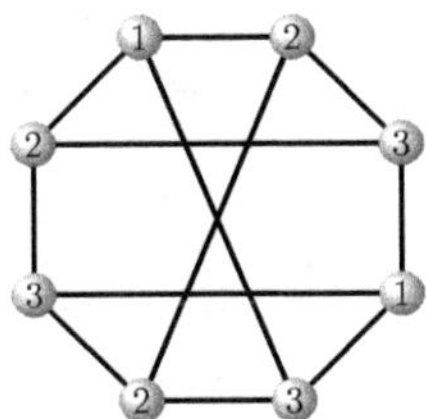

(15)(a) The polynomial $2\lambda^4 - 3\lambda^3 + 2\lambda^2 - \lambda$ is not monic.

  (d) The sum of the coefficient of the polynomial $\lambda^4 - 3\lambda^3 + 2\lambda^2 - 5\lambda$ is not zero.

(17) Recall that the order of G is the degree of $P_G(\lambda)$, its size is the opposite of the coefficient of λ^{n-1}, its chromatic number is the smallest integer k for which $P_G(k) \neq 0$. Finally, the number of connected components of G is the smallest exponent of λ appearing in $P_G(\lambda)$.

(b) $n = 9$, $m = -(-10) = 10$, $\chi(G) = 2$ and $\omega = 1$.

(21) Construct the graph G with products $C_1, \ldots, C_8$ as vertices, and edges joining products which cannot be transported in the same truck at the same time.

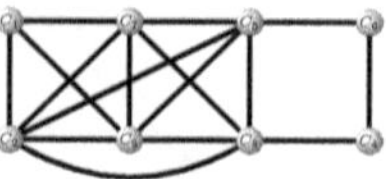

Then the minimum number of trucks needed to transport all 8 products is the chromatic number of the graph G. Note that G contains a copy of K_5 shown in bold in the diagram:

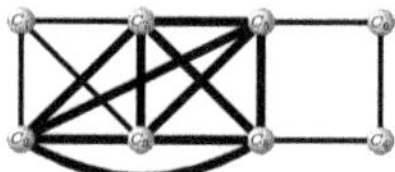

This shows in particular that $\chi(K_5) = 5 \leq \chi(G)$. On the other hand, Brooks Theorem tells us (since G is not a complete graph nor a cycle of odd length) that $\chi(G) \leq \Delta = 5$. We conclude that $\chi(G) = 5$ and this is the minimal number of trucks needed to transport all these products at the same time. The following is a possible five-coloring of the graph. Color i corresponds to a truck T_i:

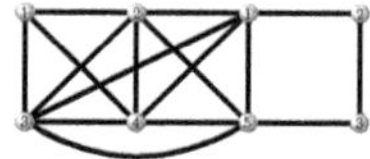

So truck 1 can transport products C_1 and C_5, truck 2 can transport products C_6 and C_7, truck 3 can transport products C_2 and C_4, truck 4 can transport product C_3 and finally truck 5 can transport product C_8.

11.2

(1) • G_3 is a forest as it has no cycles, but it is not a tree since it is disconnected.

     • G_5 is a tree as it is connected with no cycles. In particular, G_5 is a forest.

     • G_7 has a cycle (*dejihd*) so it is neither.

(5) Let v be a node in T different from the root r. Assume that there exist two distinct parents u and w of v. Then (u, v) and (w, v) are directed edges in T. Let P_u, P_w be the unique paths from r to u and from r to w respectively. Then P_u followed by (u, v) then (v, w) and finally P_w forms a cycle at the root: a contradiction to the fact that T is a tree. We conclude that v has a unique parent.

(6)(c) Node n is the parent of o.

  (e) The height of T is 5.

  (g) The descendants of node n are: $o, p, q, r, x, y, z, u, v, w$.

  (i) k and o are two nodes at level three which are not siblings.

(8) Let u, v be two siblings in a rooted tree T and let w be their unique parent. Let $l(u)$, $l(v)$ and $l(w)$ be the heights of u, v and w respectively. Then $l(u) = l(v) = l(w) + 1$.

(12)(b) Balanced but not full:

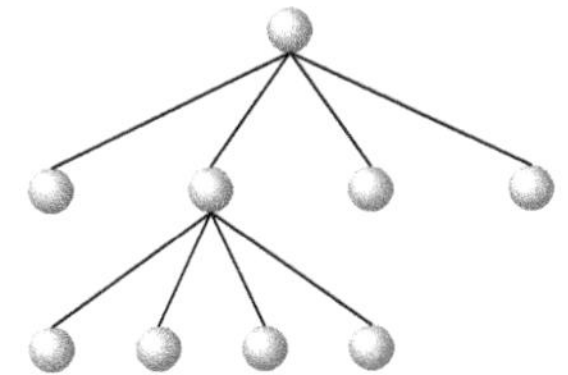

  (e) Complete but not full of height 3:

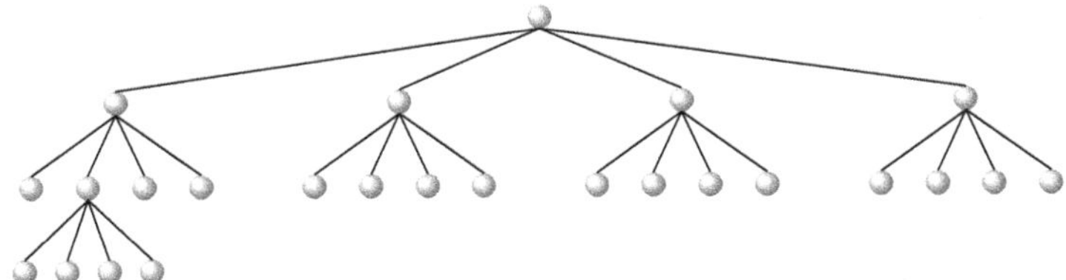

(14) Let G be a connected graph. If G is a tree and $e = \{u, v\}$ is an edge in G, then e is the only path between u and v in G. Removing G would then disconnect G. Conversely, assume that every edge of G is a bridge. If G is not a tree then it must have a cycle C. let u, v be two adjacent vertices on C. Removing edge $\{u, v\}$ will not disconnect G since there is another path from u to v using the other vertices of the cycle C. This is a contradiction. So G has no cycle and hence it is a tree.

11.3

(2) Since T is complete, the total number of nodes in T is $n = mi + 1 = 5i + 1$. The number of internal nodes in T is then $i = \frac{256-1}{5} = 51$. The number of leaf nodes is $l = n - i = 256 - 51 = 105$.

(5) If n is the number of nodes in the tree, then $\frac{4^5-1}{4-1} < n \leq \frac{4^6-1}{4-1} \Leftrightarrow 342 \leq n \leq 1365$.

(8)(a) This cannot be the degree sequence of a tree since the sequence does not have any vertex of degree 1.

(b) This is tree: a star with central vertex connected to six other nodes:

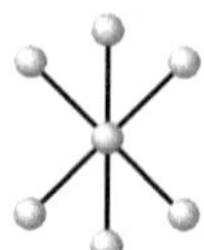

(d) This cannot be the degree sequence of a tree since the number of edges is $\frac{4+4+3+3+3+2+2+1+1+1+1+1+1+1}{2} = 14$ which is the same as the number of vertices. The number of edges in a tree is one less than the number of nodes.

(9)(a) The total number of nodes in T is $n = \frac{3^9-1}{3-1} = 9841$.

(b) The number of edges in T is one less that the number of nodes: $9841 - 1 = 9840$.

(c) The number of leaves in T is $3^8 = 6561$.

(d) The number of internal nodes in T is $9841 - 6561 = 3280$.

(e) The number of nodes at level 5 in $3^5 = 243$.

(12) Let x be the number of nodes of degree 4 in the tree. Then the tree has a total of $9 + 2 + 1 + x = 12 + x$ nodes. On the other hand, we know that the number of edges in the tree is one less than the number of nodes: $(12 + x) - 1 = 11 + x$. By the handshaking lemma, we then have that $(9 \times 1) + (2 \times 2) + (1 \times 3) + (x \times 4) = 2(11 + x)$. Solving for x gives $x = 3$.

(13) Let n, i, and l be the number of nodes, internal vertices and leaf nodes, respectively. The number of leaf nodes is $l = n - i = ki + 1 - i = (k-1)i + 1$ or $l - 1 = (k-1)i$. We conclude that $45 = (k-1)i$. Since $45 = 5 \cdot 3^2$, there are three ways to write 45 as a product of two positive integers: $1 \cdot 45 = 3 \cdot 15 = 5 \cdot 9$. Since T has height 3, $i \geq 3$. On the other hand, $l \leq k^3$ which means that $k^3 \geq 46$ or $k \geq 4$. Since neither $k - 1$ nor i can take the value 1, the pair $(i, k-1)$ can be either one of $(3, 15)$, $(5, 9)$ or $(15, 3)$.

- A tree with $i = 3$ and $k = 16$ exists. For example, the tree with 16 nodes on level 1, of which 15 are leaves; 16 nodes on level 2, of which 15 are leaves, and 16 nodes on level 3, all leaves.
- A tree with $i = 5$ and $k = 10$ exists. For example, the tree with 10 nodes on level 1, of which 7 are leaves; 30 nodes on level 2, of which 29 are leaves, and 10 nodes on level 3, all leaves.

- A tree with $i = 15$ and $k = 4$ also exists. For example, the tree with 4 nodes on level 1, all internal; 16 nodes on level 2, of which 6 are leaves, and 40 nodes on level 3, all leaves.

We conclude that the possible values for k are $4, 10$ and 16.

(17) Let T be a complete and balanced tree of height h with n nodes. Let T' be the tree obtained from T by adding m children to every leaf node at level $h - 1$, if any. So T' is a full tree of height h having at least as many nodes as T. We know that the number of nodes in T' is $\frac{m^{h+1}-1}{m-1}$, so $n \le \frac{m^{h+1}-1}{m-1}$. Let T'' be the tree obtained from T by removing all leaf nodes at level h together with their incident edges. T'' is an m-ary full tree of height $h - 1$. We know that the number of nodes in T'' is $\frac{m^h-1}{m-1}$. Since T is of height h, it must have at least one leaf at level h and consequently T has more leaf nodes than T''. We then get that $\frac{m^h-1}{m-1} < n$. Putting things together, we get that $\frac{m^h-1}{m-1} < n \le \frac{m^{h+1}-1}{m-1}$.

(18) Let T be a tree with n nodes. If $n = 1$, then the tree consists of only one node and no edges. In this case, $P_T(\lambda) = \lambda = \lambda^1(\lambda - 1)^0$ and the result is true. Let $n \ge 2$ and assume that the result is true for any tree of order less than or equal to $n - 1$. Let T be a tree of order n, v be a node in T of degree 1 (we know that at least two such nodes exist in T by) and let e be the edge incident to v. The graph $T - e$ is a disconnected graph consisting of two components: a tree T' of order $n-1$ and a single node. By the induction hypothesis and the product rule of counting: $P_{T \setminus e}(\lambda) = \lambda\left(\lambda(\lambda - 1)^{n-2}\right) = \lambda^2(\lambda - 1)^{n-2}$. The graph T/e is a tree of order $n - 1$. By the induction hypothesis, $P_{T/e}(\lambda) = \lambda(\lambda - 1)^{n-2}$. Using the Cancellation-Contraction property:

$$
\begin{aligned}
P_T(\lambda) &= P_{T \setminus e}(\lambda) - P_{T/e}(\lambda) \\
&= \lambda^2(\lambda - 1)^{n-2} - \lambda(\lambda - 1)^{n-2} \\
&= \lambda(\lambda - 1)^{n-2}(\lambda - 1) \\
&= \lambda(\lambda - 1)^{n-1}
\end{aligned}
$$

as required. The result is true by the induction principle. In particular, the chromatic number of a tree is 2.

11.4

(1)(a) For T_1:

    (i) *abdce*

    (ii) *dbeca*

    (iii) *dbace*

(b) For T_2:

    (i) *abdce*

    (ii) *dbeca*

    (iii) *bdaec*

(c) For T_3:

    (i) *abdce*

    (ii) *dbeca*

    (iii) *dbaec*

(2) For T_4: (i) *abdecfghjkilm* (ii) *debfjkhlmigca*; (iii) *dbeafcjhkglim*.

(4) For the tree T_2, the universal address labeling of the nodes is as follows.

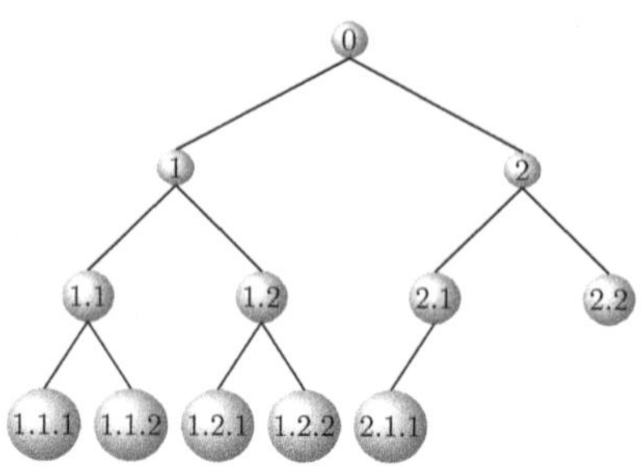

The total lexicographic ordering of the nodes induced by that labeling is $0 < 1 < 1.1 < 1.1.1 < 1.1.2 < 1.2 < 1.2.1 < 1.2.2 < 2 < 2.1 < 2.1.1 < 2.2$

(5) For the tree T_1, the universal address labeling of the nodes is as follows.

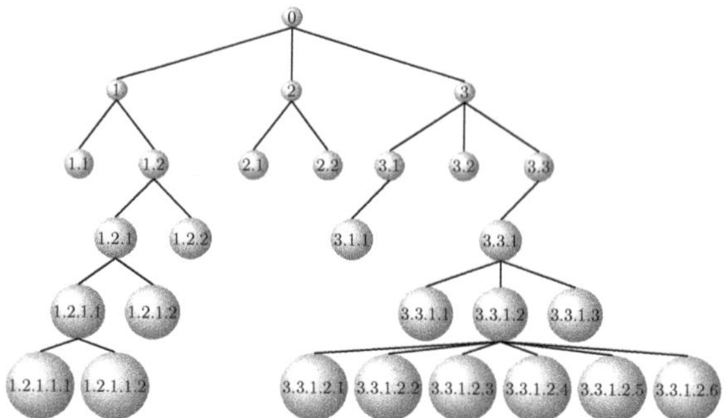

The total lexicographic ordering on the nodes is the following: $0 < 1 < 1.1 < 1.2 < 1.2.1 < 1.2.1.1 < 1.2.1.1.1 < 1.2.1.1.2 < 1.2.1.2 < 1.2.2 < 2 < 2.1 < 2.2 < 3 < 3.1 < 3.1.1 < 3.2 < 3.3 < 3.3.1 < 3.3.1.1 < 3.3.1.2 < 3.3.1.2.1 < 3.3.1.2.2 < 3.3.1.2.3 < 3.3.1.2.4 < 3.3.1.2.5 < 3.3.1.2.6 < 3.3.1.3$

(6)(a) the level of v is 5.

(c) the number of siblings of v is 6.

(d) the address of the parent node of v is 4.4.4.5.

(g) the smallest possible number of nodes of T is $1+4+4+4+5+7 = 25$.

(8)(a) Starting with the first word: *capital* at the root. Since the next word *binary* is smaller than *capital* in the sense that it appears before in the dictionary, we move to the left subtree and create a node for *binary*. The next word in the list is *mathematics* which is larger than the root, so we move to the right and create a node for that word and son on. The full binary search tree for the list is the following:

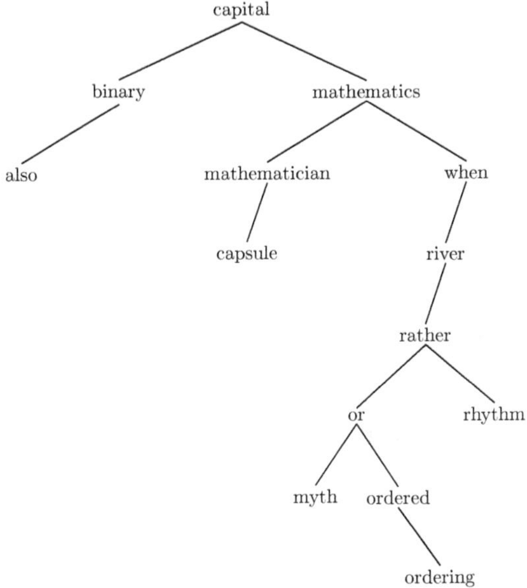

(b) Traversing T in inorder fashion produces the following order on the words in the set: *also, binary, capital, capsule, mathematician, mathematics, myth, or, ordered, ordering, rather, rhythm, river, when* which is precisely the natural order in which these words appear in the dictionary.

(c) The path followed in T to access the word *rhythm* is:

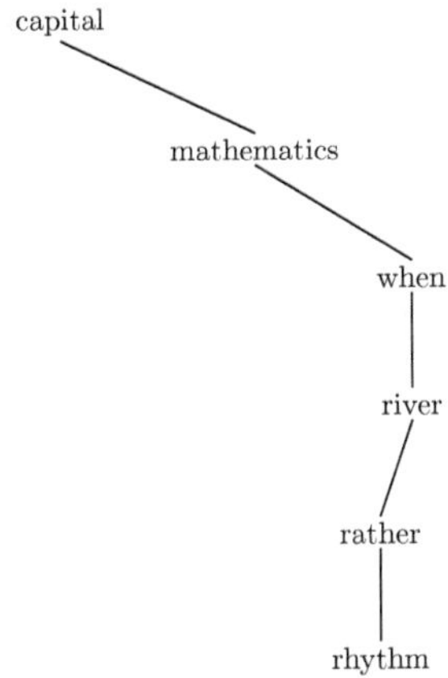

(10)(a) The binary search tree corresponding to the list of monomials with respect to the graded lexicographic ordering is the following.

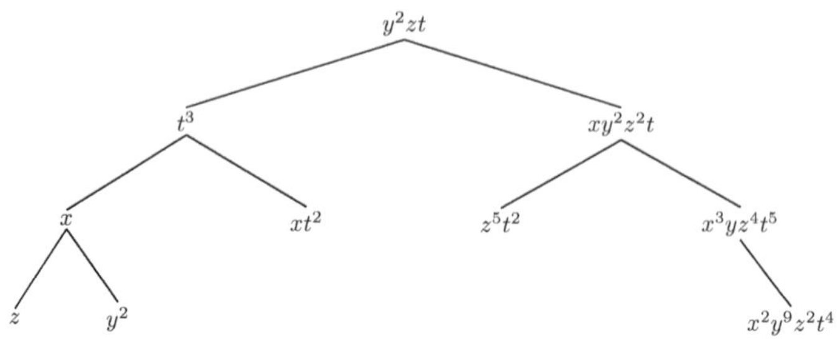

(b) Traversing T in inorder fashion gives the following order of the monomials in S (in graded lexicographic ordering): $z < x < y^2 < t^3 < xt^2 < y^2zt < z^5t^2 < xy^2z^2t < x^3yz^4t^5 < x^2y^9z^2t^4$ which is the same as the graded lexicographic order on S.

11.5

(2) Look at every saturated hydrocarbon as a tree where every carbon atom is an internal node and every hydrogen atom is a leaf node. Let k be the number of hydrogen atoms, then the handshaking lemma tells us that $4n + k = 2(k + n - 1)$ since the number of edges is $k + n - 1$. We conclude that $k = 2n + 2$.

(4)(b) $a - \left(\frac{2}{b} + 3\right)$:

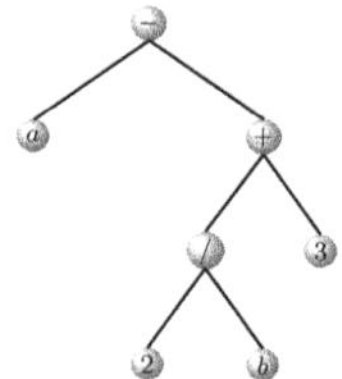

(c) $a^2 + b^2 - \left(\frac{2}{b} + \frac{a}{2}\right) + 3$:

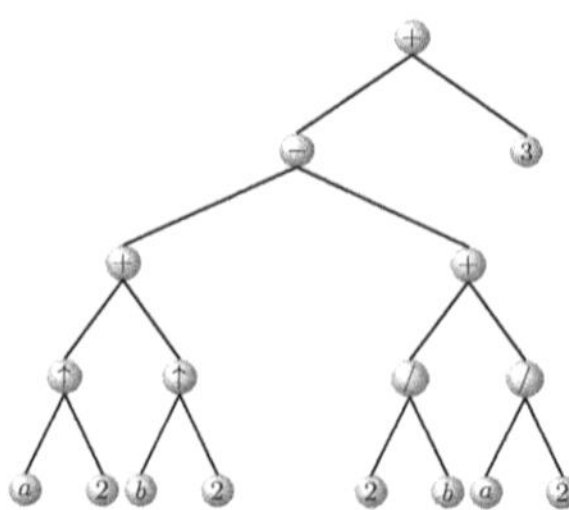

(5) Traversing the tree in inorder fashion gives the expression:
$$\left(\frac{36}{4\times\frac{18}{6}}\right)^{\frac{36}{3\times(2+4)}} = 9.$$

(7) We read the expression from right to left. When we encounter an operator, we perform the corresponding operation using the two operands immediately to the right of the operator: (a) 4; (d) 32.

(8) We read the expression from left to right. When we encounter an operator, we perform the corresponding operation using the two operands immediately to the left of the operator: (b) 32; (c) 15.

(9)(b) The ordered rooted binary tree corresponding the formula is:

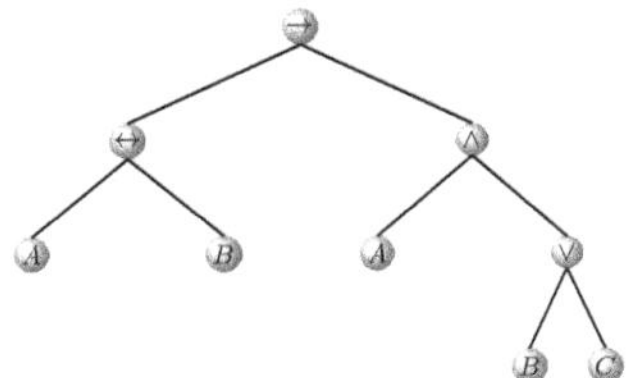

To write ϕ using prefix notation, we traverse the above tree in preorder: $\to\leftrightarrow A B \wedge A \vee B C$. To write ϕ using infix notation, we traverse the above tree in inorder: $A \leftrightarrow B \to A \wedge B \vee C$. To write ϕ using postfix notation, we traverse the above tree in postorder: $A B \leftrightarrow A B C \vee \wedge \to$.

(13) Following the procedures seen in this section:

(a)(3) $\Delta + \circ * \Delta 549182 = 9$

(b)(2) $13154 \circ \Diamond \Delta + = 4$

(14) Since the code is prefix-free, we know tat the characters correspond to leaf nodes in the tree. Since the tree is full of height 5, it has $2^5 = 32$ leaf nodes. Hence there are 32 characters in the alphabet.

(15) Note first that α cannot be 0 since otherwise the code for A would be a prefix for that of B. So $\alpha = 1$. If $\beta = 1$, B would be a prefix for E so $\beta = 0$. The possibilities for pair (γ, κ) are $(0,0)$, $(1,0)$, $(0,1)$ and $(1,1)$. The first two pairs are not possible as they make A a prefix of D. The third pair is not possible as it makes D a prefix for B. Only $\gamma = \kappa = 1$ works.

(17)(a) Starting at the root of the tree and following the branches to each leaf node we get the following binary code: $A \to 0000$, $B \to 0001$, $C \to 001$, $D \to 010$, $E \to 011$, $F \to 100$, $G \to 101$, $H \to 1100$, $I \to 1101$, $J \to 111$.

(b) 11011110000001010100110001001000111011101000100010111100011

(c) Start at the root and follow the bits from left to right. Record characters as you go through the bits: (i) BGA; (ii) BFAJEH; (iii) JBGAHFEJJ.

11.6

(1) Note that the answers may vary. Below, we give one possibility for a spanning tree for each of DFS and BSF.

- G_1:

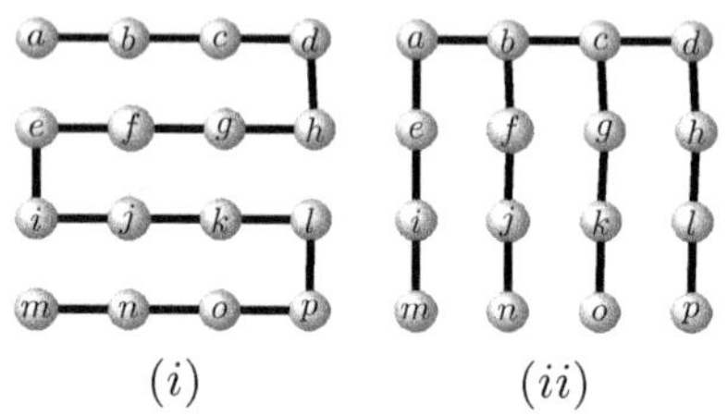

(i)   (ii)

- G_3:

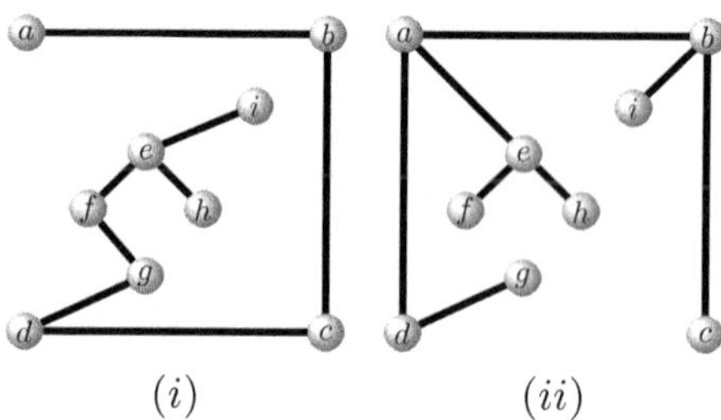

(i)   (ii)

(3) If T is a spanning tree of G, then T has the same order, say n, as G. We also know that T must have $n-1$ edges. If e is an edge of G which is not in T, then the subgraph $T+e$ of G obtained by adding e to T is not a tree since it is of order n but of size n as well.

(6)(a) In the following graph, each city (labeled by its first letter) is a vertex and each route between two cities is an edge labeled with the distance between the two cities.

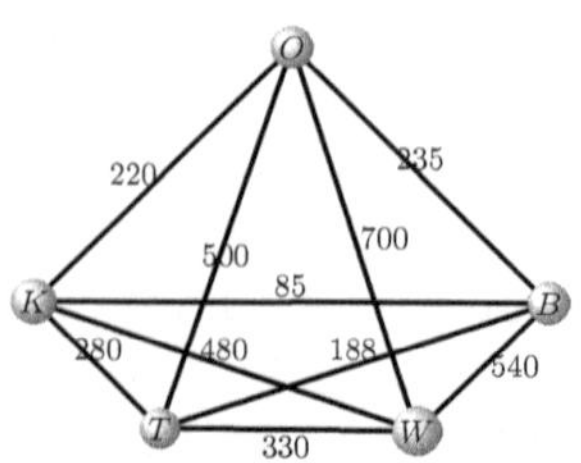

(b) Using Kruskal's algorithm. The following tables shows the edges of G sorted in a non-decreasing fashion according to their weights. The third row indicates which edge is included in the minimal spanning tree (marked with a ✓) and which edge is skipped (marked with an ×) by Kruskal's algorithm.

$\{B,K\}$	$\{B,T\}$	$\{O,K\}$	$\{O,B\}$	$\{K,T\}$
85	188	220	235	280
✓	✓	✓	×	×

$\{T,W\}$	$\{K,W\}$	$\{O,T\}$	$\{B,W\}$	$\{O,W\}$
330	480	500	540	700
✓	×	×	×	×

The minimal cost of the trip is $0.42(85+188+220+330) = 345.66\$$.

(9) Label the points as follows $O = (0,0)$, $A = (0,2)$, $B = (2,0)$, $C = (4,3)$, $D = (1,2)$, $E = (2,3)$, $F = (4,1)$. Joining each pair of points with a straight line gives the following graph:

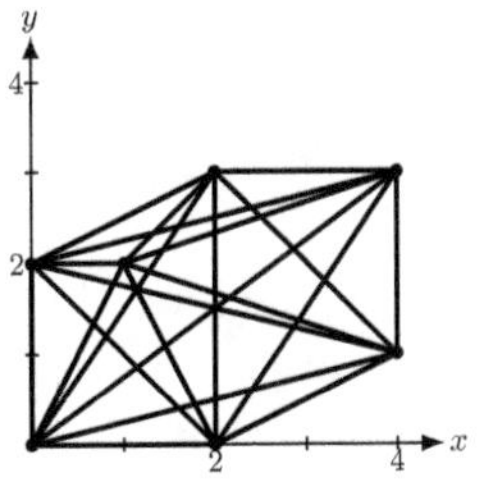

We consider a weight function w on the graph given by: $w(e)$ is the distance between the two end points of edge e. The following table shows the edges (first column), their weights (second column) in a non-decreasing order and whether or not the edge is used in Kruskal's algorithm (third column):

$\{A,D\}$	1	✓
$\{D,E\}$	$\sqrt{2}$	✓
$\{O,A\}$	2	✓
$\{O,B\}$	2	✓
$\{C,E\}$	2	✓
$\{C,F\}$	2	✓
$\{A,E\}$	$\sqrt{5}$	×
$\{A,F\}$	$\sqrt{5}$	×
$\{B,D\}$	$\sqrt{5}$	×
$\{O,D\}$	$\sqrt{5}$	×

$\{B,E\}$	$\sqrt{5}$	×
$\{B,F\}$	$\sqrt{5}$	×
$\{A,B\}$	$2\sqrt{2}$	×
$\{E,F\}$	$2\sqrt{2}$	×
$\{D,F\}$	$\sqrt{10}$	×
$\{C,D\}$	$\sqrt{10}$	×
$\{B,C\}$	$\sqrt{13}$	×
$\{O,E\}$	$\sqrt{13}$	×
$\{A,C\}$	$\sqrt{17}$	×
$\{O,F\}$	$\sqrt{17}$	×
$\{O,C\}$	5	×

The shortest path joining all seven point is the following (minimal spanning tree):

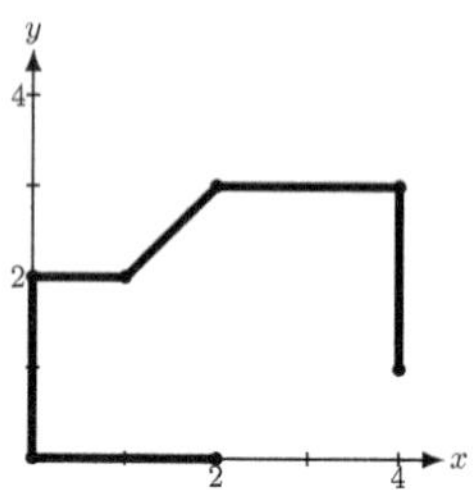

The minimal total length is $9 + \sqrt{2}$.

(11) We prove first that G' is connected. Assume that the cycle C is given by the list of vertices $(v_0, v_1, \ldots, v_k = v_0)$ and that edge e is $\{v_i, v_{i+1}\}$ for some $i \in \{0, \ldots, k-1\}$. For any pair u, v of vertices of G, we know that there exists a path P from u to v in G (since G is connected). If e is not part of P, then P is a path from u to v in G'. If e is part of P, write $P = (u, \ldots, \alpha, v_i, v_{i+1}, \beta, \ldots, v)$. Then $P' = (u, \ldots, \alpha, v_i, v_{i-1}, \ldots, v_0, v_k, v_{k-1}, \ldots, v_{i+1}, \beta, \ldots, v)$ is a path from u to v in G'. We conclude that G' is connected. Next, let T be a minimal spanning tree of G', we need to prove that T is a minimal spanning tree of G. Note first that T is a spanning tree of G since G and G' have the same set of vertices. Suppose by contradiction that T is not a minimal spanning tree for G. Then there exists a spanning tree T' of G such that $w(T') < w(T)$. Edge e must be in T' since otherwise, T' would be a tree in G' and we would have $w(T') \geq w(T)$. Remove edge e from T' and then add an edge e' of the cycle C which was not already on T' (such an edge must exist since a tree cannot contain a cycle). The resulting graph is a spanning tree T'' in G' (since it does not contain e and it still goes through all vertices). Moreover, $w(T'') = w(T') - w(e) + w(e') < w(T') < w(T)$ since $w(e) > w(e')$. This is a contradiction to the fact that T is a minimal spanning tree in G'. We conclude that T is a minimal spanning tree in G.

Index

Printed in the USA
CPSIA information can be obtained
at www.ICGtesting.com
LVHW010759290324
775682LV00001B/5